Form and Function in Birds

Form and Function in Birds

Volume 3

Edited by

A. S. KING

Department of Veterinary Anatomy,
University of Liverpool,
Liverpool,
England

J. McLELLAND

Department of Anatomy,
Royal (Dick) School of
Veterinary Studies, University
of Edinburgh, Scotland

1985

ACADEMIC PRESS

(Harcourt Brace Jovanovich, Publishers)

London · Orlando · San Diego · New York
Toronto · Montreal · Sydney · Tokyo

ACADEMIC PRESS INC. (LONDON) LTD.
24–28 Oval Road
LONDON NW1 7DX

United States Edition published by
ACADEMIC PRESS, INC.
Orlando, Florida 32887

British Library Cataloguing in Publication Data

Form and function in birds.–Vol. 3.
1. Birds–Physiology
I. King, A. S. II. McLelland, J.
598.2'1 QL693

Library of Congress Cataloging in Publication Data
(Revised for volume 3)
Main entry under title:

Form and function in birds.

Includes bibliographies and index.
1. Birds–Anatomy–Collected works. 2. Birds–Physiology–Collected works. I. King, Anthony Stuart. II. McLelland, J. (John)
QL697.F58 598.2'4 79–50523
ISBN 0–12–407503–7 (v. 3)

PRINTED IN THE UNITED STATES OF AMERICA

85 86 87 88 9 8 7 6 5 4 3 2 1

To E. C. Amoroso, biologist and teacher, counsellor and friend.

''When I look back at the three or four choices in my life which have been decisive, I find that, at the time I made them, I had very little sense of the seriousness of what I was doing and only later did I discover what had seemed an unimportant brook was, in fact, a Rubicon.''

W. H. Auden

Contents

Contributors

BETSY G. BANG, *Marine Biological Laboratory, Woods Hole, Massachusetts 02543, U.S.A.* (p. 195)

HERMAN BERKHOUDT, *Department of Morphology, Zoological Laboratory, Rijksuniversiteit Leiden, 2300 RA Leiden, The Netherlands* (p. 463)

A. BUBIEŃ-WALUSZEWSKA, *Katedra Anatomii Zwierzat Akademia Rolnicza, Wydzial Weterynaryjny, 51-631 Wrocław, Poland* (p. 149)

K.-M. GOTTSCHALDT, *Department of Neurobiology, Max Planck Institute for Biophysical Chemistry, D-3400 Göttingen, Federal Republic of Germany*[1] (p. 375)

JOAN A. HARDY, *Department of Biochemical Pathology, University College London, London WC1E 6JJ, England* (p. 1)

ROLAND KÜHNE, *Department of Biological Sciences, City of London Polytechnic, London E1 7NT, England*[2] (p. 227)

BRIAN LEWIS, *Department of Biological Sciences, City of London Polytechnic, London E1 7NT, England* (p. 227)

GRAHAM R. MARTIN, *Departments of Zoology and Comparative Physiology and Department of Extramural Studies, University of Birmingham, Birmingham B15 2TT, England* (p. 311)

ROBERT J. RAIKOW, *Department of Biological Sciences, University of Pittsburgh, Pittsburgh, Pennsylvania 15260, U.S.A.* (p. 57)

CATHERINE A. SMITH, *Department of Otolaryngology, Medical School, Oregon Health Sciences University, Portland, Oregon 97201, U.S.A.* (p. 273)

R. I. C. SPEARMAN, *Department of Biochemical Pathology, University College London, London WC1E 6JJ, England* (p. 1)

BERNICE M. WENZEL, *Department of Physiology, School of Medicine, University of California at Los Angeles, Los Angeles, California 90024, U.S.A.* (p. 195)

[1]Deceased. Requests for offprints should go to C. Vahle-Hinz, Physiology Institute, University of Würzburg, D-8700 Würzburg, Federal Republic of Germany.

[2]Present address: Institut für Zoologie der TU, D-8046 Garching, Federal Republic of Germany.

Preface

As in the first two volumes of this series, the principal theme of Volume 3 has been the adaptive advantage conferred on birds by their specialized structure. At the risk of being labelled a teleologist, one could express this objective as the study of the design requirements of birds. If, or rather when, we have strayed from our theme, it is because the many rapid advances in avian anatomy have seemed insistently to demand the marshalling of this knowledge in our pages before it becomes dispersed, or even lost, in the literature. As before, our authors have generously submitted to the somewhat austere editing which arises from the pursuit of our objectives, and we are once again grateful for their forbearance and cooperation. Additional and sometimes more painful sacrifices have been required of our authors through our continued rigorous adoption of the *Nomina Anatomica Avium,* but there is compensation in clearer international scientific communication. As in Volumes 1 and 2 we have utilized the scientific taxonomic nomenclature of J. J. Morony, W. J. Bock and J. Farrand (published in 1975 by the Department of Ornithology of the American Museum of Natural History, New York). The English vernacular names of birds are again those of E. S. Gruson (*Check List of the Birds of the World,* Collins, London, 1976). The names of the common laboratory and domestic birds refer to the following species: duck or *Anas,* domestic forms of *Anas platyrhynchos;* goose or *Anser,* domestic forms of *Anser anser;* pigeon or *Columba,* domestic forms of *Columba livia;* turkey, domestic forms of *Meleagris gallopavo;* chicken, domestic fowl or *Gallus,* domestic forms of *Gallus gallus;* quail, domestic forms of the genus *Coturnix.*

The preparation of this volume has been beset by personal difficulties of all kinds, experienced by our authors. Several withdrew during the later stage of the work through the pressure of other commitments. Others encountered health problems, which included minor abdominal surgery, two fractured arms, a broken leg and a heart attack. We recalled the curse of the pharaohs and advised our authors to take out additional insurance. This frivolity, however, was dispelled by the tragic death of Kay Gottschaldt in a skiing accident. It was a saddening end to our labours, but we hope that Kay's chapter will serve as a memorial to his life's work.

The fourth and final volume of this series is expected to follow a year after Volume 3 and will be devoted almost exclusively to the respiratory system. The only chapter deviating from this subject will be allocated to the central nervous system, its late appearance in these volumes being due to the serious illness of its principal author.

We wish to thank the staff of Academic Press Inc. (London) Ltd. for helping us to produce this book.

June, 1985

A. S. King
and
J. McLelland

List of Abbreviations

The following abbreviations of Latin terms have been used, the plural form being in brackets (as in the *Nomina Anatomica Avium,* Academic Press, New York and London, 1980).

A. (Aa.) = Arteria
Ant. = Anterior
Artc. (Artcc.) = Articulatio
Caud. = Caudalis
Cran. = Cranialis
G. (Gg.) = Ganglion
Gl. (Gll.) = Glandula
Lat. = Lateralis
Lig. (Ligg.) = Ligamentum
M. (Mm.) = Musculus
Maj. = Major
Med. = Medialis
Min. = Minor
N. (Nn.) = Nervus
Nuc. = Nucleus
Post. = Posterior
Proc. (Procc.) = Processus
R. (Rr.) = Ramus
Rdx. (Rdxx.) = Radix
Sut. (Sutt.) = Sutura
Synd. (Syndd.) = Syndesmosis
Synos. (Synoss.) = Synostosis
Tr. (Trr.) = Tractus
V. (Vv.) = Vena
Vas 1. (Vasa 1.) = Vas lymphaticum

1
Integument

R. I. C. SPEARMAN AND JOAN A. HARDY

Department of Biochemical Pathology, University College London, London England

---

CONTENTS

I. Introduction

The integument constitutes the boundary which separates the living cells of the body, with their delicately adjusted physico-chemical requirements, from the hostile conditions of the outside world. Paradoxically the integument must protect the body from contact with the environment whilst at the same time, in its role as a major sensory organ, receiving information from the environment. Free nerve endings and neuroreceptors in the skin inform the brain about warmth, cold, touch, pressure and pain. This information is partly protective in function, since it enables the bird to respond to both adverse and beneficial physical stimuli by appropriate reactions. In this way the integrity of the body surface is maintained and effects on internal homeostatic processes are minimised.

The requirements of neurosensation limit the possible thickness of the protective dead outer keratinized layer of the epidermis. This stratified layer of cornified cells has achieved its greatest thickness in reptiles, as exemplified by the thick plates of the tortoise carapace and the scales of crocodiles, lizards and snakes (Spearman, 1966, 1973). In these animals cutaneous neurosensation is mainly restricted to specialised sensillae, such as the heat-sensitive spots on the heads of snakes and the tactile spines on the scales of some lizards (Bellairs, 1969). Movement of these spines in lizards stimulates nerve endings at their bases beneath the thick scales, and the phylogenetically unrelated feathers of birds (as well as the hairs of mammals) have a similar tactile function.

The avian epidermis has developed only a thin horny layer, since the physical protection in plumage-covered sites is well provided by the keratinized feathers, which are also epidermal derivatives. The horny layer in birds is thinner than that of reptiles for two reasons. First, the much more complex skeletal muscular movements of birds, like those of mammals, necessitate a pliable integument (Spearman, 1966). Second, all inessential body weight has to be reduced to permit flight.

The still air held both within and around feathers affords efficient thermal insulation and, compared with the scales of reptiles, makes the plumage a much lighter protective covering. The evolution of homeothermy in birds has necessitated the development of a nearly continuous layer of plumage over the skin surface, and so, as in mammals, the thermoregulatory role has increased the importance of the skin to the body functions as a whole (Stephens, 1973).

II. Covering skin

A. *Dermis and subcutaneous layer*

The dermis is a layer of connective tissue which lies underneath the epidermis (Fig. 1.1). In fixed histological preparations it consists essentially of superficial and deep layers (*stratum superficiale, stratum profundum*), the deep layer being further subdivided into loose and compact layers (*stratum laxum, stratum compactum*). The junction of the dermis and subcutaneous tissue is marked by a layer of elastic connective tissue, the elastic lamina (*lamina elastica*). The junction of the dermis and epidermis is generally flat, but in the beak it is undulating with deep rete ridges (Lucas and Stettenheim, 1972). The microscopic structure of the dermis shows great species variation (Lucas and Stettenheim, 1972). In the superficial layer the scattered fibroblasts are more numerous and the connective tissue less dense than in the deep layer. The superficial compact layer of the deep dermis is made up of a dense feltwork of collagen and elastic fibres. Ultrastructurally, the fibrils of collagen are generally arranged horizontally, with a few orientated

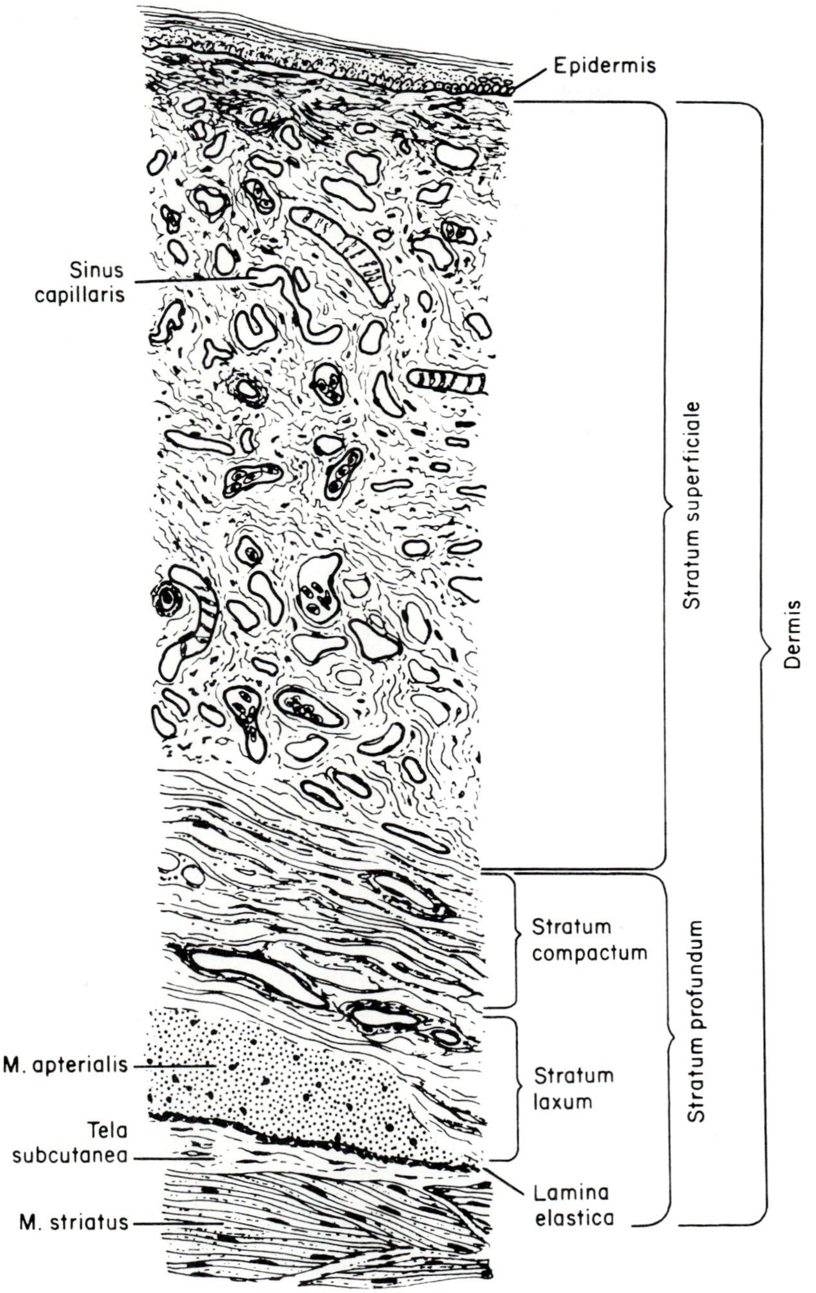

Fig. 1.1. Microscopic section of the dermis and epidermis of the skin of *Gallus*. From Lucas (1979).

vertically to the skin surface; this differs from the dermis of mammals in which the collagen is more haphazardly arranged (Matoltsy, 1969). In the deeper loose layer the relatively loosely-arranged connective tissue is interspersed with adipose cells (Spearman, 1982b) and lymphatic vessels. In *Gallus* there is more fat in both the dermis and subcutaneous tissue in feathered sites of the body than in bare areas of skin.

In life the avian dermis is highly pliable, particularly where the wings join the body and in the neck. Evidence from the mammalian dermis (Jarrett, 1974; Spearman, 1974) suggests that the collagen and mucosubstances are chemically closely linked in the form of a watery fibrous gel which readily allows movement through it of migrant macrophages, leucocytes and even nematode larvae. The ultrastructure of avian collagen resembles that of mammals. Associated with the collagen are acid glycoproteins, and in the chicken the proportion of glycoprotein varies from site to site (Fleischmajer and Krol, 1970). In addition to scattered elastin fibres, the dermis also contains thicker elastin fibres condensed into tendons. Each tendon (*tendo elasticus*) joins together end-to-end the smooth muscles of the dermis (*mm. nonstriati dermatis*), including the feather muscles (*mm. pennales*), which subtend feathers, and the apterial muscles (*mm. apteriales*), which interconnect feather tracts.

Within the subcutaneous tissue (*tela subcutanea*) (Fig. 1.1), superficial and deep fasciae (*fascia superficialis, fascia profunda*) can be distinguished. The subcutaneous tissue is basically made up of haphazardly-distributed collagen fibres which are occasionally arranged into thicker bundles. The superficial fascia is often a fat storage organ in which numerous adipose cells are aggregated into fat bodies (*corpora adiposa*). The adipose cells are especially numerous in aquatic species such as ducks, geese and swans (Anatidae), gulls and terns (Laridae) and skuas (Stercorariidae) (Spearman, 1982b). They may be either multivesicular (*lipocytus multivesicularis*) or univesicular (*lipocytus univesicularis*). For a description of the fat bodies in *Gallus* see Liebelt and Eastlick (1954). Within the subcutaneous tissue are the striated subcutaneous muscles (*mm. subcutanei*), which attach to the skeleton or the somatic muscles as well as to the overlying skin. Collagen fibres anchor the subcutaneous tissue to the fascia covering the musculature or occasionally, as in the beak, more firmly to the periosteum of the bone (Lucas and Stettenheim, 1972).

Arteries and veins supply large blood capillary plexuses just beneath the epidermis and in the deep dermis, and some capillaries enter the dermal papillae of feather follicles (Lucas and Stettenheim, 1972). In cold weather whole plexuses are shut down, thus reducing heat loss from the body. In the comb and wattles of the domestic fowl the vascularisation of the superficial layer of the dermis is especially rich and gives these organs their characteristic red colour.

Cutaneous nerve endings (*terminationes nervorum in cute*) are present in both the dermis and epidermis. The sensory nerves of the dermis are distributed in subepidermal deep plexuses (Lucas and Stettenheim, 1972). Motor sympathetic fibres course with these cutaneous nerves and innervate the adipose cells and the dermal muscles. Free nerve endings which are sensory in function penetrate between the epidermal basal cells. Several different types of multicellular neuroreceptor organs occur in the dermis (Quilliam and Armstrong, 1963; Malinovsky, 1968; and see K.-M Gottschaldt in Chapter 8 of this volume).

B. Epidermis

1. Histology

The *epidermis* is derived in the chicken embryo from a single cell layer of ectoderm. At about 40 to 45 hours of development this divides into an inner cuboidal germinal layer and an outer *periderm*. The periderm functions as the protective covering of the embryo. A few days before hatching its cells die, become keratinized, and are sloughed. The adult epidermis is derived from the cuboidal layer and is several cells thick at the time of hatching (Stettenheim, 1972). It consists basically of a deep layer of living cells, the *stratum germinativum,* and a superficial horny layer, the *stratum corneum* (Fig. 1.2).

Within the *stratum germinativum* can be distinguished three layers. At the base of the epithelium is a germinal layer of small dividing cells, the basal layer (*stratum basale*). The basal layer is separated from the dermis by the basal membrane (*membrana basalis*). Cells derived from the basal layer are pushed upwards and increase in volume to become the intermediate layer (*stratum intermedium*), which is homologous to the *stratum spinosum* or prickle cell layer of mammals. The epidermal cells of the intermediate layer are held together by desmosomal junctions. In the intervals between the desmosomes the plasma membranes of adjacent cells are separated by spaces containing tissue fluid in which nutrients are transported to the epidermal cells from the dermal blood capillaries. Fine tonofibrils converge on thickenings of the plasma membranes beneath the desmosomes, the desmosomal plaques (Spearman, 1973, 1982a). Towards the skin surface the uppermost one or two cell layers of the epidermis become flattened to form the transitional layer (*stratum transitivum*). This layer has the same position in the epidermis as the *stratum granulosum* of mammals. Whilst darkly-stained keratohyalin granules are present in the cytoplasm of the mammalian granular layer after staining by haematoxylin, no such granules were seen by Cane and Spearman (1967) and Spearman (1969) with the light microscope in either the chicken or the King Penguin (*Aptenodytes patagonica*), and this seems to be true of birds in general (Lucas and Stettenheim, 1972). Nevertheless, in electron micrographs of the epidermis of the chicken, small granules of keratohyalin can be seen around the nucleus of the transitional cells (Fig. 1.3), and these are similar to the keratohyalin granules in mammals (Matoltsy, 1969). Mammalian keratohyalin granules are aggregates of sub-particles containing a mixture of proteins, phospholipid, calcium and occasionally ribosomes. They appear to be chemical coacervates of various substances precipitated into the cytoplasm under unstable physico-chemical conditions in the cells (Spearman and Hardy, 1974). Keratohyalin does not occur in fish and amphibia (Spearman, 1966), but in reptiles ultrastructural evidence for the granules has been found in

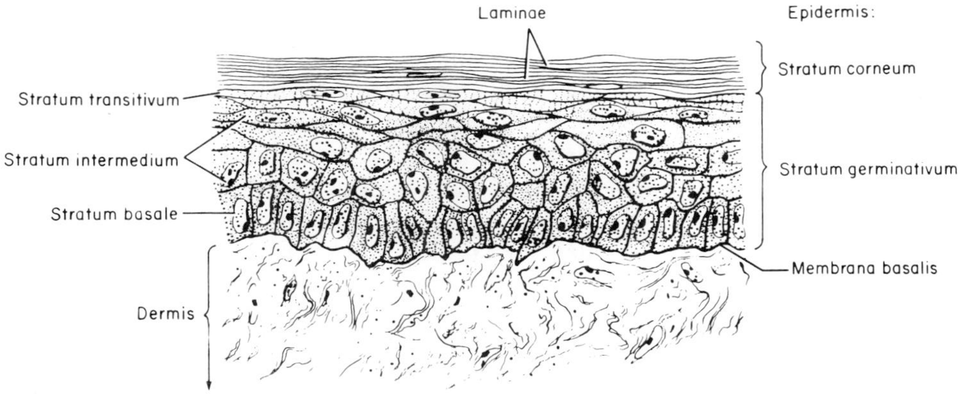

Fig. 1.2. Microscopic section of the dermis and epidermis of the skin of *Gallus*. From Lucas (1979).

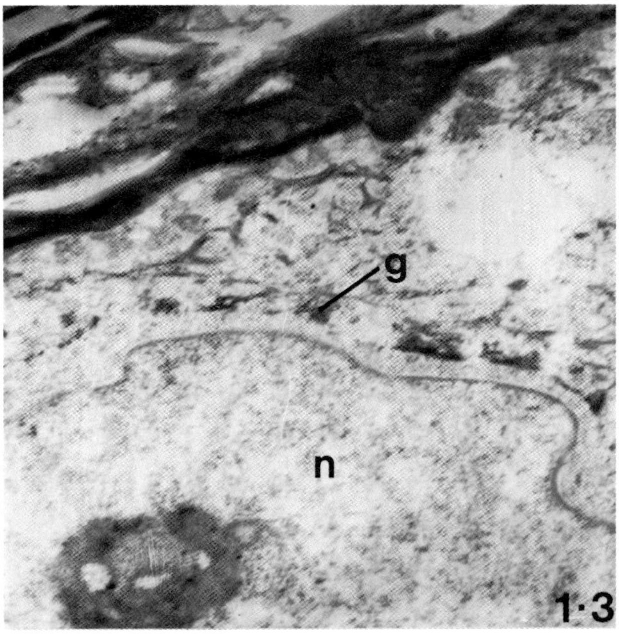

Fig. 1.3. Electron micrograph of the transitional and horny layers of the epidermis of the pectoral region of the adult domestic fowl. n, Nucleus of transitional cell; g, keratohyalin granule in transitional cell; k, keratin of *stratum corneum*. Fixed in glutaraldehyde, post-fixed in osmium and stained with phosphotungstic acid. ×20,000.

crocodile epidermis (Spearman, 1966; Parakkal and Alexander, 1972). Keratohyalin granules are seen in mammals by light microscopy, probably because calcium bound to phospholipid is stained by haematoxylin, whereas by electron microscopy it is the proteins which are shown (Jarrett and Hardy, 1957; Spearman and Hardy, 1974; Spearman, 1982a). Differences in composition in birds could therefore explain why avian keratohyalin is not stained by haematoxylin. It is of interest that eosinophilic granules, which are probably a form of keratohyalin, occur in the tongue papillae of many mammals (Spearman, 1980a). The most superficial flattened living cells constitute the zone where the final stages of keratinization occur (Cane and Spearman, 1967). Above this layer of transitional cells, death occurs suddenly with the formation of the horny *stratum corneum,* which is several cells thick.

The epidermis in the domestic fowl is thinnest in the featherless sites where the *stratum germinativum* and *stratum corneum* are both two or three cells thick. It is thickest in the *podotheca,* the non-feathered scaled integument of the foot.

Lucas and Stettenheim (1972, pp. 67, 595–598) describe four main types of scale, including large scales or scutes (*scuta*), intermediate-sized scales or *scutella,* small scales or *reticula,* and minute scales or *cancella.* (Note that in the *Nomina Anatomica Avium* the term "scuta" refers to all types of scale whilst "scutella" is used for small scales. No specific terms are provided for reticulate or cancellate scales.) Examples of the different patterns of arrangement of scales are shown in Fig. 1.4. Scutes are best developed on the dorsal surfaces of the tarsometatarsus and digits. On the sides and plantar surfaces of the toes the scales are often reticulate and in some species, including the Brown Pelican (*Pelecanus occidentalis*) and the Black-necked Stilt (*Himantopus mexicanus*), all the scales on the tarsometatarsus are reticulate. In many birds, including for example *Gallus* and the Pileated Woodpecker (*Drocopus pileatus*), the tarsometatarsus is both scutellate and reticulate. Cancellate scales occur on the webs of aquatic species. Usually scales do not overlap (non-imbricated scales), but overlapping of tar-

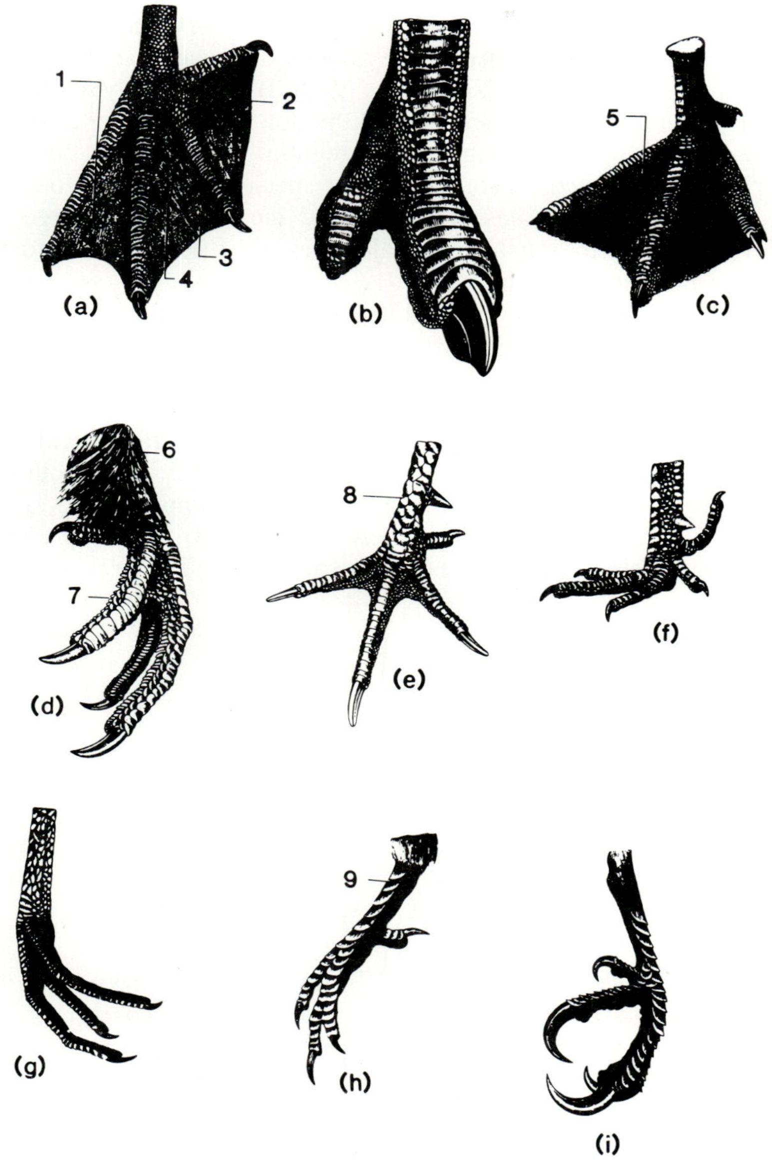

Fig. 1.4. Examples of the patterns of arrangements of scales in the foot. (a) Brown Pelican (*Pelecanus occidentalis*). (b) Ostrich (*Struthio camelus*). (c) *Anas*. (d) Ruffed Grouse (*Bonasa umbellus*). (e) *Gallus*. (f) *Gallus* (five-toed breed). (g) American Avocet (*Himantopus mexicanus*). (h) a kingfisher (*Ceryle* spp.) (i) Pileated Woodpecker (*Dryocopus pileatus*). 1, Lateral interdigital web; 2, medial interdigital web; 3, intermediate interdigital web; 4, cancellate scutes; 5, cancellate scutes; 6, feathered tarsometatarsus; 7, marginal fringe; 8, non-imbricated scutes; 9, imbricated scutes. In (g) the tarsometatarsus is reticulate. In (e), (f), (h) and (i) the tarsometatarsus is scutellate and reticulate. From Lucas and Stettenheim (1972).

sometatarsal and digital scales (imbricated scales) occurs in kingfishers (Alcedinidae) and woodpeckers (Picidae). For a detailed review of scales and scutellation see von Boetticker (1929).

At feathered sites of the body the desmosomal junctional complexes between the flat surfaces of adjacent cells in the horny layer appear to be weaker than the lateral junctions, since they tend to break down leaving spaces between the cells. This is in contrast to the tarsometatarsal scales of the domestic fowl in which the cells remain more firmly attached to one another above and below, just as in the homologous scales of reptiles (Spearman, 1966; Brush and Wyld, 1980). In the Budgerigar (*Melopsittacus undulatus*) and the domestic fowl the tarsometatarsal scales are interspersed by flexible hinge regions in which the horny layer is relatively loosely arranged. In some passerine species such as the House Sparrow (*Passer domesticus*) the scales are small and the hinge areas large. In most passerines, however, the tarsometatarsal scutes are fused together to form the "booted" or holothecal condition.

On the plantar surface of the foot the epidermis is specialised as pads to withstand compression, forming a thick horny layer of reticulate scales many cells in depth (Cane and Spearman, 1967). In *Gallus,* a metatarsal pad (*pulvinus metatarsalis*) occurs at the distal end of the tarsometatarsus and at the bases of digits II–IV, whilst several digital pads (*pulvini digitales*) are present on each of the main digits. The pads in *Columba* are shown in Fig. 1.5. In the young of a few species including woodpeckers (Picidae), toucans (Ramphastidae) and barbets (Capitonidae), a hypotarsal pad (*pulvinus hypotarsalis*) is also present and is moulted when the bird leaves the nest (Harrison, 1964). The large amount of free

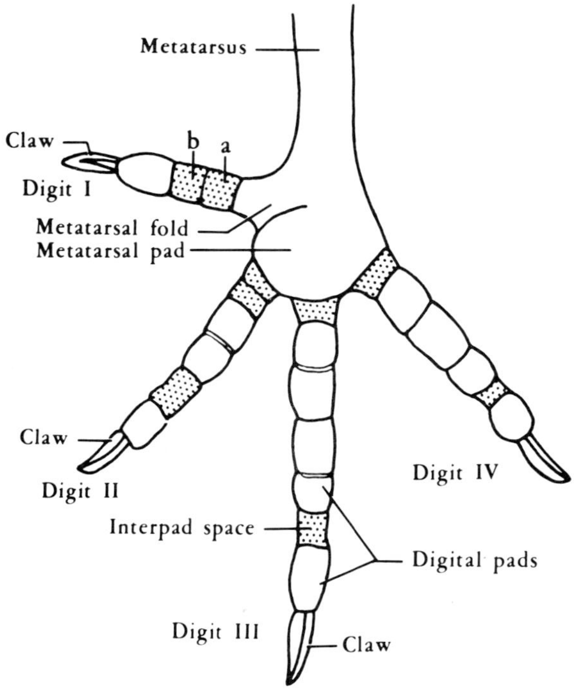

Fig. 1.5. The plantar surface of the foot of *Columba*. The metatarsus corresponds to the tarsometatarsus of the *Nomina Anatomica Avium*. From Lucas and Stettenheim (1972).

lipid in the avian horny layer, and particularly in the digital pads, probably makes the tissue spongy and resilient under the tread. These reticulate scales appear unrelated to the tarsometatarsal scales, either histologically (Cane and Spearman, 1967) or in chemical composition (Brush and Wyld, 1980).

2. *Keratinization*

The sudden death of the epidermal cells in the upper part of the *stratum germinativum,* with the formation of the horny layer, involves the process of keratinization, which combines three distinct mechanisms. These are (a) the synthesis in the intermediate cells of different pre-keratin proteins which, just prior to death, are bonded together by disulphide bonds, hydrogen bonds and polar linkages to form the keratin complex; (b) the formation at about the same time of a thickened envelope beneath the plasma membrane; and (c) the cytolysis by cytoplasmic hydrolytic enzymes of cell organelles, including the nucleus (Spearman, 1977a; Wrench *et al.,* 1980).

The amino acid composition of avian epidermal proteins has been examined by Brush (1974, 1978, 1980a,b), who found differences between the proteins in the feathers, the beak, the claws and the tarsometatarsal scales. Many of these proteins are pre-keratins and keratins, and their differences are indicative of the activity of the various keratin genes. In the horny structures mentioned before, pleated fibrils are present but these have not been found in the horny layer of the feathered sites that contain α helical keratins (Rudall, 1947; Spearman, 1966). There is no evidence that the tonofibrils are transformed into keratin, as is sometimes suggested, since keratin fibrils are produced independently by synthesis on polyribosomes. In the *stratum corneum* a layer of keratin is formed in the peripheral cytoplasm, but this cannot occur in the centres of the cells due to the presence of lipids, the formation of which is discussed in the next section. Keratin fibrils are rarely observed in the horny layer of the chicken by electron microscopy, in contrast to the mass of fibrils seen in mammalian horny cells. The reasons for this are not clear (Wrench *et al.,* 1980).

The formation of the thickened envelope beneath the plasma membrane occurs concurrently with keratin bonding. Whilst little is known about the envelope in birds, in mammals it is produced by extension of the desmosomal plaques (Matoltsy, 1980; Spearman, 1982a), and in these regions the protein structure of the envelope, although similar to keratin, is much more resistant to alkaline degradation. Formation of a thickened envelope occurs in all keratinized cells throughout the animal kingdom (Spearman, 1977a). Together with keratin it confers mechanical strength on the horny layer.

Cytolysis is most active in the epidermal transitional cells just before cell death, and this zone is rich in lysosomal non-specific esterase and acid phosphatase as in the mammalian granular layer (Cane and Spearman, 1967). However, in marked contrast to mammalian epidermis, in which there is little activity of these enzymes below the level of the granular layer, in the chicken epidermis quite strong activity occurs in the basal and intermediate layers (Cane and Spearman, 1967). The reasons for this are not clear. However, it may be related to the lipogenic activity of avian epidermis, since the germinal cells of the uropygial gland of birds and the mammalian sebaceous glands, which are also lipogenic, are similarly rich in non-specific esterase and acid phosphatase (Wrench *et al.*, 1980). In mammals, hydrolytic enzymes, including proteolases, are stored in lysosomes after their synthesis and then released at a later stage of cell maturation, but in the avian epidermis some of the enzymes are released soon after they are formed.

3. *Lipogenesis*

Keratin-bound phospholipids occur throughout the *stratum corneum* in birds and are synthesised in the basal and intermediate cells (Cane and Spearman, 1967; Lavker, 1975).

The presence of fat globules in avian epidermal cells was first shown by Varićak (1938). Since then neutral lipids have been found in the epidermis in a wide variety of species (Lucas and Stettenheim, 1972; Stettenheim, 1972), and the epidermis of birds has been shown to be both keratogenic and strongly lipogenic (Freinkel, 1972; Flaxman, 1972; Lavker, 1975; Wrench *et al.*, 1980). This is in contrast to mammalian and most reptilian epidermal cells, which are keratogenic and only undergo lipogenesis with respect to a few complex lipids (Wrench *et al.*, 1980). Amongst reptiles only tortoises (*Chelonia*) may possibly accumulate lipid droplets in the epidermis, although to a much smaller extent than in birds (Matoltsy and Huszar, 1972), and lipid droplets have not been found in either fish or amphibian epidermis.

Routine histological methods for the light microscope fail to reveal free fats which are extracted during tissue processing (Spearman, 1966, 1969; Cane and Spearman, 1967). It seems likely, however, that the vacuoles observed in the superficial epidermal cells of *Gallus* by Cane and Spearman (1967) were due to fat extraction (Spearman, 1983). Epidermal fat droplets have been demonstrated by light microscopy in the avian epidermis using frozen sections and fat stains by a number of workers including Lucas (1968, 1970, 1980), Shah and Menon (1972) and Shah *et al.* (1977). In ultrastructural studies vacuoles have been observed in the epidermis of late chicken embryos, and Parakkal and Matoltsy (1968) suggested that these were lipid droplets. In the epidermis of the newly-hatched chick Matoltsy (1969) found that small lipid droplets were apparently secreted by the Golgi apparatus of the basal cells and that these droplets later coalesced to form larger globules within the intermediate cells (Fig. 1.6). Lavker (1975), using a technique employing the uptake of labelled fatty acids and lipid analysis, demonstrated that chicken epidermal cells are able to synthesise both neutral fats and phospholipids. The transitional cells of the epidermis of newly-hatched domestic fowl were observed by Matoltsy (1969) to contain centrally-situated lipid globules, the nucleus being flattened and surrounded by the fine granules of keratohyalin.

The resemblance of the lipogenic avian epidermis to the mammalian sebaceous glands was noted by Matoltsy (1969) and by Freinkel (1972), who termed the epidermal cells "kerato-sebocytes." Wrench *et al.* (1980) adopted the term "sebo-keratocytes," since keratinization is phylogenetically the more primitive condition. The resemblance of the avian epidermis to the large sebaceous uropygial gland of birds was noted by Menon *et al.* (1981). The steady reduction

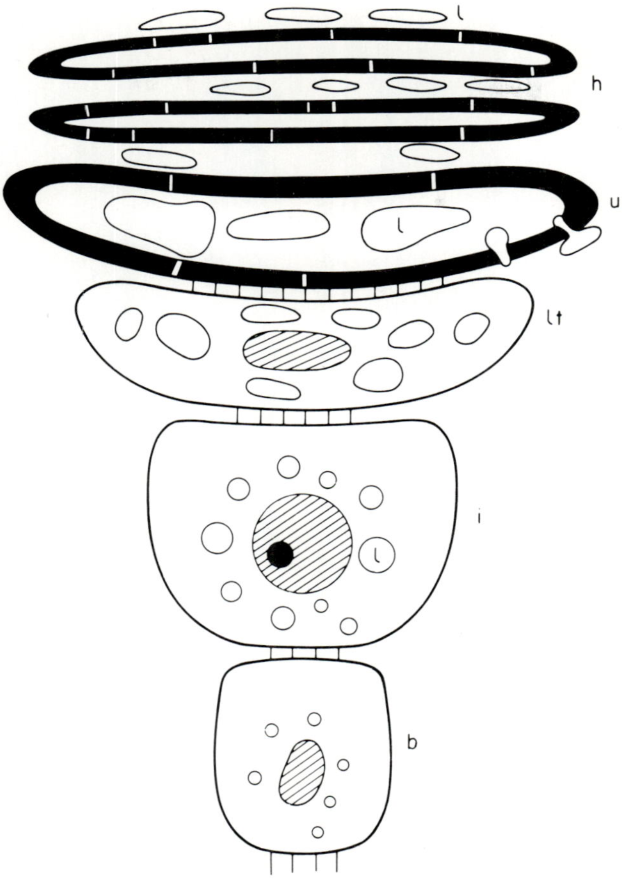

Fig. 1.6. Tier of cells in the epidermis of *Gallus*. b, Basal cell; i, intermediate cell; lt, lower transitional cell; ut, upper transitional cell; h, horny cell; l, lipid droplets. The lipid droplets diffuse out of the transitional cells through pores in the keratinized shell depicted in black. From Spearman (1983).

in the amount of lipid in the cells of the horny layer from the transitional region to the outer surface suggests that leakage occurs from the cells within the cornified layer, although lipid is not actively secreted into the narrow intercellular pathway between the living epidermal basal and intermediate cells (Matoltsy, 1969; Wrench *et al.*, 1980).

In an ultrastructural study Menon *et al.* (1981) showed that pores exist in the cornified cells of the rictus, and that lipids appear to pass out of the cells through these holes to the skin surface (Fig. 1.6). The secretions of both the uropygial gland (Jacob, 1978) and the epidermis, therefore, coat the plumage with a thin lipid film, the significance of which will be discussed in Section II,C,2.

Whilst the earliest ultrastructural investigations of the epidermis of the domestic fowl were all on young chicks, Menon's study was based on the adult. In order to determine whether any changes occur in the adult epidermis we have compared the ultrastructure of the epidermis in chick and adult domestic fowl. In both groups lipid droplets appeared first in the basal cells (Fig. 1.7), and increased in number and coalesced to form larger globules in the intermediate layer (Fig. 1.8).

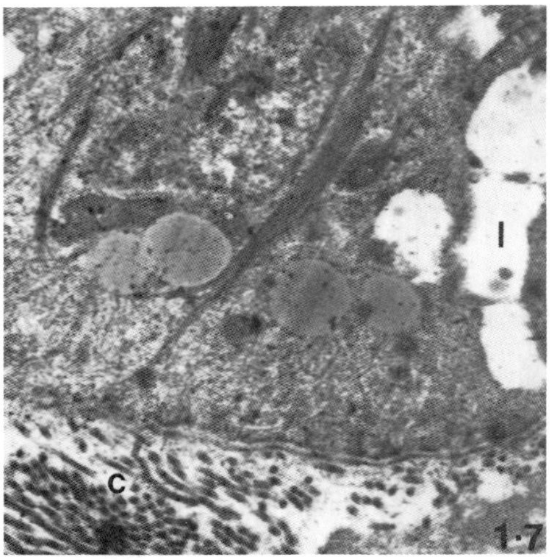

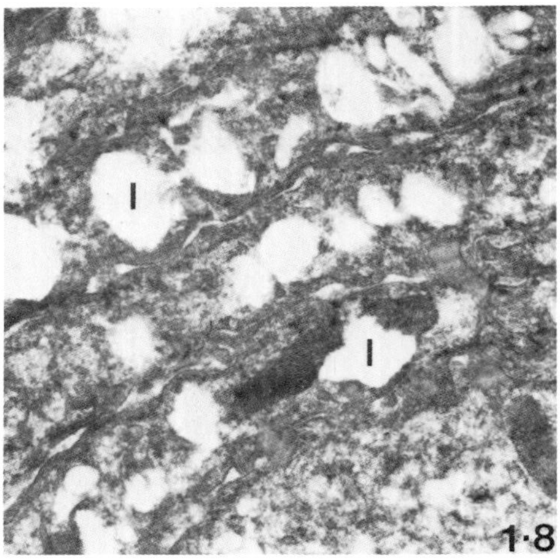

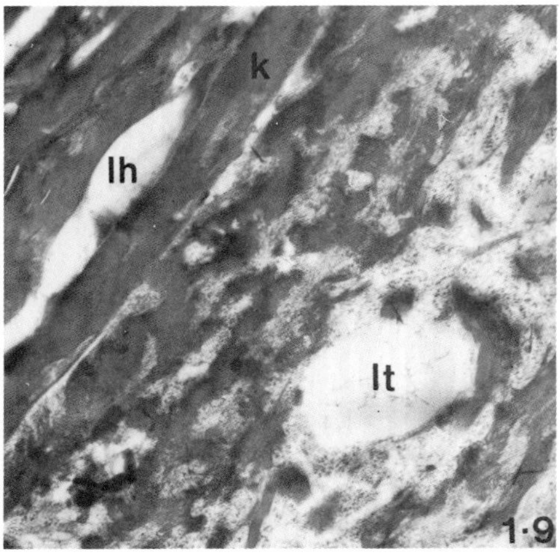

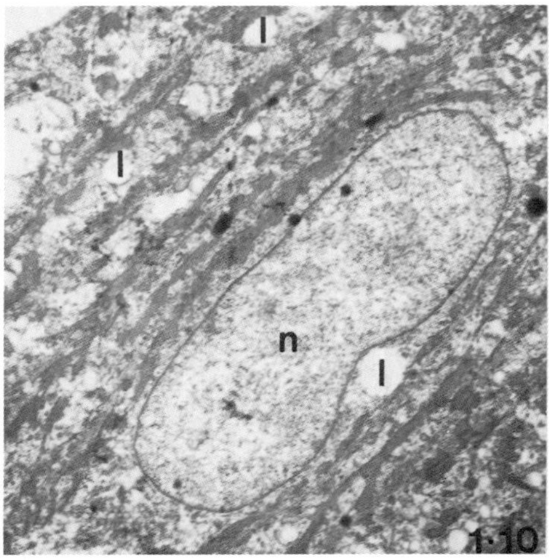

Fig. 1.7. Electron micrograph of the basal layer of the epidermis of the tarsometatarsus of the immature domestic fowl. l, Lipid globules in basal cell; c, collagen of dermis. Fixed in glutaraldehyde, post-fixed in osmium and stained with phosphotungstic acid. ×15,000.

Fig. 1.8. Electron micrograph of the intermediate layer of the epidermis of a digital pad of the adult domestic fowl. l, Lipid globules in the lower intermediate cells. Fixed in glutaraldehyde, post-fixed in osmium and stained with phosphotungstic acid. ×15,000.

Fig. 1.9. Electron micrograph of the transitional and horny layers of the epidermis of a digital pad of an adult domestic fowl. lh, Lipid globule in horny cell; lt, lipid globule in transitional cell; k, keratin. Fixed in glutaraldehyde, post-fixed in osmium and stained with phosphotungstic acid. ×20,000.

Fig. 1.10. Electron micrograph of the transitional layer of the epidermis of the tarsometatarsus of an immature domestic fowl. l, Lipid globules in transitional cell; n, nucleus of transitional cell. Fixed in glutaraldehyde, post-fixed in osmium and stained with phosphotungstic acid. ×10,000.

The quantity of lipid in the transitional cells varied from site to site. The greatest amount occurred in the interdigital webs and in the footpads (Fig. 1.9), as well as in the rictus as shown by Menon *et al.* (1981); moderate quantities were observed in the costal region and the comb; the least amount was found in the epidermis of the interramal region and the tarsometatarsus (Fig. 1.10). No differences were detected between chick skin and adult skin from the same sites. In all sites the

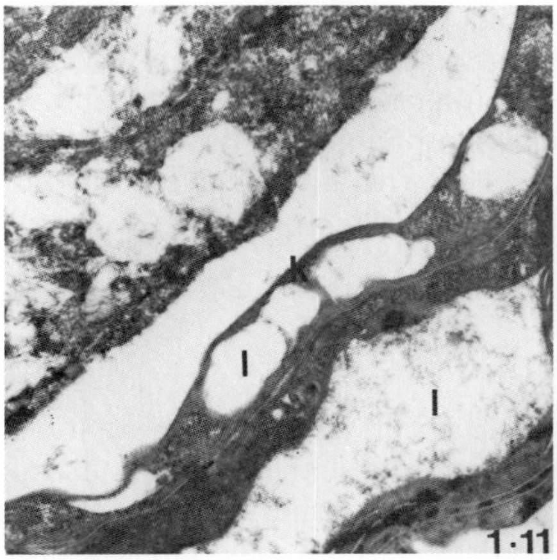

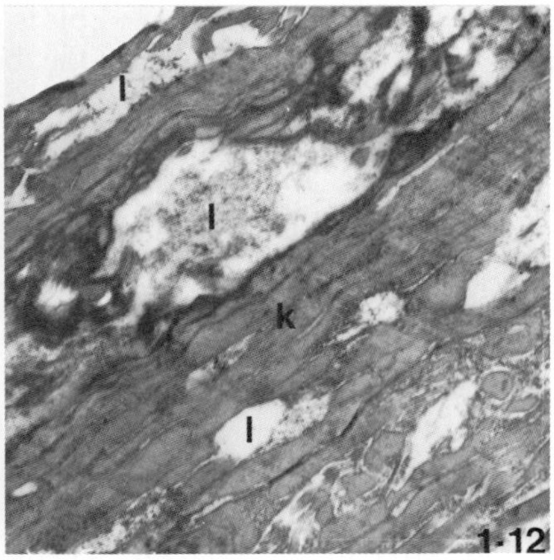

Fig. 1.11. Electron micrograph of the horny layer of the epidermis of the comb of an adult domestic fowl. l, Lipid in horny cells. Fixed in glutaraldehyde, post-fixed in osmium and stained with phosphotungstic acid. ×15,000.

Fig. 1.12. Electron micrograph of the horny layer of the epidermis of a digital pad of an immature domestic fowl. k, Keratin shells of horny cells; l, intracellular and extracellular lipid globules. Fixed in glutaraldehyde, post-fixed in osmium and stained with phosphotungstic acid. ×20,000.

horny cells which had just undergone keratinization at the base of the cornified layer contained centrally-placed elongated lipid globules (Fig. 1.11), which were largest in the interdigital webs and the reticulate scales of the toes.

Above the first layer of the *stratum corneum* the cells were more flattened and contained less lipid, with very little at all in the most superficial squames which were about to be sloughed. Some evidence was seen to suggest that lipid passes into the intercellular spaces from the horny cells (Fig. 1.12), and this supports the findings of Menon *et al.* (1981). At relatively high resolution the lipid globules, which are lightly stained by osmium, have a speckled appearance (Fig. 1.13).

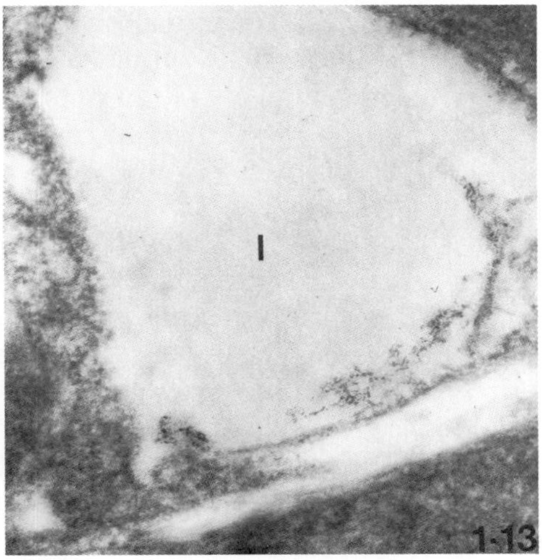

Fig. 1.13. Electron micrograph of an epidermal cell of an adult domestic fowl. l, Lipid globule which is stained faintly by osmium. Fixed in glutaraldehyde, post-fixed in osmium and stained with phosphotungstic acid. ×60,000.

The *stratum corneum* in reptiles, birds and mammals acts as a passive barrier to water movement into and out of the body (Spearman, 1973). However, this barrier is not complete and merely slows down the water flow. The main hydrophobic substances are probably keratin-bound phospholipids and other free lipids. Keratin itself takes up water vapour in equilibrium with atmospheric humidity, and so is not a good barrier material.

4. Shedding of the horny layer

The loss of horny cells from the skin surface occurs in two ways. First, small groups of cells, or individual cells, lose their desmosomal connections, as in mammalian desquamation. They are then shed amongst the plumage, and since they are coated with lipids they help to spread the sebaceous material. This type of shedding occurs in the loosely-arranged horny layer of the skin that is covered by plumage. In the second method, seen in the more compacted sites of the tarsometatarsal scales and the digits (but not in the reticulate scales), a layer of horny cells is sloughed periodically in the form of relatively large fragments. It has been shown at these sites that at intervals a single layer of unkeratinized cells enters the *stratum corneum* and later functions as a fission plane (Cane and Spearman, 1967). This mode of sloughing may be associated with feather moult. Desquamation of groups of cells in smaller fragments can occur at the same site, which confuses the picture. Periodic sloughing also occurs with the horny growths on the head of the Puffin (*Fratercula arctica*), and the claws of grouse (Tetraoninae), as well as in abnormal horny growths (Voitkevich, 1966). One is reminded of the sloughing mechanism in lizards, snakes and amphibia (Spearman, 1973), although the process in birds is actually quite different (Cane and Spearman, 1967).

By its continual renewal, the *stratum corneum* provides an effective barrier against invasion of the body by parasitic microorganisms, although many arthropod ectoparasites are able to pierce the horny layer by means of their powerful jaws (Spearman, 1980b). The horny layer on its underlying bed of resilient dermis also gives mechanical protection to the body surface as well as to the skeletal musculature underneath.

C. Uropygial gland

The cutaneous glands (*glandulae cutaneae*) of birds include the small wax-secreting auricular glands (*gll. auriculares*) in the external wall of the acoustic meatus, the small mucus-secreting vent glands (*gll. venti*) and the large uropygial or preen gland (*gl.uropygialis*), which opens by a papilla (*papilla uropygialis*) at the base of the tail (Elder, 1954). The lipid and waxy sebum which the uropygial gland secretes is coated on the beak and transferred to the plumage in preening. The gland, which is derived in the embryo from the epidermis, corresponds in function to the mammalian sebaceous gland, although the two types of gland are not necessarily homologous. Nevertheless, small, oily sebaceous-like glands have been described in crocodile skin, so a common origin could exist (Bellairs, 1969).

The uropygial gland is always visible to the naked eye. When very large, as in pelicans (Pelecanidae), it is about the size of a chicken's egg. It is very small in storks (Ciconiidae) and herons (Ardeidae). Only a few species lack a uropygial gland, notably the Ostrich (*Struthio camelus*) and other ratites, and some strains of domesticated doves (Columbidae) (Voitkevich, 1966). The uropygial gland is best developed in aquatic species, in which the sebum, together with the epidermal lipids, plays an important role in waterproofing the plumage (Spearman, 1980b).

1. Histological description

The uropygial gland (Fig. 1.14) is a lobed structure, the two lobes being separated by a connective tissue septum (*septum interlobare*) continuous with the gland capsule (*capsula gl. uropygialis*). Each lobe (*lobus gl. uropygialis*) is lined by numerous secretory alveoli which open into a central cavity. The alveoli are separated by thin connective tissue trabeculae. The central cavity of a lobe is drained by a duct (*ductus gl. uropygialis*) which has an opening (*porus ductus*

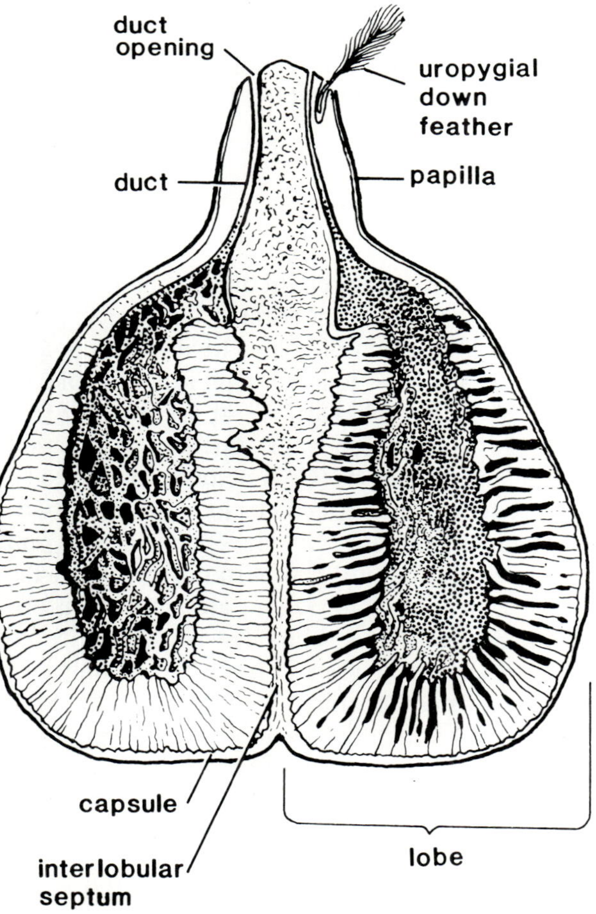

Fig. 1.14. Section of the uropygial gland of *Gallus* cut parallel to the long axis of the gland showing the right and left lobes. From Lucas (1979).

uropygialis) at the apex of the papilla. The papilla is devoid of feathers apart from a ring of down feathers (*circulus uropygialis*) close to the apex. However, these down feathers are reported to be absent in some species (Chandler, 1916). The uropygial gland, like mammalian sebaceous glands, has a holocrine mode of secretion, so that the glandular cells break down completely and liberate their products into the duct. In microscopic structure the secretory tubules have a germinal basal layer, the cells of which divide to form the secretory cells. The latter have a prominent Golgi apparatus, which is generally associated with lipogenesis, and lipid droplets accumulate in the cytoplasm as in the epidermis. Finally, these globules coalesce and cytolytic changes lead to the complete breakdown and loss of both cell organelles and plasma membrane. Keratinization does not occur in the secretory portion of the duct, but nearer the papilla a thin horny layer is formed, although the duct cells still undergo lipogenesis as in the epidermis generally. There is therefore a gradation between the unkeratinized glandular region and the fully-keratinized epidermis around the papilla in which complete cell autolysis is prevented from occuring by the keratin shells (Menon *et al.*, 1981).

The histochemistry of the uropygial gland shows that the secretory cells are rich in acid phosphatase (Cater and Lawrie, 1950), similar to intermediate epidermal cells and the cells of mammalian sebaceous glands. The secretory alveoli are rich in glycogen, which presumably forms a reserve for glucose metabolism required in lipogenesis.

2. *Sebum*

The chemical composition of the sebum of the uropygial gland has been extensively examined by Jacob (1978) in a number of species. The important hydrophobic constituents are long-chain aliphatic waxes and complex homologues of branched fatty acids and alcohols, together with saponficable and non-saponificable lipids (Haahti *et al.*, 1964; Nikkari, 1974; Jacob, 1978).

The secretion appears to be more water-repellent in aquatic birds. Only some 7% of the plumage sebum in pigeons (Columbidae) appears to be derived from the uropygial gland, as suggested by differences in plumage grease composition, and presumably the remainder is produced by the epidermis. In contrast, in the Puffin (*Fratercula arctica*), which requires a waterproof plumage, the compositions of plumage grease and uropygial gland sebum are similar (Jacob, 1978). Unfortunately the composition of epidermal sebum has not been adequately analysed, but in the domestic fowl it appears to contain more fats than uropygial gland sebum. Nothing is known about its composition in sea birds. In the Marabou (*Leptoptilos crumeniferus*), cholesterol which occurs in plumage grease is not present in the uropygial gland secretion (Jacob and Pomeroy, 1979) and so is probably derived from the epidermis. Mammalian epidermal cells secrete cholesterol, but in the sebaceous glands synthesis stops before cholesterol is formed (Wrench *et al.*, 1980).

In addition to waterproofing the feathers by providing a thin layer of sebum over the plumage, preening helps to prevent the feather keratin from drying out and becoming brittle. Nevertheless, at the end of the period just before the moult the contour feathers begin to look ragged, due to weathering when ultraviolet light weakens the keratin structure, especially in parts lacking protective melanin.

The sebum of the uropygial gland has important anti-microbial properties, as shown by Jacob (1978), and both anti-fungal and anti-bacterial constituents have been isolated. These play an important role in inhibiting growth of micro-organisms in the humid conditions of the plumage, although a few non-pathogenic bacteria thrive on the lipid secretion of the epidermis (Spearman, 1980b). The chances of skin infection are therefore lessened. Perhaps significantly, there are few infectious diseases of the intact skin of birds, and those that do occur mainly affect domestic poultry in the tropics (Lee *et al.*, 1945; Spearman, 1980b).

Birds can probably manage with only one large sebaceous gland because, unlike that of mammals, the whole epidermis is in effect a sebaceous gland, so that the secretory products of both the uropygial gland and epidermis become mixed in the plumage (Wrench *et al.*, 1980).

III. Plumage

Feathering is the most characteristic feature of the class Aves, and this section examines the structure of the plumage in relation to its functions. Feathers (*pennae*) are morphogenetically the most complex appendages produced by the epidermis in any vertebrate. Thus, in order to form the vane of a flight feather, adjacent barbules on the barbs have to hook together precisely, and any deviation in growth coordination results in non-aligned down-like barbs, making flight impossible. Not surprisingly numerous genetic mutants occur in the domestic fowl in which growth of the feathers is disordered (Spearman, 1966, 1971), but in wild birds such mutants do not survive because of both impaired flight and poor thermal insulation.

A. *Feather follicle*

1. *Phylogeny*

The possible evolution of feathers from the horny scales of reptiles is discussed in Chapter 1, by King and King, Vol. 1, p. 19. Although there is no fossil evidence for feathers in any group of extinct reptiles, there is a strong suggestion that feathers were derived from reptilian scales. Thus, the projecting overlapping scales of *Longisquama* (Bakker, 1975), though splayed out, show some similarities to the early developmental stage of the feather, the so-called pinfeather, in which the feather is contained within a sheath of *stratum corneum*. Whilst Regal (1975) suggested that the precursor of feathers could have been reptilian scales with frayed edges, not surprisingly no such scales have been found, since the horny layer of the reptilian scale is homologous to the smooth *stratum corneum* forming the horny sheath of the immature feather and not to the feather itself. The latter is formed by keratinization of the epidermal cells in the *stratum intermedium*. It would therefore be more useful to look for intact pinfeathers in fossil reptiles. The first feathers were probably down feathers required for thermal insulation.

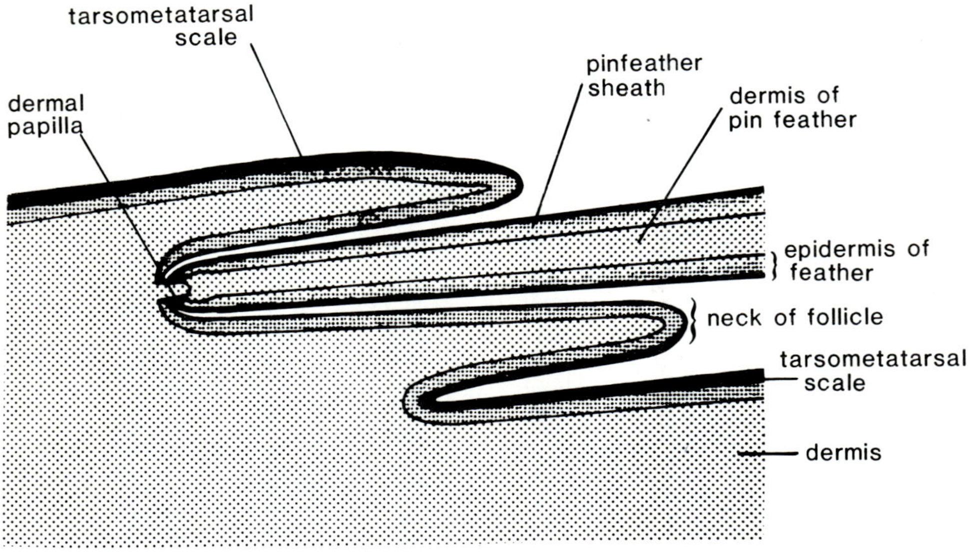

Fig. 1.15. The tarsometatarsal scales of an immature domestic fowl showing the development of an apical pinfeather. The thick *stratum corneum* of the scale is continuous with that of the feather sheath.

Birds retain tarsometatarsal scales on the legs, and these are probably homologous to the scales of reptiles (Spearman, 1966; Brush and Wyld, 1980). Support for the theory that feathers were derived from scales comes from breeds of the domestic fowl in which down feathers grow out from the apices of the tarsometatarsal scales in young chicks (Fig. 1.15), but not from the interscale hinge regions (Rawles, 1960). These feathers are lost when the juvenile plumage is produced except in scales adjacent to the feathered regions of the legs. The proximal portion is then a scale and the distal portion a feather. The feather is produced in a separate follicle at the scale apex, but the horny layer of the scale and the sheath of the pinfeather are continuous through the follicle. This suggests that scales and feathers are linked in morphogenesis, and that feathers were probably derived from scales. In contrast, morphogenetic evidence from living mammals suggests that hairs arose in the hinge regions between scales, as occurs in the tails of rodents (Spearman, 1964).

The inductive effects of the dermis on epidermal morphology have been studied by Sengel (1976) and Sengel *et al.* (1980) in recombinants of chick dermis and lizard epidermis grown on the chorioallantoic membrane of the chick egg to overcome the problem of immunological rejection. In these experiments the dorsal dermis taken from a normally-feathered site of the late chick embryo caused the normal scale buds of the lizard epidermis to form in a hexagonal array typical of chick feather buds. Furthermore, when chick tarsometatarsal dermis was combined with the lizard epidermis, the scale buds were arrayed like chick leg scales. Again, this circumstantial evidence supports the theory that feathers were evolved from reptilian scales.

2. *Embryonic development*

Feathers are tough, keratinized cellular derivatives of the epidermis and are formed in follicles that penetrate into the deep dermis. Each feather germ arises

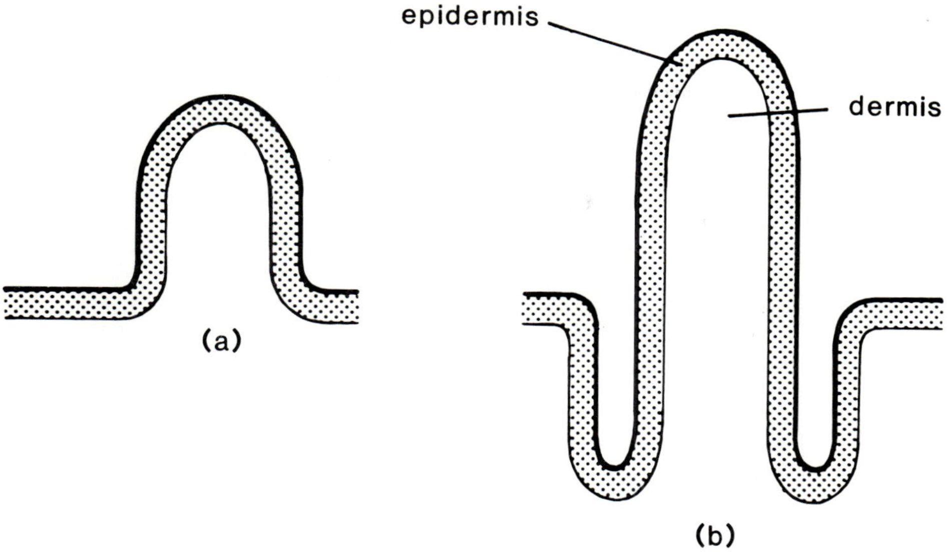

Fig. 1.16. The feather follicles in the embryo arise (a) by an outpushing of the epidermis and dermis, and (b) by an invagination of the base into the dermis.

during embryonic development as an outpushing of the epidermis enclosing a core of dermis (Fig. 1.16) (Espinasse, 1939; Spearman, 1966). Later the base invaginates into the dermis to form the follicle. When the follicle has reached its maximum length, cell division, which is at first active throughout the developing follicular epidermis, becomes localised to a collar at the base of the follicle, and this constitutes a germinal region of dividing cells responsible for further feather growth. Intermediate cells forming the barbs, barbules and rachis undergo keratinization, but the cells between them normally do not. Failing to form keratin, they die and are autolysed. In order to produce the complex feather structure, the pattern of the feather must be genetically imprinted on the follicular epidermis so that only certain cells are keratinized (Stettenheim, 1972).

3. *Adult follicle*

The mature follicle (*folliculus*) is a cylindrical pit (Fig. 1.17) closely fitting the *calamus*. It consists of a neck (*collum*), a cavity (*cavitas*) in which the calamus lies, and a wall (*paries*) formed by dermis and horny epidermis.

During feather growth the central core of dermis in the follicle is cut off from the rest of the dermis by a succession of pulp caps, each pulp cap (*galerus pulposus*) being derived by ingrowth across the collagenous pulp of follicular epidermal cells which become keratinized. The connective tissue which they cut off dries up, leaving cavities surrounded by the keratinized calamus (Lucas and Stettenheim, 1972). After the feather erupts the cavity of the calamus opens to the exterior through the distal umbilicus. When a growing feather is plucked, the soft pulp in the lower part of the calamus is seen as a projecting papilla on the skin surface, but in a feather about to be moulted the cavity of the calamus is filled with air and dried collagen. Enclosing the dermal core is the follicular epidermis, the basal layer of which is relatively inactive during the later stages of feather growth except in the formation of the pulp caps. The feather-formative region of

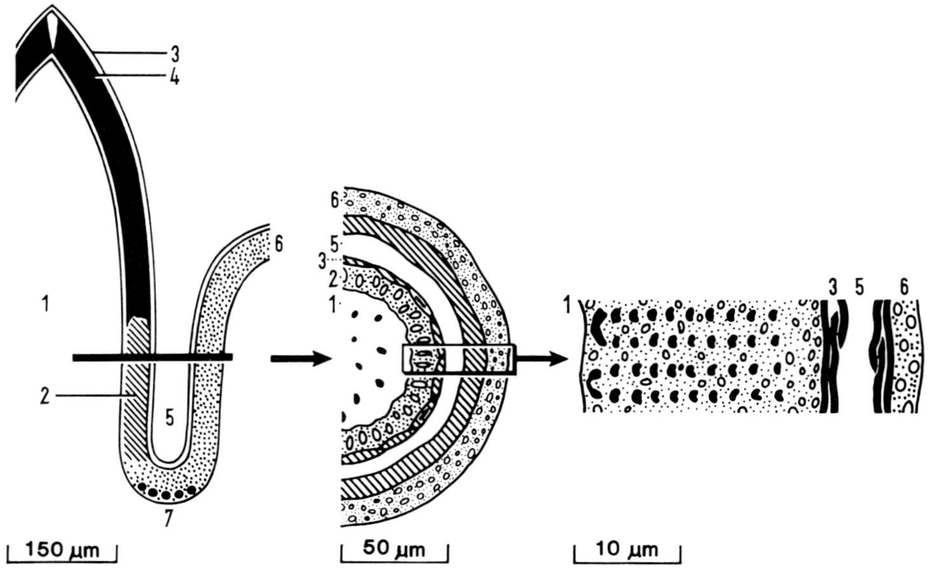

Fig. 1.17. Vertical and horizontal sections through a growing feather follicle. 1, Dermal core; 2, RNA and SH zone; 3, feather sheath; 4, feather; 5, invagination space; 6, outer sheath; 7, matrix. The matrix cells (7) in the collar constitute the germinal region of feather growth. From Cane and Spearman (1967).

the follicular intermediate layer is several cells in depth, and changes in these cells during feather formation will be described later (Section III,A,4). Surrounding the intermediate layer is the *stratum corneum* of the feather papilla, which constitutes the horny feather sheath. Again, in the early stages of growth this is formed from the underlying basal layer, but during feather growth it is derived, like the feather, by lateral division from the collar. The derivation of epidermis by lateral cell proliferation is rare in vertebrates except during rapid growth in young animals and in wound healing. The feather while still in its sheath is called the pinfeather. Later, probably because of the differential growth of sheath and feather germinal cells, it splits open to reveal the new feather.

In the early stages of follicular growth the wall of the follicle is formed from the underlying basal layer, and at this stage the cavity of the follicle is most apparent and can be probed (Lucas and Stettenheim, 1972). Later increase in follicular length is due to proliferation from the collar (Fig. 1.18), and probably for this reason the epidermis of the follicle and the feather are not always separated, strands of cells crossing from one horny layer to the other (Stettenheim, 1972). This is to be expected if the two horny layers are derived from adjacent germinal cells of the collar. In consequence, in the later stages of growth evidence for the original invagination is largely obliterated.

The kinetics of feather growth have been studied autoradiographically. In the germinal region of the collar the full thickness of the epidermis is composed of dividing cells. During feather growth, movement of cells by pressure of proliferation is into the concentric layers of the cylindrical follicle. At the end of feather growth, upward movement of cells ceases and cells from the basal layer migrate inwards forming caps in the dermal core.

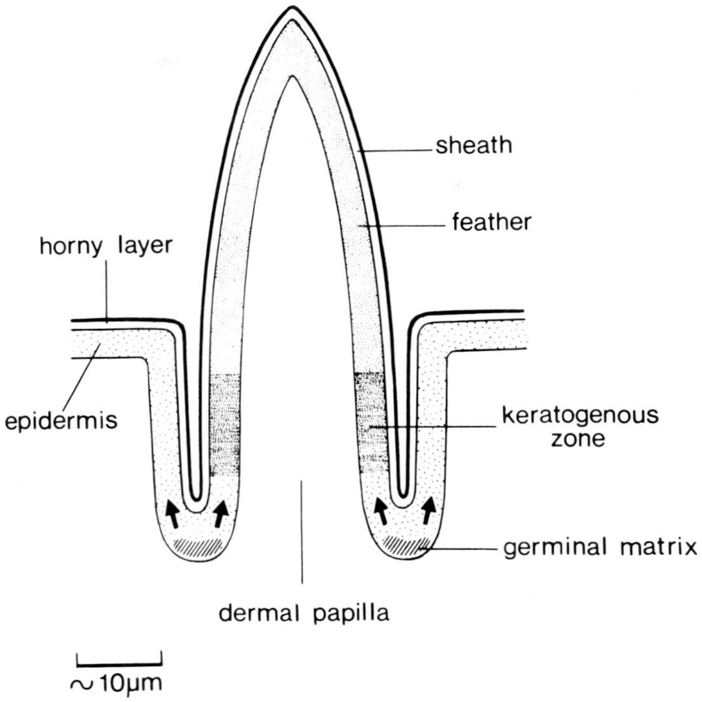

Fig. 1.18. Vertical section through a growing pinfeather showing the direction of cell proliferation from the germinal matrix of the collar. From Spearman (1973).

4. *Feather formation*

As cells are pushed upwards from the germinal matrix of the collar they cease to divide, but they undergo intense protein synthesis as well as increasing in size and changing in shape. In consequence, the cells of the formative layer of the feather are very rich in RNA (Fig. 1.17) and contain numerous polyribosomes (Cane and Spearman, 1967; Spearman, 1971; Stettenheim, 1972). Later the feather cells show an increase in the thiol SH-containing amino acid cystine (Cane and Spearman, 1967), which is subsequently oxidised to the diamino acid cystine in the formation of keratin disulphide bonds. This contrasts with the hair follicle, in which RNA synthesis ends sometime before keratin bonding is completed and the remaining RNA is broken down enzymatically prior to hardening of the hair (Spearman, 1977b,c). In a growing feather, intense protein synthesis continues right up to the time of keratin bonding and cell death, and the keratinized feather retains some RNA.

The morphogenesis of a feather involves changes in both the cell shape and the distribution of the desmosomal junctional complexes. These complexes between feather cells and non-feather cells break down, probably because the un-keratinized cells are autolysed by hydrolytic enzymes released from the lysosomes. In feather cells the hydrolytic enzymes break down organelles, but the cell structure is supported by a keratin framework so that it remains intact (Spearman, 1966; Lucas and Stettenheim, 1972). There is, however, less complete cytolysis than in hair formation. The morphogenetic changes occur in sequence. First, the barbs are laid down mid-dorsally in the form of barb ridges (Figs. 1.19, 20). These later extend downwards and ventrally to join the rachis ridge in the mid-ventral region. New barbs are laid down in sequence from the feather tip to its base near the germ, so that by the time distal barbs have keratinized, proximal barb regions within the follicle still contain living cells (Lillie, 1942; Bell and Thathachari, 1963). This suggests that changes in each parallel barb ridge may organise changes in the succeeding barb regions. Eventually all the barbs are formed and the barbules appear as cells which project laterally from the barbs. It is possible that, very occasionally, individual cells between the barbs keratinize to form a powder on the mature feather which is discussed later (see Section III,B,1,d). After the formation of the barbs and rachis, the remaining intermediate

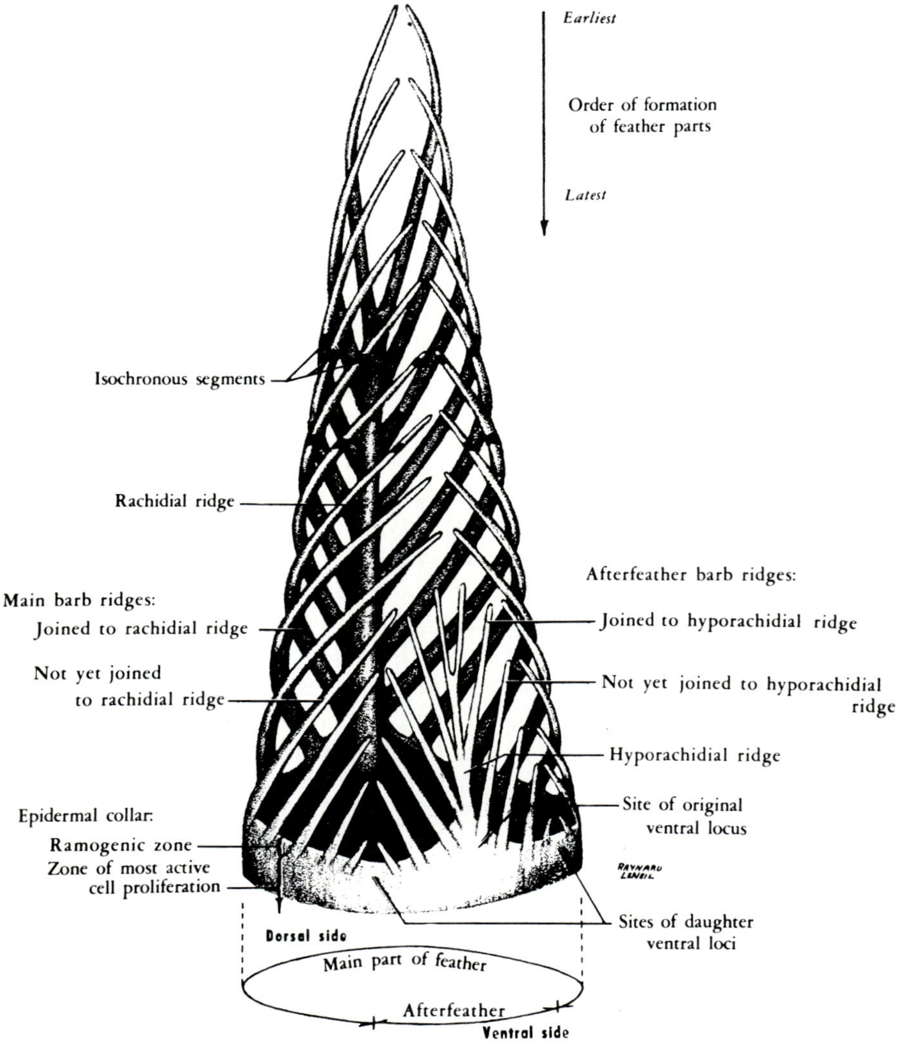

Fig. 1.19. Model of a growing contour feather showing the arrangement of the barbs. The sheath and the pulp have been left out. Isochronous segments are portions of the rachis and the barbs that were formed at the same time. From Lucas and Stettenheim (1972).

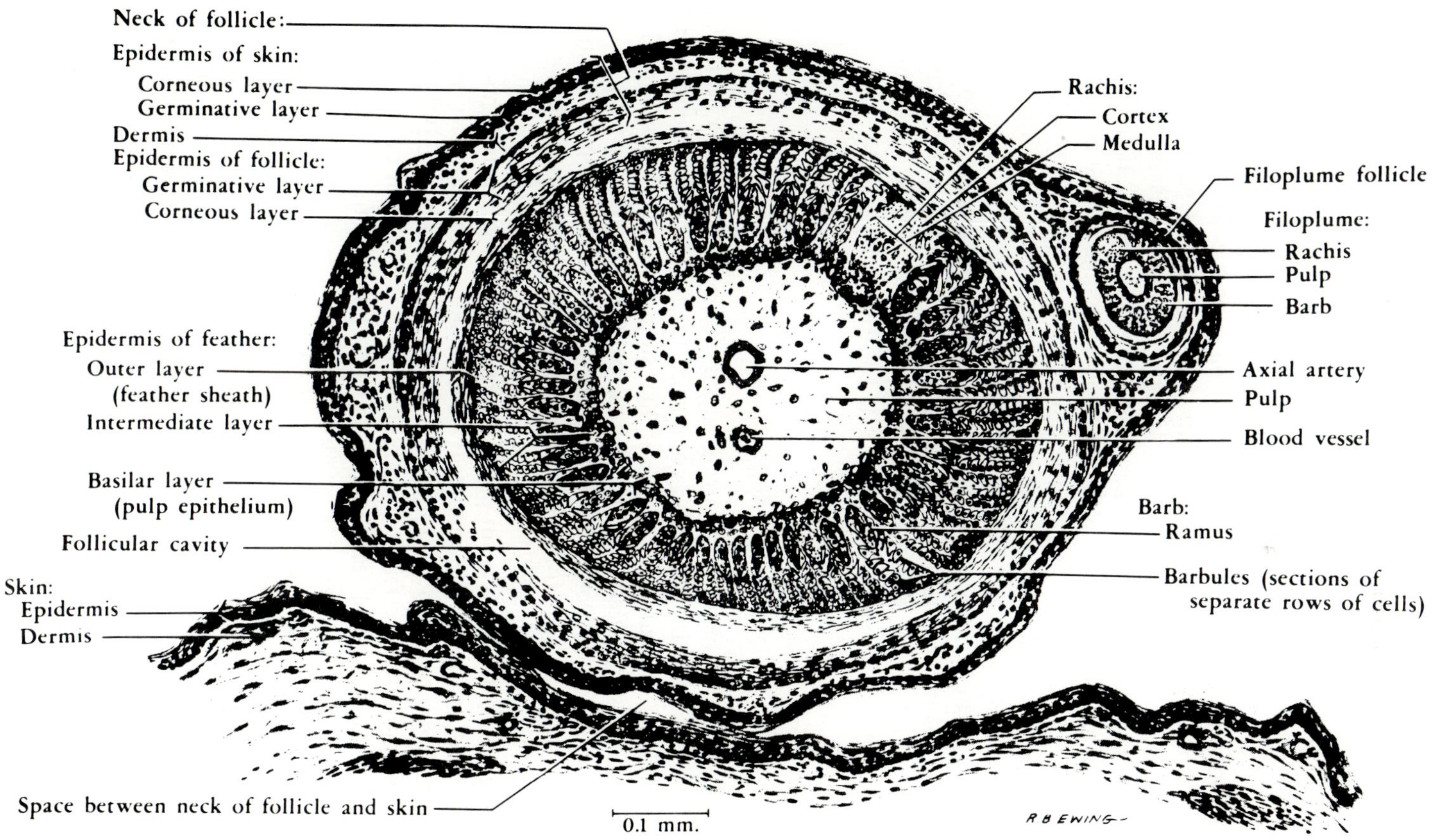

Fig. 1.20. Transverse section through regenerating chicken contour and filoplume feathers. Both feathers have completed about a quarter of their growth. The contour feather is cut just below the neck of the follicle. From Lucas and Stettenheim (1972).

cells in the feather-formative zone keratinize to form the tubular calamus, which remains in the follicle and anchors the feather in place until it is moulted. After the feather sheath splits open the barbs unfurl and, in contour feathers, the barbules on adjacent barbs interlock. In certain genetic mutants of the fowl the barbules fail to form properly (Jones and Morgan, 1956; Krimm, 1960; Spearman, 1971).

In the early stages of growth the dermal core of the follicle contains fibroblasts, collagen and blood capillaries. Around the base of the follicle, dermal cells, which are probably fibroblasts, are grouped as the dermal papilla. These cells have an inductive influence on the epidermal germinal cells of the collar and determine both follicle formation and growth (Sengel, 1976). If the cells of the dermal papilla are destroyed the follicle atrophies, whilst if they are transplanted into the dermis of a featherless area a new follicle is formed by the epidermis (Cohen, 1965). The inductive mechanism resembles that of the dermal papilla of the hair follicle (Spearman, 1973).

At the end of the feather growth the follicular pulp is occluded by the successive pulp caps in the calamus and later by ingrowth of the collar. Dermal cells in the pulp either die or migrate downwards into the dermis, while the blood capillaries collapse and shrivel.

5. *Cyclical changes*

The feather follicles of birds go through cycles of growth (anagen), followed by a rest period (telogen) in which the completed feather remains anchored in the follicle and the germinal cells of the collar stop dividing (Spearman, 1973). Renewed anagen is signalled by activity in the cells of both the dermal papilla and the germinal matrix. In contrast to hair growth, in which there is a stage of follicular atrophy (catagen) following anagen, no such degenerative stage is seen in the feather follicle (Fig. 1.21).

The telogen feather remains in place for many months. During this time the calamus is held in the follicle mainly by the tension of the non-striated feather muscles which run between the follicles (Lucas and Stettenheim, 1972; Hodges, 1974). In the condition of fright moult in chickens, relaxation of these muscles results in premature loss of feathers (Lucas and Stettenheim, 1972).

At the start of the new anagen period the germinal cells of the collar once again commence division immediately beneath the old feather, in contrast to a hair follicle in which renewed hair growth occurs at the side of the old hair. This is a major disadvantage in plumage replacement, as it results in the new pinfeather growing up underneath the old feather, which is loosened and finally pushed out in the process (Fig. 1.21) (Watson, 1963a,b). Moult in birds is therefore a passive process dependent on renewed anagen in the next feather generation. However, the old feather attached to the top of the new feather sheath is lost long before growth of its replacement is completed. In consequence, if all the feathers in a region are moulted simultaneously, temporary baldness occurs, which cannot happen in hair growth. Normally, feather moult occurs over a much more prolonged period than hair moult, so that neighbouring follicles do not moult at the same time (Palmer, 1972) and baldness does not occur. Nevertheless, in

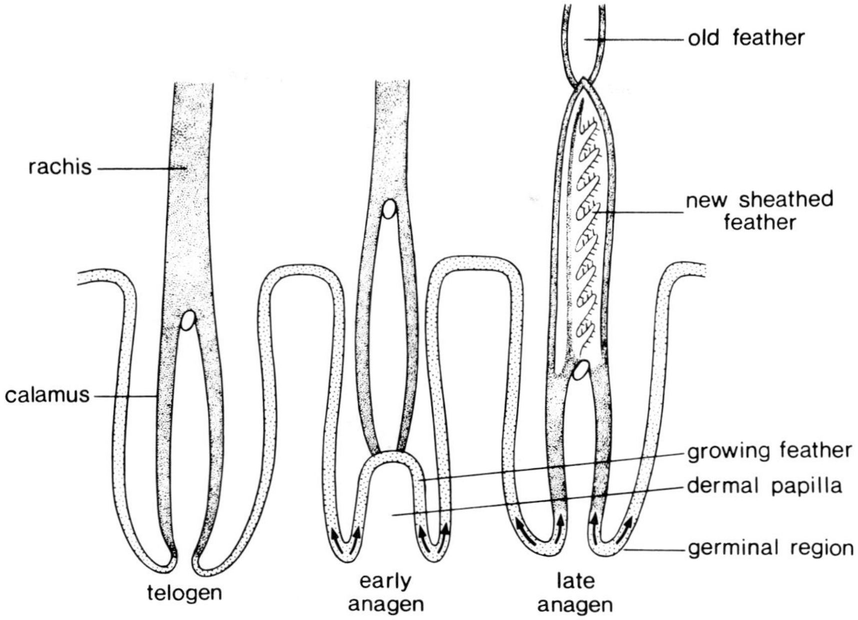

Fig. 1.21. Stages of feather follicular growth and moult. From Spearman (1973).

flamingos (Phoenicopteridae), ducks (Anatidae) and some other water birds, all the wing flight feathers are moulted at the same time so that the birds are temporarily unable to fly. Similarly when diving petrels (Pelecanoididae) moult their flight feathers, they become temporarily entirely aquatic, using their wings as paddles like penguins (Spheniscidae) for underwater swimming. Indeed, a derivation of penguins from petrel-like birds has been suggested.

B. Structure of feathers

1. Morphological appearance

There are seven basic types of feather: contour feather, semiplume, down feather, powder down feather, hypopenna, filoplume and bristle.

(a) Contour feathers. The surface plumage of adult birds is made up of contour feathers (*pennae contourae*) (Chandler, 1916; Hosker, 1936). These are characterised by a long central, relatively rigid stalk or *rachis,* from either side of which arise a series of parallel branches or barbs (*barbae*) forming a sheet or vane (*vexillum*) (Figs. 1.22, 23). The junction of the rachis and calamus is marked by a pit, the distal umbilicus (*umbilicus distalis*). The rachis and calamus together form the *scapus*. The opening at the embedded tip of the calamus is the proximal umbilicus (*umbilicalis proximalis*). A detailed terminology has been provided for the barbs. The shaft of each barb is flattened, forming a dorsal and a ventral crest (*crista dorsalis, crista ventralis*) (Fig. 1.24). In some species the ventral crest is expanded as a shelf, the *tegmen,* from the free edge of which arise many *villi.* Close to the attachment of the barb to the rachis are proximal and distal ridges

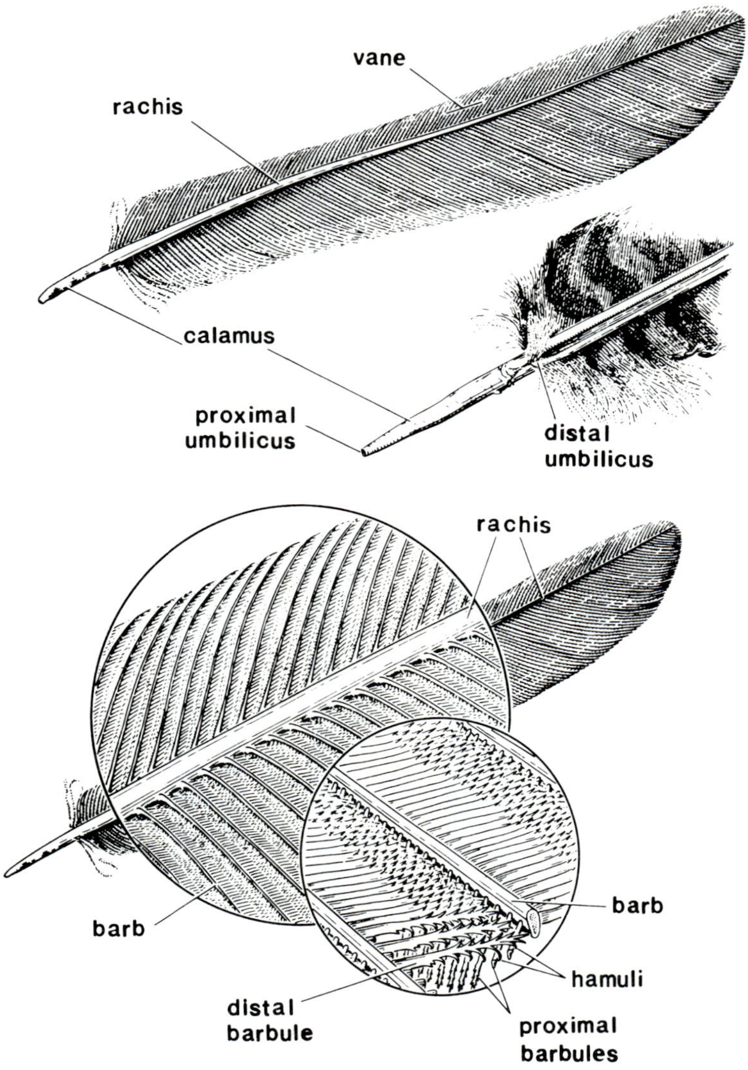

Fig. 1.22. The structure of a flight feather. From "Fundamentals of Ornithology," Van Tyne and Berger. Copyright © 1971 by John Wiley & Sons, Inc. Reprinted by permission of John Wiley & Sons, Inc.

(*ruga proximalis, ruga distalis*). The base of each barb is known as the petiole (*petiolus*), and there may be a notch (*incisura barbae*) immediately distal to it. Projecting from each side of the barb are barbules (*barbulae*) forming small vanes or vanules (Figs. 1.22, 23, 24). The vanule nearest the base of the feather is the proximal vanule (*vexillum barbae proximalis*), whilst that nearest the free tip or *apex* of the feather is the distal vanule (*vexillum barbae distalis*). The distal part of the distal barbule, the *pennula,* carries processes, the dorsal and ventral cilia (*cilium dorsale, cilium ventrale*) with hooklets (*hamuli*). These interlock with the dorsal flange (*arcus dorsalis*) on the dorsal edge of the base (*basis*) of the proximal barbule (Figs. 1.25, 26). A number of different shapes and forms of barbule have been described including pennaceous, plumaceous, a friction type, rachidial barbules and a stylet type.

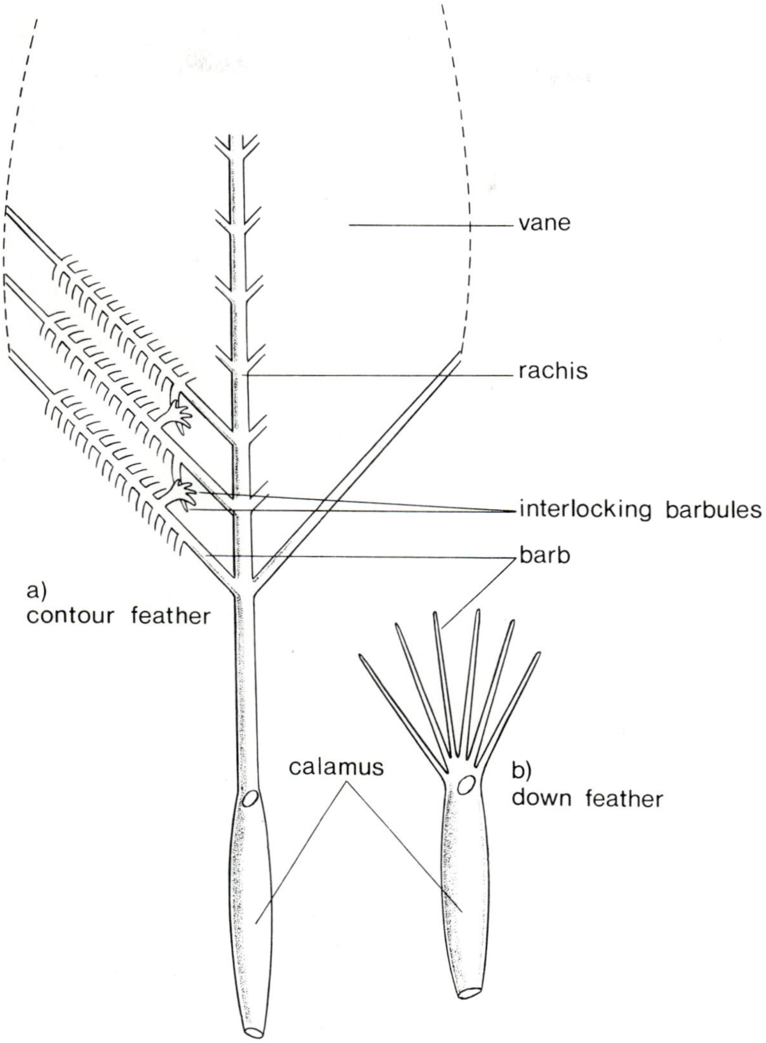

Fig. 1.23. (a) Contour feather and (b) chick down feather. From Spearman (1973).

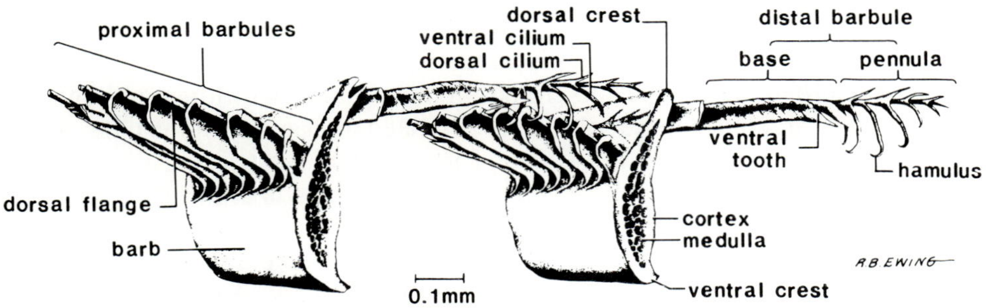

Fig. 1.24. Portions of two barbs from the pennaceous part of the vane of a chicken remix showing the interlocking of proximal and distal barbules. From Lucas and Stettenheim (1972).

Fig. 1.25. Scanning electron micrograph of a contour feather of a Common Murre (*Uria aalge*) showing the rachis (R), the barbs (B) and the interlocking barbules (H). ×110. Reproduced by courtesy of Dr. J. V. Beer.

The vane structure just described is most fully developed in the flight feathers, including the remiges in the wing and the rectrices in the tail. The barbs at the base of the contour feathers are not hooked together (Fig. 1.27), and this part of the feather is known as the fluffy or plumaceous part (*pars plumacea*). The remainder of the feather in which the barbs are united has a firm or pennaceous texture (Fig. 1.27) and forms the pennaceous part (*pars pennacea*).

The structure of the vane results from a well co-ordinated differential growth in the feather collar, so that more cells are proliferated on one side of the tube, which produces a long rachis (Lillie, 1942). The rachis is constructed of a central *medulla* of large spherical cells containing air vacuoles and an outer *cortex* of spindle-shaped cells. The cytoplasm of the cortical cells is filled with keratin. The keratin fibrils are aligned along the length of the spindle, and the spindle-shaped cells themselves are aligned along the length of the rachis (Auber, 1955; Auber and Appleyard, 1955). These cells also contain air vacuoles, melanin and other

Fig. 1.26. Scanning electron micrograph of a contour feather of a Common Murre (*Uria aalge*) showing the hooked distal barbules interlocking with the grooved proximal barbules. ×2000. Reproduced by courtesy of Dr. J. V. Beer.

pigments. The cortex provides the mechanical strength of the feather. On the outside of the feather a single layer of flattened cuticular cells has been described, but this has not been found in all feathers. The barbs are similar in structure to the rachis though smaller in size, but the barbules are made up of only two or three non-vacuolated horny cells.

(b) Semiplumes. Semiplumes (*semiplumae*) are plumaceous feathers which occur along the margins of the contour feather tracts as well as singly within the tracts and in the apteria. The rachis of the semiplumes is longer than the longest barb, and this distinguishes it from the down feather (Fig. 1.28b).

(c) Down feathers. Down feathers (*plumae*) have an entirely plumaceous vane, the rachis being shorter than the longest barb or absent altogether (Figs. 1.23, 28c). As well as occurring in the young chick, they form a thermoregulatory undercoat in the adults of many temperate and arctic species, especially water birds. Down feathers may be evenly distributed throughout the body as in pen-

guins (Spheniscidae) and ducks (Anatidae), unevenly distributed as in owls (Strigidae) and gulls (Laridae), restricted to pterylae as in tinamous (Tinamidae), and confined to apteria as in herons (Ardeidae) and most galliform species.

(d) Powder down feathers. Powder down feathers (*pulviplumae*) occur in certain groups of birds including pigeons (Columbidae), storks (Ciconiidae), herons (Ardeidae), toucans (Ramphastidae), parrots (Psittacidae) and bowerbirds (Ptilomorhynchidae). In some species, such as herons (Ardeidae), the powder down feathers are concentrated in patches. They grow continuously, and the barbs at the tips constantly break off forming a fine talc-like powder that coats the plumage (Welty, 1979). This may be supplemented by cells between the barbs of the developing feathers. The advantages of powder down to all the diverse families of birds in which it occurs is not established. However, it does alter the plumage tint, and herons have been shown to remove fish slime from their plumage after a meal by mopping it up with the powder (Austin, 1961). Powder down distribution is important in taxonomy.

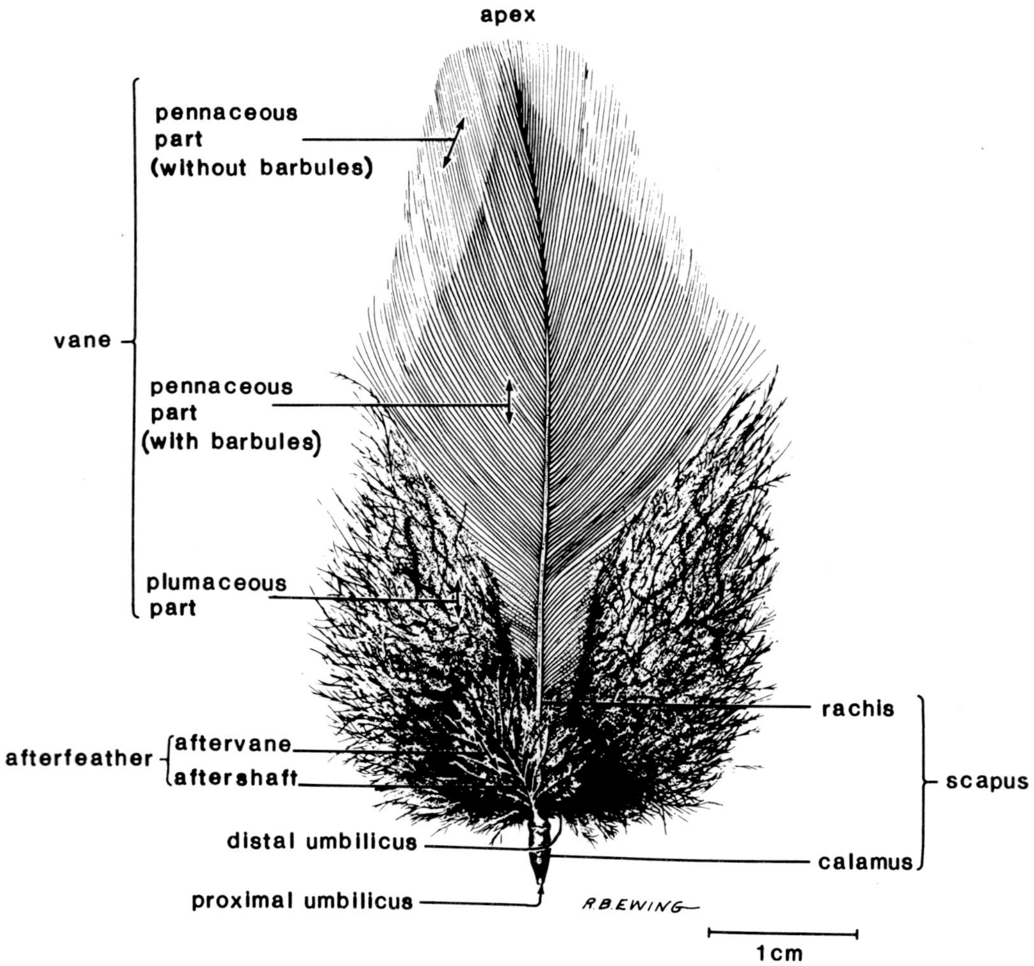

Fig. 1.27. The parts of a chicken contour feather. From Lucas and Stettenheim (1972).

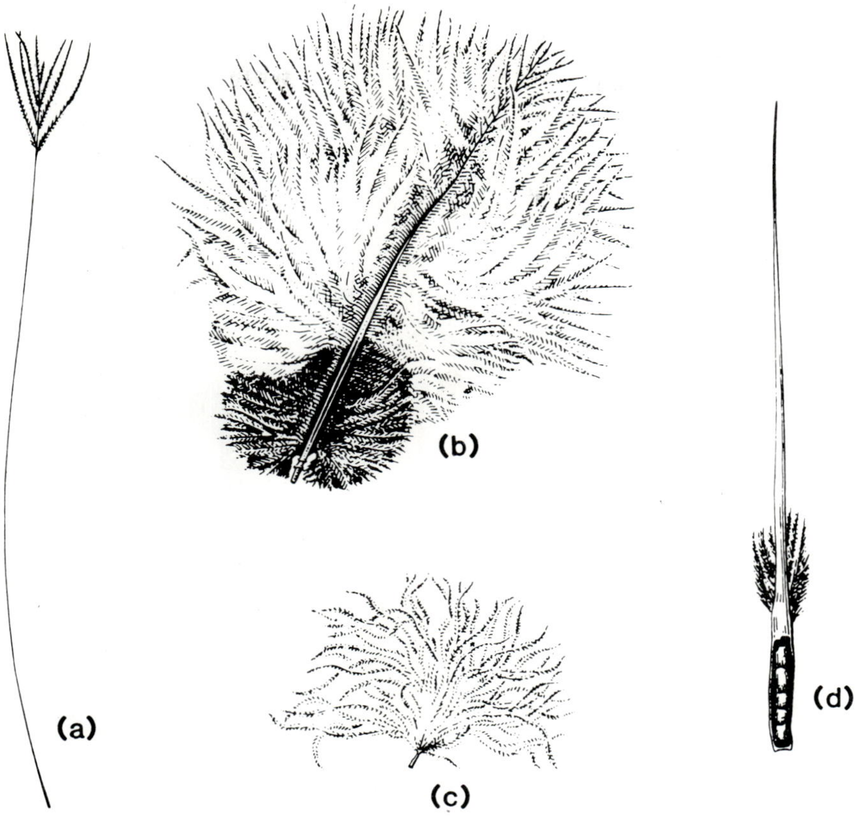

Fig. 1.28. Feathers. (a) Filoplume, *Lophura edwardsi* (×2). (b) Semiplume, a Curassow (×0.75). (c) Down feather, *Grus* sp. (×1). (d) Bristle, *Circus* sp. (×16). From "Fundamentals of Ornithology," Van Tyne and Berger. Copyright © 1971 by John Wiley & Sons, Inc. Reprinted by permission of John Wiley & Sons, Inc.

(e) Hypopennae. Projecting from the distal umbilicus of plumaceous and pennaceous feathers is a small feather with unhooked barbules, the *hypopenna* or afterfeather (Fig. 1.29). In the Emu (*Dromaius novaehollandiae*) and in cassowaries (Casuariidae) the afterfeather is as long as the main feather. The parts of the afterfeather include the *hyporachis,* the *hypovexillum* and the umbilical barbs (*barbae umbilicales*).

(f) Filoplumes. Filoplumes (*filoplumae*) have a long hair-like rachis with a bunch of flexible barbs at the tip (Fig. 1.28a). They are present in all birds except the Ostrich (*Struthio camelus*), the Emu (*Dromaius novaehollandiae*) and cassowaries (Casuariidae). Filoplumes are the only feathers which do not have muscles attached to the follicles. They are thought to be concerned with the sensory input for controlling the position of larger feathers.

(g) Bristles. Bristle feathers (*setae*) have a stiff rachis, and barbs which are generally restricted to the proximal end (Fig. 1.28d). Semibristles (*semisetae*) have barbs along most of the length of the rachis. In the majority of birds bristles occur around the eyes (*cilia palpebrarum*). *Setae nariales* are bristles situated in the loral region of some birds, such as owls (Strigidae) and barbets (Capitonidae).

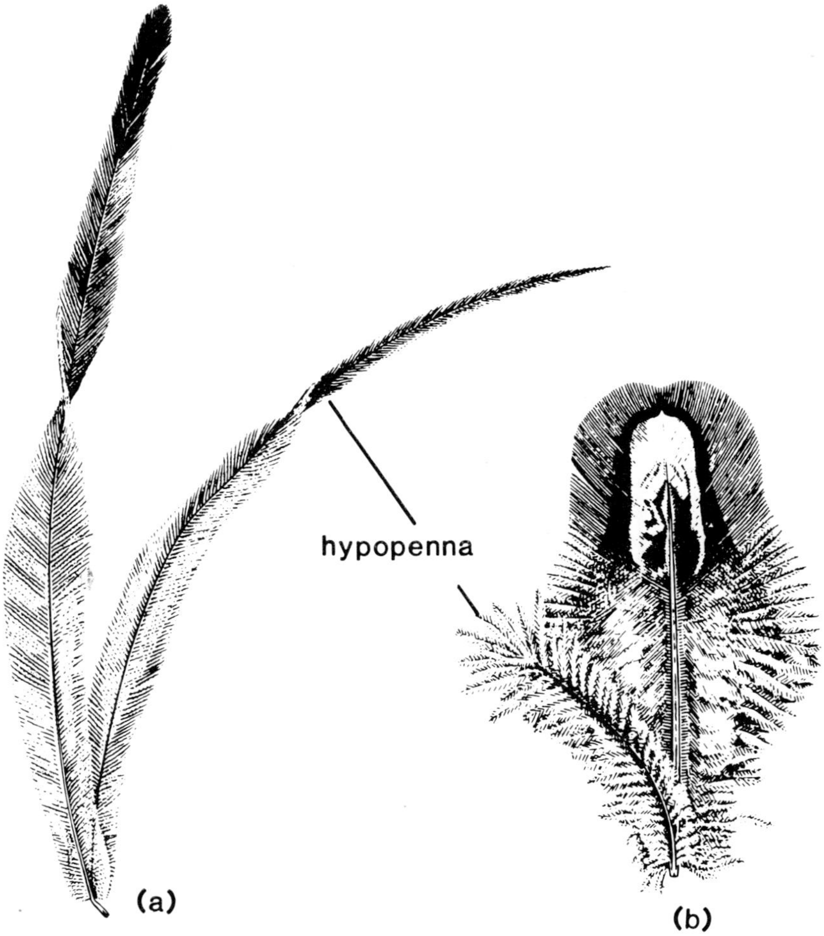

Fig. 1.29. Feathers will a well-developed hypopenna. (a) Emu (*Dromaius novaehollandiae*) (×0.75). (b) *Phasianus* sp., a back feather (×1). From "Fundamentals of Ornithology," Van Tyne and Berger. Copyright © 1971 by John Wiley & Sons, Inc. Reprinted by permission of John Wiley & Sons, Inc.

2. *Keratins*

Feather cells contain a mixture of proteins (Fraser *et al.*, 1972; Brush, 1974, 1978) which are bonded together at the time of cell death to form the durable keratin complex, which is resistant to hydrolytic enzymes and bacterial decay. Feathers possess the pleated form of keratin, which is not found in mammals but occurs in the squamate reptilian scales. The pleated sheets of feather keratin are twisted to form β filaments. In feathers there do not appear to be the separate fibrous and matrix proteins which occur in hair keratins. In feather keratins a cystine-rich zone occurs in the peripheral region of each filament and is sometimes regarded as a cystine-rich matrix, but it is joined to the pleated sheets, probably as side chains (Fraser and MacRae, 1980).

Fibrils, some 4 nm in diameter and made up of several parallel keratin filaments, are aligned along the axis of the feather cell cytoplasm. These fibril bundles are bound to each other through their side chains. The heterogeneous proteins can be separated by procedures involving reduction of disulphide bonds, and have been found to have a common N-terminal amino acid, acetyl serine (Brush, 1978, 1980b).

The mixture of proteins in feather keratins varies from site to site in the same species. There are also species differences in the proteins, which are useful in taxonomy (Brush, 1980a,b). The mixture of filamentous proteins and randomly-orientated side chains provides a composite structure, which in engineering terms is known to provide great strength with elasticity (Fraser *et al.*, 1972), as in the man-made composite material, fibre glass/resin.

Feathers differ considerably in their mechanical properties from hairs, and the α fibrils in hairs and β fibrils in feathers are mainly responsible for these differences. Astbury and Marwick (1932) found that hairs and feathers differed in tensile properties, the latter being much less extensible before rupture of the keratin than hairs. Contour feathers have to be more rigid than the hairs of the mammalian coat, because otherwise it would be impossible to maintain the form of the vane, and flight could not have evolved.

C. Pterylae and apteria

The beak, eyes and feet are generally free of feathers. Plumage is also usually absent in the tarsometatarsal region, as in the domestic fowl and passerine species, although it is feathered in many birds of prey and some grouse (Tetraoninae). The head and neck are bare in vultures (Cathartidae, Accipitridae) and the Marabou (*Leptoptilos crumeniferus*), which plunge their heads into carrion in order to feed, as well as in the Ostrich (*Struthio camelus*) and the Sacred Ibis (*Threskiornis aethiopicus*), in which the head and neck are used to release body heat in thermoregulation.

In the plumage-covered regions of most birds the contour feathers are confined to discrete tracts or *pterylae* which are separated by extensive featherless spaces, *apteria* (Lucas and Stettenheim, 1972) (Fig. 1.30). Nevertheless these bare areas are always covered by the contour feathers of adjacent pterylae. Whilst an apterium can have no contour feathers, semiplumes are usually present and sometimes there are also down feathers. The Ostrich (*Struthio camelus*) and other ratite birds lack apteria, and the same is true possibly of some marine species (Lucas and Stettenheim, 1972).

The feather follicles do not all develop in each pteryla until some time after hatching in many species, so that in the chicks the pterylae are relatively smaller than in the adults and the apteria larger. In contrast, only a few species show follicular regression as in the Sacred Ibis (*Threskiornis aethiopicus*), in which feathers are gradually lost from the head and neck, the skin of these regions in the adult bird being bare and black.

The number of pterylae in birds varies. Nearly 70 have been identified in the domestic fowl (Lucas and Stettenheim, 1972), whilst those in passerines were defined by Clench (1970). Separate names are necessary because of the physio-

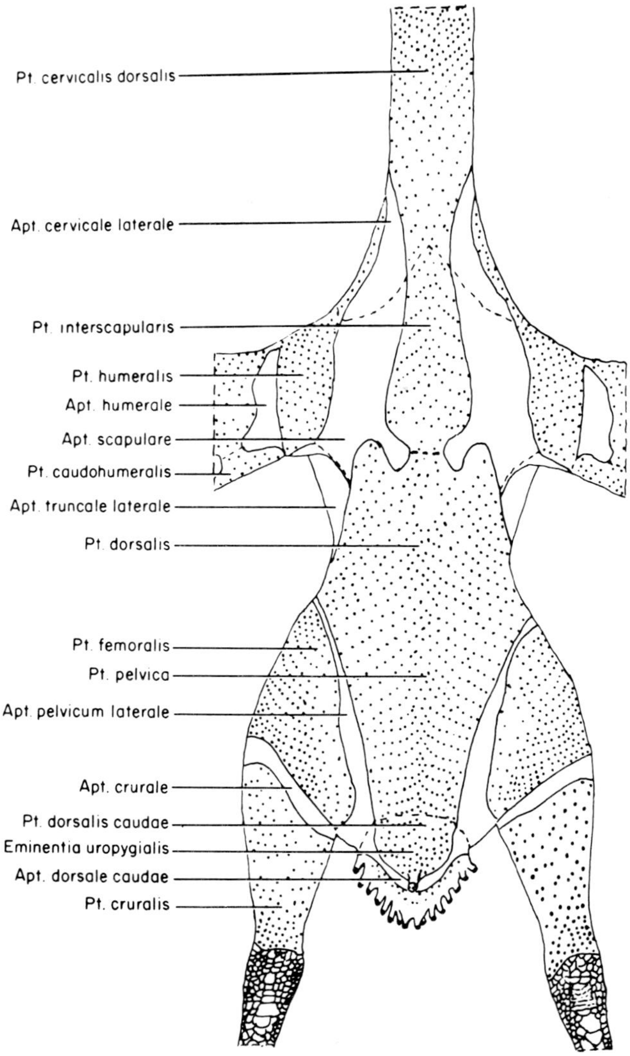

Fig. 1.30. Dorsal pterylae and apteria of *Gallus*. Apt., Apterium; Pt., pteryla. From Lucas (1979).

logical importance of pterylae as centres of plumage growth, which is not true of mammalian hair tracts (Voitkevich, 1966). The tracts in birds listed by Lucas (1979) include head (*pterylae capitales*), spinal (*pterylae spinales*), lateral trunk (*pteryla trunci lateralis*), ventral (*pterylae ventrales*), tail (*pterylae caudae*), alar (*pterylae alae*) and pelvic limb (*pterylae membri pelvici*).

Apteria were probably evolved after development of contour feathers and provide an effective covering without overcrowding. If the entire surface were covered by feathers, it would provide unnecessary extra weight in flight. The apteria listed by Lucas (1979) include head (*apteria capitalia*), spinal (*apteria spinalia*), ventral (*apteria ventralia*), lateral (*apteria lateralia*), tail (*apteria caudae*), alar (*apteria alae*) and pelvic limb (*apteria membri pelvici*).

D. Plumage renewal and moult

1. Pattern of moult

Each pteryla is a separate physiological moulting unit, with renewed feather growth commencing in the centre and then spreading to other follicles around it (Voitkevich, 1966). In each species there is a genetically-endowed plan of feather moult, so that the sequences of plumage replacement in different pterylae are bilaterally co-ordinated. However, in some moults certain pterylae may miss a moult, which complicates the picture (Palmer, 1972).

2. Types of moult

After two or three moults of the chick and juvenile plumages, these are replaced by the definitive adult plumage. As each moult is dependent on the growth of a new feather generation (Fig. 1.21), it is named by reference to the next generation of plumage (Palmer, 1972). Thus, the first adult feather generation is termed basic 1 and the moult which precedes it is a pre-basic moult. Whilst there is a general sequence of moult from head to tail, it is more complicated than the hair growth waves of mammals, because each pteryla functions as a separate growth centre.

Feathers over the head, body and tail are frequently replaced in the first basic moult, but loss of wing feathers is often offset in time so that a second moult (alternate 1) and even a third moult (supplemental 1) may occur in other sites before the flight feathers are lost by their basic 1 moult. Consequently, the number of feather generations which take place per year is greater in some sites than in others. This never occurs in mammalian hair growth, in which the same number of follicular generations occurs in all sites throughout the skin (Johnson, 1977a,b; Spearman, 1977b,c). Only when all the feathers on the bird have been replaced is the cycle considered complete. Each moult takes about 2 months. The cycle generally takes a year, but it can be longer.

In many small passerine birds such as the Common Starling (*Sturnus vulgaris*) there is only one moult in each cycle over the entire body, and this is completed in late spring after the breeding season has started. Others, such as the Blackbird (*Turdus merula*), have the main moult in the spring with an alternate generation of body plumage in late summer. A very few birds, notably the Rock Ptarmigan (*Lagopus mutus*), have three complete moults per year associated with changes in plumage coloration, and even fewer birds have four moults. Prior to incubation of the eggs most species moult all the feathers over a patch on the breast of increased vascularity—the incubation patch (*area incubationis*)—so that body heat can be readily transferred to the eggs. This occurs in both cock and hen birds which incubate their eggs. Amongst species which do not develop an incubation patch are cormorants (Phalacrocoracidae) and gannets (Sulidae).

3. Control of plumage replacement

Control of feather growth, and indirectly of moult, has been studied most extensively in the domestic fowl and pigeon (Voitkevich, 1966; Spearman,

1971), in which there is a definite involvement of sex hormones. Oestrogen promotes ovulation but retards feather growth. Testosterone, in contrast, has little effect on feather formation, although gonadectomy in both sexes leads to a prolonged basic moult. Thyroxine initiates growth in most adult feather follicles. However, growth of the adult remiges and the entire plumage in the young chick are independent of hormones (Voitkevich, 1966). With these exceptions, moult in the adult chicken is triggered by the level of thyroxine and oestrogen in the circulation. In turn, production of these hormones is controlled by gonadotrophic hormones from the adenohypophysis. Increasing or decreasing the periods of daylight falling on the retina causes neural stimulation of the hypothalamus, which determines the rate of secretion of hypophysial gonadotrophins (Spearman, 1971). Therefore, both egg production and moult are controlled by the changes in the duration of daylight occurring in spring and autumn. The experimental control of moult has been extensively investigated in the domestic fowl, since moulting and egg production cannot occur together. In battery cages shielded from sunlight, the laying birds receive alternating periods of artificial illumination and darkness, the durations of which are carefully measured. This is designed to limit the amount of moulting and to increase egg production.

A photoperiodic control of the breeding cycle and of skin changes is seen in many vertebrates (Spearman, 1973), but although birds are generally dependent in this way for the initiation of egg laying and the start of the breeding season in both sexes, photoperiodic control of plumage replacement through gonadotrophins, oestrogen and thyroxine is rare in wild birds, except pigeons (Columbidae) and ducks (Anatidae) (Payne, 1972). In the duck, removal of the ovary results in a male type of moult, and in all species in which sex hormonal control occurs there is a difference in the moult pattern in males and females. The cock birds (ducks and chickens) without the inhibitory action of oestrogens nevertheless do not moult continuously, but are possibly triggered into activity by increased levels of circulatory thyroxine. Yet in the male birds of these species the follicles are inhibited by pharmacological doses of oestrogen (Voitkevich, 1966).

The House Sparrow (*Passer domesticus*) is an example of a bird in which plumage replacement is insensitive to oestrogens, gonadectomy producing no effect on moult in either sex. Whilst it is possible that the follicles are sensitive to increased circulatory levels of thyroxine and that a photoperiodic control of moult could therefore still occur, this has not been demonstrated. It is as if the entire plumage were like the wing feathers of a chicken or the plumage in the young chick, which are insensitive to oestrogen. Indeed, the absence of sex hormonal involvement in the control of the immature and juvenile plumage replacement in birds suggests that sensitivity to oestrogen in some species is a late evolutionary development. In mammals, photoperiodic control of hair growth occurs in some species, but many others, such as rats, mice and humans, show no such response, and probably this mechanism evolved several times independently (Spearman, 1977b,c; Johnson, 1977a,b).

The exact control of moult remains to be determined in species such as the House Sparrow (*Passer domesticus*) as well as in immature birds (Payne, 1972), but it is probably due to a local rhythm of growth and rest in the skin with some influence from thyroxine levels. It is possible that follicles are responsive to the thyroid hormone only in certain stages of growth.

Moult of the wing feathers in pigeons (Columbidae) is delayed if nerves to the region are cut (Voitkevich, 1966). This suggests that some trophic action of nerves may ensure the synchrony of feather growth on the two wings, which is essential for flight and could not be achieved by hormonal control. However, severing nerves affects the circulation, and so far the crucial experiment to show that there is a direct action of nerves on the follicles has not been carried out. The work of Singer (1980) indicates that a protein which stimulates epidermal cell division is released in amphibia from both sensory and motor nerve endings, and this is probably also true of birds. Bilateral control of feather growth could therefore be explained in this way. Moulting of feathers in different species was examined extensively by Stresemann and Stresemann (1966). Lowering of ambient temperature and alteration of wet and dry seasons determine moults in some species (Welty, 1979).

IV. Functions of Feathers

A. *Thermal Insulation*

The evolution of homeothermy in birds is discussed in Chapter 1, by King and King, Vol. 1, pp. 12–18. The earliest use of feathers was probably for thermal insulation, and the most important insulatory feather is the down feather. Down occurs in a wide variety of newly-hatched birds including galliform, anseriform, charadriiform, columbiform and passeriform species. Other birds which are naked at hatching but soon grow some down include pelicans (Pelecanidae), frigate birds (Fregatidae) and cormorants (Phalacrocoracidae) (Lucas, 1979). Adult down is well developed in the Anseriformes and other water birds, such as penguins (Spheniscidae). Adult down also occurs in pelecaniform, psittaciform, strigiform and coraciiform species. In contrast, adult down occurs sparsely or is absent in ratites, such as the Ostrich (*Struthio camelus*), and in passerines, pigeons (Columbidae) and hummingbirds (Trochilidae). In species without adult down the function of thermal insulation is taken over by the fluffy hypopennae and by the unlinked proximal barbs of the contour feathers.

In cold weather small perching birds fluff themselves by elevating their contour feathers through the action of the feather muscles, which increases the depth of the thermo-insulatory air. Buntings of the temperate genus *Emberiza* (Andrew, 1956) respond to an increase in body temperature by panting and by flattening their feathers against the body, which reduces the volume of air in the plumage. They also raise their wings and so increase the surface area for heat loss. In tropical and sub-tropical species, such as for example the Ostrich (*Struthio camelus*), the contour feathers can be elevated to break the imbrication and thus increase the loss of air by convection.

The very good insulatory properties of down are demonstrated in the Common Eider (*Somateria mollisima*), which was once brought near to extinction by down collectors for quilts, although it is now protected. The adult birds pluck down from their breast plumage to line their exposed rocky nests, and it is this high-quality nest down which was so eagerly sought, with the result that the rearing of young birds was almost stopped.

TABLE I. **Variations in skin and core temperatures in the Gray Jay (*Perisoreus canadensis*) at different ambient temperatures (°C).**[a]

Ambient	Skin	Core
−50.0	+36.0	+40.0
+10.0	+39.0	+41.0
+20.0	+39.5	+41.5

[a] From Spearman (1980b), after Veghte (1964).

The down feathers trap a layer of air against the body surface, and the imbricated contour feather vanes delay the loss of this warm air by convection. Air has low thermal conductivity and provides efficient insulation so long as exchange of air with the ambient environment is minimised (Spearman, 1980b).

The body temperature of birds has a mean value of 41°C with little variation in the range 40–43°C (Dawson and Hudson, 1970). This is slightly higher than in similar-sized mammals, as shown by telemetry readings (Schallenberg *et al.*, 1974). In the domestic fowl the daytime temperature under the breast feathers is 38.9°C, but only 34°C over the exposed comb and wattles (Freeman, 1971). The variation in body temperatures with different ambient temperatures in the Gray Jay (*Perisoreus canadensis*) (Veghte, 1964), is shown in Table I. Such little variation in the surface temperature of the skin at widely different ambient temperatures demonstrates the high thermo-insulatory value of the plumage, which is much more efficient in this respect than mammalian fur. For instance, chickens are much more tolerant of cold than rabbits of the same size (Horvath *et al.*, 1948).

Dried skins are not suitable for measuring the insulatory value of plumage, because the feathers are not erected by feather muscles as they are in the living bird (Scholander *et al.*, 1950). If the air in plumage is replaced by water, the insulatory value is lost.

B. Tactile sensation

Feathers are important tactile sensory organs, and according to Dreyfuss (1937) the follicular wall is richly supplied with sensory nerve endings. Borodulina (1966) described numerous nerve endings in filoplumes, the hair-like feathers which occur between the contour feathers in some species. Probably the contour feather follicles are also supplied with nerves which are stimulated by movement of the feather rachis. See Chapter 8 of this volume by K.-M. Gottschaldt.

C. Adaptations for flight

1. Feather structure

The flight feathers are specialised contour feathers each of which has a fairly rigid rachis and a well-developed vane on either side. Adjacent feather vanes overlap, the vane lying on top forming the external vane (*vexillum externum*) and that underneath the internal vane (*vexillum internum*). The most important feathers for flight in the wings are the remiges (Fig. 1.31), comprising usually 9–11 primary remiges (*remiges primarii*) and 9–18 or so secondary remiges (*remiges secondarii*). The primary flight feathers arise from the manus, the secondaries from the caudal side of the forearm. In some groups of birds there is a space or *diastema* between the fourth and sixth secondaries. Such wings are termed "diastataxic" in contrast to the "eutaxic" wing in which the diastema is absent. From two to about seven alular remiges (*remiges alulares*) attach to the alular digit and form the so-called bastard wing, which protrudes from the leading edge of the main wing. The alular remiges are important in aerodynamics and are discussed in Section IV,C,3. Distal to the alular remiges, both the leading and trailing edges of the wing as well as the tip are formed by the primary feathers. Since these make up about a third of the wing length, birds such as flamingos (Phoenicopteridae) on ornamental lakes may have them clipped to prevent flight. The secondary feathers form the remainder of the trailing edge of the wing and can also be clipped to reduce the wing area.

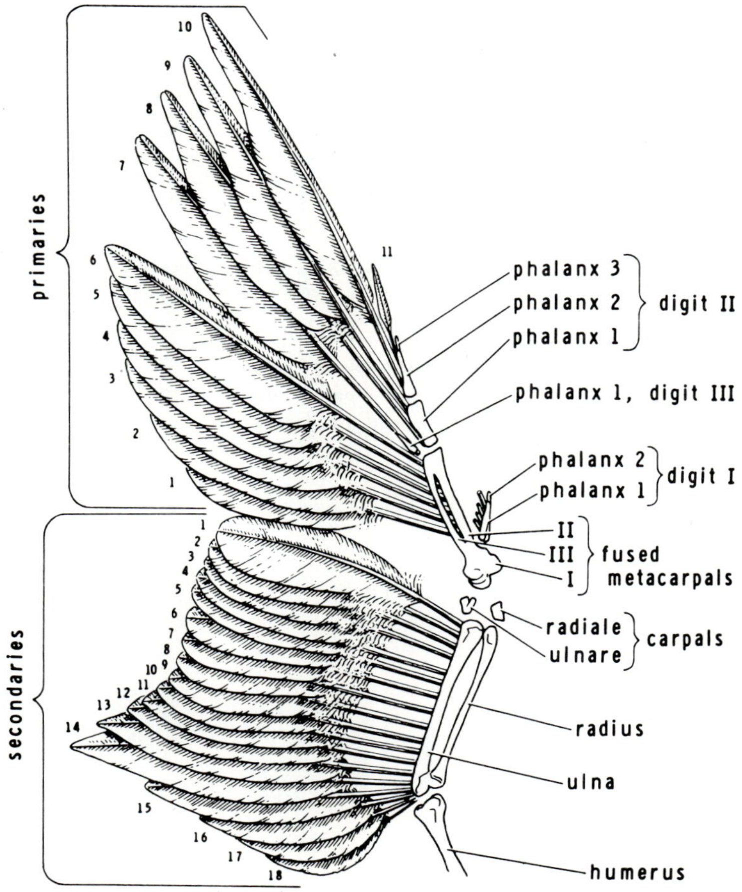

Fig. 1.31. Ventral view of the wing bones and flight feathers. The metacarpus and digits I, II and III correspond to the carpometacarpus and alular, major and minor digits of the *Nomina Anatomica Avium*. From "Fundamentals of Ornithology," Van Tyne and Berger Copyright © 1971 by John Wiley & Sons, Inc. Reprinted by permission of John Wiley & Sons, Inc.

The leading edge of the wing proximal to the alular remiges is formed by rows of small contour feathers called coverts (*tectrices*), and these overlie both the dorsal and ventral surfaces.

The tail feathers or *rectrices* are large contour feathers which can be fanned out by the feather muscles. They have a balancing action in flight and also function as an air brake when landing.

In some species (e.g. *Columba*) part of the vane of the flight feathers of the wing and tail is specialised to form a zone of friction (*zona impedens*) in which the barbules are modified to hold overlapping feathers in place next to one another (Fig. 1.32).

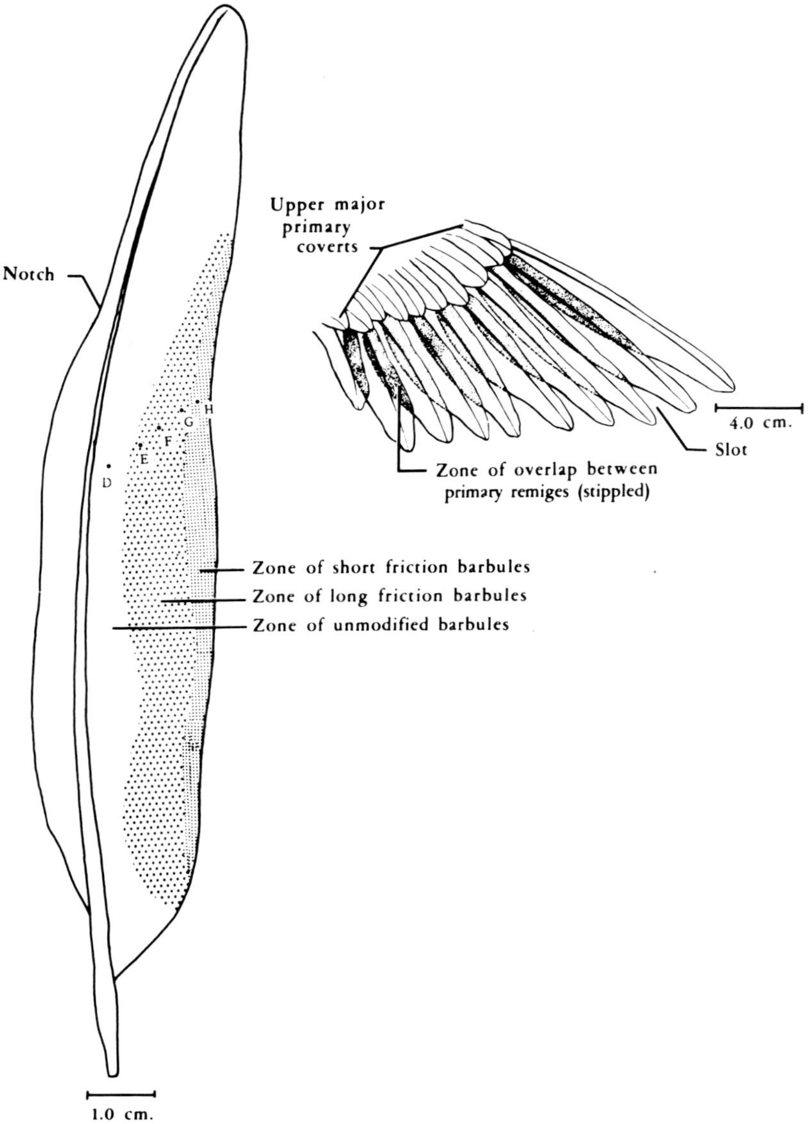

Fig. 1.32. Wing feathers of *Columba*. On the left, primary remix 8. Friction barbules are one of six types of barbules which differ in shape and form. On the right, a dorsal view of the wing tip. From Lucas and Stettenheim (1972).

The vanes of the secondary flight feathers are generally of equal width along their length, and viewed from the ventral side they overlap each other in a step-like manner from the wing tip to the body of the bird. The primary feathers have narrow vanes facing the leading edge of the wing, as do those on the bastard wing. Both sets of feathers sometimes have a notch (*incisura vexilli*) on the distal part of the vane (Fig. 1.32) which is particularly noticeable in hawks, vultures and eagles (Cathartidae, Accipitridae) and in some swans and ducks (Anatidae) (Smith, 1945; Humphrey and Clark, 1964). Birds with short and broad (low aspect ratio) wings such as the Corvidae also have well-developed notches. Aerodynamically these feathers are of great importance, because when the feathers are spread out under pressure of the air, partial separation of the overlapping vanes causes slots to appear at the sites of the indentations (Fig. 1.32). These slots are arranged along the distal leading edge of the wing and function as an anti-stalling device similar to the slots placed on some aeroplane wings.

2. *Functions of the wing feathers in flight*

The essential element which made bird flight possible was the evolution of the feather vane with its interlocking barbules (Brown, 1961, 1963; Barlee, 1964). Birds of the open sky such as swifts (Apodidae) have long, narrow (high aspect ratio) wings, and the wing span is increased by long primary feathers. Woodland birds, which need to be more manoeuvrable to fly between the branches of trees, have shorter, broader (low aspect ratio) wings with shorter primaries. Within this group are the many perching passerine species, pheasants (Phasianidae) and owls (Strigidae). The undersurface of the wings of owls has a velvet-like pile produced by filamentous processes of the barbules, and this muffles the sound of the beating wings when the owl swoops on its prey (Pycraft, 1922). The vanes on the leading edges of the primaries have unlinked barbs, which also helps to reduce sound. "As light as a feather" has been proverbial since the time of Galileo; of course this lightness is partly due to the extended vanes of the feather forming an aerofoil as well as to its mechanical structure, since the amount of strengthening keratin present is just sufficient to combine adequate rigidity with enough flexibility.

The numerous air vacuoles in the feather cells also reduce weight (Auber, 1955; Auber and Appleyard, 1955). For its strength and lightness the plumage-covered wing is more efficient than any man-made aerofoil.

3. *Aerodynamics of the avian wing*

When spread out the wings of most birds are convex on the upper dorsal side, which is desirable in an aerofoil (Storer, 1948; Brown, 1963; Welty, 1979), and the primary feathers can be slightly rotated in their follicles, the importance of which is considered in the paragraph after the next. When a cambered wing is moved through the air at the normal flight angle, the leading edge displaces molecules of air above and beneath it (Ruppell, 1977; Schlichting *et al.*, 1979). As the molecules are deflected over the curved upper surface they are accelerated, which leaves an area of low pressure above the wing. Thus, according to Daniel Bernoulli's theorem, any increase in velocity of air will reduce its pressure. This has the effect of sucking the wing upwards, and it is perhaps not generally realised that this area of reduced pressure above the wing's leading edge is very much more important in providing lift than compression of air beneath the wing, which many people imagine is why a bird or aeroplane stays in the air. In fact, 70% of the lift is provided by the area of low pressure above the wing and only 30% from compression under the wing.

The angle of attack of the wing is defined as the angle between the wing chord (the line from leading to trailing edges) and the flight path. If this angle is increased by lowering the trailing edge, two changes occur which eventually lead to stalling and, if uncorrected, to descent of the aerofoil (Schlichting *et al.*, 1979). The first change is that drag is increased, which results in a slowing of the air speed. Secondly, eddying of air occurs over the trailing edge (Fig. 1.33), and this impedes the passage of air over the leading edge, resulting in deceleration and a loss of the important zone of reduced pressure above the wing. Stalling normally ensues as a result of the reduced air speed and is of course a desirable feature on landing. The likelihood of stalling at a particular angle of attack is greatly reduced by the presence of slots along the distal part of the leading edge and behind the

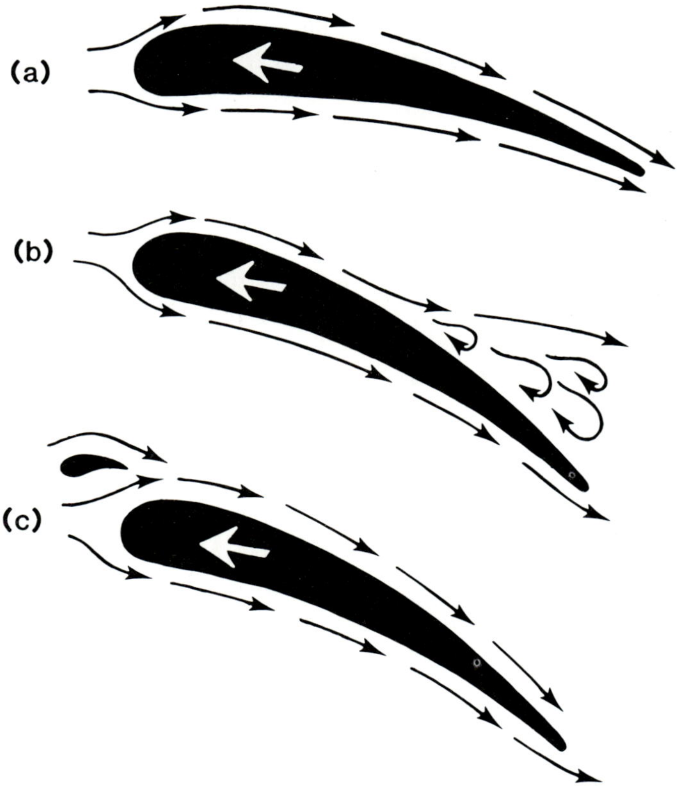

Fig. 1.33. Wing showing deflection of air molcules above and below the leading edge. A zone of low pressure is created above the cambered leading edge in normal horizontal flight (a). During stalling (b), eddying occurs over the trailing edge, impeding flow over the leading edge. In (c) the slot on the leading edge increases the air flow over the wing and reduces eddying.

bastard wing (Fig. 1.33), where stalling is most likely to occur (Graham, 1932; Urry and Urry, 1970). Thus, in the slow-flying but highly manoeuvrable Californian Condor (*Gymnogyps californianus*), multiple slots occupy 40% of the wing area. These slots in both aeroplanes and birds greatly increase the volume of air passing over the cambered wing, either through them or over the leading edge, so that eddying is hindered over the trailing edge (Fig. 1.33) (Handley-Page, 1921). In consequence, stalling occurs at a higher angle of attack than when slots are not present as in the fast-flying Wandering Albatross (*Diomedea exulans*).

On the downstroke of the wing in normal flight, the wing tip has a dangerously high angle of attack. However, stalling is prevented because the primary feathers become slightly separated under the pressure of air and are then rotated upwards and backwards (Brown, 1963), possibly in part because of the feather muscles. As a result each primary feather vane then has a reduced angle of attack and functions as a separate aerofoil. This action of the primaries is shown well in pheasants (Phasianidae) and other game birds.

In order to become airborne and remain in the air, a bird or aircraft has to gain sufficient speed to provide a lift force equal to its weight. In the beating wings of a bird the lift force is largely provided by the coverts and secondaries, whilst forward motion is provided by the primaries at the wing tips.

At high speeds the drag on the wing is proportional to its area times the speed squared, so that with the high speeds of modern aircraft very small wings are desirable. Slow flight, however, requires large wings. The aspect ratio is inversely proportional to drag, and so high aspect ratio wings are useful to birds such as the Wandering Albatross (*Diomedea exulans*) which spend long periods in the air. High aspect ratio wings also have the further advantage of being the best aerofoils for gliding and soaring. The Wandering Albatross has the largest wingspan of any bird but is relatively underpowered and designed as a high-speed sailplane. Although most sea birds have high aspect ratio wings, those of most land species have a low aspect ratio. Exceptions are the swifts (Apodidae), which have the highest aspect ratio wings of any overland flyers (Savile, 1950). In order to maintain high airspeeds in their pursuit of insects, swifts are streamlined with short legs and the wing beats are powered by exceptionally well-developed muscles.

Wing loading varies with different species. Weight increases as the cube of the linear dimensions, but wing area increases as the square. Therefore, large birds such as swans (Anatidae) would be expected to have massive wings. However, in these birds they are relatively small, because though useful for soaring birds with very slow wing beats, large wings are too cumbersome to beat rapidly in normal flight. When wing loading is expressed in kilograms per square meter, the Goldcrest (*Regulus regulus*), a very small bird, has a value of 1, crows (*Corvidae*) 3, some ducks (Anatidae) 10, and swans (Anatidae) 20.

The rapid flapping of wings in flight is a complex process, the proximal part of the wing moving less than the tip, beating downwards and forwards, and then more rapidly upwards and backwards. In hummingbirds (Trochilidae) (Savile, 1950), the hovering process is produced by pointed, uncambered wings without slots, which move mainly in the horizontal plane where they function as variable-pitch propellers. The power required for hovering flight is considerable, so that the pectoral musculature makes up 30% of the body weight.

Small birds take off by simply jumping into the air, but large, heavy birds such as the albatross (Diomedeidae), condor (Cathartidae) and swan (Anatidae) need to taxi for some distance before they become airborne. When birds land they lose speed by increasing the angle of attack and stalling, whilst fanning the tail and directing the rectrices downwards as an air brake. The wings set up far less drag than man-made aerofoils, and the flight musculature is a very efficient machine. One factor in reducing drag is the way flight feathers are tilted on the upstroke of the wing so that air passes between the vanes, but on the downstroke the vanes are held together and air cannot pass through. These features have combined to make birds masters of the air.

D. *Adaptations for diving*

The imbrication of contour feathers lessens drag in diving birds just as it does in flight. Hydrophobic lipids and waxes, derived from the uropygial gland and the epidermis, coat the plumage and help to make it waterproof, which is particularly important in aquatic species. When a duck or swan (Anatidae) sits on the water it is in fact buoyed up by the air in its plumage, and so in diving ducks such as the Common Pochard (*Aythya ferina*), diving is an active process dependent on movements of the webbed feet. Once the bird has finished feeding, it rises to the surface under its own buoyancy. The Tufted Duck (*Aythya fuligula*) keeps its wings closed whilst diving, but another duck, the White-winged Scoter (*Melanitta fusca*), partly spreads its wings with the alulae projecting as balancers. The plumage in penguins (Spheniscidae) is particularly densely packed, which improves both thermal insulation and waterproofing for underwater swimming.

Birds such as the Gannet (*Morus bassanus*) which enter the water from a considerable height have their wings half-closed to lessen drag as they hit the surface. The upper parts of the breast and neck in the Gannet and other pelecaniform species contain subcutaneous diverticula from the cervical air sacs (Groebbels, 1932; King, 1966, 1979), and these cushion the impact of the bird against the water. The Brown Pelican (*Pelecanus occidentalis*) impacts with such force that fish are stunned for up to 2 m beneath the surface and can then be caught more easily.

Large diving birds do not use their wings fully open in swimming under water, because the drag would be too great. The flightless penguins (Spheniscidae) without remiges, and diving petrels (Pelecanoididae) use their wings as paddles and employ the feet for steering. When the Razorbill (*Alca torda*) uses its wings as paddles they are only half open. The Dipper (*Cinclus cinclus*), in contrast, has been filmed using its wings fully open under water (Harrison, 1978).

E. *Display*

Many birds utilise the plumage in courtship, in natural bonding of pairs and in mating, or in territorial display. This is sometimes achieved by distinctive coloration in the cock birds, as in some passerine finches. However, in a number of species some of the feathers are also structurally modified for display purposes, and these display mechanisms are often dependent on erection of the feathers by feather muscles (Smith, 1945; Dorst, 1974). The supreme example of this is seen in the birds of paradise (Paradisaeidae) of New Guinea. In some species the tail coverts in the males have been converted into long, brightly-coloured, lace-like plumes which extend beyond the length of the tail and during display are trailed over the back in a fan-like spread. Some of the back feathers are also plume-like. The rectrices are either elongated filaments or ornate plumes of great length.

Adaptation of the tail feathers is also seen in the Australian lyrebirds (Menuridae), in which the long tail has relatively broad outer lyre-like feathers together with lace-like plumes which are laid over the back. In the Peacock (*Pavo cristatus*) the elongated, brilliantly-coloured tail coverts are without interlocking barbules and so appear ragged. In the male domestic turkey the prominent rectrices are fanned out in display.

Display adornment is generally confined to the tail, since the wing feathers cannot be tampered with so readily in flying birds without affecting aerodynamics. Instances of wing display are seen in the male Standard-winged Nightjar (*Macrodipteryx longipennis*), in which the ninth primary is 0.5 m long, and in the male Pennant-winged Nightjar (*Semieophorus vexillarius*), in which the seventh, eighth and ninth primaries are elongated (Harrison, 1978). In the male King of Saxony Bird of Paradise (*Pteridophora alberti*) two long slender plumes arise from the nape that are more than twice the length of the body. Feathers on the head are erected in some birds such as the Goldcrest (*Regulus regulus*) and the Crested Tit (*Parus cristatus*). In the Great-crested Grebe (*Podiceps cristatus*) both males and females have black ear tufts and chestnut and black frills on the sides of the head associated with an elaborate courtship water dance (Huxley, 1914), which is followed by a mating display. In the Ruff (*Philomachus pugnox*), a wading bird of fenlands, the neck in the males contains enlarged contour feathers. The male cocks-of-the-rock (*Rupicola* spp.), a brilliant red- or orange-coloured genus from tropical America, has a prominent crest reaching from the tip of the beak to the nape which is used in display.

A number of species have erectile crests in members of both sexes which they raise in display when excited or angry. The best example of this is seen in cockatoos (*Cacatua* spp.) and the related Cockatiel (*Nymphicus hollandicus*), but it is also found in the Hoopoe (*Upupa epops*), which has a black-tipped crest. The Lapwing (*Vanellus* spp.), a genus of plover, includes a number of species from Europe and Africa in which the slender erectile crest is larger in cock birds than in the hens.

F. Coloration

1. Plumage and skin pigments

Black, brown and yellow coloration is produced by the pigment melanin, which is synthesised in the melanocytes. These cells, which have long dendritic processes, lie just beneath the epidermal basal layer in many mammals but enter between the intermediate cells in birds. They are derived from the embryonic neural crest. The pigment is synthesised and deposited in cytoplasmic organelles, the melanosomes, which pass into the dendrites as they mature. In mammals, it has been shown that the epidermal basal cells phagocytose the tips of the dendrites, so that the pigmented melanosomes enter the epidermal cytoplasm where they become melanin granules (Spearman, 1982a). The similar histological appearances of these cells in birds, and the deposition of melanin granules in the feather follicle cells, suggest that the same process occurs in birds. Melanin exists in two forms: black to brown eumelanin, and yellow to reddish-brown phaeomelanin (Riley, 1977). Eumelanin is a stable and insoluble derivative of the amino acid tyrosine, being a homopolymer of indole-5,6-quinone (Mason, 1967; Nicolaus, 1968). Phaeomelanins are modified from eumelanins and contain the sulphydryl amino acid cystine as well as some other substances (Prota, 1980). Avian melanins have been studied mainly in the domestic fowl (Rawles, 1960; Lucas and Stettenheim, 1972). Melanin granules are deposited in the feather cells as well as in the surface epidermis of exposed sites such as the feet, tarsometatarsus and beak, and of the neck in the Sacred Ibis (*Threskiornis aethiopicus*). The unfeathered comb and wattles in the domestic fowl are unpigmented.

Carotenoids include a variety of different red- to yellow-coloured pigments, carotenes and xanthophils, which occur dissolved in fat globules in the feather cells (Brush, 1978, 1981). Carotenes are related to vitamin A (retinol) and confer the yellow colour to animal fats as well as the red colour to carrot juice. Carotenoids enter the body in the diet and are first metabolised and chemically altered, often to a different colour, before being taken up by the follicle cells (Fox and Vevers, 1960). Examples of these colours are seen in the yellow of the Canary (*Serinus canaria*), the bright red and yellow of the Eurasian Goldfinch (*Carduelis cardeulis*), and the brilliant orange or red of cock-of-the-rock species (Cotingidae). The altered pigments are deposited in the plumage of the different pterylae, and in the absence of the dietary carotenoids the feathers become drab. Thus, flamingos (Phoenicopteridae) and the Scarlet Ibis (*Eudocimus ruber*) have to be fed in captivity on special diets in the absence of their natural foods, if they are to maintain their brilliant plumage colours. With other birds, such as the Canary (*Serinus canaria*), the precursors for pigment formation are more widely available in seeds, and so special diets are not required. Nevertheless, by adding red cayenne pepper to the diet the plumage colours can be intensified. Carotenoids are responsible for the bright yellow colour of the feet and tarsometatarsus of the domestic Aylesbury duck.

Avian carotenoid metabolism has been studied in most detail in passerine finches, flamingos (Phoenicopteridae) and the Scarlet Tanager (*Piranga olivacea*). In the latter, seasonal colour differences are due to alterations in the extent of oxidation or reduction of xanthophils (Brush, 1967, 1981). Carotenoids are taken up into the epidermal cells by pinocytosis, a form of ultrastructural phagocytosis. Distinct xanthophores (lipophores) as occur in fish, amphibia and reptiles, are not found in birds.

Occasionally a red pigmentation is present which is not due to carotenoids but to a copper-containing pigment, porphyrin. The best examples of this are the bright red remiges and head feathers in certain turacos (Musophagidae) of southern Africa. Porphyrin pigment is removable in a dilute ammonia solution, in contrast to carotenoids, which require fat solvents (Fox and Vevers, 1960; Thiel, 1968). Porphyrins are chemically related to the iron-containing pigment of the blood, haemoglobin. In the human skin disease porphyria, porphyrins are a product of abnormal metabolism, but they occur normally in many marine invertebrates and in avian egg shells. Porphyrins sensitise living cells to sunlight so that they are damaged or killed, but in the dead feather they are unable to produce this adverse effect, as the living cells which take up the pigment are protected from sunlight by the plumage.

Different colour-producing systems may occur in the same plumage cells, or they may be found in different parts of a feather, or in different feathers within a pteryla, or in different pterylae.

2. *Physical coloration*

Iridescence which occurs in hummingbirds (Trochilidae) and starlings (Sturnidae) is due to interference of light reflected from the upper and lower surfaces of melanosomes in the feather barbules, which are possibly filled with air (Greenewalt *et al.*, 1960). Fine sculpturing of the feather surface may also produce this effect (Auber, 1957a).

The vivid white of the Mute Swan (*Cygnus olor*), which occurs in the absence of pigments, is due to the reflection and refraction of light from the surfaces of large air vacuoles in the feather cells which have interfaces longer than the wavelength of light (Fox and Vevers, 1960).

Blue pigments in feathers are rare (Lucas and Stettenheim, 1972), and the bright blue of the blue Budgerigar (*Melopsittacus undulatus*) is due to the physical scattering of light on particles less than 0.6 μm across (shorter than the wavelength of light) contained in the feather cells. Probably these small particles are unpigmented melanosomes. This blue colour (Tyndall effect) is produced in the same manner as the blue of the sky and in the blue grotto of Capri, that is, by the physical scattering of sunlight on small particles (Auber, 1957b).

3. *Mixed coloration*

When different colour-producing systems occur in the same feather cells, the resultant colour is modified. The best example of this is the colour green, which in birds is mainly due to a combination of yellow carotenoids and Tyndall blue (Auber, 1957; Fox and Vevers, 1960). Green pigments which are carotenes occur in some pigeons (Columbidae) and ducks and geese (Anatidae), and turacos (Musophagidae) have a porphyrin turacoverdin in their plumage, but these are rarities in the avian world. Most green feathers will turn blue if treated with a fat solvent, due to loss of yellow carotenoids and retention of the Tyndall effect. In order to permit these colour mechanisms the feather keratin has to be translucent to light.

4. *Colour patterning*

Colour patterning depends on the distribution of different colours in different pterylae and sometimes in different parts of the same feather. Colour patterns in the domestic fowl sometimes show sexual dimorphism, and males are more highly coloured in 19 of 85 species of British perching birds (Passeriformes), a phenomenon which is associated with territorial display. The bright colours of many birds as compared with mammals are associated with colour vision in the class Aves, whilst the Mammalia, apart from primates, are mostly colour blind.

Colour variation (plumage polymorphism) apart from sexual dimorphism occurs in many birds and is more common in birds of prey. Amongst other birds, the Parasitic Jaeger (*Stercoarius parasiticus*) has both light and dark forms. The genetics of plumage polymorphism has been discussed by Parkin (1980).

Differences in colour patterning also occur during the life of the bird. Thus, immature gulls (Laridae) are mottled greyish brown, in contrast to the white and black or grey seen in adults of several species. The young of the Mute Swan (*Cygnus olor*) are a sooty grey in their first plumage. Sometimes patterns change in the breeding plumage as in the Black-headed Gull (*Larus ridibundus*), in which a dark hood appears in summer and is generally lost in winter. A few birds such as the Rock Ptarmigan (*Lagopus mutus*) have white winter plumage. This species is relatively unusual in also having a different plumage pattern in each plumage generation during a single year.

5. *Functions of plumage coloration*

The bright coloration of many birds is associated with display. Relatively few birds rely on cryptic camouflage coloration; these are mainly ground-nesting species, such as larks (Alaudidae), the Eurasian Nightjar (*Caprimulgus europaeus*) and game birds such as the Willow Ptarmigan (*Lagopus lagopus*), which, though richly coloured, is well hidden among the varied shades of the rough moor. The mottled brown Tawny Owl (*Strix aluco*) is also well camouflaged against the bark of a tree when sleeping during the day, sitting bolt upright on a branch with its side pressed against the trunk. Small birds generally freeze in trees or bushes in the presence of birds of prey which can detect movement from a great height but easily miss still prey even when vividly coloured. Alternatively, particularly during nesting, small birds mob an intruder en masse, and even the smallest birds, the hummingbirds (Trochilidae), are capable of driving off hawks (Accipitridae) (Harrison, 1978).

V. Horny beak

The horny beak or bill (*rhamphotheca*) is a hard, tough keratinized epidermal structure in birds of prey as well as in seed and grass eaters, but in ducks (Anatidae) and other water birds which use their beaks to strain food from mud in ponds and streams, it is much thinner and softer (Mayaud, 1950). The beak grows continuously throughout life and is normally worn down by eating or biting hard objects, as in parrots (Psittacidae). The avian beak performs many of the functions of the mammalian forelimb, for example in eating, nest building and grooming. Its shape in different species reflects the functional requirements of the bird, shown well in Darwin's finches of the Galapagos Islands. For an account of the different shapes of bills, see Chapter 3, by McLelland, Vol. 1, p. 84.

The rhamphotheca is separated from the bones of the upper and lower jaws by a thin, fibrous dermis. The histological structure varies to some extent with different types of beaks (Lüdicke, 1933; Witschi and Woods, 1936; Kingsbury *et al.*, 1953; Lucas and Stettenheim, 1972). The dermo-epidermal junction in the chicken has deep ridges. The rhamphotheca is a modified horny layer consisting of layers of flattened keratin-filled cells (Frenkel and Gillespie, 1976) firmly held together by cell junctions which do not break down. When the beak is considered as a whole, its strength depends on the particular arrangement conferred by a layer of keratin on a firm bed of bone, the epidermis being bound to the periosteum by collagen fibres. As well as being keratinized, the epidermal cells of the beak are mineralised, calcium phosphate being deposited as hydroxyapatite crystals in the cytoplasm between the keratin proteins (Pautard, 1963, 1964). The egg tooth which hatching chicks use to break the egg shell is composed of horny cells containing hydroxyapatite and so is not a true tooth. In petrels (Procellariidae), the rhamphotheca is formed of segmented plates, and in puffins (Alcidae) the 7–9 segments are regularly moulted. For discussion of the morphology of the anseriform nail, the bill tip organ and the innervation of the beak in general, see Chapter 8 of this volume by K.-M. Gottschaldt.

VI. Spurs

Spurs (*calcaria*) consist of a bony core enclosed by a horny sheath. A metatarsal spur (*calcar metatarsale*) occurs in chickens, pheasants (Phasianidae) and turkeys (Meleagrididae), in which it is well developed in the cocks and usually small or absent in the hens. In Peacock-Pheasants (*Polypectron* spp.), each leg has four spurs. A carpal or metacarpal spur (*calcar carpale, calcar metacarpale*) is found in the Spur-winged Goose (*Plectropterus gambensis*), the Australian Spur-winged Plover (*Vanellus novaehollandiae*), and in jacanas (Jacanidae) and Sheathbills (Chionididae).

VII. Claws

Claws (*ungues*) resemble the rhamphotheca in being strongly keratinized and in possessing hydroxyapatite crystals (Pautard, 1963, 1964). They occur on the digits of the foot (*ungues digitorum pedis*) and in a few groups on the digits of the manus (*ungues digitorum manus*). In its formation the avian claw is like a reptilian claw in that it fits like a pointed thimble over the germinal epidermis at the tip of the digit (Biedermann, 1928; Lucas and Stettenheim, 1972). The hardest part of the toe claw is the dorsal plate (*scutum dorsale*), which is strongly keratinized and calcified. The plantar plate (*scutum plantare*), which is absent in the human nail and mammalian claw (Spearman, 1978), is relatively soft.

The growth of the claw is greatest in the dorsal plate, so that curved talons are produced for perching on branches as in passerines, for climbing trees as in woodpeckers (Picidae), and for grasping prey as in owls (Strigidae) and falcons (Falconidae). Diving birds such as ducks (Anatidae) and gannets (Sulidae), and swimming birds such as gulls (Laridae), have webbed or partly-webbed feet and smaller claws. Whilst the claw in mammals is composed of chemically distinct layers, this remains to be confirmed in birds (Spearman, 1978).

The oldest known bird, *Archaeopteryx,* found as fossils in the Jurassic limestone, had claws on its three wing digits. However, a functional claw on the major wing digit of living birds (*unguis digiti majoris*) only occurs in the young of the Hoatzin (*Opisthocomus hoatzin*), which lives in the Amazonian forests and uses the wing claws to clamber around the branches of trees, the claws being lost when the bird reaches adulthood. In a number of orders including Gaviiformes, Ciconiiformes, Anseriformes, Falconiformes, Galliformes, Gruiformes, Charadriiformes and Strigiformes, a claw is sometimes found on the alular digit (*unguis digiti alularis*) (Fisher, 1940).

VIII. Combs, wattles, rictuses and ear lobes

The comb (*crista carnosa*) is an unfeathered flap or appendage situated on the crown of the head, whilst the wattles (*paleae*) are folds of bare skin at the sides of the head (Lucas and Stettenheim, 1972). They are both well developed in the domestic fowl, in turkeys and in cassowaries (Casuaridae). Wattles also occur in jacanas (Jacanidae), cotingas (Cotingidae), starlings (Sturnidae), plovers (Charadriidae) and wattle-birds (Meliphagiidae). A comb is present in the male Andean Condor (*Vultur gryphus*). Over the comb and wattles, the epidermis, including the horny layer, is very thin. The red coloration is due to the extensive blood supply and the presence of a superficial capillary network (Lucas, 1970). The comb and wattles are mainly secondary sexual appendages used in display, but they are possibly useful in thermoregulation by dissipation of heat from their large surface area.

The *rictus* is the well-vascularised, sometimes brightly-coloured, bare, triangular area of skin at the angle of the mouth. It has maxillary and mandibular components.

The ear lobe (*lobus auricularis*) is a bare fold of well-vascularised red or white skin which in many breeds of chicken hangs down ventral to the external ear opening and overlies the caudal angle of the jaw.

IX. Thermoregulation

Many small wild birds survive very low temperatures so long as food is available, when mammals of the same size, such as rodents, hibernate or live a completely subterranean existence. The House Wren (*Troglodytes aedon*), which shows a behavioural pattern rare among birds, spends the colder months in holes in walls and trees but still has to come out to feed each day. Ambient temperatures around and under the feet can reach well below zero in arctic and antarctic species, such as penguins (Spheniscidae) and skuas (Stercorariidae), and yet the surface tissues in these exposed sites do not freeze. This appears to be because the cutaneous vasculature, though largely shut down to conserve body heat, is opened up sufficiently to warm the exposed skin of the feet. It has been shown that the skin is warmed appreciably by increased blood flow through the dermis in cold-acclimatized chickens (Chaffe and Roberts, 1971). In geese (Anatidae) from arctic regions the blood is shunted through arterioles straight to venules in the feet, thus increasing the flow of warm blood, which prevents the skin from freezing (Welty, 1979). For short periods domestic chicks can tolerate an ambient temperature as low as −50°C, and pigeons (Columbidae) can tolerate −100°C (Freeman, 1971). Birds are therefore very adaptable in their thermoregulation. At both hot and cold survivable ambient temperatures many birds become rapidly acclimatized to the new conditions. Birds tolerate high ambient temperatures better when the air is dry and with a reasonable movement, but for the chicken this advantage is lost at wind speeds over 150 m/min (Siegel and Drury, 1968). An environmental temperature of 43°C is tolerated for a short time.

Heat production in birds is mainly by oxidation of body fat and skeletal muscle proteins, and small birds such as the House Sparrow (*Passer domesticus*) are reduced significantly in weight after exposure to a cold winter's night, but this is rapidly replenished when the birds feed during the day (Newton, 1969; Barnett, 1970). Subcutaneous fat in water birds is important in conserving core heat. Shivering by rapid contractions of skeletal muscles is also a major source of body heat in birds, particularly at night in temperate regions (Steen and Enger, 1957; Chaffe and Roberts, 1971). However, in cold-acclimatized chickens shivering occurs less frequently than in unacclimatized birds (Freeman, 1971).

Cooling of the body under hot ambient conditions is achieved by panting with the beak open. Indeed, birds are much more efficient than mammals such as dogs in their panting mechanism (Jarrett and Morimoto, 1978). The latent heat of evaporation of water from the saliva cools the blood circulating through the oral mucosa, which is important because birds do not have sweat glands.

The slow but steady loss of water by evaporation at the skin surface by transpiration through the skin, which occurs in all terrestrial vertebrates, has a slight cooling effect through evaporation (Lee and Schmidt-Nielsen, 1971), but it does not appear to be of major importance in thermoregulation in birds. The regulatory system of birds, although efficient, bears no comparison with the eccrine sweating mechanism of humans, the most adaptive of any vertebrate species in thermoregulation. The importance of the skin surface in releasing excess body heat is therefore less in birds than in most mammals, and it is controversial whether the small area of bare skin of the comb, wattles and legs is sufficient for thermoregulatory purposes, although the bare head, neck, thighs and legs of the Ostrich (*Struthio camelus*) are important in this respect. It has been suggested that the larger comb size of the cockerel might be responsible for the greater heat tolerance of males, but a more efficient panting mechanism seems likely (Freeman, 1971).

Birds evolved from reptiles a similar and probably homologous hypothalamic thermostatic centre for thermoregulation as in mammals, although reptiles being poikilothermic use behavioural means to regulate their body temperature (Bligh, 1976). Neurosensors supply this centre with information on body temperature, but whilst sensors in mammalian skin are important in controlling core temperature, the importance of peripheral control in birds is uncertain. Freeman (1971) suggested that, since the first sustained thermoregulatory response occurs after the escape of the chick from the eggshell, peripheral control through the skin may be important, but sensors in the oral mucosa could also explain this. However, Randall (1943) found that shivering occurred in chickens before the core temperature was lowered in response to cold ambient conditions, which suggests that thermal sensors do occur in the skin. Despite this, Woods and Whittow (1974) believed that control of the body temperature through the skin is relatively unimportant in most birds compared with mammals, in which it is highly important.

For further discussion of the integument and thermoregulation see Chapter 5, by West *et al.*, Vol. 2, p. 320.

Acknowledgements

We are grateful to Dr. R. D. Hodges and to Professor R. Bellairs for specimens of chicken skin for electron microscopy. We also wish to thank Dr. R. Wrench for helpful discussions on epidermal lipogenesis, and Dr. J. V. Beer for his scanning electron micrographs.

References

Andrew, R. J. (1956). Some remarks on behaviour in conflict situations. *Br. J. Anim. Behav.* **4,** 41–45.

Astbury, W. T. and Marwick, T. C. (1932). X-ray interpretation of the molecular structure of feather keratin. *Nature, Lond.* **130,** 309.

Auber, L. (1955). Cortex and medulla of bird feathers. *Nature, Lond.* **176,** 1218–1219.

Auber, L. (1957a). The distribution of structural colours and unusual pigments in the class Aves. *Ibis* **99,** 463–476.

Auber, L. (1957b). The structures producing non-iridescent blue colour in bird feathers. *Proc. zool. Soc. Lond.* **129,** 455–486.

Auber, L. and Appleyard, M. H. (1955). The structure of the feathers in *chlorophanes* and *iridophanes* (Coerebidae). *Ibis* **97,** 252–258.

Austin, O. L. (1961). "Birds of the World." Hamlyn, Feltham.

Bakker, R. T. (1975). Dinosaur renaissance. *Scient. Am.* **232,** 58–78.

Barlee, J. (1964). Flight. *In* "A New Dictionary of Birds" (A. Landsborough Thomson, Ed.). Nelson, London.

Barnett, L. B. (1970). Seasonal changes in temperature acclimatization of the House Sparrow *Passer domesticus. Comp. Biochem. Physiol.* **33,** 559–578.

Bell, E. and Thathachari, Y. T. (1963). Development of feather keratin during embryogenesis of the chick. *J. Cell Biol.* **16,** 215–233.

Bellairs, A. d'A. (1969). "The Life of Reptiles," Vol. 2. Weidenfeld and Nicholson, London.

Biedermann, W. (1928). Vergleichende Physiologie des Integuments der Wirbeltiere. *Ergebn. Biol.* **3,** 354–541.

Bligh, J. (1976). Temperature regulation. *In* "Environmental Physiology of Animals" (J. Bligh, J. L. Cloudsley-Thompson and A. G. MacDonald, Eds.). Blackwell, Oxford.

Borodulina, T. L. (1966). The innervation of the filoplumes. *In* "Mechanisms of Flight and Orientation of Birds" (S. E. Kleinenberg, Ed.). Akad. Nauk S.S.S.R. Inst. Morphologie Zhivotin, Moscow.

Brown, R. H. J. (1961). Flight. *In* "Biology and Comparative Physiology of Birds" (A. J. Marshall, Ed.), Vol. 2. Academic Press, New York and London.

Brown, R. H. J. (1963). The flight of birds. *Biol. Rev.* **38,** 460–489.

Brush, A. H. (1967). Pigmentation in the Scarlet Tanager *Piranga olivacea. Condor* **69,** 549–559.

Brush, A. H. (1974). Feather keratin analysis of subunit heterogeneity. *Comp. Biochem. Physiol.* **48B,** 661–670.

Brush, A. H. (1978). Feather keratins and avian pigmentation. *In* "Chemical Zoology" (A. H. Brush, Ed.), Vol. 2. Academic Press, New York and London.

Brush, A. H. (1980a). Patterns in the amino acid composition of avian epidermal proteins. *Auk* **97,** 742–753.

Brush, A. H. (1980b). Chemical heterogeneity in keratin proteins of avian epidermal structures. *In* "The Skin of Vertebrates" (R. I. C. Spearman and P. A. Riley, Eds.), Symp. Linn. Soc. Lond., No. 9. Academic Press, London and New York.

Brush, A. H. (1981). Carotenoids in wild and captive birds. *In* "Carotenoids aa Colorants and Vitamin A Precursors" (J. C. Bauernliend, Ed.). Academic Press, New York and London.

Brush, A. H. and Wyld, J. A. (1980). Molecular correlates of morphological differentiation. *J. exp. Zool.* **212,** 153–157.

Cane, A. K. and Spearman, R. I. C. (1967). A histochemical study of keratinization in the domestic fowl *(Gallus gallus). J. Zool., Lond.* **153,** 337–352.

Cater, D. B. and Lawrie, N. R. (1950). Some histochemical and biochemical observations on the preen gland. *J. Physiol., Lond.* **111,** 231–243.

Chaffe, R. R. J. and Roberts, J. C. (1971). Temperature acclimatization in birds and mammals. *A. Rev. Physiol.* **33,** 155–202.

Chandler, D. C. (1916). A study of the structure of feathers with reference to their taxonomic significance. *Univ. Calif. Publ. Zool.* **13,** 243–446.

Clench, M. H. (1970). Variability in body pteryolysis with special reference to the genus *Passer. Auk* **87,** 650–691.

Cohen, J. (1965). The dermal papilla. *In* "Biology of the Skin and Hair Growth" (A. G. Lyne and B. F. Short, Eds.). Angus and Robertson, Sydney.

Dawson, W. R. and Hudson, J. W. (1970). Birds. *In* "Comparative Physiology of Thermal Regulation" (G. C. Whittow, Ed.). Academic Press, New York and London.

Dorst, J. (1974). "The Life of Birds." Columbia University Press, New York.

Dreyfuss, A. (1937). L'innervation de la plume. *Archs. Zool. exp. gén.* **79,** 30–42.

Elder, W. H. (1954). The oil gland of birds. *Wilson Bull.* **66,** 6–31.

Espinasse, P. G. (1939). The developmental anatomy of the Brown Leghorn breast feather and reactions to oestrone. *Proc. zool. Soc. Lond.* **109,** 247–288.

Fisher, H. I. (1940). The occurence of vestigial claws on the wings of birds. *Am. Midl. Nat.* **23,** 234–243.

Flaxman, B. A. (1972). Cell differentiation and its control in the vertebrate epidermis. *Am. Zool.* **12,** 13–25.

Fleischmajer, R. and Krol, S. (1970). Dermal specificity. *J. invest. Derm.* **54,** 472–478.

Fox, M. H. and Vevers, H. G. (1960). "The Nature of Animal Colours." Macmillan, London.

Fraser, R. D. B. and MacRae, T. P. (1980). Current views on the keratin complex. *In* "The Skin of Vertebrates" (R. I. C. Spearman, and P. A. Riley, Eds.), Symp. Linn. Soc. Lond., No. 9. Academic Press, London and New York.

Fraser, R. D. B., MacRae, T. P. and Rogers, G. E. (1972). "Keratins, their Composition, Structure and Biosynthesis." Thomas, Springfield, Illinois.

Freeman, B. M. (1971). Body temperature and thermal regulation. *In* "Physiology and Biochemistry of the Domestic Fowl" (D. J. Bell and B. M. Freeman, Eds.), Vol. 2. Academic Press, London and New York.

Freinkel, R. K. (1972). Lipogenesis in epidermal differentiation of embryonic chick. *J. invest. Derm.* **59,** 332–338.

Frenkel, M. J. and Gillespie, J. M. (1976). The proteins of the keratin component of birds' beaks. *Aust. J. biol. Sci.* **29,** 467–479.

Graham, R. R. (1932). Slots in the wings of birds. *J. R. aeronaut. Soc.* **36,** 598–600.

Greenewalt, C. H., Brandt, W. and Friel, D. D. (1960). Iridescent colors of the humming bird feathers. *J. Opt. Soc. Am.* **50,** 1005–1016.

Groebbels, F. (1932). "Der Vogel. Bau, Funktion, Lebenserscheinung, Einpassung," Vol. 1. G. Borntraeger, Berlin.

Haahti, E., Lagerspetz, K., Nikkari, J. and Fales, H. M. (1964). Lipids of the uropygial gland of birds. *Comp. Biochem. Physiol.* **12,** 435–437.

Handley,-Page, F. (1921). The Handley-Page wing. *J. R. aeronaut. Soc.* **25,** 263–289.

Harrison, C. J. O. (1978). "Bird Families of the World." Elsevier, Phaidon, Oxford.

Harrison, J. M. H. (1964). Moult. *In* "A New Dictionary of Birds" (A. Thomson, Ed.). Nelson, London.

Hodges, R. D. (1974). "The Histology of the Fowl." Academic Press, London and New York.

Horvath, S. M., Folk, G. E., Craig, F. M. and Fleischmann, W. (1948). Survival times of various warm blooded animals in extreme cold. *Science, Washington, D.C.* **107,** 171–172.

Hosker, A. (1936). Studies on the epidermal structures of birds. *Phil. Trans. R. Soc. B.* **226,** 143–188.

Humphrey, P. S. and Clark, G. A., Jr. (1964). The anatomy of the water fowl. *In* "The Waterfowl of the World" (J. Delacour. Ed.), Vol. 4. Country Life, London.

Huxley, J. S. (1914). The courtship habits of the Great-crested Grebe *(Podiceps cristatus). Proc. zool. Soc. Lond.* **2,** 491–562.

Jacob, J. (1978). Uropygial gland secretions and feather waxes. *In* "Chemical Zoology" (A. H. Brush, Ed.), Vol. X. Academic Press, New York and London.

Jacob, J. and Pomeroy, D. E. (1979). The feather lipids of the Marabou stork *(Leptoptilus crumeniferus). Comp. Biochem. Physiol.* **64B,** 301–303.
Jarrett, A. (1974). The dermis and dendrocytes. *In* "The Physiology and Pathophysiology of the Skin" (A. Jarrett, Ed.), Vol. 3. Academic Press, London and New York.
Jarrett, A. and Hardy, J. A. (1957). The value of alcohol for fixation of the skin. *Stain Technol.* **32,** 225–233.
Jarrett, A. and Morimoto, T. (1978). Heat exchange between animals and their environment. *In* "The Physiology and Pathophysiology of the Skin" (A. Jarrett, Ed.), Vol. 5. Academic Press, London and New York.
Johnson, E. (1977a). The control of hair growth. *In* "The Physiology and Pathophysiology of the Skin" (A. Jarrett, Ed.), Vol. 4. Academic Press, London and New York.
Johnson, E. (1977b). Environmental effects on the hair follicle. *In* "The Physiology and Pathophysiology of the Skin" (A. Jarrett, Ed.), Vol. 4. Academic Press, London and New York.
Jones, D. J. and Morgan, W. (1956). Woolly feathering in the fowl. *J. Hered.* **47,** 137–141.
King, A. S. (1966). Structural and functional aspects of the avian lungs and air sacs. *In* "International Review of General and Experimental Zoology" (W. J. L. Felts and R. J. Harrison, Eds.), Vol. II. Academic Press, New York and London.
King, A. S. (1979). Systema respiratorium. *In* "Nomina Anatomica Avium" (J. J. Baumel, A. S. King, A. M. Lucas, J. E. Breazille and H. E. Evans, Eds.). Academic Press, London and New York.
Kingsbury, J. W., Allen, V. G. and Rotheram, B. A. (1953). The biological structure of the beak of the chick. *Anat. Rec.* **116,** 95–115.
Krimm, S. (1960). The structure of frizzle mutant feather keratin. *J. molec. Biol.* **2,** 247–249.
Lavker, R. M. (1975). Lipid synthesis in chick epidermis. *J. invest. Derm.* **65,** 93–101.
Lee, D. H., Robinson, K. W., Yeates, N. T. M. and Scott, M. R. T. (1945). Poultry husbandry in hot climates. Experimental enquiries. *Poult. Sci.* **24,** 195–207
Lee, P. and Schmidt-Nielsen, K. (1971). Respiratory and cutaneous evaporation in the Zebra Finch. Effect on water balance. *Am. J. Physiol.* **220,** 1598–1609.
Liebelt, R. A. and Eastlick, H. L. (1954). The organlike nature of the subcutaneous fat bodies in the chicken. *Poult. Sci.* **33,** 169–179.
Lillie, F. R. (1942). On the development of feathers. *Biol. Rev.* **17,** 247–266.
Lucas, A. M. (1968). Lipid secretion in the avian epidermis. *Anat Rec.* **160,** 386–387.
Lucas, A. M. (1970). Avian functional anatomic problems. *Fedn. Proc. Fedn. Am. Socs. exp. Biol.* **29,** 1641–1648.
Lucas, A. M. (1979). Integumentum commune. *In* "Nomina Anatomica Avium" (J. J. Baumel, A. S. King, A. M. Lucas, J. F. Breazile and H. E. Evans, Eds.). Academic Press, London and New York.
Lucas, A. M. (1980). Lipid secretion by avian epidermis. *In* "The Skin of Vertebrates" (R. I. C. Spearman and P. A. Riley, Eds.), Symp. Linn. Soc. Lond., No. 9. Academic Press, London and New York.
Lucas, A. M. and Stettenheim, P. R. (1972). "Avian Anatomy. Integument," Agriculture Handbook 362. U.S. Department of Agriculture, Washington, D.C.
Lüdicke, M. (1933). Wachstum and Abnutzung des Vogelschnabels. *Zool. Jb.* **57,** 465–533.
Malinovský, L. (1968). Types of sensory corpuscles common to mammals and birds. *Folia morph., Prague* **16,** 67–73.
Mason, H. G. (1967). The structure of melanin. *In* "Advances in Biology of Skin" (W. Montagna and F. Hu, Eds.), Vol. 8. Pergamon Press, Oxford.
Matoltsy, A. G. (1969). Keratinization of the avian epidermis. An ultrastructural study of the newborn chick skin. *J. Ultrastruct. Res.* **29,** 438–458.
Matoltsy, A. G. (1980). Structure and function of the mammalian epidermal horny layer *In* "The Skin of Vertebrates" (R. I. C. Spearman and P. A. Riley, Eds.), Symp. Linn. Soc. Lond., No. 9. Academic Press, London and New York.
Matoltsy, A. G. and Huszar, T. (1972). Keratinization of the reptilian epidermis: an ultrastructural study of turtle skin. *J. Ultrastruct. Res.* **38,** 87–101.
Mayaud, N. (1950). Téguments et phanères. *In* "Traité de Zoologie" (P. P. Grassé, Ed.), Vol. 15. Masson, Paris.
Menon, G. K., Aggarwal, S. K. and Lucas, A. M. (1981). Evidence for the holocrine nature of lipid secretion by avian epidermal cells: A histological and fine structural study of rictus, toe web and the uropygial gland. *J. Morph.* **167,** 185–199.
Newton, L. (1969). Winter fattening in the Bullfinch. *Physiol. Zool.* **42,** 96–107.
Nicolaus, R. A. (1968). "Melanins." Herrmann, Paris.

Nikkari, T. (1974). Comparative chemistry of sebum. *J. invest. Derm.* **62,** 257–267.

Palmer, R. S. (1972). Patterns of moulting. *In* "Avian Biology" (D. S. Farner and J. R. King, Eds.), Vol. 2. Academic Press, New York and London.

Parakkal, P. F. and Alexander, N. J. (1972). "Keratinization. A Survey of Vertebrate Epithelia." Academic Press, London and New York.

Parakkal, P. F. and Matoltsy, A. G. (1968). An electron microscope study of developing chick skin. *J. Ultrastruct. Res.* **23,** 403.

Parkin, D. T. (1980). The ecological genetics of plumage polymorphism in birds. *In* "The Skin of Vertebrates" (R. I. C. Spearman and P. A. Riley, Eds.), Symp. Linn. Soc. Lond., No. 9. Academic Press, London and New York.

Pautard, F. G. E. (1963). Mineralisation of keratin and its comparison with normal matrix. *Nature, Lond.* **199,** 531.

Pautard, F. G. E. (1964). Calcification of keratin. *In* "Progress in the Biological Sciences in Relation to Dermatology" (A. Rook and R. H. Champion, Eds.), Vol. 2. Cambridge University Press, London.

Payne, R. B. (1972). Mechanisms of control of moult. *In* "Avian Biology" (D. S. Farner and J. R. King, Eds.), Vol. 2. Academic Press, New York and London.

Prota, G. (1980). Recent advances in the chemistry of melanogenesis in mammals. *J. invest. Derm.* **75,** 122–127.

Pycraft, W. P. (1922). "Birds in Flight." Gay and Hancock, London.

Quilliam, T. A. and Armstrong, J. (1963). Mechanoreceptors. *Endeavour* **22,** 55–60.

Randall, W. C. (1943). Factors influencing the temperature regulation of birds. *Am. J. Physiol.* **139,** 56–63.

Rawles, M. E. (1960). The integumentary system. *In* "The Biology and Comparative Physiology of Birds" (A. J. Marshall, Ed.), Vol. 1. Academic Press, New York and London.

Regal, P. J. (1975). The evolutionary origin of feathers. *Q. Rev. Biol.* **50,** 35–66.

Riley, P. A. (1977). The mechanism of melanogenesis. *In* "Comparative Biology of Skin" (R. I. C. Spearman, Ed.), Symp. zool. Soc. Lond., No. 39. Academic Press, London and New York.

Rudall, K. M. (1947). X-ray studies of the distribution of protein chain types in the vertebrate epidermis. *Biochim. Biophys. Acta* **1,** 549–562.

Ruppell, G. (1977). "Bird Flight." Van Nostrand, Reinhold, New York.

Savile, D. B. O. (1950). The flight mechanism of swifts and humming birds. *Auk* **67,** 499–504.

Schallenberg, R. J., Whittow, G. C. and Smith, R. M. (1974). Body temperature in the nestling Red-footed Booby. *Condor* **76,** 476–478.

Schlichting, H., Truckenbrodt, E. and Ramm, H. J. (1979). "Aerodynamics of the Airplane." McGraw-Hill, New York.

Scholander, P. F., Hock, R., Walters, V. and Irving, L. (1950). Adaptation to cold in arctic and tropical mammals and birds in relation to body temperature, insulation, and basal metabolism. *Biol. Bull., Woods Hole, Mass.* **99,** 259–271.

Sengel, P. (1976). "Morphogenesis of Skin." Cambridge University Press, London.

Sengel, P., Dhouailly, D. and Mauger, A. (1980). Region-specific determination of epidermal differentiation in amniotes. *In* "The Skin of Vertebrates" (R. I. C. Spearman and P. A. Riley, Eds.), Symp. Linn. Soc. Lond., No. 9. Academic Press, London and New York.

Shah, R. V. and Menon, G. K. (1972). Histochemical studies on pigeon definitive feathers during post-hatching: induced and regenerative modes of development: lipids, lipase and hydroxybutrate dehydrogenase. *Pavo* **10,** 30–42.

Shah, R. V., Menon, G. K., Desai, J. H. and Jani, M. B. (1977). Feather loss from the capital tracts of painted storks related to growth and maturity I. Histophysiological changes and lipoid secretion in the integument. *J. Anim. Morph. Physiol.* **24,** 99–107.

Siegel, H. S. and Drury, M. L. N. (1968). Physiological responses of chickens to variations in air temperature and velocity. *Poult. Sci.* **47,** 1120–1127.

Singer, M. (1980). The influence of nerves in amphibian regeneration with special emphasis on the relationship between nerves and epidermis. *In* "The Skin of Vertebrates" (R. I. C. Spearman and P. A. Riley, Eds.), Symp. Linn. Soc. Lond., No. 9. Academic Press, London and New York.

Smith, S. (1945). "How to Study Birds." Collins, London.

Spearman, R. I. C. (1964). The evolution of mammalian keratinized structures. *In* "The Mammalian Epidermis and its Derivatives" (F. J. Ebling, Ed.), Symp. Zool. Soc. Lond., No. 12. Academic Press, London and New York.

Spearman, R. I. C. (1966). The keratinization of epidermal scales, feathers and hairs. *Biol. Rev.* **41,** 59–96.

Spearman, R. I. C. (1969). The epidermis and feather follicles of the King Penguin *Aptenodytes patagonicus (Aves). Z. Morph. Okol. Tiere* **64,** 361–372.

Spearman, R. I. C. (1971). The integumentary system. *In* "Physiology and Biochemistry of the Domestic Fowl" (D. J. Bell and B. M. Freeman, Eds.), Vol. 2. Academic Press, London and New York.

Spearman, R. I. C. (1973). "The Integument." Cambridge University Press, London.

Spearman, R. I. C. (1974). The comparative biology of collagenous tissues. *In* "The Physiology and Pathophysiology of the Skin" (A. Jarrett, Ed.), Vol. 3. Academic Press, London and New York.

Spearman, R. I. C. (1977a). Keratins and keratinization. *In* "Comparative Biology of Skin" (R. I. C. Spearman, Ed.), Symp. Zool. Soc. Lond., No. 39. Acadmic Press, London and New York.

Spearman, R. I. C. (1977b). Biochemistry of hair formation and the chemistry of hair keratin. *In* "The Physiology and Pathophysiology of the Skin" (A. Jarrett, Ed.), Vol. 4. Academic Press, London and New York.

Spearman, R. I. C. (1977c). The genetics of hair growth and coloration. *In* "The Physiology and Pathophysiology of the Skin" (A. Jarrett, Ed.), Vol. 4. Academic Press, London and New York.

Spearman, R. I. C. (1978). The physiology of the nail. *In* "The Physiology and Pathophysiology of the Skin" (A. Jarrett, Ed.), Vol. 5. Academic Press, London and New York.

Spearman, R. I. C. (1980a). The tongue. *In* "The Physiology and Pathophysiology of the Skin" (A. Jarrett, Ed.), Vol. 6. Academic Press, London and New York.

Spearman, R. I. C. (1980b). The avian skin in relation to surface ecology. *Proc. R. Soc. Edinb.* **79B,** 57–74.

Spearman, R. I. C. (1982a). The biochemistry of skin disease. *Molec. Aspects Med.* **5,** 63–126.

Spearman, R. I. C. (1982b). Structure and function of subcutaneous tissue. *In* "The Physiology and Pathophysiology of the Skin" (A. Jarrett, Ed.), Vol. 7. Academic Press, London and New York.

Spearman, R. I. C. (1983). The integumentary system. *In* "Physiology and Biochemistry of the Domestic Fowl" (B. M. Freeman, Ed.), Vol. 4. Academic Press, London and New York.

Spearman, R. I. C. and Hardy, J. A. (1974). Some ultrastructural observations on keratohyalin granules of guinea pig epidermis. *Arch. Derm. Forsch.* **250,** 149–158.

Steen, J. and Enger, P. S. (1957). Muscular heat production in pigeons during exposure to cold. *Am. J. Physiol.* **191,** 157–158.

Stephens, E. D. (1973). The evolution of endothermy. *J. theor. Biol.* **38,** 597–611.

Stettenheim, P. R. (1972). The integument. *In* "Avian Biology" (D. S. Farner and J. R. King, Eds.), Vol. 2. Academic Press, New York and London.

Storer, J. H. (1948). The flight of birds. *Bull. Cranbrook Inst. Sci., Bloomfield Hills, Mich.* **28,** 1–94.

Stresemann, E. and Stresemann, V. (1966). Die Mauser der Vögel. *J. Ornithol.* **107,** 1–445.

Thiel, H. (1968). Die Porphyrine des Vogelfeder Untersuchungen uber ihre Hertrienst und Einlagerung. *Zool. Jb. Syst.* **95,** 147.

Urry, D. and Urry, K. (1970). "Flying Birds." Vernon and Yates, London.

Van Tyne, J. and Berger, A. J. (1971). "Fundamentals of Ornithology." Wiley, New York.

Varićak, R. S. (1938). Neues über Auftreten und Bedeutung von Fettsubstanzen in der Geflügelhaut. *Z. mikrosk.-anat. Forsch.* **44,** 119–130.

Veghte, J. H. (1964). Thermal and metabolic responses of the Gray Jay to cold stress. *Physiol. Zöl.* **37,** 316–328.

Voitkevich, A. A. (1966). "The Feathers and Plumage of Birds." Sedgwick and Jackson, London.

von Boetticker, H. (1929). Morphologische und phylogenetische Studien über die hornige Fussbekleidung der Vögel. *Jena. Z. Naturw.* **64,** 377–448.

Watson, G. E. (1963a). Feather replacement in birds. *Science, Washington, D.C.* **139,** 50–51.

Watson, G. E. (1963b). The mechanism of feather replacement during natural moult. *Auk* **80,** 486–495.

Welty, J. C. (1979). "The Life of Birds," 2nd edn. Saunders, Philadelphia.

Witschi, E. and Woods, R. P. (1936). The bill of the sparrow as an indicator for the male sex hormone structural basis. *J. exp. Zool.* **3,** 445–459.

Woods, J. J. and Whittow, C. G. (1974). The role of central and peripheral temperature change in the regulation of thermal polypnea in the chicken. *Life Sci.* **14,** 199–206.

Wrench, R., Hardy, J. A. and Spearman, R. I. C. (1980). Sebokeratocytes of avian epidermis with mammalian comparisons. *In* "The Skin of Vertebrates" (R. I. C. Spearman and P. A. Riley, Eds.), Symp. Linn. Soc. Lond., No. 9. Academic Press, London and New York.

2
Locomotor system

ROBERT J. RAIKOW

Department of Biological Sciences, University of Pittsburgh, Pittsburgh, Pennsylvania, U.S.A.

CONTENTS

I. Introduction

The purpose of this chapter is to provide a basic introduction to the functional gross anatomy of the avian limbs and tail in relation to locomotion. Descriptive anatomy of the skeletal and muscular systems must form the basis of any functional analysis, and an enormous literature exists on this subject. However, the present account does not attempt to give detailed descriptions of the bones and muscles, but only an outline and tabulated summary of the information which is necessary for discussions of function. For a general introduction to the avian skeleton and its literature see Bellairs and Jenkin (1960) and Feduccia (1975). For a similar coverage of the muscular system see Berger (1960), George and Berger (1966) and Vanden Berge (1975). The literature of avian anatomy has a long and confusing history with respect to nomenclature, but the publication of a standardised anatomical terminology for birds, the *Nomina Anatomica Avium* (Baumel *et al.*, 1979), is bringing order into the field, and the general adoption of this terminology is strongly recommended.

In a review of the avian skeleto-muscular system, Bock (1974) noted that most students of avian functional anatomy do not make use of the latest insights into muscular physiology and skeleto-muscular biomechanics. He discussed a number of techniques that should prove useful, including more precise application of physical principles and terminology, the use of free-body diagrams in force analysis and the recognition of the significance of muscle fibre architecture in relation to muscle activity. To this we should add electromyography, the direct recording of muscular activity in living animals. Unfortunately, little has changed in the subsequent decade. The exciting approaches to functional anatomy being used today, especially with fish and mammals, have not often been applied to the study of avian locomotion [but see Goslow (1972) and Goldspink *et al.* (1978)]. In this chapter, therefore, the analysis of skeleto-muscular activity is based largely on inferences from dissection. By the "action" of a muscle is meant the movement that the muscle will bring about as inferred from its anatomical relationships, including its origin, insertion and the form of the intervening joints. Inevitably this can give only an estimate of the actual roles that muscles play when interacting in groups, an assertion of what is potentially possible rather than a description of observed muscle activity. Perhaps one may hope that this chapter will provide a foundation and an inspiration for new and more physiologically orientated studies.

II. Forelimb

A. *Skeleton and joints*

1. Sternum and thoracic girdle

The *sternum* and the bones of the girdle of the thoracic limb (*ossa cinguli membri thoracici*) (Fig. 2.1) serve as the sites of origin of the muscles that control the movements of the wing. The girdle is composed of the coracoid (*coracoideum*), *scapula* and clavicle (*clavicula*), right and left clavicles usually being fused into a *furcula*. The coracoid articulates with the cranial margin of the sternum and is directed craniodorsolaterally, diverging so that its dorsal end lies lateral or lateroventral to the most caudal cervical vertebra. The scapula articulates with the coracoid, its blade extending caudally over the vertebral ribs to approximately the cranial end of the ilium. At the lateral side of their connection, coracoid and scapula together contribute to the formation of the glenoid cavity (*cavitas glenoidalis*), which receives the head of the humerus. At its dorsal end, the *epicleideum,* the clavicle articulates with the coracoid and often with the

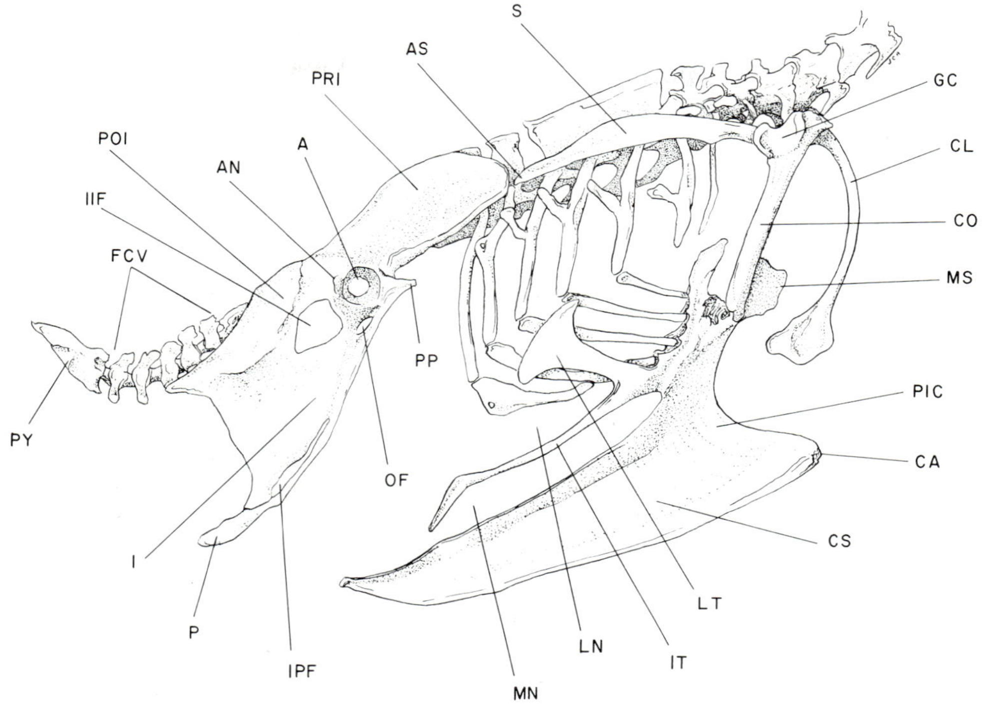

Fig. 2.1. The trunk skeleton of *Gallus*. A, Acetabulum; AN, antirochanter; AS, apex of scapula; CA, carinal apex; CL, clavicle; CO, coracoid; CS, sternal carina; FCV, free caudal vertebrae; GC, glenoid cavity; I, ischium; IIF, ilioischiadic foramen; IPF, ischiopubic fenestra; IT, intermediate trabecula; LN, lateral notch; LT, lateral trabecula; MN, medial notch; MS, *rostrum sterni* (manubrial spine); OF, obturator foramen; P, pubis; PIC, *pila carinae*; POI, postacetabular ilium; PP, preacetabular tubercle (pectineal process); PRI, preacetabular ilium; PY, pygostyle; S, blade of scapula.

scapula as well. Between the juncture of the scapula, coracoid and clavicle is a passage, the triosseal canal (*canalis triosseus*), through which the tendon of the *m. supracoracoideus* passes. In some birds this passage lies between the coracoid and scapula only, or may be formed within the coracoid alone (Baumel, 1979a).

The body of the sternum (*corpus sterni*) is a curved, dorsally concave plate of bone that forms the floor of the thorax. In most birds (carinates) a bony plate, the keel (*carina sterni*), projects ventrally from the midline. It is deepest at or near its cranial end, becoming progressively shallower caudally. The cranial margin (*margo cranialis*) of the keel is commonly strengthened by a thickened column, the *pila carinae,* the ventral margin (*margo ventralis*) of the keel often projecting cranially beyond this as the *apex carinae*. The keel serves as an important part of the area of origin of the principal flight muscles, *m. pectoralis* and *m. supracoracoideus,* and its size is proportional to theirs. In the most highly-derived flightless birds (ratites) the great reduction of the flight muscles is accompanied by a reduction in the overall size of the sternum and by loss of the keel. The keel was also absent in the flightless cretaceous diver *Hesperornis*.

At the cranial margin of the body of the sternum is the coracoidal articular sulcus (*sulcus articularis coracoideus*), with which the blade-like ventral end of the coracoid articulates. Cranially in the midline the sternum of many birds has bony projections, the ventral *spina externa* and less often the dorsal *spina interna*; these serve as attachments for the connective tissue sternocoracoclavicular membranes (*membranae sternocoracoclaviculares*) and have often been used as taxonomic characters. Together they constitute the *rostrum sterni* or manubrial spine, and in *Gallus* they are fused together. Along the lateral margins of the sternum are a series of articular surfaces (*facies articulares costales*) for the sternal ribs.

The caudal margin (*margo caudalis*) of the sternum is extremely variable in shape. It may be squared off or rounded, and may be intact or possess openings. There may be one or two openings on each side; they are then identified as being lateral or medial. The openings may be entirely surrounded by bone, the caudal margin remaining intact, in which case they are termed fenestrae (*fenestra lateralis, fenestra medialis*). Alternatively, the openings may perforate the sternal margin and are then distinguished as notches (*incisura lateralis, incisura medialis*). When a large fenestra perforates the sternal margin only slightly the condition is really intermediate, and the distinction between notch and fenestra becomes trivial. The spaces are closed by connective tissue membranes, and they essentially represent areas in which the sternum fails to ossify. In many cases they are shallow indentations, but in some forms (e.g. *Gallus*) they extend most of the length of the sternum. In general, weak (or occasional) fliers have extensive openings, while strong fliers do not. The openings are bounded by trabeculae (*trabecula lateralis, trabecula intermedia, trabecula mediana*).

The shape of the caudal margin of the sternum has been used in taxonomic studies. In a detailed review of the situation in suboscine passerines, Heimerdinger and Ames (1967) defined the following six types, which form a continuum of gradations: *type 1,* no notches or fenestrae present; *type 2,* a pair of lateral fenestrae; *type 3,* a pair of lateral notches; *type 4,* a pair of lateral notches plus small medial fenestrae; *type 5,* a pair of lateral notches plus large medial fenestrae; *type 6,* a pair of lateral notches and a pair of medial notches (the "four-notched" sternum). They found considerable individual variation and warned that the condition in a given species should only be characterised on the basis of a large series of specimens. They also noted that birds having the same type may nevertheless vary in other details. Woodpeckers (Picidae) and toucans (Ramphastidae) are type 6, but have shorter trabeculae than passerines. Barbets (Capitonidae) and a jacamar (Galbulidae) have longer and thinner trabeculae, but with wide caudal flanges which are unusual in passerines. From this it may be concluded that caution is advisable in assuming that similar sternal types necessarily indicate relationship, and the possibility of convergence must be kept in mind. Nevertheless, sternal configuration may be of value in taxonomy when used with care (Ames *et al.*, 1968).

The functional significance of these variations is uncertain. Lindsay (1885) suggested that the shape of the caudal border of the sternum is influenced by the two sets of muscles that attach there, namely the pectorals and the abdominal muscles. She concluded that, among non-passerines, strong fliers like gulls (Laridae) have long, broad sterna with short trabeculae, while such ground birds as tinamous (Tinamidae) and galliforms have short sterna with long, thin processes for the attachment of the abdominal muscles. The same generalisation (with exceptions) also serves for passerines (Heimerdinger and Ames, 1967).

Some correlation between closure of the caudal sternal margin and tree-trunk climbing habits was found by Feduccia (1972) in a number of passeriform, piciform and coraciiform birds. He suggested that an increased mass of abdominal musculature might help in holding the bird against the tree.

2. *Forelimb skeleton*

Descriptions of the wing bones (*ossa alae*) are given by Bellairs and Jenkin (1960), George and Berger (1966, pp. 13–15). Feduccia (1975) and Baumel (1979a). The present brief account will emphasise aspects not covered in those works. The forelimb skeleton of *Gallus* is shown in Fig. 2.2.

The humeral head (*caput humeri*) articulates with the glenoid cavity. At the proximal end of the *humerus* are several surface features for muscle attachment, most notably the pectoral crest (*crista pectoralis*) for insertion of the main flight muscle, the *m. pectoralis*. On the caudal surface of the proximal end of the humerus is an excavation, the pneumotricipital fossa (*fossa pneumotricipitalis*), which varies greatly in different forms, sometimes being double. Bock (1962) analysed this variation in passerine birds and concluded that the double fossa had evolved from a single condition. Bock felt that a failure to understand the functional significance of the pneumotricipital fossa and its variations weakened its value as a taxonomic character. Distally the humerus articulates with the *radius* and *ulna,* the latter being the stouter. These bones in turn articulate distally with the two free carpal elements, the radial carpal bone (*os carpi radiale*) and the ulnar carpal bone (*os carpi ulnare*).

The manus is supported by a complex element, the *carpometacarpus,* unique to birds. The carpometacarpus is formed during embryonic development by the fusion of several separate carpal and metacarpal anlagen. It provides a rigid platform for support of the primary flight feathers. The evolution of this structure by the coalescence of separate ancestral elements is a familiar story, based on a comparison of embryological data with the adult structure in *Archaeopteryx* and

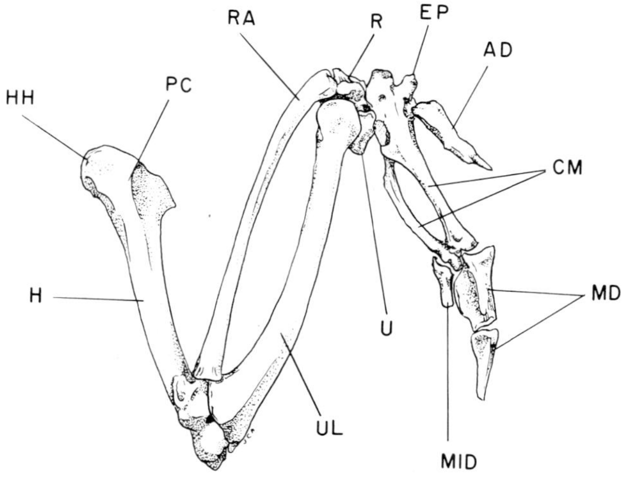

Fig. 2.2. Dorsal aspect of the forelimb skeleton of *Gallus*. AD, Alular digit; CM, carpometacarpus; EP, extensor process; H, humerus; HH, head of humerus; MD, major digit; MID, minor digit; PC, pectoral crest; R, radial carpal bone; RA, radius; U, ulnar carpal bone; UL, ulna.

reptiles. The ancestral condition is seen in *Archaeopteryx,* in which only three digits are present, as in modern birds. These are supported by three metacarpals, which are not co-ossified (Ostrom, 1976a). There are three carpal elements. In modern birds the three metacarpals (along with the rudimentary anlage of a fourth) arise separately in the embryo and then fuse together along with the distal carpal anlagen, to form the unified adult carpometacarpus. The evolutionary origin of this element has thus proceeded by a modification of embryonic development. The digits, at the same time, have undergone reduction in size and number of phalanges.

This story, however, is not as clear-cut as textbook accounts might suggest. There is considerable disagreement about the number and identity (homologies) of the embryonic structures going into the formation of the carpus and carpometacarpus. Hinchliffe (1977) reviewed the earlier studies of Montagna (1945), Holmgren (1955) and others, who investigated development through stained sections, identifying and delineating structures formed by the condensation and chondrification of the mesenchyme. One of the problems associated with this technique is that the borders of separate elements are not as distinct in the specimens as the outline drawings would suggest, and various accounts of the developing avian carpus claim that the number of pre-cartilaginous condensations range from as few as 4 to as many as 13. Hinchliffe (1977) re-studied the developing wing using autoradiography, marking chondroitin sulfate synthesis in limb buds. He found a simpler pattern of development in the carpus than reported by previous workers, with only four elements present. Montagna (1945) and Holmgren (1955) recognized more elements, interpreted as recapitulations of an ancestral tetrapod condition. In contrast, Hinchliffe (1977) argued that the evolution of the carpus has reached a more advanced stage in modern birds.

Another controversy has centred on the identity of the digits in the avian manus. Since only three are present, which two have been lost? Embryologists (Montagna, 1945; Holmgren, 1955) have considered the remaining digits to be the second, third and fourth. Hinchliffe (1977) accepted this but stressed the weakness of the supporting evidence. Others have argued that it is the first, second and third ancestral digits that are retained. Ostrom (1976a) noted that the phalangeal formula (2, 3, 4) of *Archaeopteryx* supports the latter view. He pointed out that the loss of the first and fifth digits would require the additional assumption that each remaining digit had also lost a single phalanx. The question cannot be considered as settled. As a practical solution to the problem, the structures have been re-named in the *Nomina Anatomica Avium* (Baumel, 1979a; Lucas, 1979) in a way that avoids the use of numbers. The terms are as follows: alular digit (*digitus alularis*) and alular metacarpal bone (*os metacarpale alulare*), major digit (*digitus major*) and major metacarpal bone (*os metacarpale majus*), and minor digit (*digitus minor*) and minor metacarpal bone (*os metacarpale minus*).

B. Muscles and movements

The forelimb muscles and their origins, insertions and actions are listed in Table I. Detailed descriptions of forelimb muscles are given in such works as

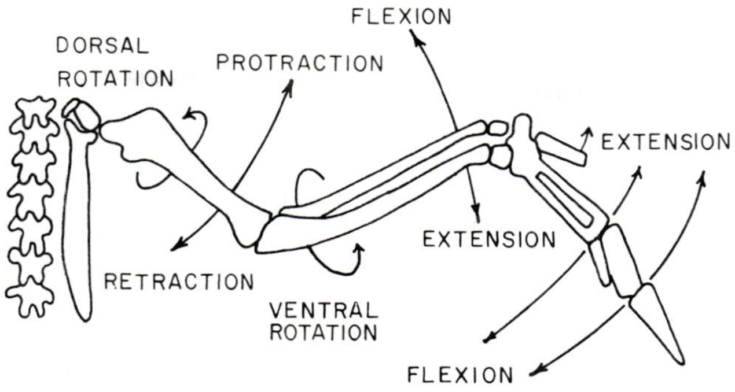

Fig. 2.3. Terminology describing the movements of the various parts of the avian forelimb. In addition, the limb may be elevated or depressed (see text).

Hudson *et al.* (1969, 1972), Vanden Berge (1970), Raikow (1977) and Rosser (1980); George and Berger (1966, pp. 288–379) review the earlier literature and provide detailed descriptions and comparisons of the muscles. Consequently in the present chapter only a general account of muscle structure will be given, with little consideration of the many variations that occur.

Movements of the forelimb (Fig. 2.3) are defined in relation to the standard anatomical position for birds, with the wing outstretched laterally as in gliding flight so that the wing surfaces are directed dorsally and ventrally (Lucas and Komárek, 1979). Movements were determined by observation of living birds, by manipulation of fresh and preserved specimens, and by reference to the literature. At any joint the potential range of movement is constrained by the arrangement of the joint surfaces. The arthrology of the forelimb is discussed by Sy (1936) and Baumel (1979b), and will not be considered to any extent in this account.

1. Movements of the arm

Movement of the humerus at the shoulder joint (Fig. 2.3) shows the greatest degree of mobility in the wing, and of course movements here affect the entire limb. Dorsal movement is elevation and ventral movement is depression. These are the principal movements of the upstroke and downstroke in flight. Movement cranially is protraction and movement caudally is retraction. Finally, movement around the axis of the wing itself provides rotation, which may be dorsal rotation, elevating the leading edge of the wing, or ventral rotation, depressing the leading edge. These movements arise from the rotation of the head of the humerus in the glenoid cavity.

(a) Depression. The downstroke of the wing in flight is brought about almost exclusively by the *m. pectoralis* (Figs. 2.6, 9). The *m. pectoralis* also gives rise to some small slips contributing to the support of the patagium or inserting into the skin (see later). As the main propulsive organ of flight, the *m. pectoralis* is the largest of the avian muscles, right and left muscles sometimes totalling one-fourth of the body weight. It arises mainly from the ventral part of the sternal carina and from the clavicle; additional areas of origin may include the sternal processes and membranes, the coracoid bone, the ribs, and the interosseous membranes of the

Table 1. **The forelimb musculature of birds.**

Name[a]	Origin[b]	Insertion	Action[c]
M. rhomboideus superficialis	Spinous processes of thoracic vertebrae	Dorsal margin of scapula	Draws scapula dorsocaudomedially
M. rhomboideus profundus	Spinous processes of thoracic vertebrae	Dorsal margin of scapula	Draws scapula dorsocraniomedially
M. serratus superficialis pars cranialis (serratus anterior)	Lateral surfaces of ribs	Ventral margin of scapula	Draws scapula caudoventrally
M. serratus superficialis pars caudalis (serratus posterior)	Lateral surfaces of ribs	Apex of scapula	Draws scapula ventrally
M. serratus superficialis pars metapatagialis (serratus metapatagialis)	Lateral surfaces of ribs	Deep surface of skin near elbow	Tenses metapatagium
M. serratus profundus	Cervical vertebrae or ribs	Costal surface of scapula	Draws scapula cranially
M. scapulohumeralis cranialis (proscapulohumeralis)	Scapula	Humerus	Retracts humerus
M. scapulohumeralis caudalis (dorsalis scapulae)	Scapula	Humerus	Retracts and elevates humerus
M. subscapularis	Scapula	Humerus	Retracts humerus
M. subcoracoideus	Coracoid and scapula	Humerus	Depresses and rotates humerus dorsally
M. coracobrachialis cranialis (coracobrachialis anterior)	Coracoid	Humerus	Protracts and depresses humerus
M. coracobrachialis caudalis (coracobrachialis posterior)	Coracoid	Humerus	Depresses and rotates humerus
M. sternocoracoideus	Sternum, ribs	Caudoventral end of coracoid	Respiratory movements
M. pectoralis	Sternum, clavicle, ribs, coracoid	Pectoral crest of humerus	Depresses humerus
M. supracoracoideus	Sternum and coracoid	Humerus	Elevates humerus
M. latissimus dorsi	Cervical and thoracic vertebrae	Humerus	Retracts and elevates humerus
M. tensor propatagialis pars longa (tensor patagii longus)	Clavicle, coracoid, scapula	Variously on carpometacarpus, radial carpal, alular digit	Extends manus, alular digit; flexes elbow; tenses propatagium
M. tensor propatagialis pars brevis (tensor patagii brevis)	Clavicle, coracoid, scapula	Belly of *m. extensor metacarpi radialis*	Flexes elbow; tenses propatagium
M. deltoideus major	Clavicle and scapula	Humerus	Elevates and retracts humerus
M. deltoideus minor	Coracoid, clavicle, scapula	Humerus	Protracts and elevates humerus
M. triceps brachii	Scapula and humerus	Ulna	Extends elbow
M. biceps brachii	Coracoid and humerus	Radius and ulna	Flexes forearm

M. expansor secundariorum	Humerus, humeroulnar trochlea, pectoral girdle or muscles	Follicles of proximal secondary remiges	Draws secondaries proximally
M. brachialis	Distal end of humerus	Proximal end of ulna	Flexes elbow
M. pronator superficialis	Distal end of humerus	Shaft of radius	Rotates forearm ventrally
M. pronator profundus	Distal end of humerus	Shaft of radius	Rotates forearm ventrally
M. flexor carpi ulnaris	Distal end of humerus	Ulnar carpal	Flexes wrist
M. flexor digitorum superficialis	Humerus	Major digit	Flexes wrist and extends major digit
M. flexor digitorum profundus	Ulna	Major digit	Flexes wrist, extends and depresses major digit
M. extensor metacarpi radialis	Humerus	Carpometacarpus	Extends wrist
M. extensor digitorum communis	Humerus	Alular digit, major digit	Flexes alular digit and manus, elevates manus, extends and elevates major digit
M. extensor metacarpi ulnaris	Distal humerus, proximal ulna	Carpometacarpus	Flexes wrist
M. extensor longus alulae (extensor pollicis longus)	Radius and/or ulna	Carpometacarpus	Extends wrist
M. extensor longus digiti majoris (extensor indicis longus)	Radius	Major digit	Extends manus and major digit
M. supinator	Distal humerus	Shaft of radius	Rotates forearm dorsally
M. ectepicondylo-ulnaris (anconeus)	Lateral surface of distal humerus	Ulna	Elevates forearm
M. entepicondylo-ulnaris	Medial surface of distal humerus	Ulna	Rotates forearm dorsally
M. ulnometacarpalis doralis (ulnimetacarpalis dorsalis)	Ulna	Carpometacarpus	Flexes wrist; in some species pulls primaries medially
M. ulnometacarpalis ventralis (ulnimetacarpalis ventralis)	Ulna	Carpometacarpus	Flexes and pronates wrist
M. interosseus dorsalis	Carpometacarpus	Distal phalanx of major digit	Extends and elevates major digit
M. interosseus ventralis (interosseus palmaris)	Carpometacarpus	Distal phalanx of major digit	Flexes and elevates major digit
M. extensor brevis alulae (extensor pollicis brevis)	Carpometacarpus	Alular digit	Extends and raises alular digit
M. abductor alulae (abductor pollicis)	Tendon of insertion of *m. extensor metacarpi radialis*	Alular digit	Extends and depresses alular digit

(continued)

TABLE 1. ***(Continued)***

Name[a]	Origin[b]	Insertion	Action[c]
M. flexor alulae (*flexor pollicis*)	Carpometacarpus	Alular digit	Flexes and depresses alular digit
M. adductor alulae (*adductor pollicis*)	Carpometacarpus	Alular digit	Flexes alular digit
M. abductor digiti majoris (*abductor indicis*)	Carpometacarpus	Proximal phalanx of major digit	Extends and depresses major digit
M. flexor digiti minoris (*flexor digiti III*)	Carpometacarpus	Minor digit	Flexes minor and major digits

[a] Names of muscles are from the *Nomina Anatomica Avium* (Baumel *et al.*, 1979). If the *N.A.A.* name differs from that used by George and Berger (1966), their term is given in parentheses.

[b] The origin and insertion given are typical for birds generally, but many exceptions and variations exist.

[c] The major action of the muscle as inferred from anatomical relations (see text).

thoracic girdle. In loons (Gaviidae), grebes (Podicipedidae), ducks (Anatidae), tinamous (Tinamidae), and some galliform birds the left and right pectorals arise in part from a midline septum extending ventrally from the carina (George and Berger, 1966, p. 307). In such forms the size of the carina will not provide an accurate estimate of pectoral muscle size, with which it is often taken to be closely correlated. The fibres of the *m. pectoralis pars thoracicus* converge cranoidorsally to insert on the pectoral crest at the cranial surface of the proximal end of the humerus. The fibre architecture of the muscle is complex, and considerable variation apparently exists. In some birds the muscle has two or three distinct layers (George and Berger, 1966, pp. 307–308; Vanden Berge, 1970). Even in the majority of birds, which have only a single layer, the internal arrangement of the fibres is not readily characterised in simple textbook terms. This

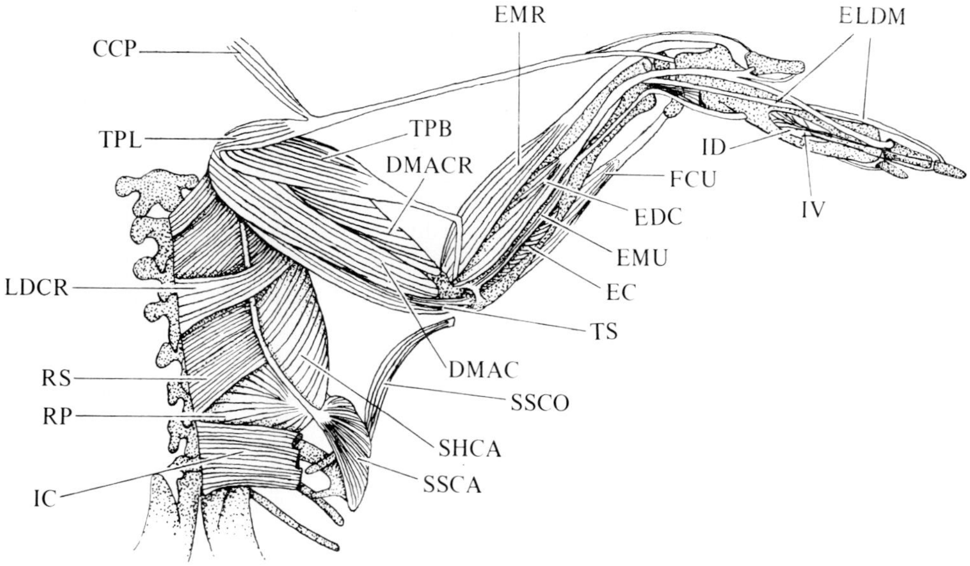

Fig. 2.4. Superficial dorsal muscles of the forelimb of a Hawaiian honeycreeper, *Loxops virens*. From Raikow (1977). Abbreviations for Figs. 2.4–12: ABA, *m. abductor alulae*; ADA, *m. adductor alulae*; ADM, *m. abductor digiti majoris*; B, *m. brachialis*; BB, *m. biceps brachii*; CC, *m. coracobrachialis caudalis*; CCP, *m. cucullaris capitis pars propatagialis*; CCR, *m. coracobrachilis cranialis*; DMA, *m. deltoideus major*; DMAC, *m. deltoideus major caudalis*; DMACR, *m. deltoideus major cranialis*; DMI, *m. deltoideus minor*; EBA, *m. extensor brevis alulae*; EC, *m. ectepicondylo-ulnaris*; EDC, *m. extensor digitorum communis*; ELA, *m. extensor longus alulae*; ELDM, *m. extensor longus digiti majoris*; EMR, *m. extensor metacarpi radialis*; EMU, *m. extensor metacarpi ulnaris*; ES, *m. expansor secundariorum*; FA, *m. flexor alulae*; FCU, *m. flexor carpi ulnaris*; FDM, *m. flexor digiti minoris*; FDP, *m. flexor digitorum profundus*; FDS, *m. flexor digitorum superficialis*; IC, *m. iliotibialis cranialis*; ID, *m. interosseus dorsalis*; IV, *m. interosseus ventralis*; LDCA, *m. latissimus dorsi pars caudalis*; LDCR, *m. latissimus dorsi pars cranialis*; P, *m. pectoralis*; PP, *m. pronator profundus*; PPB, *m. pectoralis pars propatagialis brevis*; PPL, *m. pectoralis pars propatagialis longus*; PS, *m. pronator superficialis*; RP, *m. rhomboideus profundus*; RS, *m. rhomboideus superficialis*; SBC, *m. subcoracoideus*; SBC D. Hd., *m. subcoracoideus* dorsal head; SBC V. Hd., *m. subcoracoideus* ventral head; SBS, *m. subscapularis*; SBS MED., *m. subscapularis* medial head; SC, *m. supracoracoideus*; SHCA, *m. scapulohumeralis caudalis*; SHCR, *m. scapulohumeralis cranialis*; SP, *m. serratus profundus*; SSCA, *m. serratus superficialis pars caudalis*; SSCO, *m. serratus superficialis pars metapatagialis*; SSCR, *m. serratus superficialis pars cranialis*; ST, *m. sternocoracoideus*; SU, *m. supinator*; TH, *m. triceps brachii* (humerotriceps); TPB, *m. tensor propatagialis pars brevis*; TPL, *m. tensor propatagialis pars longa*; TS, *m. triceps brachii (scapulotriceps)*; UD, *m. ulnometacarpalis dorsalis*; UV, *m. ulnometacarpalis ventralis*.

structure may be in part a consequence of the general proportions of the body, to which this large muscle makes a major contribution. Probably it is also functionally important in relation to the bird's mode of flight, but this requires investigation. The *m. pectoralis* is given minor assistance in wing depression by the much smaller *m. coracobrachialis caudalis* and perhaps also by the *m. subcoracoideus* (Figs. 2.7, 8, 9).

(b) Elevation. The largest elevator is the *m. supracoracoideus,* the smaller and deeper of the two flight muscles (Figs. 2.5, 7, 9). It arises from the dorsal part of the carina and adjacent body of the sternum, and sometimes also in part from the coracoid bone, coracoclavicular membrane and even the clavicle. Typically it is a single, bipennate muscle, but in some birds is of multiple structure (George and Berger, 1966, p. 313). Its tendon passes dorsally through the triosseal canal and turns laterally to insert on the dorsal side of the humerus. Thus, the muscle elevates the humerus despite its ventral position. Ostrom (1976b) hypothesised that the antecedent of this muscle in ancestral birds must have depressed the wing, and that its function became reversed with the elongation and modification of the coracoid, which is quite short in *Archaeopteryx.*

The humerus is also elevated by several dorsal muscles, of which the most important is the large *deltoideus major* (Figs. 2.4, 9), which arises on the clavicle, scapula, and sometimes adjacent structures such as the *os humoroscapulare.* It usually has two parts, which fan out to fleshy insertions on the dorsal side of the humerus. It is aided by the small *deltoideus minor* (Figs. 2.5, 9).

Several muscles whose main action is retraction of the humerus also contribute to its elevation; these include the *scapulohumeralis caudalis* and the *latissimus dorsi pars cranialis* and *latissimus dorsi pars caudalis,* when present (Figs. 2.4, 9). Ostrom (1976b) suggested that the small *coracobrachialis cranialis* (''ante-

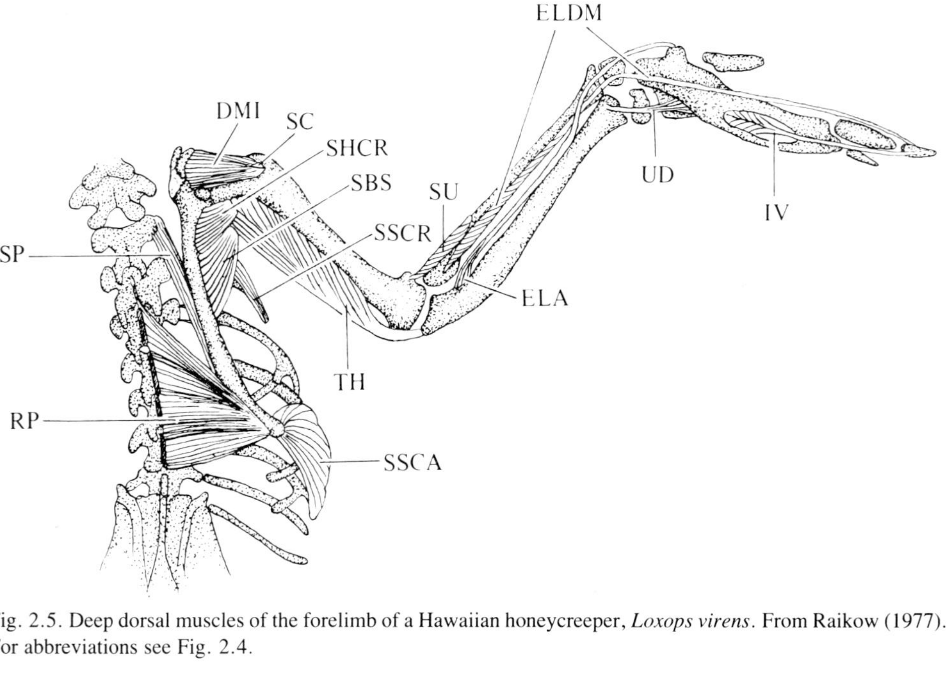

Fig. 2.5. Deep dorsal muscles of the forelimb of a Hawaiian honeycreeper, *Loxops virens*. From Raikow (1977). For abbreviations see Fig. 2.4.

rior''), whose main function is protraction, also aids in elevation (Fig. 2.9). However, Fisher (1946) considered that it has a minor action in depression instead, and my observation of the muscle in *Columba* accords with this. In sum, elevation of the wing is accomplished by one ventral muscle and several dorsal muscles, the latter also contributing variously to protraction or retraction according to the direction from which they approach the humerus.

(*c*) *Protraction.* The movement of swinging the humerus forward, as when the wing is being unfolded, is a relatively weak action provided by two small muscles in the region of the shoulder joint. The *m. coracobrachialis cranialis* (Fig. 2.9) is well developed in many non-passerines but is often reduced to a vestigial state or even lost in some passerines, as in most of the New World nine-primaried oscines (Raikow, 1978). The *m. deltoideus minor* varies considerably in its site of origin from one or more of the three bones (coracoid, scapula, clavicle) around the triosseal canal on the dorsal side of the shoulder joint, and inserts on the dorsal surface of the proximal part of the humerus, thus contributing also to elevation (Figs. 2.5, 9).

(*d*) *Retraction.* The humerous is swung caudally by several muscles, some of considerable size. This activity involves not only folding the wing, but fixing the humerus against the body when the wing is held folded, that is, a postural activity. The muscles involved include the *mm. latissimus dorsi* (both parts), the *scapulohumeralis caudalis,* the *scapulohumeralis cranialis,* the *subscapularis,* the *subcoracoideus* and the *deltoideus major.* They arise from various points around the thoracic girdle to converge on the proximal end of the humerus from

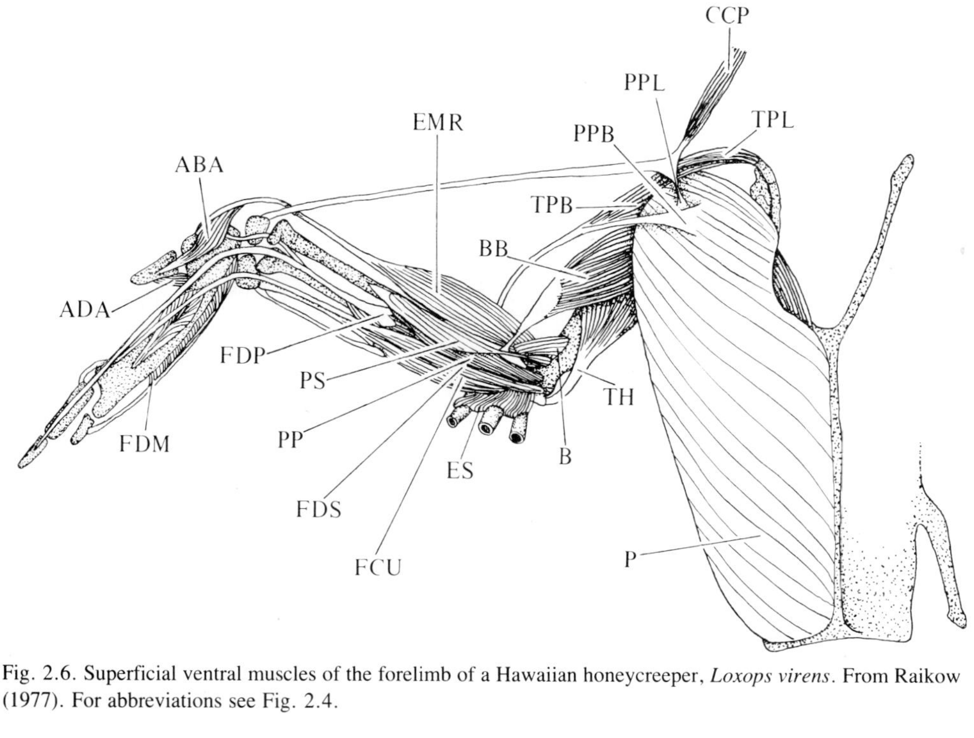

Fig. 2.6. Superficial ventral muscles of the forelimb of a Hawaiian honeycreeper, *Loxops virens.* From Raikow (1977). For abbreviations see Fig. 2.4.

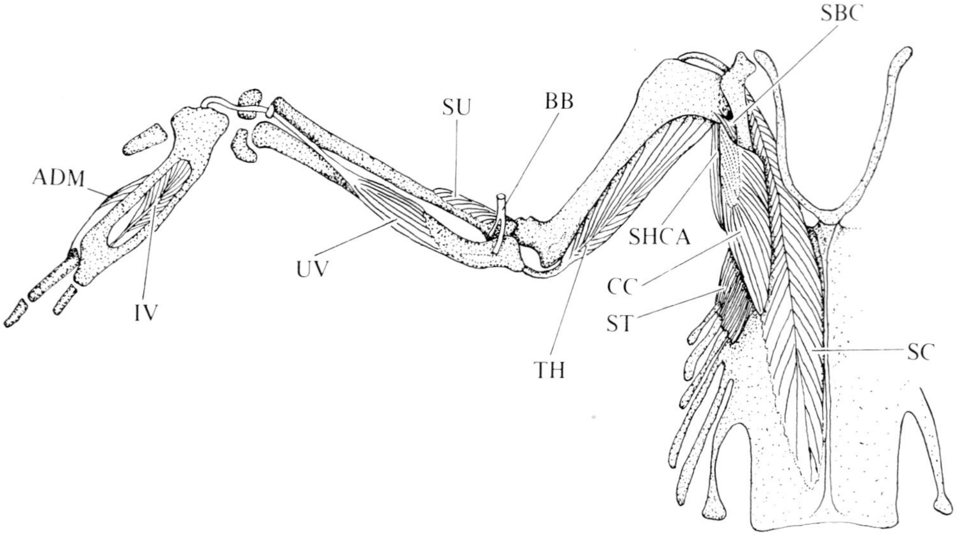

Fig. 2.7. Deep ventral muscles of the forelimb of a Hawaiian honeycreeper, *Loxops virens*. From Raikow (1977). For abbreviations see Fig. 2.4.

various angles (Figs. 2.4, 5, 7–9). The *m. latissimus dorsi pars caudalis* is absent in many birds.

(e) Rotation. Rotational movements of the humerus are probably mostly secondary results of the contraction of muscles whose lines of action function principally in one of the previously-mentioned movements. Dorsal rotation is accomplished by the *mm. supracoracoideus, coracobrachialis caudalis, deltoideus major* and *deltoideus minor*. The *mm. scapulohumeralis caudalis* and *pectoralis* provide ventral rotation, perhaps assisted by other muscles.

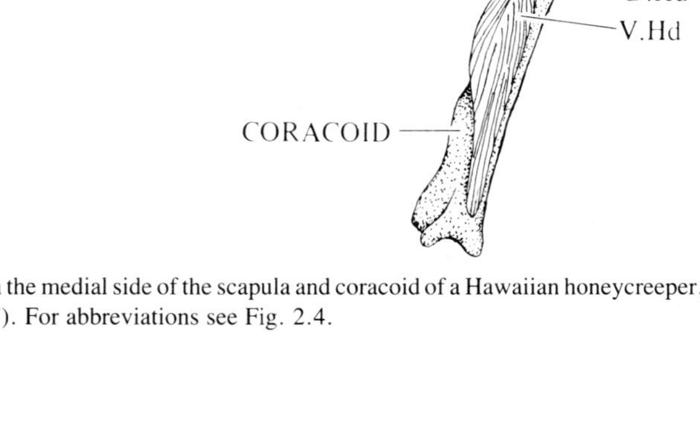

Fig. 2.8. Muscles on the medial side of the scapula and coracoid of a Hawaiian honeycreeper, *Psittirostra cantans*. From Raikow (1977). For abbreviations see Fig. 2.4.

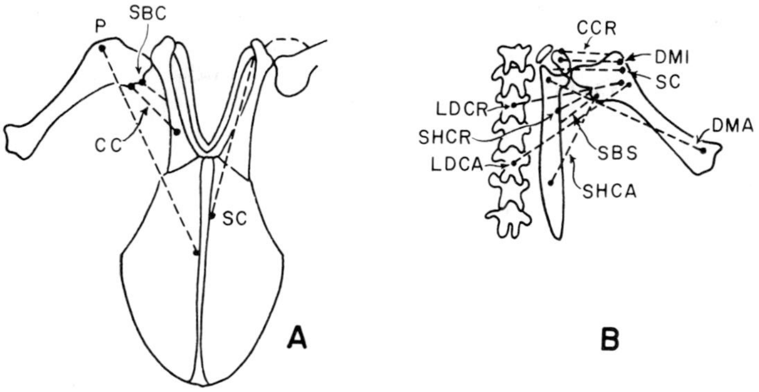

Fig. 2.9. Lines of action of the principal muscles providing for movements of the humerus. Lines of action (dashed lines) are drawn between the approximate anatomical centers of the origin and insertion of each muscle. (A) Ventral and (B) dorsal views of the pectoral girdle and humerus. See text for discussion. For abbreviations see Fig. 2.4.

2. *Movements of the forearm*

When the wing is spread, movement at the elbow (Fig. 2.3) is mostly limited to the plane of the wing, involving motion of the forearm relative to the arm. In extension the angle on the cranial side of the wing between the arm and forearm is enlarged, while in flexion it is reduced. The forearm may also be elevated or depressed relative to the arm, or rotated about its long axis. When the wing is spread the arrangement of the joint surfaces tends to lock the elbow in a way that minimises these movements, presumably stabilising the wing surface against the

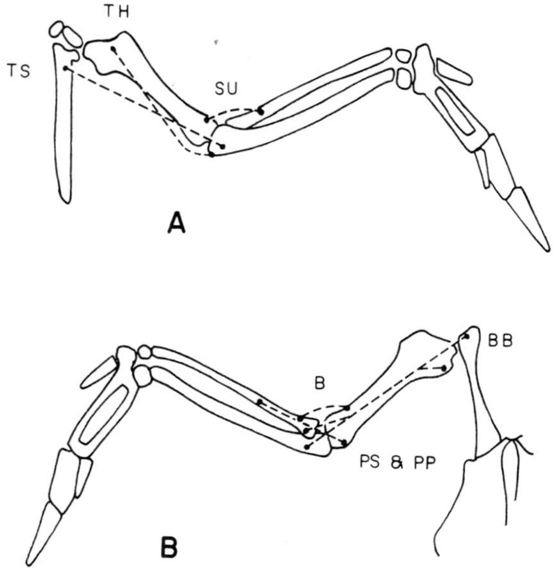

Fig. 2.10. Lines of action of muscles providing for movements of the forearm. (A) Dorsal view showing muscles that provide extension and dorsal rotation. (B) Ventral view showing muscles that provide flexion and ventral rotation. See text for discussion. For abbreviations see Fig. 2.4.

effects of air resistance during flight. When the elbow joint is flexed, however, movement at the joint becomes freer and the forearm rotates with its distal end turning ventrally as the wing is folded.

(a) Extension. The elbow joint is extended by two large muscles, the *mm. scapulotriceps* and *humerotriceps* (Figs. 2.4–7, 10). These originate on the scapula and humerus, respectively, pass down the caudal face of the humerus, and insert on or near the olecranon process of the ulna. The two muscles together constitute the *m. triceps brachii*: in some birds they are entirely separate, while in others they are connected (George and Berger, 1966, p. 335). In many birds a sesamoid (*os sesamoideum m. scapulotricipitis*) lies in the tendon of insertion of the *m. scapulotriceps.*

(b) Flexion. The main flexor of the elbow joint is the *m. biceps brachii,* which in size and position is the antagonist of the *triceps brachii* (Figs. 2.6, 7, 10). Although it has a double origin, from the coracoid and humerus, and inserts by a divided tendon onto the radius and ulna, it consists in most birds of a single, undivided belly. It is assisted by the *m. brachialis,* a short, wide muscle passing from the distal end of the humerus to the proximal end of the ulna (Figs. 2.6, 10). These two would seem to be the chief muscles directly concerned with flexion, but other muscles with different principal functions probably also contribute to flexion. In this category is the patagial muscle, the *m. tensor propatagialis,* and such forearm muscles as the *extensor metacarpi radialis* and *extensor digitorum communis* (Figs. 2.4, 6, 11, 12). It is impossible on the basis of descriptive

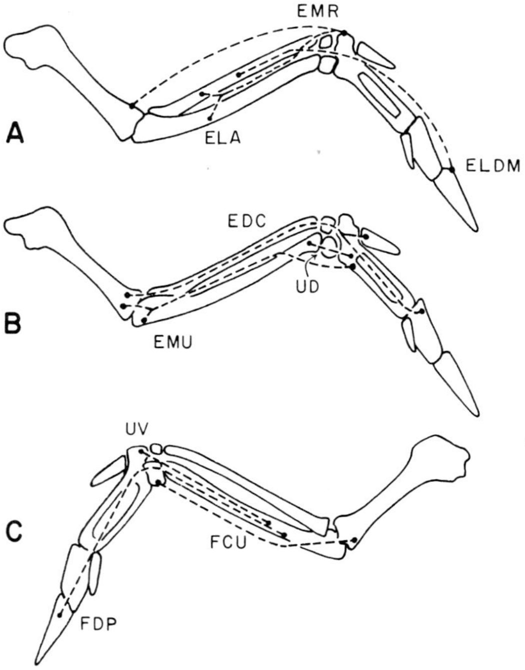

Fig. 2.11. Lines of action of muscles providing for movements of the manus. (A) Dorsal view showing muscles providing for extension. (B) Dorsal muscles involved in flexion. (C) Ventral muscles involved in flexion. See text for discussion. For abbreviations see Fig. 2.4.

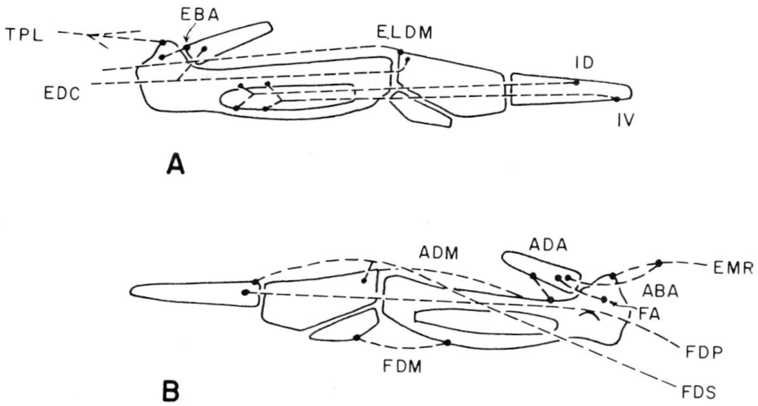

Fig. 2.12. (A) Dorsal and (B) ventral views of the manus showing lines of action of muscles providing for movements of the digits. Muscles that arise in the forearm are shown only as they pass to their insertions within the manus. See text for discussion. For abbreviations see Fig. 2.4.

anatomy alone to assess the roles of these muscles in elbow flexion, except to note the possibility of their contribution.

(c) Rotation. Rotation about the long axis of the forearm is achieved by raising or lowering the distal ends of the radius and ulna relative to each other, so that a twisting effect results. Dorsal rotation or supination involves raising the distal end of the radius and is accomplished by the *m. supinator,* which passes from the distal end of the humerus to the craniodorsal surface of the radius (Figs. 2.5, 7, 10). The opposite motion, ventral rotation or pronation, is provided by *mm. pronator superficialis* and *pronator profundus,* two closely associated muscles that pass from the distal end of the humerus to the caudoventral surface of the radius (Figs. 2.6, 10). In some passerines the *pronator profundus* has become divided into two largely separate muscles, one inserting more proximally and one more distally on the radius. Presumably these bellies can act independently of each other, which may increase the functional versatility of the muscle (Raikow, 1978).

3. Movements of the manus

Movement at the carpal joint (Fig. 2.3) involves motion of the manus relative to the forearm, and is constrained by the shapes of the joint surfaces at the articulations of the carpometacarpus, the carpal bones, and the radius and ulna. When the wing is spread, free movement is largely limited to the plane of the wing's surface. Motion of the manus cranially is extension of the carpal joint; this enlarges the angle between the manus and forearm on the caudal side of the wing. The opposite movement is flexion. Elevation, depression and rotation of the manus relative to the forearm are of more limited extent, and no muscles have these as their primary functions.

(a) Extension. Extending the carpal joint unfolds the distal part of the wing by spreading the primaries. The principal extensor is the *m. extensor metacarpi radialis,* which arises from the distal end of the humerus, passes distally as the

most cranial muscle of the forearm, and inserts by a tendon on the extensor process (*processus extensorius*) of the carpometacarpus (Figs. 2.4, 6, 11, 12). It is assisted by the smaller *m. extensor longus alulae,* which arises from the radius, ulna or both, passes distally between these elements, and inserts on the extensor process either adjacent to or in common with the tendon of the previous muscle (Figs. 2.5, 11). The manus is also extended by the *m. extensor longus digiti majoris,* which passes from the radius to the major digit: it thus extends the digit on the carpometacarpus as well as extending the manus as a whole (Figs. 2.4, 5, 11, 12). The *m. tensor propatagialis pars longa* also plays a minor role in extension (Fig. 2.4).

(b) Flexion. Several muscles contribute to flexion of the carpal joint, some directly, and others as secondary effects of their action on the digits. The largest flexor is the *m. flexor carpi ulnaris,* which forms the caudoventral margin of the forearm musculature (Figs. 2.4, 6, 11). It arises by a stout tendon from the distal end of the humerus, its passage around the elbow being guided by a ligamentous loop, the humeroulnar trochlea (*trochlea humeroulnaris*), whose considerable variation in birds has been analysed by Bentz and Zusi (1982). The muscle has two bellies, one of which has insertions on the bases of the distal secondaries. Both bellies, however, insert on the ulnar carpal bone. A dorsal counterpart to the previous muscle is the *m. extensor metacarpi ulnaris,* which arises from the distal end of the humerus and adjacent proximal end of the ulna, and inserts on the carpometacarpus (Figs. 2.4, 11). Although it functions as a flexor, it is named "extensor" because it appears to be derived from the primitive extensor group of muscles (George and Berger, 1966, p. 360). Additional direct flexors are the short *ulnometacarpalis dorsalis,* which passes from the distal end of the ulna to the carpometacarpalis (Figs. 2.5, 11), and the longer *ulnometacarpalis ventralis* (Figs. 2.7, 11). Two muscles that insert on the digits also contribute to flexion of the manus. These are the *flexor digitorum profundus* on the ventral side of the limb (Figs. 2.6, 11, 12) and the *extensor digitorum communis* on the dorsal side (Figs. 2.4, 11, 12). Fisher (1946) reported that the latter muscle acts as an automatic extensor of the manus when the forearm is extended, but Spring (1971) lists it as a flexor of the manus. The author's manipulation of a fresh specimen of pigeon supports the latter view.

4. Movements of the digits

Although the digits in birds are only three in number and their mobility is severely restricted by ligamentous connections, their movements are nonetheless important in flight. Detailed comparative accounts of their muscles are given by Stegmann (1978). The alular digit supports the alula, which forms a leading edge slot important in controlling lift during slow flight, taking off and landing. It is the most mobile digit, its principal movements being extension and flexion, by which the alula is raised away from the wing or appressed to it. Movement in other planes is more limited, but there is some capacity for elevation and depression.

The alular digit is controlled by as many as four small muscles intrinsic to the manus. The *m. abductor alulae* is unusual in that it arises from the tendon of another muscle, the *m. extensor metacarpi radialis,* the large extensor of the manus (Figs. 2.6, 12). Thus, in addition to its extension and depression of the alular digit, it may also act as a link connecting the *extensor metacarpi radialis* to the alula. The *abductor alulae* and *flexor alulae* (Fig. 2.12) are on the ventral side of the manus and thus depress the alula along with their extension and flexion actions. The *extensor brevis alulae* (Fig. 2.12) on the dorsal surface elevates as well as extends the alula. The *adductor alulae* appears only to flex the digit (Figs. 2.6, 12). The alula is also flexed and perhaps elevated by a branch of the forearm muscle, the *m. extensor digitorum communis,* and in some birds also receives an extensor branch of the *m. tensor propatagialis pars longa.* Passerine birds lack the *flexor alulae* and *extensor brevis alulae* (George and Berger, 1966; Raikow, 1977).

The major and minor digits move together, the minor digit acting mainly as a support for the major digit. These digits may be extended or flexed, which has the effect of spreading or folding the primary feathers. Because of the length of the primaries, a small movement of these digits can greatly alter the surface area of the wing, bringing about important adjustments during flight. The major digit may also rotate about its long axis, raising or lowering the leading edge of the primaries, or may be elevated or depressed relative to the carpometacarpus. These movements are highly restricted in extent, but no doubt of importance in subtle adjustments of the wing feathers.

Extension of the major and minor digits is brought about by muscles both extrinsic and intrinsic to the manus. The former include the *mm. extensor longus digiti majoris* and *extensor digitorum communis* (long branch) on the dorsal side of the manus (Figs. 2.4, 5, 11, 12), and the *mm. flexor digitorum superficialis* and *flexor digitorum profundus* ventrally (Figs. 2.6, 11, 12). Intrinsic extensors are the *mm. abductor digiti majoris* (Figs. 2.7, 12) and *interosseus dorsalis* (Figs. 2.4, 12). The digital complex is flexed by two small intrinsic muscles, the *mm. interosseus ventralis* (Figs. 2.4, 5, 7, 12) and *flexor digiti minoris* (Figs. 2.6, 12). A larger number of muscles, and a considerably larger mass of muscular tissue, are devoted to extension than to flexion. This is because the elastic properties of the forelimb ligaments draw the manus and its digits into a flexed (wing folded) position in the absence of muscular activity; the wing folds passively (as in death) unless actively extended.

Small degrees of elevation and depression of the major digit contribute to tilting of the plane of the primaries when the wing is extended, and are important in flight. These movements are contributed by some of the muscles already discussed: the *extensor digitorum communis* and the *interosseus* muscles elevate the major digit, while the *flexor digitorum profundus* and the *abductor digiti majoris* depress it.

5. *Movement and stabilisation of the thoracic girdle*

Movements of the thoracic girdle relative to the trunk are permitted to a limited degree by the articulation of the coracoid with the sternum, and the connections of the coracoid, scapula and clavicle at the shoulder. It is exeedingly difficult, however, to infer the functional roles of the various muscles. We do not know to what degree the mobility of the girdle involves contributions to the movements of the wings in flight, or alternatively, the bracing of the girdle to resist motions generated by the flight muscles. Such workers as Fisher (1946) and Owre (1967) have made suggestions as to the roles of the various muscles, but these are likely to be even more speculative than similar inferences about the limb muscles. With these reservations a few comments on muscular action may be offered (see Table I).

The *m. sternocoracoideus* (Fig. 2.7) binds the coracoid to the sternum and/or the rib cage. It may be more important as a respiratory muscle in expanding the rib cage than in relation to the forelimb. Indeed, Vanden Berge (1979, p. 196) questioned whether it has any functional role in wing movement.

The remaining muscles of the thoracic girdle insert principally on the scapula, but the movements of this bone affect the girdle as a whole. There are two muscle groups. The *rhomboideus* muscles (Figs. 2.4, 5) pass from the dorsal midline laterally to the scapula, while the muscles of the *serratus* group (Figs. 2.4, 5, 8) pass to the scapula from various positions on the rib cage or the most caudal of the cervical vertebrae. The scapula is drawn dorsally by the two *rhomboideus* muscles, and ventrally by both parts of the *serratus superficialis,* which are really separate muscles. The scapula is drawn cranially by the *m. serratus profundus,* perhaps assisted slightly by the *m. rhomboideus profundus.* It is drawn caudally by the *m. serratus superficialis pars cranialis,* with some assistance from the *m. rhomboideus superficialis.* These are the general movements to which each muscle contributes. Individually, however, a muscle may contribute to movement in more than one of these arbitrary directions.

6. *Support of the patagia*

(a) The propatagium. The surface area of the wing is greatly increased by a triangular fold of skin, the *propatagium,* which stretches between the shoulder and wrist. The leading edge of the propatagium is supported internally by the tendon of the *m. tensor propatagialis pars longa,* whose small fleshy belly lies superficially on the dorsal side of the shoulder, arising from the head of the clavicle and sometimes the adjacent coracoid and scapula (Fig. 2.4). Distally the tendon usually inserts on the extensor process of the carpometacarpus, and sometimes on adjacent structures such as the radial carpal or the alular digit. In some birds this tendon is joined by the *pars propatagialis* of the *m. cucullaris capitis,* a longitudinal muscular sheet in the neck. Additional support is often provided by the *m. pectoralis pars propatagialis longus,* which arises from the surface of the *m. pectoralis* near its insertion, and may be either a muscular slip or merely tendinous (Fig. 2.6). These structures provide a proximal pull, or at least resist distal tension along the tendon, and thus assist the *m. tensor propatagialis pars longa* in tensing the leading edge of the propatagium.

The central part of the propatagium is supported internally by another tendon, that of the *m. tensor propatagialis pars brevis* (Fig. 2.4). This muscle arises just caudal to *m. tensor propatagialis pars longa* in many birds, including passerines. In many non-passerines, however, the two muscles are fused into a single belly, which is therefore sometimes called the *m. tensor propatagialis longus et brevis.*

The brevis tendon passes distally to the forearm, and inserts on the belly or tendon of origin of the *m. extensor metacarpi radialis*. Near its origin this tendon is often supported by a muscular or tendinous slip from the surface of the *m. pectoralis*. This slip, the *m. pectoralis pars propatagialis brevis* (Fig. 2.6), may arise separately or in common with the *pars longa*. There is a great deal of variation in these patagial tendons, which have been reviewed in detail by George and Berger (1966, pp. 317–322).

The propatagium in some non-passerine birds has an additional muscular component, the biceps slip (*m. biceps brachii pars propatagialis*), which arises usually from the belly of the *m. biceps brachii,* and inserts on the tendon of the *m. tensor propatagialis pars longa,* helping that belly to draw the tendon proximally.

Although many authors have described the patagial muscles, they have principally concentrated on their use as taxonomic characters (Van Tyne and Berger, 1976, pp. 562–563) and little attention has been paid to their functions. It appears clear that the overall role of these muscles is to maintain the patagium in a more or less tensed condition so as to resist the deforming effects of air pressure during flight, and thereby to maintain the surface area of the wing. Because the wing changes shape constantly during the flapping cycle, as well as in spreading and folding, a fixed or rigid structure will not serve. Instead a mechanism is needed that will tense and relax the patagium as the shape of the wing changes during flapping flight, and along with the varying degrees of extension during gliding and soaring.

(b) The metapatagium. The surface of the wing is also enlarged caudally by a fold of skin, the *metapatagium,* between the trunk and elbow. This is smaller and simpler than the propatagium. It is apparently tensed by a small muscle, the *m. serratus superficialis pars metapatagialis,* which passes from the lateral surfaces of the ribs to insert under the skin near the elbow (Fig. 2.4).

7. *Movements of the remiges*

When a bird closes its wings the large flight feathers (remiges) fold together neatly and without interference. The primaries, which arise from the manus, slip under the secondaries, which are attached to the ulna. The manus is flexed against the forearm, and the major and minor digits are also flexed so that the shafts of the primaries lie parallel to each other and their vanes overlap. The distal secondaries swing inward, each overlying its proximal neighbor ventrally. When the wing is spread, however, the manus and the major and minor digits are extended and the remiges are unfurled, their shafts diverging so that the overlapping vanes separate. This process, like spreading and folding a hand of playing cards, is familiar to anyone who has handled a living or freshly-killed specimen. The co-ordination of the movements of the remiges depends on the action of specialised muscles and ligaments that together form a functional system unique to birds and basic to the design of the forelimb as a wing. Some functional aspects were studied by Sy (1936) and earlier workers, but except for the description of the alar ligaments by Baumel (1979b), there does not seem to have been any current study of the system. The following account is based mainly on the dissections by the author of

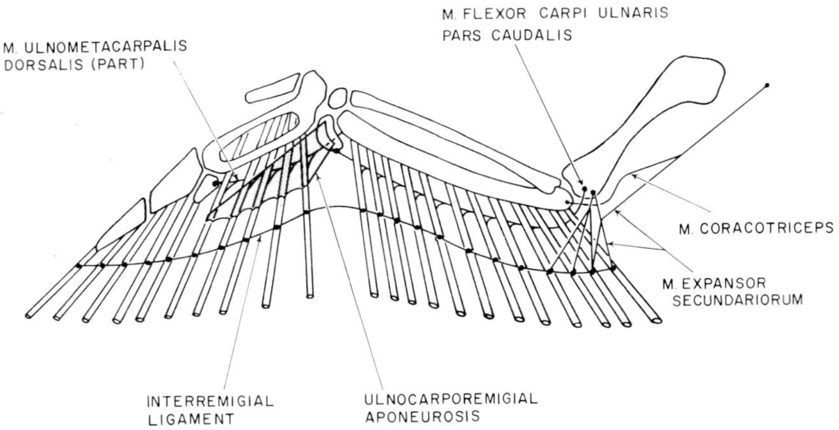

Fig. 2.13. Diagrammatic representation of the principal muscular and ligamentous mechanisms involved in coordinating the spreading and folding of the remiges in *Columba*. Ventral view. See text for discussion.

Columba and the House Sparrow (*Passer domesticus*). The individual functional components are considered first, followed by a discussion of the mechanism (Fig. 2.13).

(a) Interremigial ligament. The elastic interremigial ligament (*ligamentum elasticum interremigiale*) lies in the caudal edge of the skinfold of the forearm (*postpatagium*) and manus. It is anchored distally to the base of the outermost primary, and passes proximally, interconnecting the primaries and secondaries. At the elbow it is closely associated with the insertion of the *m. expansor secundariorum* (see later).

(b) Ulnocarporemigial aponeurosis. The ulnocarporemigial aponeurosis (*aponeurosis ulnocarporemigialis*) is a sheet of dense connective tissue which arises primarily from the ulnar carpal on the ventral side of the wing, and fans out to insert on the bases of several of the proximal primaries. In *Columba* it is a continuous sheet containing several thicker bands of fibers passing to the feathers, with thinner areas between. The fibrous bands are more distinctly separate in the House Sparrow (*Passer domesticus*).

(c) M. ulnometacarpalis dorsalis. The main attachments of the *ulnometacarpalis dorsalis* muscle are on the distal end of the ulna and the proximal end of the carpometacarpus. As noted before, it serves as a flexor of the manus. In passerines this appears to be the extent of its function (although many families have not been examined). In non-passerines there are often multiple insertions on the bases of several proximal primaries (George and Berger, 1966, p. 371). A separate belly in *Columba* passes distally to insert on the minor digit; along the way it has multiple insertions on the metacarpal primaries. Fisher (1946) described this also in cathartid vultures under the name "*m. flexor metacarpi posterior.*"

(d) M. flexor carpi ulnaris pars caudalis. The *m. flexor carpi ulnaris pars caudalis,* along with the *pars cranialis,* passes from the humerus to the ulnar carpal bone, and flexes the manus as described earlier. In addition it gives off a series of stout tendinous insertions, the *digitationes remigiales,* to the bases of all the secondaries apart from a few of the most proximal ones.

(e) M. expansor secundariorum. The *m. expansor secundariorum* is composed of smooth muscle fibres, and thus perhaps has its evolutionary origin as part of the cutaneous musculature. It is present in most birds and shows considerable diversity, which has been reviewed by Berger (1956) and George and Berger (1966, pp. 338–344). In passerines and many non-passerines it arises from the distal end of the humerus and associated ligaments, and fans out to insert on the bases of several of the most proximal secondaries. In many non-passerines there is a second origin via a long tendon that arises in various ways from the thoracic girdle and its muscles.

(f) M. coracotriceps. The *m. coracotriceps* is a tiny striated muscle that is present in some non-passerines, including *Columba,* but is sometimes reduced to a vestigial state (George and Berger, 1966, pp. 336–337). It arises by a tendon that branches from the long tendon of the *m. expansor secundariorum,* and inserts at the elbow in association with the insertion of the *m. triceps brachii.* Little is known about its structural variation amongst birds.

(g) The mechanism. The secondaries are movably attached to the ulna so that they can swing proximally or distally with ease. The more proximal primaries are likewise capable of such movement, but this mobility decreases distally, and the outer primary, the calamus of which is firmly fixed in a groove on the major digit, is virtually immobile. When the manus and major digit are extended, the outer primary is thereby swung outward. The next several primaries follow, in part because of their relatively-fixed attachments to the digits and carpometacarpus, and in part by the pull of the interremigial ligament. This ligament also transmits the pull to the secondaries, which are likewise spread distally. Thus, the interremigial ligament, probably aided to some degree by the skinfold itself, as well as the *m. flexor carpi ulnaris pars caudalis,* co-ordinates the unfolding of the remiges. If it were inelastic the feathers would turn together and not spread apart, but because it stretches they not only swing over but diverge. This can be verified by cutting the ligament in a fresh specimen, which hinders the effect.

Other structures counteract this effect by exerting passive or active tension on the calami in a proximal direction, so as to resist distal movement or to bring about proximal rotation. The ulnocarporemigial aponeurosis resists the lateral movement of the metacarpal (proximal) primaries; this effect is greatest on the most proximal primary and least on the most distal one involved. As a result, when the manus is extended the primaries do not merely swing outward in a group, but fan out as their shafts diverge from their initially parallel orientation. This effect is easily demonstrated by manipulating a fresh specimen that has been plucked; in a small bird like the House Sparrow (*Passer domesticus*) it is often unnecessary even to skin the wing because the silvery ulnocarporemigial aponeurosis is easily visible through the skin. In a bird of this size the interaction of the interremigial ligament and the ulnocarporemigeal aponeurosis is evidently sufficient to spread the primaries. In others, like *Columba,* the effect is aided by the *m. ulnometacarpalis dorsalis,* the contraction of which adds an active force to the passive effect of the ulnocarporemigeal aponeurosis.

For the secondaries the *m. flexor carpi ulnaris pars caudalis* and the *m. expansor secundariorum* provide the proximal tension to resist or counteract the outward pull of the interremigeal ligament. The former muscle affects the more distal secondaries. The latter controls the most proximal group directly, but via the interremigeal ligament must also draw the entire set of remiges proximally. The variations in the *m. expansor secundariorum* must be functionally related to the force needed to bring about this action. The long, tendinous connection to the thoracic girdle in many birds may act as a passive supplement to the contraction of the *m. expansor secundariorum,* especially when the shoulder joint is extended and the tendon tensed. Possibly the *m. coracotriceps* extends the range of positions in which this is effective by shortening so that the tendon is taut when the shoulder joint is less fully extended.

The closing of the remigeal fan when the wing is being folded appears to be a more passive side-effect of the flexion of the limb segments, perhaps assisted by contraction of the three muscles discussed previously. The foregoing explanation for the wing spreading and folding mechanism is based on a passive consideration of anatomical relationships and manipulation of specimens, and is clearly incomplete. The roles of various lesser connective tissue structures not discussed already, and the actual patterns of muscular activity remain to be elucidated. The time is ripe for a modern functional anatomical study of this mechanism, including its gross structural, histological and physiological aspects.

C. Locomotor specialisations

1. Adaptive variations in flight

Birds show many different modes of flight associated with different ways of life. Flapping flight, soaring over land or water, diving, hovering, and so forth are all correlated with specific morphological characteristics. Particular attention has been paid to wing shape (Savile, 1957), wing tip slotting and wing loading in relation to aerodynamics. Other factors considered important include the relative weight of the flight muscles as an estimation of their strength, and the relative weight of the heart as a predictor of endurance. Hartman (1961) and Greenewalt (1962) compiled quantitative data on these factors. A general theory of avian flight would require an integration of morphology, physiology and aerodynamics. Although much progress has been made in this direction, our understanding of the biological basis for adaptive variation in avian flight is still largely at a descriptive and speculative level, and lacks a fully comprehensive experimental foundation. A detailed review of the topic is outside the scope of this chapter, but is readily available in a number of works including Gray (1968), Herzog (1968), Rüppell (1977), Norberg (1981) and Rayner (1981).

2. Wing-propelled diving

Mechanisms of underwater propulsion have evolved a number of times in different groups of birds. Foot-propelled diving will be discussed later. Wing-propelled diving, often (though somewhat loosely) termed "underwater flight," has arisen independently in penguins (Spheniscidae), auks (Alcidae), and diving petrels (Pelecanoididae), and in the extinct family Plotopteridae. The penguins are probably the most highly-specialised wing-propelled divers, with wings greatly modified into reduced, relatively rigid paddles suited to the density of water. The wing skeleton is dorsoventrally flattened, and its joints allow little movement between bones. The wing musculature is profoundly modified (Schreiweis, 1982). The *propatagialis* is present as a single belly, its tendon extending along the cranial margin of the humerus and radius, the patagium being absent. The *m. brachialis* inserts on the radius rather than the ulna because of the extreme wing flattening, but shows an occasional vestigial ulnar attachment. Many wing muscles are small or vestigial. Schreiweis (1982) found that eight muscles were replaced by tendons (more accurately, perhaps, termed ligaments), although in some individual birds a number of tendons had a few vestigial muscle fibres. A dozen wing muscles are entirely lost, except as occasional vestiges. Only nine muscles distal to the shoulder were found by Schreiweis (1982) to have fleshy fibres.

The wing area is reduced compared to flying birds, and is covered with small feathers; the large remiges of flying birds are lacking. Clark and Bemis (1979) studied underwater swimming in penguins by videotape and motion picture analysis of birds in a large tank. They showed that penguins produce thrust during both phases of the wingbeat cycle, elevation and depression, and not just during depression as in most aerial flight. In association with this, the chief elevator, the *m. supracoracoideus,* is unusually large relative to the *m. pectoralis*.

Flightless wing-propelled divers also evolved in the order Pelecaniformes. The Plotopteridae of Oligocene and Miocene times have been studied most extensively by Olson (1980), who stressed their convergent similarities with penguins.

Flightless auks are known from Miocene to recent times, but except for the recently extinct Great Auk (*Pinguinus impennis*), extant representatives of the family use their wings for both aerial and underwater flight. Storer (1960, 1971; see also Feduccia, 1980, pp. 74–77) presented a theory for the evolution of wing-propelled diving that postulated three successive stages. In the first, wings are used for aerial flight only, in the second, for both aerial and sub-marine flight, and in the third, for sub-marine flight only. Storer regarded the second stage as a "compromise" adaptation because a smaller wing is more effective underwater, a larger wing in air. In this sense the second stage may be seen as a relatively inefficient intermediate condition between two adaptive peaks. Nevertheless, the flightless auks are all extinct, while the less-specialised "compromisers" of today are numerous and successful. Spring (1971) analysed the underwater flight of two species of auks and found the wing movements to be complex, and anyone who watches auks swimming underwater cannot help but be impressed by their manoeuvrability. Comparing Spring's description of wing movements in auks with that of Clark and Bemis (1979) for penguins, one gets the impression that it would be simplistic to regard auks as "incomplete penguins," or to assume that they are necessarily evolving toward flightlessness. The adaptive value of a whole life-style may be enhanced by the avoidance of extreme specialisation in one particular direction.

TABLE II. **Flightless and nearly flightless birds.**

Taxa	Time of occurrence	Distribution	References[a]
Order Paleognathiformes			de Beer (1956)
Casuariidae (cassowaries)	Pleistocene–recent	Australia, New Guinea	
Dromaiidae (emus)	Pleistocene–recent	Australia	
Dromornithidae[b]	Miocene–Pleistocene (recent?)	Australia	Rich (1980)
Apterygidae (kiwis)	Pleistocene–recent	New Zealand	
Dinornithidae (moas)[b]	Pleistocene–recent	New Zealand	Cracraft (1976a,b)
Aepyornithidae (elephant-birds)[b]	Pleistocene–recent	Madagascar	
Struthionidae (ostriches)	Pliocene–recent	Europe, Asia, Africa	
Rheidae (rheas)	Eocene–recent	South America	Raikow (1969) Bruning (1974)
Order Gruiformes			
Diatrymidae[b]	Paleocene–Eocene	Europe, North America	
Phorusrhacidae[b]	Oligocene–Pleistocene	South and North America	
Bathornithidae[b]	Eocene–Miocene	North America	Cracraft (1973)
Mesitornithidae (mesites)[c]	Recent	Madagascar	Glenny and Friedmann (1954)
Rhynochetos jubatus (Kagu; Rhynochetidae)	Recent	New Caledonia	
Rallidae (rails)[d]	Oligocene (possibly Eocene)–recent	Widespread	Olson (1973, 1977a)
Order Anseriformes: Anatidae			
Thambetochen chauliodus[b]	Pleistocene	Hawaii	Olson and Wetmore (1976)
Cnemiornis calcitrans[b]	Pleistocene	New Zealand	
Chendytes lawi[b]	Pleistocene–recent	North America	Morejohn (1976)
Tachyeres pteneres, *T. leucocephalus*, *T. brachypterus*, *T. patachonicus*[e] (steamer ducks)	Recent	South America, Falkland Islands	Humphrey and Livezey (1982)
Anas auklandica (Chestnut Teal)	Recent	Aukland and Campbell Islands, New Zealand	Weller (1980)
Order Columbiformes: Columbidae			
Raphus cucullatus (Dodo)[b]	Recent	Mauritius Island	Hutchinson (1954)
Pezophaps solitarius (solitaire)[b]	Recent	Rodriguez Island	Hutchinson (1954)
Order Strigiformes: Strigidae			
Ornimegalonyx oteroi[b,f]	Pleistocene	Cuba	Arredondo (1976)

Order Psittaciformes: Psittacidae			
Strigops habroptilus (Kakapo)	Recent	New Zealand	
Order Ciconiiformes: Threskiornithidae			
Apteribis glenos[b]	Pleistocene	Hawaii	Olson and Wetmore (1976)
Xenicibis xympithecus[b]	Late Quaternary	Jamaica	Olson and Steadman (1977)
Order Pelecaniformes			
Nannopterum harrisi (Galapagos Flightless Cormorant: Phalacrocoracidae)	Recent	Galapagos Islands	
Plotopteridae[b]	Oligocene–Miocene	North America, Japan	Olson (1980)
Order Podicipediformes: Podicipedidae			
Rollandia micropterum (Short-winged Grebe)	Recent	South America	
Podilymbus gigas (Atitlan Grebe)[c]	Recent	Central America	
Order Hesperornithiformes[b]	Cretaceous	Europe, North and South America	Martin and Tate (1976), Martin (1980)
Order Sphenisciformes: Spheniscidae (penguins)	Eocene–recent	Southern hemisphere	Simpson (1946), Clark and Bemis (1979)
Order Charadriiformes: Alcidae			
Pinguinis (great auks)[b]	Pliocene–recent	North America, North Atlantic Islands	Olson (1977b)
Mancallinae (Lukas auks)[b]	Miocene–Pleistocene	North America	Howard (1976)
Order Passeriformes			
Menura novaehollandiae, M. alberti (lyrebirds, Menuridae)[c]	Recent	Australia	Sibley (1974), Smith (1976)
Atrichornis rufescens, A. clamosus (scrub-birds: Atrichornithidae)[c]	Recent	Australia	Sibley (1974), Smith (1976)
Zeledonia coronata (Wren-Thrush: Parulidae)[c]	Recent	Central America	Hunt (1971), Raikow (1978)
Xenicus lyalli (Stephen Island Wren: Acanthisittidae)[b,f]	Recent	New Zealand	

[a] For information on many flightless and nearly flightless birds see Greenway (1967) and Feduccia (1980, Chapter 6).
[b] Extinct.
[c] Nearly flightless or very reluctant to fly.
[d] The number of flightless rails (Rallidae) is too large to list here. See Olson (1973) and Ripley (1977).
[e] Some individuals flightless. See text.
[f] Uncertain whether nearly flightless or completely so.

3. *Flightlessness*

(a) Morphological changes. Because the basic organisation of the avian body is so closely associated with its adaptation for flight, a fightless bird is almost a contradiction in terms. Yet the frequency and diversity of groups in which flightlessness has arisen (Table II) suggests that the ability to fly may be a luxury the costs of which have often exceeded its advantages. Flightlessness is associated with various modes of life, but two patterns stand out. Most commonly it occurs in small birds living on islands that lack predators, as in the numerous flightless rails (Rallidae) and waterfowl (Anatidae). Alternatively, it often occurs in continental birds that have achieved extreme body size when freed from the weight constraints of flight. The ratites and such extinct carinates as the diatrymas and phorusrhascids exemplify this trend. Whatever its ecological context, flightlessness is usually associated with a common pattern of morphological changes having reduction of the forelimb as the basic theme. At its extreme this involves a diminution in the size of the skeleton of the wing and thoracic girdle, along with the flight muscles, especially the *mm. pectoralis* and *supracoracoideus*. The sternum is reduced, and the keel becomes shallow or is lost entirely. The clavicles may entirely disappear as in ratites, or their distal portions may be reduced to ligaments (Glenny and Friedmann, 1954) or disappear completely apart from the epicleideum. The angle between the scapula and coracoid increases from acute to obtuse, and the two bones may fuse into a single scapulocoracoid. Feduccia (1980) has reviewed these changes in numerous groups and considers that the selective advantage of the complex modification lies in the saving of energy that would otherwise be needed to grow, maintain and power the full flight apparatus. The similar pattern of changes in diverse flightless groups suggests a common mechanism, and Feduccia (1980) develops the discussion of Olson (1973) that flightlessness evolves through neoteny, a modification of ontogeny. Changes in the relative growth rates of different parts of the body can bring about great modifications in the body proportions of the adult. Olson (1973) analysed the evolution of island rails, and showed that they evolved from flying ancestors by retarding the development of the forelimb relative to the rest of the body. In essence, the adults reach full size and maturity while retaining sub-adult body proportions.

General accounts of flightlessness have been provided by Olson (1973), Feduccia (1980) and Diamond (1981). Some aspects not emphasised by those authors are discussed next.

(b) Passerine birds. Considering that passerines constitute well over half the living species of birds, it is remarkable that flightlessness is so rare in this group; indeed there is apparently no passerine species known with certainty to be completely flightless. The leading candidate is the Stephen Island Wren (*Xenicus lyalli*) of the endemic New Zealand family Acanthisittidae. This species inhabited a small island lacking predators, where it lived on the ground among the rocks and was never seen to fly (Gilliard, 1958, p. 289). Unfortunately it was exterminated by a domestic cat in 1894, so we will never know whether it retained any residual ability to fly.

The Wren-Thrush (*Zeledonia coronata*) is a Central American oscine related to the New World warblers (Parulidae) (Raikow, 1978). It feeds on the ground or in low, dense vegetation. Hunt (1971) reported that it makes only short flights (usually under 10 m) by gliding or fluttering downhill from a perch, and he considered that it was incapable of sustained flights.

Australia is the home of two small families of nearly flightless passerines. The lyrebirds (Menuridae) include two species in a single genus, and the scrub-birds (Atrichornithidae) also comprise a pair of congeneric species. Neither group is completely flightless. Lyrebirds can leap and flutter up to a low branch, and may work their way up through a tree in this manner, and then glide down. Scrub-birds are less capable, but can make low flights of a few meters. Specimens of the Superb Lyrebird (*Menura novaehollandiae*) and the Noisy Scrub-Bird (*Atrichornis clamosus*) have been dissected by the author, and both show structural changes associated with the reduction of flight, although these are more advanced in the scrub-bird. In both, the flight muscles are reduced in size. In five North American oscines dissected for comparison, the *m. pectoralis* constituted an average of 12.64% of the body weight, *m. supracoracoideus* 1.42% (bilateral values). In contrast the *pectoralis* muscles in *Menura* accounted for only 5.64% of the body weight, and the *supracoracoideus* muscles only 0.64%. In *Atrichornis,* muscle reduction has gone farther, the corresponding values being 3.71% (*m. pectoralis*) and 0.38% (*m. supracoracoidenus*) (Fig. 2.14). The forelimb appears to be slightly shortened in *Menura,* while in *Atrichornis* it is distinctly shortened. Most notably, as first reported by Garrod (1876), the clavicles in *Atrichornis* lack the distal portions of their shafts; they are therefore not in contact ventrally and so there is no furcula as such (Fig. 2.15). This is presumably associated with the great reduction of the *m. pectoralis,* part of which normally arises from the clavicle. *Menura* retains an intact furcula (Fig. 2.15).

Why is flightlessness so rare in passerines? There is no obvious answer, but

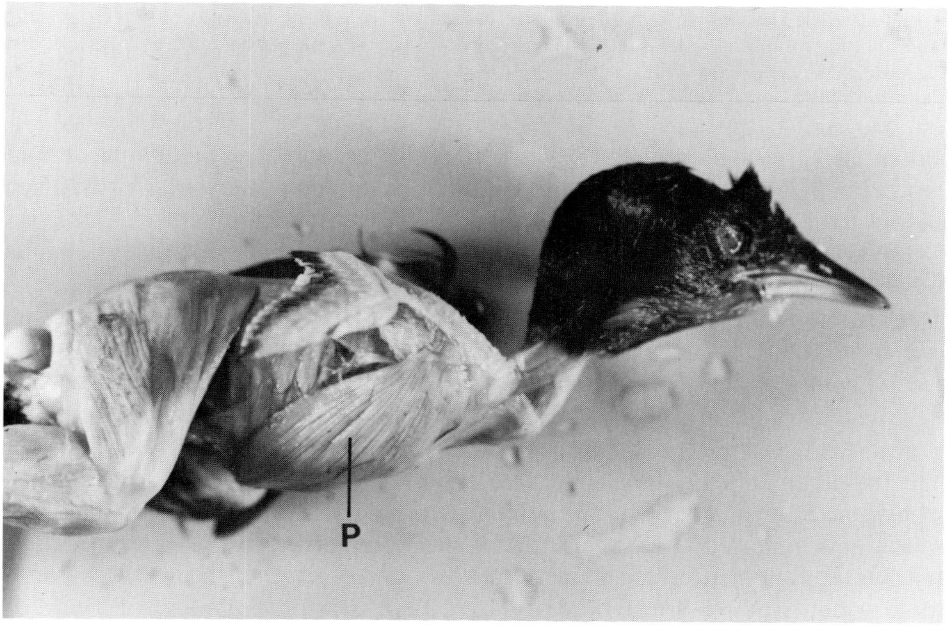

Fig. 2.14. Skinned specimen of the almost flightless Noisy Scrub-Bird (*Atrichornis clamosus*) showing the greatly-reduced *m. pectoralis* (P).

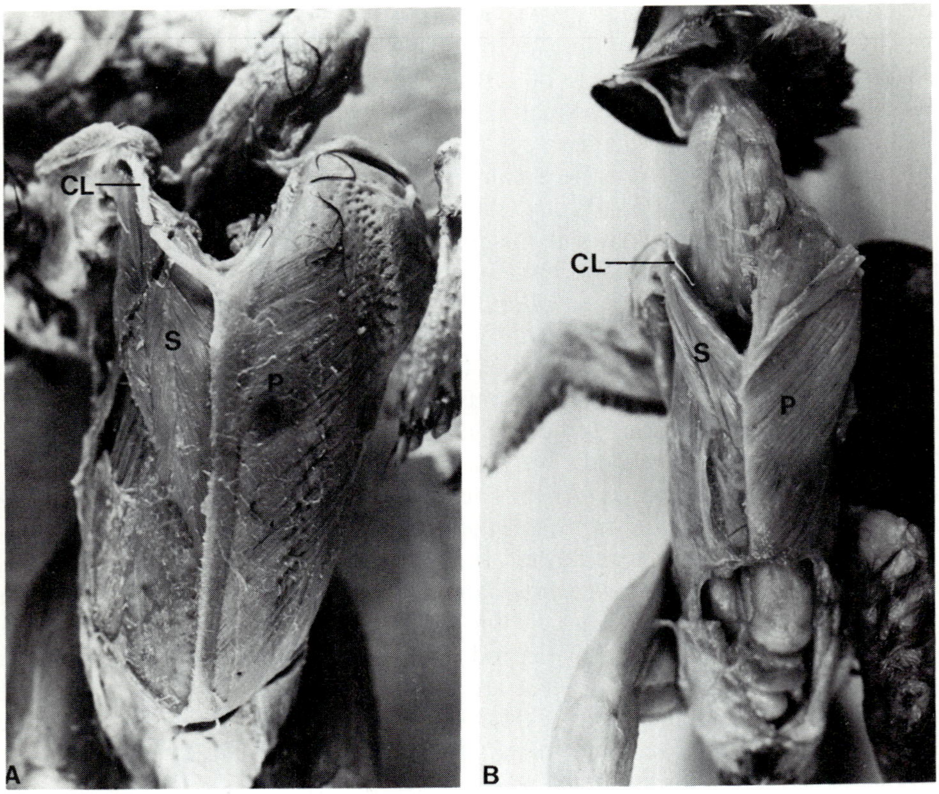

Fig. 2.15. Ventral views of specimens of (A) a Superb Lyrebird (*Menura novaehollandiae*) and (B) a Noisy Scrub-Bird (*Atrichornis clamosus*). The *m. pectoralis* has been removed on one side. In *Menura* a furcula is present, since the clavicles are complete and meet ventrally. (The clavicle in *Menura* was bisected during dissection.) In *Atrichornis,* however, the clavicles are reduced to vestiges and do not meet. CL, Clavicle; P, *m. pectoralis*; S, *m. supracoracoideus*.

perhaps it is because the majority are arboreal in habits, so that flight remains important. Flightlessness usually arises in either aquatic or terrestrial birds. There are no really aquatic passerines, but it is probably significant that the few flightless or near-flightless passerines just discussed are of terrestrial habits. Nevertheless, most ground-dwelling passerines have not become flightless. In any event it is apparent that when they do evolve towards flightlessness passerines exhibit structural changes comparable to those in non-passerines.

(c) Steamer ducks. To understand fully how birds become flightless, it would seem necessary to study a species just at the time that the event is occurring. The scrub-birds described earlier come close to this, and demonstrate that considerable reduction of the flight apparatus can precede the complete loss of flight behaviour. Still, they are not quite flightless. In contrast, Humphrey and Livezey (1982) have reported on a species that appears just now to be making the transition. Of the four species of steamer duck (*Tachyeres* spp.) of the southern part of South America, three are flightless, whilst *Tachyeres patachonicus* of Tierra del Fuego, Patagonia, and the Falkland Islands is known as the Flying Steamer Duck for obvious reasons. Previous workers had noted the occasional reluctance or

even apparent inability of some Flying Steamer Ducks to take off, and this was confirmed by Humphrey and Livezey. These workers measured the wing loadings of the steamer ducks and found that the wing loading of *Tachyeres patachonicus* is the highest recorded for flying birds, whilst that for the flightless steamer ducks is even higher. Using the estimate of Meunier (1951), that avian flight becomes impossible with wing loadings of 2.5 g/cm^2 and greater, they concluded that many adult male Flying Steamer Ducks are unable to fly, and this would explain why several of the male Flying Steamer Ducks which they collected did not fly when evading pursuit. Furthermore, they estimated that many males that have not quite reached the threshold value could be rendered flightless by the additional weight of a full meal.

These findings suggest a scenario for the evolution of flightlessness in steamer ducks that may also be applicable to other birds. Apparently the birds undergo evolutionary changes in absolute size or in relative proportions (or both) that lead to increasing wing loadings as body weight increases relative to wing area. Although these changes may be gradual and continuous, there is a threshold value at which flight becomes impossible. This threshold is not necessarily reached by all members of a population at the same time; for some period the species may be polymorphic with respect to flightlessness because of intra-populational variation in the parameters that determine wing loading. Perhaps as time passes and body weight increases faster than wing area, an ever-increasing proportion of the population will become flightless, until at last no flying individuals remain.

III. Hindlimb

A. *Skeleton and joints*

1. Synsacrum and pelvic girdle

The *synsacrum* is a rigid unit formed by the fusion of thoracic, sacral and caudal vertebrae, and by having an attachment to the pelvic girdle functions as the base of support for the body on the hindlimbs. In *Archaeopteryx* the pelvis was relatively small and the attachment of the ilium to the vertebral column appears to have been rather loose and to have involved only about six vertebrae (de Beer, 1954, p. 26). This number may be more than doubled in modern birds, as the ilium has become extended both cranially and caudally to establish connections with many more vertebrae. The more extensive iliovertebral connection is presumably correlated in part with bipedalism, and in part by the need for a broadened area of contact to absorb the shock of landing. It may also be a fortuitous side-effect of an extension of the ilium initially associated with an enlargement of the hindlimb muscles that arise from the pelvis. All of these factors are probably functionally interrelated.

There has been much discussion about the exact identity of the various vertebrae comprising the synsacrum. Cranially there are one to three "thoracic" vertebrae bearing ribs that approach or reach the sternum, and which are overlapped dorsally by the cranial end of the ilium. These are followed by several vertebrae having massive transverse processes that abut firmly against the ilium, and which are often divided into dorsal and ventral parts. Next there follows a series of vertebrae having only delicate dorsal bars, perhaps corresponding to the lumbar region. Two or three vertebrae follow that have been considered "true" sacrals, having distinct though usually slender dorsal and ventral transverse processes. The remaining vertebrae are believed to be fused caudals, because they resemble the free caudal vertebrae that continue past the end of the synsacrum. Beyond this general sequence, attempts to homologise the highly-modified synsacral vertebrae with the more distinct regions in other tetrapods have not reached a generally-accepted conclusion (Bellairs and Jenkin, 1960). Jollie (1977) considered the attempt "artificial."

Descriptive accounts of the bones of the girdle and of the pelvic limb (*ossa cinguli membri pelvici*) are given by Gadow (1896, pp. 860–863), Beddard (1898), Boas (1933), Bellairs and Jenkin (1960) and Feduccia (1975). The pelvis is formed by the ilium, ischium and pubis (Fig. 2.1), which arise as separate ossifications in the embryo but subsequently fuse together into a single unit, the pelvic bone (*os coxae*). Except in the Ostrich (*Struthio camelus*), which has a pubic symphysis, the pelvis is entirely open ventrally, a presumed adaptation to egg-laying. The *ilium* is the largest of the three elements, and joins with the synsacrum along its medial surface as already noted. Cranial to the acetabulum the ilium forms an elongated plate, the preacetabular wing (*ala preacetabularis ilii*), the concave surface and margins of which form the site of origin of the iliotrochanteric muscle group. The medial margin of the preacetabular ilium, the dorsal iliac crest (*crista iliaca dorsalis*) lies close to the dorsal crest of the synsacrum, right and left iliac crests and the synsacral crest, sometimes fusing in the midline to form the iliosynsacral crest (*crista iliosynsacralis*). The postacetabular ilium is more convex dorsally, and together with the synsacrum gives origin dorsally and ventrally to the levator and depressor muscles of the tail. The lateral surface of the postacetabular ilium is oriented vertically.

The *ischium* arises at the caudal margin of the acetabulum and extends caudally ventral to the ilium. Late in the development of most birds the ilium and ischium fuse in the caudal half of the pelvis. Between this and a short ilioischiadic contact of the dorsocaudal margin of the acetabulum, the large ilioischiadic foramen (*foramen ilioschiadicum*) remains open.

The *pubis* forms the ventral edge of the acetabulum and extends caudally as a narrow rod ventral and parallel to the ischium. Caudoventral to the acetabulum the pubis and ischium are separated by a small obturator foramen (*foramen obturatum*), and more caudally by the ischiopubic window (*fenestra ischiopubica*). The pubis and ischium are often fused between these two openings, but sometimes the two spaces are continuous. The broad surface of the post-acetabular pelvis serves as the site of origin of several retractor muscles of the femur and flexors of the knee joint.

The *acetabulum* is a cup-shaped depression that holds the head of the femur. Its floor is incomplete, forming the acetabular foramen (*foramen acetabuli*), which is closed by the acetabular membrane (*membrana acetabuli*). At the dorsal or dorsocaudal margin of the acetabulum an elevation, the *antitrochanter,* articulates with the trochanter of the femur. In many birds a bony process, the preacetabular tubercle (*tuberculum preacetabulare*), extends cranially from the cranioventral margin of the acetabulum. In most birds it is formed from the ilium, but in some species from the pubis. It serves as the site of origin of the *ambiens* muscle of the thigh.

Adaptive variations in the pelvis of birds have been discussed by Dabelow (1925), Boas (1933), Miller (1937), Fisher (1946), Wilcox (1952), Owre (1967), Raikow (1970) and others. Major variations occur in relative proportions, orientation and fusion of elements, and surface features. A comprehensive review of the subject is beyond the scope of this chapter, but a few examples may be mentioned. Aquatic birds show a narrowing of the pelvis and an elongation of the postacetabular region as adapations for swimming. In contrast, birds of prey often have a shortened and downcurved postacetabular region. A detailed and illustrated account of the pelvis in birds of prey, which is more taxonomic than functional, has been provided by Jollie (1977).

Despite many variations, the pelvic girdle in carinates generally adheres to a common pattern. In contrast, the ratites show a remarkable diversity of unusual and even bizarre pelvic features, which are still not clearly understood from the standpoint either of function or evolution (Beddard, 1898; Pycraft, 1901; Glutz von Blotzheim, 1960; Friant, 1968; Cracraft, 1974). In ratites the fusion of the ilium and ischium either does not occur at all, or else occurs only slightly or in a way different from that in carinates. The ilioischiadic foramen is therefore quite large and usually open caudally. A foramen which is open like this otherwise occurs among living birds only in the tinamous (Tinamidae), which are considered close relatives of the ratites (Cracraft, 1974). An open foramen is characteristic of such ancient birds as *Hesperornis* and *Ichthyornis* of Cretaceous times, and might therefore be considered a primitive condition in the ratites. However, Cracraft (1974) noted that these birds are not closely related to the ratites, and preferred the hypothesis that a large ilioischiadic foramen is a derived character. Inasmuch as fusion occurs late in development, with the adult pelvis of some ratites (e.g. kiwis, Apterygidae) closely resembling the embryonic pelvis of carinates (Olson, 1973), it may be that the condition in ratites is neotenic.

Beyond this general similarity there is great diversity in the ratite pelvis (Fig. 2.16). The moas (Dinornithiformes) of New Zealand and the elephant-birds (Aepyornithiformes) of Madagascar were the largest ratites, and their pelvic girdles are strikingly similar in form (Andrews, 1897; Archey, 1941; Oliver, 1949). The acetabula appear to be more widely separated than in other ratites, perhaps as an aid to stability. This, a moderate lateral expansion of the pelvis, and the structure of the hindlimb (Storer, 1971) suggest that they were graviportal rather than cursorial like the other ratites, and that their movements must have been more ponderous and deliberate. The postacetabular width exceeds the preacetabular, so that in dorsal view the pelvis resembles that of most carinates. Laterally, however, the postacetabular ilium is typically ratite, short and shallow, and not connected to the ischium caudally. Among cursorial ratites the pelvis of the Emu (*Dromaius novaehollandiae*), cassowaries (Casuariidae) and kiwis (Apterygidae) is similar in that the dorsal margin is convex and the ilioischiadic foramen open, although in older emus and cassowaries it may be closed by a narrow bony connection. All show postacetabular compression, especially that of kiwis. Indeed, the entire kiwi skeleton, including the rib cage and sternum, is laterally compressed, a condition accentuated by the reduced wings. This is apparently an adaptation for their dense forest habitat, where lateral compression

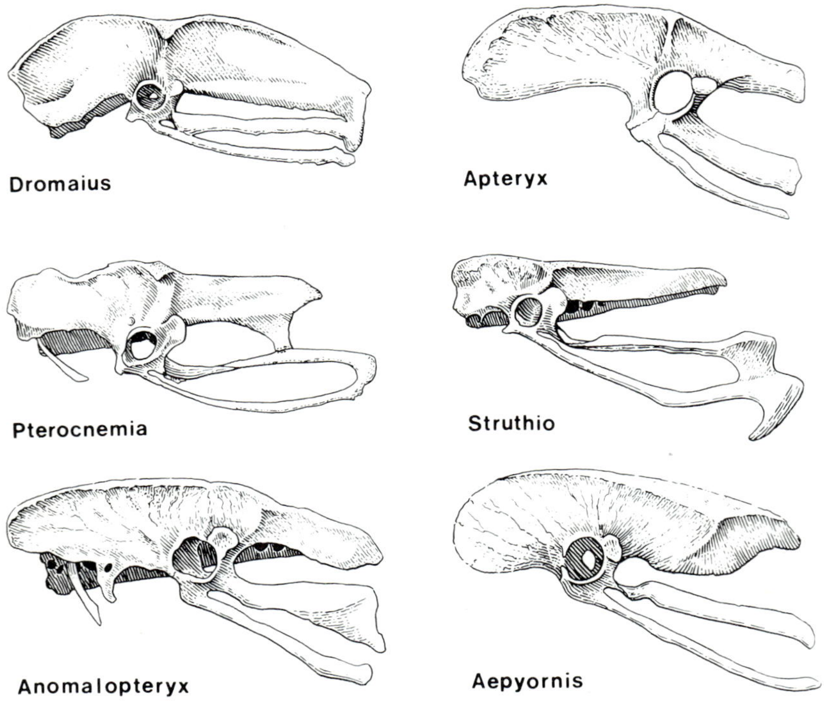

Fig. 2.16. Lateral views of the pelvic girdles of six species of ratites; the Emu (*Dromaius novaehollandiae*); the Brown Kiwi (*Apteryx australis*); the Lesser Rhea (*Pterocnemia pennata*); the Ostrich (*Struthio camelus*); a Moa (*Anomalopteryx didiformis*); and an elephantbird (*Aepyornis hildebrandti*). *Anomalopteryx* is after Archey (1941) and *Aepyronis* is after Andrews (1897).

must help them to manoeuvre through the vegetation, and for the nonrestricted spaces of their burrows (Gilliard, 1958, p. 26).

In rheas (Rheidae) the postacetabular region is unique among birds. The ischia converge medially to join and run parallel side-by-side, forming a bony shelf, the ischiadic symphysis (*symphysis ischiadica*) in the midline that separates the kidneys from the rest of the viscera. The functional significance of this is unknown. The postacetabular ilium is joined sub-terminally to the ischium by a unique bony column. The ischium and pubis fuse caudally, together forming an ellipse, the caudal extremity of which projects beyond the end of the ilium.

In the Ostrich (*Struthio camelus*) the pubis extends beyond the caudal end of the ischium, which is fused to it by a flat sheet of bone. The end of the pubis curves ventromedially and cranially to meet its counterpart in the midline, forming the only pubic symphysis in birds. This has been regarded as a primitive reptilian character, but more likely it is a secondary adaptation to large size. When the Ostrich sits on the ground its weight is supported laterally by the heels, cranially by the sternum, and caudally by the pubic symphysis. Because of the very short sternum, this appears necessary to prevent compression of the viscera. According to Gadow (1896, p. 862), the pubic symphysis of the Ostrich acts to support the weight of the viscera.

What is the significance of the diversity of pelvic structure in the ratites? Inasmuch as ratites are mostly very large birds, the obvious suggestion is that the ratite pelvis is modified for weight-bearing on a large scale. Because of the diversity of its structure, Feduccia (1980, pp. 133–135) argued against a close inter-relationship of the ratites. In contrast, Cracraft (1974) proposed a transformation pattern for the evolution of the pelvis in the context of his hypothesis of ratite monophyly. In either case the diversity requires explanation. One possibility is that, as ratites evolved through a combination of neoteny and increasing body size, the neotenic pelvis became modified for large size in different ways in the different groups. This suggests that whatever the phyletic relationships of the various groups, they underwent most of their increase in size independently of each other, after divergence from their common ancestors. In any case we still lack an understanding of the biomechanical functions of these pelvic girdles, the analysis of which would require study of the pelvic musculature as well as the osteology.

2. *Hindlimb skeleton*

A detailed description of the hindlimb skeleton will not be given here as it is adequately treated elsewhere (Gadow, 1896, pp. 863–867; Stolpe, 1932; Bellairs and Jenkin, 1960, pp. 261–265; Cracraft, 1971; Feduccia, 1975; Baumel, 1979a,b). Furthermore, many of the references listed in this section that discuss particular groups also include osteological analyses.

As compared to the primitive tetrapod hindlimb, that of birds (Fig. 2.17) is highly modified in adaptation to bipedal cursorial locomotion, to perching, and to leaping from the substrate in takeoff and absorbing the shock of landing. The orientation of the limb is mainly vertical (except, as will be described, in the most specialised divers), the *femur* articulating with the acetabulum by an angulated head (*caput femoris*). The tarsometatarsus is elongated and the stance digitigrade. Movement at the hip, knee and intertarsal ("ankle") joints is mainly in a roughly parasagittal plane, with limited rotational and even more restricted adduction–abduction capabilities. This pattern is not fundamentally avian, however, but was already developed to a considerable degree in the theropod ancestors of birds. Birds have evolved further by additional fusion of elements, giving the superficially and misleadingly simple appearance of the limb skeleton in adults.

The main bone of the *crus* is the *tibiotarsus,* which is formed by the fusion of the tibia with the proximal row of tarsal bones (*ossa tarsi proximalia*). The *fibula* articulates with the femur proximally, but distally it narrows to a fine splint. In the foot (*pes*) the *tarsometatarsus* is formed through fusion of the second, third and fourth metatarsals (*ossa metatarsalia II–IV*) and the distal tarsal bones (*ossa tarsi distalia*). In birds having a first digit, the short first metatarsal bone (*os metatarsale I*) is joined more or less securely to the tarsometatarsus, but is reduced to its distal portion and is not in contact with the tarsus. The "ankle" joint thus passes between the two rows of tarsals and is an intertarsal articulation (*articulatio intertarsalis*). The general phalangeal formula of birds is as follows: the first digit or *hallux,* two phalanges; the second digit (*digitus secundus*), three phalanges; the third digit (*digitus tertius*), four phalanges; and the fourth digit (*digitus quartus*), five phalanges. The phalangeal bones are referred to as proximal, intermediate and distal phalanges (*phalanges proximales, phalanges intermedia, phalanges distales*). The distal phalanges carry the keratinized claws.

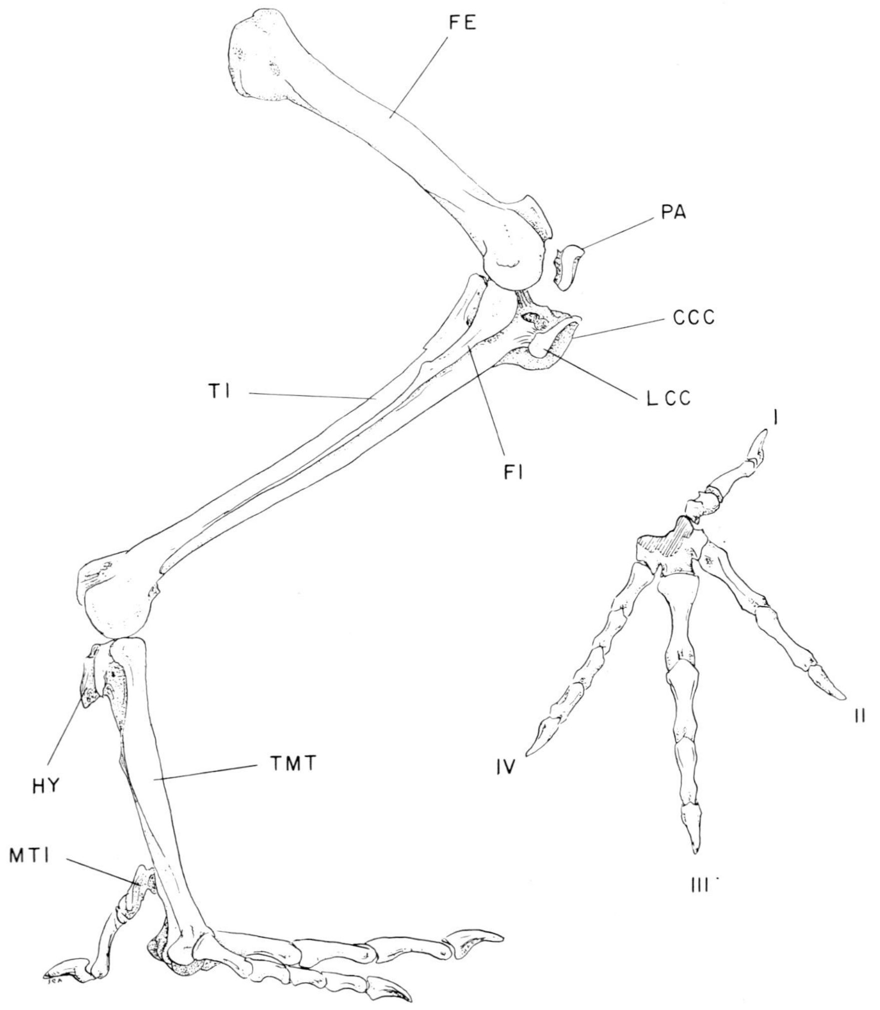

Fig. 2.17. Hindlimb skeleton of *Gallus*. CCC, Cranial cnemial crest; FE, femur; FI, fibula; HY, hypotarsus; LCC, lateral cnemial crest; MTI, metatarsal I: PA, patella; TI, tibiotarsus; TMT, tarsometatarsus.

The evolution of the avian hindlimb is revealed through a correlated study of fossils (Ostrom, 1976a) and embryonic development [see Romanoff (1960) for a summary of the literature, and Lansdown (1970) for an account of the embryology in *Coturnix*]. Ostrom analysed the hindlimb skeleton of *Archaeopteryx* and showed that in many ways it is intermediate between the conditions in therapods and modern birds. The rather small hallux is reversed in position, giving the anisodactyl configuration (see later). Metatarsals II, III and IV are well developed, and metatarsal I is reduced. One specimen has a small splinter of bone passing from the tarsal region distally along metatarsal IV. This may be a vestige of metatarsal V as seen in some therapods, but which makes only a transient appearance in the embyronic development of modern birds. The main metatarsals (II, III, IV) are distinguishable in *Archaeopteryx* but are neither separated nor displaced in any of the specimens, so it is uncertain whether they were fused together. In modern birds they arise as separate elements but fuse so completely that for most of the length of the tarsometatarsus the original boundaries are not discernible at maturity. The distal articular surfaces for the digits, however, are separate. The tarsal bones of *Archaeopteryx* are separately disinguishable but firmly attached to the ends of the long bones, so that an intertarsal joint is present. The hindlimb skeleton of *Archaeopteryx* resembles the foetal skeleton of modern birds, the skeleton of modern birds having evolved beyond that condition by the further fusion of embryonically separate bones. Ostrom considered that as a whole the hindlimb skeleton of *Aracheopteryx* is more bird-like than theropod-like, especially in its slender form and relative proportions.

B. Muscles and movements

The hindlimb muscles and their origins, insertions and actions are listed in Table III. Detailed descriptions of the muscles and their variations among birds have been given by Owre (1967), Klemm (1969), Raikow (1970, 1976), Vanden Berge (1970, 1975), Cracraft (1971), Hudson *et al.* (1972) and Berman and Raikow (1982); George and Berger (1966, pp. 380–464) reviewed the earlier literature and provided a comparative account. Consequently, in the present chapter only a brief overview of anatomical relations will be given without considering the many variations and exceptions that occur.

1. Movements of the thigh

Because the femur is the most proximal segment of the leg, its movements affect the entire limb (Fig. 2.18). Protraction is the movement of the distal end of the femur cranially, while retraction is its movement caudally. Adduction is movement of the distal end of the femur toward the ventral midline, whereas abduction is its movement away from the ventral midline. Rotation is movement of the femur along an axis that passes from the head of the femur to its distal end. This axis does not coincide with that of the femoral shaft, because the head of the femur diverges at an angle from the shaft. Medial rotation is a movement that turns the lateral surface of the femur craniomedially, whilst lateral rotation turns it caudomedially. These terms correspond to the inward and outward rotation of Raikow (1970).

Most of the muscles involved in movement of the femur are one-joint muscles that arise on the pelvic girdle or adjacent vertebrae and insert on the femur itself, but important roles may also be played by two-joint muscles that pass from the pelvis to the tibiotarsus or fibula.

(a) Protraction. The main femoral protractor is the *m. iliotibialis cranialis* (Figs. 2.19, 24, 28C). In *Columba* this muscle contains mostly twitch fibres (Cracraft, 1971). Although it is also capable of extending the tibiotarsus, Cracraft's motion analysis of walking in *Columba* suggested that this is of minor importance. Vanden Berge (1976) described a new muscle, the *m. iliotibialis medialis,* lying medial and running approximately parallel to the *m. iliotibialis cranialis* in flamingos (Phoenicopteridae), and Olson and Feduccia (1980) re-

TABLE III. **The hindlimb musculature of birds.**

Name[a]	Origin[b]	Insertion	Action[c]
M. iliotibialis cranialis (*sartorius*)	Dorsal iliac crest	Medial edge of proximal end of tibiotarsus	Protracts femur, extends tibiotarsus
M. iliotibialis lateralis (*iliotibialis*)	Dorsal and dorsolateral iliac crests	Patella and proximal end of tibiotarsus via patellar tendon	Abducts femur; slightly extends tibiotarsus
M. iliotibialis medialis[d]	Dorsal iliac crest	Medial aspect of tibiotarsus at base of cranial cnemial crest	Inwardly (medially) rotates tibiotarsus?
M. iliofibularis (*biceps femoris*)	Dorsal and dorsolateral iliac crests, mainly caudal to acetabulum	Fibula	Flexes crus; retracts femur
M. ambiens	Pectineal process of ilium or pubis	Knee joint or tendon of origin of one or more digital flexors	Assists digital flexors; possibly adducts thigh
M. iliotrochantericus caudalis (*iliotrochantericus posterior*)	Preacetabular ilium	Lateral surface of head of femur	Uncertain; retracts or protracts femur (see text)
M. iliotrochantericus cranialis (*iliotrochantericus anterior*)	Lateral margin of preactabular wing of ilium	Lateral surface of femur	Protracts and inwardly rotates femur
M. iliotrochantericus medius	Lateral margin of preacetabular wing of ilium	Lateral surface of femur	Protracts and inwardly rotates femur
M. iliofemoralis externus (*gluteus medius et minimus*)	Ilium dorsal to acetabulum	Lateral surface of femur	Abducts femur
M. iliofemoralis internus (*iliacus*)	Lateral margin of preacetabular wing of ilium	Medial surface of femur	Adducts, protracts, and outwardly rotates femur
M. femorotibialis externus	Lateral surface of femur	Proximal end of tibiotarsus	Extends crus
M. femorotibialis medius	Cranial surface of femur	Proximal end of tibiotarsus (via patella)	Extends crus
M. femorotibialis internus	Medial surface of femur	Medial side of proximal end of tibiotarsus	Inwardly rotates tibiotarsus
M. flexor cruris lateralis (*semitendinosus*)	Caudal end of ilium and adjacent caudal vertebrae	Femur (*pars accessoria*) and tibiotarus	Retracts femur, flexes crus
M. flexor cruris medialis (*semimembranosus*)	Caudolateral surface of wing of ischium	Tibiotarsus	Flexes crus, retracts femur
M. caudofemoralis (*piriformis pars caudofemoralis*)	Pygostyle	Femur	Depresses tail, retracts femur (?)
M. iliofemoralis (*piriformis pars iliofemoralis*)	Lateral surface of ilium	Femur	Retracts femur
M. ischiofemoralis	Lateral surface of ischium	Femur	Outwardly rotates and retracts femur
M. obturatorius lateralis (*obturator externus*)	Lateral surface of pelvis caudoventral to acetabulum	Caudal surface of head of femur	Outwardly rotates femur

M. obturatorius medialis (obturator internus)	Medial surface of body of pubis and wing of ischium	Dorsolateral edge of head of femur	Outwardly rotates and abducts (?) femur
M. pubo-ischio-femoralis (adductor longus et brevis)	Caudoventral margin of pelvis	Distal part of caudal surface of femur	Retracts femur
M. tibialis cranialis (tibialis anterior)	Cranial surface of tibiotarsus	Dorsal surface of proximal part of tarsometatarsus	Flexes ankle
M. extensor digitorum longus	Cranial surface of tibiotarsus	Dorsal surface of ungual and other phalanges of digits II, III, and IV	Extends digits II, III, and IV; flexes ankle
M. fibularis longus (peroneus longus)	Proximal end of tibiotarsus	Proximolateral corner of tibial cartilage, tendon of *flexor perforatus digiti III*	Extends ankle, assists *flexor perforatus digiti III*
M. fibularis brevis (peroneus brevis)	Craniolateral surface of fibula and tibiotarsus	Lateral surface of proximal part of tarsometatarsus	Inwardly rotates tarsometatarsus
M. gastrocnemius	Distal end of femur and proximal end of tibiotarsus	Hypotarsus and body of tarsometatarsus	Extends ankle
M. plantaris	Caudomedial surface of proximal end of tibiotarsus	Proximomedial corner of tibial cartilage	Extends ankle
M. popliteus	Head of fibula	Caudal surface of proximal part of tibiotarsus	Uncertain, see text
M. flexor perforans et perforatus digiti II	Femur; sometimes also tibiotarsus or fibula	Plantar surface of digit II	Flexes digit II
M. flexor perforans et perforatus digiti III	Tibiotarsus, femur and/or fibula	Plantar surface of digit III	Flexes digit III
M. flexor perforatus digiti II	Femur; sometimes also *ambiens* tendon	Plantar surface of proximal phalanx of digit II	Flexes digit II
M. flexor perforatus digiti III	Femur; sometimes also *ambiens* tendon or fibula	Plantar surface of first and sometimes second phalanx of digit III	Flexes digit III
M. flexor perforatus digiti IV	Femur; sometimes also *ambiens* tendon or fibula	Plantar surface of digit IV	Flexes digit IV
M. flexor hallucis longus	Caudal surface of distal end of femur; sometimes also fibula	Plantar surface of ungual phalanx of hallux	Flexes hallux; extends tarsometatarsus
M. flexor digitorum longus	Caudal surfaces of proximal parts of tibiotarsus and fibula; sometimes also femur	Plantar surfaces of ungual and other phalanges of digits II, III, and IV	Flexes digits II, III, and IV simultaneously; extends tarsometatarsus
M. extensor hallucis longis	Craniomedial surface of tarsometatarsus	Dorsal surface of base of ungual phalanx of hallux	Extends hallux
M. flexor hallucis brevis	Hypotarsus and caudomedial surface of tarsometatarsus	Base of proximal phalanx of hallux	Flexes hallux

(*continued*)

TABLE III. ***(Continued)***

Name[a]	Origin[b]	Insertion	Action[c]
M. abductor digiti II	Dorsomedial surface of tarsometatarsus	Dorsomedial surface of base of proximal phalanx of digit II	Abducts digit II
M. adductor digiti II	Plantar surface of tarsometatarsus	Lateral surface of base of proximal phalanx of digit II	Adducts digit II
M. extensor proprius digiti III	Dorsal surface of tarsometatarsus	Dorsal surface of base of proximal phalanx of digit III	Extends digit III
M. extensor brevis digiti III	Dorsal surface of tarsometatarsus	Dorsal surface of base of proximal phalanx of digit III	Extends digit III
M. extensor brevis digiti IV	Craniolateral surface of tarsometatarsus	Medial surface of base of proximal phalanx of digit IV	Extends and adducts digit IV
M. extensor proprius digiti IV[e]	Dorsal surface of belly of *m. extensor brevis digiti IV*	Dorsal surface of proximal part of proximal phalanx of digit IV	Extends digit IV
M. abductor digiti IV	Lateroplantar surface of tarsometatarsus	Lateral surface of proximal phalanx of digit IV	Abducts digit IV
M. adductor digiti IV	Plantar surface of tarsometatarsus	Medial surface of base of proximal phalanx of digit IV	Adducts digit IV
M. lumbricalis	Tendon of insertion of *m. flexor digitorum longus*; sometimes also *flexor hallucis longus*	Joint pulley of digit III and/or IV	Draws deep flexor tendon(s) distally

[a] Names of muscles are from the *Nomina Anatomica Avium* (Baumel *et al.*, 1979). If the *N.A.A.* name differs from that used by George and Berger(1966), their term is given in parentheses.
[b] The origin and insertion given are typical for birds generally, but many exceptions and variations exist.
[c] The major action of the muscle as inferred from anatomical relations (see text).
[d] Not listed in George and Berger (1966). See Vanden Berge (1976) and Olson and Feduccia (1980).
[e] Not listed in George and Berger (1966). See Berman and Raikow (1982).

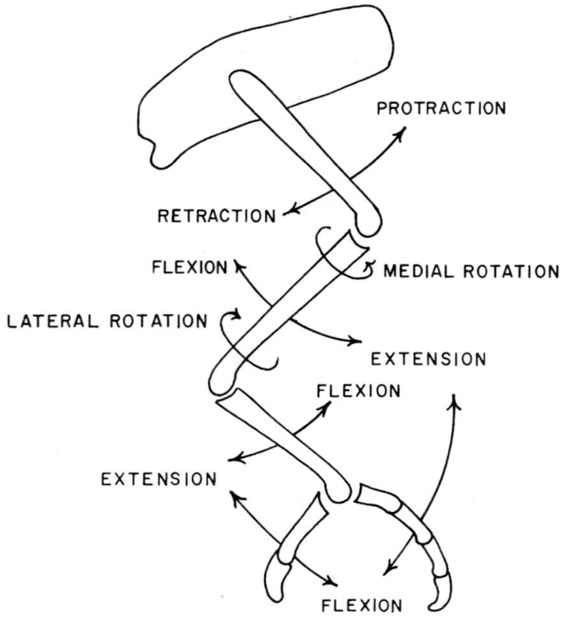

Fig. 2.18. The terminology describing the movements of the various parts of the avian hindlimb.

ported the same muscle in the Banded Stilt (*Cladorhynchus leucocephalus*). It probably acts in conjunction with the *m. iliotibialis cranialis*, but neither paper discussed its function.

The *mm. iliotrochantericus cranialis* (Fig. 2.28C), *iliotrochantericus medius* (Fig. 2.28C) or the combined *iliotrochantericus cranialis et medius* (Figs. 2.20, 21, 24), as well as the *m. iliofemoralis internus* (Figs. 2.25, 28C), are all possible protractors, but their short lever arms and small size suggest that they must be only minor synergists of the *m. iliotibialis cranialis*, and their significant functions probably lie more in femoral rotation.

The large size of the *m. iliotrochantericus caudalis* (Figs. 2.20, 28C) suggests that it is an important muscle, yet its action is uncertain. Most authors have considered it to be a femoral protractor (Miller, 1937; Fisher, 1946; Berger, 1952; Raikow, 1970). Klemm (1969) emphasised its role in medial rotation and adduction, while Raikow (1970) considered joint stabilisation to be its primary function. Cracraft (1971) argued that this muscle produces retraction rather than protraction, at least in *Columba*. The problem is that the muscle inserts so close to the level of the hip joint that its effect is difficult to infer, especially in preserved specimens in which stiffened joints interfere with manipulation. Electromyography is needed to clarify the action of this muscle.

(b) Retraction. Retractor muscles are involved not only in retractive motion but also in postural support, where they resist over-protraction of the femur. In swimming birds these muscles must resist the femoral protraction that would otherwise follow upon extension of the foot paddle. Protraction also occurs in landing, and this must be resisted by retractors acting as shock absorbers. The chief retractors are one-joint muscles that insert on the femur (Fig. 2.28B), but their

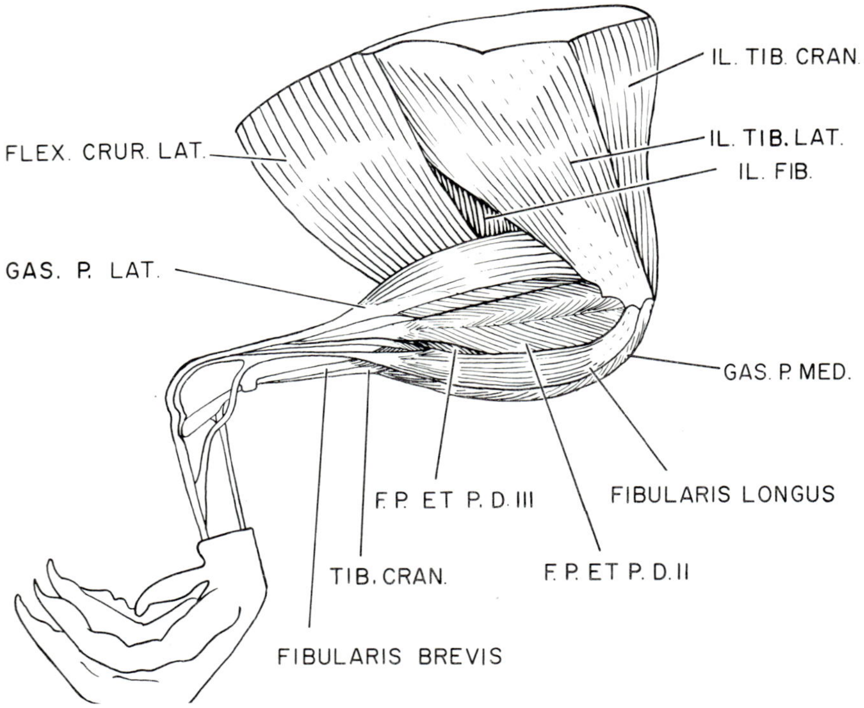

Fig. 2.19. First layer of muscles on the lateral side of the hindlimb of *Anas*. After Raikow (1970). Abbreviations for Figs. 19–27: ABD. DIG. II, *m. abductor digiti II*; CAUD. FEM., *m. caudofemoralis*; EXT. BREV. DIG. IV, *m. extensor brevis digiti IV*; EXT. DIG. L., *m. extensor digitorum longus*; EXT. HAL. L., *m. flexor hallucis longus*; FEM. TIB. EXT. ET MED., *m. femorotibialis externus et medius*; FEM. TIB. INT., *m. femorotibialis internus*; F. DIG. L., *m. flexor digitorum longus*; F. HAL. BREV., *m. flexor hallucis brevis*; F. HAL. L., *m. flexor hallucis longus*; FIB. BREV., *m. fibularis brevis*; FLEX. CRUR. LAT., *m. flexor cruris lateralis*; FLEX. CRUR. MED., *m. flexor cruris medialis*; F.P.D. II, *m. flexor perforatus digiti II*; F.P.D. III, *m. flexor perforatus digiti III*; F.P.D. IV, *m. flexor perforatus digiti IV*; F.P. ET P.D. II, *m. flexor perforans et perforatus digiti II*; F.P. ET P.D. III, *m. flexor perforans et perforatus digiti III*; GAS. P(ARS) INTERMED., *m. gastrocnemius pars intermedia*; GAS. P. LAT., *m. gastrocnemius pars lateralis*; GAS. P. MED., *m. gastrocnemius pars medialis*; IL. FEM. EXT., *m. iliofemoralis externus*; IL. FIB., *m. iliofibularis*; IL. TIB. CRAN., *m. iliotibialis cranialis*; IL. TIB. LAT., *m. iliotibialis lateralis*; IL. TROC. CAUD., *m. iliotrochantericus caudalis*; IL. TROC(H). CRAN. ET MED., *m. iliotrochantericus cranialis et medius*; OBT. LAT., *m. obturatorius lateralis*; OBT. MED., *m. obturatorius medialis*; PIF. *m. pubo-ischio-femoralis*; PIF PARS LAT., *m. pub-ischio-femoralis pars lateralis*; PIF PARS MED., *m. pubo-ischio-femoralis pars medialis*; TIB. CRAN., *m. tibialis cranialis*.

roles vary. The *m. ischiofemoralis* (Figs. 2.21, 22, 28B), with its numerous short fibres of limited excursion, probably functions mostly as a rotator in a postural role. The *m. caudofemoralis* (Figs. 2.21, 28B) is probably most active in tail movement. The main retractors are the *mm. pubo-ischio-femoralis* (Figs. 2.20–22, 24, 25, 28B) and *iliofemoralis* (Figs. 2.20, 28B). If Cracraft (1971) is correct in his aforementioned interpretation of the *m. iliotrochantericus caudalis,* that muscle too is an important retractor. On the basis of the relative numbers of twitch and tonus fibres, Cracraft (1971) postulated that the *m. pubo-ischio-femoralis* is

---

Fig. 2.21. Third layer of muscles on the lateral side of the hindlimb of *Anas*. After Raikow (1970). For abbreviations see Fig. 2.19.

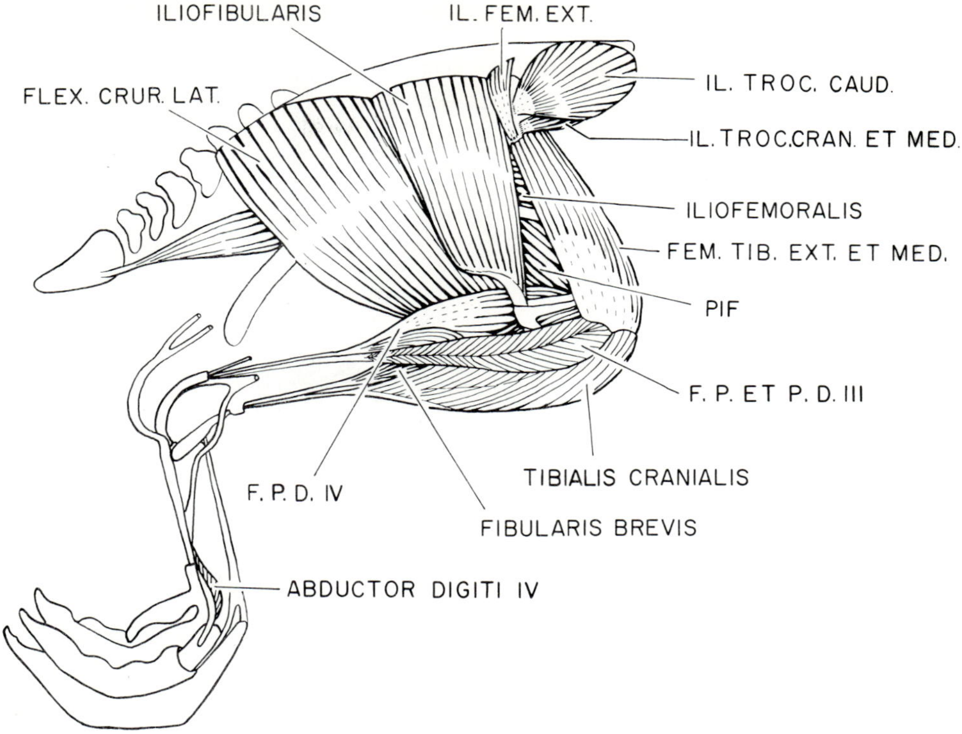

Fig. 2.20. Second layer of muscles on the lateral side the hindlimb of *Anas*. After Raikow (1970). For abbreviations see Fig. 2.19.

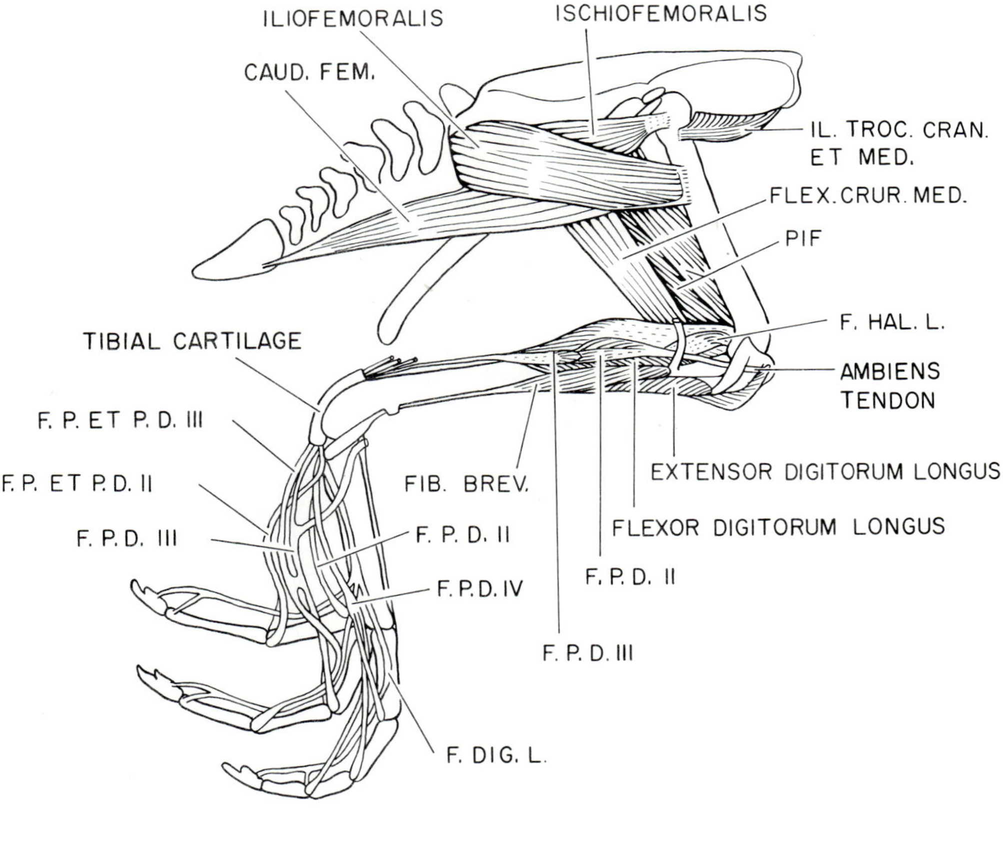

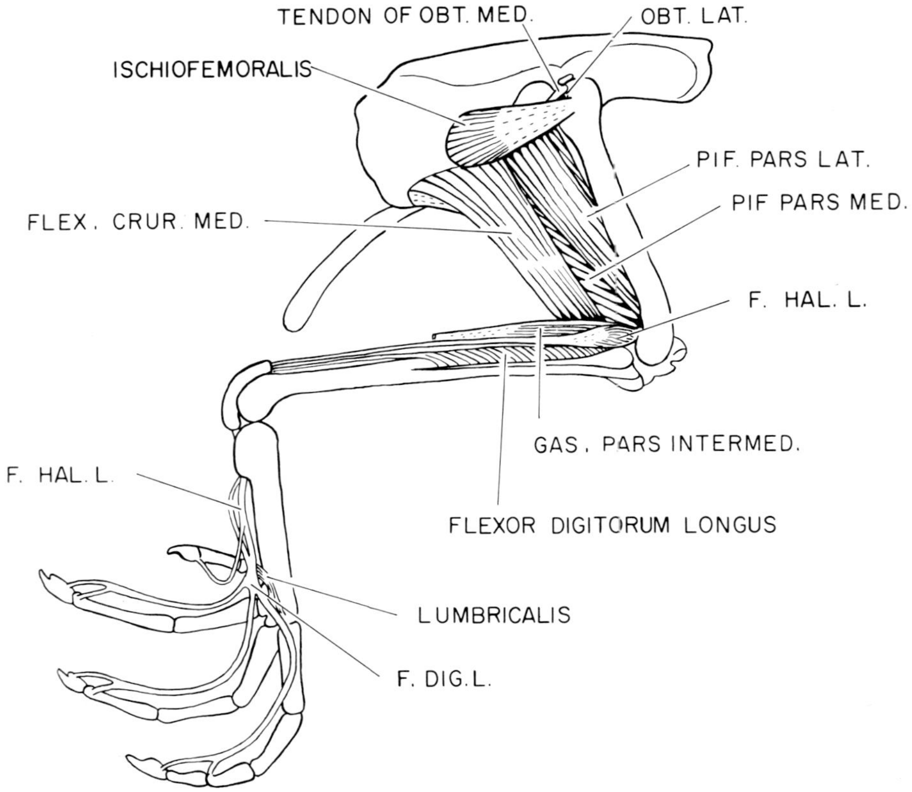

Fig. 2.22. Fourth layer of muscles on the lateral side of the hindlimb of *Anas*. After Raikow (1970). For abbreviations see Fig. 2.19.

primarily postural in function, and that the *m. iliotrochantericus caudalis* is mainly active in movement.

Femoral retraction is also a potential role of certain two-joint muscles that pass from the pelvic girdle to the tibiotarsus (the *mm. flexor cruris medialis* and *flexor cruris lateralis*) or to the fibula (*m. iliofibularis*) (Fig. 2.28F). The main functions of these muscles are probably as flexors of the crus, but their possible contribution

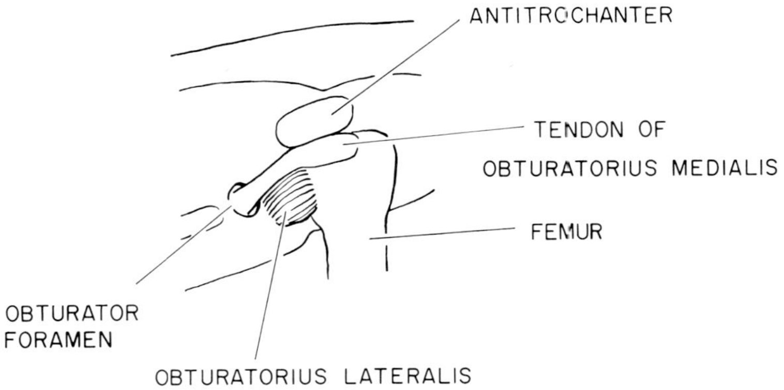

Fig. 2.23. The *m. obturatorius lateralis* in *Anas*. After Raikow (1970).

to protraction must also be considered. The action of two-joint muscles cannot be inferred with any confidence by indirect means, because the effects that they have on the two joints that lie between their attachments can vary according to the actions of other muscles. Such muscles may hold one or the other joint steady so that the muscle in question acts on one joint only. Alternatively, many combinations of relative movement at the two joints are possible. The action of such muscles will be determined with certainty only when electromyographic studies are done.

Cracraft (1971) believed that the *m. iliofibularis* (Figs. 2.19, 20, 28F) (his *biceps femoris*) is principally active in flexing the crus. However, the tendon of insertion passes through a tendinous loop, the loop of the *iliofibularis* muscle (*ansa m. iliofibularis*) that arises from the caudolateral surface of the distal end of the femur, so the muscle must exert a protracting effect directly on the femur. The *m. flexor cruris medialis* (Figs. 2.21, 22, 25, 28F), on the other hand, inserts directly on the tibiotarsus.

The *m. flexor cruris lateralis* (Figs. 2.19, 20, 24, 28F) is especially hard to interpret functionally because of its variation. It arises from the caudodorsal region of the pelvic girdle and sometimes also partly from the first few caudal vertebrae. The belly passes distally toward the crus. In most birds this portion, the *pars pelvica* (Fig. 2.28F, FCLP), ends at a connective tissue inscription or raphe, which is joined by another fleshy belly, the *pars accessoria* (Fig. 2.28F, FCLA), from the distal part of the caudal surface of the femur. The raphe continues

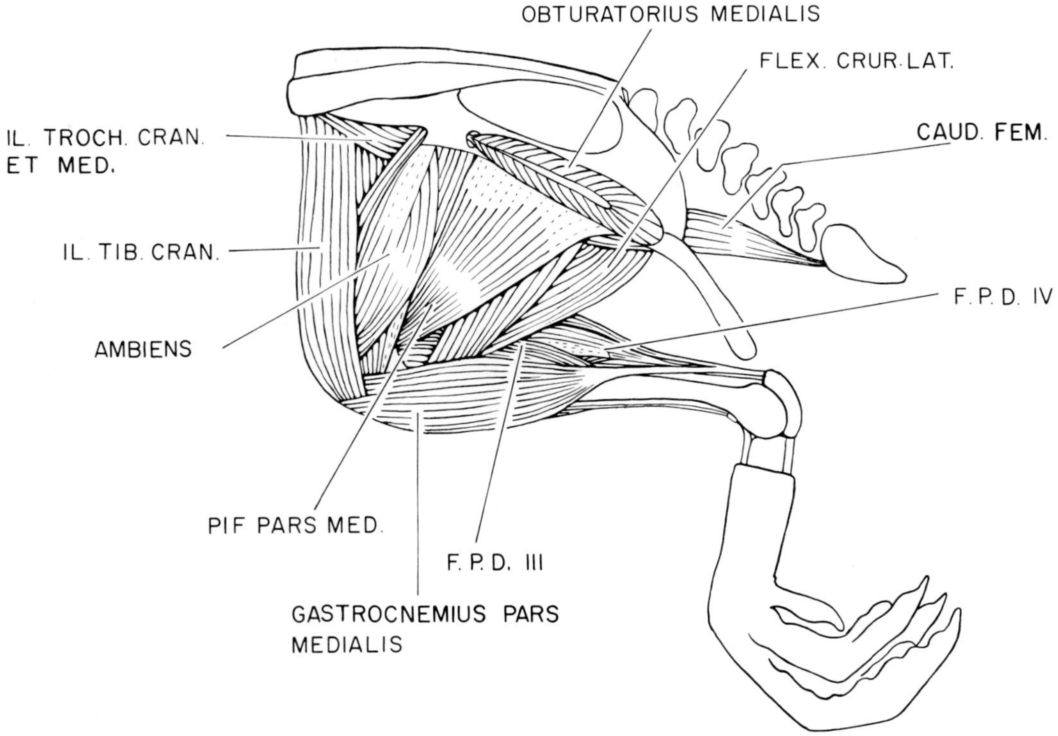

Fig. 2.24. First layer of muscles on the medial side of the hindlimb of *Anas*. After Raikow (1970). For abbreviations see Fig. 2.19.

beyond the connection of the two bellies and has a variety of interconnections with other muscles in various birds. Thus, the *m. flexor cruris lateralis* in this form may be considered as passing from the pelvis to the femur, and therefore as acting as a femoral retractor. The totality of its action is complicated, however, because it also has an indirect connection to the tibiotarsus via the extension of the raphe, which may insert on that bone directly or in combination with the tendon of the *m. flexor cruris medialis*. Thus, it may function simultaneously in protraction and flexion, depending on the action of other muscles. In some birds, however (including *Anas* as illustrated), the *pars accessoria* is lacking, and the *m. flexor cruris lateralis* inserts directly on the tibiotarsus, usually in conjunction with the *m. flexor cruris medialis*. In this case it seems to be primarily a tibial flexor, with any protraction a secondary effect. George and Berger (1966, p. 403) list the groups in which the *pars accessoria* is present or absent. Occasionally this varies within a family (e.g. Picidae), but in general the accessory part is present in land birds and absent in water birds. Although there are exceptions to this rule, the concordance with habit is so close as to suggest a functional correspondence. One possibility is that in land birds the muscle may function as a femoral retractor in a postural role to counteract the effect of gravity.

(c) Abduction and adduction. Abductive and adductive movements are limited in birds. In *Columba* Cracraft (1971) found that the antitrochanter–trochanter articulation effectively prevents abduction when the femur is fully protracted, while the hip joint ligaments also largely prevent adduction. With retraction the

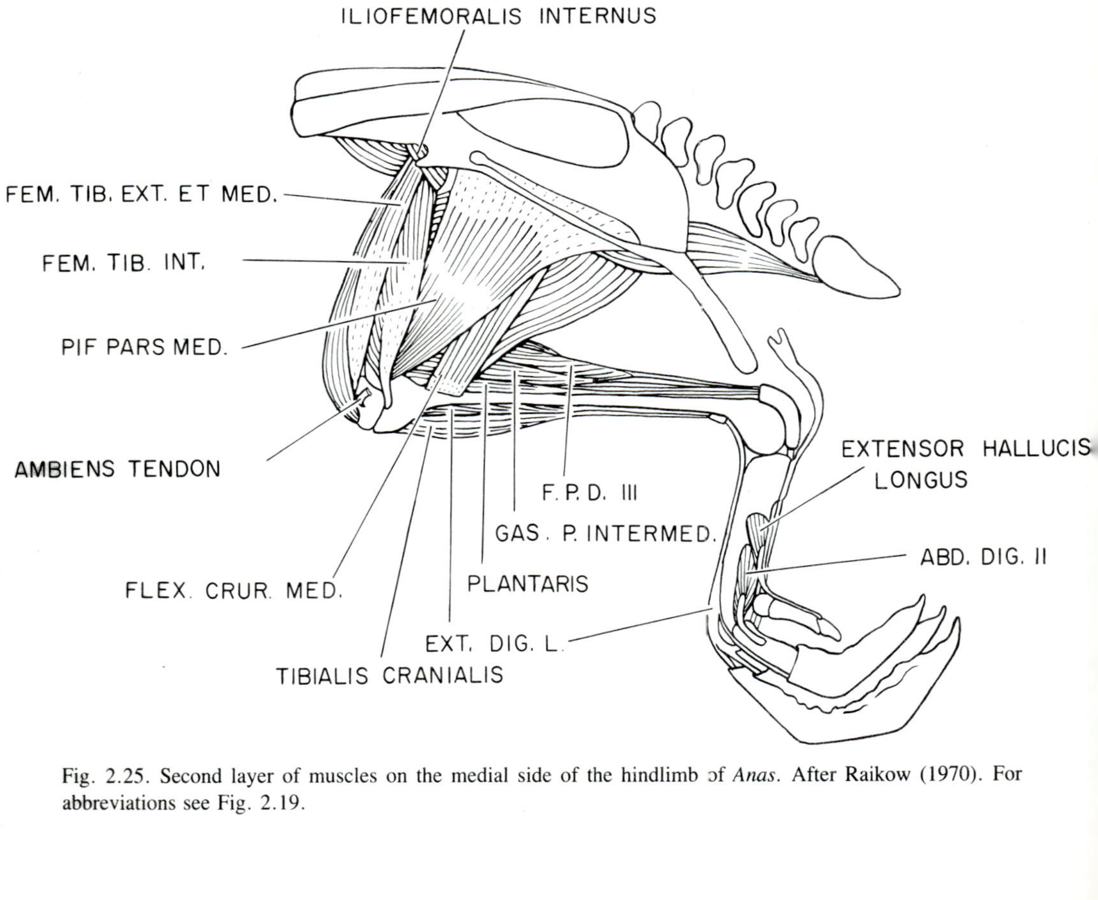

Fig. 2.25. Second layer of muscles on the medial side of the hindlimb of *Anas*. After Raikow (1970). For abbreviations see Fig. 2.19.

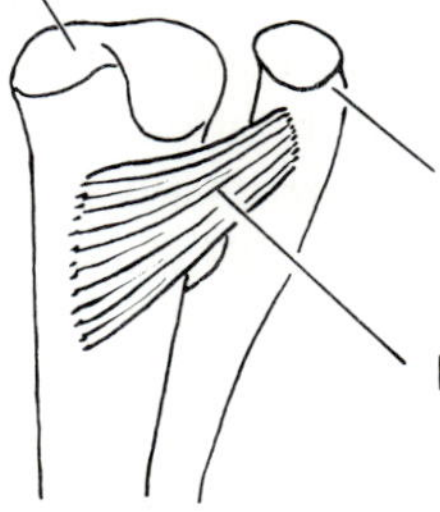

Fig. 2.26. The *m. popliteus* on the caudal surface of the proximal parts of the tibiotarsus and fibula of *Anas*. From Raikow (1970).

trochanter moves cranial to the antitrochanter, allowing up to a 35° arc of movement. However, Cracraft concluded that this probably exceeds the range of movement attained in life. In most birds the condition is probably similar, with the hip joint structure holding the leg mainly in a vertical position. Some foot-propelled diving birds, like loons (Gaviidae) and grebes (Podicipedidae), hold the femur horizontally. Because this position is largely fixed, these birds cannot adduct the limb into the position used by terrestrial birds, and this contributes to their clumsiness on land.

The only muscles in a position to abduct the limb are the *mm. iliotibialis lateralis* (Figs. 2.19, 28E) and *iliofemoralis externus* (Figs. 2.20, 28E). The former is weak in most birds, parts of it often being lost completely. The *m. iliofemoralis externus* is a small muscle, and its abductive powers must be quite feeble given the relatively large mass of the limb. It may function more to press the head of the femur into the acetabulum than to abduct the femur. It is absent in many birds, especially smaller forms (George and Berger, 1966, p. 303, list species possessing and lacking this muscle), but it sometimes reappears as a developmental anomaly or even as a re-established feature (Raikow, 1975; Raikow *et al.*, 1979). The mousebirds (Coliidae) are an exception to the general rule, having the *iliofemoralis externus* enormously developed at the expense of the adjacent *iliotrochantericus caudalis* (Berman and Raikow, 1982). The insertion is on the caudolateral surface of the trochanter of the femur, giving the muscle a role in retraction as well as abduction. This unusual condition is functionally associated with the remarkable repertory of acrobatic movements and postures of the mousebirds (see later).

There are no muscles specifically positioned for adduction. Muscles passing from the pelvis to the medial side of the thigh, such as the *mm. ambiens, iliofemoralis internus,* or even *iliotibialis cranialis* and *flexor cruris medialis,* could provide adduction through a minor inward component of their principal actions.

(d) Rotation. Rotation is provided by muscles that arise on the pelvic girdle and insert at the proximal end of the femur (Fig. 2.28A). The results of their action include the rotation of the limb during the recovery phase of walking, and perhaps more significantly, the rotation of the body around the hip joint, bringing it in over the supporting leg while the opposite leg is in the air during recovery. This

imparts a waddling motion most apparent in such short-legged birds as ducks (Anatidae). These muscular forces are also important in joint stabilisation and postural support. Inward rotation is imparted by the three iliotrochanteric muscles, of which the *m. iliotrochantericus caudalis* (Figs. 2.20, 28A) is by far the largest and most important. Lateral rotation is provided by the *mm. ischiofemoralis* (Figs. 2.21, 22, 28A), *obturatorius medialis* (Figs. 2.22–24, 28A), *obturatorius lateralis* (Figs. 2.22, 23, 28A) and *iliofemoralis internus* (Figs. 2.25, 28A). The latter two muscles are so small, however, that their contribution must verge on insignificance.

2. *Movements of the crus*

The most important movements of the crus (Fig. 2.18) are its swinging motions craniocaudally in a roughly parasagittal plane. Movement of the distal end cranially is extension of the knee joint, whilst movement caudally is flexion of the knee. In addition the crus may turn around its long axis: movement of the lateral surface craniomedially is medial rotation, whilst movement caudomedially is lateral rotation, as for the femur. The crus may also undergo abduction and adduction or lateral–medial swing (Cracraft, 1971) relative to the femur.

On the caudal aspect of the knee in some birds is a very small muscle, the *m. popliteus,* which passes from the fibular head to the tibiotarsus (Fig. 2.26). Various authors have speculated as to its possible function, usually in relation to joint action, but no convincing explanation has emerged (Cracraft, 1971).

(a) Extension. The principal extensor of the knee joint is a large muscle mass that arises along the lateral and cranial surfaces of the shaft of the femur. The lateral part, the *m. femorotibialis externus,* often has distinct outer (*pars proximalis*) and inner (*pars distalis*) parts. The cranial portion of the complex is the *m. femorotibialis medius.* The two muscles are more or less fused along their border, sometimes so intimately that the entire mass is given the single name *m. femorotibialis externus et medius* (Raikow 1970, p. 32). The muscle fibres converge to form the bulk of a tendon, which contains the *patella* and inserts on the cranial edge of the tibiotarsus (Figs. 2.20, 25, 28D). Cracraft (1971) showed that different groups of fibres are functionally significant at different degrees of extension. Lesser roles in extension of the knee joint are played by the *mm. iliotibialis lateralis* (Figs. 2.19, 28D) and *iliotibialis cranialis* (Figs. 2.19, 24, 28D). The latter muscle appears to function primarily in protraction of the femur, but if the femur is fixed the muscle is in a position to extend the knee joint.

(b) Flexion. Flexion of the knee joint is principally accomplished by the *mm. iliofibularis, flexor cruris medialis* and *flexor cruris lateralis,* which were discussed earlier (Fig. 2.28F). The *m. gastrocnemius* (discussed later) may contribute to this movement if the intertarsal joint is prevented from being extended. Medial rotation of the tibiotarsus is accomplished by the *m. femorotibialis internus* (Fig. 2.25), the insertion of which on the medial side of the head of the tibiotarsus is too far caudal to allow it to assist the other femorotibialis components in extension. There are no muscles with a specific function of lateral rotation, adduction or abduction; these movements must arise as secondary effects of other muscles. For example, the lateral insertion of the *mm. iliofibularis* and

iliotibialis lateralis presumably permit them to contribute to outward rotation in addition to their other actions. Similarly, the *m. femorotibialis internus* may contribute to adduction, especially when the knee joint is extended.

3. *Movements of the foot*

Movements of the foot as a whole take place by the motion of the tarsometatarsus against the tibiotarsus at the intertarsal joint (Fig. 2.18). The chief movement is cranial-caudal in orientation. Movement of the distal end of the tarsometatarsus cranially is flexion of the joint, whereas the opposing caudal movement is extension of the joint. Medial and lateral rotation likewise occur. The former is apparently brought about by the *m. fibularis brevis,* but there are no muscles specifically positioned for lateral rotation. Cracraft (1971) also noted the capacity for lateral–medial swing and a limited degree of translation, but functional studies have generally not considered these movements.

(a) Flexion. The primary flexor of the ankle is the *m. tibialis cranialis,* which arises from the cranial surface of the tibiotarsus (Figs. 2.19, 20, 25, 27, 29). At the distal end of the tibiotarsus the tendon of the *tibialis cranialis* passes under a fibrous loop, the transverse ligament (*retinaculum extensorium tibiotarsi*) before inserting on the proximal part of the dorsal surface of the tarsometatarsus. This muscle in *Columba* has numerous long twitch fibres adapting it to flex the intertarsal joint rapidly through a wide arc of joint movement (Cracraft, 1971). This is the only muscle which has flexion of the intertarsal joint as its principal function, but it is probably assisted at times by the *m. extensor digitorum longus* (Figs. 2.21, 25, 29). The *m. extensor digitorum longus* arises from the cranial surface of the tibiotarsus deep to the *m. tibialis cranialis,* and is the deepest muscle on the

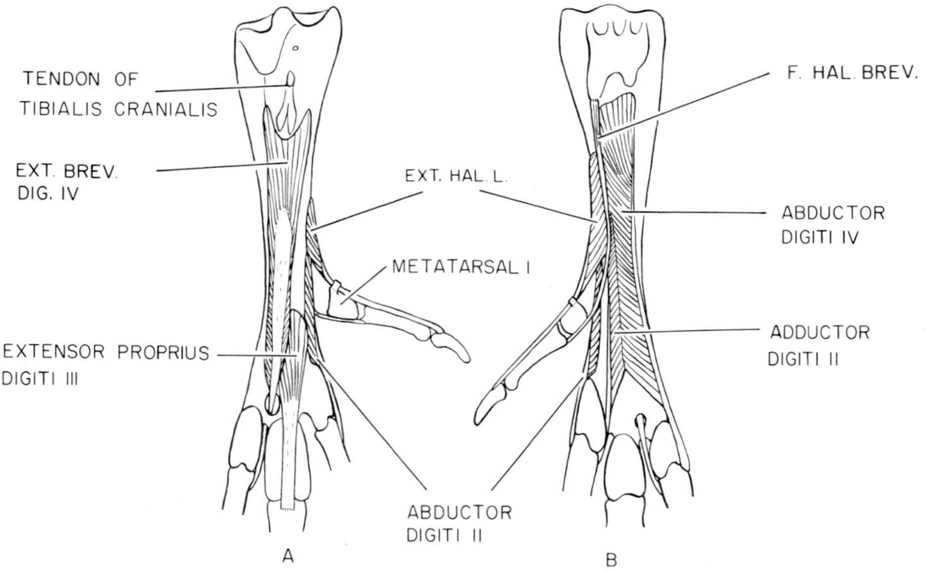

Fig. 2.27. Intrinsic muscles of the foot of *Anas*. (A) Cranial surface; (B) plantar surface. After Raikow (1970). For abbreviations see Fig. 2.19.

cranial surface of the crus. Its tendon passes across the intertarsal joint and continues distally along the dorsal surface of the tarsometatarsus. Near the distal end of the tarsometatarsus the tendon gives rise to branches that continue over the dorsal surfaces of digits II, III and IV, inserting on the distal phalanx of each, and also on the other phalanges by branches displaying various patterns in the different taxa. Although it is considered primarily as an extensor of those digits, this muscle will also flex the ankle joint.

(b) Extension. The principal extensor of the ankle is the *m. gastrocnemius,* which is the most superficial muscle on the caudal surface of the crus (Figs. 2.19, 22, 24, 25, 29). It has three separate bellies. The *pars lateralis (externa)* lies on the lateral surface and arises from the femur; the *pars intermedia* lies deeper in the crural musculature and also arises from the femur; the *pars medialis (interna)* lies on the medial surface of the crus and originates from the head of the tibiotarsus. The *pars intermedia* is small, but the other bellies are generally quite large. The

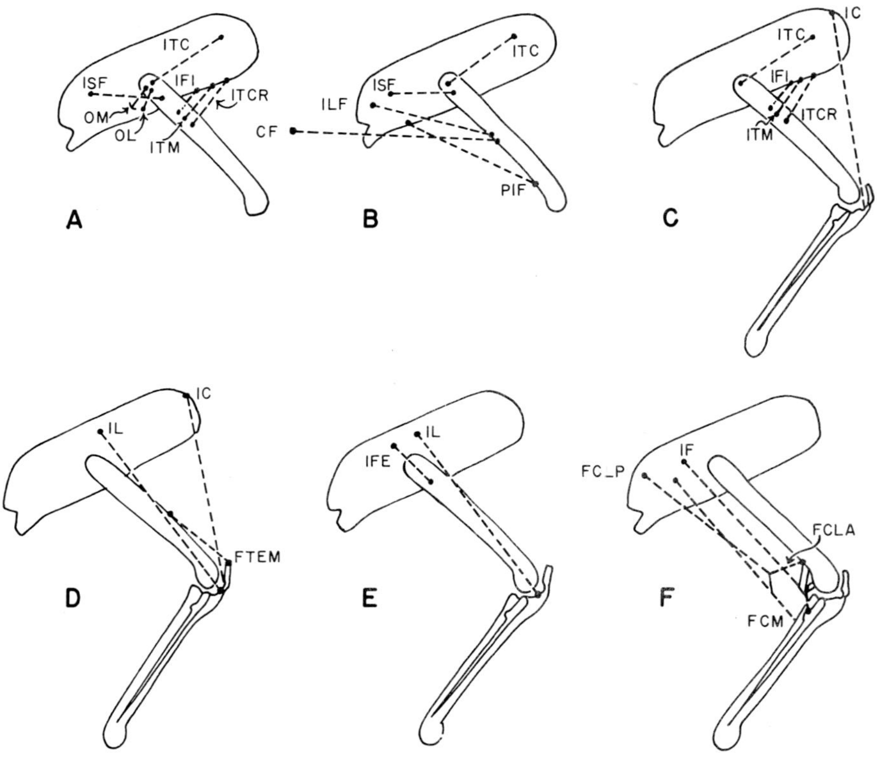

Fig. 2.28. Lines of action of the principal muscles involved in various movements of the hindlimb. (A) Rotation of the femur; (B) retraction of the femur; (C) protraction of the femur; (D) extension of the tibiotarsus; (E) abduction of the limb; (F) flexion of the tibiotarsus. CF, *m. caudofemoralis*; FCLA, *m. flexor cruris lateralis pars accessoria*; FCLP, *m. flexor cruris lateralis pars pelvica*; FCM, *m. flexor cruris medialis*; FTEM, *mm. femorotibialis externus* and *femorotibialis medius*; IC, *m. iliotibialis cranialis*; IF, *m. iliofibularis*; IFE, *m. iliofemoralis externus*; IFI, *m. iliofemoralis internus*; IL, *m. iliotibialis lateralis*; ILF, *m. iliofemoralis*; ISF, *m. ischiofemoralis*; ITC, *m. iliotrochantericus caudalis*; ITCR, *m. iliotrochantericus cranialis*; ITM, *m. iliotrachantericus medius*; OL, *m. obturatorius lateralis*; OM, *m. obturatorius medialis*; PIF, *m. pubo-ischio-femoralis*.

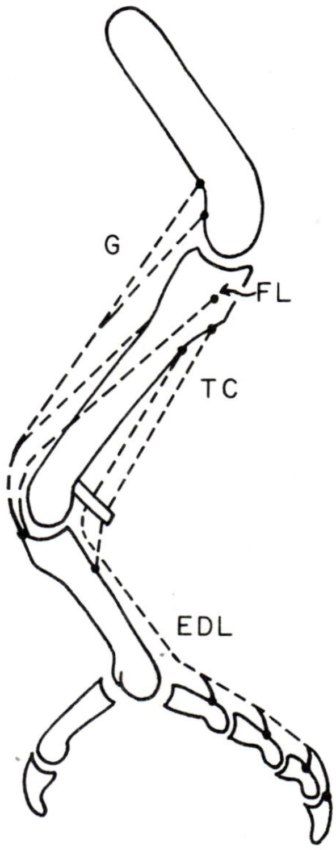

Fig. 2.29. Lines of action of the principal muscles involved in the movements of the foot. EDL, *m. extensor digitorum longus*; FL, *m. fibularis longus*; G, *m. gastrocnemius*; TC, *m. tibialis cranialis*.

three bellies contribute to a single tendon of insertion that passes over a depression on the surface of the tibial cartilage and inserts on the *hypotarsus* (a process on the plantar surface at the proximal end of the tarsometatarsus) (Fig. 2.17) and sometimes along the shaft of the tarsometatarsus. Cracraft (1971) reported that, in *Columba,* the *pars lateralis* is primarily a twitch muscle and therefore probably functions to extend the intertarsal joint during locomotion. The other heads, in contrast, contain mostly tonus fibers, and therefore probably act to maintain posture by resisting flexion.

The *m. fibularis longus* (Figs. 2.19, 29), which inserts on the lateral side of the tibial cartilage (*cartilago tibialis,* the fibrocartilaginous block on the caudal surface of the trochlea at the distal end of the tibiotarsus), assists in extension of the intertarsal joint. The *m. plantaris* (Fig. 2.25) similarly inserts on the proximomedial corner of the tibial cartilage, and presumably does the same. Fisher (1946) suggested that the function of the *m. plantaris* is to draw the tibial cartilage proximally so as to prevent "crushing" of the tendons passing through it, but Cracraft (1971) questioned this. In any event the *m. plantaris* is a small muscle and is absent in many birds.

The flexors of the toes (see later) also pass around the plantar surface of the intertarsal joint. If flexion of the toes is resisted by their extensors, then these muscles would also serve as ankle extensors.

4. Movements of the toes

The system of muscles to the toes is extremely complex because of the number of toes, the variety of their movements and the diversity of adaptive specialisations in different groups of birds. The basic movements of the toes (Fig. 2.18) are flexion and extension. Flexion of the toes brings about the grasping action of the foot; extension is the releasing movement. In addition the toes show varying degrees of abduction and adduction, which are defined in terms of the limb itself rather than the body. Thus, in an anisodactyl foot adduction of digit II is movement laterally toward digit III, while abduction is movement medially away from digit III. For digit IV, however, adduction is movement medially toward digit III, and abduction is movement laterally away from it.

The muscles may be better understood by considering them in groups according to both their positions and their actions. With regard to position, the toe muscles are divided into intrinsic and extrinsic groups. Muscles of the intrinsic group lie entirely within the foot, arising from the tarsometatarsus and inserting on the phalanges. The extrinsic toe muscles lie in the crus, arising from the femur, tibiotarsus or fibula. They send their long tendons distally around the intertarsal joint and down the tarsometatarsus to their phalangeal insertions. Alternatively, the toe muscles may be classified functionally into two opposing groups. The extensor system includes those muscles involved in releasing the grip, including the extensors proper and a few small intrinsic adductors and abductors. The flexor system includes those muscles that bring about the grasping movement. More strength is required for grasping than for letting go. Thus, the flexor system contains mostly larger extrinsic muscles, and the extensor system mostly smaller intrinsic muscles (Table IV).

(a) Extension. The only extrinsic extensor muscle is the *m. extensor digitorum longus,* which was already described. It inserts on digits II, III and IV, and extends them all simultaneously (Fig. 2.30A). In addition there are a number of intrinsic muscles, each of which inserts on one digit only (Figs. 2.20, 22, 25, 27, 28C). The capability of a species to extend the digits individually depends on how many of the intrinsic muscles are present. No birds possess the full set of muscles listed in Table IV, and there is a tendency for them to be lost as birds get smaller, or in species which are generally considered to be more highly advanced. Information on the presence or absence of the intrinsic muscles in the various groups of birds is given by George and Berger (1966). Similar data are provided for the Coraciiformes by Maurer and Raikow (1981) and for the Piciformes by Swierczewski and Raikow (1981). In the Passeriformes the intrinsic muscles to digits II, III and IV are almost entirely lost (see later).

Fig. 2.30. Muscles of the digits in birds. Intrinsic muscles of digits II–IV (Fig. 2.27) are excluded. (A) The only extrinsic extensor is the *m. extensor digitorum longus* (EDL), the antagonist of which is the *m. flexor digitorum longus* (FDL). The chief flexor of the hallux is the extrinsic *m. flexor hallucis longus* (FHL). In some birds the tendons of the *flexor digitorum longus* and the *flexor hallucis longus* are connected by a vinculum (V). (B) Tendons of insertion of the extrinsic superficial flexor muscles of digits II–IV. FPD2, *m. flexor perforatus digiti II*; FPPD2, *m. flexor perforans et perforatus digiti II*; FPD3, *m. flexor perforatus digiti III*; FPPD3, *m. flexor perforans et perforatus digiti III*; FPD4, *m. flexor perforatus digiti IV*. (C) Intrinsic muscles of the hallux. EHL, *m. extensor hallucis longus*; FHB, *m. flexor hallucis brevis*. From Raikow (1982).

TABLE IV. **Muscles of the toes in birds.**

	Intrinsic muscles	Extrinsic muscles[a]
I. Extensor system		
Digit I	*Extensor hallucis longus*	
Digit II	*Adductor digiti II*	[*Extensor digitorum longus*]
	Abductor digiti II	
Digit III	*Extensor proprius digiti III*	[*Extensor digitorum longus*]
	Extensor brevis digiti III	
Digit IV	*Extensor brevis digiti IV*	[*Extensor digitorum longus*]
	Extensor proprius digiti IV	
	Adductor digiti IV	
	Abductor digiti IV	
II. Flexor system		
Digit I	*Flexor hallucis brevis*	*Flexor hallucis longus*
Digit II		*Flexor perforatus digiti II*
		Flexor perforans et perforatus digiti II
		[*Flexor digitorum longus*]
Digit III		*Flexor perforatus digiti III*
		Flexor perforans et perforatus digiti III
		[*Flexor digitorum longus*]
Digit IV		*Flexor perforatus digiti IV*
		[*Flexor digitorum longus*]

[a] Bracketed muscles have multiple insertions.

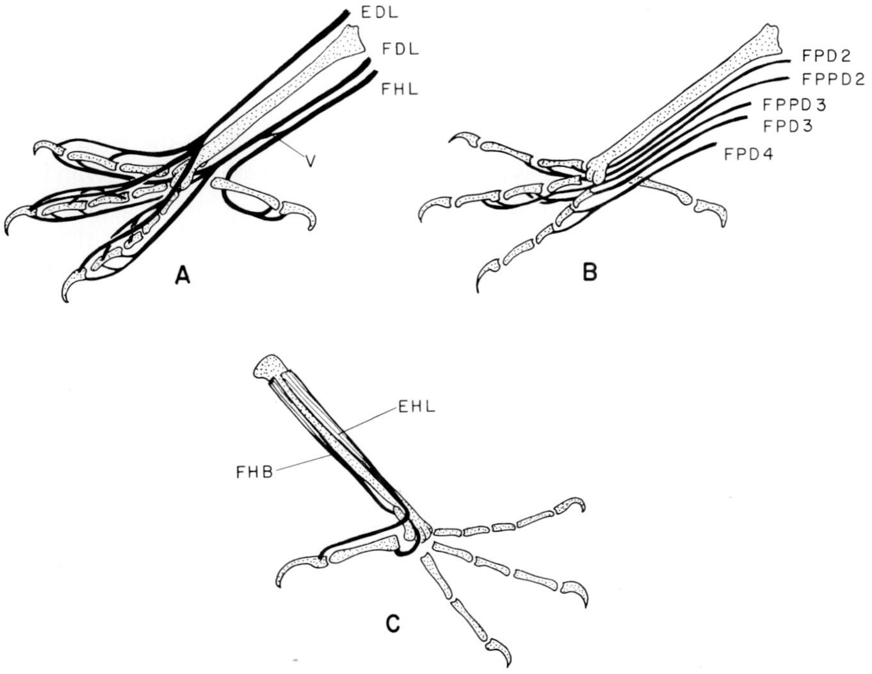

Extension of digit I is performed by the intrinsic *m. extensor hallucis longus* (Figs. 2.27, 30C). This is retained in passerines, no doubt because there is no extrinsic extensor of the hallux in birds (Raikow, 1982). It is assisted in some birds by an elastic automatic extensor ligament (*ligamentum elastica extensoria digitorum*), which holds the distal phalanx extended when the *flexor hallucis longus* is relaxed (George and Berger, 1966, p. 455). The muscle is absent in some birds in which the hallux is lost or vestigial (George and Berger, 1966, p. 447; Klemm, 1969).

(b) Flexion. The flexors of the toes include one intrinsic muscle and several extrinsic muscles (Table IV). For convenience of study the latter may be divided into the superficial flexors and the deep flexors. The superficial flexors arise from the caudodistal end of the femur or from the proximal tibiotarsus or fibula, or the ligaments of the knee region. They lie in the lateral and caudal region of the crus deep to the *m. gastrocnemius* and superficial to the deep flexors. Their tendons pass distally through the tibial cartilage, and down the plantar surface of the tarsometatarsus, where they lie superficial to the deep plantar tendons. They continue out along the plantar surfaces of the digits to their insertions. These muscles each flex only one digit. There is much variation in the superficial flexors (George and Berger, 1966), but a general picture may be given without concern for details and variations. The typical pattern of insertions of the tendons of the superficial flexors is shown in Fig. 2.30B.

Digit II is supplied by two superficial flexors. The *m. flexor perforatus digiti II* inserts at the base of the first phalanx and thus flexes the whole digit at its articulation. The *m. flexor perforans et perforatus digiti II* inserts farther out along the plantar surface of the digit, flexing the second phalanx on the first phalanx as well as contributing to flexion at the articulation. Digit III is similarly provided with two superficial flexors, the *m. flexor perforatus digiti III* inserting more proximally and the *m. flexor perforans et perforatus digiti III* more distally. Digit IV, in constrast, has only one superficial flexor, *m. flexor perforatus digiti IV*. The hallux has no superficial flexors.

The most superficial muscle of the thigh in many birds is the *m. ambiens*. It arises from the pectineal process (*proc. pectinealis*) of the ilium and passes distally, giving rise to a tendon that overlies the lateral side of the crus by penetrating through or deep to the patellar ligament or the patella itself (Figs. 2.24, 25). The tendon passes medial to the iliofibularis tendon and then gives rise to part of one or more of the superficial flexor muscles of digits II, III and IV. In a few species, however, the tendon merely terminates in the knee joint capsule. It is difficult to infer the action of the *ambiens* muscle given its complex relations, but it appears likely that its principal role in most cases is to assist the digital flexors. George and Berger (1966, pp. 418–421) discuss its occurrence and variation among birds.

The two deep flexors are the deepest muscles on the caudal side of the crus, lying between the tibiotarsus and the superficial flexors. There is considerable variation in these muscles (George and Berger, 1966). In general the *m. flexor digitorum longus* is the deepest muscle on the caudal surface of the crus, arising from the proximal ends of the fibula and tibiotarsus, and sometimes also a femoral head (Figs. 2.21, 22, 30A). The *flexor hallucis longus* typically lies superficial to the *flexor digitorum longus*, with an origin from the distal end of the femur (Figs.

2.21, 22, 30A). The tendons of these muscles pass around the intertarsal joint and down the plantar surface of the tarsometatarsus, where they lie deep to the tendons of the superficial flexors. The tendons of *flexor digitorum longus* and *flexor hallucis longus* are known as the "deep plantar tendons" and show considerable variety in their interconnections with each other and in their distribution to the digits. This diversity will be explored later, after the basic arrangement has been described (Fig. 2.30A). The *flexor hallucis longus* sends its tendon to the hallux, where it inserts on the flexor tubercle (*tuberculum flexorium*) at the base of the plantar surface of the distal phalanx, with a secondary insertion on the proximal phalanx. The muscle thus flexes the hallux on metatarsal I, and the distal phalanx on the proximal phalanx. It is assisted by the *m. flexor hallucis brevis,* the only intrinsic flexor of the toes (Fig. 2.30C). This muscle is usually small, though it may be enlarged in birds with a strong grip (Raikow *et al.*, 1980). Digits II, III and IV are flexed by the *flexor digitorum longus,* the tendon of which divides into three branches, each branch inserting on the distal phalanx of its digit, with various secondary insertions more proximally (Fig. 2.30A).

Thus, between the superficial and deep flexor groups, each digit is provided with flexor tendons from two or three different muscles, which insert at various points along the length of the digit and bring about the curvature of the toe as it is flexed. The *flexor digitorum longus* provides co-ordinated, simultaneous flexion of the three forward toes whilst the superficial flexors provide individual movements.

The *m. lumbricalis* (Fig. 2.22) is a very small muscle that arises from the distal end of the tendon of the *m. flexor digitorum longus* (sometimes also from the *flexor hallucis longus*) and passes distally to insert on the base of one or two of the forward toes. Given its small size (often apparently vestigial), its function is difficult to predict. Perhaps it provides for some delicate adjustment of the *flexor digitorum longus* tendon. George and Berger (1966, p. 464) suggest that it is often absent in birds, but it is easily missed in dissection unless a stain is used. The author has found it in many passerines, whilst Swierczewski (1977) observed it throughout the Piciformes, and Maurer (1977) in many of the Coraciiformes. It is also present in the Coliidae (Berman and Raikow, 1982).

C. The foot

The diversity of avian feet has long stood as a textbook example of adaptive variation, and a considerable terminology has arisen to describe it. Several authors (Coues, 1890, pp. 190–195; Harrison, 1964, p. 431) have identified three general types of foot amongst birds: (1) the *perching* or *insessorial* type (including the *raptorial* or *prey-grasping* subtype); (2) the *walking* and *running* (*cursorial*) type, which includes the *wadding* or *grallatorial* subtype; and (3) the *swimming* or *natatorial* type. As the discussions of these authors make clear, however, this is not a classification of structure or even function, but one of behavior. Thus, distinctly different results have been attained by the evolution of alternative solutions to the common problems presented by the behavioural choices.

Several terms used in describing the avian foot may be introduced at this point (see also Table V). An *incumbent* hallux is articulated at the same level as the other toes. If it is attached further proximally it is *elevated*. The hallux may be *large, reduced, vestigial* etc. in relation to the other toes. Progressive elevation, reduction and even loss of the hallux often occur in birds that lack a grasping usage of the foot. Digits II and IV are described as being *reversed* when they are directed cadually. A piece of skin stretched between adjacent toes is a *web*. When there is little or no webbing between the toes, the toes are said to be *cleft* or *free*. Cleft toes have the capacity for independent movement. If webbing extends for up to half (or less) the length of the toes, the condition is *semipalmate*. Although the semipalmate condition occurs both in wading birds and in some non-passerine forms, such as galliform and falconiform species, for some reason the term is usually applied only to wading birds. In the *palmate* foot the webbing extends for more than half the length of the toes. This term is restricted to birds in which the hallux is not included in the webbed portion. In contrast, pelecaniform species have webbing that extends between all four toes, giving the *totipalmate* condition. Webbing has arisen in the course of evolution through a modification of the ontogeny of the foot, which initially develops in most tetrapods as a flattened paddle at the end of the rudimentary limb. After the axes of the digits are established the areas between them undergo localised cell death, leading to the digits becoming separated. Failure of this process results in the retention of living tissue between the digits in the form of webbing. Lobation is the development of paddle-like outgrowths of the skin along the margins of the toes, as in phalaropes (Scolopacidae) and grebes (Podicipedidae). Webbing and lobation are alternative swimming adaptations that increase the paddle area of the foot.

1. Foot types

The feet of birds may be classified according to the number and arrangement of the toes. The types described here follow general usage, but an attempt has been made to clarify the definitions. As in any sort of classification there are ambiguities where one type grades into another, and also because some birds have the ability to shift the positions of certain toes. The occurrence of different foot types is listed in Tables V, VI and VIII, and the arrangements themselves are illustrated in Fig. 2.31.

(a) Anisodactyl. The hallux is directed caudally, and the second, third and fourth toes are directed cranially. This is believed to be the ancestral type of foot in birds, from which the other types evolved. Its origin was discussed by Bock and Miller (1959). Accepting the then-current view that birds evolved from pseudosuchian thecodont reptiles, which had five cranially-directed toes, they hypothesised that the evolution of the avian foot involved the loss of digit V for unknown reasons, and the reversal of the hallux to a caudal position as an adaptation for perching in trees. *Archaeopteryx* the earliest known bird, had an anisodactyl foot. Whilst Bock and Miller noted that the hallux in *Archaeopteryx* is small and somewhat elevated, which would make it a relatively poor perching foot, they accepted the idea that *Archaeopteryx* was arboreal. They suggested that if ancestral birds had had an efficient perching foot from the start, there would have been no need for the other types of perching feet to develop. John Ostrom has since championed the view that birds evolved from coelurosaurian dinosaurs,

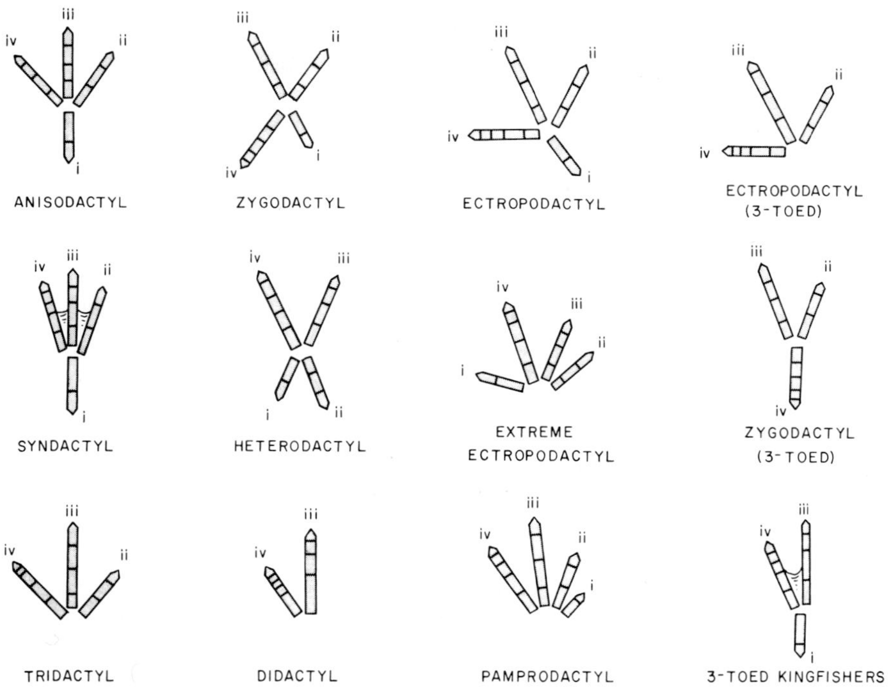

Fig. 2.31. Diagrammatic representation of the classifications of the feet of birds, according to the number and arrangement of the digits. See text for discussion.

and that *Archaeopteryx* was cursorial rather than arboreal. Ostrom (1974) showed that the foot structure, including the toe arrangement and proportions, is similar in these reptiles to the condition in *Archaeopteryx*. Analysing many data besides foot structure, he concluded that the advanced conditions shared by *Archaeopteryx* and coelurosaurs are due to common ancestry rather than to convergence. In addition, anisodactyly is by far the most common configuration in birds (Tables V and VI), occurring in a wide variety of forms of many adaptive types, whereas other arrangements are generally limited to one or a few groups where they are associated with particular functional specialisations. For

TABLE V. **Characteristics of the feet in birds.**

Order and family	Type of foot	Hallux	Type of deep plantar tendons
Paleognathiformes			
Tinamidae	Anisodactyl or tridactyl	Small and elevated or lacking	II or IV
Struthionidae	Didactyl	Lacking	IV[a]
Rheidae	Tridactyl	Lacking	IV
Casuariidae	Tridactyl	Lacking	IV
Dromaiidae	Tridactyl	Lacking	IV
Apterygidae	Anisodactyl	Small, elevated	II

(*continued*)

TABLE V. (*Continued*)

Order and family	Type of foot	Hallux	Type of deep plantar tendons
Podicipediformes			
Podicipedidae	Anisodactyl, lobate, claws flattened	Small, elevated	IV
Gaviiformes			
Gaviidae	Anisodactyl, palmate	Small, elevated	IV
Sphenisciformes			
Spheniscidae	Anisodactyl, palmate	Small, elevated, directed medially	II
Procellariiformes			
Diomedeidae	Anisodactyl or tridactyl, palmate	Rudimentary or lacking	IV
Procellariidae	Anisodactyl, palmate	Minute, elevated	IV
Hydrobatidae	Anisodactyl, palmate	Minute, elevated	IV
Pelecanoididae	Tridactyl, palmate	Lacking	IV
Pelecaniformes			
Phaethontidae	Totipalmate	Small, directed medially	IV
Sulidae	Totipalmate	Incumbent, directed medially	I
Phalacrocoracidae	Totipalmate	Incumbent, directed medially	I
Anhingidae	Totipalmate	Incumbent, directed medially	I
Pelecanidae	Totipalmate	Incumbent, directed medially	
Fregatidae	Totipalmate, webbing reduced	Incumbent, directed medially	Va
Ciconiiformes			
Ardeidae	Anisodactyl	Large, incumbent	I or VII
Balaenicipitidae	Anisodactyl	Large, incumbent	I
Scopidae	Anisodactyl	Large, incumbent	I[b]
Ciconiidae	Anisodactyl, toes webbed basally	Large, slightly elevated	I
Threskiornithidae	Anisodactyl, toes webbed basally	Large, slightly elevated	I
Phoenicopteridae	Anisodactyl or tridactyl, palmate	Small and elevated or lacking	IV
Anseriformes			
Anhimidae	Anisodactyl, semipalmate	Large, incumbent	II
Anatidae	Anisodactyl, palmate[c]	Small, elevated, sometimes lobed	II
Falconiformes[d]			
Cathartidae	Anisodactyl	Elevated	Va
Sagittariidae	Anisodactyl, toes short, partly webbed	Small, incumbent	III
Accipitridae	Anisodactyl, raptorial	Large, incumbent	III
Pandionidae	Semizygodactyl, raptorial	Large, incumbent	XI[e]
Falconidae	Anisodactyl, raptorial	Large, incumbent	III
Galliformes			
Megapodiidae	Anisodactyl	Large, incumbent	I
Cracidae	Anisodactyl	Incumbent	I
Phasianidae	Anisodactyl	Elevated	I
Opisthocomidae	Anisodactyl	Large, incumbent	I
Gruiformes			
Turnicidae	Tridactyl	Lacking	IV
Pedionomidae	Anisodactyl	Small, elevated	

(*continued*)

TABLE V. (*Continued*)

Order and family	Type of foot	Hallux	Type of deep plantar tendons
Gruidae	Anisodactyl	Small, elevated	I
Aramidae	Anisodactyl	Slightly elevated	I
Rallidae	Anisodactyl[f]	Small, elevated	I
Psophiidae	Anisodactyl	Small	I
Heliornithidae	Anisodactyl, lobate	Small	II[g] or X[h]
Eurypygidae	Anisodactyl	Small, elevated	I
Rhynochetidae	Anisodactyl	Small, elevated	I
Cariamidae	Anisodactyl	Small, elevated	IV
Otididae	Tridactyl	Lacking	
Mesitornithidae	Anisodactyl	Large, incumbent	I
Charadriiformes[i]			
Jacanidae	Anisodactyl; toes and claws elongate	Large, incumbent	IV[j]
Rostratulidae	Anisodactyl	Small, elevated	I
Haematopodidae	Tridactyl	Lacking	
Scolopacidae	Anisodactyl or tridactyl,[k] lobate, semipalmate	Small, elevated[k]	I
Recurvirostridae	Anisodactyl or tridactyl, palmate or semipalmate	Vestigial or lacking	
Charadriidae	Anisodactyl or tridactyl	Vestigial or lacking	
Dromadidae	Anisodactyl	Small, elevated	I
Burhinidae	Tridactyl, semipalmate	Lacking	
Glareolidae	Anisodactyl	Present or lacking	I
Thinocoridae	Anisodactyl	Small, elevated	
Chionididae	Anisodactyl, semipalmate	Small, elevated	I
Stercorariidae	Anisodactyl, palmate	Small	IV
Laridae	Anisodactyl or tridactyl, palmate	Small or vestigial	I or IV
Rhynchopidae	Anisodactyl, palmate	Small, slightly elevated	IV
Alcidae	Anisodactyl or tridactyl, palmate	Vestigial or lacking	IV
Columbiformes			
Pteroclididae	Anisodactyl or tridactyl	Small or absent	IV[l]
Columbidae	Anisodactyl	Incumbent or slightly elevated	I
Psittaciformes			
Psittacidae	Zygodactyl[m]	Incumbent	I
Cuculiformes			
Musophagidae	Semizygodactyl	Large, incumbent	I
Cuculidae	Zygodactyl	Large, incumbent	I
Strigiformes			
Tytonidae	Semizygodactyl, raptorial	Large, incumbent	I
Strigidae	Semizygodactyl, raptorial	Large, incumbent	I
Caprimulgiformes			
Steatornithidae	Pamprodactyl	Incumbent	Va[n]
Podargidae	Anisodactyl	Incumbent	Va[n]
Nyctibiidae	Syndactyl	Incumbent	See text
Aegothelidae	Anisodactyl	Incumbent	I[n]
Caprimulgidae	Anisodactyl	Small, elevated	Va

(*continued*)

TABLE V. ***(Continued)***

Order and family	Type of foot	Hallux	Type of deep plantar tendons
Apodiformes			
Apodidae	Anisodactyl, pamprodactyl	Incumbent	Va
Trochilidae	Anisodactyl	Large, incumbent	I[l]
Coliiformes			
Coliidae	Semizygodactyl[o]	Large, incumbent	Va
Coraciiformes			
Trogonidae	Heterodactyl	Incumbent	VIII
Coraciidae	Anisodactyl[p]	Incumbent	Va
Brachypteraciidae	Anisodactyl	Small	
Leptosomidae	Semizygodactyl	Incumbent	Va
Bucerotidae	Syndactyl	Incumbent	Va
Phoeniculidae	Syndactyl	Large, incumbent	Vd[q]
Upupidae	Syndactyl	Incumbent	Vd or VII[q]
Todidae	Syndactyl	Incumbent	Vb
Momotidae	Syndactyl	Incumbent	Vb
Meropidae	Syndactyl	Incumbent	Vb
Alcedinidae	Syndactyl[r]	Incumbent	Vb[q]
Piciformes			
Bucconidae	Zygodactyl	Incumbent	VI
Galbulidae	Zygodactyl	Incumbent or lacking	VI
Capitonidae	Zygodactyl	Incumbent	VI
Ramphastidae	Zygodactyl	Incumbent	VI
Indicatoridae	Zygodactyl	Incumbent	VI
Picidae	Zygodactyl[s]	Incumbent or lacking	VI
Passeriformes	Anisodactyl or syndactyl	Incumbent, large	VII or I[t]

[a] Modified by loss of branch to digit II (lacking).
[b] Unusual form (see Beddard 1898, p. 421).
[c] Semipalmate in *Anseranas semipalmata* and *Branta sandvicensis*.
[d] Jollie (1977) gives an exceptionally-detailed description.
[e] Gadow (1896). George and Berger (1966, p. 448) discuss only types I–VIII and list *Pandion* under type V.
[f] Lobate in *Fulica*.
[g] *Podica*.
[h] *Heliornis*.
[i] Ridgway (1919, p. 1) lists the order Charadriiformes as type I, but tridactyl forms are more likely to be type IV.
[j] Ridgway (1919, p. 5) describes a type IV (without a number), that is, lacking a branch to the hallux. This is questionable because the hallux is well developed in the Jacanidae.
[k] Hallux lacking in *Calidris alba*.
[l] See text.
[m] Pamprodactyl in *Micropsitta* (Bock and Miller, 1959).
[n] These are my interpretations of the descriptions given by Hoff (1966).
[o] An extreme form; see text.
[p] Verges on being syndactyl.
[q] Variable, see text.
[r] Digit II lacking in two genera; see text.
[s] Adjustable to ectropodactyl in some species; see text.
[t] Eurylaimidae and some Philepittidae.

TABLE VI. **Numbers of non-passerine families having each type of foot**[a]

Foot type	Number of families[b]
Anisodactyl	61
Tridactyl	16
Zygodactyl	8
Syndactyl	8
Semizygodactyl	6
Totipalmate	6
Pamprodactyl	2
Heterodactyl	1

[a] A family that includes species with different foot types is counted separately under each type.
[b] Data from Table V.

these reasons we may conclude that anisodactyly is the primitive condition in birds.

(b) Tridactyl. The hallux is absent, and digits II, III and IV are directed cranially. Although tridactyly has not been included in most lists, it is the second most common arrangement, is distinct from anisodactyly, and thus merits recognition. Tridactyly is common in cursorial birds (Table V) and has clearly evolved from the anisodactyl condition by the loss of the hallux in groups that do not perch or have other grasping functions. Intermediate stages in some birds suggest that the loss of the hallux involves gradual size reduction and frequently its movement from an incumbent position to an elevated position. In some groups of birds there are species that retain a reduced, elevated hallux, while other species have lost the digit entirely, such as flamingos (Phoenicopterdiae) and tinamous (Tinamidae) (see Table V). There are also changes associated with the hallucial muscles, especially in the transfer of the *flexor hallucis longus* tendon to the forward digits by its fusion with the *flexor digitorum longus* tendon (see later). Presumably the hallux is expendable under these conditions, yet many cursorial birds retain it. In large birds like moas the hallux is of good size and incumbent, presumably helping to support the bird's weight. In some small birds the hallux may play a role in support on the ground, as suggested by the elongated and straightened claw which occurs in terrestrial passerines like longspurs (*Calcarius* spp.) and larks (*Alaudidae*). Although the usual tridactyl condition involves the loss of the hallux, a few birds have retained the hallux and lost other toes (see Fig. 2.31). Technically this is not tridactyly.

(c) Didactyl. Amongst living birds only the Ostrich (*Struthio camelus*) has a two-toed foot, having lost digits I and II. Digit III is the longer of the two remaining toes, both of which are relatively short. This condition is often regarded as a running adaptation analogous to the toe reduction in hoofed mammals. Reduction and loss of the inner toe also occurred in the fossil Gruiform family Ergilornithidae (Feduccia, 1980, pp. 126–127).

(d) Syndactyl. In this type of foot the second, third and fourth toes are directed cranially, and are variously joined at their bases by common skin sheaths. The hallux is directed caudally and is usually long. According to Bock and Miller (1959, p. 30), fusion occurs between the second, third and fourth toes, whilst Van

Tyne and Berger (1976) state that only two toes must be coalesced. The latter definition is probably the most generally accepted and is used in the present account. Bock and Miller (1959) consider syndactyly to be an adaptation for perching, in that the binding of the toes at their bases holds them parallel to each other so that they are held at right angles to the branch, which assures that all of the force of flexion is directed against the branch and none is lost by being directed laterally. Clark (1981) found a high (though not perfect) correlation between toe fusion and perching habits in oscines.

The syndactyl and anisodactyl conditions intergrade so closely that the distinction between them is sometimes arbitrary. The problem is simply that a tendency for basal fusion of the forward toes in anisodactyl forms has often occurred independently in different groups. For example, in the Coraciidae the second and third toes are united basally (Van Tyne and Berger, 1976, p. 692). This is a modest beginning of syndactyly, but the condition is poorly developed compared to most coraciiform families. Perhaps impressed by this intra-ordinal difference, Bock and Miller (1959) consider the family to be anisodactyl rather than syndactyl as sometimes stated. A similar problem occurs in the Passeriformes. Although this order is typically described as being anisodactyl, several families including both sub-oscine (Eurylaimidae, Dendrocolaptidae, Furnariidae, Formicariidae and Pipridae) and oscine (Hirundinidae, Troglodytidae, Mimidae, Bombycillidae, Vangidae and Vireonidae) species show varying degrees of adherence of the forward toes (Van Tyne and Berger, 1976). In a relatively comprehensive survey, Clark (1981) also found varying degrees of toe fusion in many oscine families, a wide range occurring between genera, and within a genus. Given this situation it is impossible to characterise most oscine groups accurately as being anisodactyl or syndactyl without being vague or arbitrary.

(e) Heterodactyl. In heterodactyly the first and second digits are directed caudally, whilst the third and fourth are directed cranially in opposition to them. This arrangement occurs only in trogons (Trogonidae) and is considered to be a perching specialisation. Bock and Miller (1959, p. 35) believe that the ancestral avian foot is anisodactyl with a weak hallux, and that the reversal of the second digit in the heterodactyl foot is to compensate for the weak first digit. The effect is similar to the zygodactyl arrangement (see later) but with different pairs of toes in opposition. Both the zygodactyl and heterodactyl arrangements are sometimes termed isodactyly, because they involve equal numbers of opposing toes. The deep flexor tendons of trogons (Trogonidae) are uniquely modified in association with the reversal of the second digit (see later).

(f) Zygodactyl. The second and third digits are directed cranially in opposition to the first and fourth in zygodactyly, which occurs in the Piciformes, Cuculiformes, and Psittaciformes. The hallux in the zygodactyl foot is usually short, and it has been lost in a few piciforms, including the Three-toed Jacamar (*Jacamaralcyon tridactyla*), several woodpeckers in the genera *Picoides, Dinopium* and *Gecinulus,* and piculets (*Sasia* spp.). The evolution of the zygodactyl foot has involved the rotation of digit IV from a cranial to a caudal orientation. Probably zygodactyly evolved independently more than once, and in some forms the fourth toe can be shifted forward (see the ectropodactyl foot and the semi-zygodactyl foot later). The zygodactyl foot is adapted for grasping, especially in perching. Bock and Miller (1959) point out that many authors incorrectly consider

zygodactyly to be a climbing (scansorial) adaptation, although most zygodactyl birds are not scansorial. In addition many scansorial birds are not zygodactyl, such as woodcreepers (Dendrocolaptidae), nuthatches (Sittidae), creepers (Certhiidae), the Black-and-White Warbler (*Mniotilta varia*), the Hawaiian creepers (*Paroreomyza*) and others. Bock and Miller clearly showed that most woodpeckers (Picidae) do not use the zygodactyl configuration while climbing vertically, but become instead ectropodactyl (see later). They pointed out that zygodactyly is mechnically inappropriate for climbing. Although artists frequently depict woodpeckers climbing with their feet in the incorrect zygodactyl position, this is probably because woodpeckers appear zygodactyl in museum specimens and when they perch.

(g) Ectropodactyl. Bock and Miller (1959) used this term to describe the position of the toes in most woodpeckers when climbing on vertical surfaces. The second and third toes are directed cranially, while the fourth toe is not directed caudally but instead is swung laterally or even cranially. The hallux is usually held caudally, but in the extreme form of this configuration it is directed cranially, in which case it lies on the lateral side of the foot, beside digit IV. In the ectropodactyl configuration the second, third and fourth digits, and sometimes the hallux, suspend the bird from the tree against the pull of gravity.

(h) Semizygodactyl. This term refers to birds that can rotate the fourth digit cranially, giving an anisodactyl configuration, or caudally to give a zygodactyl arrangement. Presumably this ability enhances the functional potential of the foot. Birds usually described as semizygodactyl include the Osprey (*Pandion haliaetus*), turacos (Musophagidae), owls (Strigidae) and the Kirombo Coural (*Leptosomus discolor*). Mousebirds (Coliidae) can shift both the first and fourth digits, giving them the capability of being anisodactyl, zygodactyl or pamprodactyl at will (Berman and Raikow, 1982). There does not seem to be a term for the latter condition, which goes beyond semizygodactyly as found in the other forms mentioned, but it is simpler to include them here than to coin a new term for one example.

(i) Pamprodactyl. As usually defined, pamprodactyly is the condition in which all four toes are directed cranially. It is used by birds that cling to vertical surfaces, the four toes combining to suspend the bird against the pull of gravity. There is some ambiguity in terminology, however. Most books simply define pamprodactyly as a condition in which all four toes point forward, without mention of their relative positions. However, two such arrangements are known. The usual arrangement is for the hallux to lie on the medial side of the foot next to digit II. This has been described in certain swifts (Apodidae), in mousebirds (Coliidae) at times, in the Oilbird (*Steatornis caripensis*) and in one parrot genus (*Micropsitta*). In contrast, in the Ivory-billed Woodpecker (*Campephilus principalis*) the hallux lies on the lateral side of the foot beside digit IV, the lateral position. Whilst Bock and Miller (1959) included the medial position of the hallux in their definition of pamprodactyl, this would not be pamprodactyl in a strict sense but an extreme ectropodactyl configuration. Perhaps by not calling this arrangement pamprodactyl, the different positions of the hallux in the two arrangements are emphasised.

The pamprodactyl condition of swifts has been re-assessed by Charles T. Collins (1983), who has handled living specimens of many supposedly pam-

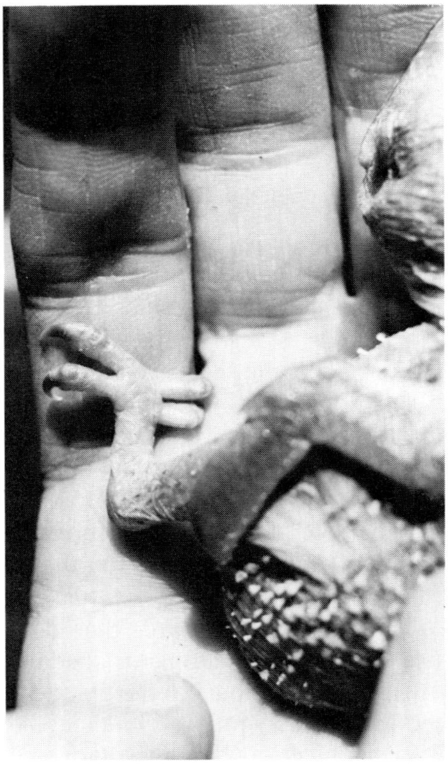

Fig. 2.32. Left foot of the nestling Alpine Swift (*Apus melba*) showing the lateral grasping conformation of the toes. See text for discussion. Photo courtesy of Charles T. Collins.

prodactylous species. He found that these swifts have a unique grasping mechanism in which digits I and II are spread medially, opposing digits III and IV, which are directed laterally. This lateral grasping movement (Fig. 2.32) is effective in holding the soft nesting material of plant floss and feathers, or palm fronds. It is intriguingly convergent to the lateral grasping specialisations in chameleons among reptiles, and in some phalangeroid mammals such as the Koala (*Phascolarctos cinereus*). Collins did note, however, that the foot assumes a pamprodactyl arrangement when relaxed, and this arrangement is seen in museum specimens.

(j) Other variations. Some birds have feet which are not easily defined according to the preceding classification. In the totipalmate condition of the Pelecaniformes the hallux is directed medially while the other digits point cranially, apparently because of the presence of a full set of interdigital webs. There is no specific term for this configuration, but none is really needed because the term totipalmate adequately identifies the pelecaniform foot. Penguins (Spheniscidae) also have the hallux reduced and somewhat medially directed. This may be considered anisodactyl in a broad sense, as it clearly is not one of the other major types, and is probably best regarded as a degenerate condition.

In a few birds there are losses of toes other than the hallux. The otherwise obscure Three-toed Parrotbill (*Paradoxornis paradoxus*) of China has achieved a measure of fame as the only passerine in which digit IV is reduced to a vestige.

Similarly a few kingfishers (Alcedinidae) in the genera *Alcedo* and *Ceyx* have lost digit II (Fig. 2.31), although other species in each genus retain it (Delacour, 1951). There is no need for special terms to cover these cases.

2. *The deep plantar tendons*

Variation in the arrangement and interconnections of the tendons of the deep digital flexor muscles, the *mm, flexor digitorum longus* and *flexor hallucis longus,* was apparently first noted by Sundevall (1835). Garrod (1875) greatly expanded these observations, describing various new configurations. Gadow and Selenka (1891), and later Gadow (1896, pp. 614–618) formalised the study of the tendons by numbering distinctive types and listing groups in which they occur. Various other authors have examined the tendons, sometimes referring to them as Gadow's types, sometimes merely describing them. For reviews see George and Berger (1966, pp. 442–449) and Sibley and Ahlquist (1972, pp. 17–20).

The tendon arrangements were initially investigated as taxonomic characters, and despite their obvious functional importance, they have seldom been considered from the latter viewpoint. Sibley and Ahlquist (1972) question the importance of the tendons in taxonomy, and without doubt they do not have the significance that Garrod and Gadow would have liked. Nevertheless some value in taxonomy may be derived from the tendons when they are studied by modern methods of analysis (Swierczewski and Raikow, 1981; Maurer and and Raikow, 1981). An analysis of the variations in the tendons in terms of their functional significance could be advantageous, not only in the classification of birds but also in the interpretation of the biological significance of their evolutionary history.

As in any attempt to classify structural variations there are limitations and exceptions to the system, and as tendon types inter-grade the classification of them is sometimes arbitrary. On the other hand, arrangements with distinctive differences may be classified together, because they fulfill the requirements of a definition while varying in features that the definition ignores. In many cases the tendon types attributed to particular groups are based on a few species (sometimes only on one species), and the family may well possess unsuspected diversity. Furthermore, some birds, such as for example the Nyctibiidae, do not fit into the established patterns (Hoff, 1966, p. 82). In most groups, notable exceptions being the passerines, the *flexor hallucis longus* tendon is connected to the *flexor digitorum longus* tendon by a fibrous band, the vinculum (*vinculum tendinum flexorum*) details of which show considerable variation. Eight main types of relationship between the tendons can be recognised and are described next (see also Fig. 2.33 and Tables V, VII and VIII).

(a) Type I. The *flexor digitorum longus* tendon divides into three, sending branches to digits II, III and IV, while the *flexor hallucis longus* tendon supplies digit I. Although a vinculum is present, the *flexor hallucis longus* tendon remains well developed as it passes to the hallux. Because of the arrangement of the vinculum, the *flexor hallucis longus* can assist the *flexor digitorum longus* but not vice versa (Garrod, 1875). There is much structural variation within this basic arrangement, Hudson *et al.* (1959), for example, describing intra-generic differences in the structure of the vinculum in the Galliformes. Functionally, the

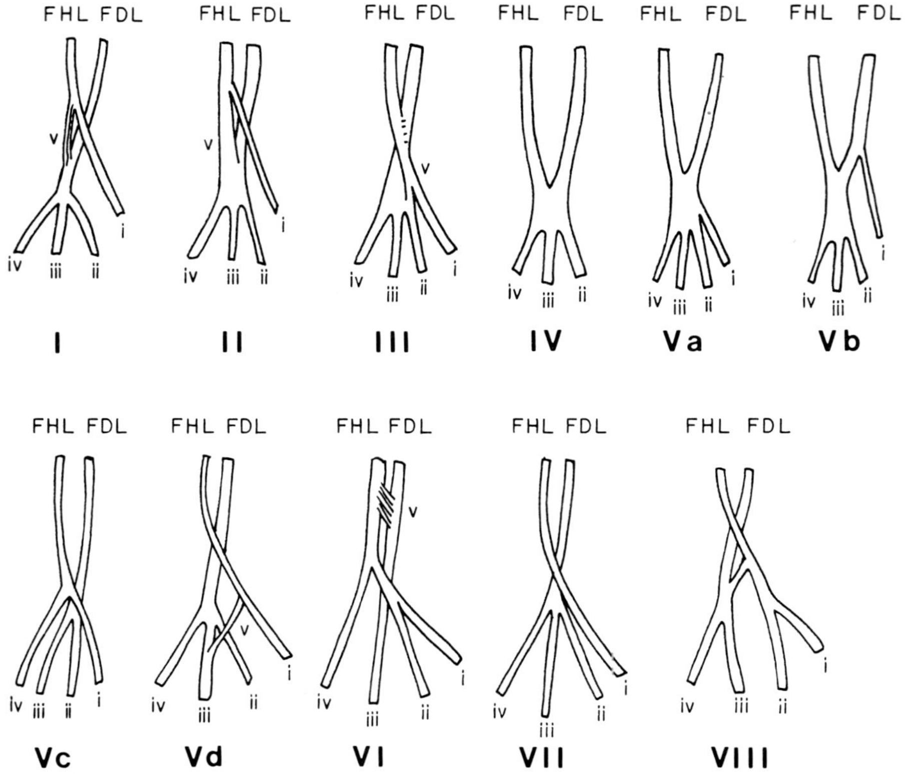

Fig. 2.33. The major types of arrangement (types I–VIII) of the deep plantar tendons in birds. FDL, Tendon of *m. flexor digitorum longus*; FHL, tendon of *m. flexor hallucis longus*; v, vinculum. No example of the type Vc arrangement is known to exist. After Gadow (1896).

TABLE VII. **Numbers of non-passerine families having each tendon arrangement.**[a]

Tendon types	Number of families[b]
I	36
IV	21
Va	10
II	6
VI	6
Vb	4
III	3
Vd	2
VII	2
VIII	1

[a] A few families are excluded. See footnote to Table VIII.

[b] Data from Table V.

TABLE VIII. **Numbers of non-passerine families having various combinations of foot types and tendon arrangements.**[a]

Foot type	Tendon type									
	I	II	III	IV	Va	Vb	Vd	VI	VII	VIII
Anisodactyl	28	6	3	13	5	—	—	—	1	—
Tridactyl	—	—	—	10	—	—	—	—	—	—
Zygodactyl	2	—	—	—	—	—	—	6	—	—
Syndactyl	—	—	—	—	1	4	2	—	1	—
Semizygodactyl	3	—	—	—	2	—	—	—	—	—
Totipalmate	3	—	—	1	1	—	—	—	—	—
Pamprodactyl	—	—	—	—	2	—	—	—	—	—
Heterodactyl	—	—	—	—	—	—	—	—	—	1

[a] A few families are excluded. The Ostrich (Struthionidae) has a didactyl foot with a modified type IV tendon arrangement having only two branches. The Nyctibiidae have a unique tendon arrangement (Hoff, 1966). Data for the Otididae, Apodidae and most families of the Charadrii are lacking or ambiguous. Where a family includes species with different combinations, all are listed.

type I arrangement allows strong flexion of the hallux in birds in which that digit is well developed, while the vinculum presumably co-ordinates flexion of the hallux with that of the forward toes.

Type I is the most common condition in non-passerine birds (Tables V, VII and VIII) and is usually associated with the anisodactyl foot, but occasionally with the zygodactyl or semizygodactyl foot (Table VIII). A detailed description of the arrangement is given by Owre (1967) in cormorants (Phalacrocoracidae) and anhingas (Anhingidae). Although type I is the typical arrangement in herons (Ardeidae), Vanden Berge (1970) confirmed earlier reports that in bitterns (*Ixobrychus* and *Botaurus*) the vinculum is absent, giving a type VII condition (see later). Among passeriform species the type I arrangement occurs only in the Eurylaimidae and Philepittidae.

(b) Type II. Type II is similar to type I but the vinculum is stouter, being the main continuation of the *flexor hallucis longus* tendon, and the tendon to the hallux is relatively feeble. It occurs in anisodactyl birds with a reduced hallux (Tables V, VII and VIII). A detailed description of the evolutionary changes of the tendons in tinamous (Tinamidae) from the type II to the type IV arrangements was provided by Hudson *et al.* (1972). In *Nothoprocta* and *Nothura* the tendons are connected by a short, wide vinculum, and the arrangement is clearly type II. In *Tinamus* and *Crypturellus* the tendons are fused together. In *Eudromia* and *Tinamotis* the tendons are fused together and the hallux is absent, giving a clear-cut type IV arrangement (see later). The tinamous illustrate nicely the transition from one tendon arrangement to another in association with evolutionary changes in foot structure.

In the Musk Duck (*Biziura lobata*), which is the most highly-specialised diver in the family Anatidae (Raikow, 1970), the tendon to the hallux has a fine insertion on the plantar side of the proximal phalanx rather than on the distal phalanx which is its usual insertion (Raikow, 1970, p. 37). This is probably a degenerate condition associated with the extreme aquatic specialisation and loss of cursorial abilities in the Musk Duck.

(c) Type III. The *flexor hallucis longus* and *flexor digitorum longus* tendons are more or less fused together for much of the length of the tarsometatarsus, and the vinculum arises at a relatively distal level and inserts on the branch of the *flexor digitorum longus* tendon that passes to digit II. Type III is associated only with an anisodactyl foot, but more specifically, it is limited to the Falconiformes in which it occurs among hawks (Accipitridae), falcons (Falconidae) and the Secretary Bird (*Sagittarius serpentarius*). It is not present, however, in the New World vultures (Cathartidae) and in the Osprey (*Pandion haliaetus*). For detailed descriptions see Hudson (1948) and Jollie (1977).

(d) Type IV. The *flexor hallucis longus* and *flexor digitorum longus* tendons fuse together completely in type IV, rather than crossing one another, the combined tendon dividing into three to supply digits II, III, and IV. Even though a hallux may be present it is not supplied with a tendon. Type IV is characteristic of birds in which the hallux is small or absent as in anisodactyl and tridactyl types of foot (Table VIII). In the Ostrich (*Struthio camelus*) the conjoined tendons divide into two branches instead of the usual three, since digit II has been lost (Garrod, 1875). The arrangement in sandgrouse (Pteroclididae) appears to be uncertain. While George and Berger (1966, p. 448) list *Pterocles* as type IV, Gadow (1896) described it under type I as well as under type IV. Furthermore, although Gadow (1882) reported that the *m. flexor hallucis longus* is absent in *Pterocles,* the author observed it in a specimen of the Black-faced Sandgrouse (*Pterocles decoratus*) in which the tendon arrangement was type IV.

(e) Type V. Gadow (1896, p. 616) divided the type V arrangement into four subtypes (types Va, Vb, Vc and Vd), although not all authors agree with this analysis (George and Berger, 1966, p. 448).

In type Va the *flexor hallucis longus* and *flexor digitorum longus* tendons do not cross but fuse together completely, the combined tendon giving off four branches, one to each digit. Whilst the four branches may arise together at the distal end of the tarsometatarsus, typically the branch to the hallus splits off first. In the Magnificient Frigatebird (*Fregata manificens*) the tendons for digits I and II are fused for a short distance proximally (Hudson, 1937). In contrast to Gadow (1896), a type Va arrangement is depicted for penguins (Spheniscidae) by Grunert (1906).

Type Vb is similar to type Va, but the tendon to the hallux is unusual in arising from the *flexor digitorum longus* tendon, proximal to the point where it joins the *flexor hallucis longus*. Since the *flexor hallucis longus* is the normal muscle of the hallux, type Vb must have evolved from a type Va condition. It is associated with a syndactyl foot in four coraciiform families (Table VIII), including the todies (Todidae), motmots (Momotidae), bee-eaters (Meropidae) and some kingfishers (Alcedinidae), an assemblage that it helps to define as a clade (Maurer and Raikow, 1981).

Although a type Vc arrangement (Fig. 2.33) was described by Gadow (1896, p. 616) in hummingbirds (Trochilidae), it is now well established that the relationship of the tendons in this group is a variation of the type I arrangement (Lucas, 1895; Gadow, 1895, Zusi and Bentz, 1984). Since no other evidence for type Vc can be found in the literature, it seems likely that the arrangement does not occur in any bird.

In the type Vd arrangement of Gadow (1896, p. 618), the vinculum connects the *flexor hallucis longus* tendon to the *flexor digitorum longus* branch to digit III. According to Gadow, the arrangement is characteristic of the Hoopoe (*Upupa epops*) and wood-hoopoes (although it is uncertain to which species of wood-hoopoe he is referring). Sundevall (1835), however, reported that in *Upupa* the *flexor hallucis longus* and *flexor digitorum longus* tendons are entirely separate (see type VII, to follow), and this was confirmed by Garrod (1875). The absence of a connection between the tendons in *Upupa* is also reported by George and Berger (1966, p. 446), who found a branch of the *flexor hallucis longus* tendon inserting on the trochlea for digit II. Although Maurer (1977) agrees with the description by George and Berger for *Upupa,* he did find the type Vd condition as reported by Gadow in the White-headed Wood-Hoopoe (*Phoeniculus bollei*). In another wood-hoopoe which he examined, the Scimitar-Bill (*Rhinopomastus cyanomelas*), the vinculum ended in the branch of the *flexor digitorum longus* to digit IV as well as that to digit III. The conflict in these reports may be due in part to differences in dissection technique by various investigators, or perhaps may reflect individual variation in the specimens.

(f) Type VI. The *flexor hallucis longus* tendon bifurcates near the distal end of the tarsometatarsus, one branch going to digit IV whilst the other branch divides again to supply digits I and II. The *flexor digitorum longus* tendon is not branched and supplies only digit III. The *flexor hallucis longus* thus appears to have acquired the tendons to digits II and IV that are normally provided by the *flexor digitorum longus.* A vinculum is present. The most detailed and comprehensive account of this arrangement was given by Swierczewski (1977), who greatly extended the original description of Garrod (1875) and subsequent workers. Swierczewski (1977) confirmed in dissections of 45 species representing 31 genera that the type VI arrangement is without exception characteristic of the six families of the order Piciformes, and that it is an important derived character arguing for monophyly of the order (Swierczewski and Raikow, 1981). The only notable variation observed by Swierczewski occurred in the Three-toed Jacamar (*Jacamaralcyon tridactyla*) and the Black-backed Three-toed Woodpecker (*Picoides arcticus*), in which there is no hallux and therefore no hallucial tendon.

The functional significance of the type VI arrangement is probably associated with zygodactyly, but this is not a necessity, since the Cuculidae and Psittacidae, both of them large and diverse families of zygodactyl birds, have a type I arrangement.

(g) Type VII. The *flexor hallucis longus* tendon supplies only the hallux, whilst the *flexor digitorum longus* tendon divides into three to send branches to digits II, III and IV. There is no vinculum, and the two muscles thus operate entirely independently of each other. Type VII is found mainly in the Passeriformes, in which it occurs in association with both anisodactyl and syndactyl feet. Exceptions are the Eurylaimidae and Philepittidae, in which the relationship is type I.

(h) Type VIII. The *flexor digitorum longus* tendon bifurcates to insert on digits III and IV, whilst the *flexor hallucis longus* tendon bifurcates to supply digits I and II. Type VIII is found only in the trogons (Trogonidae) (Garrod, 1875; and

verified by George and Berger, 1966, p. 449; Maurer, 1977; Raikow, unpublished data). Functionally, it is clearly correlated with the heterodactyl foot, which is also unique to the trogons. By this arrangement the *flexor digitorum longus* flexes the two forward toes, whilst the *flexor hallucis longus* flexes the backward toes and, via the vinculum, also apparently assists the *flexor digitorum longus*. Gadow (1896, p. 618) suggested that the type VIII arrangement is derived from a type Va ancestral condition by a splitting and separation of the tendons (see also Maurer and Raikow, 1981).

3. Relationship between foot types and tendon types

The different types of feet and tendons have been used for over a century as higher-level taxonomic characters, yet even though both include suites of structural variations associated with the functional specialisations of the foot, there has been little consideration of how they inter-relate. What is the relative frequency of occurrence of the different foot and tendon types among the families of birds? What foot types occur together with what tendon types, and what is the frequency of these combinations? What combinations do not occur amongst birds, and why are they absent? Such questions will be examined briefly in this section.

Most passerines are anisodactyl, some are syndactyl, and some as already noted are difficult to classify because of their intermediate form. All but eurylaimids and philepittids have the type VII tendons. Therefore, despite the large number of passerine families, there is little variation in these characters within the order. The remainder of this discussion will focus on non-passerine birds, which are less numerous as species but much more diverse in their structural variety.

Ignoring a few foot types which are of rare occurrence (e.g. didactyl), or are merely positional variants rather than fixed postures (e.g. ectropodactyl), eight general foot types will be considered. Table VI shows that these types are not evenly distributed, but that most non-passerine families are of one type (anisodactyl) and that the remaining types are of relatively infrequent occurrence. Again, excluding a few tendon arrangements of rare occurrence or ambiguous distinctiveness, 10 basic types are shown in Table VII with the numbers of families that possess them. As before, a few variations predominate, though not as much as in the foot types.

Given these 8 foot types and 10 tendon types, Table VIII shows all the possible combinations that could theoretically occur. Only about one-fourth of the possible combinations have been reported in the literature (Table V), and it is clear that a small number of combinations predominate, with most being of infrequent occurrence. Why are so many combinations absent? Some of them are functionally incongruous; one would not expect a combination of syndactyly with type IV tendons, for example, because the elongated hallux, which opposes the forward toes in grasping, would be deprived of its principal flexor. Yet, most of the absent combinations are not functionally inappropriate in this way, at least to the extent that one can estimate from a mental picture alone. It may be that many potentially functional combinations simply never evolved in any known group. The preponderance of a few combinations suggests that certain ones may evolve more readily than others, although this speculation would be difficult to test.

Ironically, however, the second most frequent combination of foot and tendon arrangements among the families of non-passerine birds is one that appears to make no sense from a functional point of view. This is the combination of anisodactyly with type IV tendons. It is functionally incongruous, because there are four toes but only three branches of the deep plantar tendons; there is no provision for the hallux. (There are few data on the small flexor of the hallux, the *m. flexor hallucis brevis,* but this is at best a feeble muscle that could never compensate for the loss of the deep plantar connection.) Although this combination is meaningless from a design standpoint, there is a simple historical explanation, namely that this combination is an intermediate stage in the loss of the hallux.

The most common non-passerine condition is the combination of anisodactyly with type I tendons (Table VIII, Fig. 2.34A). This provides for the separate flexion of the forward three toes by the *flexor digitorum longus* and of the hallux by the *flexor hallucis longus.* The vinculum is presumed to aid the grasping movement in some way, but the nature of this functional connection is uncertain. Possibly the *flexor hallucis longus* simply assists the *flexor digitorum longus* in flexing the forward toes, the *flexor digitorum longus* having the harder task, so to speak, because it must deal with three toes instead of one. Another possibility is that the *flexor hallucis longus* alone may flex all four toes weakly, being joined by the *flexor digitorum longus* when additional strength is needed. The vinculum may also serve to co-ordinate the flexion of the two sets of toes. These possibilities are not mutually exclusive, but only electromyographic studies are likely to tell us what the muscles actually do. In any event, the vinculum is not indispensable, as seen by its loss in bitterns (Ardeidae) and passerines, leading to type VII. The combination of anisodactyly with type I tendons commonly occurs in various walking and perching birds, and is possibly the ancestral avian condition. It is widespread, for example in Ciconiiformes, Galliformes, Gruiformes and Charadriiformes (Table V).

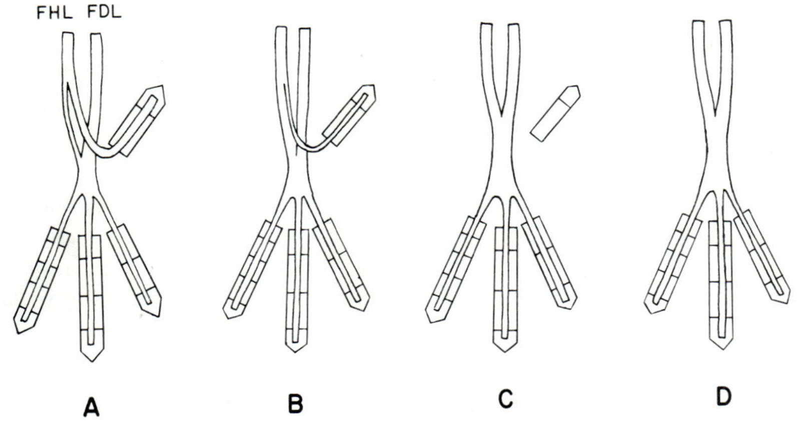

Fig. 2.34. Stages in the evolution of the tridactyl foot from an initial anisodactyl condition. (A) Anisodactyl foot with type I tendons, associated with a strong hallux. FDL, *Flexor digitorum longus*; FHL, *flexor hallucis longus*. (B) Anisodactyly with type II tendons, associated with a reduced hallux. (C) Anisodactyl foot with type IV tendons, associated with a vestigial hallux. (D) Tridactyl condition with type IV tendons; hallux lost.

From the foregoing we may postulate a series of stages leading to the loss of the hallux in groups that do not use the foot for grasping. This speculative scenario will account for most non-passerine families. The hallux becomes reduced in size, and is often elevated so that it no longer touches the ground. At some point, size reduction and elevation will combine to render the digit non-functional (vestigial). At the same time the tendons evolve from the type I to the type II arrangement (Fig. 2.34B), the contractile strength of the *flexor hallucis longus* being transferred from the hallux to the forward toes. Eventually the diminishing connection to the hallux is lost entirely (Fig. 2.34C), the *flexor hallucis longus* having shifted its action entirely to the forward toes, thus giving the type IV tendon configuration. (In a few cases the belly of the *flexor hallucis longus* also becomes reduced.) Eventually the hallux itself is lost (Fig. 2.34D), giving the common and functionally meaningful combination of tridactyly with the type IV tendons.

Thus, the kind of avian foot having an anisodactyl toe arrangement with type IV tendons (Fig. 2.34C) is explainable as a stage in the evolutionary loss of the hallux. In almost all cases the digit itself remains after its tendon has been lost; once function is abandoned, different parts of the mechanism may evolve at different rates. In an evolving structure that maintains a functional role one would expect the various components to evolve together in an integrated manner so that functional capability is retained as structure changes. Such integration would result from the co-ordination of developmental mechanisms such as tissue interactions. In the case of a structure evolving towards loss, however, such constraints might cease to operate once a stage of functionlessness has been reached. Possibly the relative rates of loss of different components are related to the relative complexity of their developmental bases. In the present case the hallux almost always remains after the tendon has been lost. The opposite arrangement, retention of a vestigial tendon after loss of the digit, has been reported only once to the knowledge of the author, in a Rhinoceros Auklet (*Cerorhinca monocerata*) (Vanden Berge, 1970).

Obligate zygodactyly occurs in association with type I tendons in the Cuculiformes and Psittaciformes, and with type VI in the Piciformes. Although the tendon patterns are different, the effect is similar, namely that one muscle provides simultaneous flexion of opposing toes, giving a co-ordinated grasping movement. In Cuculiformes and Psittaciformes the *flexor digitorum longus* flexes the backward-directed digit IV along with the forward-directed II and III, whilst in Piciformes the *flexor hallucis longus* flexes the backward-directed IV and I and the forward-directed II. In all of these cases digit IV appears to be the principal backward digit, being longer than digit I. Reduction and loss of the hallux in some Piciformes has already been noted. Berger (1960) reported that the *flexor hallucis longus* belly is reduced in *Tauraco* (Musophagidae) and in the cuculid *Centropus*. The latter also lacks a tendon to the hallux, although there are compensatory modifications of other structures. These observations support the idea of Bock and Miller (1959) that the reversed fourth toe in zygodactyl birds evolved to compensate for a primitively weak hallux.

The syndactyl configuration occurs in association with several tendon types. In each case, however, it is functionally associated with only one habit, perching, thus showing less diversity than other arrangements. The frequent occurrence of syndactyly suggests that it is advantageous among perchers whatever their plantar tendon arrangement may be.

Semizygodactyl feet may have either type I (Musophagidae, Tytonidae, Strigidae) or type Va (Pandionidae, Coliidae, Leptosomatidae) tendon arrangements. The two types are similar, sharing strong tendons to all four toes and a close interconnection between the *flexor hallicus longus* and *flexor digitorum longus* tendons. Semizygodactyl feet have the ability to shift toe positions in association with different activities, but the common factor in all cases is probably the capability of simultaneous flexion of all four toes in any arrangement. Pamprodactyl feet occur in association with the same two tendon arrangements. Although the range of movement is more restricted, the common factor of importance again is probably the ability for strong simultaneous flexion of all four toes. Heterodactyly, as already noted, occurs only in association with type VIII tendons as part of the unique perching specialisation of the trogons (Trogonidae).

D. Terrestrial locomotion

Birds adapted for walking or running on the ground show fewer specific adaptations than those of arboreal or aquatic habits. Most terrestrial birds have either anisodactyl or tridactyl feet. The anisodactyl condition is primitive for birds, but an opposing hallux is of no particular use in the absence of a grasping function, and the hallux commonly undergoes elevation, reduction and loss, with numerous species showing various intermediate stages.

Birds that walk on solid ground, like pheasants and grouse (Phasianidae), generally have toes of moderate length at best, while large forms have relatively smaller toes, minimising contact with the ground. The Ostrich (*Struthio camelus*), with only two short toes, is the closest approach among birds to the extreme adaptations of hoofed mammals, although Storer (1971) considered that the problems of balance related to their bipedal stance prevented birds from developing the most extreme cursorial adaptations found in quadrupedal mammals.

Birds that walk on soft substrates have adaptations for spreading their weight over a larger surface area to prevent them from sinking. Those that wade on muddy or sandy bottoms may have elongated toes as in herons (Ardeidae), or webbing as in avocets (Recurvirostridae). Rails (Rallidae) have long toes and claws. The ultimate development of this kind is seen in the Jacanidae, which walk on floating lily pads with the aid of bizarrely-elongated toes and claws. Ptarmigan (*Lagopus*) and sandgrouse (*Syrrhaptes*) have densely-feathered feet for walking on snow and sand, respectively. Storer (1971) discusses these and other adaptations for terrestrial locomotion.

E. Arboreal locomotion

1. Perching

The basic adaptation for perching is a strong grip, which requires the provision of opposing sets of toes set at the same level (hallux incumbent) so as to encircle the perch. Bock and Miller (1959) hypothesised that the ancestral avian foot was anisodactyl with a weak hallux, as seen in *Archaeopteryx,* and was not suitable for powerful grasping. From this condition several types of perching feet developed in different groups, including heterodactyl, syndactyl and zygodactyl types, as well as improved anisodactyl arrangements. These foot types, and their functional relationships to the plantar tendons, were discussed earlier. In this section two very different solutions to the problems of adapting to an arboreal habitat are considered.

(a) The passerine birds. The Passeriformes are commonly known as "the perching birds," a misnomer because they are not all arboreal, and because many non-passerines also perch. Nevertheless, the perching habit is sufficiently common within this largest order of birds that one might expect to find specialisations in the hindlimb underlying the function. Although such adaptations are indeed present, they are not of a profoundly complex nature. Passerines seem to have evolved a structurally simple and straightforward perching foot, much in contrast to the almost baroque complexity of the mousebirds discussed later.

In passerines the hallux is large and incumbent, and the deep plantar tendons are separate, except in eurylaimids and Philepitta, in which the vinculum is weak and probably of little functional significance. The *flexor hallucis longus* is large, and the lack of a vinculum means that no part of its contractile force is diverted to the forward toes. Thus, the hallux has the size, strength and independence of action to oppose the forward three toes effectively.

One of the most distinctive characteristics of the passerines is the loss of most of the intrinsic foot muscles of the forward toes (Raikow, 1982). Except for the occasional retention of a vestigial *abductor digiti IV,* passerines have lost the set of small muscles that provide for independent action of the forward toes in releasing the grip, retaining only the ability to extend these toes simultaneously through the action of the *m. extensor digitorum longus.* Apparently the subtlety of individual extension, adduction and abduction, which might be useful to some birds in climbing over irregularly-shaped surfaces or in manipulating objects, is not needed by passerines. Although the toes are deeply cleft in most passerines, there are trends towards the development of syndactyly in some of the more arboreal groups as noted before.

A mechanism for locking the toes in the flexed position during perching has been described by Schaffer (1903). Rows of large cells on the plantar surfaces of the flexor tendons lock against ridges on the inner sides of the toe pads. These structures are readily seen in dissection and have often been noted by later writers, but do not seem to have been investigated since Schaffer's work. This would seem to be an excellent subject for new research using modern techniques such as scanning electron microscopy.

(b) The mousebirds. The colies or mousebirds of Africa are a small group of six species in the genus *Colius* and are sufficiently unusual that their closest relatives have not been determined, for which reason they are placed in their own order and family (Coliiformes, Coliidae). They have the remarkable ability to reverse both the first and fourth digits, so that their feet can be shifted at will between anisodactyl, pamprodactyl and zygodactyl configurations, and the right and left feet of an individual bird may be held in different arrangements at the same time. Mousebirds creep about in the bushes with their bodies parallel to the branch.

They seldom perch in a conventional manner, but suspend their body between widespread legs or hang by one or both feet like swifts (Apodidae). They also use their feet to hold food and for fighting. Along with their description of the hindlimb muscles in captive birds, Berman and Raikow (1982) included detailed observations on the movements of these birds. They point out that the hanging postures so common to mousebirds require returning the body to an upright posture. Modifications of the thigh musculature for this include enlargement of the *iliofemoralis externus,* and an expansion of the insertion of the *flexor cruris lateralis,* both of these muscles being involved in femoral retraction. In addition the *pubo-ischio-femoralis* has an accessory belly not reported in other birds.

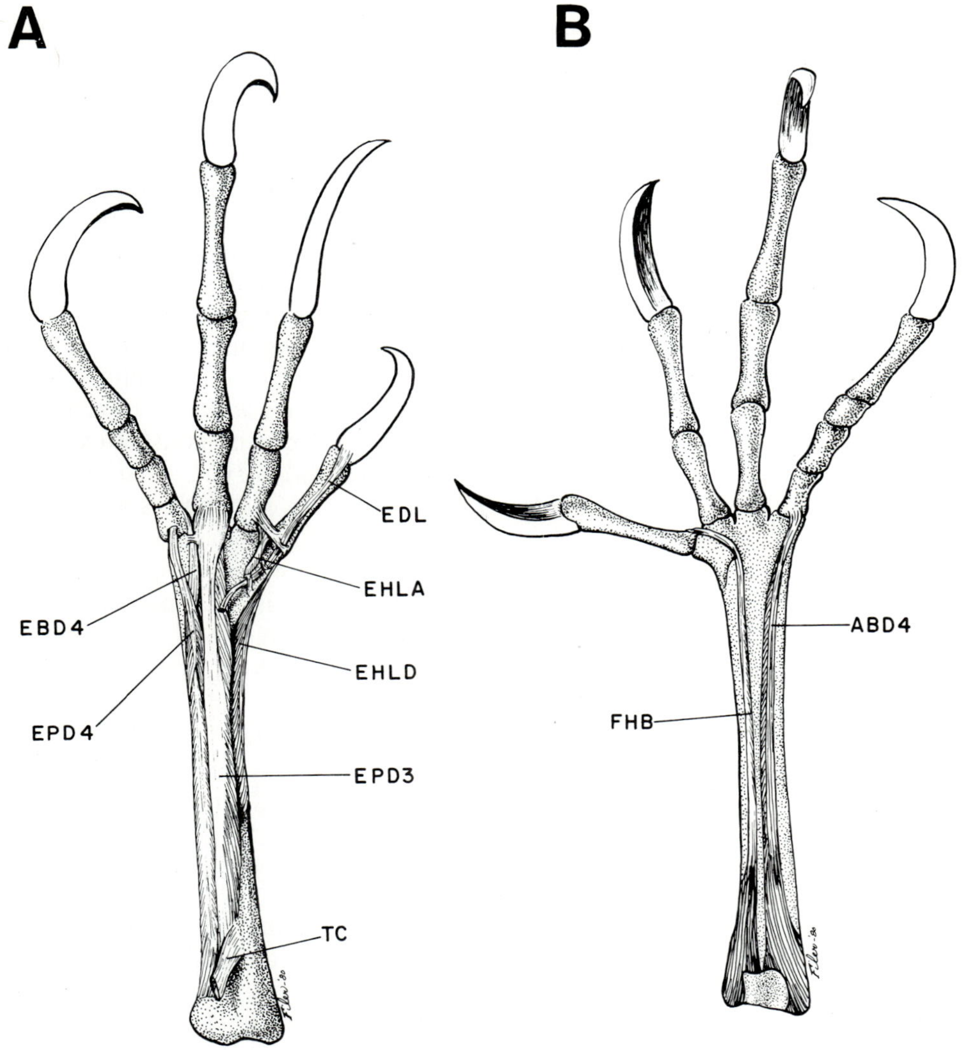

Fig. 2.35. Intrinsic muscles of the foot of the Speckled Mousebird (*Colius striatus*). (A) Dorsal view, (B) plantar view. ABD4, *m. abductor digiti IV*; EBD4, *m. extensor brevis digiti IV*; EDL, branch tendon of insertion of *m. extensor digitorum longus*; EHLA, *m. extensor hallucis longus accessorius*; EHLD, *m. extensor hallucis longus pars distalis*; EPD3, *m. extensor proprius digiti III*; EPD4, *m. extensor proprius digiti IV*; FHB, *m. flexor hallucis brevis*; TC, tendon of insertion of *m. tibialis cranialis*. From Berman and Raikow (1982).

An especially curious modification in mousebirds is that the *m. extensor digitorum longus* sends a tendon to the hallux as well as to the second, third and fourth digits, a condition otherwise known only in parrots (Psittacidae). This permits co-ordinated, simultaneous extension of all four toes. Extension of the hallux in most birds is provided by the *m. extensor hallucis longus pars proximalis,* but this muscle is absent in mousebirds. The probable explanation for this peculiarity is that the belly of the *pars proximalis* was lost, but its tendon was transferred to that of the *m. extensor digitorum longus* in a manner comparable to similar changes noted earlier in the deep flexor tendons.

In marked contrast to the condition in passerines, the foot of mousebirds has a set of well-developed intrinsic muscles, including two (the *extensor proprius digiti IV* and *extensor hallucis longus accessorius*) that are not known in other birds (Fig. 2.35). The ability to shift the first and fourth digits forward or back in changing toe configurations is provided by these intrinsic muscles. The *extensor hallucis longus* in some birds includes two bellies. The *pars proximalis* extends the digit as described earlier. The *pars distalis,* when present, is usually a very small muscle that inserts on the dorsal surface of the base of the proximal phalanx, generally just becoming incorporated into the joint capsule. In *Colius,* however, the *pars proximalis* (as noted earlier) is absent, but the *pars distalis* is unusually large and its insertion has shifted to the medial surface, where its contraction brings about caudal rotation of the hallux. Cranial rotation of the hallux is performed by the large *flexor hallucis brevis,* the insertion of which has shifted from the usual plantar position to the lateral surface of the digit. Cranial rotation of digit IV is performed by the *m. extensor brevis digiti IV,* the insertion of which has shifted medially. It may have been replaced in its usual extension function by the new muscle, the *m. extensor proprius digiti IV*. Caudal rotation is the function of the *m. abductor digiti IV,* the insertion of which is also shifted slightly from its more typical position.

Despite the complexity of the locomotor behavior of mousebirds and its underlying morphological basis, this group of birds, which is known at least from Miocene times (Ballman, 1969), can hardly be considered to be evolutionarily successful compared to the Passeriformes. Perhaps the very complexity of its specialised hindlimb mechanism has precluded the diversification made possible by the relatively simple passerine limb, but many other factors are probably also involved.

2. *Climbing*

Several groups of birds have independently evolved the habit of climbing on vertical tree trunks, generally in order to feed on the arthropods that live there. Mechanisms are required that will hold the bird close to the trunk against the effect of gravity. A brief review was given by Storer (1971). Climbing birds include forms which do not use the tail as a support, such as the nuthatches (Sittidae), the Black-and-White Warbler (*Mniotilta varia*), and Hawaiian Creepers (Paroreomyza). These birds often climb both upward and downward, and the claw of the hallux is long and curved to support the bird during head-first descent. In other groups the tail is pressed against the tree as a brace, and such birds generally climb upward only. They include the treecreepers (Certhiidae), woodcreepers (Dendrocolaptidae), some ovenbirds (Furnariidae) and the woodpeckers (Picini), but not their sister group the piculets (Picumnini). In tail-supported birds the pygostyle and free caudal vertebrae are enlarged to support the rectrices, which are stiffened by one means or another (Richardson, 1942). Attempts to analyse the functional anatomy of climbing birds have resulted in suggestions for certain common features, such as sharply-curved claws and shortened legs, but a general structural plan for climbers has not emerged (Burt, 1930; Stolpe, 1932; Richardson, 1942; Feduccia, 1972; Orenstein, 1977).

The woodpeckers (Picidae) are probably the most highly-specialised climbers and have been the most thoroughly studied (see Spring, 1965, for an analysis of film of their climbing and pecking behaviour). The evolutionary history of the foot in this group was traced by Bock and Miller (1959), who showed (as discussed earlier) that zygodactyly is not an adaptation for climbing, but in the woodpeckers developed into an ectropodactyl condition. They hypothesised a morphological sequence beginning with the primitive condition in the wryneck (*Jynx* spp.) and continuing through the piculet *Picumnus*, to *Colaptes*. A "short-hallux line" is traced from *Colaptes* to *Picoides*, in which some species lose the hallux altogether. A "long-hallux line" is traced from *Colaptes* through *Dryocopus* to *Campephilus*, in which the lengthened hallux remains functional in an extreme ectropodactyl foot. Bock and Miller (1959) did not consider this pattern to form an evolutionary sequence. However, Swierczewski and Raikow (1981) reconstructed the phylogeny of the woodpeckers via a cladogram based mainly on limb muscle anatomy, and found a similar pattern.

The oxpeckers (*Buphagus* spp.) are African starlings which climb, not trees, but over the hides of large mammals in search of parasites. Borecky (1977) dissected the Red-billed Oxpecker (*Buphagus erythrorhynchus*), along with members of 19 other genera of Sturnidae, and found that the hindlimb muscles of *Buphagus* are generally similar to those of other starlings, except that in *Buphagus* the *m. flexor digitorum longus* has a femoral head (not present in other species) and the *m. flexor hallucis brevis* is extremely enlarged. Both these modifications of the musculature of oxpeckers must increase the grasping strength of the feet and are adaptations to their unusual foraging substrate.

F. Aquatic locomotion

Many groups of birds have evolved aquatic locomotion independently, probably passing to various degrees through progressive stages of wading, swimming and diving. Common specialisations include (a) a widening of the body, which promotes stability on the water's surface, (b) dense plumage, providing insulation and buoyancy, and (c) a large uropygial gland for waterproofing the plumage. The general specialisation of the locomotor system is the development of either webbing or lobation to convert the foot into a paddle. Diving birds may be divided into two groups according to whether they are wing-propelled (discussed earlier) or foot-propelled.

Foot-propelled divers include the following groups arranged in their sequence corresponding roughly to increasing specialisation; some ducks (Anatidae), anhingas (Anhingidae) and cormorants (Phalacrocoracidae), grebes (Podicipedidae), loons (Gaviidae) and the toothed divers of Cretaceous times (Hesperornithes). Several studies have discussed the structural and functional mechanisms of these birds, and a comparison between them shows considerable convergence toward a common pattern that evidently represents the most effective design for a foot-propelled diver (Dabelow, 1925; Miller, 1937; Owre, 1967; Raikow, 1970; Stolpe, 1932, 1935; Storer, 1960; Wilcox, 1952).

In diving birds the pelvic girdle, especially the postacetabular portion, becomes narrowed. This is accomplished in part by a narrowing of the synsacrum and in part by lateral compression of the pelvic bones. Storer (1956) traced this process in fossil and extant loons (Gaviidae), and Raikow (1970) recorded it in ducks (Anatidae). Since specialised diving birds do not walk on land and therefore do not require a wide separation of the legs, this narrowing contributes to streamlining by merging the mass of the legs with the trunk, giving a flattened, teardrop shape to the body. In addition, the postacetabular part of the pelvis becomes elongated relative to the preacetabular part. This appears to correlate with the large size and increased mechanical advantage of *mm. iliofibularis, flexor cruris lateralis* and *flexor cruris medialis,* which hold the tibiotarsus parallel to the body axis against the extending effect of the propulsive stroke of the foot.

In diving birds the femur is short and stout. It lies permanently in a highly-abducted position compared to terrestrial birds, extending laterally perpendicular to the body. The head of the femur articulates with the acetabulum, and the trochanter with the vertically-oriented antitrochanter dorsal to the acetabulum. Thus, the femur is limited to protraction and retraction in a plane parallel to the surface of the water.

The elongated tibiotarsus lies parallel to the vertebral column. It is held close to the body and is bound in by the skin so that it cannot be extended significantly, but is limited largely to fore-and-aft movements with some rotation about the knee. The tibial axis is extended cranial to the knee as a device for increasing the area of origin of the crural muscles, although this has been accomplished differently in the various groups. In loons (Gaviidae) the tibiotarsus is elongated, the patella being reduced to a vestige. In grebes (Podicipedidae) both the cnemial crest of the tibiotarsus and the patella are enlarged. The patella alone provides the additional area of attachment in anhingas (Anhingidae), more so in cormorants (Phalacrocoracidae), and also in *Hesperornis.* A similar enlargement of the patella was traced by the author in a series of stifftail ducks (Oxyurini) showing progressively increased diving specialisations (Raikow, 1970, pp. 20–21).

The end of the tibiotarsus extends from the body mass at the level of the tail, so that the tarsometatarsus projects from the body and can move freely at, above or below the level of the body when the bird is underwater. This allows the propulsive stroke to be used also for steering. The tarsometatarsus in grebes (Podicipedidae), loons (Gaviidae), and *Hesperornis* is compressed so as to minimise water resistance during the recovery stroke. The phalanges are also narrow in these birds.

The functions of the hindlimb muscles in diving birds have been analysed in detail, on the basis of inferences from anatomical structure, by Wilcox (1952) in the Common Loon (*Gavia immer*), Owre (1967, pp. 99–104) in the Anhinga (*Anhinga anhinga*) and the Double-crested Cormorant (*Phalacrocorax auritus*), and Raikow (1970, pp. 39–48) in the stifftail ducks (Oxyurini). There are many differences, but in general it appears that the thigh muscles act primarily to hold the thigh and crus relatively immobile as a base for the principal swimming movements provided by the foot. The power stroke is extension of the foot, and in these divers the main extensor, the *m. gastrocnemius,* is greatly enlarged. It is aided by the peroneal muscles and the digital flexors. In the recovery stroke the foot is brought forward with the toes flexed and abducted, closing the paddle so as to minimise water resistance. The main muscle active here is the *tibialis cranialis*. Increased size of the crural muscles relative to the thigh muscles involves the extension of their area of origin cranial to the knee as noted earlier, and an increase in the length and girth of the muscle bellies. Progressive stages in the evolution of these and other changes are documented in the stifftail ducks (Raikow, 1970).

A few other factors associated with foot-propelled diving may be noted briefly. The evolution of diving habits in ducks (Anatidae) is accompanied by an increased area of webbing relative to body weight. This ratio (the paddle index) has been measured in various North American ducks by Raikow (1973). The smallest paddle (with a paddle index value of 0.48) was found in the American Wigeon (*Anas americana*), which is relatively terrestrial in habit. Four other species in the same genus, but with more aquatic habits, had a value of 0.77. Diving ducks in general had larger paddles, ranging from 0.91 in the Lesser Scaup (*Aythya affinis*) to a high of 1.01 in the Ruddy Duck (*Oxyura jamaicensis*). The eiders (*Somateria* spp.) were exceptions in that three species had values ranging from 0.73 to 0.79, and this is associated with their ground-nesting habits. In general, diving ducks have more caudally-placed feet, larger paddles, and reduced walking ability, but the pattern varies with their overall habits.

Grebes (Podicipedidae) have lobed rather than webbed feet, and use a different method of swimming. During the propulsion stroke the foot is extended with the surface area of the paddle perpendicular to the plane of movement so as to maximise thrust. In the recovery stroke, however, the toes cannot be folded together as in webbed forms because of the relative rigidity of the lobes. Instead the grebe rotates the toes as it flexes them, bringing them forward lateral side first, with the medial lobes trailing and the lateral lobes folded under the toe (Stolpe, 1935; Storer, 1960).

IV. Tail

A. Structure and functions

The avian tail is a cluster of feathers, the rectrices, that fans out from the caudal end of the body, its base being covered by shorter feathers, coverts, dorsally and ventrally. The feathered appendage has no counterpart in other vertebrates. Movements of the feathered tail are brought about by a series of mobile caudal vertebrae and their muscles. The tail is enormously varied in size and shape. It is considered to be short when it is shorter than the trunk, and long when it exceeds the length of the trunk, but these terms are inadequate because the tail varies from extremes of great length to virtual non-existence. This rectrices are paired, most birds having 12 (six pairs), but the number varies from at least 4 to 32 (Van Tyne and Berger, 1976, p. 137). In a few species the most conspicuous tail feathers are not the rectrices but the exceptionally well-developed upper tail coverts. Thus, in the male peafowl (*Pavo cristatus*) the rectrices, which are of normal size, support the enlarged coverts and are hidden by them when viewed from in front. Likewise the gorgeous train of the male Resplendent Quetzal (*Pharomachrus mocinno*) is composed of upper tail coverts. The ornamental tail plumes of the male Ostrich (*Struthio camelus*) are also thought to be upper coverts because of their position, although it is difficult to be certain in the absence of distinguishable rectrices.

The extent of variation in the rectrices defies description, ranging from complete absence of a discernible tail (e.g. most ratites) to the bizarre extremes of pheasants (Phasianidae), lyrebirds (Menuridae), birds of paradise (Paradisaeidae) and many others, in which the evolution of the tail as a display mechanism has apparently been driven by selection for both conspicuousness and uniqueness. Even within the range of "normal" tails, however, the variation is considerable. The shape of the tail has often been used as an aid in taxonomy and identification (Coues, 1890, pp. 173–175; Pettingill, 1970, pp. 155–156).

Birds use the tail in several ways. It may serve in flight as a lift-generating surface, as a rudder for steering, and as an air brake, the latter being most important in landing. The tail is used as a rudder underwater by some diving birds, and as a balancing organ on the ground or in trees. Some birds use it as a support in climbing trees, and its specialisation for social displays has already been noted. There appears to be no general treatise on the functional anatomy of the tail in relation to its diverse roles in the lives of birds, but some studies of its function and role in different groups are outlined next. A brief review of the structure and functions of the tail in birds is provided by Baumel (1983).

B. The caudal skeleton

The most cranial of the caudal vertebrae (*vertebrae caudales*) are incorporated into the synsacrum (see earlier) and are not functionally part of the support of the tail. The remaining free caudal vertebrae extend caudally from the synsacrum, terminating in the pygostyle (*pygostylus*), to which the innermost rectrices are attached. The free caudal vertebrae have a complete vertebral arch (*arcus vertebrae*) and, in most birds, a well-developed transverse process (*proc. transversus*), from the dorsal and ventral surfaces of which important caudal muscles arise (see later). Since the *zygopophyses* tend towards reduction, the articulation between the caudal vertebrae is principally via a fibrocartilaginous intervertebral disk (*discus intervertebralis*) connecting the vertebral bodies. For a comparative account of the vertebrae see van Oort (1905).

The pygostyle forms the terminal part of the vertebral column. It is a compound bone formed by the fusion of several embyronically separate vertebral anlagen, the number involved varying from 4 to 10 in different species. Coues (1890, p. 210) offers the intriguing speculation that the number of vertebrae combined in the pygostyle should equal the number of pairs of rectrices, since primitively (presumably as suggested by *Archaeopteryx*) each caudal vertebra supported one pair. The degree of development of the pygostyle is closely correlated with the size of the rectrices. It is quite large in most flying birds, but is reduced or indistinguishable in many flightless or weakly-flying forms, for example most ratites, *Hesperornis* and tinamous (Tinamidae). The Ostrich (*Struthio camelus*) has a small pygostyle, which de Beer (1956) considered to be evidence that the Ostrich has evolved from flying ancestors. In its well-developed form the pygostyle has a cranial articular surface (*facies articularis cranialis*) for attachment to the last free caudal vertebra. Above and below this are processes for the insertions of tendons. The body of the pygostyle may have an expanded ventral base (*basis pygostyli*) and a dorsally-directed vertical crest (*lamina pygostyli*).

Erroneously the pygostyle is sometimes referred to as the coccyx or urostyle. These structures, found in humans and anuran amphibians, respectively, are also formed by the fusion of caudal vertebrae, but are not homologous, as products of vertebral fusion, to the condition in birds.

C. Evolution of the tail

The shortened tail seen in living birds does not characterise the class Aves as a whole but is a derived condition that arose after the origin of birds. In the earliest birds, as exemplified by *Archaeopteryx,* there was no pygostyle, and the caudal vertebral column consisted of a long series of 20 to 23 separate vertebrae (Ostrom, 1976a). The pairs of rectrices projected laterally from all but the first five vertebrae and were attached by ligaments (de Beer, 1954, p. 37). The tail was therefore an elongated, flattened structure that no doubt could be raised, lowered and tilted, but there was no rectricial fan as in modern birds. Maynard Smith (1952) argued that, in evolving flight, animals pass from an early condition of aerodynamic stability to a more advanced condition of instability. The latter provides greater manoeuvrability but depends on the perfection of the neurosensory system. He suggested that the elongated, flattened tail of *Archaeopteryx* provided an effective stabilising surface. The subsequent evolution of the tail in birds therefore must have followed on the improvement of the neurosensory system. The most important changes were the truncation of the caudal vertebral column, the formation of the pygostyle, the separation of most rectrices from direct attachment to vertebrae, and the evolution of the rectricial bulbs (Baumel, 1983). This process has been analysed from an embryological point of view by Steiner (1938).

D. Muscles and movements

The movements of the tail are accomplished in part by movements of the caudal vertebral column, which moves the tail as a whole, and in part by movements of the feathers themselves, which bring about the spreading and folding of the tail. The caudal muscles (*musculi caudae*) are summarised in Table IX and illustrated in Figs. 2.36 and 37.

TABLE IX. **The caudal musculature of birds.**

Name	Origin	Insertion	Action
M. levator caudae (*levator coccygis* + *levator caudae*)	Dorsal surface of synsacrum, adjacent ilium and free caudal vertebrae	Spinous processes of free caudal vertebrae, pygostyle and bases of rectrices	Elevates tail, also contributes to rotation
Mm. interspinales	Spinous processes of caudal vertebrae	Spinous processes of adjacent caudal vertebrae (including pygostyle)	Elevate tail
M. depressor caudae (*lateralis coccygis*)	Caudoventral surface of postacetabular wing of ilium, ventral surfaces of transverse processes of free caudal and fused caudal vertebrae	Ventral surfaces of free caudal vertebrae and pygostyle, bases of some rectrices	Depresses tail, possibly aids in tail rotation
M. lateralis caudae	Transverse processes of cranial free caudal vertebrae, sometimes adjacent ilium	Dorsal surfaces of rectrices, especially outer ones	Abducts rectrices (spreads tail)
M. pubocaudalis externus (*depressor caudae*)	Caudal margin of pelvis, mainly pubis	Ventral surface of outer rectrices	Depresses tail
M. pubocaudalis internus (*depressor coccygis*)	Caudal end of shaft of pubis, sometimes also ischium and ilium	Ventral margin of pygostyle and bases of inner rectrices	Depresses tail
M. caudofemoralis (*piriformis pars caudofemoralis*)	Base of pygostyle	Caudal surface of shaft of femur	Depresses tail
M. bulbi rectricium	Pygostyle	Bases of rectrices and fibrous partitions of rectricial bulb	Abducts rectrices (spreads tail) and stabilises rectrices
M. adductor rectricium	Bases of rectrices	Bases of adjacent rectrices	Adducts rectrices (folds spread tail)

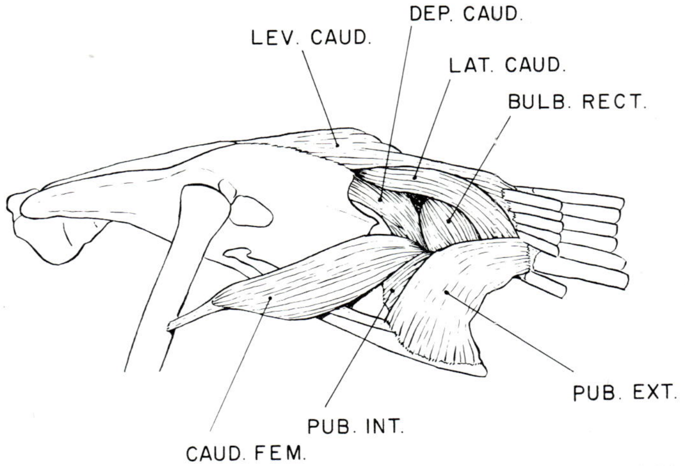

Fig. 2.36. Lateral view of the caudal musculature of *Columba*. Abbreviations for Figs. 2.36 and 37: BULB. RECT., *m. bulbi rectricium*; CAUD. FEM., *m. caudofemoralis*; DEP. CAUD., *m. depressor caudae*; INT., *mm. interspinales*; LAT. CAUD., *m. lateralis caudae*; LEV. CAUD., *m. levator caudae*; LIG. ELAST., *ligamentum elasticum interrectriciale*; PUB. EXT., *m. pubocaudalis externus*; PUB. INT., *m. pubocaudalis internus*.

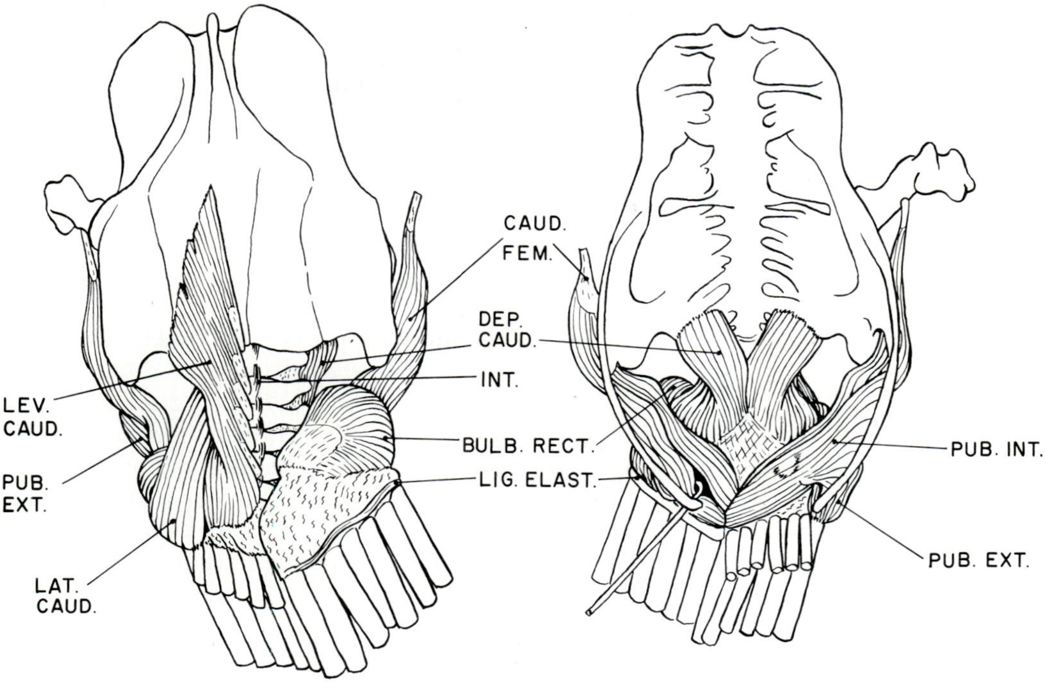

Fig. 2.37. Dorsal (left) and ventral (right) views of the caudal musculature of *Columba*. For abbreviations see Fig. 2.36.

1. Movements of the tail as a whole

In the sagittal plane the tail may be raised (elevation) or lowered (depression). Elevation of the tail is accomplished by a large dorsal muscle, *m. levator caudae,* which arises from the dorsal surfaces of the synsacrum, ilium and free caudal vertebrae. It is a complex muscle giving rise to numerous tendons that insert on the spinous processes of the free caudals, on the pygostyle and over the dorsal bases of the rectrices. A minor contribution to elevation is also provided by a series of small muscles, the *mm. interspinales,* which pass between the spinous processes of the free caudal vertebrae as well as extending to the pygostyle.

Depression of the tail involves a greater number and mass of muscles, perhaps because of the importance of depression (along with spreading of the tail) as an aerial brake. The *m. depressor caudae* is the ventral counterpart of the *m. levator caudae,* both being modified axial muscles. The *m. depressor caudae* arises from the ventral surfaces of the pelvic girdle and the free caudal vertebrae, and inserts more distally on the free caudals, the pygostyle and the ventral bases of the rectrices. The tail is also depressed by two muscles that arise on the caudal margin of the pubis and sometimes the adjacent pelvic surfaces. The *m. pubocaudalis externus* passes dorsally to insert on the ventral surfaces of the outer rectrices, thus drawing the outer base of the tail ventrally. In contrast, the *m. pubocaudalis internus,* which originates just deep to the former muscle, passes more medially to insert on the pygostyle and the inner rectrices. Another muscle that apparently depresses the tail is the *m. caudofemoralis,* which passes cranially from the pygostyle to the femur. According to convention the pygostyle would be considered its origin and the femur its insertion, but Fisher (1946) suggested that the relative massiveness of the hindlimb compared to the tail indicates that the tail is the more likely functional insertion. To test this hypothesis Fisher (1957) incapacitated the *caudofemoralis* in pigeons (Columbidae) by removing pieces from each end of the muscle, but found no increase in landing forces, although there was a slight decrease in tail depression. He concluded that any function the muscle has in moving the tail is so slight that other muscles can easily compensate for it.

In the horizontal plane the tail as a whole may be deflected to one side by the unilateral contraction of the above-mentioned muscles, perhaps assisted by the *m. lateralis caudae* (see later). In the transverse plane the tail may be rotated or tilted, with the rectrices of one side raised and those of the other side lowered. This would be accomplished by the simultaneous contraction of the *lateralis caudae* (and perhaps the *levator caudae*) on one side, and some combination of the depressor muscles, including the *depressor caudae,* of the other side. Complex motions of the whole tail in three dimensions may thus be visualised as resulting from the selective actions of the caudal muscles in various unilateral and bilateral combinations.

2. *Spreading and folding*

Spreading and folding of the tail occur independently of the movements of the vertebral column. The tail consists of two sets of rectrices, one on each side, with the medial rectrix of each side bound to the pygostyle by connective tissue fibers. The remainder of the rectrices are not attached directly to the skeleton, but their follicles are interconnected by a complex connective tissue matrix, the rectricial bulb, which has been analysed by Baumel (1971). Within the bulb the base of each rectrix is suspended in a connective tissue septum, and the spaces between adjacent septa are filled with fat. This arrangement allows considerable mobility to the individual rectrices during tail spreading and folding. The bulb is enveloped by a thin sheet of muscle, the *m. bulbi rectricium,* which helps anchor it to the pygostyle. The bases of the rectrices are, additionally, interconnected by an elastic interrectricial ligament (*lig. elasticum interrectriciale*), so that when the outer rectrix is drawn laterally the others are pulled out also, maintaining their spacing. This tail spreading is accomplished by the *m. lateralis caudae,* aided by the *m. bulbi rectricium.* The tail is folded by the elastic shortening of the ligament, which is stretched when the tail is spread. This action is assisted by the *m. adductor rectricium,* which is formed by a series of small muscles interconnecting the bases of rectrices. Folding of the tail may be demonstrated in a fresh specimen, the spread tail returning to the closed position when released by hand. Coues (1890, p. 170) pointed out that the vanes of the members of each pair of rectrices are not quite at the same level, so that when the tail is folded they interdigitate rather than colliding. A detailed account of the structure of the rectrices in several species is given by Lucas and Stettenheim (1972, pp. 255–262).

E. Adaptive variations

In flight, the tail is primarily functional in controlling pitching movements during forward motion, but also acts as a rudder when the bird is turning (Brown, 1961). A bird with a small tail cannot turn quickly in fast flight, and manoeuvrable flight is often associated with a broad tail having stiff feathers, as in hawks (Accipitridae) and flycatchers (Tyrannidae). Fisher (1946) made a detailed study of the structure and function of the tail in New World vultures (Cathartidae). In such soarers the tail is put to more delicate use than in most birds, undergoing continual fine adjustments in area through spreading and folding, as well as delicate movements of elevation, depression and tilting. Brown (1961) noted comparable movements in soaring gulls (Laridae), and Owre (1967, p. 48) described the use of the large tail in the Anhinga (*Anhinga anhinga*) during soaring and in its "accipitrine-like" manoeuvring among trees. He noted that tail movements are less evident in the Double-crested Cormorant (*Phalacrocorax auritus*), in which the tail is relatively small. Most birds use the tail as an air brake in landing, the cormorant also using it as a water brake when alighting on the surface. The arboreal *Anhinga* uses its tail as a balancing organ and as a prop when moving or resting on its perch. Fisher (1946) noted that the most extreme tail movements are seen in preening, when the tail may be rotated at 90° from its horizontal position, and this is a common sight in other birds as well. Fisher observed a number of differences in tail use between species. For instance, the California Condor (*Gymnogyps californianus*) apparently does not use the tail as an air brake in the typical manner; in flight, however, its tail movements are more subtle and precise than those of the Turkey Vulture (*Cathartes aura*) or the Black Vulture (*Coragyps atratus*). The tail movements of *Coragyps* are more extensive and less precisely controlled, perhaps because its rectrices are relatively shorter and the tail thereby less effective as a control surface. Possibly in correlation with this, the caudal muscles are also relatively smaller. *Cathartes* appears to be intermediate in this respect. Fisher also noted that Turkey Vultures lacking most of their rectrices due to moulting or injury soared effectively in moderate air currents but had trouble maintaining their balance in stronger winds. The muscles of the larger Cathartidae (*Gymnogyps, Vultur, Sarcoramphus*) insert more distally than in other species, which he interpreted as permitting more precise and delicate movements of the rectrices and of the tail as a whole. Likewise, Owre (1967) reported differences in the caudal muscles of *Anhinga anhinga* and *Phalacrocorax auritus,* which he interpreted as providing more intricate movements in the anhinga, and stronger but more restricted motion in the cormorant. This in turn he correlated with behavioural differences. Thus, the tail muscles of *Anhinga* facilitate soaring and manoeuvrable flight, while those of *Phalacrocorax* are more suited to use in heavy flapping flight, braking, and possibly in underwater manoeuvring.

The stifftail ducks of the tribe Oxyurini show progressive specialisation for underwater swimming, ranging from the Black-headed Duck (*Heteronetta atricapilla*) of South America, through the more advanced species of the widespread genus *Oxyura,* to the Musk Duck (*Biziura lobata*) of Australia, which is the most highly-specialised diver among the Anatidae. Raikow (1970) compared these forms with the Mallard (*Anas platyrhynchos*), a dabbling duck thought to be similar to the non-diving ancestors of the Oxyurini. The stifftails owe their name to their stiffened rectrices, and use their tail both as an underwater rudder and in sexual displays. The number of freely movable caudal vertebrae is six in *Anas* and *Heteronetta,* and five or six in *Oxyura,* but is increased to eight in *Biziura.* Including also the pygostyle, the total length of the caudal skeleton (as a percentage of trunk length) is lowest in *Anas* and *Heteronetta,* significantly greater in *Oxyura,* and again much increased in *Biziura.* This progressive increase in length and vertebral number in the Oxyurini is presumably correlated with the use of the tail both as a rudder and in such displays as the ''whistle-kick'' (Johnsgard, 1966), in which the tail is raised so far over the back that the rectrices are drawn down horizontally and point directly forward. The major change found in the caudal muscles occurs in the *m. levator caudae,* the mass of which is slightly increased in *Heteronetta* and *Oxyura,* and greatly so in *Biziura.* Along with this its area of origin is expanded cranially. This is interpreted as providing greater strength and excursion for tail elevation, especially in the Musk Duck.

Acknowledgements

I would like to thank Julian J. Baumel and Philip S. Humphrey for reading and criticising the manuscript. Julian J. Baumel and Charles T. Collins very kindly allowed me to make use of their pre-publication manuscripts. The artists for the various figures were John Anderton (Figs. 2.1, 2, 17), Lee Ambrose (Figs. 2.4–8), Gene M. Christman (Fig. 2.16), William R. Filer(Fig. 2.35), and Robert J. Raikow (remaining drawings). Charles T. Collins provided Fig. 2.32. My research is supported by N.S.F. grant DEB-8010898.

References

Ames, P. L., Heimerdinger, M. A. and Warter, S. L. (1968). The anatomy and systematic position of the antpipits *Conopophaga* and *Corythopis. Yale Univ., Peabody Mus., Postilla* **114,** 1–32.

Andrews, C. W. (1897). Note on a nearly complete skeleton of *Aepyornis* from Madagascar. *Geol. Mag.* **4,** 241–250.

Archey, G. (1941). The Moa: a Study of the Dinornithiformes. *Bull. Aukland Inst. Mus.* No. 1.

Arredondo, O. (1976). The great predatory birds of the Pleistocene of Cuba. *Smithson. Contr. Paleobiol.* No. 27, 169–187.

Ballman, P. (1969). Les oiseaux Miocenes de la Grive-Saint-Alban (Isere). *Geobios* **2,** 157–204.

Baumel, J. J. (1971). Morphology of the tail apparatus in the pigeon *(Columba livia). Proc. 19th World Vet. Congr.,* Mexico City, Vol. 3, p. 849.

Baumel, J. J. (1979a). Osteologia. *In* ''Nomina Anatomica Avium'' (J. J. Baumel, A. S, King, A. M. Lucas, J. E. Breazile and H. E. Evans, Eds.). Academic Press, London and New York.

Baumel J. J. (1979b). Arthrologia. *In* ''Nomina Anatomica Avium'' (J. J. Baumel, A. S. King, A. M. Lucas, J. E. Breazile and H. E. Evans, Eds.). Academic Press, London and New York.

Baumel, J. J. (1985). Tail apparatus. *In* ''A Dictionary of Birds'' (B. Campbell and E. Lack, Eds.). T. & A. D. Poyser, Ltd. Calton, U. K.

Baumel, J. J., King, A. S., Lucas, A. M., Breazile, J. E. and Evans, H. E.. Eds. (1979). ''Nomina Anatomica Avium.'' Academic Press, London and New York.

Beddard, F. E. (1898). ''The Structure and Classification of Birds.'' Longmans, Green, London.

Bellairs, A. d'A. and Jenkin, C. R. (1960). The skeleton of birds. *In* "Biology and Comparative Physiology of Birds" (A. J. Marshall, Ed.), Vol. 1. Academic Press, New York and London.

Bentz, G. D. and Zusi, R. L. (1982). The humeroulnar pulley and its evolution in hummingbirds. *Wilson Bull.* **94,** 71–73.

Berger, A. J. (1952). The comparative functional morphology of the pelvic appendage in three genera of Cuculidae. *Am. Midl. Nat.* **47,** 513–605.

Berger, A. J. (1956). The expansor secundariorum muscle, with special reference to passerine birds. *J. Morph.* **99,** 137–168.

Berger, A. J. (1960). The musculature. *In* "Biology and Comparative Physiology of Birds" (A. J. Marshall, Ed.), Vol. 1. Academic Press, New York and London.

Berman, S. L. and Raikow, R. J. (1982). The hindlimb musculature of the mousebirds (Coliiformes). *Auk* **99,** 41–57.

Boas, J. E. V. (1933). Kreuzbein, Becken und Plexus lumbosacralis der Vögel. *Kgl. Danske Vidensk. Selsk. Skr., Naturvidensk.* **5,** 3–74.

Bock, W. J. (1962). The pneumatic fossa of the humerus in the Passeres. *Auk* **79,** 425–443.

Bock, W. J. (1974). The avian skeletomuscular system. *In* "Avian Biology" (D. S. Farner and J. R. King, Eds.), Vol. IV. Academic Press, New York and London.

Bock, W. J. and Miller, W. DeW. (1959). The scansorial foot of the woodpeckers, with comments on the evolution of perching and climbing feet in birds. *Am. Mus. Novit.* No. 1931, 1–45.

Borecky, S. R. (1977). The appendicular myology and phylogenetic relationships of the avian "corvid assemblage." Ph.D. Thesis, University of Pittsburgh.

Brown, R. H. J. (1961). Flight. *In* "Biology and Comparative Physiology of Birds" (A. J. Marshall, Ed.), Vol. 2. Academic Press, New York and London.

Bruning, D. F. (1974). Social structure and reproductive behaviour in the Greater Rhea. *Living Bird* **13,** 251–294.

Burt, W. H. (1930). Adaptive modifications in the woodpeckers. *Univ. Calif. Publ. Zool.* **32,** 455–524.

Clark, B. D. and Bemis, W. (1979). Kinematics of swimming penguins at the Detroit Zoo. *J. Zool., Lond.* **188,** 411–428.

Clark, G. A., Jr. (1981). Toe fusion in oscines. *Wilson Bull.* **93,** 67–76.

Collins, C. T. (1983). A reinterpretation of pamprodactyly in swifts: a convergent grasping mechanism in vertebrates. *Auk* **100,** 735–737.

Coues, E. (1890). "Handbook of Field and General Ornithology." Macmillan, London.

Cracraft, J. (1971). The functional morphology of the hind limb of the Domestic Pigeon, *Columba livia*. *Bull. Am. Mus. Nat. Hist.* **144,** 171–268.

Cracraft, J. (1973). Systematics and evolution of the Gruiformes (Class Aves). *Bull. Am. Mus. Nat. Hist.* **151,** 1–128.

Cracraft, J. (1974). Phylogeny and evolution of the ratite birds. *Ibis* **116,** 494–521.

Cracraft, J. (1976a). Covariation patterns in the postcranial skeleton of moas (Aves, Dinornithidae): a factor analytic study. *Paleobiology* **2,** 166–173.

Cracraft, J. (1976b). The hindlimb elements of the Moas (Aves, Dinornithidae): a multivariate assessment of size and shape. *J. Morph.* **150,** 495–526.

Dabelow, A. (1925). Die Schwimmanpassung der Vögel. Ein Beitrag zur biologischen Anatomie der Fortbewegung. *Morph. Jb.* **54,** 288–321.

de Beer, G. (1954). "Archeopteryx lithographica," Publ. No. 224. British Museum, London.

de Beer, G. (1956). The evolution of ratites. *Bull. Br. Mus. Nat. Hist., Zool.* **4,** 57–70.

Delacour, J. (1951). The significance of the number of toes in some woodpeckers and kingfishers. *Auk* **68,** 49–51.

Diamond, J. M. (1981). Flightlessness and fear of flying in island species. *Nature, Lond.* **293,** 507–508.

Feduccia, A. (1972). Variation in the posterior border of the sternum in some tree-trunk foraging birds. *Wilson Bull.* **84,** 315–328.

Feduccia, A. (1975). Aves osteology. *In* "Sisson and Grossman's The Anatomy of the Domestic Animals" (R. Getty, Ed.), 5th ed. Saunders, Philadelphia.

Feduccia, A. (1980). "The Age of Birds." Harvard University Press, Cambridge, Massachusetts.

Fisher, H. I. (1946). Adaptations and comparative anatomy of the locomotor apparatus of New World vultures. *Am. Midl. Nat.* **35,** 545–727.

Fisher, H. I. (1957). The function of *M. depressor caudae* and *M. caudofemoralis* in pigeons. *Auk* **74,** 479–486.

Friant, M. (1968). Sur les ceintures de membres des oiseaux. *Acta anat.* **69,** 262–273.

Gadow, H. (1882). On some points in the anatomy of *Pterocles,* with remarks on its systematic position. *Proc. zool. Soc. Lond.* pp. 312–332.

Gadow, H. (1895). The deep plantars in the Trochilidae. *Ibis* **1,** 299–300.

Gadow, H. (1896). Muscular system. *In* "A Dictionary of Birds" (A. Newton, Ed.). Black, London.

Gadow, H. and Selenka, E. (1891). Aves. *In* "Dr. H. G. Bronn's Klassen und Ordnungen des Thierreichs" (H. Gadow and E. Selenka, Eds.), Vol. 6, Part 4. Winter, Leipzig.

Garrod, A. H. (1875). On the disposition of the deep plantar tendons in different birds. *Proc. zool. Soc. Lond.* pp. 339–348.

Garrod, A. H. (1876). On some anatomical characters which bear upon the major divisions of the passerine birds. Pt. I. *Proc. zool. Soc. Lond.* pp. 506–519.

George, J. C. and Berger, A. J. (1966). "Avian Myology." Academic Press, New York and London.

Gilliard, E. T. (1958). "Living Birds of the World." Doubleday, Garden City, New York.

Glenny, F. H. and Friedmann, H. (1954). Reduction of the clavicles in the Mesoenatidae, with some remarks concerning the relationship of the clavicle to flight-function in birds. *Ohio J. Sci.* **54,** 111–113.

Glutz von Blotzheim, U. (1960). Zur Morphologie und Ontogenese von Schultergurtel, Sternum, und Becken von *Struthio, Rhea,* und *Dromiceius.* Ein Beitrag zur Phylogenese der Ratiten. *Proc. Int. Ornithol. Congr.* **12,** 240–251.

Goldspink, G., Mills, C. and Schmidt-Nielsen, K. (1978). Electrical activity of the pectoral muscles during gliding and flapping flight in the Herring Gull, *Larus argentatus. Experientia* **34,** 862–864.

Goslow, G. E. (1972). Adaptive mechanisms of the raptor pelvic limb. *Auk* **89,** 47–64.

Gray, J. (1968). "Animal Locomotion." Norton, New York.

Greenewalt, C. H. (1962). Dimensional relationships for flying animals. *Smithson. misc. Collns.* **144,** 1–46.

Greenway, J. C., Jr. (1967). "Extinct and Vanishing Birds of the World." Dover Publications, New York.

Grunert, P. (1906). Der Beckengürtel und die Hintern Extremitäten von Eudyptes chrysocome. Dissertation, Univ. Leipzig.

Harrison, J. G. (1964). Leg. *In* "A New Dictionary of Birds," (A. L. Thomson, Ed.). Nelson, London.

Hartman, F. A. (1961). Locomotor mechanisms of birds. *Smithson. misc. Collns.* **143,** 1–49.

Heimerdinger, M. A. and Ames, P. L. (1967). Variation in the sternal notches of suboscine passeriform birds. *Yale Univ., Peabody Mus., Postilla* **105,** 1–44.

Herzog, K. (1968). "Anatomie und Flugbiologie der Vögel." Fischer, Stuttgart.

Hinchliffe, J. R. (1977). The chondrogenic pattern in chick limb morphogenesis: A problem of development and evolution. *In* "Vertebrate Limb and Somite Morphogenesis" (D. A. Ede, J. R. Hinchliffe and M. Balls, Eds.). Cambridge University Press, London.

Hoff, K. M. (1966). A comparative study of the appendicular muscles of strigiformes and caprimulgiformes. Ph.D. Thesis, Washington State University.

Holmgren, N. (1955). Studies on the phylogeny of birds. *Acta zool., Stockh.* **36,** 243–328.

Howard, H. (1976). A new species of flightless auk from the Miocene of California (Alcidae:Mancallinae). *Smithson. Contr. Paleobiol.* No. 27, 141–146.

Hudson, G. E. (1937). Studies on the muscles of the pelvic appendage in birds. *Am. Midl. Nat.* **18,** 1–108.

Hudson, G. E. (1948). Studies on the muscles of the pelvic appendage in birds II: the heterogeneous order Falconiformes. *Am. Midl. Nat.* **39,** 102–127.

Hudson, G. E., Hoff, K. M., Vanden Berge, J. and Trivette, E. C. (1969). A numerical study of the wing and leg muscles of Lari and Alcae. *Ibis* **111,** 459–524.

Hudson, G. E., Lanzillotti, P. J. and Edwards, G. D. (1959). Muscles of the pelvic limb in galliform birds. *Am. Midl. Nat.* **61,** 1–67.

Hudson, G. E., Schreiweis, D. O. and Wang, S. Y. C. (1972). A numerical study of the wing and leg muscles of tinamous (Tinamidae). *NW. Sci.* **46,** 207–255.

Humphrey, P. S. and Livezey, B. C. (1982). Flightlessness in flying Steamer-Ducks. *Auk* **99,** 368–372.

Hunt, J. H. (1971). A field study of the Wren-Thrush, *Zeledonia coronata. Auk* **88,** 1–20.

Hutchinson, G. E. (1954). Marginalia: the Dodo and the Solitaire. *Am. Scient.* **42,** 300–305.

Johnsgard, P. A. (1966). Behavior of the Australian Musk Duck and Blue-Billed Duck. *Auk* **83,** 98–110.

Jollie, M. (1977). A contribution to the morphology and phylogeny of the Falconiformes. *Evol. Theory* **3,** 1–142.

Klemm, R. D. (1969). Comparative myology of the hind limb of procellariiform birds. *Sth. Ill. Univ. Monogr. Sci. Ser.* **2,** 269 pp.

Lansdown, A. B. G. (1970). A study of the normal development of the leg skeleton in the Quail *(Coturnix coturnix japonica). J. Anat.* **106,** 147–160.

Lindsay, B. (1885). On the avian sternum. *Proc. zool. Soc. Lond.* pp. 684–716.

Lucas, A. M. (1979). *Anatomia topographica externa. In* "Nomina Anatomica Avium" (J. J. Baumel, A. S. King, J. E. Breazile and H. E. Evans, Eds.). Academic Press, London and New York.

Lucas, A. M. and Komárek, V. (1979). Termini situm et directionem, termini generales. *In* "Nomina Anatomica Avium" (J. J. Baumel, A. S. King, A. M. Lucas, J. E. Breazile and H. E. Evans, Eds.). Academic Press, London and New York.

Lucas, A. M. and Stettenheim, P. R. (1972). "Avian Anatomy. Integument," Agriculture Handbook 362. U.S. Department of Agriculture, Washington, D.C.

Lucas, F. A. (1895). The deep plantars in the Trochilidae. *Ibis* **1,** 298–299.

Martin, L. D. (1980). Foot-propelled diving birds of the Mesozoic. *Acta 17th Congr. Int. Ornithol.* Vol. 2, pp. 1237–1242.

Martin, L. D. and Tate, J., Jr. (1976). The skeleton of *Baptornis advenus* (Aves:Hesperornithiformes). *Smithson. Contr. Paleobiol.* No. 27, 35–66.

Maurer, D. R. (1977). The appendicular myology and relationships of the avian order Coraciiformes. Ph.D. Thesis, University of Pittsburgh.

Maurer, D. R. and Raikow, R. J. (1981). Appendicular myology, phylogeny, and classification of the avian order Coraciiformes (including Trogoniformes). *Ann. Carneg. Mus.* **50,** 417–434.

Maynard Smith, J. (1952). The importance of the nervous system in the evolution of animal flight. *Evolution* **6,** 127–129.

Meunier, K. (1951). Korrelation und Umkonstruktion in den Grössbenbeziehungen zwischen Vogelflügel and Vogelkörper. *Biol. Gen.* **19,** 403–443.

Miller, A. H. (1937). Structural modifications in the Hawaiian Goose (*Nesochen sandvicensis*). A study in adaptive evolution. *Univ. Calif. Publ. Zool.* **42,** 1–80.

Montagna, W. (1945). A re-investigation of the development of the wing of the fowl. *J. Morph.* **76,** 87–113.

Morejohn, G. V. (1976). Evidence of the survival to recent times of the extinct flightless duck *Chendytes lawi* Miller. *Smithson. Contr. Paleobiol.* No. 27, 207–211.

Norberg, U. M. (1981). Flight, morphology and the ecological niche in some birds and bats. *In* "Vertebrate Locomotion" (M. H. Day, Ed.), Symp. zool. Soc. Lond., No. 48. Academic Press, London and New York.

Oliver, W. R. B. (1949). "The Moas of New Zealand and Australia," Dominion Mus. Bull. No. 15. Wellington, N.Z.

Olson, S. L. (1973). Evolution of the rails of the South Atlantic Islands. *Smithson. Contr. Zool.* No. 152, 1–53.

Olson, S. L. (1977a). A synopsis of the fossil Rallidae. *In* "Rails of the World" (S. D. Ripley, Ed.). Godine, Boston.

Olson, S. L. (1977b). A great auk, *Pinguinis,* from the Pliocene of North Carolina. *Proc. biol. Soc. Wash.* **90,** 690–697.

Olson, S. L. (1980). A new genus of penguin-like pelecaniform bird from the Oligocene of Washington (Pelecaniformes: Plotopteridae). *Los Angeles Co. Mus., Contr. Sci.* **330,** 51–57.

Olson, S. L. and Feduccia, A. (1980). Relationships and evolution of flamingos (Aves: Phoenicopteridae). *Smithson. Contr. Zool.* No. 316, 1–73.

Olson, S. L. and Steadman, D. W. (1977). A new genus of flightless ibis (Threskiornithidae) and other fossil birds from cave deposits in Jamaica. *Proc. biol. Soc. Wash.* **90,** 447–457.

Olson, S. L. and Wetmore, A. (1976). Preliminary diagnosis of two genera of birds from Pleistocene deposits in the Hawaiian Islands. *Proc. biol. Soc. Wash.* **89,** 247–258.

Orenstein, R. (1977). Morphological adaptations for bark foraging in the Australian treecreepers (Aves: Climacteridae). Ph.D. Thesis, University of Michigan.

Ostrom, J. (1974). *Archeopteryx* and the origin of flight. *Q. Rev. Biol.* **49,** 27–47.

Ostrom, J. (1976a). *Archeopteryx* and the origin of birds. *Biol. J. Linn. Soc.* **8,** 91–182.

Ostrom, J. (1976b). Some hypothetical anatomical stages in the evolution of avian flight. *Smithson. Contr. Paleobiol.* No. 27, 1–21.

Owre, O. T. (1967). Adaptations for locomotion and feeding in the Anhinga and the Double-crested Cormorant. *Ornithol. Monogr.* No. 6.

Pettinghill, O. S., Jr. (1970). "Ornithology in Laboratory and Field." Burgess, Minneapolis.

Pycraft, W. P. (1901). On the morphology and phylogeny of the Paleognathae (Ratitae and Crypturi) and Neognathae (Carinatae). *Trans. zool. Soc., Lond.* **15,** 149–290.

Raikow, R. J. (1969). Sexual and agonistic behavior of the Common Rhea. *Wilson Bull.* **81,** 196–206.

Raikow, R. J. (1970). Evolution of diving adaptations in the stifftail ducks. *Univ. Calif. Publ. Zool.* **94,** 1–52.

Raikow, R. J. (1973). Locomotor mechanisms in North American ducks. *Wilson Bull.* **85,** 295–307.

Raikow, R. J. (1975). The evolutionary reappearance of ancestral muscles as developmental anomalies in two species of birds. *Condor* **77,** 514–517.

Raikow, R. J. (1976). Pelvic appendage myology of the Hawaiian Honeycreepers. (Drepanididae). *Auk* **93,** 774–792.

Raikow, R. J. (1977). Pectoral appendage myology of the Hawaiian honeycreepers (Drepanididae). *Auk* **94,** 331–342.

Raikow, R. J. (1978). Appendicular myology and relationships of the New World nine-primaried oscines (Aves: Passeriformes). *Bull. Carneg. Mus. Nat. Hist.* **7,** 1–43.

Raikow, R. J. (1982). Monophyly of the passeriformes: Test of a phylogenetic hypothesis. *Auk* **99,** 431–445.

Raikow, R. J., Borecky, S. R. and Berman, S. L. (1979). The evolutionary reestablishment of a lost ancestral muscle in the bowerbird assemblage. *Condor* **81,** 203–206.

Raikow, R. J., Polumbo, P. J. and Borecky, S. R. (1980). Appendicular myology and relationships of the shrikes (Aves: Passeriformes: Laniidae). *Ann. Carneg. Mus.* **49,** 131–152.

Rayner, J. M. V. (1981). Flight adaptations in vertebrates. *In* "Vertebrate Locomotion" (M. H. Day, Ed.), Symp. Zool. Soc. Lond., No. 48. Academic Press, London and New York.

Rich, P. V. (1980). The Australian Dromornithidae: A group of extinct large ratites. *Los Angeles Co. Mus., Contr. Sci.* **330,** 93–103.

Richardson, F. (1942). Adaptive modifications for tree-trunk foraging in birds. *Univ. Calif. Publ. Zool.* **46,** 317–368.

Ridgway, R. (1919). The Birds of North and Middle America, Part VIII. *Bull. U.S. natn. Mus.* No. 50.

Ripley, S. D. (1977). "Rails of the World." Godine, Boston.

Romanoff, A. L. (1960). "The Avian Embryo." Macmillan, New York.

Rosser, B. W. C. (1980). The wing muscles of the American coot *(Fulica americana Gmelin). Can. J. Zool.* **58,** 1758–1773.

Rüppell, G. (1977). "Bird Flight." Van Nostrand, Reinhold, New York.

Savile, D. B. O. (1957). Adaptive evolution in the avian wing. *Evolution* **11,** 212–224.

Schaffer, J. (1903). Über die Sperrvorrichtung an den Zehen der Vögel. *Z. Wiss. Zoöl.* **73,** 377–428.

Schreiweis, D. O. (1982). A comparative study of the appendicular musculature of penguins (Aves: Sphenisciformes). *Smithson. Contr. Zool.* No. 341, 1–46.

Sibley, C. G. (1974). The relationships of the lyrebirds. *Ibis.* **74,** 65–79.

Sibley, C. G. and Ahlquist, J. E. (1972). A comparative study of the egg white proteins of non-passerine birds. *Bull. Peabody Mus. Nat. Hist., Yale Univ.* **39,** 1–276.

Simpson, G. G. (1946). Fossil penguins. *Bull. Am. Mus. Nat. Hist.* **87,** 1–99.

Smith, G. T. (1976). Ecological and behavioural comparisons between the Atrichornithidae and Menuridae. *Proc. Int. Ornithol. Congr.* **16,** 125–136.

Spring, L. (1965). Climbing and pecking adaptations in some North American woodpeckers. *Condor* **67,** 457–488.

Spring, L. (1971). A comparison of functional and morphological adaptations in the Common Murre (*Uria aalge*) and Thick-billed Murre (*Uria lomvia*). *Condor* **73,** 1–27.

Stegmann, B. C. (1978). Relationships of the superorders Alectoromorphae and Charadriomorphae (Aves): a comparative study of the avian hand. *Publ. Nuttall Ornithol. Club* **17,** 1–119.

Steiner, H. (1938). Der "Archeopteryx"-Schwanz der Vogelembryonen. *Vierteljahrschr. Naturf. Ges. Zürich Beibl.* **83,** 279–300.

Stolpe, M. (1932). Physiologisch-anatomische Untersuchungen über die hintere Extremität der Vögel. *J. Ornithol., Lpz.* **80,** 161–247.

Stolpe, M. (1935). *Colymbus, Hesperornis, Podiceps*; ein vergleich ihrer hinteren Extremitat. *J. Ornithol., Lpz.* **83,** 115–128.

Storer, R. W. (1956). The fossil loon, *Colymboides minutus. Condor* **58,** 413–426.

Storer, R. W. (1960). Evolution in the diving birds. *Proc. Int. Ornithol. Congr.* **12,** 694–707.

Storer, R. W. (1971). Adaptive radiation of birds. *In* "Avian Biology" (D. S. Farner and J. R. King, Eds.), Vol. I. Academic Press, New York and London.

Sundevall, C. J. (1835). Ornithologiskt system. *Kgl. Svenska Vetenskapsakad. (Stockh.), Hand.* pp. 43–130.

Swierczewski, E. V. (1977). The hindlimb myology and phylogenetic relationships of the Piciformes (Class Aves.). Ph.D. Thesis, University of Pittsburgh.

Swierczewski, E. V. and Raikow, R. J. (1981). Hindlimb morphology, phylogeny, and classification of the Piciformes. *Auk* **98,** 466–480.
Sy, M. (1936). Funktionell-anatomische Untersuchungen am Vogelflügel. *J. Ornithol., Lpz.* **84,** 199–296.
Vanden Berge, J. C. (1970). A comparative study of the appendicular musculature of the Order Ciconiiformes. *Am. Midl. Nat.* **84,** 289–364.
Vanden Berge, J. C. (1975). Aves myology. *In* "Sisson and Grossman's The Anatomy of the Domestic Animals" (R. Getty, Ed.), 5th edn. Saunders, Philadelphia.
Vanden Berge, J. C. (1976). *M. iliotibialis medialis* and a review of the *M. iliotibialis complex* in flamingos. *Auk* **93,** 429–433.
Vanden Berge, J. C. (1979). Myologia. *In* "Nomina Anatomica Avium" (J. J. Baumel, A. S. King, A. M. Lucas, J. E. Breazile and H. E. Evans, Eds.). Academic Press, London and New York.
van Oort, E. D. (1905). Beitrag zur Osteologie des Vogelschwanzes. *Tijdschr. Ned. dierk. Vereen.* **9,** 1–144.
Van Tyne, J. and Berger, A. J. (1976). "Fundamentals of Ornithology," 2nd edn. Wiley, New York.
Weller, M. W. (1980). "The Island Waterfowl." Iowa State University Press, Ames.
Wilcox, H. H. (1952). The pelvic musculature of the loon, *Gavia immer*. *Am. Midl. Nat.* **48,** 513–573.

Supplementary Readings

Berman, S. L. (1984). The hindlimb musculature of the White-fronted Amazon (*Amazona albifrons,* Psittaciformes). *Auk* **101,** 74–92.
Bolen, E. G. and Rylander, M. K. (1983). Whistling-ducks: Zoogeography, ecology, anatomy. *Spec. Publ., Mus., Texas Tech Univ.* No. 20.
Campbell, K. E., Jr. and Tonni, E. P. (1983). Size and locomotion in teratorns (Aves: Teratornithidae). *Auk* **100,** 390–403.
Hall, B. K. (1984). Developmental mechanisms underlying the formation of atavisms. *Biol. Rev.* **59,** 89–124.
Hinchliffe, J. R. and Gumpel-Pinot, M. (1983). Experimental analysis of avian limb morphogenesis. *Current Ornithol.* **1,** 293–327.
McGowan, C, (1979). The hind limb musculature of the Brown Kiwi, *Apteryx australis mantelli. J. Morph.* **160,** 33–74.
McGowan, C. (1982). The wing musculature of the Brown Kiwi *Apteryx australis mantelli* and its bearing on ratite affinities. *J. Zool., Lond.* **197,** 173–219.
Olson, S. L. (1982). The distribution of fused phalanges of the inner toe in the Accipitridae. *Bull. Br. Ornithol. Club* **102,** 8–12.
Rosser, B. W. C., Secoy, D. M. and Riegert, P. W. (1982). The leg muscles of the American Coot *(Fulica americana Gmelin). Can. J. Zool.* **60,** 1236–1256.
Simpson, S. F. (1983). The flight mechanism of the pigeon *Columba livia* during take-off. *J. Zool., Lond.* **200,** 435–443.
Vanden Berge, J. C. (1982). Notes on the myology of the pelvic limb in kiwi (*Apteryx*) and in other birds. *Auk* **99,** 309–315.
Weinstein, G. N., Anderson, C., and Steeves, J. D. (1984). Functional characterization of limb muscles involved in locomotion in the Canada goose, *Branta canadensis. Can. J. Zool.* **62,** 1596–1604.
Zusi, R. L. and Bentz, G. D. (1978). The appendicular myology of the Labrador Duck *(Camptorhynchus labradorius). Condor* **80,** 407–418.
Zusi, R. L. and Bentz, G. D. (1982). Variation of a muscle in hummingbirds and swifts and its systematic implications. *Proc. biol. Soc. Wash.* **95,** 412–420.
Zusi, R. L. and Bentz, G. D. (1984). Myology of the Purple-throated Carib (*Eulampis jugularis*) and other hummingbirds (Aves: Trochilidae). *Smithson. Contr. Zool.* No. 385, 1–70.

3
Somatic peripheral nerves

A. Bubień-Waluszewska

Katedra Anatomii Zwierzat, Akademia Rolnicza, Wydzial Weterynaryjny, Wrocław, Poland

---

CONTENTS

I. General introduction

The somatic innervation of the avian body is mediated by the peripheral nervous system, consisting of the cranial and spinal divisions joined, respectively, to the brain and spinal cord. The somatic structures supplied by the individual peripheral nerves develop either from the corresponding branchial arches or from somites, and constitute mainly the striated muscles and skin of the head, neck, trunk, wings, legs and tail. A typical spinal nerve gives rise to two principal branches: a dorsal branch (*ramus dorsalis*) and a ventral branch (*ramus ventralis*). In each segment of the trunk there is a clear regional distribution of these two great branches, the dorsal ramus supplying the epaxial aspect of the body and the ventral ramus being distributed mainly to the hypaxial components of the body. The limbs are innervated largely by peripheral nerves which arise from a series of plexuses of spinal nerves (*plexus nervorum spinalium*), notably the brachial, lumbar and sacral plexuses; the plexuses themselves arise from the ventral rami of spinal nerves.

Many peripheral nerves of birds run alongside satellite vessels, forming neuro-vascular triads; numerous very small vasa nervorum and nervi vasorum are also present. Within the appendages these neuro-vascular systems are usually located on the flexor aspect, which gives them greater protection against trauma.

For a survey of the autonomic peripheral nerves and the cranial nerves of birds, see Chapter 7, by A. R. Akester, in Vol. 1, and Chapter 7, by the present author, in Vol. 2.

II. Spinal nerves

The spinal nerves of birds vary in number much more from species to species than those of mammals, and this is a reflexion of the relatively greater metameric variability of the avian axial skeleton. It is in fact very difficult to establish reliably the boundaries between the main regions of the avian vertebral column (Gadow, 1891; Boas, 1929, 1933). This inherently complex problem is worsened by the existence in the literature of several different anatomical criteria for identifying the segments in the various transitional zones of the vertebral column. Therefore, whilst the total number of spinal nerves in certain species of birds may be more-or-less readily agreed in the ornithological literature, authors differ about the number of spinal nerves belonging to each region of the spinal cord. This difficulty affects the recognition of those nerves which are homologous to the cervical and thoracic nerves of mammals. The *Nomina Anatomica Avium* (Baumel, 1979a) defines the first thoracic vertebra as the cranial-most vertebra with a complete rib, that is, a rib with both a vertebral and a sternal part, the sternal part articulating directly or indirectly with the sternum; the spinal nerve emerging immediately caudal to this vertebra is then regarded as the first thoracic spinal nerve. This somewhat arbitrary criterion for identifying the first thoracic vertebra has been established in the literature for a long time; it was advocated by Fürbringer (1888), subsequently adopted by Gadow (1896) and Beddard (1898), and employed by many authors since then, including Bellairs and Jenkin (1960), Lucas and Stettenheim (1972) and King and McLelland (1984). However, various other alternatives have been suggested (see Fürbringer, 1888, for an analysis of the early literature), and it was subsequently proposed by Gadow (1933) that the cervical vertebrae should be identified by the ankylosis of their ribs, whereas the thoracic vertebrae should be recognised by the synovial articulations by which their ribs are carried. It is the author's judgement that this principle is correct from the point of view of comparative anatomy. Having used it previously (Waluszewska-Bubień and Pospieszny, 1982a) I would prefer to use it here also. However, the definition of the *Nomina Anatomica Avium* is adopted in this chapter in order to be consistent with other chapters in this present series of volumes. It is even more difficult to distinguish between the lumbar, sacral and caudal nerves. To simplify this problem, in the *Nomina Anatomica Avium* Breazile and Yasuda (1979) have adopted the principle (previously utilised for example by Jungherr, 1969) of collectively naming all the nerves of the pelvic region as synsacral nerves (*nervi synsacrales*), a device which obviates the necessity of identifying the boundaries between the groups of segments incorporated in the synsacrum; this procedure has also been adopted here.

Because it is so difficult to establish a *regional* enumeration of the spinal nerves in birds, many investigators have used the simpler method of counting the paired nerves in sequence, starting at the base of the skull; thus, the first spinal nerve emerges between the occipital bone and the atlas, and in most birds the last spinal nerve escapes between the last free caudal vertebra and the pygostyle (Huber, 1936; Baumel, 1958, 1975; Breazile and Yasuda, 1979). The total number of spinal nerves is 38 in *Columba* (Huber, 1936) and 51 in *Struthio* (Streeter, 1904). Since there are over 60 vertebrae in swans (*Cygnus*) (Bellairs and Jenkin, 1960), more than 51 spinal nerves may be expected in this group. However, Romanoff (1960) noted that the total number of spinal nerves may not necessarily correspond with the total number of vertebrae, since several of the most caudal of the *nervi caudales* may be absent. Moreover, according to Stresemann (1934), metameric nerves are present in the region of the pygostyle in some species but are absent in others including *Gallus* and *Columba*. Intra-specific variations in the numbers of neural segments have also been reported (Fürbringer, 1879, 1888; Gadow, 1891).

A. *General organisation of spinal nerves*

Each spinal nerve is connected with its corresponding segment of the spinal cord by two separate roots, the ventral root and the dorsal root. The two roots differ from each other functionally, according to the principle of Bell-Magendie. Thus, the dorsal root consists mainly of general somatic afferent and visceral afferent fibres, and therefore constitutes the sensory component of the spinal nerve; the ventral root is composed of somatic efferent and visceral efferent fibers and is therefore essentially the motor component. It is possible, however, that in the highest vertebrates this segregation of afferent fibres into the dorsal root and efferent fibres into the ventral root is not absolute, since there are several claims for the presence of afferent fibres in the ventral roots and efferent fibres in the dorsal roots (Davenport and Ranson, 1931; Windle and Orr, 1934; Huber, 1936; Young and Zuckermann, 1936; Coggeshall *et al.*, 1975; see also Akester, 1979). The presence of afferent fibres in the ventral roots appears to be well documented, at least in mammals, but the evidence for efferent fibres in the dorsal roots of birds seems less convincing. Anyway, each spinal nerve (except for the first and the last, which are mainly motor) is essentially a mixed nerve in birds as in the majority of vertebrates (Horst, 1934).

Typically, each dorsal root (*radix dorsalis*) possesses a dorsal root ganglion (*ganglion radiculare dorsale*) (Fig. 3.1). The first two cervical spinal nerves in the adult bird, however, are an exception to this rule (see p. 159). A dorsal root ganglion consists of pseudounipolar nerve cells, many of which are somatic sensory in function; some dorsal root ganglia also contain visceral sensory cell bodies. The central axonal processes of these neurons form the delicate filamentous rootlets (*fila radicularia*) which make up the dorsal root of the spinal nerve. In contrast to the ventral root, the filaments of the dorsal root are assembled into several relatively large fascicles (Fig. 3.2), as is usual in the higher vertebrates (Kappers *et al.*, 1936). The dorsal root ganglia of the cervical and thoracic spinal nerves generally lie within the intervertebral foramina, whereas those of the synsacral and caudal nerves are situated just outside the intervertebral foramina. In *Gallus* the dorsal root ganglia of the thoracic and cranial synsacral regions fuse

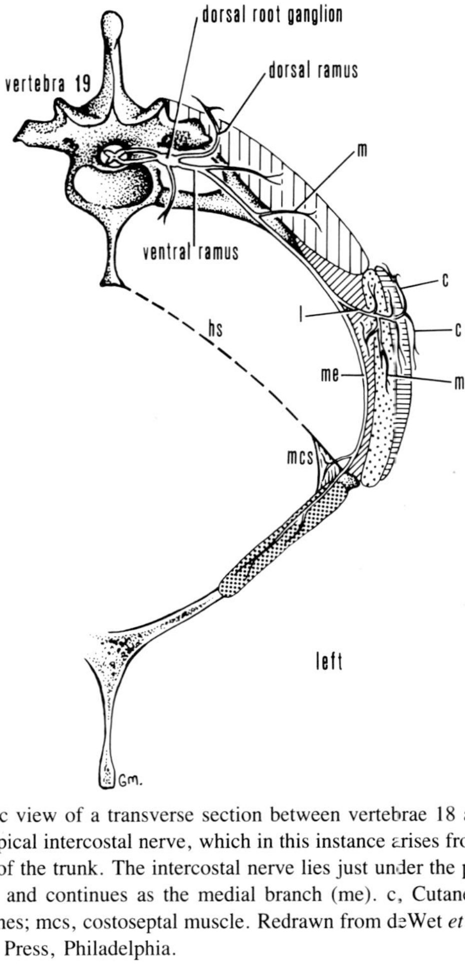

Fig. 3.1. Semidiagrammatic view of a transverse section between vertebrae 18 and 19 of *Gallus* to show the course and branches of a typical intercostal nerve, which in this instance arises from the ventral ramus of spinal nerve XIX on the left side of the trunk. The intercostal nerve lies just under the parietal pleura (not shown). It forms its lateral branch (l) and continues as the medial branch (me). c, Cutaneous branches; hs, horizontal septum; m, muscular branches; mcs, costoseptal muscle. Redrawn from deWet *et al.* (1967), by courtesy of the author and Wistar Institute Press, Philadelphia.

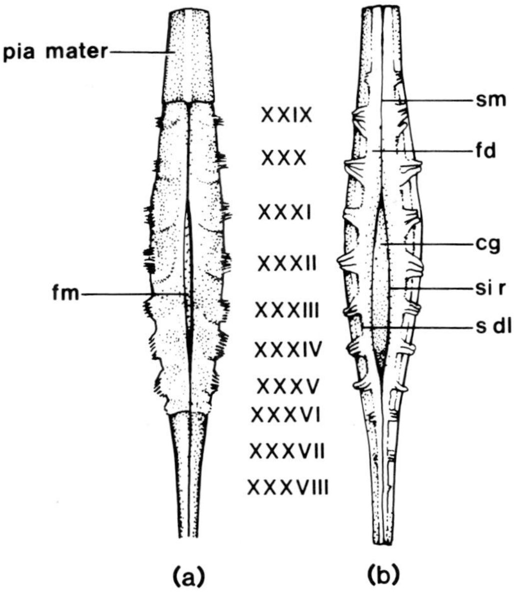

Fig. 3.2. (a) Ventral and (b) dorsal view of the lumbosacral enlargement of the spinal cord of the Ostrich (*Struthio camelus*). XXIX–XXXVIII, Spinal nerves. The ventral roots and dorsal roots increase greatly in size in the region of the enlargement. The ventral roots are formed by numerous fine rootlets, whilst the dorsal roots consist of a few larger bundles. cg, *Corpus gelatinosa*; fd, dorsal funiculus; fm, median fissure; s dl, dorsolateral sulcus; si r, rhomboidal sinus; sm, median sulcus. Based on Streeter (1904).

directly with the corresponding paravertebral ganglia of the sympathetic chain (Baumel, 1975).

The filamentous rootlets of the ventral root (*radix ventralis*) are formed essentially by the peripheral axonal processes of somatic motor neurons with their cell bodies in the ventral horn, and by the preganglionic axons of visceral motor neurons with their cell bodies in the *nucleus intermedius medullae spinalis,* or column of Terni (Huber, 1936; Baumel, 1975; Akester, 1979). Unlike those of the dorsal roots (see earlier), the ventral rootlets leave the spinal cord in many minute bundles (Fig. 3.2), as in the higher vertebrates generally. The number of filaments in the dorsal root is generally similar to the number in the ventral root, but in a typical spinal nerve the total number of nerve fibres within all of its dorsal root filaments together outnumber the total in its ventral filaments, and this too seems to be the rule in the higher vertebrates (Clara, 1959).

The number of fibres and the actual size of the roots vary segmentally, depending on the functional importance of the region which they innervate. Thus, the ventral roots contain the most fibres and have the greatest dimensions in the caudal cervical and the first thoracic segments of strong-flying birds, but the ventral roots are particularly well developed in the synsacral region of ratites (Fig. 3.2); these variations reflect the different degree of development of the respective limbs in those two groups of birds (Fürbringer, 1879, 1888). The dorsal roots of the synsacral region of the Ostrich (*Struthio camelus*) (Fig. 3.2) are also correspondingly well developed (Streeter, 1904).

In adult mammals the spinal cord is much shorter than the neural canal. Consequently the dorsal and ventral roots of the cervical, thoracic, lumbar, sacral and caudal nerves travel for progressively-increasing distances within the neural canal before emerging, finally forming the cauda equina. In birds the spinal cord is virtually the same length as the neural canal. Consequently the origins of the spinal nerves correspond quite closely to their related vertebrae, and there is no true cauda equina. However, at least in *Gallus* the nerve roots of the caudal region of the spinal cord do descend within the neural canal for about the length of one vertebra (Baumel, 1975).

B. The main divisions of a spinal nerve

The trunk of each spinal nerve, formed from the fusion of its ventral and dorsal roots, emerges from the vertebral canal and divides into the following three main branches: (1) the relatively small meningeal branch (*ramus meningeus*), which re-enters the neural canal to supply the spinal meninges, (2) the dorsal branch (*ramus dorsalis*) (Fig. 3.1), and (3) the ventral branch (*ramus ventralis*) (Fig. 3.1). True *rami communicantes* joining the spinal nerves to the paravertebral trunk (*truncus paravertebralis*) of the thoracolumbar (sympathetic) division of the nervous system are mostly not present in the thoracic and cranial synsacral regions of birds. This is because the segmental paravertebral ganglia in these regions are actually fused with the corresponding dorsal root ganglia (see p. 152). However, in *Gallus* and *Columba* (and probably in other avian species) the more caudal of the synsacral ganglia and the caudal ganglia themselves are closely connected to the ventral rami of their related spinal nerves (Fig. 3.20) by distinct rami communicantes (Baumel, 1975). By means of either these rami or the fusion of the dorsal root ganglia with the segmental paravertebral ganglion, each spinal nerve receives an input of postganglionic autonomic efferent fibres.

C. General distribution of peripheral branches

The peripheral branches of spinal nerves typically form muscular and cutaneous nerves. The cutaneous nerves are composed mainly of afferent fibres but also contain autonomic efferent fibres such as postganglionic vasomotor fibres and postganglionic fibres innervating the smooth muscles of the skin. The muscular nerves obviously contain many efferent fibres but also include afferent fibres arising for example from muscle spindles and tendon organs. (For a review of the structure and function of avian muscle spindles and tendon organs see Chapter 8, by K.-M. Gottschaldt, of this volume.)

A typical muscular nerve distributes branches to a group of muscles sharing a common function. Muscular nerves may also give rise to periosteal and articular rami. Striated muscle fibres which are adapted for fast, intermittent contractions (mainly, but not exclusively, white fibres) are typically innervated by a single axonal branch. On the other hand, striated muscle fibres capable of slow, sustained contraction (typically red fibres) are usually innervated by multiple axonal branches (Bolton, 1976). The cutaneous branches emerge subcutaneously on the exterior of the superficial fascia. The exact topographical sites where these nerves emerge has been established only in the domestic fowl, by Yasuda (1964). The branches of peripheral nerves are generally distributed to a definite region and with a characteristic sequence of arborization, but individual variation and bilateral asymmetry occur (Baumel, 1958).

D. Microscopic structure of peripheral nerves

The microscopic structure and functional differentiation of the peripheral nerves of birds are similar to those of other vetrebrates (Young, 1951; Clara, 1959; Bennett, 1974). Thus, an avian peripheral nerve consists of a number of bundles which vary in the size, composition and function of their nerve fibres. The bundles are enclosed within an epineurium consisting of loose, irregular connective tissue with collagen fibres and elastic fibres; blood vessels are also contained within the epineurium, the peripheral nerves of birds being comparatively richly vascularised. Each individual bundle of nerve fibres is surrounded by a perineurial sheath composed of concentric lamellae formed by flattened cells (Fig. 3.3); as in mammals, the cells of adjacent lamellae are attached to each other by tight junctions (D. Z. King, personal communication) (Fig. 3.3). The perineurium is pierced by small branches of the epineurial blood vessels, which then form extensive vascular networks between the nerve fibres. Efferent and afferent fibres are mingled together within a bundle. Fibres tend to change their course within a nerve, passing from one bundle to another and thereby forming internal "plexuses" within a nerve. The individual nerve fibres within a bundle are separated from each other by endoneurial connective tissue (Fig. 3.3).

The ultrastructure of the nerve fibre itself is essentially the same as in mammals. Myelinated nerve fibres possess a myelin sheath formed by the spiral growth of the Schwann cell around the axon cylinder. An unmyelinated fibre is an axon cylinder individually suspended by a mesaxon within a Schwann cell; often several such axons are suspended within the same Schwann cell each by its own mesaxon (Fig. 3.3). As in mammals, the axon cylinder contains neurotubules and neurofilaments. Smooth endoplasmic reticulum, vesicles and mitochondria are seen occasionally. Ribosomes are typically absent, a useful feature for distinguishing axonal profiles from circular profiles of other cell processes in electron micrographs. Axoplasm appears to be in a state of continuous movement, transport from the cell body taking place simultaneously in a slow phase of only a few millimetres per day and a fast phase perhaps 50–100 times faster. The effector endings of efferent nerve fibres are characterised ultrastructurally by large accumulations of pale or dense-cored vesicles, and by the presence of moderate numbers of mitochondria. In some tissues such as striated muscle and the carotid body, the site of synaptic transmission may show electron-dense thickenings, with relatively narrow presynaptic and relatively wide postsynaptic regions. The differing widths of these dense regions, and the accumulation of vesicles against the presynaptic membrane, have been interpreted as indicators of the direction of transmission, although autoradiographic studies show that such interpretations can be misleading at least in the carotid body (Taha and King, 1983). The receptor endings of afferent axons tend to be enlarged and packed with

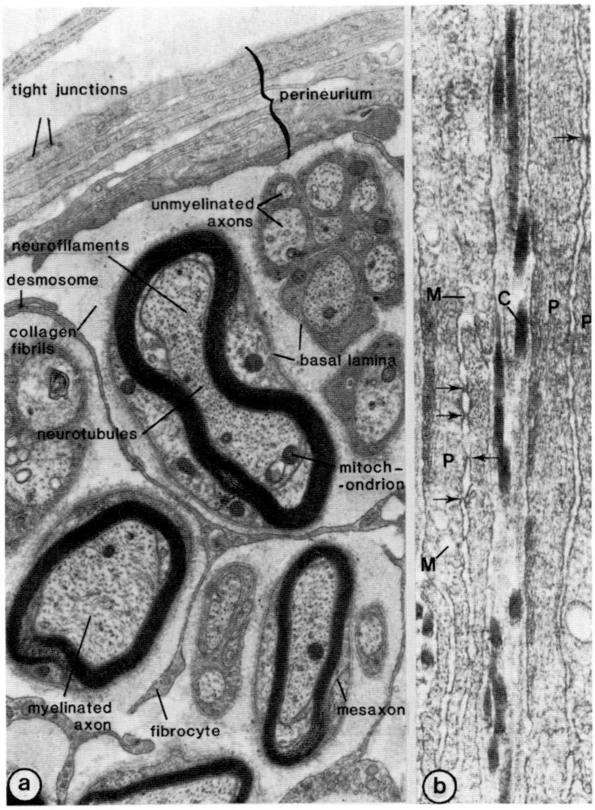

Fig. 3.3. (a) Electron micrograph of a small nerve arising from the brachial plexus of *Gallus,* cut in transverse section. The perineurial cells are arranged in flattened concentric lamellae, some of which are attached to each other by tight junctions. The collagen fibrils belong to the endoneurium. Several unmyelinated and myelinated axons are shown. The axon cylinders contain an occasional mitochondrion, and numerous neurofilaments and neurotubules. ×18,000. (b) Electron micrograph of part of the perineurium of the same nerve. P, Perineurial cells; M, micropinocytotic vesicles; C, collagen fibrils. Arrows, tight junctions. ×75,000. From D. Z. King (unpublished observations).

mitochondria; some mechanoreceptor endings are surrounded by complex laminae (see the account by Gottschaldt in Chapter 8 of this volume).

E. Dorsal rami of spinal nerves

As already stated, the peripheral nerves of the neck, trunk and limbs are formed by the derivatives of the dorsal and ventral rami of the spinal nerves (Figs. 3.1, 8). Because of the weak development of the avian epaxial musculature, the dorsal rami (*rami dorsales*) of birds are relatively much more slender than those of mammals. The feeble development of the epaxial musculature in birds is particularly evident in the trunk, where it is associated with (1) the reduction in mobility which results in many species from the ossification of thoracic ligaments and tendons, so forming the fused vertebrae (Fig. 3.4) constituting the notarium (Baumel, 1979a), and especially with (2) the characteristic rigidity of the fused synsacral vertebrae of birds. In the neck, however, the epaxial muscles and the dorsal rami which innervate them are relatively well developed (Fig. 3.5); the extreme mobility of the cervical part of the vertebral column is just as typical of the avian body as is the rigidity of the other vertebral regions. Thus, the iliocostalis–longissimus system is well represented in the avian neck by the *mm. cervicalis ascendens* and *thoracicus ascendens,* and rotation is possible through the action of these muscles in combination with the *mm. longus colli dorsalis* and *intertransversarius.* Most of the dorsal rami of the cervical spinal nerves (Fig. 3.6) constitute independent nerves, which proceed to bifurcate directly into medial and lateral branches. The first three cervical dorsal rami are an exception. After these nerves have pierced between the *rectus* and *obliquus* muscles, their *medial* branches combine to form the dorsal cervical plexus (*plexus cervicalis dorsalis*), which is also known as the suboccipital plexus (Nishi, 1938). The lateral branches of all of the cervical nerves run directly to their epaxial cervical muscles and the adjacent skin (Watanabe, 1961; Yasuda, 1964).

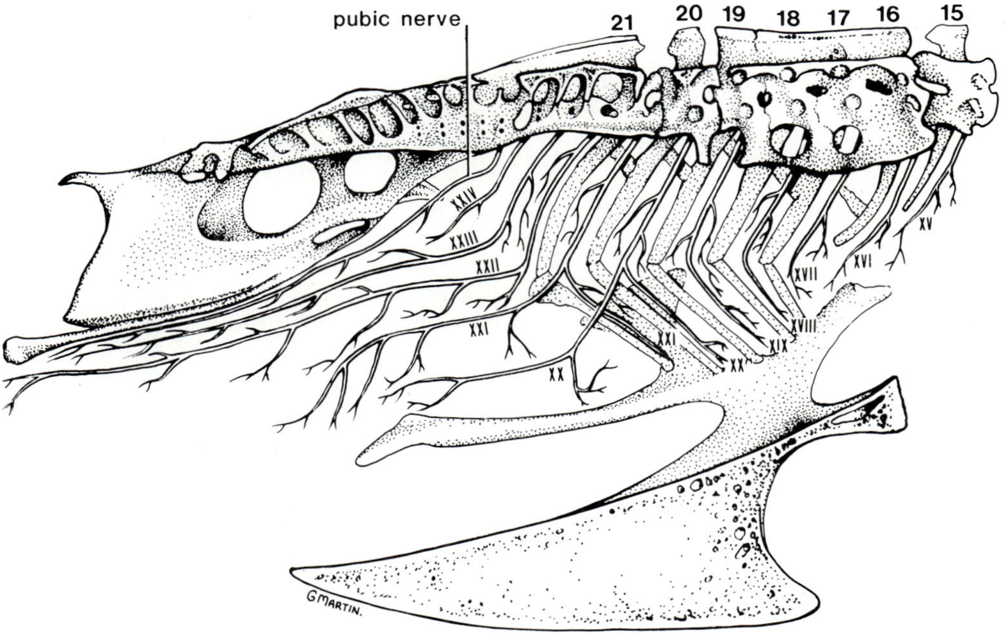

Fig. 3.4. Semidiagrammatic medial view of the principal nerves of the trunk of *Gallus,* seen from the right side. 15–21, Vertebrae 15–21, inclusive. According to the criteria of the *Nomina Anatomica Avium* (see p. 150), in this specimen vertebra 17 is the first thoracic vertebra, since it is the first vertebra to carry a rib which articulates with the sternum. (In other specimens of this species it may be vertebra 18). The fused vertebrae 16–19 constitute the notarium in this specimen of *Gallus.* As defined by the *Nomina Anatomica Avium,* the most cranial vertebra in the notarium (16) is the last cervical vertebra. XV–XXIV, Ventral rami of spinal nerves. In this specimen XVII is the last cervical spinal nerve, and XXII is the first synsacral nerve. The components of those spinal nerves that lie between the ribs are the intercostal nerves. The four intercostal nerves formed by spinal nerves XVIII–XXI are bounded both cranially and caudally by a rib which articulates with the sternum. Nerves XVII and XVI, which lie caudal to the two freely movable cervical ribs (articulating with vertebrae 16 and 15 here) are also referred to as intercostal nerves. Redrawn from deWet *et al.* (1967), by courtesy of the author and the Wistar Institute Press, Philadelphia.

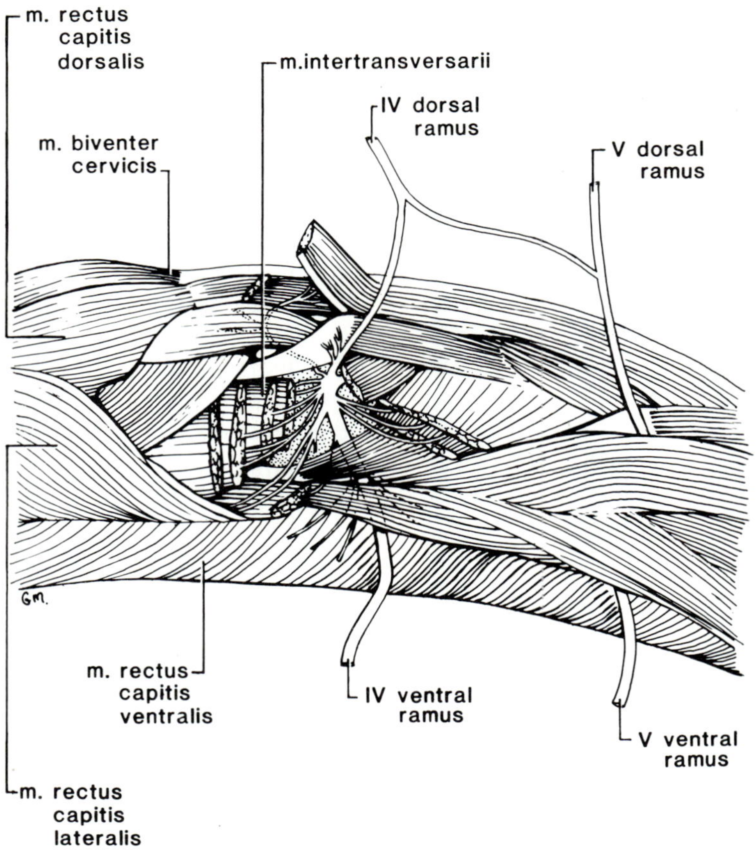

Fig. 3.5. Left lateral view of the dorsal and ventral rami of the fourth (IV) and fifth (V) cervical spinal nerves, and their branches innervating the epaxial and hypaxial musculature of *Gallus*. Redrawn from T. Watanabe (1961), by courtesy of the author and the *Japanese Journal of Veterinary Science*.

The dorsal rami in the pelvic region are especially poorly developed due to the extreme reduction of the dorsospinal musculature, and are restricted mainly to vasomotor and sensory components.

F. Ventral rami of spinal nerves

The ventral rami of the spinal nerves are generally much larger than the dorsal rami. In some segments of the neuraxis they form massive plexuses which innervate the limbs. In other regions, such as the upper cervical and lower thoracic segments, the ventral rami preserve their metameric arrangement and greater morphological independence, and supply the lateral and ventral walls of the neck and trunk, for instance via the intercostal nerves. Like the dorsal rami, these segmentally-distributed ventral rami eventually divide into medial and lateral branches (Fig. 3.1), both of which contain efferent as well as afferent fibres. The lateral branches give off muscular nerves and then innervate the skin of the lateral surface of the neck and trunk. The ventral aspect of the skin of the neck and trunk is innervated mainly by cutaneous nerves from the medial branches (Yasuda, 1964).

III. The neck

The cervical nerves in birds are much more variable in number than those in mammals, ranging from 10 pairs in species with very short necks as in some of the Oscines and swifts (Apodidae) to 26 in long-necked birds such as swans (*Cygnus*). Compared with other higher vertebrates, there seems to be also a greater intra-specific variation of avian cervical nerves, which appears to arise from the loss of segments or from the interpolation of extra segments during the development of the neuraxis (Fürbringer, 1879). The neuromeres are closely correlated with the development of the myotomes (Horst, 1934). The cervical nerves in birds leave the vertebral canal *cranial* to their corresponding segment, as in higher vertebrates generally, via the intervertebral foramina. The first cervical nerve emerges from the vertebral canal between the occipital bone and the atlas, and the last escapes caudal to the last cervical vertebra.

A detailed analysis of the cervical nerves in the domestic fowl and their distribution was made by Watanabe (1961). The first two cervical nerves of adult birds differ from the rest of the cervical series in being mainly motor in function, since the first nerve does not possess a dorsal root ganglion and the second has only a vestigial ganglion (at least in some species, including *Gallus*). Consequently the dorsal rami of the first two cervical nerves fail to form cutaneous nerves (Watanabe, 1961). The ramifications of the dorsal rami of the first two cervical nerves participate in the innervation of muscles such as the *biventer servicis, rectus capitis lateralis, complexus, splenius capitis* and *rectus capitis dorsalis,* principally via branches of the dorsal cervical plexus (see p. 157). The afferent components of the third and fourth cervical nerves supply the skin of the occiput (Yasuda, 1964).

Most of the ventral ramus of the first cervical spinal nerve (Baumel, 1975; Breazile and Yasuda, 1979), and perhaps a part of the second one also (Gadow, 1891), combine with the hypoglossal nerve (*n. hypoglossus*) to form the hypoglossocervical nerve (*n. hypoglossocervicalis*); these communicating pathways appear to be the homologue of the plexus cervicalis of other vertebrates. The cervical elements of the hypoglossocervical nerve are believed to continue mainly within the slender filaments of its descending cervical ramus (*r. cervicalis descendens*) to supply the *m. sternotrachealis,* thus being reminiscent of the descending ramus of the hypoglossal nerve in some mammals. The ventral ramus of the third cervical nerve communicates via the *connexus cum n. faciali* with the cervical ramus of the hyomandibular branch of the facial nerve (Yasuda, 1964; Baumel, 1975; Bubień-Waluszewska, 1981), thus forming the *nervus cutaneus colli* (Fig. 3.6), which supplies the cranial part of the *m. constrictor colli* (Yasuda, 1964). The ventral rami of the first 10 or more cervical nerves are much more slender than those at the caudal end of the series which contribute to the formation of the brachial plexus.

The cutaneous nerves in *Gallus* were thoroughly studied by Yasuda (1964). They are distributed to the subcutaneous musculature and then to the integument

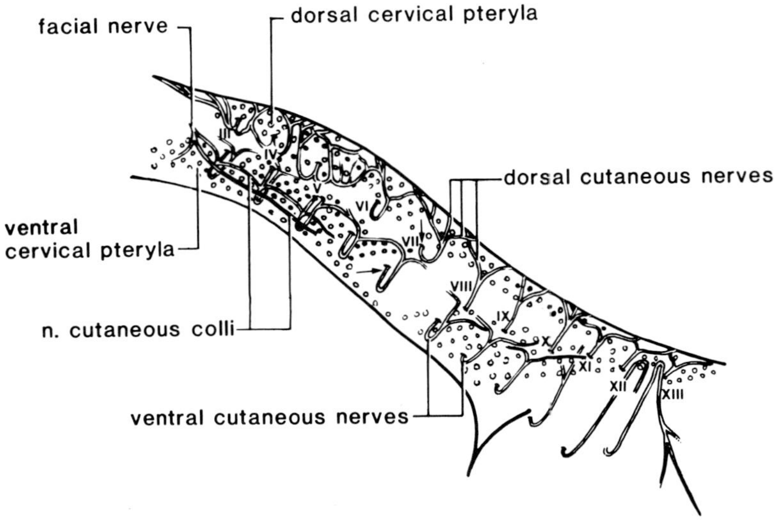

Fig. 3.6. Left lateral view of the cutaneous nerves of the neck of *Gallus*. III–XIII, Spinal nerves each represented by a dorsal and a ventral ramus. The dorsal and ventral rami of nerve VII are indicated by an upper and a lower arrow, respectively. Redrawn from M. Yasuda (1964), by courtesy of the author and the *Japanese Journal of Veterinary Science*.

itself and its organs (Stammer, 1961). The first cervical cutaneous nerves (*nervi cutanei cervicales*) arise from the third and immediately successive cervical spinal nerves (Watanabe, 1961; Yasuda, 1964). Each of these cervical spinal nerves forms a dorsal cervical cutaneous nerve (*n. cutaneus cervicalis dorsalis*) from its dorsal ramus, and a ventral cervical cutaneous nerve (*n. cutaneus cervicalis ventralis*) from its ventral ramus. The areas of skin supplied by these successive pairs of dorsal and ventral cervical cutaneous nerves take the form of bands surrounding the neck one after another (Fig. 3.6), like those of the trunk (Kaiser, 1924).

The innervation of the various components of the skin in birds was described by Lucas and Stettenheim (1972). It includes the nerves of the follicular wall, the papilla and pulp of the feather, and also the feather muscles. Complex laminated sensory endings are found throughout much of the integument, and have been commonly known as lamellar corpuscles (Malinovsky, 1967, 1968). These structures are present in the skin of the apteria, of the feather tracts and of the feet. Within the tracts they are situated beside the follicles of contour feathers, though not always at every follicle (Lucas and Stettenheim, 1972). For further discussion of these cutaneous mechanoreceptors see Gottschaldt (Chapter 8 of this volume).

IV. The trunk

In the earlier literature the "thoracic nerves" in birds are said to number from 4 to 5 pairs as in some vultures and Columbidae, up to 9 to 10 in *Struthio,* and up to 11 pairs in swans (*Cygnus*); these figures reflect the numbers of thoracic vertebrae, which are estimated by Bellairs and Jenkin (1960) to range from about 5 to 10 in the various species. However, statements in the literature about the numbers of thoracic spinal nerves are only approximate guides, since the identity of the first thoracic nerve is controversial (see p. 150), and there is no certainty at all about the number of thoracic vertebrae in the synsacrum. Unlike the cervical nerves, the thoracic spinal nerves leave the vertebral canal caudal to their related vertebra and hence caudal also to their related rib.

The thoracic spinal nerves retain an obvious segmental arrangement. Their ventral rami form intercostal nerves (*nn. intercostales*) which run in the intercostal spaces (Figs. 3.1, 4) accompanied by the corresponding vein and artery, forming together an anatomical triad. Typically the intercostal nerves lie essentially just beneath the parietal pleura, along the caudal border of the corresponding rib. Each nerve passes a considerable distance down its intercostal space and then forms a deep (medial) branch which innervates muscles (Fig. 3.1), and a superficial (lateral) branch which innervates muscles and also supplies skin (deWet *et al.*, 1967); the medial branch is essentially the direct continuation of the main trunk. The innervated muscles are those of the thoracic and abdominal walls. The muscles of the thoracic wall include the *levatores costarum* and the *scalenus,* the internal and external intercostal muscles, the costosternal muscle, and also the costoseptal muscle (*m. costoseptalis*) (deWet *et al.*, 1967). The last named is the striated muscle that is attached to the horizontal septum (*septum horizontale*) and in the earlier literature was often referred to as the "diaphragm"; since there is no structure in any bird which is anatomically, embryologically or physiologically comparable to the mammalian diaphragm, the term diaphragm (and hence also phrenic nerve) should be avoided altogether in birds (King and Molony, 1971). Branches of intercostal nerves are distributed to all of the muscles of the abdominal wall, that is, the *rectus abdominis, transversus abdominis,* and external and internal oblique abdominal muscles.

The first few synsacral nerves also supply the trunk. Thus, the first two synsacral nerves (nerves XXII and XXIII in Fig. 3.4) distribute numerous muscular and cutaneous rami to the abdominal muscles and feather tracts in *Gallus,* and the third synsacral nerve (nerve XXIV) gives off a few similar branches (Yasuda, 1964; deWet *et al.*, 1967). In this species and in *Columba* the lumbar plexus itself (or the intrapelvic part of the medial femoral cutaneous nerve) forms the pubic nerve (*n. pubicus*); in the Great Bustard (*Otis tarda*) this nerve arises from the second synsacral nerve as in Fig. 3.21 (Waluszewska-Bubień and Pospieszny, 1982b). The pubic nerve runs a long distance caudally, parallel with the pubis (Figs. 3.4, 20), distributing cutaneous branches and muscular branches to the abdominal muscles (Baumel, 1975) in the region between the two pubic bones; this nerve was previously known as the ilioinguinal nerve (Chomiak, 1950; Buchholz, 1959–1960). For a detailed description of the nerves of the trunk in *Gallus* see Yasuda (1960a).

V. General arrangement of the great plexuses

The formation of the great nerve plexuses by the ventral rami of a series of spinal nerves occurs at the segmental levels of the limbs, as is characteristic of vertebrates adapted for terrestial locomotion (Fürbringer, 1879). These plexuses are created by the exchange of nerve fibres between the ventral rami of those spinal nerves that belong developmentally to the somites forming the muscles, bones and skin of the limb. The individual ventral rami constitute the roots of origin of the plexus (*radices plexus*). The roots join each other to form two or three short trunks of the plexus (*trunci plexus*), in which there is an exchange of fibres. The trunks may then recombine into great fascicles (*fasciculi plexus*); each fascicle gives rise to the named peripheral nerves, as in the brachial plexus (Figs. 3.7–10). Alternatively, the trunks themselves may directly form peripheral nerves. In this way a peripheral nerve, derived from either a trunk or its fasciculus (fascicle), may contain fibres from two or more roots of the plexus (Baumel, 1958). The points of origin of the individual peripheral nerves from a plexus, as well as their secondary branches, generally have more-or-less definite topographical characteristics in a given species; however, intra-specific variations from the typical pattern occur at least in the brachial plexus, and Baumel (1958) has pointed out the consequent hazards of generalising about the species anatomy of the plexus from dissections of only one or two specimens.

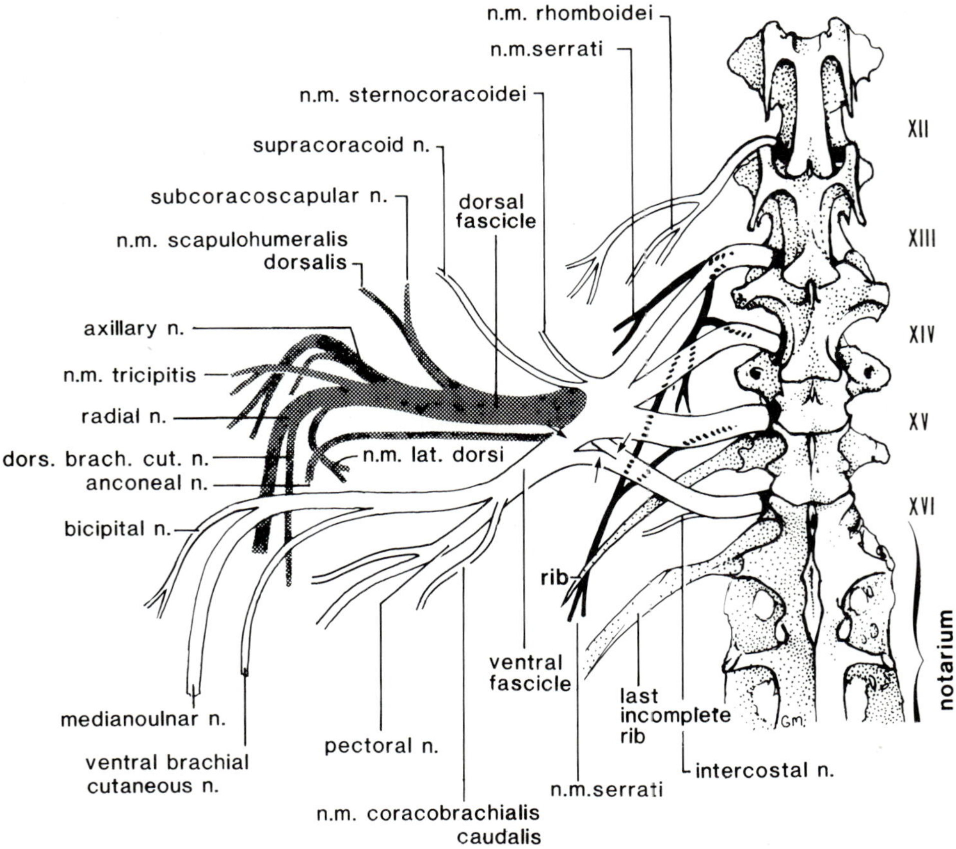

Fig. 3.7. Ventral view diagram of the right brachial plexus of *Gallus*. The accessory plexus is shown in black, and the dorsal components arising from the main plexus are hatched. XII–XVI, Ventral rami of the five spinal nerves forming the roots of the plexus (*radices plexus*). The four main roots (XIII–XVI) unite to form three trunks (*trunci plexus*) indicated by arrows, in which there is an exchange of fibres. Abbreviations: dors. brach. cut. n., dorsal brachial cutaneous nerve; lat. dorsi, *latissimus dorsi*; m., muscle; n., nerve. Redrawn from J. J. Baumel (1975), by courtesy of the author and Saunders Co., Philadelphia.

VI. The wing

A. *Brachial plexus*

1. The roots of the brachial plexus

The roots of the avian brachial plexus (*radices plexus*) generally arise from the ventral rami of four or five spinal nerves. According to Gadow's summary (1896) of Fürbringer's remarkably extensive data (see Table I), the smallest number is three (in *Bucorvus abyssinicus*) and the greatest is six (in *Charadrius* and *Columba*). In *Gallus* (Fig. 3.7) the relevant spinal nerves are most commonly XIII–XVI, inclusive (Baumel, 1975), and in some specimens XVII also (Gadow, 1891; Ohmori *et al.*, 1982). In *Columba* (Fig. 3.8) nerves X (or XI)–XV, inclusive, typically contribute (Gadow, 1891; Baumel, cited in Breazile and Yasuda, 1979). In the gruiform Great Bustard (*Otis tarda*), Waluszewska-Bubień and Pospieszny (1982a) observed contributions from nerves XIII–XVII, inclusive (Fig. 3.9); Fürbringer reported that nerves XIV–XVII formed the plexus in this species. Nerves XV–XVIII contribute in the Whooping Crane (*Grus americana*) (Fisher and Goodman, 1955). From a survey of the brachial plexus of a good number of specimens of the passeriform Purple Martin (*Progne subis*), Baumel (1958) found that nerves XII–XV invariably contributed and that nerve XI did so in about half the specimens (Fig. 3.10). In the closely-related *Hirundo rustica,* six spinal nerves were found to form the plexus, that is, XI–XVI (Buri, 1900). Spinal nerves XI–XV also formed the plexus in the passeriform Eurasian Jay (*Garrulus glandarius*) (Fürbringer, 1879). On the basis of the earlier literature and especially the numerous observations by Fürbringer (see Table I), Gadow (1891, 1896) confirmed that in many Passeriformes the nerves concerned are indeed XII–XV. He also confirmed the contributions of XIII to XVII in *Gallus* and of XI to XV (usually) in *Columba,* and reported XV to XVIII in *Anas.* He concluded that the known craniocaudal extremes occurred in the Common Swift (*Apus apus*) (X–XIV) and in *Cygnus* (XXII–XXVI). In general the middle nerves of those that contribute to the plexus are the largest, particularly in short-necked species; in long-necked birds the caudal-most roots seem to be predominant (Gadow, 1891).

Baumel (1985) discovered the presence, hitherto completely overlooked, of a ligamentous apparatus for supporting the roots of the brachial plexus. It consists of short and long ligaments (*ligamenta suspensoria nervorum*) which arise from the vertebrae and blend with the epineurium of the ventral surface of the three largest roots of the plexus (e.g. roots XII–XIV in *Columba*). In principle there are a pair of short and a pair of long ligaments for each of these nerves. The two ligaments of each pair arise, respectively, from the cranial and caudal aspects of the border of the intervertebral foramen (i.e. on each side of their nerve root) and then converge on the root to give it a V-shaped suspensory apparatus. The short ligaments attach to the root at the level of the dorsal root ganglion, and the long roots attach just central to where the plexus trunks are formed. The long ligaments arise from small tubercles on the vertebral body, or in the case of vertebra 12 in *Columba* from its short costal process. These ligaments were found in *Columba, Gallus,* the passerine *Progne subis,* a nighthawk (*Chordeiles minor*), a grebe (*Podilymbus podiceps*), a hawk (*Circus cyaneus*), an owl (*Strix varia*), a tern

Table I. **The contributions of spinal nerves to the brachial plexus.**[a]

Species	Spinal nerves[b]																
	X	XI	XII	XIII	XIV	XV	XVI	XVII	XVIII	XIX	XX	XXI	XXII	XXIII	XXIV	XXV	XXVI
Struthio camelus								+	+	+	+	+					
Rhea americana						+	+	+	+								
Casuarius casuarius							+	+	+	+	+						
Apteryx australis				+	+	+	+	*									
Spheniscus demersus				+	+	+	+										
Alca torda				+	+	+	+										
Gavia arctica				+	+	+	+										
Podiceps cristatus								+	+	+	+	+	*				
Larus ridibundus																	
Juvenile			+	+	+	+	+										
Adult				+	+	+	+										
Fulmarus glacialis				+	+	+	+										
Puffinus spp.				+	+	+	+										
Phalacrocorax carbo							+	+	+	+	+	*					
Pelecanus rufescens						+	+	+	+								
Tadorna cristata						+	+	+	+								
Anser anser						+	+	+	+	+							
Anser anser							+	+	+	+							
Anser anser							+	+	+	+*	+						
Anser anser								+	+	+	+						
Cygnus atratus													+	+	+	+	+
Phoenicopterus ruber								+	+	+	+						
Ciconia ciconia						+	+	+	+								
Threskiornis aethiopicus					+	+	+	+	+								
Grus canadenis								+	+	+	+*	+					
Otis tarda					+	+	+	+									
Charadrius spp.			+	+	+	+	+*	+									
Numenius phaeopus				+	+	+	+*	+									
Crypturellus noctivagus					+	+	+	+	*								
Talegalla spp.				+	+	+	+	*									
Argusianus argus				+	+	+	+	+									

Gallus domesticus				+	+	+	+	*			
Gallus domesticus				+	+	+	+	+			
Numida meleagris				+	+	+	+	+			
Opisthocomus hoazin							+	+	+	+	*
Goura cristata		+	+	+	+	+	*				
Columba livia	+	+	+	+	+	+					
Columba livia		+	+	+	+	*					
Columba livia		+	+	+	+	+					
Columba livia		+	+	+	+	+	*				
Columba livia			+	+	+	+	*				
Psittacus erithacus		+	+	+	+*	+					
Sagittarius serpentarius				+	+	+	+				
Aquila audax			+	+	+	+*	+				
Aquila audax			+	+	+	+	+				
Falco tinnunculus			+	+	+	+	+				
Buteo buteo			+	+	+	+*	+				
Accipiter nisus			+	+	+	+	+				
Pandion haliaetus			+	+	+	+	+				
Ketupa spp.		+	+	+	+	+					
Glaucidium passerinum		+	+	+	+	+					
Otus spp.			+	+	+	+					
Cuculus canorus		+	+	+	+	*					
Harpactes spp.			+	+	+	+	*				
Caprimulgus europaeus		+	+	+	+						
Caprimulgus europaeus			+	+	+*	+					
Podargus spp.		+	+	+	+	*					
Podargus spp.		+	+	+	+	+					
Eurystomus orientalis			+	+	+	+					
Momotus mexicanus			+	+	+	+	*				
Todus spp.			+	+	+	+	*				
Merops apiaster			+	+	+	+					
Buceros spp.			+	+	+	+					
Bucorvus abyssinicus				+	+	+	*				
Pelargopsis spp.			+	+	+	+	*				

(*continued*)

TABLE I. ***(Continued)***

Species	Spinal nerves[b]																
	X	XI	XII	XIII	XIV	XV	XVI	XVII	XVIII	XIX	XX	XXI	XXII	XXIII	XXIV	XXV	XXVI
Apus apus	+	+	+	+	+												
Megalaima spp.			+	+	+	+											
Ramphastos spp.			+	+	+	+											
Picus spp.		+	+	+	+	+											
Picus spp.			+	+	+	+											
Gecinulus viridis		+	+	+	+	+											
Gecinulus viridis			+	+	+	+											
Urocissa spp.			+	+	+	+											
Garrulus glandarius		+	+	+	+	+											
Garrulus glandarius			+	+	+	+											
Corvus corone			+	+	+	+											
Turdus pilaris			+	+	+	+											
Fringilla spp.	+	+	+	+	+												
Emberiza citrinella			+	+	+	+											
Alauda arvensis			+	+	+	+											
Lanius spp.			+	+	+	+											
Upupa epops		+	+	+	+												

[a] Observed by Fürbringer (1888) and cited by Gadow (1891).

[b] Asterisks indicate the last cervical spinal nerve, emerging between the last cervical vertebra and the first thoracic vertebra. Fürbringer defined the first thoracic vertebra as the first vertebra to carry a rib articulating with the sternum, that is, the same definition as that used in *Nomina Anatomica Avium* and in the present chapter.

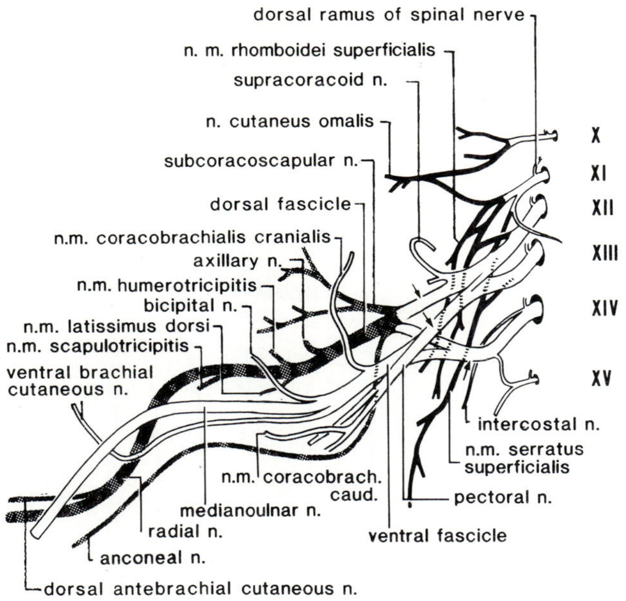

Fig. 3.8. Ventral view diagram of the right brachial plexus of *Columba livia*. The accessory brachial plexus is shown in black, and the dorsal components arising from the main plexus are hatched. X–XV, Ventral rami of the six spinal nerves that form the roots of the plexus (*radices plexus*). The roots unite and form three short trunks (arrows), the *trunci plexus,* in which there is an exchange of fibres. Redrawn from J. J. Baumel (cited in Breazile and Yasuda, 1979), by courtesy of the author.

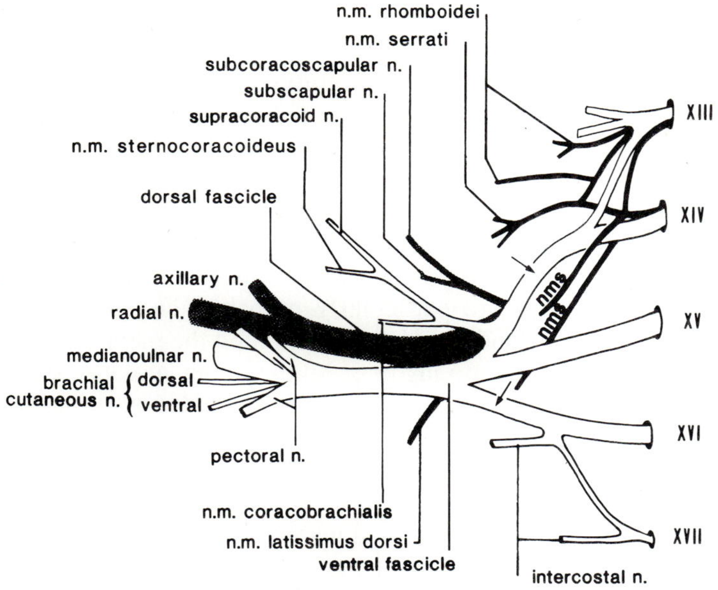

Fig. 3.9. Ventral view diagram of the right brachial plexus of the gruiform Great Bustard (*Otis tarda*). The accessory plexus is shown in black, and the dorsal components arising from the main plexus are hatched. XIII–XVII, Ventral rami of the five spinal nerves that form the roots of the plexus (*radices plexus*). These unite, forming two short trunks (arrows), the *trunci plexus,* in which fibres are exchanged. nms, Nerves to serrati muscles. From A. Waluszewska-Bubień (unpublished observations).

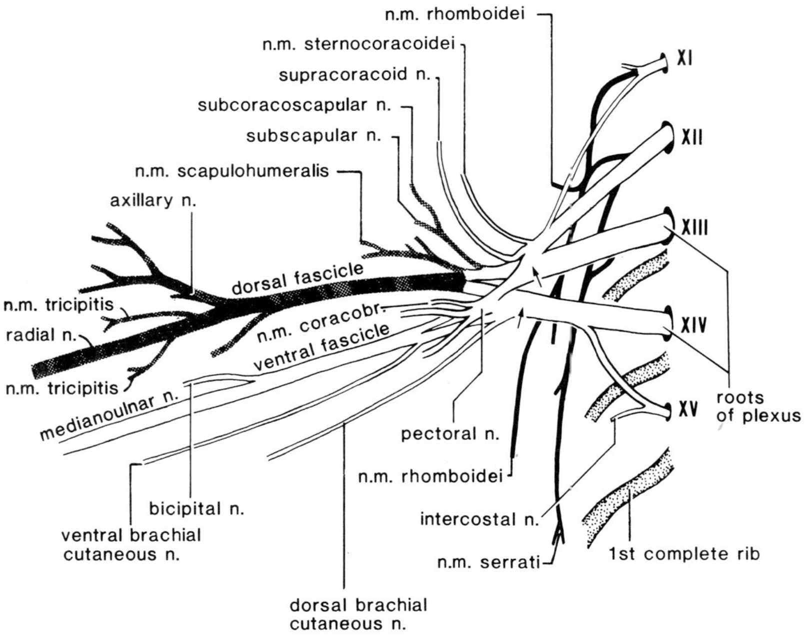

Fig. 3.10. Ventral view diagram of the right brachial plexus, as typical for the passeriform Purple Martin (*Progne subis*). The accessory plexus is shown in black, and the dorsal components arising from the main plexus are hatched. XI–XV, Ventral rami of the five spinal nerves forming the roots of the plexus (*radices plexus*). These unite, forming two short trunks (arrows), the *trunci plexus,* in which fibres are exchanged. n.m. coracobr., Nerve to the *m. coracobrachialis*. Redrawn from J. J. Baumel (1958), by courtesy of the author and S. Karger, Basel.

(*Gygis alba*), a toucan (*Pteroglossus torquatus*) and the psittaciform *Melopsittacus undulatus*. The ligaments were larger in the bigger birds. All of these species possessed the short ligaments, and all except the nighthawk and the grebe had the long ligaments as well. It therefore seems likely that suspensory ligaments of the brachial plexus roots are widespread among Aves. No such ligaments were found on the roots of the lumbosacral plexus of *Columba*. Baumel suggested that flight must impose tensile forces on the plexus roots which have to be buffered by the ligaments. Such forces are particularly probable during the extreme wing movements of takeoff and landing. Unusual strain on the plexus might also occur when captors strike their prey, diving birds hit the water, or the extended wings of soaring birds are buffeted by gusts of wind. The extended tubercles to which the long ligaments are attached suggest, however, that traction on these ligaments is a routine, everyday event. The suspensory ligaments appear therefore to be adaptations to protect the spinal cord against the rigours of flight.

Gadow (1896) interpreted all the spinal nerves which constitute the roots of the brachial plexus as being cervical, and the same view was taken by Portmann (1950) and Portmann and Stingelin (1961). Baumel (1958) too regarded all the nerves in *Progne subis* as being cervical. Other authors, however, have considered the caudal-most one or two nerves to be thoracic, as in *Garrulus glandarius, Picus viridis, Columba, Gallus, Anser* and *Otis tarda* (Fürbringer, 1879;

Moskoff, 1929; Stresemann, 1934; Dozsa, 1939; Buchholz, 1959–1960; Waluszewska-Bubień and Pospieszny, 1982a), and in a number of other species (see Table I). Differing methods of distinguishing between cervical and thoracic vertebrae (see p. 150) must, however, be considered when evaluating these statements. Many of these authors, and especially Baumel (1958), have also reported intra-specific variations. Asymmetry of the left and right sides of the plexus is often as evident as variation between individuals of a species (Gadow, 1891; Fisher and Goodman, 1955; Baumel, 1958; Waluszewska-Bubień and Pospieszny, 1982a), but in general such differences are minor (Baumel, 1958); examples are variations in the level at which the roots of the plexus unite or the trunks diverge, and differences in the levels of origin of branches or in their pattern of distribution.

Despite the extreme variability of the vertebral column and all the uncertainties as to whether the roots of the plexus belong to cervical or thoracic spinal nerves, the basic pattern of the roots seems to be surprisingly regular in birds generally.

2. *The trunks of the brachial plexus*

Having given short motor branches directly to the cervical muscles, the roots then unite to form trunks (*trunci plexus*), which are short and more-or-less difficult to define. There appear to be two trunks in *Grus americana* (Fisher and Goodman, 1955), *Progne subis* (Baumel, 1958), and *Otis tarda* (Waluszewska-Bubień and Pospieszny, 1982a) (arrows in Figs. 3.9, 10); in *Gallus* (Baumel, 1975) and *Columba livia* (Baumel, cited in Breazile and Yasuda, 1979) there seem to be three (arrows in Figs. 3.7, 8). Each trunk tends to divide into dorsal and ventral divisions. The trunks of the plexus and their dorsal and ventral divisions lie medial to the cranial extremity of the scapula, being closely related topographically to the jugular vein and carotid artery. Baumel's observations (1958) indicate that on the whole the cranial roots appear to supply more dorsal than ventral fibres, and vice versa. According to Harris (1904, 1939), this is a general characteristic of the brachial plexuses of the higher vertebrates. It has been suggested that the ventral and dorsal divisions are likely to be homologous with the medial and lateral branches, respectively, of the intercostal nerves (Patterson, 1887; Durward, 1951). It should be noted that here, and in the subsequent account of the peripheral nerves, the wing is regarded as being in the extended position so that its surfaces are dorsal and ventral (rather than lateral and medial as in the folded position), this conforming to the *Nomina Anatomica Avium*.

The dorsal divisions which arise from the more cranial trunk or trunks recombine with the dorsal divisions of the more caudally-situated trunk or trunks to produce a dorsal fascicle (*fasciculus dorsalis*); likewise the ventral divisions unite to form a ventral fascicle (*fasciculus ventralis*). The peripheral nerves which arise from the dorsal fascicle supply principally the musculature and the integument of the dorsal or extensor surface of the wing; the nerves formed by the ventral fascicle innervate the ventral or flexor aspect of the wing. The quite difficult distinctions between roots, trunks and fasciculi are analysed by Baumel (1975, p. 2038).

3. Localisation of motoneurons within the spinal cord

The retrograde degeneration experiments of Ohmori *et al.* (1982) have localised within the spinal cord the motoneurons of the wing muscles of *Gallus* (Fig. 3.11). All of them are sited in lamina 9 of the ventral horn. This lamina occupies the most ventrolateral aspect of the ventral horn and consists of several groups of nerve cells forming elongated craniocaudal columns. The lamina begins to appear at the cranial end of segment XIII of the spinal cord, increases progressively in size reaching a maximum at segments XIV and XV of the spinal cord, and then decreases gradually to fade out in the cranial region of segment XVII. There is one other column of nerve cells in the ventral horn, lamina 8, which is placed medial to lamina 9. No motoneurons of the wing are situated in lamina 8. The detailed analysis by Ohmori *et al.* (1982) extends even to the localisation of individual muscles. The muscles in the proximal region of the limb (e.g. muscles acting on the shoulder joint, and also the triceps and biceps muscles which act on both the shoulder and elbow joints) are mainly innervated by motoneurons in the more cranial segments, especially XIII–XV; the motoneurons of the muscles of the elbow and more distal joints are sited mainly in the more caudal segments, particularly XV and XVI (Fig. 3.11a). Ohmori *et al.* also

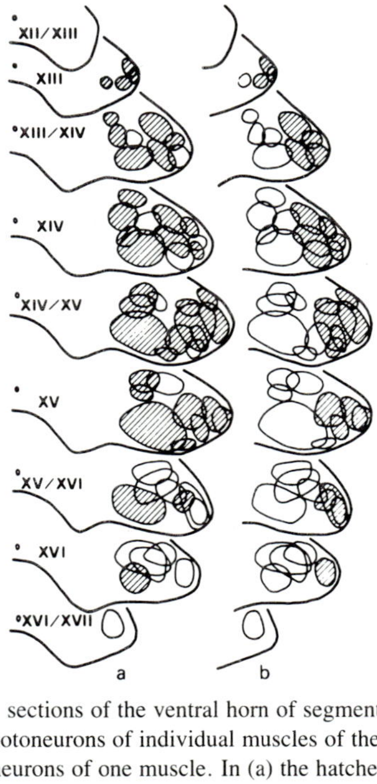

Fig. 3.11. Diagrammatic transverse sections of the ventral horn of segments XII–XVII in *Gallus,* showing the sites within lamina 9 wherein the motoneurons of individual muscles of the wing are sited. Each encircled area represents the location of the motoneurons of one muscle. In (a) the hatched areas belong to muscles which act on the shoulder joint, and also to *m. triceps brachii* and *m. biceps brachii,* which act on both the shoulder and the elbow joint; the motoneurons of these muscles tend to lie in the more cranial segments of the ventral horn. The open areas represent muscles which act on the elbow and even more distal joints; these motoneurons tend to be located in the more caudal segments. In (b) the hatched areas, which are mostly lateral in position, represent motoneurons which contribute to the *dorsal* fascicle of the brachial plexus. The open areas, mainly located medially, are the sites of motoneurons which contribute to the ventral fascicle of the plexus. These results were obtained by retrograde degeneration experiments. From Ohmori *et al.* (1982), by courtesy of the authors and Paul Parey Verlag, Berlin.

showed that the motoneurons of the radial nerve, and hence of the great dorsal fascicle, are placed in the lateral part of the whole length of lamina 9 (Fig. 3.11b); those of the medianoulnar nerve, and therefore of the great ventral fascicle, are located all along the medial region of lamina 9 (Fig. 3.11b). These latter observations on the sources of the dorsal and ventral fibres of the plexus appear to argue against Baumel's suggestion (p. 169) that the cranial roots of the plexus appear to supply predominantly dorsal fibres, and vice versa.

B. Accessory brachial plexus

The concept of an accessory brachial plexus (*plexus brachialis accessorius*), as opposed to the main brachial plexus, was introduced by Fürbringer (1879) under the mame Nebenplexus. It refers to a network of small nerves which arises from the ventral rami of the two, three or four spinal nerves that form the cranial roots of the main brachial plexus. In Figs. 3.7–10 it is shaded black. In some birds (*Anas, Buteo* and *Garrulus*) the most caudal roots of the main plexus also contribute (Gadow, 1891). The detailed form of the network is extremely variable (Gadow, 1891; Baumel, 1958), even though the spinal nerves forming it tend to be rather constant (Baumel, 1958). Apparently in some species the accessory plexus may be absent altogether, as in *Struthio* and *Phalacrocorax* (Gadow, 1891).

The accessory plexus and the peripheral nerves which it forms lie dorsal to the main plexus (Figs. 3.7–10). This topographical location can be correlated with the distribution of its branches to the group of dorsally-placed muscles that originate mainly on the axial skeleton and insert on the scapula, that is, the *mm. rhomboidalis superficialis* and *thomboidalis profundus* and the *mm. serrati.* These muscles appear to be mainly involved in stabilising the scapula, whilst other muscles which arise from the scapula and insert distally are contributing to the locomotory movements of the wing (Vanden Berge, 1975). The *serratus "dorsalis" muscle* has furthermore been shown electromyographically by Kadono *et al.* (1963) to be also an inspiratory muscle in *Gallus.* Branches also supply the adjacent skin via the cutaneous scapular nerve (*n. cutaneus omalis*) (Fig. 3.8).

C. Peripheral nerves arising from the brachial plexus

The following survey of the peripheral nerves which arise from the main plexus is based mainly on studies of *Progne subis* (Baumel, 1958), *Gallus* (Buchholz, 1959–1960; Yasuda, 1960b; Baumel, 1975), *Columba* (Baumel, cited in Breazile and and Yasuda, 1979) and *Otis tarda* (Waluszewska-Bubień and Pospieszny, 1982a), and the general works of Fürbringer (1879, 1888), Gadow (1891) and Breazile and Yasuda (1979). Of these accounts that of Baumel (1975), albeit confined to a single species (*Gallus*), is so detailed and clear that it constitutes the principal source. For a survey of the actions of the muscles innervated by the nerves of the brachial plexus, reference should be made to Section II,B of Chapter 2, by R. J, Raikow, in this volume.

In essence, the nerves can be divided into ventral and dorsal components, and this classification is adopted throughout the following account.

1. Nerves of the shoulder region

(a) Ventral components. The massive pectoral nerves, controlling the all-important muscles of the downstroke of the wing, arise either from the ventral divisions of the most caudal trunk or trunks (Figs. 3.8, 10) or from the ventral fascicle itself (Figs. 3.7, 9). The pectoral nerve usually forms cranial and caudal pectoral nerves (*n. pectoralis cranialis, n. pectoralis caudalis*) which supply muscular rami to the related components of the pectoral muscle. The cranial pectoral nerve also has branches which innervate the integument of the pectoral region (*nn. cutanei pectorales*); the caudal pectoral nerve forms branches (*nn. cutanei abdominales*) which innervate an area of abdominal skin (Breazile and Yasuda, 1979). The supracoracoid nerve (*n. supracoracoideus*), also important for flight because of its control of the upstroke of the wing by the supracoracoid muscle, arises from the cranial aspect of the most cranial trunk of the main plexus, generally from the ventral division of that trunk (Figs. 3.7–10). The nerves to the sternocoracoid muscle (*n. m. sternocoracoideus*) and the nerves to the coracobrachial muscles (*nn. mm. coracobrachiales*) arise either from the ventral division of the most cranial trunk or from the ventral fascicle itself (Figs. 3.7, 8, 10), or from the supracoracoid nerve (Fig. 3.9).

(b) Dorsal components. The subcoracoscapular nerve (*n. subcoracoscapularis*) and the subscapular nerve (*n. subscapularis*) innervate the subcoracoid and subscapular muscles. These nerves and the nerve to the scapulohumeral muscle (*n. m. scapulohumeralis*) generally arise dorsally and cranially, either from the dorsal division of the cranial trunk of the plexus or from the dorsal fascicle itself (all these nerves are shown in Fig. 3.10, and some of them in Figs. 3.7–3.9).

2. The ventral fascicle and its principal nerves

The nerves which arise from the ventral fascicle supply the flexor muscles of the wing and the integument of the ventral aspect of the wing (i.e. the underside of the extended wing). They are mainly derivatives of the medianoulnar nerve, which is the direct continuation of the great ventral fascicle of the plexus (Figs. 3.7–10).

(a) Branches in the brachial region. The ventral brachial cutaneous nerve (*n. cutaneus brachialis ventralis*) generally arises from the ventral fascicle (Figs. 3.7–10) and is distributed to the skin of the shoulder region and to the ventral surface of the brachial part of the propatagium. In *Progne subis* and *Otis tarda* not only the ventral but also the dorsal brachial cutaneous nerve (*n. cutaneus brachialis dorsalis*) appear to arise from the ventral fascicle or medianoulnar nerve (Figs. 3.9, 10), but this nerve is typically considered to be a branch of the radial nerve (Breazile and Yasuda, 1979). According to Gadow (1891) it is distributed to the skin on the dorsal surface of the brachium and metapatagium, overlapping the sensory field of the cutaneous axillary nerve, and it also innervates the smooth muscle of the *expansor secundariorum*.* The bicipital nerve (*n.*

*Baumel (1975) and Vanden Berge (1975) state that the *expansor secundariorum* is supplied by the anconeal nerve; see p. 175.

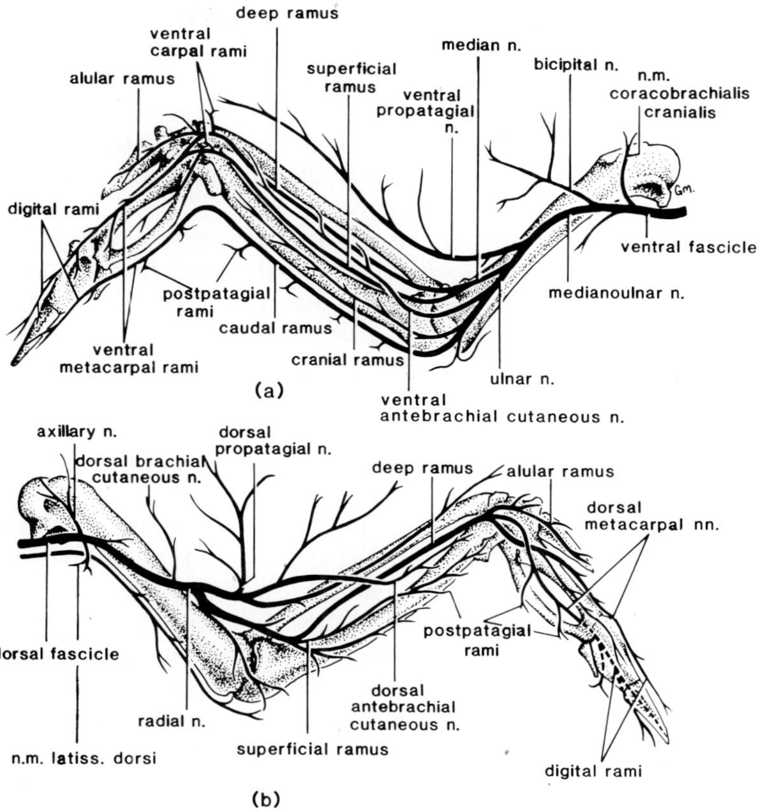

Fig. 3.12. Diagram of the innervation of the right wing in *Columba livia*. (a) Ventral view; (b) dorsal view. latiss. dorsi, *latissimus dorsi*. Redrawn from J. J. Baumel (cited in Breazile and Yasuda, 1979), by courtesy of the author.

bicipitalis) is a substantial branch arising from the ventral fascicle or medianoulnar nerve (Figs. 3.7, 8, 10, 12a). It innervates the *m. biceps brachii,* including its *pars propatagialis*; the *m. biceps brachii* is the main flexor of the elbow joint (see Raikow, Chapter 2 of this volume). This nerve also distributes cutaneous fibres ventrally towards the leading edge of the propatagium (*rami propatagiales*). As it approaches the elbow (Fig. 3.12a), the medianoulnar nerve forms the ventral propatagial nerve (*n. propatagialis ventralis*), which distributes cutaneous branches to the skin on the ventral surface of the lateral region of the propatagium; this nerve is often a branch of the median nerve (Breazile and Yasuda, 1979). Just proximal to the elbow joint, the medianoulnar nerve divides into the ulnar and median nerves (Fig. 3.12a).

(b) Ulnar nerve. The ulnar nerve (*n. ulnaris*) supplies motor nerve fibres to the flexor musculature of the carpus and to some of the muscles of the manus; it also distributes sensory fibres to the ventral aspect of the follicles and adjacent skin of the primary and secondary flight feathers and to the joints of the elbow, wrist and hand (Baumel, 1975). Just above the elbow the ulnar nerve forms the cubital cutaneous nerve (*n. cutaneus cubiti*), and then gives off muscular rami to the *m. flexor carpi ulnaris* (Baumel, 1975). At the elbow the ulnar nerve divides into its

cranial and caudal rami. The cranial ramus (*r. cranialis*), which is the smaller of the two, forms no branches in the forearm and ends in three metacarpal rami which innervate the skin and joints of the manus. The caudal ramus (*r. caudalis*) gives off postpatagial rami (*rr. postpatagiales*) and culminates in ventral metacarpal rami (*rr. metacarpales ventrales*) and digital rami (*rr. digitales*) (Fig. 3.12a). The latter contribute to the sensory innervation of the skin and joints of the manus.

There are differing accounts of the motor innervation of the muscles of the forearm and manus by the ulnar nerve. For example, according to Baumel (1975), in *Gallus* the main trunk of the ulnar nerve supplies the *m. flexor carpi ulnaris*, and in the manus the cranial ramus innervates the *mm. interosseous ventralis* and *flexor digiti minoris*. Following Buchholz (1959–1960), Vanden Berge (1975) agreed about the *mm. interosseus ventralis* and *flexor digiti*; however, he allocated the *m. flexor carpi ulnaris* to the median nerve. Yasuda (1960b) attributed only the "m. flexor digiti quarti longus," "m. expansor primarium" and the m. entepicondylo-ulnaris to the ulnar nerve, and again allotted the *m. flexor carpi ulnaris* to the median nerve. According to Gadow (1891), the ulnar nerve innervates the *m. flexor carpi ulnaris*, and also the *mm. flexor digitorum superficialis, interosseus dorsalis, abductor digiti majoris, flexor alulae* and *abductor alulae*.

(c) Median nerve. The majority of the flexor (ventral) muscles of the carpus and digits, and several muscles of the manus, are innervated by the median nerve (*n. medianus*); it also supplies sensory fibres to the skin of the ventral aspect of the forearm and manus including the ventral surface of the propatagium near the carpus, and to the ventral aspects of the joints of the elbow, carpus and digits (Baumel, 1975). Just proximal to the elbow it sends motor branches to the *m. brachialis* and the *mm. pronator superficialis* and *pronator profundus* (Yasuda, 1960b; Baumel, 1975). Here it also forms the ventral propatagial nerve (*n. propatagialis ventralis*), if this nerve has not already arisen from the medianoulnar nerve. At the elbow the median nerve divides into superficial and deep rami. According to Yasuda (1960b) and Baumel (1975), the smaller superficial ramus (*r. superficialis*) innervates the *mm. flexor digitorum superficialis* and *flexor digitorum profundus*. The deep ramus (*r. profundus*) runs along the caudal border of the radius (Fig. 3.12a), close to the deep ramus of the radial nerve; at this level the deep ramus of the median nerve supplies the *m. ulnometacarpalis ventralis* (Baumel, 1975; Vanden Berge, 1975). This is not mentioned by Yasuda (1960b). In the manus the deep ramus also innervates the *mm. abductor digiti majoris* and *flexor alulae* (Baumel, 1975; Vanden Berge, 1975). Yasuda (1960b) noted in principle a distribution to intrinsic muscles of the manus. The deep ramus finally forms an alular ramus (*r. alularis*), ventral carpal rami (*rr. carpales ventrales*), ventral metacarpal nerves (*nn. metacarpales ventrales*), postpatagial rami (*rr. postpatagiales*) and digital rami (*rr. digitales*) (Fig. 3.12a). These terminal branches distribute sensory fibres to the joints and skin of the manus (Baumel, 1975).

3. The dorsal fascicle and its principal nerves

The motor components of these nerves elevate the humerus (deltoid muscles and latissimus dorsi muscles; Vanden Berge, 1975) and extend the elbow, carpal and digital joints. Their sensory components innervate the integument of the dorsal surface of the wing. Most of these nerves arise from the radial nerve itself, which is the direct continuation of the dorsal fascicle (Fig. 3.12b).

(a) Branches in the brachial region. The axillary nerve (*n. axillaris*) is the largest branch of the dorsal fascicle (Figs. 3.7–10, 12b). Its function is to supply muscular, articular and cutaneous branches to the dorsal shoulder region (Baumel, 1975). After leaving the axilla it innervates the *mm. deltoideus major* and *deltoideus minor* and distributes some of its terminal branches in the *m. tensor propatagialis*. Its sensory fibres include articular branches to the shoulder joint and the large cutaneous axillary nerve (*n. cutaneous axillaris*). The latter innervates the skin of the dorsal region of the shoulder and proximal region of the arm (Baumel, 1975) and the adjacent part of the propatagium (Ribbing, 1938). The anconeal nerve (*n. anconealis*) generally arises from the dorsal fascicle (Figs. 3.7, 8) and innervates the *m. triceps brachii* and the smooth muscle of the *expansor secundariorum* (Baumel, 1975). The dorsal fascicle also innervates the *mm. triceps brachii* (*mm. humerotriceps, scapulotriceps* in Fig. 3.8, and *m. triceps* in Figs. 3.7, 3.10); it also distributes cutaneous rami to the caudodorsal skin of the distal region of the arm and the elbow, as well as to the follicles of the proximal secondary flight feathers (Baumel, 1975). The *mm. triceps brachii* are the extensors of the elbow joint (Raikow, Chapter 2, of this volume). The dorsal fascicle furthermore supplies the latissimus dorsi muscle (*n. m. latissimus dorsi*) (Figs. 3.7–9 12b).

(b) Radial nerve. The radial nerve (*n. radialis*) (Figs. 3.7–10, 12b) is the motor nerve of the extensor muscles of the carpal and digital joints. Its cutaneous branches innervate the integument of the dorsal aspect of the propatagium, forearm and manus, including the dorsal aspect of the follicles of the primary and secondary flight feathers. In the proximal part of its course the radial nerve spirals round the humerus (Fig. 3.12b). Fürbringer noted that the deltoid muscle covers the radial nerve to a varying extent in different species, and this was considered to be of taxonomic importance (Gadow, 1891). In the region of the elbow (Fig. 3.12b) the radial nerve forms the large dorsal propatagial nerve (*n. propatagialis dorsalis*) which supplies the dorsal surface of the propatagium. The dorsal antebrachial cutaneous nerve (*n. cutaneus antebrachialis dorsalis*) arises from the radial nerve (Fig. 3.8) or from the dorsal propatagial nerve (Fig. 3.12b) and supplies the skin of the dorsal and caudal aspect of the forearm (Baumel, 1975). At the distal end of the humerus the radial nerve innervates the *m. extensor metacarpi radialis* and sends articular branches to the dorsal aspect of the elbow joint. Just distal to the elbow joint it divides into its superficial and deep rami (Fig. 3.12b). The relatively small superficial ramus (*r. superficialis*) innervates the *m. extensor digitorum communis, m. extensor metacarpi ulnaris,* and *m. ectepicondylo-ulnaris,* and ends in postpatagial rami (*rr. postpatagiales*) to the follicles and skin associated with the secondary flight feathers (Baumel, 1975). The deep ramus (*r. profundus*) runs along the caudal border of the radius (Fig. 3.12b). It innervates the *m. ulnometacarpalis dorsalis, mm. interosseus dorsalis* and *interosseus ventralis, m. extensor brevis alulae,* and *m. adductor alulae* (Baumel, 1975; Vanden Berge, 1975). Finally, the deep ramus ends by forming an alular ramus (*r. alularis*), and dorsal metacarpal nerves (*nn. metacarpales dorsales*) with their postpatagial rami (*rr. postpatagiales*) and digital rami (*rr.*

digitales) (Fig. 3.12b); these contribute nerve fibres to the skin and follicles associated with the primary flight feathers and to the digital joints. The cutaneous branches of the radial and ulnar nerves which supply the follicles of the primary and secondary flight feathers are presumed to contain afferent nerve fibres from proprioceptive nerve endings and efferent fibres which are motor to the smooth muscles (*mm. pennales*) which move the feathers (Baumel, 1975).

VII. The pelvic limb

The peripheral nerves of the pelvic limb arise from the lumbosacral plexus, which is generally considered, as in other vertebrates (Boas, 1933), to comprise two subsidiary plexuses, the lumbar plexus and the sacral plexus. Earlier writers on the avian lumbosacral plexus (Gadow, 1891; Boas, 1933) were largely preoccupied with the problem of homologising the synsacral vertebrae of birds with those of other vertebrates, especially reptiles. However, the anatomical account of the lumbosacral plexus by Boas gives clear details of many avian species, and was judged by Breazile and Yasuda (1979) to be "the most definitive work available on the description of this plexus." The following review is based on their survey of numerous species representing Rheiformes, Casuariiformes, Tinamiformes, Ciconiiformes, Anseriformes, Charadriiformes, Falconiformes, Galliformes, Gruiformes, Strigiformes and Oscines.

A. The roots of the lumbosacral plexus

One basic type of lumbosacral plexus (Fig. 3.13) prevails throughout these orders (Boas, 1933). The ventral rami of eight spinal nerves form the roots of the plexus. The first of these is the second synsacral nerve, arising between the second and third synsacral vertebrae (Fig. 3.13). This and the next two nerves (i.e. synsacral nerves 2–4) form the lumbar plexus (*plexus lumbaris*), which gives rise to two major peripheral nerves of the pelvic limb, that is, the femoral and obturator nerves. The sacral plexus (*plexus sacralis*) arises from the ventral rami of six spinal nerves. The most cranial of these (synsacral nerve 4) is also the most caudal root of the lumbar plexus; because this nerve divides to contribute one root to the lumbar and one to the sacral plexus it has been named the "furcal" nerve (Gadow, 1891; Baumel, 1975). The most caudal root of all, which is generally very thin, is synsacral nerve 9 (Fig. 3.13). The ischiadic nerve is the main peripheral nerve to arise from the sacral plexus.

The most substantial modifications of this basic pattern of roots occur at the cranial and caudal ends of the series. Variants observed by Boas include the following:

Variant 1: In individuals of some species (e.g. *Fulica atra, Porphyrio, Gallinula chloropus*) there were nine roots (although other individuals of the same species had the typical number). The extra root arose at the cranial end of the series, so that the lumbar plexus now had four roots, the roots of the sacral plexus being unchanged (Fig. 3.14). This variant was found in *Grus americana* by Fisher and Goodman (1955).

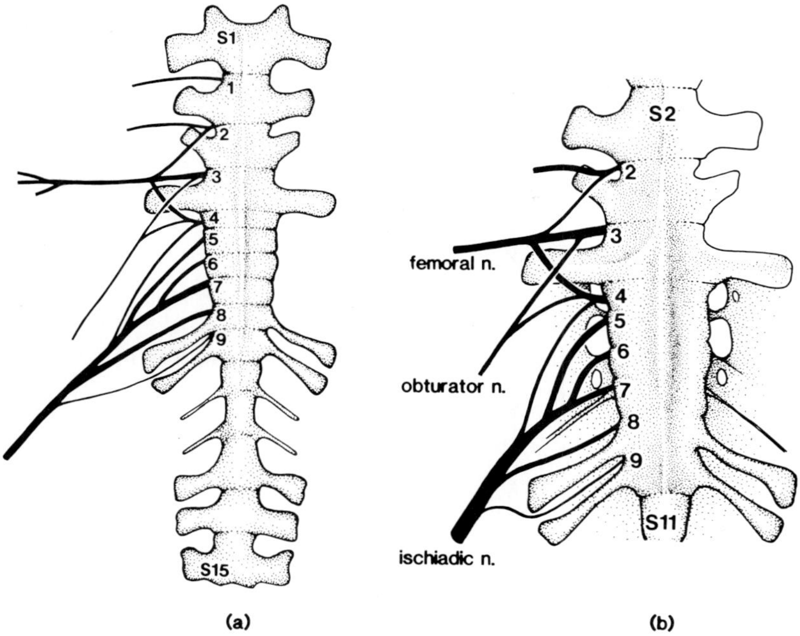

Fig. 3.13. Ventral view diagrams of the basic pattern of the roots of the lumbosacral plexus in birds generally, as established by Boas (1933) from observations on representatives of 11 orders. (a) The whole synsacrum and the lumbosacral plexus; (b) enlarged view of the same plexus. The ventral rami of the first nine synsacral spinal nerves are numbered 1–9. The lumbar plexus has three roots formed by synsacral nerves 2, 3 and 4, and the sacral plexus has six roots formed by nerves 4–9. By forking, nerve 4 contributes a root to both plexuses and is therefore known as the "furcal nerve." S1 is the first and S15 the last synsacral vertebra, but the total number of synsacral vertebrae is not consistent in birds generally. In the great majority of the 159 species examined by Boas, the total ranged between 13 and 17; higher numbers up to 22 were observed in ratites and anseriforms, and lower numbers down to 10 were found in many oscines. S2 and S11 are synsacral vertebrae 2 and 11.

Variant 2: In *Rhea* (Fig. 3.15) 10 roots were observed, there being 2 additional roots at the cranial end of the lumbar plexus, making a total of 5 for that plexus. *Dromaius* (Fig. 3.16) was similar. The extra roots, especially in *Dromaius,* were massive (Fig. 3.16), as might be expected from the great size of the spinal nerves in the region of the lumbar enlargement of the spinal cord of ratites (Fig. 3.2, and p. 153). In a drawing of another specimen of *Rhea,* Boas showed 11 roots, 2 extra roots being at the cranial end of the plexus (giving 5 roots to the lumbar plexus) and 1 extra root at the caudal end (giving 7 roots to the sacral plexus).

Variant 3: The lumbosacral plexus still had eight roots, but the lumbar plexus had an extra root at its cranial end and the sacral plexus lost one at the caudal end, as in the Common Buzzard (*Buteo buteo*) (Fig. 3.17) and the Secretary Bird (*Sagittarius serpentarius*) (Fig. 3.18). Thus, in this variant the lumbar plexus had four roots and the sacral plexus had only five. Boas observed a few species in which there were only four sacral roots. As in the brachial plexus, the general pattern of the roots of the lumbosacral plexus seems to be rather more constant than might be expected from the variability of the vertebral column.

Minor variations reported by Boas (1933) in the mid-region of the lumbosacral

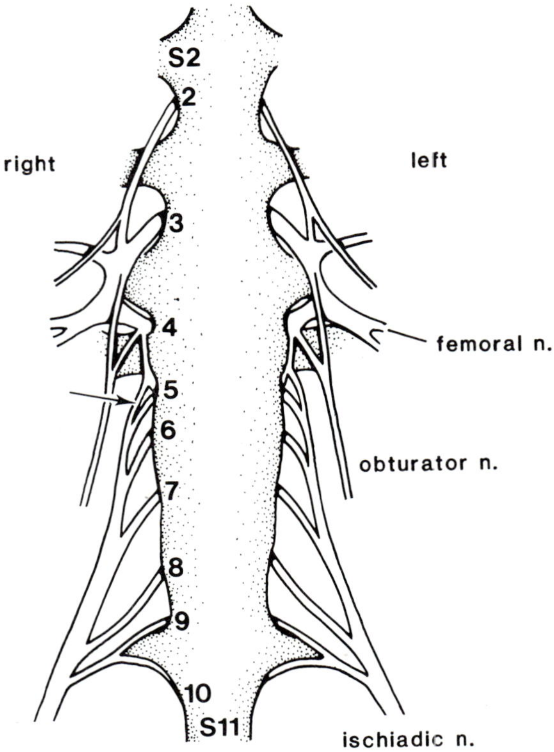

Fig. 3.14. Semidiagrammatic ventral view of the lumbosacral plexus in the Coot (*Fulica atra*). 2–10, Ventral rami of synsacral spinal nerves 2–10; S2 and S11, synsacral vertebrae 2 and 11; arrow, accessory root from synsacral nerve 4 to the sacral plexus. The lumbosacral plexus has nine roots (one more than the typical avian plexus in Fig. 3.13), the lumbar plexus having an extra root making four in all (synsacral nerves 2–5, inclusive). Synsacral nerve 5 is the furcal nerve. The sacral plexus has the typical number of (main) roots, that is, six (synsacral nerves 5–10, inclusive). Redrawn from Boas (1933).

plexus included asymmetry of the two sides (Fig. 3.18) and the presence of a slender accessory root to the sacral plexus (arrow in Figs. 3.14–18).

Further details have been mentioned by other authors. Gadow (1891) reported only two roots of the lumbar plexus in *Paradisaea, Gymnorhina, Prosthemadera, Fringilla, Ramphastos, Apus, Cacatua, Amazona, Turtur, Sula, Sterna* and *Mergus*. In some species, including *Rhea* (Fig. 3.15) and *Otis tarda,* the first root of the lumbar plexus may be formed by the first synsacral nerve (Gadow, 1891; Chomiak, 1950; Buchholz, 1959–1960; Waluszewska-Bubień and Pospieszny, 1982b). Typically, however, the ventral ramus of the first synsacral nerve is restricted to muscular and cutaneous rami to the abdominal wall and thigh (see nerve XXII, Fig. 3.4; also Fig. 3.19). Four roots of the lumbar plexus were mentioned by Gadow (1891) in *Bubo, Crax, Numida, Phasianus, Goura, Grus, Ciconia, Phoenicopterus, Uria* and *Rhea,* and by Waluszewska-Bubień and Pospieszny (1982b) in *Otis tarda*. In *Gallus* (Fig. 3.19) the lumbosacral plexus typically has the standard avian pattern of eight roots, the lumbar plexus being formed by three roots from synsacral nerves 2, 3, and 4—that is spinal nerves

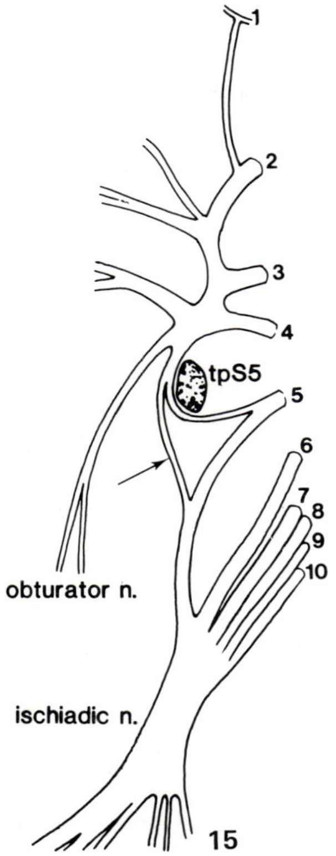

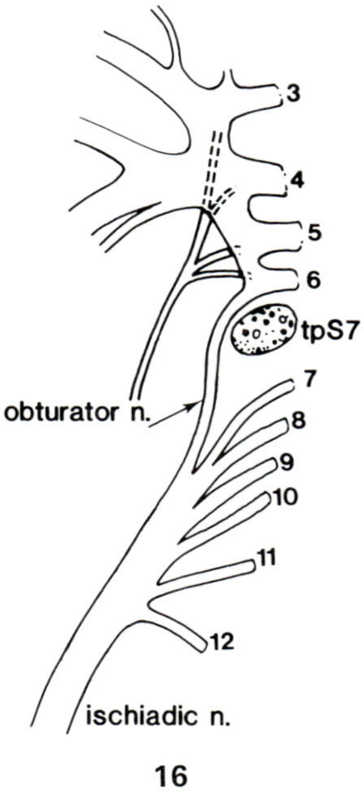

Fig. 3.15. Semidiagrammatic ventral view of the right lumbosacral plexus of *Rhea*. 1–10, Ventral rami of synsacral spinal nerves 1–10; tpS5, Lateral extremity of the transverse process of the fifth synsacral vertebra; arrow, accessory root from synsacral nerve 4 to the sacral plexus. The lumbosacral plexus has two additional roots, giving a total of 10. The two extra roots belong to the lumbar plexus, giving five altogether (synsacral nerves 1–5, inclusive), nerve 5 being the furcal nerve. The plexus is also unusual in that the *first* synsacral spinal nerve contributes a root. The sacral plexus has the typical number of six (main) roots (synsacral nerves 5–10, inclusive). Redrawn from Boas (1933).

Fig. 3.16. Semidiagrammatic ventral view of the right lumbosacral plexus of *Dromaius*. 3–12, Ventral rami of synsacral spinal nerves 3–12. tpS7, Lateral extremity of the transverse process of the seventh synsacral vertebra; arrow, accessory root from synsacral nerve 6 to the sacral plexus. The lumbosacral plexus has 10 roots—that is, two more than is usual—the two extra ones being at the cranial end of the lumbar plexus (synsacral nerves 3, 4). The lumbar plexus has four roots (synsacral nerves 3–6, inclusive). The sacral plexus has the typical number of six (main) roots (synsacral nerves 7–12, inclusive). All four roots of the lumbar plexus are relatively massive. Redrawn from Boas (1933).

XXIII–XXV (Baumel, 1975)—but four roots to the lumbar plexus have often been noted (Chomiak, 1950; Buchholz, 1959–1960). In Baumel's drawing of *Columba* (Fig. 3.20), the lumbosacral plexus has only seven roots, the sacral plexus being reduced by one (to five). According to Gadow (1891), in *Falco peregrinus* the number of roots to the sacral plexus is even further reduced (to four or even three). Gadow (1891) also confirmed the presence in some species of additional roots to the sacral plexus in *Ardea, Mergus, Scolopax, Rhea* and *Struthio,* totalling seven (as observed in one of Boas' specimens of *Rhea,* see p. 177), and in *Casuarius* there were eight roots in the sacral plexus.

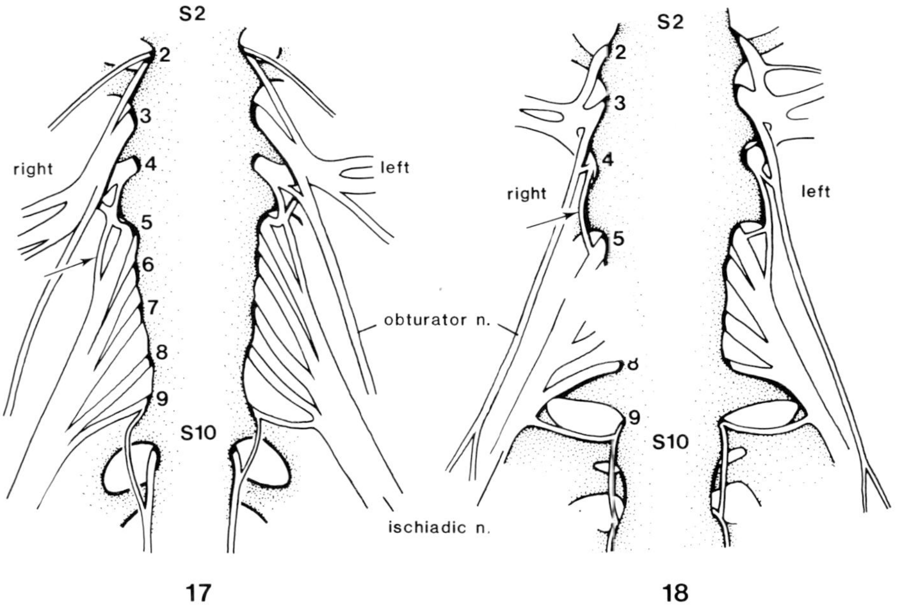

Figs. 3.17, 18. Semidiagrammatic ventral views of the lumbosacral plexuses of the Common Buzzard (*Buteo buteo*) and Secretary Bird (*Sagittarius serpentarius*), respectively. 2–9, Ventral rami of synsacral spinal nerves 2–9; S2 and S10, synsacral vertebrae 2 and 10; arrows, accessory root from synsacral nerve 4 to the sacral plexus. Nerve 5 is considered to be the furcal nerve. The lumbosacral plexus has the typical avian number of eight roots. However, the lumbar plexus has one extra root (synsacral nerve 2), giving it four in all, and the sacral plexus has one less root than usual, giving it a total of only five roots (excluding the accessory root). The middle of the lumbosacral plexus is slightly asymmetrical. Redrawn from Boas (1933).

B. The trunks of the lumbosacral plexus

The roots of the lumbar plexus tend to form two trunks (*truci plexus*), as in *Gallus* (Fig. 3.19). The two cranial roots may form a short cranial trunk; this supplies rami to the *m. iliotibialis cranialis (sartorius)* and gives off the lateral femoral cutaneous nerve (Baumel, 1975). The middle and caudal roots of the plexus may form a short caudal trunk which continues directly into the femoral nerve (Baumel, 1975). The trunks and their roots lie in contact with the dorsal surface of the cranial division of the kidney.

The roots of the sacral plexus tend to form three trunks (Baumel, 1975). A cranial trunk may arise from the fusion of the first three roots of the plexus (Figs. 3.14, 19), the first of these being the most caudal root of the lumbar plexus (the furcal nerve, p. 176). In a number of species this trunk is reinforced by the small accessory root from the cranial branch of the furcal nerve (Fig. 3.14), which was mentioned above (p. 178). A middle trunk may be formed by the independent fourth root of the plexus (Figs. 3.14, 19). A caudal trunk may be formed by the remaining two caudal roots (Figs. 3.14, 19, 21).

The aforementioned pattern of lumbosacral trunks holds for *Gallus* and *Fulica atra,* and perhaps for *Otis tarda* (Fig. 3.21), but appears not to fit the species

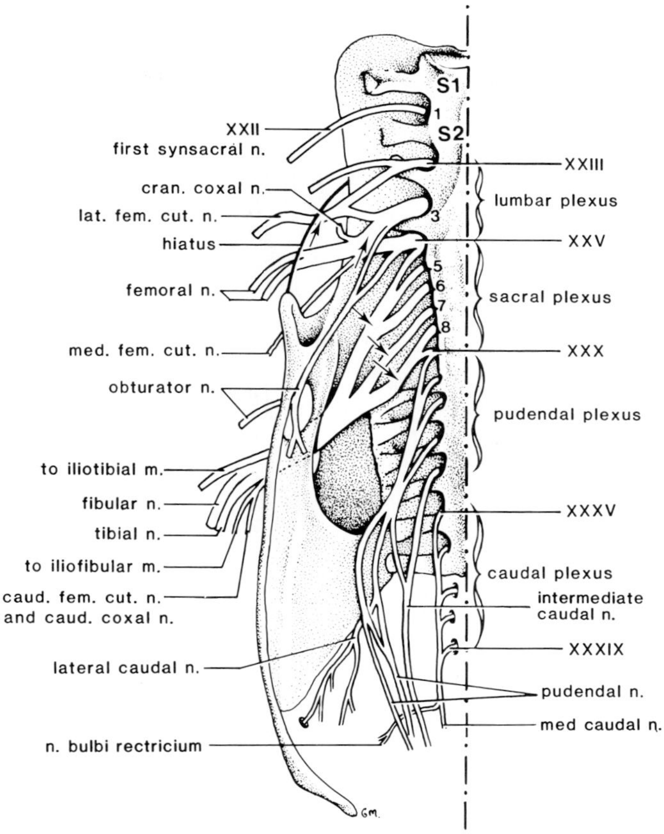

Fig. 3.19. Ventral view of the right lumbosacral, pudendal and caudal plexuses of *Gallus*. XXIII–XXXIX, Ventral rami of spinal nerves; arrows pointing cranially, trunks of lumbar plexus; arrows pointing caudally, trunks of sacral plexus; S1 and S2, first and second synsacral vertebrae (vertebrae 21 and 22 in this specimen); hiatus, aperture between the lateral border of the ilium and the abdominal muscles through which the peripheral nerves arising from the lumbar plexus escape from the pelvis. Nerves 1, 3 and 5–8 are the first to eighth synsacral nerves. The lumbosacral plexus of this species is typical of birds generally (compare Fig. 3.13). It has eight roots. The lumbar plexus has three roots and the sacral plexus six roots. Nerve XXV (the fourth synsacral nerve) divides to contribute a root to both the lumbar and the sacral plexus, and is therefore the "furcal nerve." The pudendal plexus has five roots (nerves XXX–XXXIV, inclusive). This is one more root than is typical; the extra root comes from nerve XXX, which is then termed the "bigeminal nerve," because it forks to give a root to both the sacral and the pudendal plexus. The caudal plexus is formed by five roots (nerves XXXV–XXXIX, inclusive). caud., Caudal; cran., cranial; cut., cutaneous; fem., femoral; lat., lateral; med., medial. Redrawn from J. J. Baumel (1975), by courtesy of the author and Saunders Co., Philadelphia.

shown in Figs. 3.15–18 and 20. Nor does it fit the pattern illustrated for *Grus americana* by Fisher and Goodman (1955). Horst (1934) stated that "the division into a dorsal and ventral component is more distinct in birds than in reptiles," thus indicating that dorsal and ventral divisions of the trunks should be recognisable in birds. Fisher and Goodman (1955) illustrated the bifurcation of the ischiadic nerve as forming a dorsal division (fibular nerve) and a ventral division (tibial nerve) in *Grus americana*. Further observations are required to show what patterns of trunks are applicable to birds in general.

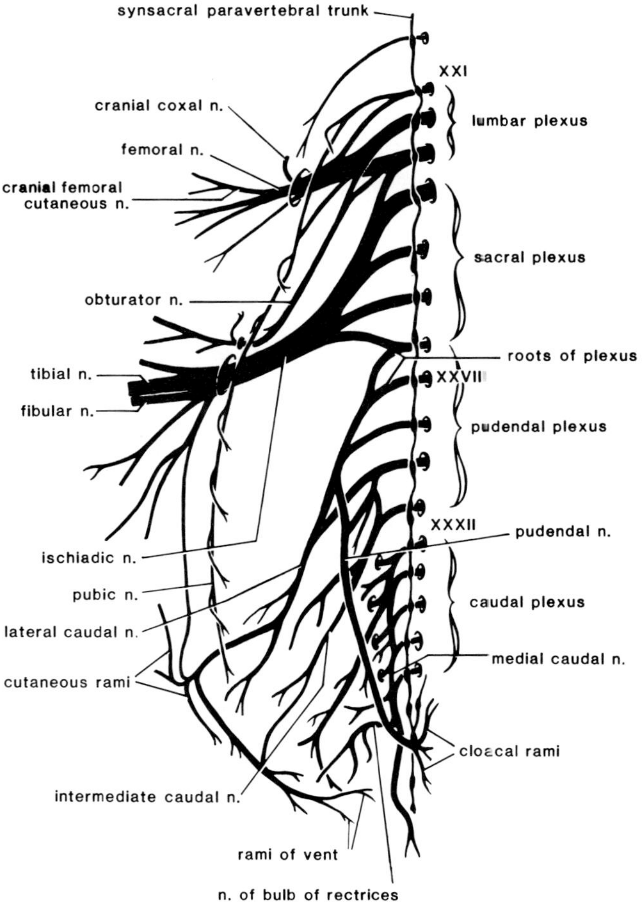

Fig. 3.20. Ventral view diagram of the right lumbosacral, pudendal and caudal plexuses of *Columba*. XXI–XXXII, Ventral rami of spinal nerves. The lumbosacral plexus in this specimen has one less root (seven altogether) than is usual in birds generally, the sacral plexus being reduced by one (to five roots). Nerve XXIII is the furcal nerve. The paravertebral ganglia are closely connected to the ventral roots of the synsacral nerves, especially in the more caudal synsacral segments. The pudendal plexus has five roots (from nerves XXVII–XXXI, inclusive), the first of these being the so-called bigeminal nerve supplying both the sacral and the pudendal plexus. The caudal plexus is also formed from five roots (from nerves XXXII–XXXVI, inclusive). Redrawn from J. J. Baumel (cited in Breazile and Yasuda, 1979), by courtesy of the author.

C. Localisation of motor neurons within the spinal cord

Ohmori *et al.* (1983) have succeeded in localising within the spinal cord the motoneurons of the muscles of the hindlimb of the domestic fowl. They are all situated in lamina 9 of the lumbosacral enlargement of the spinal cord, thus occupying most of the ventrolateral part of the ventral horn. The neurons which innervate the muscles of the hip joint and thigh region occur from segments XXIII–XXX, inclusive. In segments XXIII–XXV (which form the lumbar plexus, Fig. 3.19) they are distributed throughout the whole of lamina 9. In segments

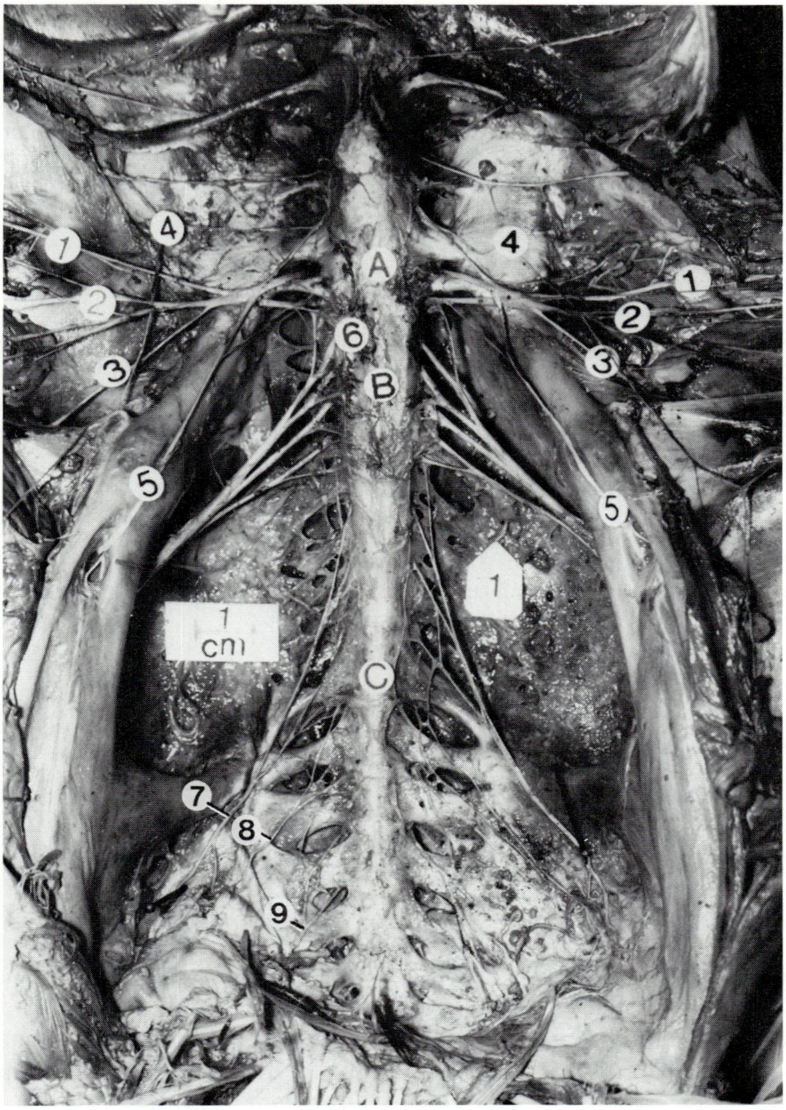

Fig. 3.21. Ventral view of the lumbosacral and pudendal plexuses of the Great Bustard (*Otis tarda*). A, Region of lumbar plexus; B, region of sacral plexus; C, region of pudendal plexus. 1, Lateral femoral cutaneous nerve; 2, femoral nerve; 3, medial femoral cutaneous nerve; 4, pubic nerve; 5, obturator nerve; 6, synsacral paravertebral trunk; 7, lateral caudal nerve; 8, intermediate caudal nerve; 9, medial caudal nerve. From Waluszewska-Bubień and Pospieszny (1982b).

XXVI–XXX, inclusive (which give rise to the sacral plexus, Fig. 3.19), they are confined to the dorsolateral regions of the lamina. Motoneurons innervating the muscles acting on the intertarsal and foot joints are restricted to segments XXVI–XXX, where they occupy the large majority of lamina 9, especially its ventromedial regions. The more ventral components of the lumbar and sacral plexuses arise from motoneurons in the medial parts of lamina 9, from segments XXIII–XXX, inclusive; the relatively dorsal components are derived from motoneurons in the lateral region of lamina 9, from segments XXIII–XXIX. These relationships between the ventral and dorsal parts of the plexuses on the one hand, and the medial and lateral regions of lamina 9 on the other, are the same as for the ventral and dorsal components of the brachial plexus (p. 171).

D. Peripheral nerves arising from the lumbosacral plexus

Substantial contributions to our knowledge of the peripheral nerves of the pelvic limb in a variety of avian species have been made by Gadow (1891), Chomiak (1950), Buchholz (1959–1960), Yasuda (1961, 1964), Baumel (1975), Breazile and Yasuda (1979) and Waluszewska-Bubień and Pospieszny (1982b). Of these, that by Baumel (1975) is again outstanding and forms the principal source in the following account, although it has the limitation of being confined to *Gallus*. Reference should be made to Section III,B of Chapter 2, by R. J. Raikow, in this volume for an account of the actions of the muscles of the pelvic limb which are innervated by the nerves arising from the lumbosacral plexus.

1. Nerves arising from the lumbar plexus

The principal branches of the lumbar plexus arise within the pelvis and escape at the lateral border of the ilium via a neurovascular hiatus bounded by the ilium and the abdominal muscles (Fig. 3.19). These nerves are destined mainly for the femoral region of the limb, and innervate no muscles distal to the knee joint; however, they do supply articular rami to the craniomedial aspect of the knee joint (*junctura genus*) and distribute cutaneous fibres as far distally as the region of the hock joint (*articulatio intertarsalis*) (Baumel, 1975).

(a) Lateral femoral cutaneous nerve. The *n. cutaneus femoralis lateralis* is a substantial nerve derived from the cranial trunk of the lumbar plexus, and hence from two plexus roots (Fig. 3.19). It supplies fibres to the integument of the craniolateral region of the thigh (Fig. 3.22), lateral region of the knee and proximal part of the leg (Baumel, 1975). Its cutaneous distribution includes the femoral feather tract (*pteryla femoralis*) (Lucas and Stettenheim, 1972). In the pelvic region it may give a connexion to the pubic nerve (Buchholz, 1959–1960). Motor fibres to the *m. iliotibialis cranialis (sartorius)* are sometimes mentioned in association with this nerve. It seems more likely that these motor filaments arise from the cranial trunk of the lumbar plexus (Baumel, 1975), or perhaps from the femoral nerve (Yasuda, 1961; Vanden Berge, 1975).

(b) Medial femoral cutaneous nerve. The caudal trunk of the lumbar plexus (and therefore two plexus roots) forms the *n. cutaneus femoralis medialis* (Figs. 3.19, 21). Baumel (1975) stated that in the region of the pelvis it innervates the *m. iliofemoralis internus (iliacus)*; according to Yasuda (1961) and Vanden Berge (1975), this muscle is suppled by the femoral nerve. After leaving the pelvis with the femoral artery and vein it passes to the medial aspect of the thigh (Fig. 3.22),

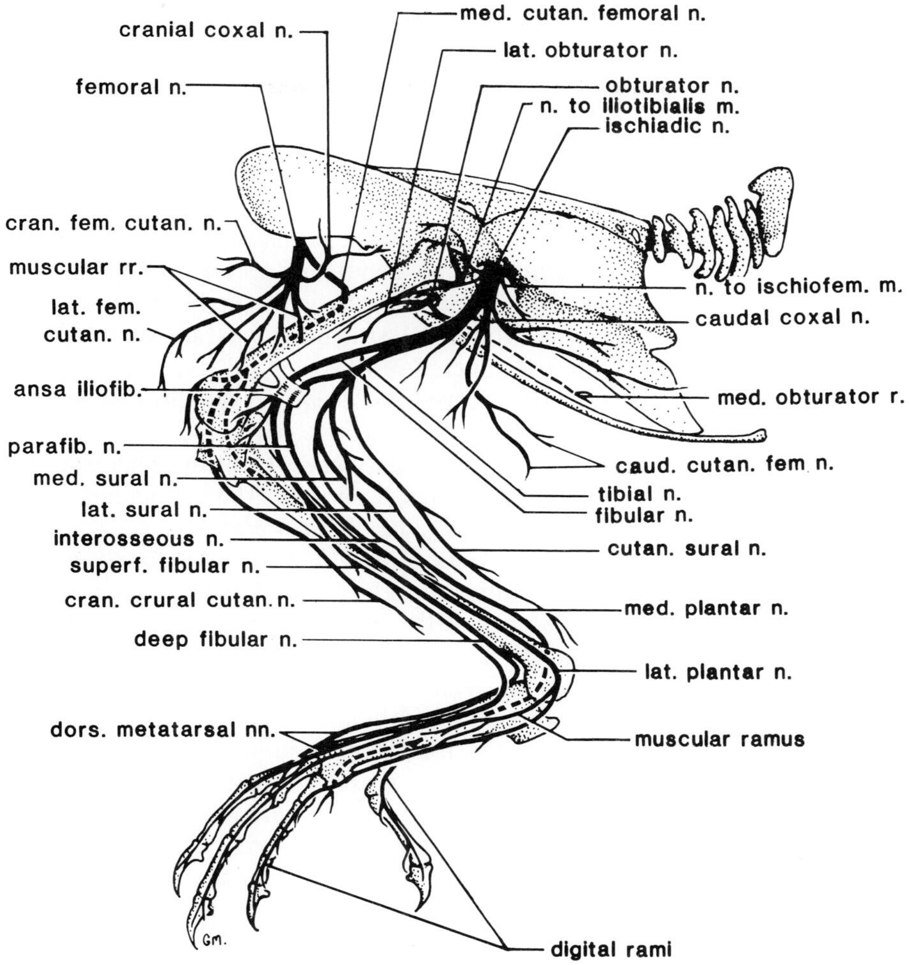

Fig. 3.22. Lateral view diagram of the nerves of the left pelvic limb of *Columba*. caud., Caudal; cran., cranial; cutan., cutaneous; dors., dorsal; fem., femoral; iliofib., *iliofibularis*; ischiofem., *ischiofemoralis*; lat., lateral; med., medial; parafib., parafibular; superf., superficial. Redrawn from J. J. Baumel (cited in Breazile and Yasuda, 1979), by courtesy of the author.

supplying the skin of the medial aspect of the knee and of the medial proximal part of the leg (tibial region). It continues along the cranial aspect of the leg as the cranial crural cutaneous nerve (*n. cutaneus cruralis cranialis*) (Fig. 3.22), which innervates the medial region of the extensive crural feather tract (Lucas and Stettenheim, 1972).

(c) Pubic nerve. The origin and distribution of the pubic nerve (*n. pubicus*) (Figs. 3.4, 20, 21) have been summarised on p. 161.

(d) Femoral nerve. The femoral nerve (*n. femoralis*) is the largest terminal branch of the lumbar plexus. In *Gallus* (Fig. 3.19) it is the direct continuation of the caudal trunk of the plexus and therefore of the second and third plexus roots.

In *Columba* (Fig. 3.20), and indeed in many other birds (e.g. Fig. 3.13), it also receives a contribution from the first root of the plexus. In the region of the hip joint it emits the cranial coxal nerve (*n. coxalis cranialis*) (Figs. 3.19, 20, 22), which innervates the muscle mass on the dorsal aspect of the preacetabular ala of the ilium. The femoral nerve also gives off the cranial femoral cutaneous nerve (*n. cutaneus femoralis cranialis*), which supplies the extensive area of skin on the cranial aspect of the thigh (Figs. 3.20, 22). The muscles innervated by the femoral nerve are the extensors of the knee joint, especially the *m. femorotibialis* and to a lesser extent the *iliotibialis muscles,* and the *m. ambiens* (Vanden Berge, 1975; Baumel, 1975). The *m. iliotibialis cranialis* is the main protractor of the femur (Raikow, Chapter 2 of this volume). Articular fibres are supplied to the patellar region of the knee joint (Baumel, 1975).

(e) Obturator nerve. The obturator nerve (*n. obturator*) typically arises directly from the caudal two roots of the lumbar plexus (Figs. 3.13, 14, 17–20). Usually the more caudal of these two roots is the furcal nerve. Contributions from three roots have been noted occasionally in *Gallus* (Chomiak, 1950; Buchholz, 1959–1960). Three roots have been reported in *Crax, Uria, Pelecanus, Cygnus, Rhea* and *Struthio,* and four in *Casuarius* (Gadow, 1891). Four were found in *Dromaius* (Fig. 3.16) by Boas (1933). Covered only by peritoneum, the nerve runs caudolaterally along the lateral border of the os coxae to the obturator foramen (Figs. 3.19, 21). Just cranial to the foramen it innervates the medial obturator muscle by the medial obturator ramus (*ramus obturatorius medialis*); this muscle appears to contribute to lateral rotation of the femur (Raikow, Chapter 2 of this volume). Within the foramen it sends the lateral obturator ramus (*ramus obturatorius lateralis*) to the lateral obturator muscle. After passing through the foramen it innervates the *m. pubo-ischio-femoralis* (adductor) (Yasuda, 1961; Vanden Berge, 1975; Baumel, 1975), which appears to be one of the main retractors of the femur (Raikow, Chapter 2 of this volume).

(f) Nerves to the iliotrochanteric muscles. One or two long nerves arise from the ventrolateral aspect of the lumbar plexus, mainly from the second root. Having traversed the lateral margin of the preacetabular ala of the ilium, these nerves escape from the pelvis and innervate the iliotrochanteric muscles. Some authors have considered these muscles to be unique to birds. On the other hand, others indicated them as homologues to the mammalian gluteal system (Vanden Berge, 1979); consequently, their nerves have sometimes been named gluteal nerves (Horst, 1934; Buchholz, 1959–1960; Lucas and Stettenheim, 1972; Schummer, 1973). However, the term gluteal is not appropriate, since the preacetabular ilium does not correspond at all to the mammalian gluteal region (Baumel, 1979b, p. 371). These muscles seem to contribute to inward rotation of the femur (Raikow, Chapter 2 of this volume).

2. *Nerves arising from the sacral plexus*

The sacral plexus is totally embedded within the tissue of the middle division of the kidney, in the cranial region of the renal fossa of the os coxae. The synsacral paravertebral trunk, which has exceptionally large segmental ganglia, runs over the ventral surface of the roots (Fig. 3.20). The roots, trunks and subsequent nerves of the sacral plexus lie more obliquely than those of the lumbar plexus (see Figs. 3.19, 21). The trunks of the plexus unite to form a very large compact bundle of nerves. At least in *Gallus* (Baumel, 1975), these nerves remain closely associated together, within a common epineurial sheath, until they leave the pelvis by passing through the ilioischiadic foramen with the ischiadic vessels. The principal component of this massive bundle is the ischiadic nerve (*n. ischiadicus*), which itself consists of the tibial and fibular nerves bound together. The main branches of the sacral plexus, all of which arise intra-pelvically, are the caudal coxal nerve, the caudal cutaneous femoral nerve, muscular rami, a connexus linking the sacral and pudendal plexuses, and the ischiadic nerve with its main derivatives, the tibial and fibular nerves.

(a) Caudal coxal nerve. The *n. coxalis caudalis* arises from the caudal aspect of the plexus (Fig. 3.19). It innervates the *m. caudo-ilio-femoralis* and also the *mm. flexor cruris lateralis* and *flexor cruris medialis,* when these are not supplied directly from the sacral plexus (Baumel, 1975).

(b) Caudal femoral cutaneous nerve. Being the main continuation of the caudal trunk of the plexus (Fig. 3.19), the caudal femoral cutaneous nerve (*n. cutaneus femoralis caudalis*) is formed by the most caudal roots of the plexus. It supplies the skin of the caudal region of the thigh (Fig. 3.22).

(c) Muscular rami. These include branches to the iliotibial (Fig. 3.22), iliofibular and iliotrochanteric muscles (Baumel, 1975), and also to the ischiofemoral muscle (Vanden Berge, 1975). The iliofibular muscle is one of the main flexors of the knee joint (Raikow, Chapter 2 of this volume).

(d) Connexus caudalis. This nerve connects the caudal coxal nerve of the sacral plexus to cutaneous rami of the lateral caudal nerve of the pudendal plexus (Baumel, 1975) (see also p. 190).

(e) Ischiadic nerve. The ischiadic nerve (*n. ischiadicus*) divides into the tibial and fibular nerves, which are adherent to each other but easily separated, at least in *Gallus,* after removal of the common epineurial sheath (Baumel, 1975). The ischiadic nerve is not only by far the largest nerve to arise from the sacral plexus, but is also the largest nerve in the avian body. In the thigh it supplies a cutaneous branch to the femoral feather tract (Lucas and Stettenheim, 1972). At levels of the thigh which vary in different birds it may give rise to the cutaneous sural nerve (*n. cutaneus suralis*), but this nerve may more typically be a branch of the tibial nerve as in Fig. 3.22 (Breazile and Yasuda, 1979). Proximal to the popliteal region the ischiadic nerve ends by separating into the tibial and fibular nerves (Fig. 3.22).

(f) Tibial nerve. The *n. tibialis* is the larger of the two terminal branches of the ischiadic nerve. The following survey of its distribution is based on Baumel's account of *Gallus* (1975). A short distance proximal to the knee joint the tibial nerve divides into the medial and lateral sural nerves, the medial nerve being the larger of the two (Fig. 3.22).

The medial sural nerve (*n. suralis medialis*) innervates the *mm. gastrocnemius, popliteus, flexor digitorum longus, flexor hallucis longus* and *plantaris* (Yasuda, 1961; Vanden Berge, 1975; Baumel, 1975). The gastrocnemius is the principal extensor of the hock joint; the digital flexors are involved in grasping (Raikow, Chapter 2 of this volume). Another branch of the medial sural nerve, the interosseous nerve (*n. interosseus*), runs parallel with the interosseous membrane between the tibia and fibula and may be associated with mechanoreceptors in the membrane (Baumel, 1975). Distal to the muscle bellies which the medial sural nerve has supplied, its main trunk continues into the medial plantar nerve (*n. plantaris medialis*) (Fig. 3.22). This latter nerve distributes cutaneous branches on the medial aspect of the hock joint, and ends in the metatarsal region, where it again supplies the integument.

The lateral sural nerve (*n. suralis lateralis*) innervates the gastrocnemius muscle, the *flexor perforans et perforatus muscle* of digits II and III, and the *flexor perforatus muscle* of digits II, III and IV (Yasuda, 1961; Vanden Berge, 1975; Baumel, 1975).

Proximal to the formation of the medial and lateral sural nerves, the main trunk of the tibial nerve gives off the parafibular nerve (*n. parafibularis*) (Fig. 3.22). This nerve accompanies the fibular nerve within a common epineural sheath at least in *Gallus,* and the two nerves pass together (Fig. 3.22) through the ligamentous pulley for the tendon of the iliofibularis muscle (*ansa m. iliofibularis*) (Baumel, 1975). In the distal part of the leg the parafibular nerve continues directly into the lateral plantar nerve (*n. plantaris lateralis*). The latter passes on the lateral aspect of the hock joint (Fig. 3.22), and distal to this level distributes muscular rami to the *mm. abductor digiti IV* (Yasuda, 1961; Vanden Berge, 1975; Baumel, 1975), *adductor digiti II* (Vanden Berge, 1975; Baumel, 1975) and *flexor hallucis brevis* (Yasuda, 1961; Vanden Berge, 1975; Baumel, 1975), and also to the *m. lumbricalis* (Vanden Berge, 1975; Baumel, 1975). The lateral plantar nerve distributes cutaneous fibres to the lateral aspect of the distal region of the leg, hock and tarsometatarsus, and to the metatarsal pad of the foot; it also supplies articular branches to the lateral side of the hock joint. Finally it continues into the lateral digital nerve of the fourth digit, supplying cutaneous fibres to the lateral aspect of the fourth toe.

(g) Fibular nerve. The *n. fibularis* passes through the ligamentous pulley for the *m. iliofibularis* (Fig. 3.22) within the same epineurial sheath as the parafibular nerve. It passes lateral to the neck of the fibula, and then ends halfway down the leg by dividing into the superficial and deep fibular nerves. The fibular nerve innervates the *mm. tibialis cranialis, extensor digitorum longus,* and *fibularis longus* (Yasuda, 1961; Vanden Berge, 1975; Baumel, 1975). The *m. tibialis cranialis* is the primary flexor of the hock joint, and is probably assisted by the *m. extensor digitorum longus,* which otherwise is primarily an extensor of the toes; the *m. fibularis longus* appears to extend the hock joint (Raikow, Chapter 2 of this volume).

The superficial fibular nerve (*n. fibularis superficialis*) descends over the dorsal aspect of the hock joint and continues into the lateral dorsal metatarsal and dorsal digital nerves (Fig. 3.22). The main trunk of the superficial fibular nerve reinforces the innervation of the *m. fibularis longus* (Yasuda, 1961; Baumel, 1975) and sends other muscular rami to the *extensor brevis digiti III* (Yasuda, 1961; Vanden Berge, 1975; Baumel, 1975) and to the *extensor brevis digiti IV* (Vanden Berge, 1975; Baumel, 1975). The sensory distribution of the superficial fibular nerve includes articular branches to the lateral aspect of the knee joint and hock joint (Baumel, 1975). The lateral dorsal metatarsal nerve supplies the podotheca of the tarsometatarsus; the digital rami innervate the adjacent sides of their toes, the metatarsal pad and the web between the bases of the toes (Baumel, 1975).

The deep fibular nerve (*n. fibularis profundus*) crosses the hock joint and proceeds distally into the medial dorsal metatarsal and dorsal digital nerves (Fig. 3.22). It innervates the *m. fibularis brevis* (Yasuda, 1961; Vanden Berge, 1975), the *m. extensor hallucis longus* (Vanden Berge, 1975; Baumel, 1975) and the *m. abductor digiti II* (Vanden Berge, 1975; Baumel, 1975). According to Yasuda (1961), this last-named muscle is innervated by a branch of the tibial nerve. The extrinsic and intrinsic extensor muscles of the digits, and the intrinsic adductors and abductors of the digits, are involved in releasing the grip of the toes (Raikow, Chapter 2 of this volume). The sensory field of the deep fibular nerve and its continuation includes articular fibres to the hock joint, and twigs to the metatarsal pad and toes (Baumel, 1975).

VIII. The tail

The peripheral nerves of the tail arise from the pudendal and caudal plexuses.

A. The roots of the pudendal plexus

The roots of the pudendal plexus (*plexus pudendus*) have apparently not been studied in a wide range of avian species. In *Gallus* (Baumel, 1975) the plexus is usually formed from the ventral rami of four spinal nerves, XXXI–XXXIV, inclusive (synsacral nerves 10–13). In some specimens there are five roots, since the last root of the sacral plexus (XXX) also contributes as in Fig. 3.19; this root therefore forks to supply both plexuses and is then known as the "bigeminal nerve" (XXX in Fig. 3.19). Some of the roots are embedded in the kidney.

There are typically five roots in *Columba* (Fig. 3.20), from spinal nerves XXVII to XXXI. In three specimens of the Great Bustard (*Otis tarda*), the number of roots varied from five to seven, ranging between spinal nerves XXXI and XXXVIII; nerve XXXI or XXXII formed the bigeminal nerve (Waluszewska-Bubień and Pospieszny, 1982b).

B. Peripheral nerves arising from the pudendal plexus

The somatic nerves arising from the pudendal plexus innervate the striated muscles of the tail and cloaca, and the skin in the region of the tail and vent. The actions of the tail muscles are discussed by Raikow in Section IV,D of Chapter 2 in this volume.

(a) Lateral caudal nerve. This nerve (*n. lateralis caudae*) arises from the cranial roots of the plexus (Figs. 3.19, 20). It accompanies the lateral caudal vessels and gradually enters the internal pelvic musculature. It innervates the muscle of the rectricial bulb, and the *mm. lateralis caudae, depressor caudae, pubocaudalis internus* and *transversus cloacae* (Vanden Berge, 1975; Baumel, 1975). Its cutaneous rami supply the integument of the ventrolateral aspect of the tail, the skin between the tail and the pubis, the ventral abdominal skin (Baumel, 1975) and the vent (Breazile and Yasuda, 1979).

(b) Intermediate caudal nerve. The intermediate caudal nerve (*n. intermedius caudae,* the caudal cutaneous nerve of Baumel, 1975) arises from the caudal two roots of the plexus (Figs. 3.19, 20). It may take the form of two or three separate

but intercommunicating and parallel filaments (Baumel, 1975). It penetrates and innervates the *m. pubocaudalis externus* (Vanden Berge, 1975; Baumel, 1975). It also innervates the sphincter cloacae muscle and distributes cutaneous branches to the skin of the dorsal lip of the vent, ventral surface of the tail (Baumel, 1975), and the ventral region of the vent (Yasuda, 1960c).

(c) Pudendal nerve. The first two or three roots of the pudendal plexus (Figs. 3.19, 20) give rise to the *n. pudendus* (Baumel, 1975), which accompanies the ureter to the dorsolateral aspect of the cloaca (Baumel, 1975). Here it forms several cloacal ganglia which connect with the rectal ganglia (Watanabe, 1972), to constitute the cloacal plexus (Baumel, 1975). The visceral distribution of the pudendal nerve emanates from the cloacal plexus in the form of (a) ureteral, oviductal and deferential rami which innervate the terminal parts of the urogenital ducts, and (b) cloacal rami which supply the phallus, paracloacal vascular body and all parts of the coprodeum, urodeum and proctodeum (Baumel, 1975; see also Chapter 2, by A. S. King, in Vol. 2).

The somatic component of the pudendal nerve is distributed in cloacal rami to the striated muscles of the cloaca and to the integument around the vent and ventral aspect of the tail.

(d) Other somatic branches. Several fine branches of the pudendal plexus supply the striated muscles of the cloaca and the ventral tail muscles. Cutaneous rami innervate the striated *m. adductor rectricium* and the smooth *mm. pennales* of the rectrices, as well as proprioceptive endings around the tail bulb, around the follicles of the rectrices and in the cloacal wall (Baumel, 1975).

(e) Connexus caudalis. As mentioned on p. 187, this nerve connects the sacral plexus with the pudendal plexus. It arises from the sacral plexus (often from the caudal coxal nerve) and communicates with a branch of the lateral caudal nerve. The combined nerve forms a cloacal ramus which innervates the distal cloacal wall and all of the cloacal musculature (Baumel, 1975).

C. Caudal plexus

In *Gallus* and *Columba* (Figs. 3.19, 20) the ventral rami of up to five spinal nerves contribute to the caudal plexus (*plexus caudalis*), namely spinal nerves XXXV–XXXIX, inclusive, in *Gallus* (Baumel, 1975), and XXXII–XXXVI, inclusive in *Columba* (Baumel, cited in Breazile and Yasuda, 1979). These roots pass caudally (Figs. 3.19, 20) along the ventral aspects of the free caudal vertebrae, forming the medial caudal nerve (*n. medialis caudae*). This nerve innervates the *m. depressor caudae* (Vanden Berge, 1975; Baumel, 1975). It also supplies the muscle of the rectricial bulb (Vanden Berge, 1975; Baumel, 1975).

The last two or three of the caudal spinal nerves may have small or even no ventral rami. Their dorsal rami supply the dorsal axial muscles including the *m. levator caudae* (Vanden Berge, 1975; Baumel, 1975). Cutaneous branches supply the skin of the dorsal aspect of the tail, and also the uropygial gland (Yasuda, 1960c; Lucas and Stettenheim, 1972; Baumel, 1975).

IX. Annex

During the last 100 years the brachial and lumbosacral plexuses have been studied in a wide variety of avian species, especially by Fürbringer, Gadow and Boas.

Some further lines of investigation have also been re-opened in a few species, for example through Baumel's emphasis on the trunks of the plexuses and their division into dorsal and ventral cords (fascicles), and these need further exploration in a wider variety of birds. The weakest aspect of current knowledge of the peripheral somatic nerves of birds appears to be their distribution to the muscles of the wing and pelvic limb. This field has been relatively intensively investigated in domestic birds, especially by Yasuda and his colleagues, but comparable studies on other species are rare. In 1974 Bock remarked that the innervation of avian muscles had been neglected and needed a searching review. This comment still holds good today.

References

Baumel, J. J. (1958). Variation in the brachial plexus of *Progne subis*. *Acta anat.* **34,** 1–34.

Baumel, J. J. (1975). Aves. Nervous system. *In* "Sisson and Grossman's The Anatomy of the Domestic Animals" (R. Getty, Ed.), Vol. 2. Saunders, Philadelphia.

Baumel, J. J. (1979a). Osteologia. *In* "Nomina Anatomica Avium" (J. J. Baumel, A. S. King, A. M. Lucas, J. E. Breazile and H. E. Evans, Eds.). Academic Press, London and New York.

Baumel, J. J. (1979b). Systema cardiovasculare. *In* "Nomina Anatomica Avium" (J. J. Baumel, A. S. King. A. M. Lucas, J. E. Breazile and H. E. Evans, Eds.). Academic Press, London and New York.

Baumel, J. J. (1985). Suspensory ligaments of nerves: an adaptation for protection of the avian spinal cord. *Anat. Histol. Embryol.* **14**, 1–5.

Beddard, F. E. (1898). "The Structure and Classification of Birds." Longmans, Green, London.

Bellairs, A. d'A. and Jenkin, C. R. (1960). The skeleton of birds. *In* "Biology and Comparative Physiology of Birds" (A. J. Marshall, Ed.), Vol. 1. Acaemic Press, New York and London.

Bennett, T. (1974). Peripheral and autonomic nervous system. *In* "Avian Biology" (D. S. Farner and J. R. King, Eds.), Vol. IV. Academic Press, New York and London.

Boas, J. E. V. (1929). Biologisch-anatomische Studien über den Hals der Vögel. *Kgl. Danske Vidensk. Selsk. Skr., Naturvidensk.* **1,** 101–222.

Boas, J. E. V. (1933). Kreuzbein, Becken und Plexus lumbosacralis der Vögel. *Kgl. Danske Vidensk. Selsk. Skr., Naturvidensk.* **5,** 1–74.

Bock, W. J. (1974). The avian skeletomuscular system. *In* "Avian Biology" (D. S. Farner and J. R. King, Eds.), Vol. 4. Academic Press, New York and London.

Bolton, T. B. (1976). Nervous system. *In* "Avian Physiology" (P. D. Sturkie, Ed.), 3rd edn. Springer-Verlag, Berlin and New York.

Breazile, J. E. and Yasuda, M. (1979). Systema nervosum peripherale. *In* "Nomina Anatomica Avium" (J. J. Baumel, A. S. King, A. M. Lucas, J. E. Breazile and H. E. Evans, Eds.). Academic Press, London and New York.

Buchholz, V. (1959–1960). Beitrag zur makroskopischen Anatomie des Armgeflechtes und der Beckennerven beim Haushuhn *(Gallus domesticus). Wiss. Z. Humboldt-Univ. Berlin, Math.-Naturw.* **9,** 515–594.

Buri, R. O. (1900). Zur Anatomie des Flügels von *Micropus melba* und einigen anderen Coracornithes, zugleich Beitrag zur Kenntnis der Systematischen Stelling der Cypselidae. *Jena Z. Med.* **33,** 361–610.

Chomiak, M. (1950). Badania nad splotem ledzwiowym i krzyzowym kury domowej. *Ann. Univ. Mariae Curie-Sklodowska, Sect. DD.* **5,** 29–45.

Clara, M. (1959). "Das Nervensystem des Menschen." Barth Verlag, Leipzig.

Coggeshall, R. E., Coulter, J. D. and Willis, W. D. (1975). Unmyelinated axons in the ventral roots of the cat lumbosacral enlargement. *J. comp. Neurol.* **153,** 39–58.

Davenport, H. A. and Ranson, S. W. (1931). Ratios of cells to fibers and of myelinated to unmyelinated fibers in spinal nerve roots. *Am. J. Anat.* **49,** 193–207.

deWet, P. D., Fedde, M. R. and Kitchell, R. L. (1967). Innervation of the respiratory muscles of *Gallus domesticus. J. Morph.* **123,** 17–34.

Dozsa, S. (1939). Das periphere Nervensystem der Hausvogel. Disseration, Budapest.

Durward, A. (1951). Peripheral nervous system. *In* "Cunningham's Test-book of Anatomy." Oxford University Press, New York.

Fisher, H. I. and Goodman, D. C. (1955). The myology of the Whooping Crane, *Grus americana. Ill. Biol. Monogr.* **24,** 1–127.

Fürbringer, M. (1879). Zum Lehre von den Umbildungen der Nervenplexus. *Morph. Jb.* **5,** 324–394.

Fürbringer, M. (1888). "Untersuchungen zur Morphologie und Systematik der Vögel, zugleich ein Beitrag zur Anastomie der Stütz-und Bewegungsorgane," Vol. 2. Van Holkema, Amsterdam.

Gadow, H. (1891). Vogel. I. Anatomischer Theil. *In* "Dr. H. G. Bronn's Klassen und Ordnungen des Thierreichs" (H. Gadow and E. Selenka, Eds.), Vol. 6, Part 4. Winter, Leipzeg.

Gadow, H. (1896). Skeleton. *In* "A Dictionary of Birds" (A. Newton, Ed.). Black, London.

Gadow, H. F. (1933). *In* "The Evolution of the Vertebral Column: a Contribution to the Study of Vertebrate Phylogeny" (J. F. Gaskell and H. L. H. H. Green, Eds.). Cambridge University Press, Cambridge.

Harris, W. (1904). The true form of the brachial plexus and its motor distribution. *J. Anat. Physiol., Lond.* **38,** 399–442.

Harris, W. (1939). "The Morphology of the Brachial Plexus." Oxford University Press, London.

Horst, J. G. (1934). Spinal nerven. *In* "Handbuch der vergleichenden Anatomie der Wirbeltiere" (L. Bolk, E. Goppert, E. Kallius and W. Lubosch, Eds.), Vol. 2. Urban and Schwarzenberg, Berlin.

Huber, J. F. (1936). Nerve roots and nuclear groups in the spinal cord of the pigeon. *J. comp. Neurol.* **65,** 43–91.

Jungherr, E. L. (1969). The neuroanatomy of the domestic fowl *(Gallus domesticus). Avian Diseases* Special Issue.

Kadono, H., Okada, T. and Ono, K. (1963). Electromyographic studies on the respiratory muscles of the chicken. *Poult. Sci.* **42,** 121–128.

Kaiser, L. (1924). L'innervation segmentale de la peau chez le pigeon *(Columba livia var. domestica). Archs. néerl. Sci.* **9,** 299–379.

Kappers, C. U. A., Huber, G. C. and Crosby, E. (1936). "The Comparative Anatomy of the Nervous System of Vertebrates, including Man." Macmillan, New York.

King, A. S. and McLelland, J. (1984). "Birds: Their Structure and Function." Baillière, Tindall, London.

King, A. S. and Molony, V. (1971). The anatomy of respiration. *In* "Physiology and Biochemistry of the Domestic Fowl" (D. J. Bell and B. M. Freeman, Eds.), Vol. 1. Academic Press, London and New York.

Lucas, A. M. and Stettenheim, P. R. (1972). "Avian Anatomy. Integument," Agriculture Handbook 362. U.S. Department of Agriculture, Washington, D.C.

Malinovsky, L. (1967). Die Nervenendkorperchen in der Haut von Vogeln und ihre Variabilitat. *Z. mikrosk.-anat. Forsch.* **77,** 279–303.

Malinovsky, L. (1968). Types of sensory corpuscles common to mammals and birds. *Folia morph., Prague* **16,** 67–73.

Moskoff, M. (1929). Der Plexus brachialis der Hausvògel. *Jb. Vet. Ned. Faf. Sofia* **5,** 300–314.

Nishi, S. (1938). Muskeln des Rumpfes. *In* "Handbuch der vergleichenden Anatomie der Wirbeltiere" (L. Bolk, E. Göppert, E. Kallius and W. Lubosch, Eds.), Vol. V. Urban and Schwarzenberg, Berlin.

Ohmori, Y., Watanabe, T. and Fujioka, T. (1982). Localization of the motoneurons innervating the forelimb muscles in the spinal cord of the domestic fowl. *Anat. Histol. Embryol.* **11,** 124–137.

Ohmori, Y., Watanabe, T. and Fujioka, T. (1984). Localization of the motoneurons innervating the hindlimb muscles in the spinal cord of the domestic fowl. *Anat. Histol. Embryol.* **13,** 141–155.

Patterson, A. M. (1887). On the fate of the muscle plate and the development of the spinal nerves and limb plexus in birds and mammals. *J. Am. Sci.* **28,** 109–129.

Portmann, A. (1950). Systeme nerveux. *In* "Traité de Zoologie, Oiseaux" (P. P. Grassé, Ed.), Vol. 15. Masson, Paris.

Portmann, A. and Stingelin, W. (1961). The central nervous system. *In* "Biology and Comparative Physiology of Birds" (A. J. Marshall, Ed.), Vol. 2. Academic Press, New York and London.

Ribbing, L. (1938). Muskelm und Nerven der Extremitaten. *In* "Handbuch vergleichenden Anatomie der Wirbeltiere" (L. Bolk, E. Göppert, E. Kallius and W. Lubosch, Eds.), Vol. V. Urban and Schwarzenberg, Berlin.

Romanoff, A. L. (1960). "The Avian Embryo," p. 305. Macmillan, New York.

Schummer, A. (1973). "Anatomie der Hausvögel." Paul Parey, Berlin.

Stammer, A. (1961). Die Nervenendorgane der Vogelhaut. *Acta biol. Acad. Sci.* **7,** 115–131.

Streeter, G. L. (1904). The structure of the spinal cord of the ostrich. *Am. J. Anat.* **111,** 1–27.

Stresemann, E. (1934). Sauropsida: Aves. *In* "Handbuch der Zoolgie" (W. Kukenthal and T. Krumbach, Eds.). de Gruyter, Berlin.

Taha, A. A. M. and King, A. S. (1983). Autoradiographic observations on the innervation of the carotid body of the domestic fowl. *Brain Res.* **266,** 193–201.

Vanden Berge, J. C. (1975). Aves myology. *In* "Sisson and Grossman's The Anatomy of the Domestic Mammals" (R. Getty, Ed.), Vol. 2. Saunders, Philadelphia.

Vanden Berge, J. C. (1979). Myologia. *In* "Nomian Anatomica Avium" (J. J. Baumel, A. S. King, A. M. Lucas, J. E. Breazile and H. E. Evans, Eds.). Academic Press, London and New York.

Waluszewska-Bubień, A. and Pospieszyn, N. (1982a). Morphometric study of the brachial plexus in Bustard (*Otis tarda* L., 1758). *Folia Morph., Warsaw* **41,** 89–102.

Waluszewska-Bubień, A. and Pospieszny, N. (1982b). Pelvic plexuses of Bustard (*Otis tarda* L., 1758). *Folia Morph., Warsaw* **41,** 253–267.

Watanabe, T. (1961). Comparative and topographical anatomy of the fowl. 8. On the distribution of the nerves in the neck of the fowl. *Jap. J. vet. Sci.* **23,** 85–94.

Watanabe, T. (1972). Comparative and topographical anatomy of the fowl. 64. Sympathetic nervous system of the fowl. Part 2. Nervus intestinalis. *Jap. J. vet. Sci.* **34,** 303–313.

Windle, W. F. and Orr, D. W. (1934). The development of behaviour in chick embryos: spinal cord structure correlated with early somatic motility. *J. comp. Neurol.* **60,** 287–307.

Yasuda, M. (1960a). Comparative and topographical anatomy of the fowl. 15. The distribution of the nerves to the trunk. *Jap. J. vet. Sci.* **22,** 434.

Yasuda, M. (1960b). Comparative and topographical anatomy of the fowl. 3. On the nervous supply of the thoracic limb in the fowl. *Jap. J. vet. Sci.* **22,** 89–101.

Yasuda, M. (1960c). Comparative and topographical anatomy of the fowl. 20. On the nervous supply of the coccygeal region. *Jap. J. vet. Sci.* **22,** 503.

Yasuda, M. (1961). Comparative and topographical anatomy of the fowl. 21. On the nervous supply of the hind-limb. *Jap. J. vet. Sci.* **23,** 145–155.

Yasuda, M. (1964). Comparative and topographical anatomy of the fowl. 22. Distribution of the cutaneous nerves of the fowl. *Jap. J. vet. Sci.* **26,** 241–254.

Young, J. Z. (1951). The size of the nerve fibres of the muscle nerves. *Proc. R. Soc. B.* **139,** 38.

Young, J. Z. and Zuckermann, S. (1936). The course of fibres in the dorsal roots of *Macaca mulatta,* the rhesus monkey. *J. Anat.* **71,** 447–457.

4

Nasal cavity and olfactory system

BETSY G. BANG AND BERNICE M. WENZEL

Marine Biological Laboratory, Woods Hole, Massachusetts, U.S.A., and Department of Physiology, School of Medicine, University of California at Los Angeles, California, U.S.A.

CONTENTS

I. Introduction

Odour perception is part of an integrated sensory system whereby a bird continually receives information that regulates its behaviour. Evolutionary pressures have induced enormous variation in the relative size, position and shape of the olfactory sense organ and the parts of the brain that sub-serve it. The avian nasal cavity is both the organ of smell and the portal of entry into the respiratory system. It has shown remarkable plasticity in adapting to these two vital functions and has developed a formidable arsenal of defence mechanisms against respiratory disease (Bang, 1971).

The peculiar value of the nasal cavity to avian biology may lie in its use as a model in which comparative nasal bioengineering, olfactory neurophysiology and olfactory behaviour can be correlated. This has not yet been done for any vertebrate species. The first sections of this chapter will introduce the avian nasal cavity as a model in which relationships between form and function can be explored. The later ones will review laboratory and field studies of the physiology and function of the avian olfactory system.

The reason that greater emphasis is placed on the olfactory than on the respiratory function of the nasal cavity is that for respiratory purposes the nasal passages could be a pair of simple, vascularised mucociliated tubes. The amount of intranasal space required by the olfactory mucous membrane, however, has directly influenced the size and structure of the nasal cavity and brain and sometimes of the bony skull.

II. History of investigation of the sense of smell in birds

Meckel's fine drawing of a goose brain (Meckel, 1816) was the first of many observations on the impressive size of the olfactory equipment of several species of bird of ancient lineage. Owen (1872) and Parker (1891) called attention to the large olfactory bulbs and conchae of *Apteryx,* and in 1906 Benham was so impressed by the kiwi's olfactory apparatus that he designed experiments in which earthworms were placed in buckets covered with 4 in. of earth (control buckets contained only earth), and the snuffling, crepuscular birds unerringly and promptly probed the worm-containing buckets and "would not even look at" the control one (Benham, 1906).

The famous but seriously-faulted negative experiments of Audubon (1834) with vultures had the positive result of stimulating a lively transatlantic controversy over the sense of smell in birds. An amusing contemporary review by Waterton (1839) was followed by Gurney's (1922) survey of the roll-call of "scenters" and "no scenters" who took sides during the next decades. Meanwhile, the large olfactory organs of certain vultures (Sells, 1837), an albatross (Owen, 1872), a fulmar (Klinckowström, 1890) and some petrels (Burne, 1908) had been noted. Miller (1942) found that marked Black-footed Albatrosses (*Diomedea nigripes*) were attracted from distances of at least 20 miles to bacon drippings poured on the ocean surface, and that the birds clearly distinguished between bacon fat and paint flux poured on the same area. The great range in sizes of olfactory bulbs among diverse species was recorded by Bumm (1883) and by the neuroanatomists Ariëns-Kappers, Huber and Crosby (1936). Craigie (1940) noted that members of the "ostrich family" that had generally similar habits, also had generally similar bulb sizes.

On the premise that birds with major olfactory equipment must use it in nature, and that "we may judge of the importance of smell to the animal according as this organ is large or small in relation to the rest of the brain" (Edinger, 1908), Strong (1911) wrote a 12-page review of the literature on the avian sense of smell, illustrated the olfactory bulbs of 15 species, and did a long series of experiments attempting to train pigeons to olfactory cues. Results were negative, as were those of Walter (1943) who used conditioned reflexes to evaluate olfactory training of pigeons, ducks, siskins and parakeets. Nonetheless, both of these experimenters were sufficiently encouraged to leave the question open to investigation by improved techniques.

Michelsen (1959) first succeeded in using operant conditioning to train common Rock Doves (*Columba livia*) to respond to the chemical odorant iso-octane. The next decade saw substantial progress regarding the sense of smell in birds. Cobb (1960a,b) not only developed a method of comparative measurement of the olfactory bulbs of different species, but as Bumm (1883) had done on a small scale, demonstrated broad correlations between bulb sizes and general habits. In a series of observations made between 1958 and 1968 Bang showed that a good many species, whose behaviour indicated responses to olfactory cues in nature, had well-developed olfactory conchae. Pooling resources, Cobb and Bang accumulated olfactory rank-order data on 151 species in 23 of the orders and showed quite consistent correlations between olfactory bulb sizes and living habits in most species (Bang and Cobb, 1968; Bang, 1971).

From 1964 to the present, Stager (1964, 1967), Tucker (1965), Grubb (1972, 1974) and Wenzel's school (Wenzel, 1967, 1968, 1971a,b, 1980; Sieck and Wenzel, 1969; Rieke and Wenzel, 1978; Macadar *et al.*, 1980; Hutchison and Wenzel, 1980a,b, 1982; Fox *et al.*, 1982, 1983; Hutchison et al., 1984) have used physiological tests and/or controlled field experiments to show that definite responses to chemical odorants are evoked in avian olfactory tissue, and that several macrosmatic species use olfactory cues in food finding and homing. Current research in this area will be discussed in Section V,E of this chapter.

III. Nasal dynamics

The structures involved in nasal dynamics are illustrated in topographical maps of the nasal cavity (*cavitas nasalis*) of the chicken in Fig. 4.1. On this basic ground plan, types of adaptation of the structures in each intranasal region will be considered in relation to function.

A. *Airstreaming*

It would be ideal to demonstrate areas of impact of inhaled and exhaled molecules by pumping coloured particles in and out of a transparent model and studying the cumulative density of staining. Lacking such a model, we have tried in Fig. 4.1b to postulate the areas of impact during inhalation by envisaging the lateral wall of a sagittally-sectioned nasal cavity, using the chicken as a general avian model. The nostril (*naris*) acts as nozzle, the rostral (*vestibular*) concha functions as an insulating baffle, and the upslant of the vestibule (vestibular region of the *Nomina Anatomica Avium*) thrusts air onto the roof, fanwise into the olfactory region, along the middle concha and into the choana. Presumably, the positioning of receptors for the most critical odorants or classes of odorants is related to the areas of primary impact during normal breathing and the changed order of impact during rapid breathing and sniffing. Exhalation would create eddy currents in certain areas and would result in secondary areas of impact.

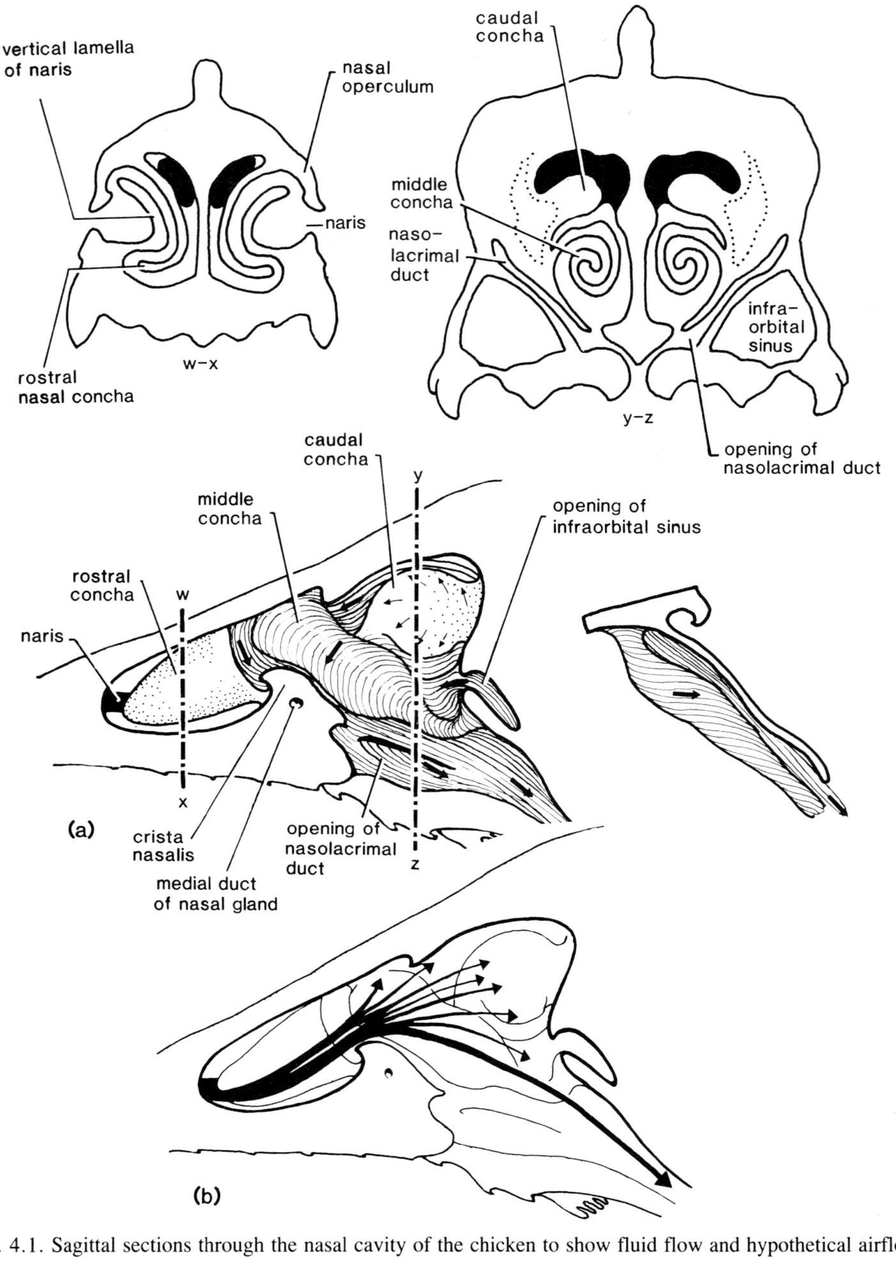

Fig. 4.1. Sagittal sections through the nasal cavity of the chicken to show fluid flow and hypothetical airflow. w–x and y–z, Transverse sections cut through the levels indicated on sagittal section (a). (a) Fluid flow: mucociliary traction pulls olfactory gland secretions from the caudal concha (arrows) to merge with those of the mucociliary blanket of the respiratory region (lines and arrows). The lines represent alternate rows of mucous acini and cilia (see Fig. 4.6), and the arrows show the directions in which the cilia transport the mucous blanket. Mucociliary blankets from the nasolacrimal duct and the infraorbital sinus also merge with these secretions and are carried into the pharynx to be swallowed. The inset shows the mucociliary flow on the inner scroll of the middle nasal concha. (b) Presumed airflow: entering via the naris, the airstream would be baffled by the vertical lamella of the naris and the rostral nasal concha, directed roofwards by the nasal crest (*crista nasalis*), then fanwise either along the roof to the caudal nasal concha or around the middle nasal concha into the pharynx, lower respiratory tract and lungs. From Bang (1971), *Acta anatomica,* **79,** supplementum 58, p. 4, by permission of S. Karger, Basel.

B. Fluid flow

Fluid secretions produced by the lacrimal complex of glands and the nasal gland (*glandula nasalis*) are discharged into the nasal cavity by way of their ducts, whilst secretions of olfactory glands and the mucous glands of the respiratory membrane (*glandulae mucosae nasalis*) are produced *in situ* and are constantly removed by mucociliary traction.

The mucous and lacrimal secretions are usually swallowed, whilst the secretions of the nasal gland invariably discharge into the nasal vestibule (*regio vestibularis*), where they may be temporarily retained or may drip from the beak. In species that lack nares, the ducts of the nasal gland discharge into the choana.

The structures involved in the interacting airstreams and fluids will be considered in each functional region: the naris, the vestibular region, the respiratory region and the olfactory region.

IV. Regional topography

A. Naris

The nares, which are often referred to as the external nares, may be open (pervious) or closed (impervious). In several Pelecaniformes (*Pelecanus, Morus,*

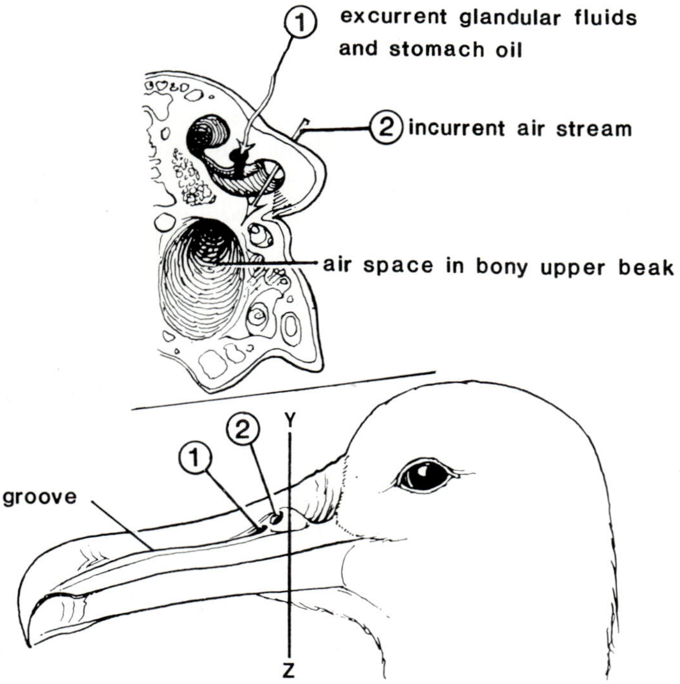

Fig. 4.2. Air and fluid flow patterns in the tube nostril of the Black-footed Albatross (*Diomedea nigripes*). Incurrent air is pulled into the short tube, while excurrent nasal gland fluids and stomach oil are discharged via a channelled opening into a shallow groove in the beak, whence fluids drip from the beak tip. The upper drawing is a transverse section through the beak at Y–Z. From Bang (1966), *Acta anatomica*, **65**, p. 403, by permission of S. Karger, Basel.

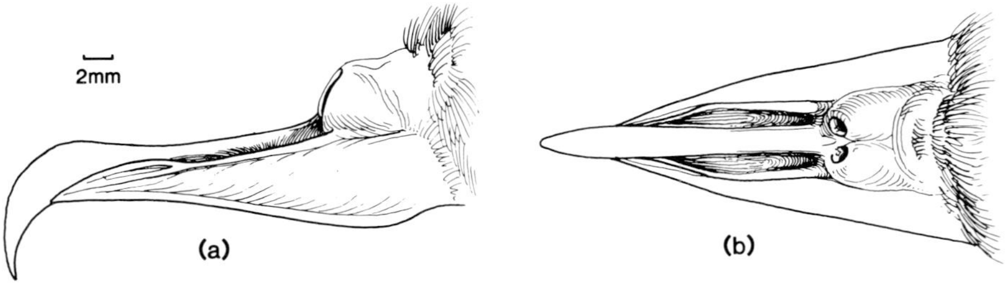

Fig. 4.3.(a) Lateral and (b) dorsal views of the beak of the Dove Prion (*Pachyptila desolata*). Excurrent fluids from the tube nostrils are channelled into a deep groove in the beak. The groove may also act as an additional fluid reservoir. From Bang (1966), *Acta anatomica,* **65,** p. 397, by permission of S. Karger, Basel.

a few Phalacrocoracidae, *Anhinga*), the nares are covered over by secondary growth of the horny beak. These species breathe by "secondary nares" at the angle of the mouth (MacDonald, 1960). The nares may be circular, oval or slit-like, may be placed dorsally (as in *Pelecandoides*), laterally (*Porphyrio*) or ventrally (*Fratercula*) on the beak, less commonly in the cere (*Columba* spp.), at the beak tip (*Apteryx*) or in tubes (Procellariiformes, *Caprimulgus*). A cornified flap, the nasal operculum (*operculum nasale*), partly covers the naris as a flexible (tapaculos, Rhinocryptidae) or rigid (*Gallus*) overhang or as an upthrust (*Jynx*). The "secondary nares" developed at the jugal margins of the upper mandible in *Morus* also have opercular flaps. There are many variations in the partial or complete coverage of the nares by soft feathers (as in *Corvus*) or rictal bristles, as in some Laniidae. Rarely, nostrils seem designed to obtain olfactory information from the rear, as in *Tockus* and *Ramphastos* in which the nostrils face caudally.

The most elegantly-engineered nostrils are in the Procellariiformes. In the albatross *Diomedia nigripes,* for example, the small tubes neatly segregate the incoming airstream from the excurrent gland fluids and stomach oil (Fig. 4.2). The latter fluids are discharged into either a shallow groove as in *D. nigripes,* or a very deep groove as in the Dove Prion (*Pachyptila desolata*) (Fig. 4.3).

B. Vestibular region

The vestibular region (*regio vestibularis*) is the most structurally variable region of the nasal cavity, is the primary determinant of intranasal airflow, and is critical for warming and vapourizing the inspired air. It is lined almost entirely with keratinized or non-keratinized squamous epithelium, beneath which lies a richly vascular bed. Cool air condenses in this blood-warmed "nasal vestibule" to provide one source of the vapour needed for olfaction and respiration.

The main source of saline vapour is the fluid of the nasal gland (*glandula nasalis*). The duct mouth(s) of the single- or bi-lobed nasal gland always discharges into the nasal vestibule at its caudal boundary; secretions accumulate in a shallow reservoir, the vestibular cisterna, where they are vapourized at inspiration to spread a moist cloud over the nasal mucous membranes (Bojsen-Möller, 1965). The *cisterna* is an avian analogue of the moist vestibular moat or Nasenspiegel of many terrestrial vertebrates. Since the nasal gland is known to be missing in *Casuarius, Opisthocomus, Anhima, Pterocles* and *Ramphastos* (Technau, 1936), these species have other, compensatory fluid sources.

The most important baffling–warming structure in many species is the rostral nasal concha (*concha nasalis rostralis*), which varies from the simple convex flap of *Gallus* to the complex grooved structure in *Merops* (Fig. 4.4a). In *Pagodroma* it is an aerodynamic torque (Bang, 1960). This very vascular baffle is much reduced in some species and is lacking entirely in a few such as *Dromaius, Gyps bengalensis* and *Pelecanus*. The rostral concha intrudes conspicuously into the nostril in a number of species. This has led to the misinterpretation that it is a special structure, a "paraseptal process," "nasal flange" (Murphy and Harper, 1921), "*lamina vestibuli*" (Wood-Jones, 1937a,b) in Procellariiformes, or a "central tubercle" (Van Tyne and Berger, 1959) in falcons. *Falco,* like the Procellariiformes, has remarkable vestibular engineering.

Looking into the round nostril of *Falco,* one indeed sees a fleshy "tubercle" which occludes much of the opening. If the naris is cut away, it is seen that the tubercle is the rostral part of a U-shaped rostral nasal concha (Technau, 1936, p. 571) which has a partial bony skeletal support (Fig. 4.5a). The smooth medial face of the same concha, facing the septum, is perforated by the mouth of the duct of the medial lobe of the nasal gland, which deposits fluids into an unusually deep cisterna below an unusually steep nasal crest (*crista nasalis*) (Fig. 4.5b).

The nasal septum forms a solid partition between the right and left nasal cavities of most birds, but in species in at least 10 orders, the septum is pierced by an open window, usually opposite the nostril. This presumably allows proximal mixing of odorant molecules and might initiate some turbulence during inspiration. The selective advantage is obscure. There seems to be no correlation between perforate septa and the habits of the species which have them (Table I). Suschkin (1905) noted a septal "fontanelle" in very young Falconiformes which closed over with age, and suggested that the ancestral progenitor, like modern Cathartidae, may have had a perforate septum.

The nasal crest (the Schwelle of the old German literature), when present, abruptly demarcates the vestibular from the respiratory region of the nasal cavity (Figs. 4.1, 5b). The medial duct of the nasal gland always runs through the crest (Fig. 4.1a). Because of its mechanical effect of forcing inspired air to strike the most rostral extension of the olfactory membrane, the nasal crest is probably most important in species in which it is exceptionally steep, such as *Rhea, Falco, Fratercula, Porphyrio, Gallus* and *Fulmarus*. The crest is reduced or lacking in some species (Bang, 1971).

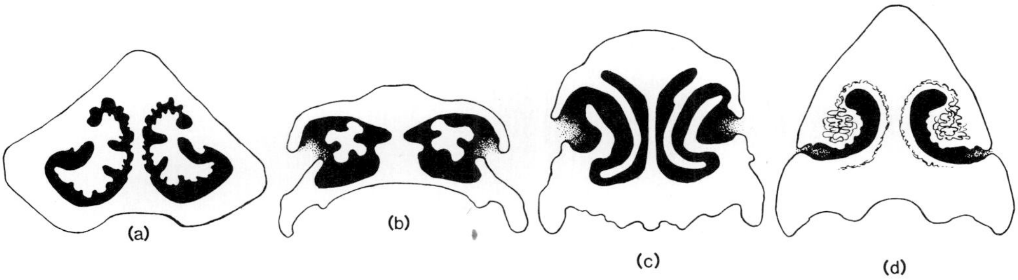

Fig. 4.4. Transverse sections through the rostral nasal concha of four unrelated species to show different shapes adapted to different requirements. (a) The insectivorous Little Green Bee-Eater (*Merops orientalis*); (b) the nectar-feeding Purple-rumped Sunbird (*Nectarinia zeylonica*); (c) the ground-living chicken (*Gallus gallus*); (d) the cave-dwelling frugivorous Oilbird (*Steatornis caripensis*). The rich sub-vascular epithelial plexus of the rostral concha is indicated only in *Steatornis*. Airways black. Not to scale.

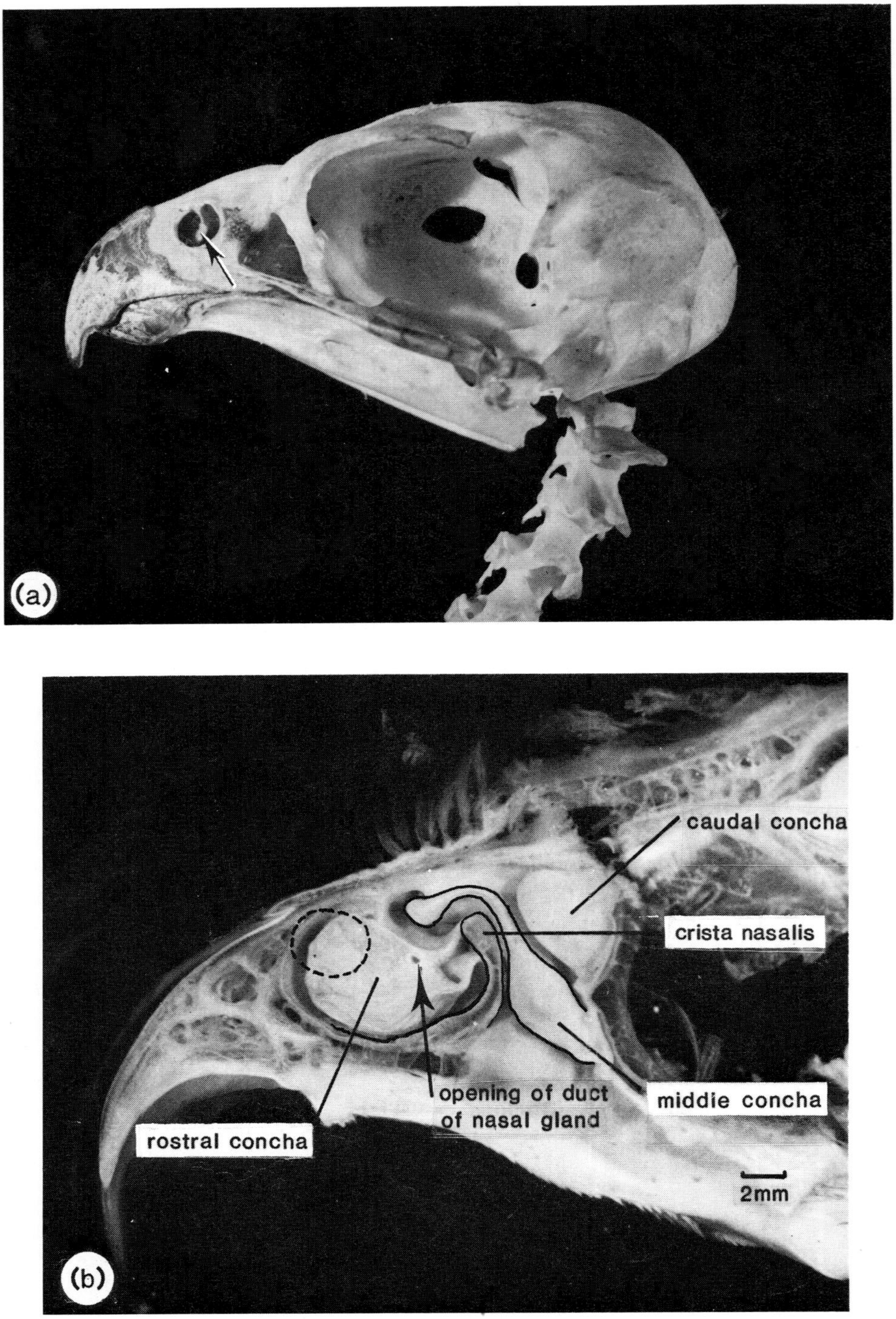

Fig. 4.5. (a) Skull of Peregrine Falcon (*Falco peregrinus*). The bony portion of the skeletal support of the rostral concha protrudes (arrow) into the nostril, lending credence to the erroneous concept of a specific "nasal tubercle." (b) Sagittal section of the nasal cavity of the Peregrine Falcon, showing the medial surface of the rostral concha. The position of the naris is outlined by dashed lines; the contours of the nasal crest and middle concha have been outlined with ink.

TABLE I. **Orders or groups with perforate nasal septa.**

Sphenisciformes	Gruiformes (except Turnicidae[b])
Gaviiformes	Charadriiformes
Podicipediformes	Passeriformes[c]
Pelecaniformes (only *Phaethon*[a])	Hirundinidae
Ciconiiformes	Motacillidae
Anseriformes	Campephagidae
Falconiformes (only Cathartidae)	Laniidae
	Cinclidae
	Zosteropidae

[a] *Phaethon* is the only member of the order Pelecaniformes that has a nasal valve.

[b] The Turnicidae are very atypical members of the Gruiformes, living much more like gallinaceous birds, which they also physically resemble.

[c] The species of Passeriformes listed are all insectivorous and all have a large, often complex *concha rostralis*.

C. Respiratory region

The outstanding characteristic of the respiratory region (*regio respiratoria*) is the ceaseless movement of the sheet of secreted mucus along genetically-determined pathways, transported by cilia which beat irreversibly in a given direction. In all terrestrial vertebrates, including birds, this function is of paramount importance to both olfaction and respiration, and is the primary means of defence against infectious disesases of the lower respiratory tract (Bang and Bang, 1958, 1967, 1969; Bang *et al.*, 1966). In *Gallus* the general rate of mucociliary transport is about 10 mm/min; piling up of mucus is avoided by having a constantly rapid rate in some areas and slower rates in others. The mucociliary flow patterns shown in the inset in Fig. 4.6 represent the directions in which individual rows of cilia transport the secretions produced by individual rows of mucous acini (Bang, 1961). The photograph shows that in this species the entire sheet moves forward as an arc, most of it into the pharynx.

In *Gallus* the mucus is produced by two types of gland: the goblet cells of the mucociliary membrane itself, and the submucosal acinar cells. Histochemical staining reveals that distinctive types of mucins are synthesised in different local areas within the respiratory region of the chicken: mixed types in the lacrimal ducts, sulphated acid types in the deeper portions of each acinus, and more neutral types in the glands surrounding the olfactory region and the luminal portions of each acinus. In the few other avian species sampled, it is clear that no two families have quite the same distribution of histochemical types of mucins. The biological significance of this characteristic distribution has not been investigated.

In the respiratory region, in rostrocaudal progression, one encounters the middle concha, the nasal valve when present, the opening of the nasolacrimal duct which discharges secretions from the gland of the nictitating membrane (*glandula membranae nictitantis,* formerly known as Harder's gland) and from the lacrimal gland (*glandula lacrimalis*) and the aperture of the infraorbital sinus (*apertura sinus infraorbitalis*). All of these are lined by mucociliated epithelium.

The middle nasal concha (*concha nasalis media*), rarely missing (as in certain Phalacrocoracidae), varies in size, shape (Fig. 4.7) and prominence; the surface

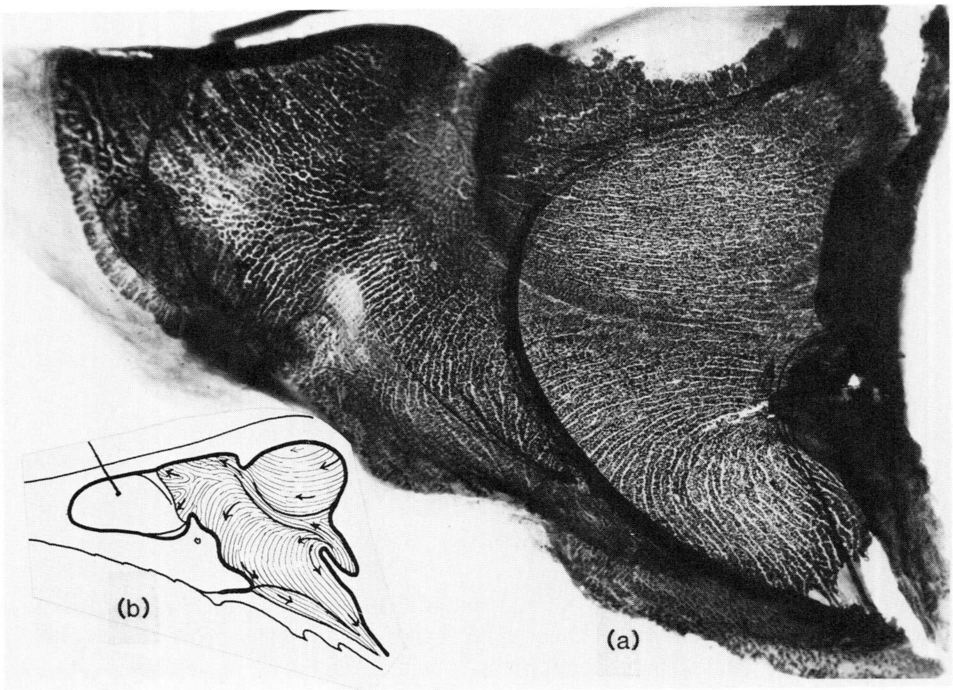

Fig. 4.6. (a) Photograph of the whole nasal septum of a chicken in which the mucous glands were selectively stained with periodic acid–Schiff stain. To show mucociliary transport, India ink was placed at the caudal limit of the mucous membrane within minutes after it was dissected out; 30 seconds later, the septum was fixed in formol alcohol to arrest ciliary movement. The specimen was then stained. It can be seen that small wisps of ink were carried to the rostral mucocutaneous junction, while most of the mucous blanket was being transported into the pharynx. The dark rows are stained mucous acini; the alternating colourless rows are rows of cilia. (b) Diagram of a sagittal section that reveals the nasal septum of a chicken, sectioned just to the left of the septum, the tip of the beak being cut off. It shows the direction of transport of the ink-laden blanket of mucus by ciliary movement along rows of cilia.

facing the airway is well vascularised, like the opposing septal wall. It is usually decurved or scrolled. As in mammals, the amount of surface area developed by the respiratory concha of each species is proportional to the surface area of olfactory membrane required for that species.

The nasal valve (*valvula nasalis*) is a delicate but strong crescent-shaped dou-

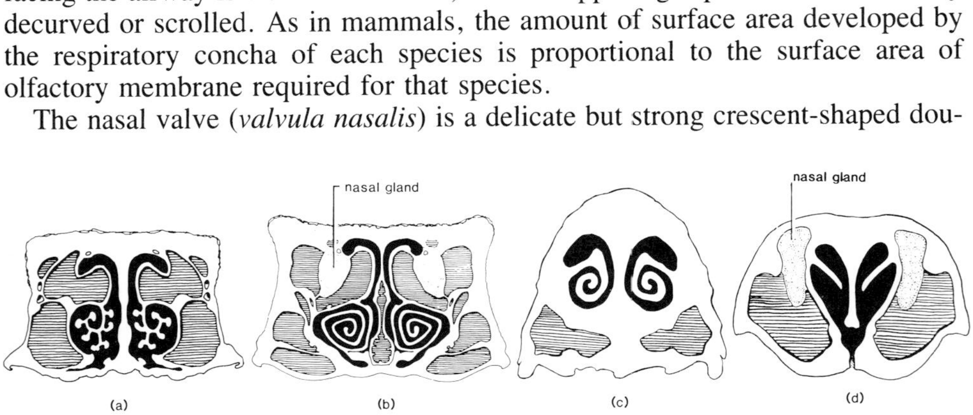

Fig. 4.7. Transverse sections through the middle concha to show variations in size and shape. (a) Emu (*Dromaius novaehollandiae*) and (b) the Greater Rhea (*Rhea americana*), which have broadly similar habitats and behaviour but have evolved quite different means of increasing the surface area of the concha. (c) The domestic chicken (*Gallus gallus*) with the most common prototype of middle concha. (d) Magnificent Frigatebird (*Fregata magnificens*). Not to scale.

ble fold of mucous membrane which forms a blind pouch. The caudal blind limit is firmly attached to the mucoperiosteum of the nasal septum or nasal roof, while the rostral open edge is free. In living birds, the forward edge of the valve lies in close apposition to the septum or the arch of the roof in its "resting" position, but when a bird dives or dips its beak in water, the valve is forced into the operative position by water pressure (Fig. 4.8) and then deflects water from the olfactory region. During diving, dipping or dabbling, the olfactory membrane is thus protected from impingement of seawater particles. The valve is present (bilaterally) in all water-feeding and diving birds with patent nostrils except *Cinclus*, and is inexplicably present in a few land species also (Table II).

Nasal valves are so prominent in Procellariiformes that Mangold suggested that they acted as sensors of air speed (Mangold, 1946). Since they are not present in falcons, swifts or other fast-flying land birds, and since they are found in all water birds and waders, their function in passive underwater protection of the olfactory membrane is more probable.

The infraorbital sinus (*sinus infraorbitalis*) opens through its aperture precisely where the ventral portion of the olfactory membrane would be desiccated by mouth breathing. The sinus is moist, the mucous glands surrounding its opening are very rich, and the mucociliary current at the opening is very rapid. These same characteristics are true of the mammalian maxillary sinus—a probable homologue. [Paulli (1928) has analysed the phenomenon of the sinus spaces in the bony skull of mammals.]

The sinus is much reduced in *Halietor pygmaeus* and is lacking in *Phalacrocorax carbo* and in *Pelecanus*. One adult Adelie Penguin (*Pygoscelis ade-*

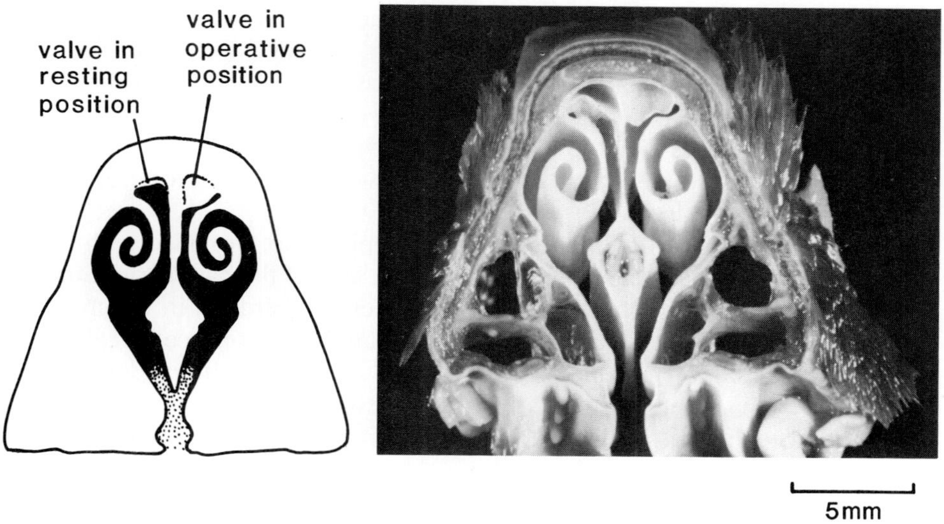

Fig. 4.8. Nasal valves of the Tasmanian Waterhen (*Tribonyx mortierii*). The photograph is of a transverse section through the nasal fossa at the level of the valves of a formalin-fixed specimen. Fixation had stiffened the normally flexible valves. The valve on the right almost completely occludes the dorsal airway as when the bird is underwater; in this position it deflects waterborne particles from the olfactory mucous membranes during feeding. The outline drawing shows the relative natural positions of the "resting" valve which lies in close apposition to the roof arch when the bird is out of water, and of the fully operative valve when the bird is feeding, diving or swimming underwater.

TABLE II. **Groups which have nasal valves.**

Sphenisciformes	Ciconiiformes
Opisthocomus	Anseriformes
Gaviiformes	Gruiformes (except *Turnix*)
Podicipediformes	Charadriiformes
Procellariiformes	Alcedinidae[a]
Phaethontidae[a]	

[a] The only members within their orders. Most of the birds in this table are diving, swimming or wading birds. Yet, kiwis and two land kingfishers which are not known to dive—the Laughing Kookaburra (*Dacelo novaeguineae*) and the Woodland Kingfisher (*Halcyon senegalensis*)—have a nasal valve as well as do the diving kingfishers.

liae) in our collection has the sinuses completely filled with fat, possibly for storage against the long fast of the breeding season?

D. Olfactory region

The Oilbird (*Steatornis caripensis*) and Turkey Vulture (*Cathartes aura*) have roughly the same total amount of olfactory equipment, but a food-seeking Oilbird probably is not motivated by the smell of carrion, nor a Turkey Vulture by ripe fruits. As Adrian (1951) has pointed out, each species selectively discriminates among only those odours relevant to its own habits. Adrian also emphasised that a species with a very small olfactory system can detect odours just as efficiently as one with a very large olfactory system, but the latter has a much bigger screen on which to discriminate more items of information. Thus, (1) the size of the surface area of the olfactory (caudal) concha (*concha nasalis caudalis*) should reflect the spectrum of discrimination that a bird is capable of in nature, (2) the behaviour of a species indicates the kinds of odours to which its receptors react, and (3) the bioengineering of the nose probably indicates where the receptors for the most vital information are located.

The olfactory mucous membrane is a neuroepithelium provided with intrinsic olfactory glands that secrete an ample fluid. The tips of the bare nerve endings of the receptor neurons are covered by this mucoserous secretion in which odorous molecules must be dissolved in order to stimulate the receptor cells. Like that of other warm-blooded vertebrates, the avian neuroepithelium develops ontogenetically some time before the adjacent respiratory epithelium, is pseudostratified and columnar, taller than the adjacent respiratory membrane, and (in most species) lacks goblet cells. Each receptor cell is surrounded by a cluster of supporting cells; each nerve cell dendrite passes between these supporting cells to reach the lumen by means of a long extension which ends in a knob from which 6–15 non-motile cilia extend to the surface, covered only by a thin mucoserous sheet (Fig. 4.9). The axons of individual cells are gathered into bundles which constitute the olfactory nerve and go to the first relay station in the central olfactory pathway, the olfactory bulb. The course of the nerve varies among species, sometimes passing through the orbit (Watanabe and Yasuda, 1968) but

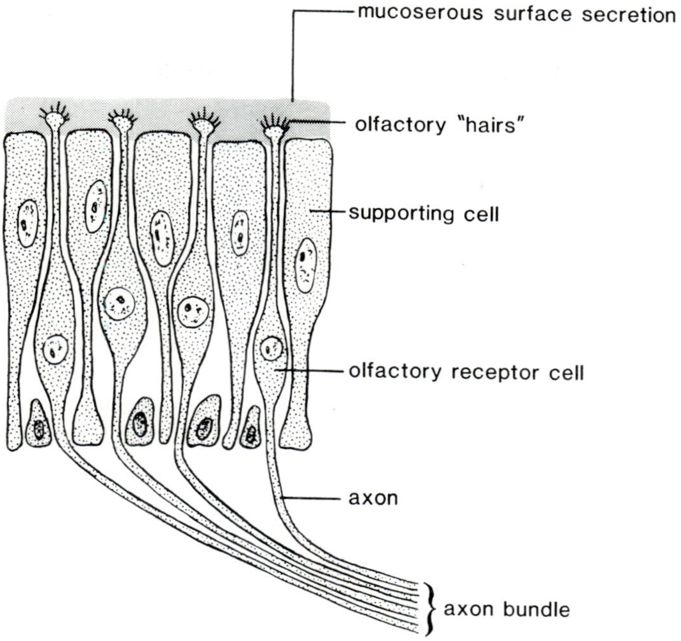

Fig. 4.9. Diagram of a generalised vertebrate olfactory mucous membrane. The avian olfactory membrane is typical, varying in depth according to the total number of nerve cells crowded into the membrane: from one to three in microsmatic species to eight or more in the Turkey Vulture (*Cathartes aura*).

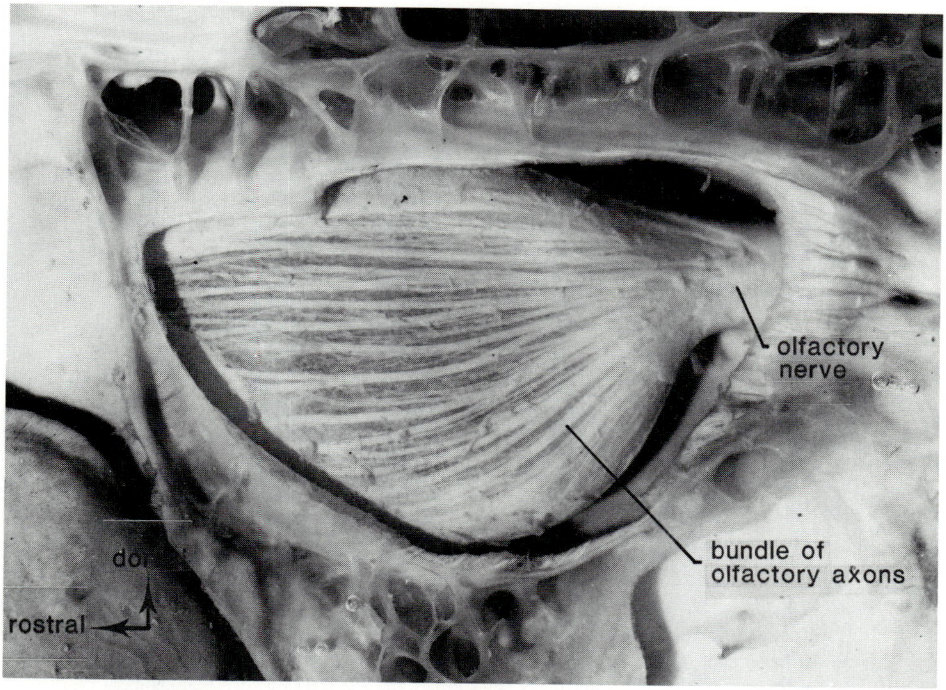

Fig. 4.10. Caudal concha of the Black-footed Albatross (*Diomedea nigripes*), from which the olfactory membrane was peeled off to show the olfactory nerve bundles. From Bang (1966), *Acta anatomica*, **65,** p. 402, by permission of S. Karger, Basel.

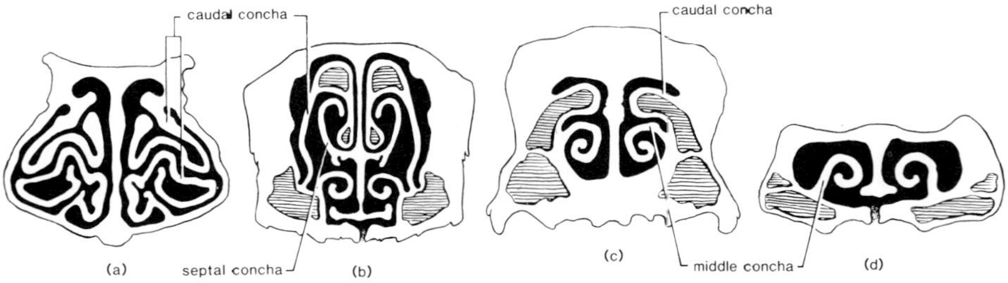

Fig. 4.11. Transverse sections through the caudal concha of four species with different behaviour and habitats. (a) The flightless, solitary, ground-living, nocturnal Brown Kiwi (*Apteryx australis*). (b) The colonial, piscivorous, highly olfactory Snow Petrel (*Pagodroma nivea*). (c) The omnivorous domestic chicken (*Gallus gallus*). (d) The cave-dwelling, echolocating, insectivorous White-bellied Swiftlet (*Collocalia esculenta*), which possesses no caudal concha but has a very rich olfactory membrane covering the roof arch of the nasal cavity. Black areas indicate the airways; horizontal lines are sinuses or air spaces in bones. Not to scale.

more typically lying in the medial interorbital space. Its connection with the olfactory bulb also takes varying patterns, breaking up into many widely-distributed small bundles in some species and remaining tightly collected in others (L. V. Hutchison and B. M. Wenzel, unpublished observations). Figure 4.10 shows the *concha caudalis* of the Black-footed Albatross after the epithelial layer was carefully stripped off to reveal the main twigs of the olfactory nerve as it leaves the mucosa.

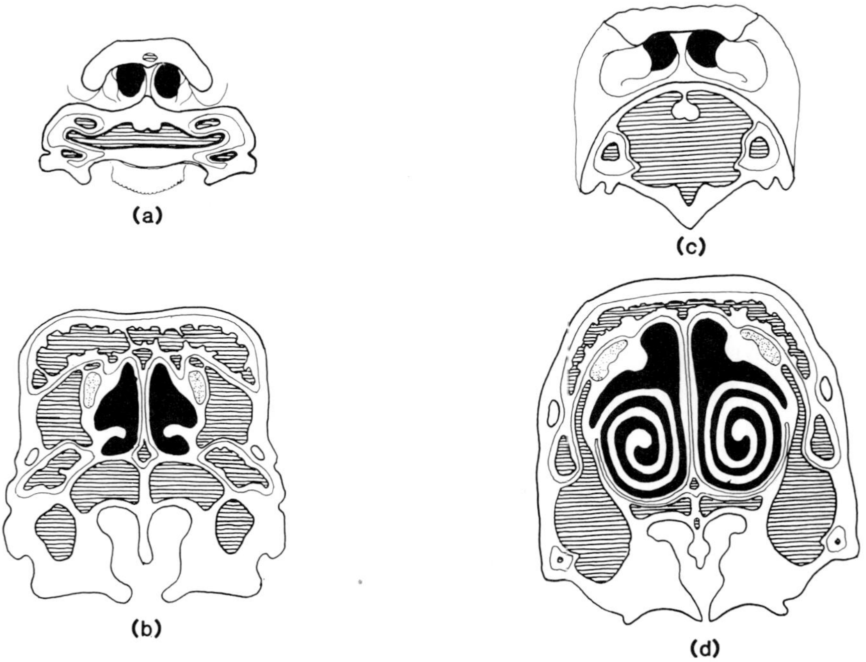

Fig. 4.12. Transverse sections through the nares (upper drawings) and through the greatest dimensions of the caudal conchae (lower drawings) of (a) and (b) the Black Vulture (*Coragyps atratus*), and (c) and (d), the Turkey Vulture (*Cathartes aura*). The airways are black, and in (b) and (d) the olfactory regions have black backgrounds; horizontal lines indicate sinuses or air spaces in the bony skull. In *Cathartes* the caudal concha is remarkable in being a scroll with two and one-half turns. Drawn to the same scale. From Bang (1971), *Acta anatomica*, **79,** supplementum 58, p. 32, by permission of S. Karger, Basel.

TABLE III. **Species listed in olfactory rank-order.**[a]

Olfactory rank-order (%)					
30–40	25–30	20–25	15–20	10–15	3–10
Snow Petrel (*Pagodroma nivea*)[b]	Dove Prion (*Pachyptila desolata*)[b]	Hoatzin (*Opisthocomus hoazin*)[b]	Koel (*Eudynamys scolopacea*)	Hoopoe (*Upupa epops*)	Common Starling (*Sturnus vulgaris*)
Brown Kiwi (*Apteryx australis*)[b]	Black-footed Albatross (*Diomedea nigripes*)[b]	Weka (*Gallirallus australis*)[b,c]	Chimney Swift (*Chaetura pelagica*)	Great Cormorant (*Phalacrocorax carbo*)	Blue-naped Mousebird (*Colius macrourus*)
Leach's Storm-Petrel (*Oceanodroma leucorhoa*)[b]		Steller's Eider (*Polysticta stelleri*)[b,c]	Greater Rhea (*Rhea americana*)	shrike (*Lanius*)[c]	Gannet (*Morus bassanus*)
Wilson's Storm-Petrel (*Oceanites oceanicus*)[b]	Turkey Vulture (*Cathartes aura*)[c]	Oilbird (*Steatornis caripensis*)	Mallard (*Anas platyrhynchos*)[b,c]	Common Snipe (*Gallinago gallinago*)[b,c]	Brown Pelican (*Pelecanus occidentalis*)
Wedge-tailed Shearwater (*Puffinus pacificus*)[b]	Cape Petrel (*Daption capense*)[b]	Purple Swamp Hen (*Porphyrio poliocephalus*)[b,c]	Short-eared Owl (*Asio flammeus*)	Indian Roller (*Coracias benghalensis*)	Red-faced Cormorant (*Phalacrocorax urile*)
Greater Shearwater (*Puffinus gravis*)[b]	Northern Fulmar (*Fulmarus glacialis*)	Ruddy-breasted Crake (*Porzana fusca*)[b,c]	Palm Swift (*Cypsiurus parvus*)	Dyal (*Copsychus saularis*)[c]	Crimson-breasted Barbet (*Megalaima haemacephala*)
	Horned Grebe (*Podiceps auritus*)	White-breasted Water-Hen (*Amaurornis phoenicurus*)[b,c]	Common Swift (*Apus apus*)	Osprey (*Pandion haliaetus*)	Common Grackle (*Quiscalus quiscula*)
	Emu (*Dromaius novaehollandiae*)	Indian Nightjar (*Caprimulgus asiaticus*)	Blue-winged Pitta (*Pitta brachyura*)	Puffin (*Fratercula arctica*)[b,c]	Black-and-White Casqued Hornbill (*Bycanistes subcylindricus*)
	Tasmanian Waterhen (*Tribonyx mortierii*)[b,c]	Little Grebe (*Tachybaptus ruficollis*)[b,c]	diving-petrel (*Pelecanoides*)[b]	Yellow Wagtail (*Motacilla flava*)[c]	Pelagic Cormorant (*Phalacrocorax pelagicus*)

(*continued*)

Table III. **(Continued)**

Olfactory rank-order (%)					
30–40	25–30	20–25	15–20	10–15	3–10
	Virginia Rail (*Rallus limicola*)[b,c]	Lapwing (*Vanellus vanellus*)[b,c]	Rock Dove (*Columba livia*)	Turkey (*Meleagris gallopavo*)	Common Flicker (*Colaptes auratus*)
	Wood Duck (*Aix sponsa*)[b,c]	Lesser Flamingo (*Phoeniconaias minor*)[b,c]	Common Coucal (*Centropus sinensis*)	Black-headed Oriole (*Oriolus xanthornus*)	Hill Myna (*Gracula religiosa*)
	Whip-Poor-Will (*Caprimulgus vociferus*)	Purple Moorhen (*Porphyrio porphyrio*)[b,c]	Chestnut-backed Shrike (*Lanius collurioides*)[c]	Common Myna (*Acridotheres tristis*)	American Robin (*Turdus migratorius*)
	Coot (*Fulica atra*)[b,c]	Yellow-footed Green Pigeon (*Treron phoenicoptera*)	Black Vulture (*Coragyps atratus*)[c]	Purple-rumped Sunbird (*Nectarinia zeylonica*)	Indian Treepie (*Dendrocitta vagabunda*)
		Yellow-billed Cuckoo (*Coccyzus americanus*)	Adelie Penguin (*Pygoscelis adeliae*)[b,c]	Bustard-Quail (*Turnix suscitator*)	Brown-headed Cowbird (*Molothrus ater*)
		Peregrine Falcon (*Falco peregrinus*)	American Woodcock (*Scolopax minor*)[b,c]	Brahminy Kite (*Haliastur indus*)	Blue Jay (*Cyanocitta cristata*)
		Gray Moorhen (*Gallinula chloropus*)[b,c]	Barn Swallow (*Hirundo rustica*)[c]	Black-headed Cuckoo-Shrike (*Coracina melanoptera*)[c]	Black-capped Chickadee (*Parus atricapillus*)[d]
		Clapper Rail (*Rallus longirostris*)[b,c]	Belted Kingfisher (*Ceryle alcyon*)[b]	Dipper (*Cinclus cinclus*)[c]	American Goldfinch (*Carduelis tristis*)[d]
		Red-billed Tropicbird (*Phaethon aethereus*)[b,c]	Herring Gull (*Larus argentatus*)[c]	Double-crested Cormorant (*Phalacrocorax auritus*)	House Sparrow (*Passer domesticus*)[d]

Black-crowned Night-Heron (*Nycticorax nycticorax*)[b,c]	Fork-tailed Drongo (*Dicrurus adsimilis*)	Red Avadavat (*Amandava amandava*)	Common Crow (*Corvus brachyrhynchos*)[d]
Common Loon (*Gavia immer*)[b,c]	Large Pied Wagtail (*Motacilla maderaspatensis*)[c]	Downy Woodpecker (*Picoides pubescens*)	Rose-breasted Grosbeak (*Pheucticus ludovicianus*)[d]
Common Hawk-Cuckoo (*Cuculus varius*)	Little Black Cormorant (*Phalacrocorax sulcirostris*)	Rose-ringed Parakeet (*Psittacula krameri*)	Purple Finch (*Carpodacus purpureus*)[d]
Green-winged Teal (*Anas crecca*)[b,c]	Blue-throated Barbet (*Megalaima asiatica*)		
American Jacana (*Jacana spinosa*)[b,c]	Red-breasted Merganser (*Mergus serrator*)[c]		
Cattle Egret (*Ardeola ibis*)[b,c]	Black Kite (*Milvus migrans*)		
	Magnificent Frigatebird (*Fregata magnificens*)		
	Common Murre (*Uria aalge*)[b,c]		
	Lapwing (*Vanellus vanellus*)[b,c]		
	Short-billed Dowitcher (*Limnodromus griseus*)[b,c]		
	domestic chicken (*Gallus gallus*)		

[a] The rank-order is obtained from Cobb's olfactory ratio, which expresses the ratio between the longest diameter of an olfactory bulb and the longest diameter of the ipsilateral cerebral hemisphere. When expressed as, say, 25% this means that the length of the bulb was 25% of the length of the hemisphere.

[b] Birds with nasal valves.

[c] Birds with perforate nasal septa.

[d] Birds with fused olfactory bulbs.

The secretions that cover the receptors must be constantly removed in order for new information to be received. The cilia of the olfactory neuroepithelial cells have no transport capability, but the olfactory gland secretions are miscible with those of the mucus of the respiratory region of the nasal cavity and are removed by traction of the surrounding respiratory cilia, as in Fig. 4.1. The receptor cells themselves are replaced about once a month (Graziadei and Monti Graziadei, 1977).

Like the rostral and middle nasal conchae, the caudal concha has evolved many shapes (Fig. 4.11). A caudal concha is lacking in a number of species (Bang, 1971), but even these species, such as the cave-dwelling White-bellied Swiftlet (*Collocalia esculenta*), have a very rich olfactory epithelium lining the arch of the roof. Examples of the divergent evolution of olfactory equipment are found in the two most common cathartid vultures in the eastern United States, the Black Vulture (*Coragyps atratus*) and the Turkey Vulture (*Cathartes aura*). These cathartids are locally sympatric in a good many areas, yet the differences in their

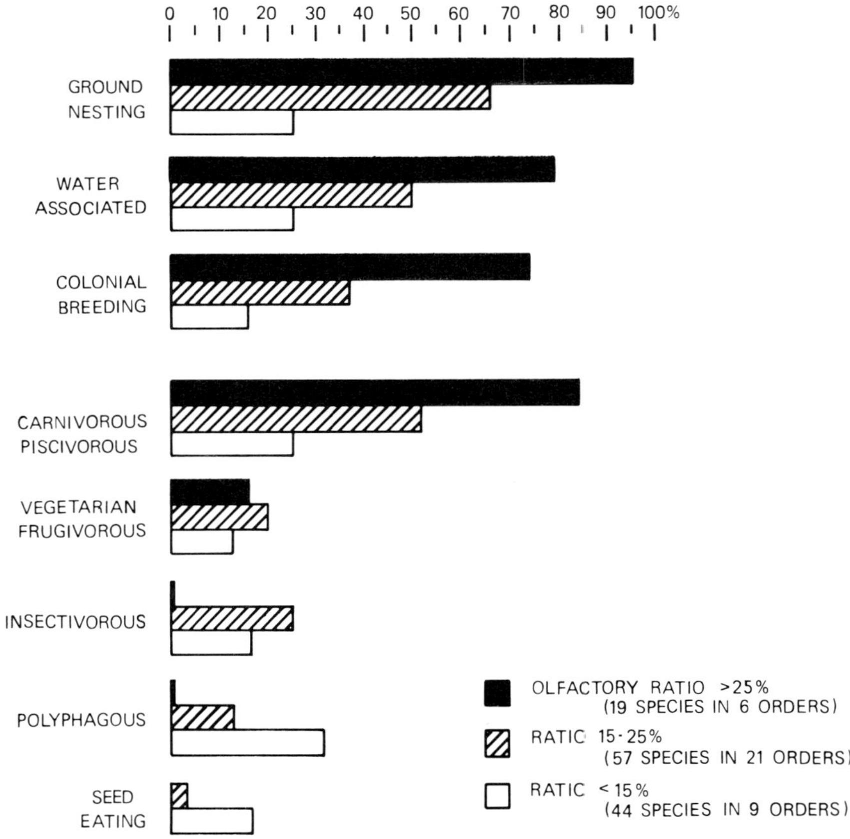

Fig. 4.13. Correlations between olfactory rank-order (based on olfactory ratio) and habits: percentage of each rank-order group which has a particular habit. The olfactory ratio expresses the ratio of the longest diameter of the olfactory bulb to the longest diameter of the ipsilateral cerebral hemisphere (Cobb, 1960b). A ratio of 15 to 25% means that the length of the olfactory bulb was 15 to 25% of the length of the ipsilateral cerebral hemisphere. As an example of interpretation, note that species with ratios over 25% are mainly ground-nesting, water-associated, colonial-breeding, and carnivorous or piscivorous birds From Bang (1971), *Acta anatomica*, **79,** supplementum 58, p. 63, by permission of S. Karger, Basel.

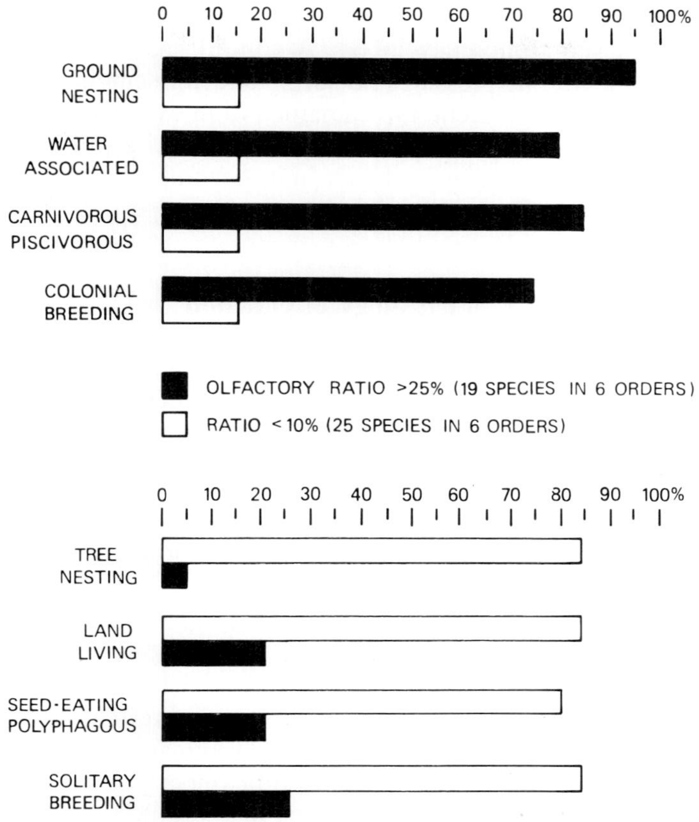

Fig. 4.14. Percentage of species which have the highest olfactory ratios compared to percentage which have the lowest ratios, correlated with particular habits. The olfactory ratio is established and interpreted as in Fig. 4.13. From Bang (1971), *Acta anatomica,* **79,** supplementum 58, p. 64, by permission of S. Karger, Basel.

hunting behaviour (Bang, 1964) seem to correlate with the significant differences in the size of their olfactory conchae (Fig. 4.12) and of their olfactory bulbs (Stager, 1964; Bang, 1971). Thus, the Turkey Vulture has a relatively much greater olfactory region and evidently hunts by olfactory cues, while the Black Vulture hunts visually.

Cobb's "olfactory ratio" (Cobb, 1960b) expresses the ratio between the longest diameter of the olfactory bulb and the longest diameter of the ipsilateral hemisphere (Table III). Olfactory ratios correlate in a general way with the relative surface area of the olfactory membrane in the nasal cavity and with habits and habitats (Figs. 4.13, 14). For example, Fig. 4.13 shows that a species in which the length of the olfactory bulb is over 25% that of the ipsilateral hemisphere is likely to be a ground-nesting, water-associated, colonial-breeding, carnivorous or piscivorous bird. On the other hand, Fig. 4.14 indicates that a species with a ratio below 10% is likely to show mainly tree-nesting, land-living, solitary-breeding, and seed-eating or polyphagous characteristics. Our lack of detailed understanding of avian evolution prevents any generalisation about phylogenetic age and olfactory ratio. Except for the cluster of passerines, a group known to be relatively recently evolved, near the bottom of the list (i.e. where the ratios are low), there is no strong correlation. Much more information is needed about when major groups and sub-groups developed. A fruitful analysis will probably be oriented towards identifying pressures favouring olfactory emphasis or de-emphasis. Stager's discussion of olfaction in cathartids is an example of this approach. He concluded that only the Turkey Vulture had retained a primary reliance on olfaction and that this was the major factor in its becoming "the most successful of all cathartine vultures in the New World today" (Stager, 1964, p. 56).

V. Physiology of olfaction

Knowledge of how the avian olfactory system works lacks the rich descriptive detail now available for the peripheral structures. This section presents the main features of olfactory neuroanatomy and neurophysiology, as well as behavioural evidence for olfactory function.

A. *Olfactory mucosa and nerve fibres*

When olfactory receptor cells are excited by an odorant, nerve impulses are initiated in their axons and the entire mucosa shows a slow potential called the electro-olfactogram (EOG). Both of these events have been recorded in several avian species (Tucker, 1965) and indicate that the avian system does not differ in general function from that of amphibians, reptiles and mammals. The nerve fibres are fine and unmyelinated so that the conduction velocity is slow, ranging well below 1.5 m/sec on the average (Macadar *et al.,* 1980). Even in species like *Passer domesticus* with very small olfactory bulbs and, presumably, correspondingly small areas of mucosa and few nerves, the fibres respond to the odour of amyl acetate and the firing rate increases with stimulus intensity (Tucker, 1965).

Neither transduction nor coding is yet understood for olfaction, so that ignorance of these processes in birds is inevitable at present. Nothing in the limited data available, however, suggests that their functioning differs in any fundamental way from that of other vertebrates.

B. *Olfactory bulb*

In all species examined, the olfactory bulb has the general vertebrate multilayered concentric structure and there is usually a ventricle (Huber and Crosby, 1929; Fox *et al.,* 1982, 1983) (Fig. 4.15). The incoming olfactory nerve fibres make up the outermost layer. Their fine branching terminals penetrate to the next, glomerular layer, where they connect with dendrites of mitral and tufted cells in basket-like arborizations called glomeruli. The perikarya of these cells are in the deeper, mitral cell layer from which their axons leave en route to many areas of the forebrain. Intervening layers of the bulb consist of small nerve cells that make complex intrabulbar connections. As mentioned before, proportional olfactory bulb size varies a great deal as well as bulb shape, and these differences affect the specific arrangement of the basic internal structure. In the small fused bulb of

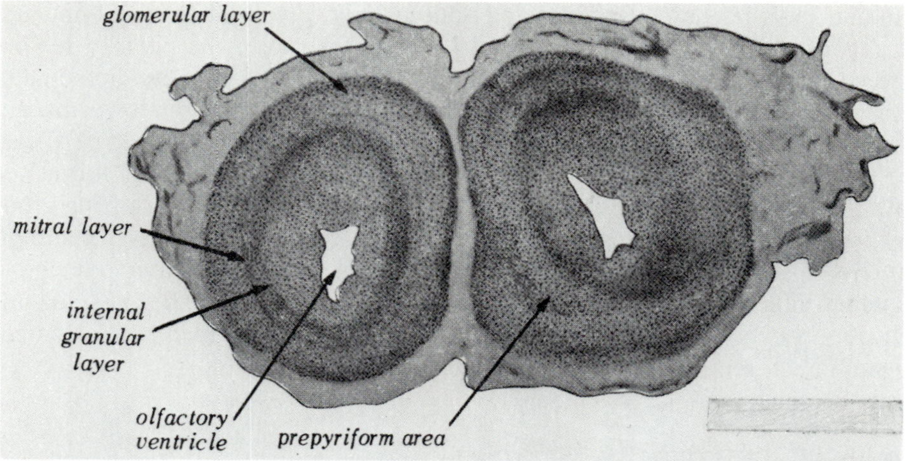

Fig. 4.15. Photomicrograph of a transverse section through the caudal portion of the olfactory bulb of the Rock Dove (*Columba livia*). The prepyriform cortex also appears on the right. From Wenzel (1983), *in* "Physiology and Behaviour of the Pigeon," M. Abs, Ed., Academic Press, London and New York.

Passer domesticus, for example, the layers are not as sharply drawn nor are they concentrically arranged, cell numbers are low, and there is no ventricle.

Not surprisingly, the avian bulb's response characteristics to olfactory nerve input also resemble those of other vertebrates. Electrical stimulation of the pigeon's olfactory nerve elicits an ipsilateral evoked potential of the monophasic

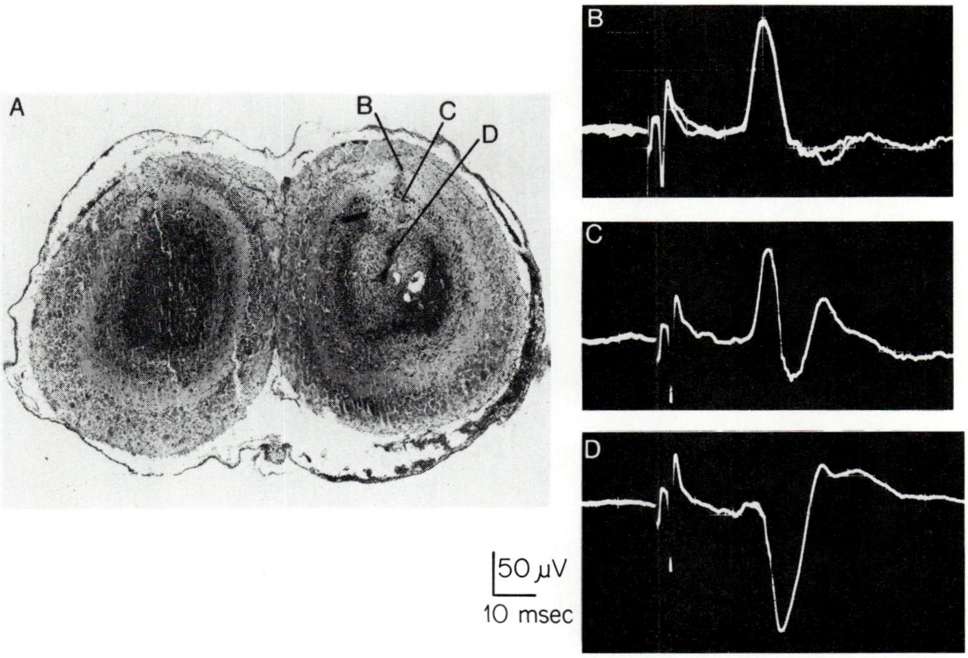

Fig. 4.16. (A) Transverse section through olfactory bulbs of the Rock Dove (*Columba livia*). Labels show points where evoked potentials were recorded after electrical stimulation of the ipsilateral olfactory nerve. (B) Potential evoked in glomerular layer, (C) in mitral cell layer and (D) in granule cell layer. From Macadar *et al.* (1980), by permission of Springer-Verlag.

form and short latency to be expected with a single synapse between stimulation and recording sites (Macadar *et al.*, 1980). Furthermore, the polarity reverses between the outermost and innermost layers in accordance with the dipole nature of the bulb (Fig. 4.16). Responses in the contralateral bulb are slower because there are no direct connections from bulb to bulb (Rieke and Wenzel, 1978), and the stimulus pulse must cross at least two more synapses. Individual mitral cells show characteristic spike responses and appear to be distributed in shifting groups of excited, inhibited and unaffected units as different odours are used for stimulation (Hutchison and Wenzel, 1980a). Multicellular electrical activity changes in frequency and amplitude as odour quality and intensity vary, and it also includes high-frequency spindle bursts associated with inspiration, which regularly occur in mammalian records (Sieck and Wenzel, 1969).

C. Higher centres

The projections from the olfactory bulb have been studied only in *Columba livia,* where they are so extensive as to suggest considerable involvement of olfaction in the bird's behaviour (Rieke and Wenzel, 1978). The avian brain has evolved separately from that of mammals, and the forebrain especially emphasises the divergence. Not only is telencephalic anatomy almost completely different in the two classes, but functional analogies are still far from understood,

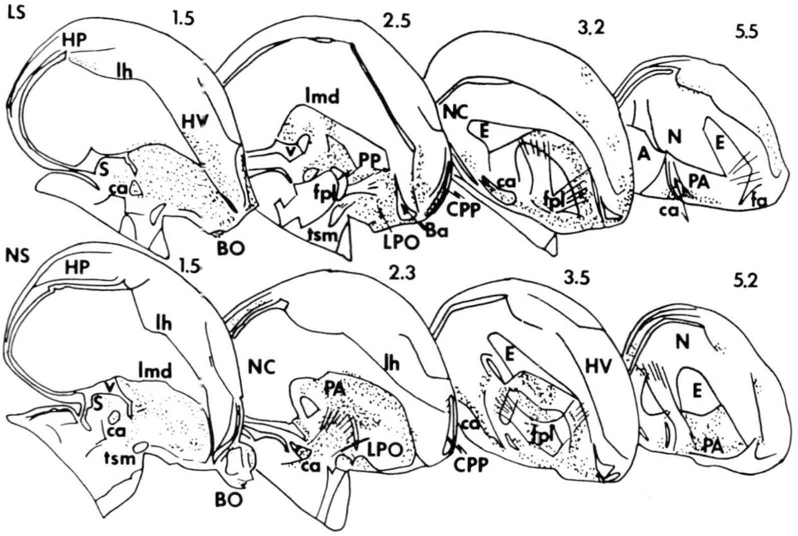

Fig. 4.17. Diagrammatic parasagittal sections through each side of the brain of a pigeon (*Columba livia*) 4 days after destruction of one olfactory bulb. The dots represent degenerating nerve fibres and indicate the areas of projection from the lesioned olfactory bulb on the same side of the brain (LS) and on the opposite side (NS). The numbers are lateral distances in millimetres from the midline. A, *archistriatum*; Ba, *nucleus basalis*; BO, olfactory bulb; ca, rostral commissure; CPP, prepyriform cortex; E, *ectostriatum*; fa, fronto-archistriatal tract; fpl, prosencephalic lateral fasciculus; HP, hippocampus; HV, ventral hyperstriatum; lh, *lamina hyperstriatica*; lmd, *lamina medullaris dorsalis*; LPO, parolfactory lobe; N, *neostriatum;* NC, caudal neostriatum; PA, *paleostriatum augmentatum*; PP, *paleostriatum primitivum*; S, septal nucleus; tsm, septomesencephalic tract; v, ventricle. From Rieke and Wenzel (1978), by permission of the Wistar Institute Press.

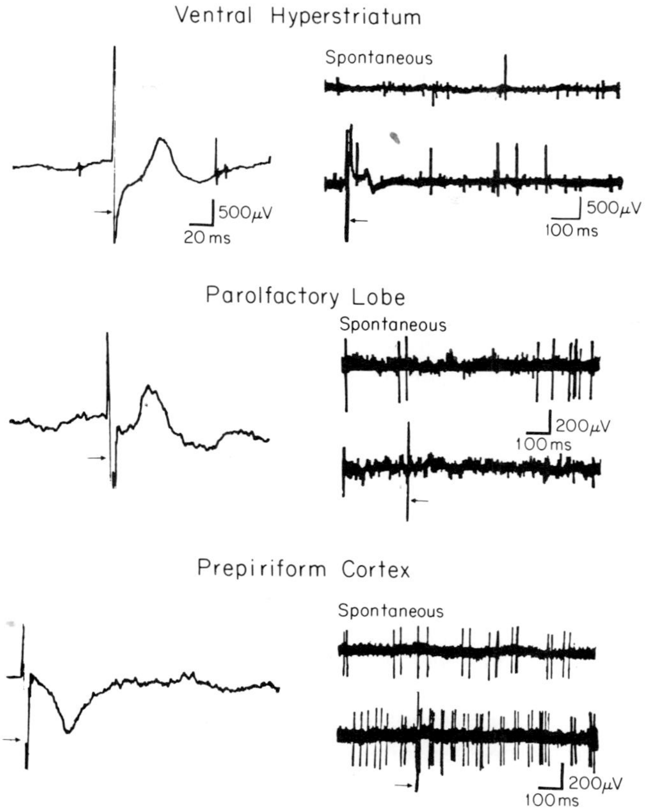

Fig. 4.18. Evoked potentials (left) and unit activity (right) in three forebrain areas of the Northern Fulmar (*Fulmarus glacialis*) after electrical stimulation (arrow) of the ipsilateral olfactory nerve. From Hutchison and Wenzel (1980a), by permission of IRL Press, Ltd.

making it difficult to compare the avian and mammalian olfactory pathways. One difference, at least in the pigeon, is the apparent absence of an anterior olfactory nucleus (Rieke and Wenzel, 1978) which serves mammals as a major relay station for signals from the bulb to other brain sites and for signals returning to the bulb. The absence of such a relay point would have implications for the system's functional style. Another difference is the lack of any projection to the putative analogue of the mammalian amygdalar complex, the *archistriatum*. Similarities include projections to the pre-pyriform cortex, to the paleostriatal complex, which is probably analogous to the basal ganglia of mammals (Karten and Dubbeldam, 1973), and both directly and indirectly to areas where other sensory modalities are also represented or where vegetative functions are partially controlled. The ipsilateral and contralateral projection fields of the pigeon's olfactory bulbs are outlined in detail in Fig. 4.17.

Functional characteristics of the various projection sites are largely unknown. Some recordings have been made from single units in these sites before and after electrical stimulation of the olfactory nerve (Hutchison and Wenzel, 1980a, 1982) (Fig. 4.18), but no differential analysis is possible at present nor can such responses yet be related to the effects of normal stimulation by odours.

D. Sensitivity

An important question in understanding any sensory system is whether it responds to stimulation of low intensity and to small changes in suprathreshold intensity. A related question is how many different odours can be perceived. Early laboratory studies with birds were designed as learning tasks using odours as critical stimuli; if the tasks were learned, the investigator concluded that the bird could smell the odours presented. Such experimental designs are useful for some problems but are unnecessarily demanding and inefficient for answering the basic question of whether a given bird can perceive a certain odour. A simpler approach is to record some continuous physiological process, such as olfactory bulb electrical activity, heart rate or respiration rate, and note whether regular changes occur whenever an odour is presented. Using this method, good evidence for odour perception has been obtained from any species examined, regardless of bulb size (Neuhaus, 1963; Wenzel, 1967; Wenzel and Sieck, 1972).

Estimates of both absolute and differential odour thresholds have been made only for pigeons and only with odours of no biological meaning for them. Operant-conditioning techniques were used in all cases. On the basis of the very limited information available, absolute thresholds of pigeons to standard laboratory stimuli range from 0.1 to 39.7 ppm and the difference threshold is about 65% (Henton, 1969; Shumake *et al.*, 1969; Stattelman *et al.*, 1975), comparable to the rat but poorer than human differential sensitivity (Davis, 1973). The pigeon's bulb: hemisphere ratio (0.18) is intermediate, and sensitivity data for birds at each extreme are very much needed. One study compared the absolute thresholds of pigeons for pentane, hexane and heptane with those of the Bobwhite (*Colinus virginianus*) (ratio unknown), Bustard-Quail (*Turnix suscitator*) (ratio 0.125) and chickens (ratio 0.15), and found that pigeons were somewhat more sensitive than the Bobwhite to all three stimuli but less sensitive than Gallus to pentane and hexane (Stattelman *et al.*, 1975). An equally urgent need is the use of stimulus odours presumed relevant to the species being tested. It may be that sensitivity to significant odours is as keen in small-bulbed birds as in those with the largest bulbs, and that greater amounts of olfactory tissue permit a greater range of functions rather than keener sensitivity (Adrian, 1951), but we do not know that at present.

E. Normal behaviour: field studies

The final term for an equation relating form to function is species-specific behaviour that is dependent on or modulated by odours. Here, especially, the experiments must be designed with great care. One of the first rules should be to select a species for study according to an appropriate hypothesis based on features of olfactory anatomy, certain behavioural patterns or anecdotal reports. Another critical consideration is to provide all of the control conditions necessary to rule out interpretations other than odour use. Lastly, the outcome should not be generalised to other species, even close relatives of the species studied. The main lesson to be learned from the early work on this topic is the erroneous extrapolation of negative results with certain species to birds in general. The evidence now suggests that odour-based behaviour is variably distributed amongst birds just as it is amongst mammals.

Most of the current research has concentrated on birds with the largest olfactory bulbs. The top dozen (Table III) includes 10 procellariiforms, the Brown Kiwi (*Apteryx australis*) and the Turkey Vulture (*Cathartes aura*); several of these have been studied in the field. Stager's (1964) monograph on the Turkey Vulture presents extensive evidence that they forage by detecting windborne carrion odour, which they track to the area of its source, and that they then find the carcass by sight. On the basis of comparing the brains of current and extinct vultures, Stager hypothesised that only the Turkey Vulture now consistently relies on odour in finding food. His experiments in several places around the world supported this hypothesis. Audubon's negative results probably were due to studying a species other than the Turkey Vulture, in addition to flaws of design and execution.

Because the Brown Kiwi's nares are at the tip of its long beak and it can be heard to sniff, it is the one bird that was usually said "to have a sense of smell." Convincing evidence for this conclusion was collected in experiments in which baited and unbaited feeding tubes were buried under 3 cm of dirt and covered with nylon screening (Wenzel, 1968, 1971a). The birds, which have poor eyesight and feed at night, always broke into the baited locations and never into an unbaited one. Other types of observations in the same study establish this bird as a macrosmat.

The procellariiforms' pre-eminence in bulb size classifies them as a group of olfactory giants (Bang, 1965, 1966). The behavioural data collected so far reinforce such status. Under seminatural conditions, the Snow Petrel (*Pagodroma nivea*) was able to find pieces of raw fish concealed in plastic cups around an Antarctic laboratory room or buried outside in the snow. A local skua had no success in the latter test (Jouventin, 1977). The most extensive work, in the eastern Pacific Ocean off California, has dealt with foraging behaviour of Sooty Shearwaters (*Puffinus griseus*) and Northern Fulmars (*Fulmarus glacialis*), both of which are attracted from downwind to odours of fish oils, squid and krill, and make investigatory manoeuvres around the releasing wick (Fig. 4.19) (Hutchison and Wenzel, 1980b; Hutchison *et al.*, 1984). Other local procellariiforms, such as the Pink-footed Shearwater (*Puffinus creatopus*), Black-footed Albatross (*Diomedea nigripes*) and Ashy Storm-Petrel (*Oceanodroma homochroa*), also approach the same stimuli. Non-procellariiforms in the same area show no such behaviour, but some, especially gulls, are readily attracted from all directions by a visual stimulus like popcorn or pieces of paper spread on the water. The smell of cod-liver oil also attracted Wilson's Storm-Petrels (*Oceanites oceanicus*), as well as Sooty Shearwaters, in the western Atlantic Ocean off Maine (Grubb, 1972). These observations were made at night and have been confirmed for the Sooty Shearwater in darkness off California, as well as for the Ashy Storm-Petrel (L. V. Hutchison, K. E. Stager and B. M. Wenzel, unpublished observations).

The older literature is rich in suggestive accounts of procellariiform reliance on odours, not only for foraging but also for components of reproductive behaviour (Wenzel, 1980). Relatively few positive results have been obtained in this area, however. It is well known that procellariiforms typically approach their breeding islands from downwind, but it is not clear whether this is a navigational or an

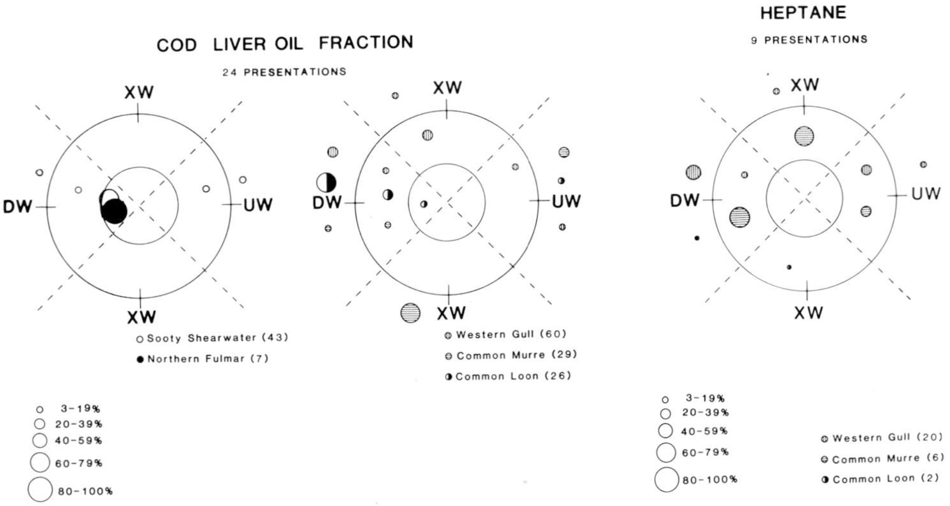

Fig. 4.19. Proportions of approaches by two procellariiforms, the Sooty Shearwater (*Puffinus griseus*) and Northern Fulmar (*Fulmarus glacialis*), and three non-procellariiforms, the Common Murre (*Uria aalge*), Western Gull (*Larus occidentalis*) and Common Loon (*Gavia immer*), to odours of the volatile fraction of cod-liver oil and to heptane as control stimulus according to wind direction and closeness of approach. The outer circle represents an approach between 10 and 50 m of the releasing wick, the inner circle of less than 10 m. DW, Approach from downwind of the wick; UW, approach from upwind; XW, approach from crosswind. Numbers in parentheses are total sightings. For example, 65% of 43 Sooty Shearwaters sighted during all tests with cod-liver oil fraction approached from downwind and flew within 10 m of the releasing wick. Of the remaining 18% that approached from downwind, some flew by at more than 50 m and some were between 10 and 50 m. The other Shearwaters sighted passed more than 10 m away from upwind. From Hutchison *et al.* (1984), by permission of the Canadian Wildlife Service.

aerodynamic aid. Grubb (1974) reported that Leach's Storm-Petrels (*Oceanodroma leucorhoa*), after landing on the island where their colony was located, walked upwind towards breeding or non-breeding burrows but moved at random if no burrow was entered. This finding has not been confirmed for the Wedge-tailed Shearwater (*Puffinus pacificus*) (Shallenberger, 1975) or Leach's Storm-Petrel (V. S. Fox, unpublished observations) on Pacific islands where no heavy forest cover darkened the island floor.

Olfaction has been implicated in pigeon homing as one of the mechanisms for orientation (Papi and Wallraff, 1982). Extensive positive evidence has come from one Italian laboratory with some support from other groups, but the issue appears complex and remains unsettled. The present interpretation is that orientation mechanisms are highly multifactorial and olfaction appears to be one component (Keeton, 1979).

Finally, a claim has been made for the ability of the Black-chinned Hummingbird (*Archilochus alexandri*) to use odour as a cue for the location of nectar (Goldsmith and Goldsmith, 1982). However, no olfactory bulb measurements have been made for any hummingbird species. The full significance of this report, therefore, is not clear until hummingbirds' locations on the scale of olfactory anatomy are known, but this finding emphasizes the importance of considering olfaction as a working modality in any avian species.

VI. Discussion

It is clear from this review that exploration of avian olfactory systems has been extremely limited. No histochemistry has been done on olfactory tissue, virtually no electron microscopy, and very little light microscopy on structures from the mucosa to higher brain centres. In addition to these great gaps in our knowledge of olfactory structures there are corresponding gaps for functional characteristics, both physiological and behavioural. Present evidence points to as much variation within Aves as across classes, at least in terms of proportional amounts of olfactory tissue, and so the range of species studied should be extended to avoid inappropriate generalisations. To characterise all birds on the basis of studying only the pigeon would be fallacious indeed.

Enough information has been gathered to suggest guidelines for future work. In anatomical and histochemical studies, it would be useful to include species with small, medium and large olfactory bulbs. Bulb size is positively correlated with the number of receptor cells in the mucosa and their axons in the olfactory nerve, and presumably with the extent of central projections as well. Physiological questions relate to the types of odours that excite the central neurons and the patterns of neural responses elicited by different stimuli. Adrian's proposal (1951) that discriminative capacity, not absolute sensitivity, increases with the number of sensory neurons predicts that the fulmar can discriminate more odours than the pigeon, for example, and such a difference should appear in neuronal response patterns. Accurate behavioural measurement of discriminative ability in a non-verbal subject requires carefully-designed procedures that usually involve training the animal under laboratory conditions. Such experiments are always challenging with undomesticated animals, if only because of the frequent difficulty of keeping them healthy during long periods in captivity.

The ultimate question is how any bird uses olfactory information. The most difficult studies are the behavioural ones, which should concentrate on natural species-specific behaviour. Among the gems of nineteenth-century natural history is an editorial letter in the May 15, 1873 issue of *Nature* citing observations by Mr. Tegetmeier on "Suppression of Scent in Pheasants": during incubation the hen protects the brood by suppressing her normal body and excretory odours so that hunting dogs pass within a few feet of her "without evincing the slightest cognizance of her proximity." On the other hand, Hoopoe chick nestlings develop enlarged uropygial glands and reserve the foul-smelling secretions until threatened by a predator, at which time they turn their rear ends toward the intruder and accurately eject a mixed stream of faeces and fatty acids at the target; the glands return to normal proportions and content when chicks are a month old (Sutter, 1946; Skead, 1950). Jones and Black (1979) have proposed that in groups of chicks the smell of blood during intra-flock pecking stimulates further aggressive pecking until social rank-order is established. Kolb (1973) suggested that a species of low-flying insectivorous bat, *Myotis myotis,* emits specific low-level (16–20 kHz) sonar pulses—"olfactory sounds"—which whirl up odorant molecules and insect prey from hidden niches in moss. Both *Steatornis caripensis* and *Collocalia esculenta* nest in very crowded colonies in total darkness; nests contain droppings and epidermal debris, possibly oil-gland marking. Could these species have developed a similar low-frequency "dust-raising" sonar for precision in terminal homing?

A century and a half after Audubon, then, our incomplete understanding of olfactory function in birds is that they, like other vertebrates, use olfactory cues in nature (Stager, 1967; Wenzel, 1968; Grubb, 1972, 1974; Hutchison and Wenzel, 1980b; Hutchison *et al.*, 1984) and can be trained to respond to olfactory cues in the laboratory (Michelsen, 1959; Henton, 1969), that olfactory thresholds for different odours vary among species (Stattelman *et al.*, 1975) and that avian olfactory bulb: hemisphere ratios differ more than 10-fold amongst species and correlate closely with olfactory membrane surface area (Bang and Cobb, 1968). Birds' uses of olfactory cues may include food finding, social integration, alarm response, defence and orientation. There is some evidence for hormonal modulation of odorant sources and of olfactory behaviour. The possibility of imprinting to odours has not been directly examined.

Field studies are urgently needed. The analysis of life habits according to bulb: hemisphere ratios (Figs. 4.13, 14) shows that relatively large bulbs are associated with ground nesting, colonial breeding, water habitat, and a diet of meat or fish; it implies that odours are especially useful under these conditions and less so under their opposites. Studies that manipulate olfactory variables and observe the natural behaviour of species deliberately chosen to represent varying patterns of life habits will help to identify the role of odours for the species involved. The experiments with certain procellariiforms described earlier are examples of such an approach. Even astute observation, with olfaction specifically in mind, during field work on other topics can be helpful. Procellariiforms had been suspected of olfactory reliance on the basis of field reports, for example, before any anatomical evidence was available (see Murphy, 1936, for abundant examples).

Among the open questions about structure and function in the avian nose are the following:

1. Are perforate and imperforate nasal septa random accidents of evolution, or is each type genetically encoded because it has survival value for a species? Would air current eddying confer some advantage on species with perforate septa? In binarual hearing there is logarithmic enhancement of sound in relation to concentration, and the effect is doubled by binaural reception. Would perforate septa "scramble" olfactory information without interfering with receptor function? Diffusion constants differ in air and water, yet both aquatic and land birds have perforate septa (Table I). As for concentration, one would speculate that in all animals receptors for the most critical odours are located at the site of initial impact: in birds this would be the rostral limit of the receptor membrane on the arch of the nasal roof.

2. Nasal valves are a vital mechanism in aquatic diving birds with open nares, but why should kiwis, kookaburras and one land kingfisher which is not known to dive (*Halcyon senegalensis*) have them? Are they hangovers from ancestral prototypes? Were the archetypal coraciiforms piscivorous divers? Bee-eaters (Tree, 1961; Took, 1963) and rollers (Reynolds, 1965) have been seen to dive occasionally, but valves have not been found in preserved specimens. Perhaps they are present in embryos. Alone of Rallidae, Button-Quails (*Turnix*) have lost the valves, and *Phaethon*, alone of Pelecaniformes, has retained both open nares and

diving valves. Suggestions that the valves may prevent dust from entering the nose, or that they function as velocity sensors in air-diving birds are untenable, for they are activated only in water as the result of water pressure.

3. The sheet of continually-renewed mucus that bathes the olfactory receptors is produced by glands that have species-specific staining characteristics. These histo-chemical properties change when internal or external environmental conditions are altered (Bang and Bang, 1969). Olfactory receptor specificity is one of the selective forces in specialisation, yet the biochemical role of olfactory gland mucus in modulating the receptor response is an area of profound ignorance.

4. Since right and left cerebral hemispheres in the human brain are known to have independent but complementary functions, might the same be true of vertebrate olfactory bulbs and secondary centres?

References

Adrian, E. D. (1951). Olfactory discrimination. *L'Année psychol.* **50,** 107–113.

Ariëns-Kappers, C. V., Huber, G. C. and Crosby, E. C. (1936). "Comparative Anatomy of the Nervous System of Vertebrates." Macmillan, New York.

Audubon, J. J. (1834). "Ornithological Biography," Vol. II, p. 33. Adam and Charles Black, Edinburgh.

Bang, B. G. (1960). Anatomical evidence for olfactory function in some species of birds. *Nature, Lond.* **188,** 547–549.

Bang, B. G. (1961). The surface pattern of the nasal mucosa and its relation to mucus flow—a study of chicken and herring gull nasal mucosae. *J. Morph.* **109,** 57–72.

Bang, B. G. (1964). The nasal organs of the Black and Turkey Vultures; a comparative study of the Cathartid species *Coragyps atratus* and *Cathartes aura septentrionalis atratus. J. Morph.* **115,** 153–184.

Bang, B. G. (1965). Anatomical adaptations for olfaction in the Snow Petrel. *Nature, Lond.* **205,** 513–515.

Bang, B. G. (1966). The olfactory apparatus of tubenosed birds (Procellariiformes). *Acta anat.* **65,** 391–415.

Bang, B. G. (1971). Functional anatomy of the olfactory system in 23 orders of birds. *Acta anat.* **79,** Suppl., 1–76.

Bang, B. G. and Bang, F. B. (1958). A comparative study of the vertebrate nasal chamber in relation to upper respiratory infections. *Bull. Johns Hopkins Hosp.* **104,** 107–149.

Bang, B. G. and Bang, F. B. (1967). Laryngotracheitis virus in chickens. *J. exp. Med.* **125,** 409–428.

Bang, B. G. and Bang, F. B. (1969). Experimentally induced changes in nasal mucus secretory systems and their effects on virus infections in chickens. *J. exp. Med.* **130,** 105–120.

Bang, B. G. and Cobb, S. (1968). The size of the olfactory bulb in 108 species of birds. *Auk* **85,** 55–61.

Bang, B. G., Bang, F. B. and Foard, M. A. (1966). Responses of upper respiratory mucosa to drugs and viral infections. *Am. Rev. resp. Dis.* **93,** 142–149.

Benham, W. B. (1906). The olfactory sense of Apteryx. *Nature, Lond.* **74,** 222–223.

Bojsen-Möller, F. (1965). The anterior nasal glands. *Int. Rhinol.* **3,** 117–127.

Bumm, A. (1883). Das Grosshirn der Vögel. *Z. wiss. Zool.* **38,** 430–467.

Burne, R. H. (1908). Exhibitions of . . . olfactory organs . . . of an albatross *Diomedea exulans. Proc. zool. Soc. Lond.* p. 65.

Cobb, S. (1960a). A note on the size of the avian olfactory bulbs. *Epilepsia* **1,** 394–402.

Cobb, S. (1960b). Observations on the comparative anatomy of the avian brain. *Perspect. Biol. Med.* **3,** 383–408.

Craigie, E. H. (1940). The cerebral cortex in Paleognathine and Neognathine birds. *J. comp. Neurol.* **73,** 179–234.

Davis, R. G. (1973). Olfactory psychophysical parameters in man, rat, dog, and pigeon. *J. comp. physiol. Psychol.* **85,** 221–232.

Edinger, L. (1908). The relations of comparative anatomy to comparative psychology. *J. comp. Neurol. Psychol.* **18,** 437–457.

Fox, V. S., Davidian, H., Hutchison, L. V. and Wenzel, B. M. (1982). Olfactory neuroanatomy of the Sooty Shearwater and Northern Fulmar. *Joint Meeting, Pacific Seabird Group–Australian Seabird Group, Honolulu, Hawaii* (abstr.).

Fox, V. S., Davidian, H., Wenzel, B. M. and Hutchison, L. V. (1983). Comparative anatomy of the olfactory bulb of the Sooty Shearwater, Northern Fulmar and Rock Dove. *Ass. Chemoreception Sci., Sarasota, Fla.* (abstr. 41).

Goldsmith, K. M. and Goldsmith, T. H. (1982). Sense of smell in the Black-chinned Hummingbird. *Condor* **84,** 237–238.

Graziadei, P. P. C. and Monti Graziadei, G. A. (1977). Continuous nerve cell renewal in the olfactory system. *In* "Handbook of Sensory Physiology" (M. Jacobson, Ed.), Vol. IX. Springer-Verlag, New York.

Grubb, T. C., Jr. (1972). Smell and foraging in shearwaters and petrels. *Nature, Lond.* **237,** 404–405.

Grubb, T. C., Jr. (1974). Olfactory navigation to the nesting burrow in Leach's petrel *Oceanodroma leucorrhoa*. *Anim. Behav.* **22,** 192–202.

Gurney, J. H. (1922). On the sense of smell possessed by birds. *Ibis* **4,** 225–253.

Henton, W. W. (1969). Conditional suppression to odorous stimuli in pigeons. *J. exp. Anal. Behav.* **12,** 175–186.

Huber, G. C. and Crosby, E. C. (1929). The nuclei and fiber paths of the avian diencephalon with consideration of telencephalic and certain mesencephalic centers and connections. *J. comp. Neurol.* **48,** 1–225.

Hutchison, L. V. and Wenzel, B. M. (1980a). Olfactory function in the Northern Fulmar (Fulmaris glacialis). *In* "Olfaction and Taste 7" (H. van der Starre, Ed.). IRL Press, London and Washington, D.C.

Hutchison, L. V. and Wenzel, B. M. (1980b). Olfactory guidance in foraging by procellariiforms. *Condor* **82,** 314–319.

Hutchison, L. V. and Wenzel, B. M. (1982). Activity of central olfactory neurons in the pigeon. *In* "Avian Navigation" (F. Papi and H. Wallraff, Eds.). Springer-Verlag, Berlin and New York.

Hutchison, L. V., Wenzel, B. M., Stager, K. E. and Tedford, B. L. (1984). Further evidence for olfactory foraging by Sooty Shearwaters and Northern Fulmars. *In* "Proceedings of the Pacific Seabird Group Symposium on Marine Birds: Feeding Ecology and Commercial Fisheries Relationships" (D. N. Nettleship, G. A. Sanger and P. F. Springer, Eds.). *Can. Wildl. Serv., Spec. Publ,* 72–77.

Jones, R. B. and Black, A. J. (1979). Behavioral responses of the domestic chick to blood. *Behav. Neurol. Biol.* **27,** 319–329.

Jouventin, P. (1977). Olfaction in Snow Petrels. *Condor* **79,** 498.

Karten, H. J. and Dubbeldam, J. L. (1973). The organization and projections of the paleostriatal complex in the pigeon (Columba livia). *J. comp. Neurol.* **148,** 61–89.

Keeton, W. T. (1979). Avian orientation and navigation. *A. Rev. Physiol.* **41,** 353–366.

Klinckowström, A. (1890). Les lobes olfactoires du *Fulmarus glacialis*. *Biol. Ogiska Förening Stockh.* **3,** 10–11.

Kolb, A. (1973). Riechverhalten und Riechlaute der Mausohrfledermaus. *Sonderdruck aus Z. Säugetierk.* **38,** 277–284.

Macadar, A. W., Rausch, L. J., Wenzel, B. M. and Hutchison, L. V. (1980). Electrophysiology of the olfactory pathway in the pigeon. *J. comp. Physiol.* **137,** 39–46.

MacDonald, J. D. (1960). Secondary external nares of the gannet. *Proc. zool. Soc. Lond.* **135,** 357–363.

Mangold, O. (1946). Die Nase der segelnden Vögel ein Organ des Stromungesinnes? *Naturwissenschaften* **33,** 19–23.

Meckel, A. (1816). Anatomie des Gehirns der Vögel. *Dt. Arch. Physiol.* **2,** 25.

Michelsen, W. J. (1959). Procedure for studying olfactory discrimination in pigeons. *Science, Wash., D.C.* **130,** 630–631.

Miller, L. (1942). Some tagging experiments with blackfooted albatrosses. *Condor* **44,** 3–9.

Murphy, R. C. (1936). "Oceanic Birds of South America," Vols 1 and 2. American Museum of Natural History, New York.

Murphy, R. C. and Harper, F. (1921). A review of the diving petrels. *Bull. Am. Mus. Nat. Hist.* **44,** 495–554.

Neuhaus, W. (1963). On the olfactory sense of birds. *In* "Olfaction and Taste" (Y. Zotterman, E.), 111–123. Macmillan, New York.

Owen, R. (1872). On *Dinornis* (part XVI). *Trans. zool. Soc. Lond.* **7,** 381–396.

Papi, F. and Wallraff, H. G., Eds. (1982). "Avian Navigation." Springer-Verlag, Berlin and New York.

Parker, T. J. (1891). Observations on the anatomy and development of *Apteryx*. *Phil. Trans. R. Soc.* **182B,** 25–134.

Paulli, S. (1928). "Beiträge zur Lehre von der Pneumaticität des Säugetierschädels." Kandrup & Wunsch, Copenhagen.

Reynolds, J. F. (1965). Bee-eaters and rollers diving into water. *E. Afr. Wildl. J.* **3,** 129–130.

Rieke, G. K. and Wenzel, B. M. (1978). Forebrain projections of the pigeon olfactory bulb. *J. Morph.* **158,** 41–55.

Sells, W. (1837). On the habits of the *Vultur aura* with notes on the dissections of two heads by R. Owen. *Proc. zool. Soc. Lond.* Part V, 33–35.

Shallenberger, R. J. (1975). Olfactory use in the Wedge-tailed Shearwater (*Puffinus pacificus*) on Manana Island, Hawaii. *In* "Olfaction and Taste 5" (D. A. Denton and J. P. Coghlan, Eds.). Academic Press, New York and London.

Shumake, S. A., Smith, J. C. and Tucker, D. (1969). Olfactory intensity-difference thresholds in the pigeon. *J. comp. physiol. Psychol.* **67,** 64–69.

Sieck, M. H. and Wenzel, B. M. (1969). Electrical activity of the olfactory bulb of the pigeon. *Electroenceph. clin. Neurophysiol.* **26,** 62–69.

Skead, C. J. (1950). A study of the African hoopoe. *Ibis* **92,** 434–463.

Stager, K. E. (1964). The role of olfaction in food location by the turkey vulture *(Cathartes aura), Los Angeles County Mus. Contr. Sci.* **81,** 1–63.

Stager, K. E. (1967). Avian olfaction. *Am. Zool.* **7,** 415–419.

Stattelman, A. J., Talbot, R. B. and Coulter, D. B. (1975). Olfactory thresholds of pigeons (*Columba livia*), quail (*Colinus virginianus*) and chickens *(Gallus gallus). Comp. Biochem. Physiol.* **50A,** 807–809.

Strong, R. M. (1911). On the olfactory organs and sense of smell in birds. *J. Morph.* **22,** 619–662.

Suschkin, P. (1905). Vergleichende Osteologie der normalen Tagraubvögel (Accipitres) und die Fragen der Klassifikation. *Nouv. Mém. Soc. impér. Nat. Moscou* **16.** Cited by Technau (1936).

Sutter, E. (1946). Das abwehrverhalten nestjunger Widehopfe. *Ornithol. Beobach.* **43,** 72–81.

Technau, G. (1936). Die Nasendruse der Vögel. *J. Ornithol.* **4,** 511–617.

Tree, A. J. (1961). Bee-eaters diving into water. *Ostrich* **32,** 188–189.

Took, J. M. E. (1963). Bee-eaters diving into water. *Ostrich* **34,** 176.

Tucker, D. (1965). Electrophysiological evidence for olfactory function in birds. *Nature, Lond.* **207,** 34–36.

Van Tyne, J. and Berger, A. J. (1959). "Fundamentals of Ornithology." Wiley, New York.

Walter, W. G. (1943). Some experiments on the sense of smell in birds. *Arch. néerl. Physiol.* **27,** 1–72.

Watanabe, T. and Yasuda, M. (1968). Peripheral course of the olfactory nerve in the fowl. *Jap. J. vet. Sci.* **30,** 275–279.

Waterton, C. (1839). "Essays on Natural History." Longmans, Orme, Brown, Green and Longmans, London.

Wenzel, B. M. (1967). Olfactory perception in birds. *In* "Olfaction and Taste 2" (T. Hayashi, Ed.). Pergamon Press, Oxford.

Wenzel, B. M. (1968). Olfactory prowess of the kiwi. *Nature, Lond.* **220,** 1133–1134.

Wenzel, B. M. (1971a). Olfactory sensation in the kiwi and other birds. *In* "Orientation: Sensory Basis" (H. E. Adler, Ed.). *Ann. N.Y. Acad. Sci.* **188,** 183–193.

Wenzel, B. M. (1971b). Olfaction in birds. *In* "Handbook of Sensory Physiology" (L. M. Beidler, Ed.), Vol. 4, Part 1. Springer-Verlag, Berlin and New York.

Wenzel, B. M. (1980). Chemoreception in seabirds. *In* "Behavior of Marine Animals: Current Perspectives in Research, Marine Birds" (J. Burger, B. L. Olla and H. E. Winn, Eds.), Vol. 4. Plenum Press, New York.

Wenzel, B. M. and Sieck, M. H. (1972). Olfactory perception and bulbar electrical activity in several avian species. *Physiol. Behav.* **9,** 287–294.

Wood Jones, F. (1937a). The olfactory organ of the Tubinares. I. *Emu* **36,** 281–286.

Wood Jones, F. (1937b). The olfactory organ of the Tubinares. III. *Emu* **37,** 128–131.

5
External and middle ears

ROLAND KÜHNE* AND BRIAN LEWIS

Department of Biological Sciences, City of London Polytechnic, London, England

---

CONTENTS

I. Introduction

Hearing is, after vision, the second most important sense in birds. It fulfills one or more of three main functions: (1) a warning function, signalling the presence of a

*Present address: Institut für Zoologie der TU, Lichtenbergstrasse 4, D-8046 Garching, Federal Republic of Germany.

predator, (2) a social function in intra-specific acoustic communication, and (3) a function in detecting prey. In addition, a few bird species have evolved the use of acoustic signals for echo location (Pye, 1980). The communicatory and prey-tracking functions have been highly evolved in some groups of birds. In fact, songbirds are the vertebrates which display the most complex auditory communication signals, and no terrestrial animal is known to reach the acuity of the hunting owl in acoustically locating a target in three-dimensional space.

To react in a biologically appropriate manner to an acoustic signal, a bird has to be able to recognise and to locate the sound source; that is, its auditory system has to provide information about both the nature and the position of the sound source. Information about the nature of the signal is contained mainly in the frequency spectrum and the time course; information about the position of the source is contained mainly in the intensity and the direction of the signal.

In the chain of events leading from the arrival of a stimulus to its central nervous representation, the external and middle ears perform the first stages of signal perception, that is, the first interaction of the stimulus with a body structure and the first part of the physical transformation of sound energy. Sensitivity, frequency range, pitch discrimination, time resolution and directionality of the auditory system are eventually the result of signal transduction in the inner ear and of central nervous processing. However, the extent of these abilities is determined by the limitations of the external and middle ears, and by their capacities for collecting sound energy and transforming it to provide optimal stimulation of the receptor organ so as to preserve all the relevant information contained in the signal. In fact, directional cues are provided only as a result of the special design of the external and/or middle ears, on each side alone, or in combination. Therefore, the highly-evolved auditory performance of some birds should be reflected in the form and function of their external and middle ears, bringing about optimisation of sensitivity and resolution. In some cases, the resulting morphological characteristics have even been used as an aid in taxonomic problems (Norberg, 1977; Saiff, 1981).

Comprehensive anatomical studies of the avian external and middle ear structures have been carried out since before the beginning of the century (for reviews of the earlier literature see Smith, 1904; Pohlman, 1921; Stellbogen, 1930; Freye-Zumpfe, 1953). However, investigations of the function of the middle ear have been directed mainly at mammals, and an understanding of the physical problems and functions of the peripheral auditory system of birds has developed only after the pioneer studies of Schwartzkopff (1949, 1950, 1955). Apart from some detailed studies of ecophysiological specialisations, particularly in owls (Norberg, 1968, 1978; Payne, 1971; Knudsen and Konishi, 1979), few biophysical and physiological investigations of the external and middle ears of birds have been carried out (Hill *et al.*, 1980; Coles *et al.*, 1980; Rosowski and Saunders, 1980; Moiseff and Konishi, 1981b.).

The avian ear is the product of a complex evolutionary process, and to understand the roles of the structures found in the external and middle ears it is necessary to study the physical problems and the functional requirements which contributed to the shaping of these structures during their evolution. Most birds hear over about the same frequency range as humans. However, due to the differences in body size and in anatomical and functional requirements (e.g. the need for higher time resolution, localisation in three dimensions rather than just in the horizontal plane, and aerodynamic disadvantages in having pinnae), different physical and physiological problems occur and different constructional solutions are to be found. In this chapter the adaptations of the external and middle ears to the specific biophysical problems of hearing in birds are outlined, in so far as these are known.

II. Biophysical aspects

The external and middle ears of vertebrates (and analogous structures in invertebrates) can be considered as accessory structures evolved to transmit acoustic energy incident on the head or body surfaces to the sensory receptor cells of the inner ear. Since the receptor cells are bathed in some form of tissue fluid, and since different media differ in their responses to sound, some such accessory structure system is necessary to transform the large velocities of movement that occur in a compressible medium such as air, to the much smaller velocities occurring in a dense medium such as water, without energy loss. The efficiency of the transforming system is reflected in the extent to which the "effective force" acting on the tympanic membrane is maintained at the inner ear.

A second major problem associated with the design of the peripheral auditory system is that, unlike the visual system, there is only a single pathway by which the airborne sound can pass from the surface of the head to the linearly-arranged receptor cells. Directional information cannot, therefore, be obtained directly from a spatially-organized grid of receptors but must depend on characteristic, spatially-dependent changes in the sound quality that are produced by the accessory structures.

The accessory structures, including the head, have finite physical dimensions and will therefore affect both the transformer function and the directionality of the ear. The point has been made, however, that only rarely can the physical effects be predicted from theory, since biological materials and conditions differ considerably from the ideal situations used to derive physical equations (Michelsen, 1983). Theory must be supplemented by experimentation, and too few experiments have been performed to allow any definitive conclusions on any aspect of outer or middle ear function in birds.

A. The middle ear as an acoustic transformer

To function as an efficient transformer, the middle ear must match the low impedance of the air to the much higher impedance of the cochlear fluids. In the absence of a transformer, much higher pressures are needed to achieve a given velocity in an incompressible medium such as water. From a consideration of air: seawater impedance ratios, Wever and Lawrence (1954) estimated the optimal transformer ratio of the middle ear to be 63. However, such calculations are based on progressive acoustic waves in an effectively infinite medium. In the audible range, the cochlea is much smaller than the wavelength and its impedance will depend on the way the fluids flow from the vestibular window to the cochlear window. As a result, cochlear impedance will be lower than the estimation of Wever and Lawrence (1954). Nevertheless, the problem of impedance matching is a fundamental one, since at an air–water interface only about 0·1% of sound of normal incidence is transmitted, the remainder being reflected. This corresponds to a transmission loss of about 30 dB. The middle ear apparatus, by acting as an acoustic impedance transformer, reduces this attenuation considerably. On the basis of the geometry of the middle ear structures, Norberg (1978) determined the ratio to be of the order of 56 in the Boreal Owl (*Aegolius funereus*), that is, close to the optimum transformation ratio.

In mammals, three principal mechanisms have been shown to contribute to the total impedance transformer ratio of the middle ear. The first of these involves the ratio of the area of the tympanic membrane to the area of the vestibular window. The concentration of the forces collected over the large area of the tympanic membrane onto the smaller area of the vestibular window results in a higher pressure. The second mechanism involves the lever action of the auditory ossicles, which decreases velocity and increases force at the scapus. Thirdly, a buckling movement of the tympanic membrane occurs during its vibrations and contributes to the velocity–force transformation.

In birds, the presence of the first mechanism, a large area ratio between the tympanic membrane and the vestibular window, is obvious. The experimental evidence for the employment of the other two mechanisms will be discussed in Section IV,D.

The impedance of the middle ear has three components that govern the transformation of the large-velocity, low-pressure meatus sound to the low-velocity, high-pressure vibrations in the *scala vestibuli,* that is, that determine the transfer function of the middle ear. At low frequencies, the main component determining middle ear impedance is the *stiffness reactance,* which expresses the elasticity of the system; the value of this component increases with stiffness and decreases with frequency. The air enclosed in the middle ear adds to the stiffness reactance, which is inversely proportional to the volume. Thus, the large air cavity of the middle ear of most birds, and its communication with several other air cavities (see p. 241, later), should result in a lowering of the impedance of the stiffness-controlled frequency region and therefore an improved transmission of these frequencies to the cochlea. These conclusions are certainly true for mammals (Møller, 1965), but too few experiments have yet been performed in birds (Saunders and Johnstone, 1972). At high frequencies, the *mass reactance* controls the system, reflecting its inertia. This value increases with the mass of the middle ear structures and with frequency. Thus, the high-frequency roll-off in auditory sensitivity is very much a reflection of the mass of the transmitting apparatus. Any means of reducing this mass, such as by trabeculation or hollowing the structures, will improve high-frequency sensitivity. At some intermediate frequency, the mass and stiffness cancel. The system is then dominated by the *resistance* or friction, which is independent of frequency. The impedance reaches a minimum, and resonance occurs. This frequency therefore usually corresponds to the region of best sensitivity (Saunders and Johnstone, 1972).

B. Sound localisation

To achieve directional information from a comparison of the binaural signal inputs, four principal cues may be used: ΔI, the difference of overall signal intensity at the two ears, ΔT, the difference in sound arrival (onset) time, $\Delta\phi$, the difference in ongoing phase, and Δs, the difference in relative intensities of certain frequency bands in a broad-banded signal (Knudsen, 1980).

Until recently our knowledge of the basic mechanisms for providing directional cues for sound localisation in terrestrial vertebrates has been based almost entirely on data obtained from mammals, and in particular, from humans. Humans rely on interaural sound pressure difference (ΔI) and on the time-of-arrival of the sound at the two ears (either onset time, ΔT, or phase difference, $\Delta\phi$).

Differences in sound pressure occur when the dimensions of the head exceed one-tenth the wavelength (λ). For the human head, a surplus pressure starts to build up above a few hundred hertz at the ear facing the source, this value increasing to a maximum of +6 dB at 5 kHz. Note that sound "shadows" at the contralateral ear do not occur until much higher frequencies are used. Clearly, the smaller the head, the higher the frequencies at which such effects begin to occur. Small mammals may overcome this problem by using high or ultrasonic frequencies and by developing large, parabolic-type receivers (the pinnae) on the head. For many other terrestrial vertebrates, including many birds, however, the frequencies used in communication have wavelengths much longer than head size. They can therefore expect little or no ΔI due to sound diffraction.

Sound travels in air with a velocity of about 340 m/sec, so that the maximum difference in time of arrival at human ears is about 0·5 msec. Long, pure tones of low frequencies (below about 1·5 kHz) produce phase-locked responses in the cochlear nerves, and the differences in the times of occurrence of responses can be computed by the central nervous system. Small mammals may find greater difficulty in using this cue, but since the latency of the neural response depends on the magnitude of the sound pressure, ΔI values may result in latency differences that are usable as time cues. The situation is very different for other small vertebrates. In small songbirds for example, with a distance of 1 cm or less between the ears, the difference in time of arrival will be less than 35 μsec and there will be no ΔI at the head.

Thus, if the ears of small birds acted as independent "mammalian-type" *pressure receivers,* minimal ΔI values would be established and, since they would also have minimal ΔT values ($\Delta\phi$), they would be faced with a very real problem of sound localisation. The solution to this problem was first realised by Autrum (1940), who pointed out that significant directionality could be achieved if the ears were *pressure difference receivers,* that is, if the sound waves could reach both the inner and outer surfaces of each tympanic membrane. The pressures and phases of the sounds would then interact at the tympanic membrane; its response would be dependent on the *net* pressure. Since the phase difference across the membrane is dependent on the angle of incidence of the sound, the displacement of the contralateral membrane could (in theory at least) fall to zero for some angles. In other words, very large differences can be created by these means, even at low frequencies, when no ΔI values are measureable outside the tympanic membranes. It is known that birds have extensive air spaces connecting their middle ears, and evidence is accumulating that these cavities transmit sounds from one ear to the other (Lewis, 1983).

In this respect, owls may be a special case (Knudsen, 1980). However, it is worth noting that little work has been done on the low-frequency, intra-specific communication calls of owls, attention having been concentrated on predator–prey interactions. The small prey (rodents, shrews), as they scuttle in the undergrowth, produce many high frequencies, and the localisation of the prey may be based on ΔI values produced by head diffraction. Further, the extreme morphological modifications of the outer ears of owls are considered to be correlated with their phenomenal localisation acuity, greatly improving both ΔI and ΔT components of the binaural cues.

III. The external ear and signal reception

Apart from some specialists in auditory localisation, birds have not evolved external sound-collecting structures, presumably because these would be aerodynamically disadvantageous. In most birds the external ear (*auris externa*) is inconspicuous, and the position of the external ear opening may be detectable only because of the slight difference in the covering feathers. However, within its physical and physiological limits, the external ear is well adapted (and in some cases extremely specialised) to fulfill the function of sound reception and the transmission of the signal to the tympanic membrane without distortion.

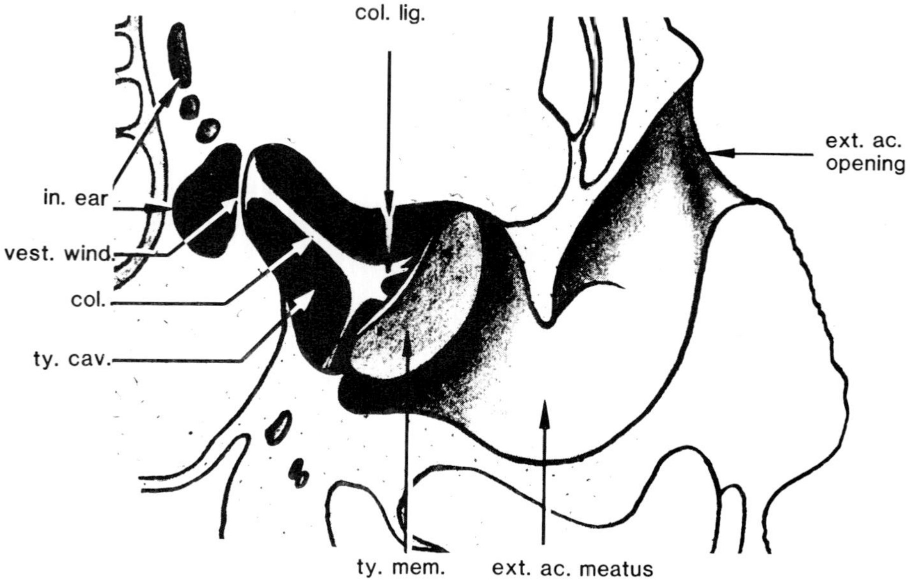

Fig. 5.1. Transverse section through the external and middle ears of the Ostrich (*Struthio camelus*) showing the shape and course of the external acoustic meatus and its relationship to the middle ear. The pharyngotympanic tube is not in the plane of the section. col., Columella; col. lig., columellosquamosal ligament; ext. ac. meatus, external acoustic meatus; ext. ac. opening, external acoustic opening; in. ear, inner ear; ty. cav., tympanic cavity; ty. mem., tympanic membrane; vest. wind., vestibular window. After Portmann (1950), courtesy of Masson, Paris.

A. *General anatomy*

The general design of the external ear is similar in most birds. As an example, the principal regions of the external ear of the Ostrich (*Struthio camelus*) and their relationships to the middle ear are shown in Fig. 5.1.

1. External acoustic opening

The external acoustic opening (*porus acusticus externus*) is located ventral and caudal to the eye, at the level of, or a little dorsal to, the gape. The shape of the ear opening may be a horizontal slit as in the Anhinga (*Anhinga anhinga*), oval or circular as in most Galliformes, or a vertical slit as in some Ciconiiformes and (most obviously) in owls (Strigiformes). The diameter of the ear opening is of the order of a few millimetres in most birds and is correlated with hearing ability or ecological specialisations (e.g. diving birds) rather than with body or head size. Only in some birds are structural specialisations found (see later).

2. Feathers

With only a few exceptions, such as casuariiforms, some vultures (Cathartidae) and ibises (Threskiornithidae), the ear opening is covered by the auricular feathers (*pennae auriculares*). Normally the feathers insert around the whole rim of the opening, it being possible to distinguish rostral and caudal groups (*tectrices auriculares rostrales, tectrices auriculares caudales*). In most birds the feathers inserting in the rostral and sometimes the ventral part of the rim are directed caudally. In diving birds, such as albatrosses (Diomedeidae) and storm-petrels (Hydrobatidae), the auricular feathers are no different from the feathers covering the rest of the head (Saiff, 1974). Usually, however, the auricular feathers are filamentous and have a slender rachis and long barbs without barbules. They protect the external acoustic meatus and the tympanic membrane, but they may also have an aerodynamic and acoustic function as suggested by Pumphrey (1948). Thus, they may ensure laminar air flow across the external ear opening during flight, reducing drag caused by turbulence at the ear opening. They may also act as a high-pass filter reducing the low-frequency wing noises which could otherwise mask important acoustic signals. The auricular feathers of the Barn Owl (*Tyto alba*) produce a sound attenuation of less than 1 dB between 3 and 20 kHz (Payne, 1971). Certainly, the external ear and the surrounding feathers are highly specialised in the Barn Owl, and few experiments have been done on other species in which the frequency ranges of important sounds extend well below 3 kHz. However, the auricular feathers do not appear to be an acoustic obstacle for biologically important sounds.

3. External acoustic meatus

Generally, the external acoustic meatus (*meatus acusticus externus*) has the form of a short tube, but its orientation and course, and as a result the structures which form its boundaries, vary. Most often, the meatus runs caudomedially or caudoventrally, an exception being the Ostrich (*Struthio camelus*), in which the orientation is rostromedial (Saiff, 1981). The meatus is often curved but can be straight as in the Great Blue Heron (*Ardea herodias*) (Saiff, 1978). In the Silver-gray Fulmar (*Fulmerus glacialoides*) the meatus is divided into an external and an internal part by the shaft of the quadrate (Saiff, 1974). The cross-sectional area of the meatus can be larger than that of the ear opening, a relatively large meatus being found in songbirds and parrots (Psittacidae) (Schwartzkopff, 1973). In specialised birds the meatus may widen to form a cavity with a trough-shaped cross-section (Iljitschew and Izwekowa, 1961). An extremely large cavity in the external ear, which communicates with air spaces in the surrounding structures, is found in many owls (Strigiformes).

Anatomically, the external acoustic meatus is defined rostrally by the quadrate and the caudal end of the lower jaw, whilst the *m. depressor mandibulae* forms the ventral and caudal walls, and the jaw musculature the dorsal wall. As the tympanic membrane is usually larger than the diameter of the meatus, and orientated obliquely to its axis, the membrane can form a considerable part of the medial, ventral and/or rostral wall. The meatus is lined with skin which differs little from the outer epidermis except for the lack of feathers and the occurrence of more glands (*glandulae meatus acustici externi*). Saiff (1974, 1978) described the skin of the meatus as wrinkled in a number of bird species, but he could not

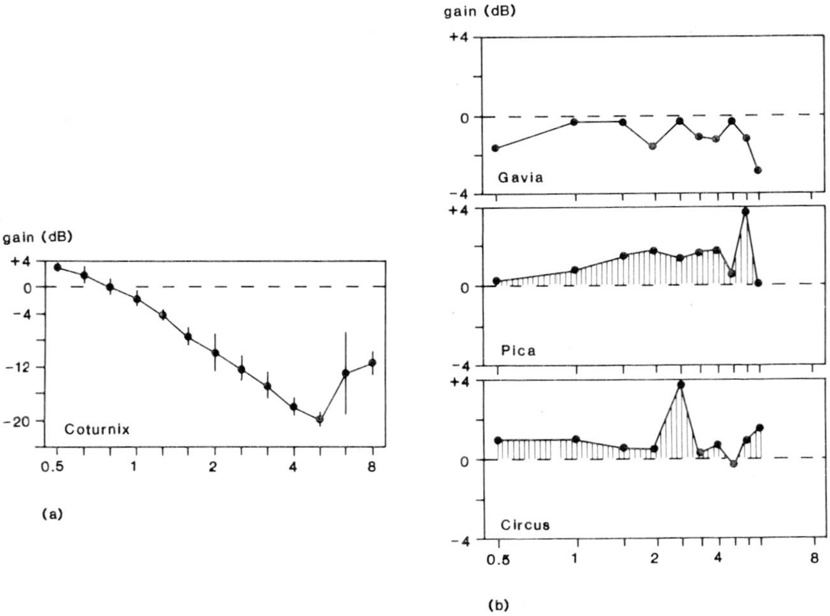

Fig. 5.2. Acoustic properties of the external acoustic meatus in some birds. (a) Sound pressure level measured near the site of the tympanic membrane with respect to the sound intensity outside the ear, using monaural stimulation in the Japanese Quail (*Coturnix japonica*). After Hill *et al.* (1980), courtesy of the Company of Biologists, Ltd. (b) Attenuation and amplification of sound measured at the site of the tympanic membrane after passing down the meatus in the Common Magpie (*Pica pica*), which has good hearing abilities, and in the Pallid Harrier (*Circus macrourus*), where some positive gain is observed for certain frequencies. The external ear of the Red-throated Loon (*Gavia stellata*), however, shows no amplification at any frequency. After Iljitschew and Izwekowa (1961), courtesy of Akademija Naūk SSSR, Moskva.

exclude preservation artifacts. In owls (Strigiformes), the skin lining the outer parts of the meatus is certainly thin, wrinkled and very moveable over the underlying skull bones (Stellbogen, 1930; Norberg, 1978).

B. Acoustic properties of the external ear

The acoustic meatus removes the tympanic membrane from the head surface to a more protected location. The membrane can therefore be larger and thinner, and hence more sensitive particularly to higher frequencies. This may be the main function of the meatus in most birds, since its diameter and length are too small (compared to the wavelengths of audible tones) either to produce major interference and/or resonance phenomena which would lead to positive gain, or confer directionality on the external ear. On the other hand, neither does the meatus greatly attenuate the sound during transmission to the tympanic membrane. In the Japanese Quail (*Coturnix japonica*), Hill *et al.* (1980) found the meatus to be a low-pass filter, with a roll-off of about 8 dB/octave above 1 kHz (Fig. 5.2a). Iljitchew and Izwekova (1961) measured the gain of the external ears in a number of avian species and found that the ears of birds with poor auditory sensitivity, such as loons (Gavidae), showed no gain; other birds with good auditory sensitivity such as the Corvidae, birds-of-prey (Fig. 5.2b), and owls (Strigiformes) in particular, showed considerable positive gain. In addition, the external ears of birds showing positive gain had some degree of inherent directionality, at least in certain frequency bands (see later).

C. Structural and functional specialisations

1. The external ear of owls

The largest and geometrically most complicated external ear is found in some owl species. Owls are acoustically the most sensitive of birds: up to 10 kHz, the auditory threshold of a Barn Owl (*Tyto alba*) matches that of the cat (Konishi, 1973). Many owls are partly, or entirely, nocturnal hunters. They use the noises produced by prey animals such as small rodents or shrews to track them, either with the support of vision or, as in the Barn Owl (*Tyto alba*), the Barred Owl (*Strix varia*) and the Long-eared Owl (*Asio otus*), entirely by auditory means (Payne and Drury, 1958).

Structural specialisations may affect the feathers and skin around the ear opening, the form of the ear opening itself and the external acoustic meatus. A remarkable specialisation is the bilateral ear asymmetry which occurs in some owls and will be described in detail in Section VI,B.

The most obvious structure associated with the external ear in owls is the facial ruff. The facial ruff is formed of modified, dense feathers which have a relatively thick rachis and reduced vanes. The ruff feathers insert directly caudal to the ear opening in the postaural skin flap. They form a concave wall which can almost encircle the face in some species, the left and right parts of the ruff being united below the beak by the feathers of the chin region. The form of the facial ruff is not very prominent in many species of the sub-family Buboninae, but is well developed in a number of Striginae. The most complete and elaborate facial ruffs are found in the Tytonidae. In the Barn Owl (*Tyto alba*) the ruff feathers insert (on each side of the head) in the postaural skin flap in a hexagonal pattern and are obviously packed as densely as is physically possible (Payne, 1971). Together they form two almost parabolic troughs in the centre of which lies the ear opening covered by the preaural skin flap (*operculum auris*). The facial ruff is covered by auricular feathers which insert rostral to the ear opening to form a facial disk which partly covers the ruff. In many owls, ruff and disk feathers are of different colour. The auricular feathers do not attenuate or reflect sound (Payne, 1971); the acoustically functional structure is the facial ruff (Fig. 5.3).

The widest diameter of the ruff is about 70 mm, which approximately corresponds to the wavelength of a 5-kHz tone. Above this frequency, the facial ruff can be expected to act as a parabolic antenna, collecting sound and focussing it onto the small area of the ear opening. It will therefore show both a gain function and an increase in directionality with frequency. Konishi (1973) calculated the sound amplification function of this parabolic antenna to be about 10 dB at 7 kHz, a frequency which falls in the range used by the owl for localisation. In biophysical measurements Payne (1971) found the ear of the Barn Owl (*Tyto alba*) to be highly directional above 8 kHz. The facial ruff appears to be essential for the Barn Owl to determine the elevation of a sound source (Knudsen and Konishi, 1979).

In the Barn Owl, the ear opening itself is almost square and relatively small, measuring about 7 mm. In other owls, the ear opening is a vertical slit which in

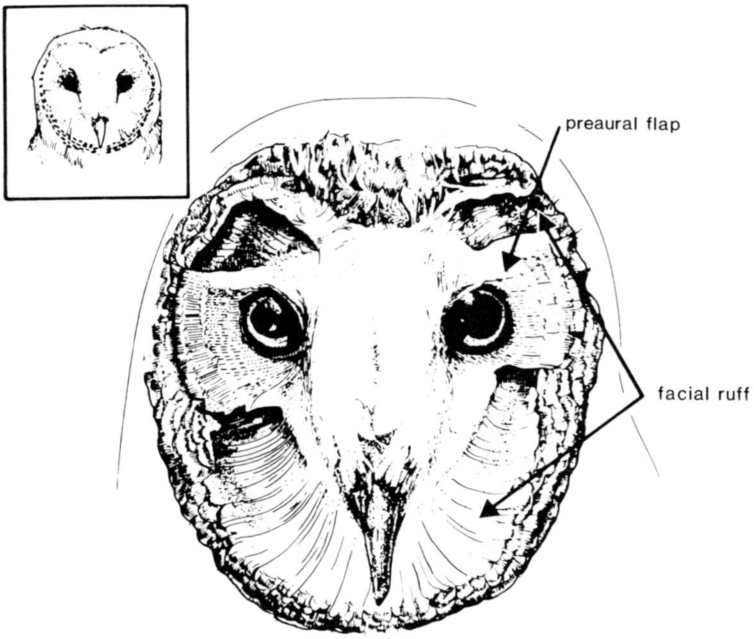

Fig. 5.3. The face of the Barn Owl (*Tyto alba*), showing the facial ruff after removal of the facial disk feathers. The inset shows the normal appearance of the owl. Note the different heights of the left and right preaural flaps. From Knudsen (1980), courtesy of Springer-Verlag, New York.

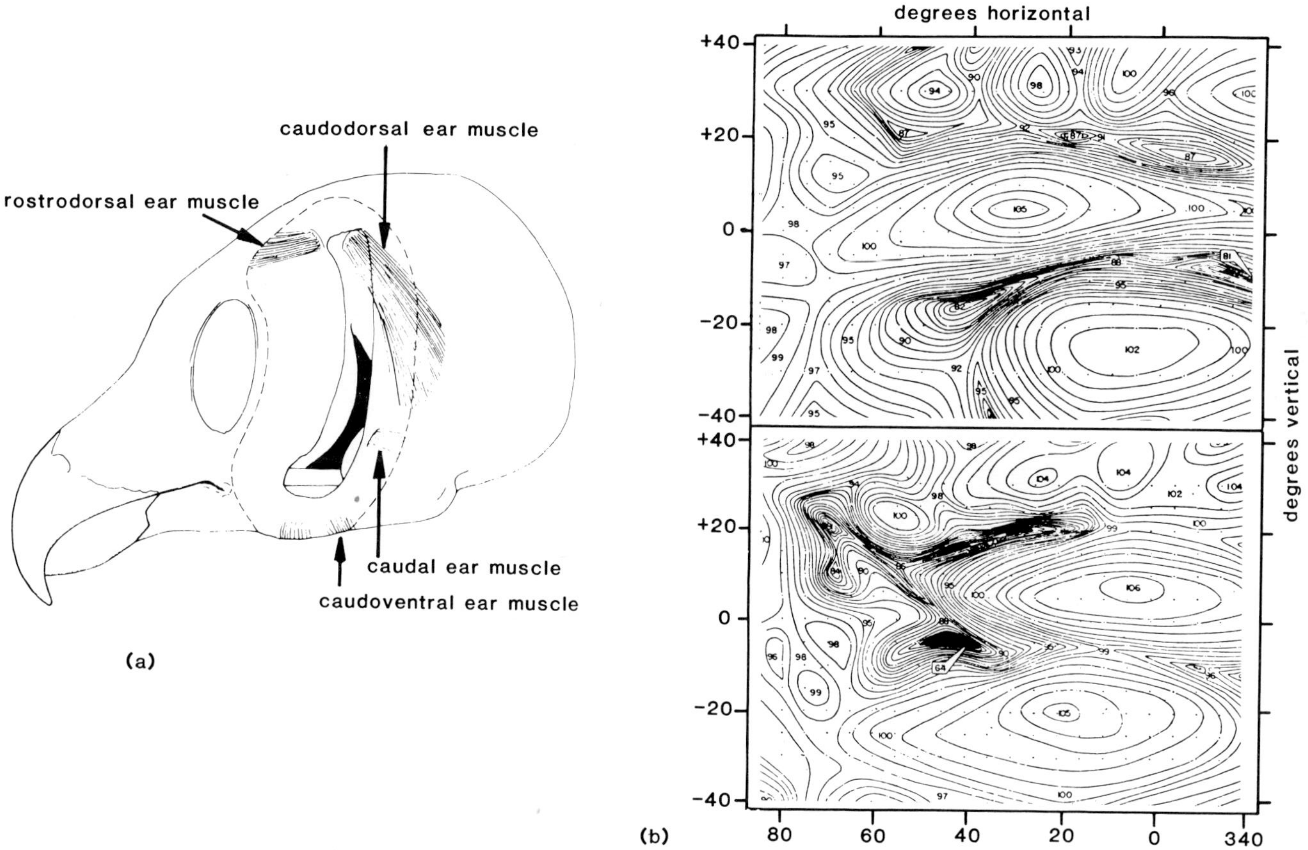

Fig. 5.4. (a) Schematic representation of the four external ear muscles in the Boreal Owl (*Aegolius funereus*). The rostrodorsal muscle acts on the preaural flap, the other three muscles on the postaural flap. The caudodorsal muscle is most important in moving the postaural flap rostrally and caudally. From Norberg (1978), courtesy of the Royal Society, London. (b) The change in directional sensitivity of the right ear of a Barn Owl (*Tyto alba*) at 13 kHz with a change in the position of the ear flap from ''resting,'' i.e. pushed rostrally (top), to ''alert'' (below). The patterns show the sound intensity measured at the site of the tympanic membrane when the sound source was moved around the owl's head to the positions indicated by the degree scales. The iso-intensity lines represent 1-dB steps. From Payne (1971), courtesy of the Company of Biologists.

extreme cases, such as the Long-eared Owl (*Asio otus*) and the Boreal Owl (*Aegolius funereus*), can be the height of the skull (Norberg, 1977). Often the ear opening is surrounded by a skin ridge or skinfold forming preaural (rostral) and/or postaural (caudal) flaps, the opercula of the ear (*opercula auris*). The pre-aural flap may entirely or partly cover the ear opening as in the Tawny Owl (*Strix aluco*) and the Long-eared Owl (*Asio otus*). The position of the flaps can be controlled by muscles. Although these muscles are found in most birds, they are very prominent in owls. Whilst numerous muscle fibres occur under the skin around the ear opening, it appears that only some of them are arranged as discrete muscle groups (Norberg, 1978). This may be one reason why the anatomical and functional descriptions of the muscles vary greatly for different birds. Thus, Stellbogen (1930) described three external ear muscles in the Tawny Owl (*Strix aluco*), whereas Norberg (1978) identified four muscles in the Boreal Owl (*Aegolius funereus*) (Fig. 5.4a), three of them acting on the post-aural skin flap. The exact positions of the aural flaps and facial ruff greatly affect the directional pattern of the ear (Schwartzkopff, 1962; Norberg, 1968; Payne, 1971) (Fig. 5.4b), but the extent to which owls acutally use this signal control maneouvre to improve localisation is not yet known.

The trend to enlarge the components of the external ear extends to the external meatus, which can be quite spacious. Further, some owls possess air cavities between the skin covering the skull and the ear flaps, that open into the superficial part of the meatus. As a result, the external ear structures and the air cavities of many owls reach sizes which correspond to half or the full wavelength of tones above about 5 kHz. The occurrence of frequency-dependent interference and resonance phenomena cannot therefore be excluded in these cases; the external ear of owls may have functions in sound localisation similar to those performed by the pinnae of mammals.

External ear specialisations for improving auditory sensitivity are not, however, restricted to owls. Although not as elaborate as those of owls, facial ruffs which may collect sound and enhance directionality are also found in some other birds. The occurrence of such structures appears to be correlated with an increase in the importance of hearing in one or other behavioural context. Examples include birds of prey such as the Pallid Harrier (*Circus macrourus*) and the Marsh Harrier (*Circus aeruginosus*) or the Eurasian Nightjar (*Caprimulgus europaeus*) (Iljitschew, 1974).

2. *The external ear of diving birds*

The middle ear of diving birds has to be protected from water and water pressure; the tympanic membrane cannot resist the change in pressure that occurs when diving to depths of up to 30 m (Kartaschew and Iljitschew, 1964). However, this protection cannot be allowed to affect too much the function of the ear when the head is not submerged, and in fact, diving birds maintain reasonable hearing abilities in air. Various mechanisms have been developed to solve this problem. In diving birds, the auricular feathers do not usually differ from the head plumage and they fulfil the same hydrodynamic and waterproofing functions (Kartaschew and Iljitschew, 1964; Saiff, 1974). Often the feathers are particularly

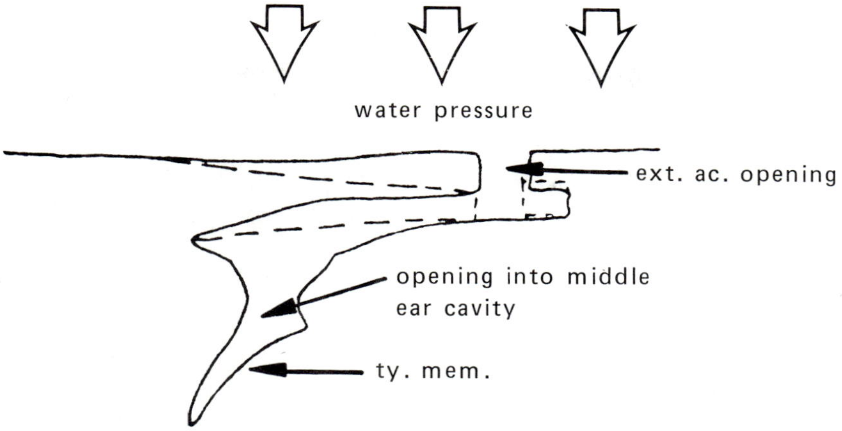

Fig. 5.5. The mechanism to "shut off" the ear in auks (Alcidae). The schematic drawing shows a horizontal section through the external ear. The external acoustic opening (ext. ac. opening) and the bony opening in the skull to the middle ear are not in line. The external ear is closed when increasing water pressure during diving compresses the external ear sac as indicated by the dotted lines. ty. mem., Tympanic membrane. From Kartaschew and Iljitschew (1964), courtesy of *Deutsches Journal für Ornithologie,* **105,** 128.

dense rostral to the ear opening and the feathers overlap, thus effectively closing the ear when pressed to the head underwater.

The ear opening in the skin of diving birds is usually smaller than the meatus diameter; it is extremely small in the Guanay Cormorant (*Phalacrocorax bougainvillii*), measuring only 0·1 mm in diameter (Saiff, 1978). Also in the cormorant, the ear opening is relatively rostral in position and the distal end of the acoustic meatus expands to produce a sac draped externally over the quadrate (Saiff, 1978). Furthermore, the external ear opening and the opening in the skull do not face each other. A similar arrangement is found in auks (Alcidae). Kartaschew and Iljitschew (1964) suggested that underwater the ear sac would be compressed by water pressure and collapse, thus shutting off the proximal part of the external ear (Fig. 5.5). In addition, in auks the ear opening can be actively closed by the *m. cucullaris,* which inserts on the sac.

All these protective mechanisms will, of course, impair the acoustic properties of the external ear to some extent. In auks, Kartaschew and Iljitschew (1964) described a number of improvements of middle ear and/or inner ear structure and function, which partly compensate for the peripheral loss in sensitivity. They also showed the close correlation between the prominence of protective mechanisms and ecology in related auk species.

IV. The middle ear and signal transformation

The function of the middle ear (*auris media*) is to transfer accoustically-induced vibrations of the tympanic membrane to the vestibular window. Since the impedance of air must be matched to that of the inner ear perilymph, this transfer must be in the form of a power transformation: the large amplitude and velocity of the vibrations of the tympanic membrane must be converted into vibrations of small

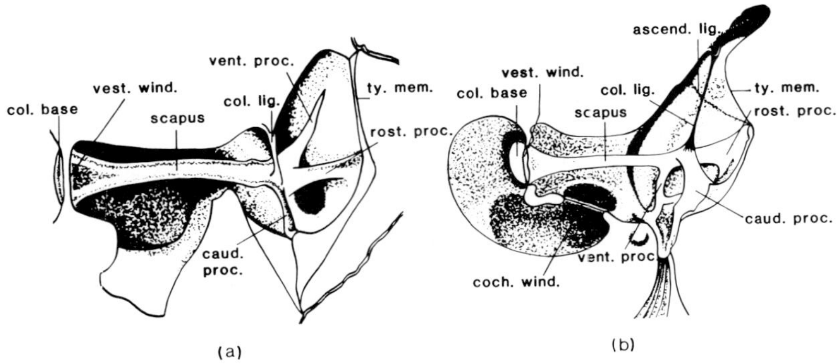

Fig. 5.6. Schematic drawings of the middle ear of (a) *Anser* [after Freye-Zumpfe (1953), courtesy of Martin-Luther-Univ. Halle Wittenberg] and (b) *Gallus* [after Pohlman (1921)]. ascend. lig., Ascending ligament; coch. wind., cochlear window; col. base, columellar base; col. lig., columellosquamosal ligament; rost. proc., vent. proc., caud. proc,, rostral, ventral and caudal processes of the extracolumellar cartilage; ty. mem., tympanic membrane; vest. wind., vestibular window.

amplitude but great force at the vestibular window. For many years, the avian middle ear with its single ossicle has been considered to be a simple version of the mammalian ear. However, studies of middle ear dynamics have shown that the middle ear apparatus of birds has its own specific and efficient mechanism for performing its function.

A. *General design of the middle ear*

The middle ear apparatus resides in the air-filled tympanic cavity (*cavitas tympanica*), which is separated from the acoustic meatus by the tympanic membrane (*membrana tympanica*) (Fig. 5.6). The tympanic cavity communicates with air spaces in the surrounding bones and, via the pharyngotympanic tube, with the oral cavity. The acoustic transfer from the tympanic membrane to the vestibular window (*fenestra vestibularis*) is mediated by the single bony ossicle, the columella (*columella auris*). The shaft or scapus of the columella (*scapus columellae*) expands proximally to form the base (*basis columellae*) or footplate, which fits into the vestibular window. The columella articulates distally with the tripod-like extracolumellar cartilage (*cartilago extracolumellaris*), which is attached to the tympanic membrane. The position of the columella is stabilised by the elastic columellosquamosal ligament (*lig. columellosqamosum*) and, to some extent, by the columellar muscle (*m. columellae*). The contact the extracolummellar cartilage makes with the tympanic membrane is strengthened by the so-called ascending ligament.

B. *The middle ear cavity*

1. *Anatomy*

The form of the tympanic cavity is irregular, usually narrowing medially. In general it is a spacious cavity, extending ventrally to the mandible, and caudally

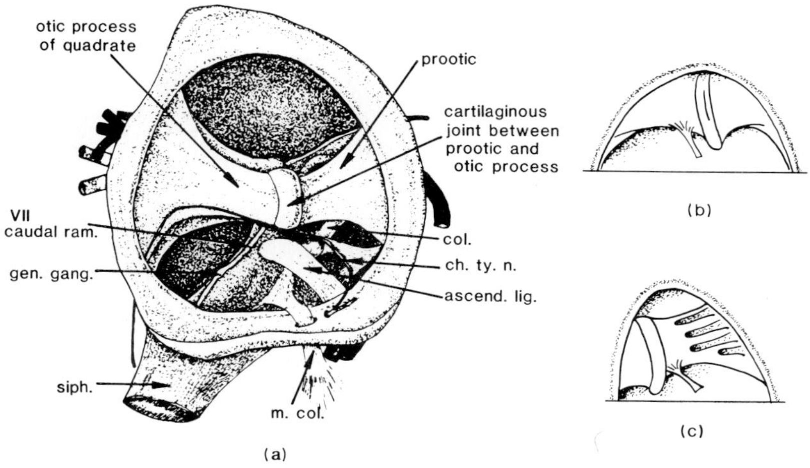

Fig. 5.7. (a). Drawing of the tympanic cavity of the Tawny Owl (*Strix aluco*) as seen from the outside after removal of the tympanic membrane, showing the ridge formed by the otic process of the quadrate and the proötic bone. ascend. lig., Ascending ligament; ch. ty. n., chorda tympani nerve; col., columella; gen. gangl., geniculate ganglion; m. col., columellar muscle; siph., siphonium; VII caud. ram., caudal ramus of facial nerve. From Stellbogen (1930). (b) and (c) Appearance of the ridge formed by the otic process of the quadrate and the proötic bone in (b) *Anser* and (c) the House Sparrow (*Passer domesticus*). After Freye-Zumpfe (1953), courtesy of Martin-Luther-Univ. Halle Wittenberg.

further than the epibranchial bone. The cavity is lined ventrolaterally by the tympanic membrane; mediodorsally by the squamosal bone; mediocaudally by the proötic, opisthotic and exoccipital bones; ventrally by the basisphenoid bone and rostrally by the orbitosphenoid bone. The proötic bone and the otic process of the quadrate form a bony ridge dorsally, which in most birds separates the tympanic cavity from an air space above it. The form of this bony ridge varies; it is perforated in some passerines (Freye-Zumpfe, 1953) (Fig. 5.7); in owls it divides the tympanic cavity into ventral and dorsal parts (Stellbogen, 1930) (Fig. 5.7).

The tympanic cavity is lined by a mucosa which also covers the inner surface of the tympanic membrane. At the attachment of the extracolumellar cartilage to the tympanic membrane and at the vestibular window, the mucosa is reflected to envelope the whole columellar complex (Stellbogen, 1930).

2. *Air cavities associated with the tympanic cavity*

The tympanic cavity communicates with the oral cavity via the pharyngotympanic tube (*tuba pharyngotympanica*). The opening of the tube (*ostium tympanicum tubae pharyngotympanicae*) is found in the rostroventral part of the cavity. The tube then runs rostrally in a ventromedial direction and at its medial end joins the tube of the opposite side to form the common pharyngotympanic tube (*tuba pharyngotympanica communis*). This opens into the dorsal part of a median chamber in the roof of the pharynx, the pharyngotympanic infundibulum (*infundibulum pharyngotympanicum*). The infundibulum communicates ventrally with the pharynx via a short median cleft (*rima infundibuli*).

The tympanic cavity also communicates with air cavities in the skull. Typ-

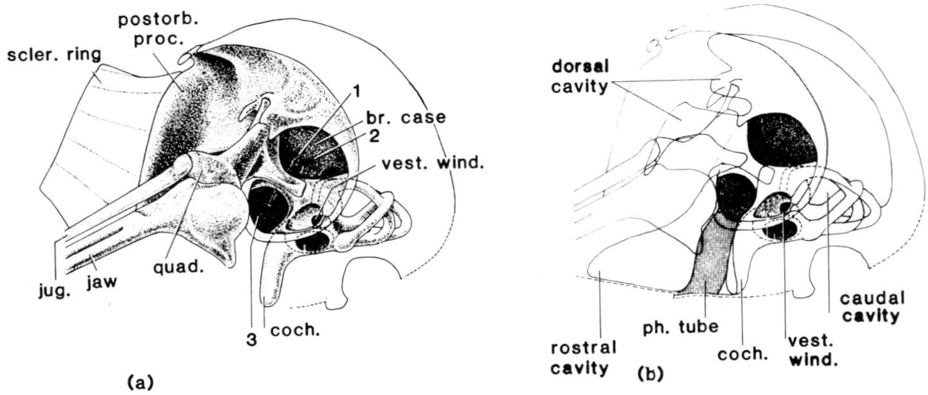

Fig. 5.8. The three air cavities associated with the cavity of the middle ear in the Boreal Owl (*Aegolius funereus*). (a) Lateroventral view of the skull with the tympanic ring in the plane of the figure. 1, Communication between the rostral and dorsal air cavities; 2, opening of the rostral air cavity into the cavity of the middle ear; 3, opening of the rostral air cavity and pharyngotympanic tube. (b) View of the skull of an owl, same orientation as in (a), to show the extension of the three air cavities. br. case, Brain case; coch., cochlea; jug., jugal arch; ph. tube, pharyngotympanic tube; postorb. proc., postorbital process; quad., quadrate bone; scler. ring, scleral ring; vest. wind., vestibular window. From Norberg (1978), courtesy of the Royal Society, London.

ically, there are three prominent air cavities (Fig. 5.8). A dorsal cavity opens into the tympanic cavity above the ridge formed by the proötic bone and the otic process of the quadrate. A rostral cavity opens near the exit of the pharyngotympanic tube and extends into the orbitosphenoid bone. A caudal cavity opens above the vestibular window. The rostral and caudal cavities usually communicate extensively with their counterparts on the other side. The tympanic cavity is also continuous with air spaces in the pneumatised quadrate and, via the siphonium (a connective tissue tube), with the lower jaw.

The middle ear region of the skull has been comprehensively studied in a number of avian orders (Saiff, 1974, 1976, 1978, 1981). Details of the middle ear cavity and the air spaces which are associated with it in owls (Strigiformes) are found in Stellbogen (1930) and Norberg (1978). The ontogenesis of the avian middle ear region was described by de Beer (1937), Crompton (1953) and Müller (1960).

3. *Acoustic aspects*

Part of the transformer efficiency, and the transfer function of the middle ear, depend on its own mechanical impedance. This is determined by friction, mass and stiffness, the stiffness component being dominant at low frequencies. The air enclosed in the middle ear cavity contributes to this stiffness component, and its effect is inversely proportional to its volume. Therefore, any enlargement of the middle ear cavity itself or any communication with the surrounding air cavities will decrease the impedance of the middle ear at low frequencies and extend the frequency sensitivity to lower ranges. Thus, the more spacious middle ear cavities and the most extensive communication with air spaces, are found in birds with good or excellent hearing abilities, such as birds-of-prey (Freye-Zumpfe, 1953) and especially owls (Stellbogen, 1930; Payne, 1971; Norberg, 1978; see also Section VI, later).

Another acoustic function of the air spaces associated with the tympanic cavity was suggested by Pumphrey (1961). The impedance of the inner ear is decreased by the opening of the cochlear window into the cavity. This enables the release of acoustic vibrations occurring in the inner ear fluid. Pumphrey noted that the air spaces are often filled with bony trabeculae (see for example Fig. 5.20) and may be constructed to act as efficient sound attenuators, thus contributing to a quick damping of air vibrations arising from the cochlear window. However, the cochlear window is at an air–water interface, and such an interface in the absence of some transformer structure reflects rather than transmits vibrations (Beranek, 1954).

C. The sound-transferring apparatus

1. The tympanic membrane

The removal of the tympanic membrane from the head surface into a stable supportive framework allows it to be larger, thinner and more delicate than if it occurred on the surface of the body. The result is an improvement in overall sensitivity and an extension of the high-frequency response (Henson, 1974). As is true for the acoustic meatus and the tympanic cavity, the size of the tympanic membrane is correlated with hearing ability as well as body or head size. Since auditory sensitivity calls for a certain absolute size of the tympanic membrane, small birds generally have large tympanic membranes in relation to body size or weight; the best relationship is again found in owls (Schwartzkopff, 1955, 1957). Areas range from a few square millimetres (e.g. 6·25 mm^2 in the House Sparrow, *Passer domesticus*) to the extensive membranes of owls (e.g. 50·6 mm^2 in the Tawny Owl, *Strix aluco*) (Freye-Zumpfe, 1953).

The most characteristic feature of the tympanic membrane of birds is that it protrudes outwards, supported by the processes of the extracolumellar cartilage, like a "tent by the centre pole" (Pohlman, 1921). Thus, the tympanic membrane has the form of a flat cone with an eccentric umbo (Fig. 5.9). It is a somewhat irregular oval, the long axis of which is vertical. Its orientation is inclined with respect to the median plane of the head, the ventral and caudal parts being more medial (see Fig. 5.1). The angle between the long axis and the median plane is between 40 and 60°; the angle between the short axis and the median plane is usually between 10 and 30° (Freye-Zumpfe, 1953).

The tympanic membrane consists of three layers. The elastic lamina propria is covered on its outer side by the epidermal lining of the meatus. This epidermis is only loosely attached to the propria (Stellbogen, 1930; Saiff, 1978), and its removal does not affect the acoustic properties of the middle ear (Schwartzkopff, 1952). The inner surface is covered by the mucosal lining of the tympanic cavity, and is very thin and firmly connected to the lamina propria to form a functional unit. The margin of the tympanic membrane is thickened by fibroelastic tissue (*margo fibroelasticus*).

The bony frame of the tympanic membrane is usually formed by parts of the

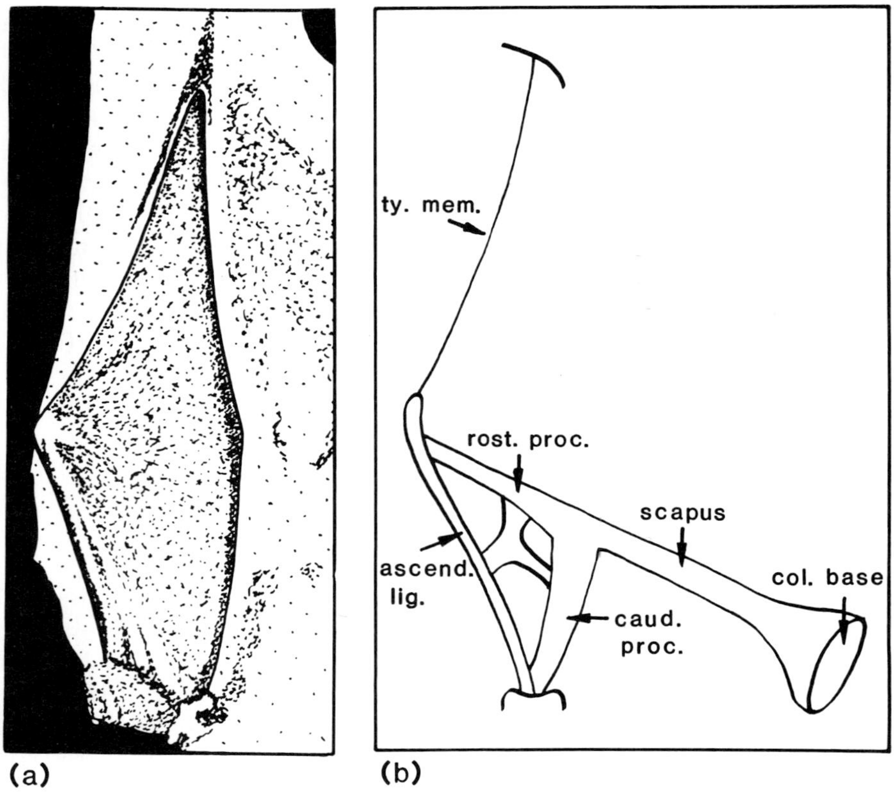

Fig. 5.9. (a) The tympanic membrane of an owl *in situ* as seen from the outside. Redrawn from a photograph by Gaudin (1968), courtesy of Almquist and Wiksell, Stockholm. The cone-shaped form of the membrane, which is typical for all birds, is caused mainly by the rostral process of the extracolumellar cartilage; this process pushes the membrane outwards and has its tip at the umbo resting on the ascending ligament, as shown in the schematic cross-section in (b). After Norberg (1978), courtesy of the Royal Society, London. ascend. lig., Ascending ligament; col. base; columellar base; rost. proc., caud. proc., rostral and caudal processes of the extracolumellar cartilage; ty. mem., tympanic membrane.

squamosal bone dorsally, the exoccipital bone caudally, the basisphenoid bone ventrally, and rostrolaterally by a strong ligament extending over the otic process of the quadrate bone near its articulation with the squamosal bone (Pohlman, 1921). Thus, the tympanic membrane is mechanically isolated from the movements of the quadrate. In a number of birds, and especially in owls (Strigiformes), the isolation of the tympanic membrane is even more effective: parts of the squamosal bone, exoccipital bone and basisphenoid bone fuse to form a bony tympanic ring in which the tympanic membrane is mounted by means of connective tissue. The ring can have a small gap in the region of the quadrate as in the Boreal Owl (*Aegolius funereus*) (Norberg, 1978) or be completely closed as in the Tawny Owl (*Strix aluco*) (Stellbogen, 1930) and the Barn Owl (*Tyto alba*) (Freye-Zumpfe, 1953). Ligaments insert near the rostral margin of the membrane and pass to the opening of the pharyngotympanic tube (Pohlman, 1921) or, as in the Boreal Owl (*Aegolius funereus*), to the tympanic ring (Norberg, 1978). Their function may be to keep the membrane taut.

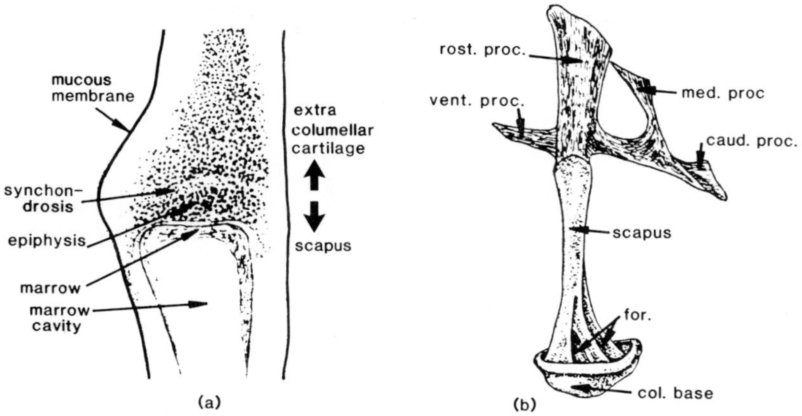

Fig. 5.10. The columella of the Tawny Owl (*Strix aluco*). (a) Longitudinal section of the synchondrosis between the scapus and the extracolumellar cartilage. (b) The main parts of the columella. col. base, Columellar base; for., foramina; rost. proc., vent. proc., caud. proc., rostral, ventral and caudal processes of the extracolumellar cartilage. From Stellbogen (1930).

2. *The columella, extracolumellar cartilage and ligaments*

The columella of birds consists of two parts, the bony shaft or scapus and the expanded proximal base or footplate. The extracolumellar cartilage is attached to the distal end of the shaft and may partly ossify with age (Gaudin, 1968). The two parts are fused by synchrondrosis (Fig. 5.10), the connection being more or less rigid in most birds, although it may be capable of elastic deformation and bending as in the Boreal Owl (*Aegolius funereus*) (Norberg, 1968). The columella forms an angle of about 50° with the plane of the tympanic ring, approaching the membrane from behind.

(a) Extracolumellar cartilage and ligaments. The main part of the extracolumellar cartilage is the rostral process (*processus rostralis*). It joins with the scapus and continues it distally along the same axis. The process is attached to the dorsal radius of the tympanic membrane, from its caudal margin to near the centre (Fig. 5.11); its tip marks the umbo of the cone of the tympanic membrane. The process does not make direct contact with the lamina propria but inserts in the tissue of the ascending ligament. This ligament has the form of a convex plate and runs from the caudal or caudoventral margin of the tympanic membrane between the ventral and caudal processes of the extracolumellar cartilage to its centre (Fig. 5.1). It possibly acts to stabilise and protect the delicate tympanic membrane, particularly at the umbo, and to transfer oscillations from a wider area of the tympanic membrane to the extracolumellar cartilage in a manner analogous to the manubrium of the mammalian malleus (Norberg, 1978). The orientation of the ascending ligament varies between different birds and may be correlated with the position of the foramen magnum (Freye-Zumpfe, 1953).

The three other parts of the extracolumellar cartilage arise from the shaft of the rostral process (Figs. 5.10, 11). The ventral process (*processus ventralis*) stretches ventrorostrally to the margin of the tympanic membrane; the caudal process (*processus caudalis*) passes in the same plane dorsocaudally. The latter is at-

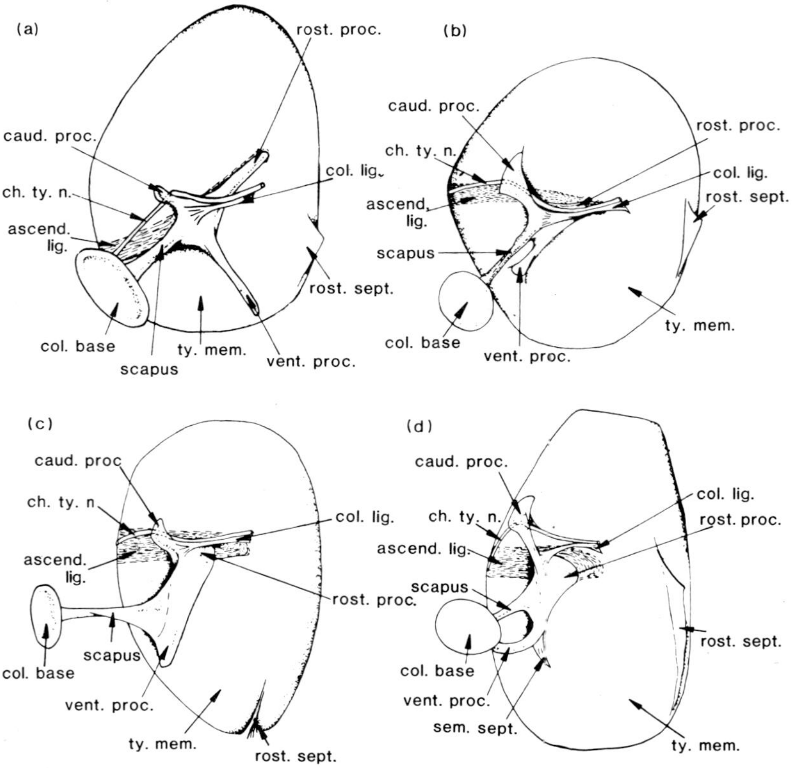

Fig. 5.11. Medial view of the columella and tympanic membrane of (a) the House Sparrow (*Passer domesticus*), (b) the Common Buzzard (*Buteo buteo*), (c) the White Stork (*Ciconia ciconia*) and (d) a gull (*Larus*). ascend. lig., Ascending ligament; ch. ty. n., chorda tympani nerve; col. base, columellar base; col. lig., columellosquamosal ligament; rost. proc., vent. proc., caud. proc., rostral, ventral and caudal processes of the extra-columellar cartilage; rost. sept., rostral septum; sem. sept., semilumar septum; ty. mem., tympanic membrane. From Freye-Zumpfe (1953), courtesy of Martin-Luther-Univ. Halle Wittenberg.

tached by a broad base to the tympanic membrane or, in some birds such as *Gallus,* to the ascending ligament (Freye-Zumpfe, 1953). Near the origin of the caudal process the columellosquamosal ligament is attached to the shaft of the rostral process. It runs at an angle of between 50 and 90° dorsorostrally to join with the connective tissue covering the bony ridge formed by the proötic bone and the otic process of the quadrate. It is thought to provide additional stabilisation of the columellar complex (Pohlman, 1921; Stellbogen, 1930; Freye-Zumpfe, 1953; Norberg, 1978). The fourth process of the extracolumellar cartilage, not prominent in all birds, is the medial process. It arises from the caudal process as a short transverse rod and attaches to the ascending ligament; the foramen thus formed has been referred to as "Huxley's foramen" (Crompton, 1953).

The shape, the relative size and the insertion on the tympanic membrane of the processes of the extracolumellar cartilage vary greatly (Fig. 5.11). They are slender in passerines but are very compact in the Falconidae and Strigidae. In these birds, the ventral and caudal processes reach to the margins of the tympanic membrane (Freye-Zumpfe, 1953). In the Boreal Owl (*Aegolius funereus*) they are attached to the bony tympanic ring rather than the membrane, forming a kind of hinge for the movement of the rostral process (Norberg, 1978). An unusual arrangement is found in the White Stork (*Ciconia ciconia*), in which the rostral and ventral processes are fused (Fig. 5.11).

(b) The scapus and base of the columella. The shape of the scapus or shaft varies among birds, a relatively slender form being found in species with good hearing abilities. The scapus is a slender, hollow cylinder which broadens at its proximal (vestibular window) end to form the base or footplate (Fig. 5.12) and at its distal end to make a stable connection with the extracolumellar cartilage. For a description of the scapus and base in 70 species see Krause (1901). The scapus and base have a typical lightweight design, with grooves and ridges and foramina which open to the hollow interior containing little marrow (Stellbogen, 1930). Thus, mass is reduced without loss in bending strength.

The margin of the base of the columella fits neatly into the vestibular window and is fixed by an annular ligament (*lig. annulare columellae*) which allows small movements of the base with respect to the rim of the window. The annular ligament is broader at the rostral margin of the base and may thus act as a hinge allowing greater displacements of the rostral part of the base (Gaudin, 1968). However, great inter-specific differences occur in the form of the base. Thus, it varies from a small trough-like broadening of the scapus as in the vulture *Gypaetus,* to the elaborate structures of some owls such as *Strix aluco* and *Strix flammea* (Fig. 5.12). In the latter species the inner surface of the base is prolonged into a hemispherical tip which again is hollow. As the margin of the base fits into the vestibular window, this tip projects into the scala vestibuli. Schwartzkopff (1955) suggested two functions for this type of columellar base. First, it provides a "mass equilibration" to reduce the forces on the hinge (i.e. the annular ligament) during level-like vibrations of the scapus. Second, the hemispherical shape prevents the occurrence of turbulence in the perilymph. This form of base is not

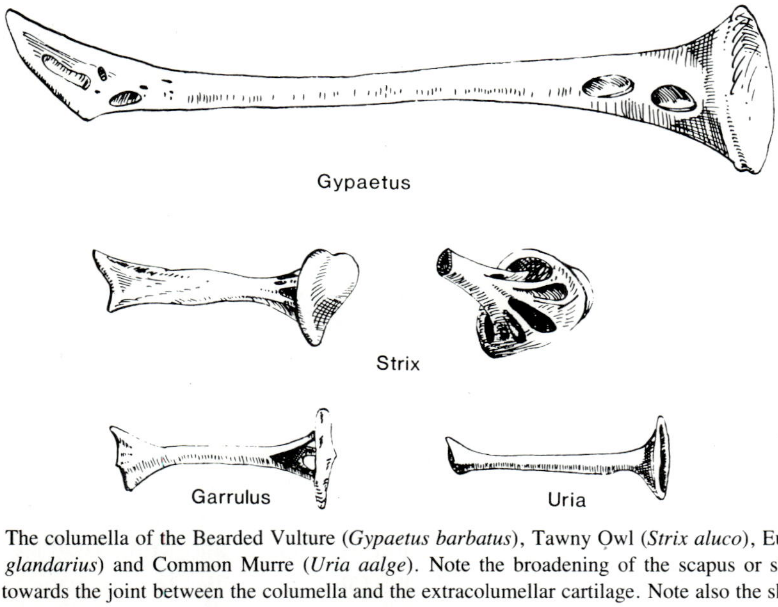

Fig. 5.12. The columella of the Bearded Vulture (*Gypaetus barbatus*), Tawny Owl (*Strix aluco*), Eurasian Jay (*Garrulus glandarius*) and Common Murre (*Uria aalge*). Note the broadening of the scapus or shaft of the columella towards the joint between the columella and the extracolumellar cartilage. Note also the shape of the base of the columella in the owl. Redrawn from Krause (1901).

found in all owls, however, *Asio otus* and *Athene noctua* displaying a more usual design (Krause, 1901).

D. Transformation ratio and transfer function

1. Area ratio

The area reduction ratio between the tympanic membrane and the base of the columella varies between 11 and 40 in the 47 species investigated by Schwartzkopff (1957). For comparison, this ratio is about 36 in the cat and 21 in humans (Wever and Lawrence, 1954). In birds the ratio shows some correlation with hearing ability, high values suggesting high sensitivity. The lowest ratios were found in newly-hatched domestic fowl (11), grebes (*Podiceps ruficollis, Podiceps cristatus*) (15) and ducks (Anatidae) (up to 20). Intermediate ratios (20–30) are possessed by songbirds, birds-of-prey, galliform species and gulls (Laridae). The highest ratios occur in owls (Strigiformes), the Long-eared Owl (*Asio otus*) having a ratio of 40. The highest ratios are mainly due to the enlargement of the tympanic membrane. However, morphological ratio determinations must be applied to physiological situations with great care. One example is discussed by Schwartzkopff (1957): because of a large tympanic membrane, gulls (Laridae) show area ratios of up to 33, but the membrane is particularly thick and stiff and may be rather insensitive. Other limiting factors involve the vibration modes of the tympanic membrane, which may cause a considerable decrease in the "effective area" at higher frequencies, as has been described for the cat by Khanna and Tonndorf (1972). The influence of such phenomena on sensitivity has not yet been studied in birds.

2. Lever ratio

The middle ear apparatus in birds is not a simple "membrane-plus-piston" array, the amplification function of which depends only on the area ratio. The geometry of the columella attachments to the tympanic membrane and vestibular window results in lever-like movements of the extracolumellar cartilage and/or scapus. From morphological considerations of the middle ear of the Boreal Owl (*Aegolius funereus*), Norberg (1978) calculated the probable movements of the tympanic membrane and columellar complex. He assumed that a rotatory movement of the extracolumellar cartilage was transformed into a piston-like movement of the scapus by elastic deformation at the extracolumellar cartilage–scapus joint. As a result, the oblique orientation of the columellar complex with respect to the tympanic membrane (and hence to the force of the membrane movement) yields a lever ratio of about 1·6 (Fig. 5.13). With an area ratio of 35, Norberg calculated the total transformer ratio for the Boreal Owl to be about 56. The lever ratio for the cat middle ear was estimated by Khanna and Tonndorf (1972) to be about 1·32 with a total ideal transformer ratio of about 65.

The actual movements of the columella and columellar cartilage may be even more complex than this. Lever-like movements of the scapus itself were noted,

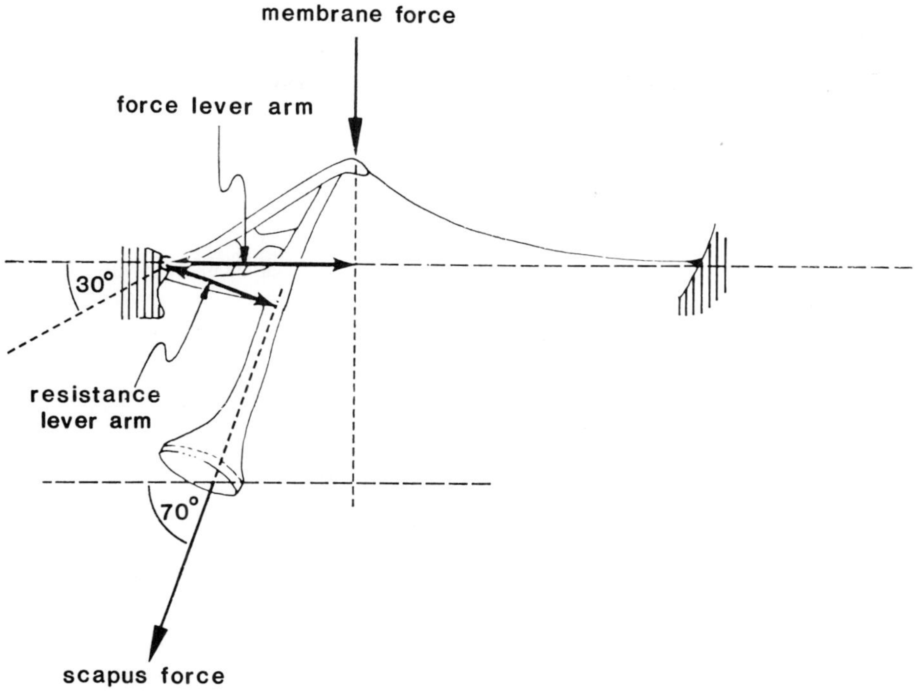

Fig. 5.13. Norberg's model of middle ear function in the Boreal Owl (*Aegolius funereus*). The schematic cross-section through the middle ear shows how parts of the extracolumellar cartilage act as lever arms and increase the force component of the vibrations of the columellar base with respect to the original vibration of the tympanic membrane. This model assumes a pure piston-like movement of the scapus itself, and Norberg calculated the lever ratio to be 1:1·6. From Norberg (1978), courtesy of the Royal Society, London.

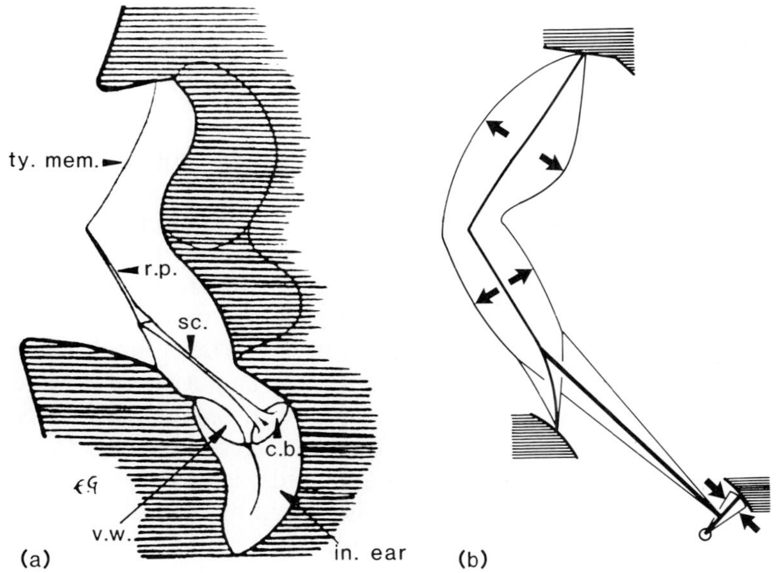

Fig. 5.14. Gaudin's model of middle ear function in birds. (a) Schematic horizontal section through the middle ear. c.b., Columellar base; in. ear, inner ear; r.p., rostral process; sc., scapus; ty. mem., tympanic membrane; v.w., vestibular window. (b) The extreme positions during the movements of the structures in (a) are indicated by the thin lines. Gaudin's model assumes a transverse movement of the columellar base and thus a high lever ratio. From Gaudin (1968), courtesy of Almquist and Wiksell, Stockholm.

not only in owls (Strigiformes) (Schwartzkopff, 1955) but also in auks (Alcidae) (Kartaschew and Iljitschew, 1964). Similar movements were observed in a number of avian species by Gaudin (1968) during experiments in which low-frequency pressure changes were applied. He concluded that the columella vibrated, not parallel to its long axis, but in a transverse mode. The hinge-like insertion of the columellar base in the vestibular window produced displacements mainly of the rostral part of the base, which vibrated like "the movements of a muscian's foot tapping as it rests on the heel" (Fig. 5.14). In this model, the two lever arms are the columella plus the tympanic membrane on one side, and the columellar base on the other. The result is a high lever ratio. However, it remains to be demonstrated that this model works under physiological conditions.

3. *Curved-membrane concept*

A small additional transformation factor may arise from the cone-shaped arrangement of the tympanic membrane. In mammals, the tympanic membrane is an inwardly-directed cone which buckles when it vibrates, so that the *amplitude* of the vibrations of the structures transmitting the sounds is less than that of the surface of the membrane, although the transmitted *force* may be greater (Khanna and Tonndorf, 1972). The curved-membrane concept may also apply to the bird's tympanic membrane (Norberg, 1978) and may account for the high lever ratio of the tympanic membrane-plus-columella of the Ringed Turtle Dove (*Streptopelia risoria*) (Saunders and Johnstone, 1972): the amplitude loss of 10 to 12 dB between tympanic membrane and columellar base corresponds to a lever ratio of about 5 (see Fig. 5.15).

4. *Transfer function*

The transfer function of the middle ear represents its frequency response, which clearly is one important factor determining the hearing range. Saunders and Johnstone (1972) measured the transfer function and the amplitude loss between the tympanic membrane and the columellar base in the Ringed Turtle Dove (*Streptopelia risoria*), using the Mossbauer technique. Although they determined only the piston-like components of the membrane and base movement, the results reasonably reflect the frequency response (Fig. 5.15). The optimal response occurs between 1 and 2 kHz where tympanic membrane amplitude is close to ($<$10 dB below) the theoretical air particle movement. At these frequencies, the response of the system is determined mainly by friction effects. At lower frequencies the response, with its more-or-less constant amplitude, is determined by elastic stiffness. The sharp, high-frequency roll-off is caused by mass-dominated transmission loss. The sharp, high-frequency roll-off is the major difference between the avian and mammalian middle ear transfer functions (Saunders and Johnstone, 1972). Up to 3 kHz, the frequency range of communication and best sensitivity for most birds, the transfer function of the avian middle ear is within 6 dB of that of the three-ossicle chain of the mammalian ear (Fig. 5.15b).

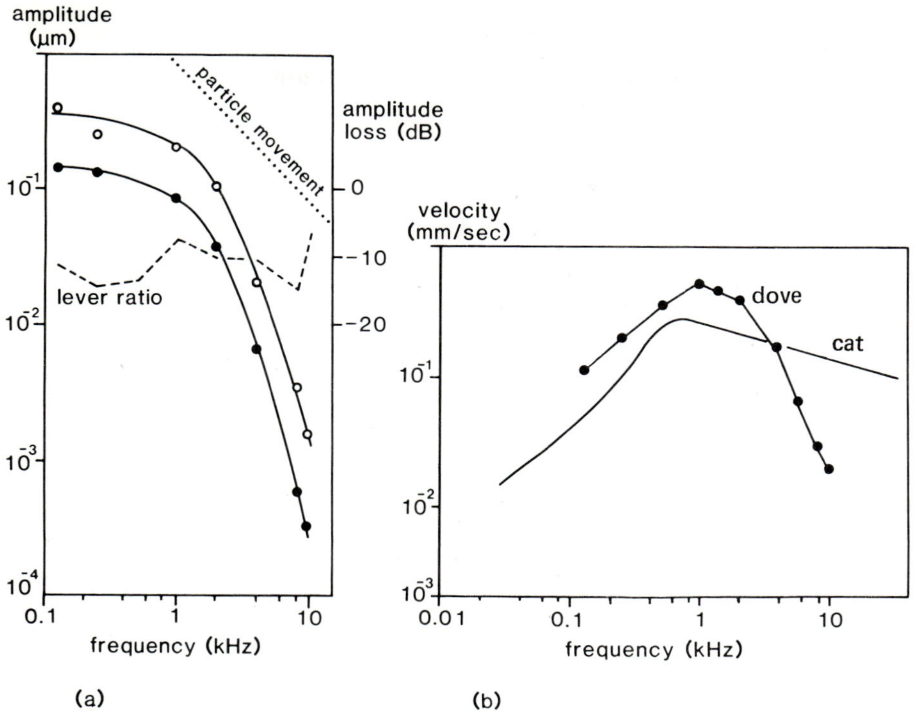

Fig. 5.15. Transfer function of the avian middle ear. (a) Middle ear frequency response for the Ringed Turtle Dove (*Streptopelia risoria*) compared to the theoretical amplitude of air particle movement at the applied sound intensity of 100 dB SPL. The amplitude of tympanic membrane (○) and columella (●) vibrations were measured and used to calculate an overall lever ratio (---, expressed in decibels) for the middle air. (b) Comparison of the frequency responses, measured as the velocity of the columellar base vibration, in the dove and cat. Both middle ears show a bandpass characteristic with a low-frequency roll-off of about 6 dB/octave; the high-frequency roll-off is greater in the bird. After Saunders and Johnstone (1972), courtesy of Almquist and Wiksell, Stockholm.

V. The middle ear muscle and signal control

Unlike mammals, birds possess only one middle ear muscle, the columellar muscle (*m. columellae*). The insertion, innervation and action of this muscle show a number of similarities to the tympanic muscle of crocodiles and the stapedius muscle of mammals (Henson, 1974). The columellar muscle has been described under various names by different authors (e.g. tympanic muscle, stapedius muscle, m. tensor tympani, m. laxor tympani), reflecting the variety of functions which have been ascribed to it. However, several studies have since suggested that the function is not a static one, providing a permanent tension on the tympanic membrane (Wada, 1924), but is used in a well-balanced, efferent signal control mechanism.

A. *Anatomy*

The contractile part of the columellar muscle is situated outside the middle ear. The muscle arises by a broad attachment from a shallow groove on the ventral

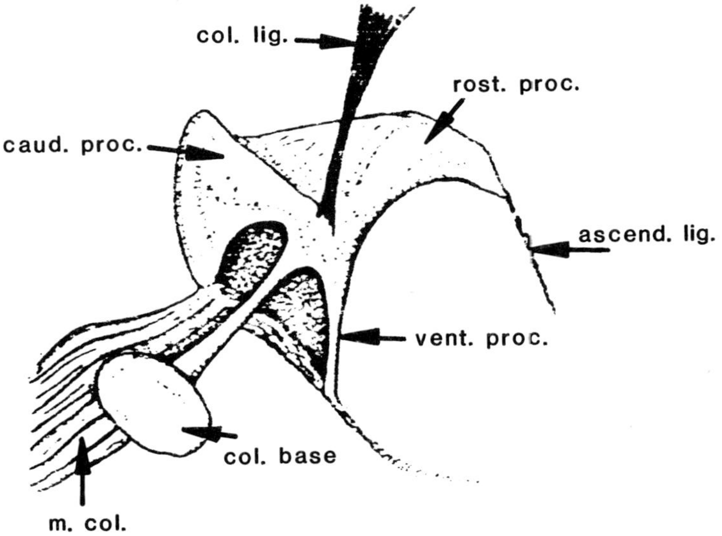

Fig. 5.16. Schematic representation of the insertions in *Gallus* of the columellar muscle (m. col.) at the margin of the tympanic membrane and on the ventral and caudal processes of the extracolumellar cartilage. ascend. lig., Ascending ligament; col. base, columellar base; col. lig., columellosquamosal ligament; rost. proc., vent. proc., caud. proc., rostral, ventral and caudal processes of the extracolumellar cartilage. From Henson (1974), courtesy of Springer-Verlag, Berlin.

surface of the basioccipital bone, lateral to the occipital condyle. The tendon enters the tympanic cavity via a foramen (*foramen m. columellae*) in its caudo-ventral wall and passes to the margin of the tympanic membrane. It then branches into three parts: one part attaches to the caudal margin of the tympanic membrane between the caudal and ventral processes of the extracolumellar cartilage (Fig. 5.16); a minor part attaches to the caudal process itself; and the major part inserts on the ventral process near the base of the caudal process. The columellar muscle is innervated by a branch of the facial nerve (Smith, 1904; Pohlman, 1921; Stellbogen, 1930; Norberg, 1978) (Fig. 5.7).

The fine structure of the columellar muscle was studied in *Gallus* by Borg *et al.* (1979). Their results indicate that the muscle fulfils all the criteria of a homogeneous fast muscle. It contains only fibres with well-developed sarcomeres and high ATPase activity; the muscle fibres are innervated individually via neuromuscular endplates which are distributed over a large area of the muscle.

B. Contraction of the columellar muscle and its effects

Borg *et al.* (1979) characterised the columellar muscle of *Gallus* as a fast-twitch muscle with a contraction and a half-relaxation time of about 20 msec each. The response of the columellar muscle to acoustic stimulation in the Long-eared Owl (*Asio otus*) was studied by Golubjewa (1972). The electromyogram she obtained consisted of single impulses of 1 to 2 msec duration, with a latency of 1·5 to 6 msec after stimulus onset. However, Borg *et al.* (1979) also suggested that the muscle showed slow-contraction properties, since repetitive stimulation resulted in summation; but, as the contraction of the muscle was measured indirectly by changes in middle ear volume and impedance, these slow components may be due to the properties of the attachments of the muscle to the tympanic membrane and extracolumellar cartilage. Contraction of the columellar muscle changes both the tension of the tympanic membrane and the position of the columella and extracolumellar cartilage. The tympanic membrane is tensed in its upper and rostral quadrants, the other quadrants relaxing. The extracolumellar cartilage is drawn caudoventrally (Pohlman, 1921). Thus, the joint between the columella and the extracolumellar cartilage is moved towards the tympanic membrane, and the base of the columella moves away from the vestibular window (Norberg, 1978). Activation of the columellar muscle should therefore decrease the transformer efficiency of the middle ear, and hence the amount of sound energy transmitted to the inner ear.

C. The middle ear muscle and auditory sensitivity

The effect of columellar muscle activity on auditory sensitivity has been studied in the Common Starling (*Sturnus vulgaris*) (Oeckinghaus and Schwartzkopff, 1975, 1983) and in the Long-eared Owl (*Asio otus*) (Golubjewa, 1972). In the starling, electrical stimulation of the muscle caused a decrease in the amplitude of cochlear microphonic potentials (CM) (Fig. 5.17). For a given frequency, activation of the muscle shifted the intensity–response curve of the CM to higher sound

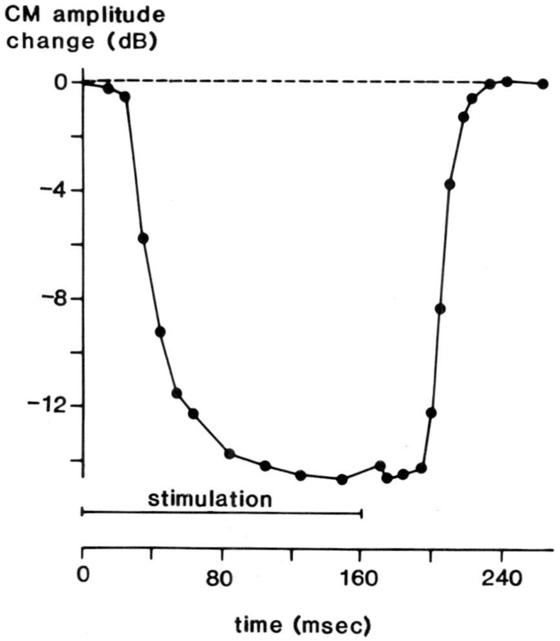

Fig. 5.17. Time course of the effect of columellar muscle activity on the amplitude of the cochlear microphonic potential (CM) in the Common Starling (*Sturnus vulgaris*). The columellar muscle was stimulated electrically during the time indicated by the horizontal bar. From Oeckinghaus and Schwartzkopff (1983), courtesy of Springer-Verlag, New York.

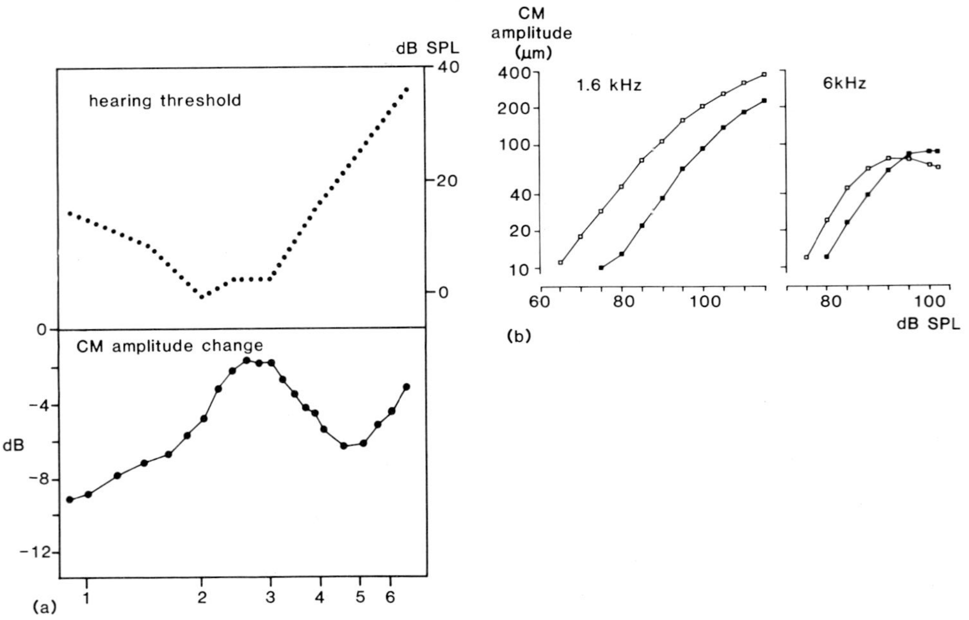

Fig. 5.18. Effects of columellar muscle activation on the cochlear microphonic potential (CM) in the Common Starling (*Sturnus vulgaris*). (a) Frequency course of the CM attenuation caused by constant electrical stimulation of the columellar muscle compared with the behavioural hearing threshold. The effect on CM is minimal in the frequency range of highest hearing sensitivity. After Kuhn *et al.* (1982), courtesy of Springer-Verlag, Berlin, and Oeckinghaus and Schwartzkopff (1983), courtesy of Springer-Verlag, New York. (b) Activation of the columellar muscle shifts the intensity response curve of CM to higher intensities, shown here at frequencies of 1·6 and 6 kHz. From Oeckinghaus and Schwartzkopff (1983), courtesy of Springer-Verlag, New York.

pressure levels (Fig. 5.18). The effect was frequency-dependent, the CM amplitude being strongly attenuated in the lower frequency range. With increasing frequency, suppression decreased to a minimum between 2 and 3 kHz and then increased again (Fig. 5.18). Maximal CM amplitude change was equivalent to 24 dB. This change in sensitivity had a time course somewhat slower, but in the same range as the contraction time of the columellar muscle demonstrated in *Gallus* by Borg *et al.* (1979). The columellar muscle thus acts as a bandpass filter. In the starling, the range of minimal suppression coincides with the range of auditory maximum sensitivity (Kuhn *et al.*, 1982), and these effects have also been demonstrated at the level of single auditory receptor fibres (Oeckinghaus, 1978).

D. The biological function of the middle ear muscle

An acoustic middle ear reflex (i.e. an activation of the middle ear muscle in response to auditory stimulation of the ear) is common in mammals. However, in birds the reflex has only been found in owls (Strigiformes), and it has been shown to be positively absent in *Gallus* and the Common Starling (*Sturnus vulgaris*). In the Long-eared Owl (*Asio otus*), this reflex is frequency-dependent and is most sensitive to sounds around 6 kHz, when it can be elicited by intensities of 50 to 60 db SPL (Golubjewa, 1972). At sufficiently high intensities, the muscle is active during the whole duration of the sound stimulus, its contraction producing a complicated and characteristic amplitude modulation of the CM response (Golubjewa, 1972). This modulation of the CM during middle ear muscle activity and an excitation of the columellar muscle by auditory stimulation was also found in the Tawny Owl (*Strix aluco*) by Oeckinghaus and Schwartzkopff (1983). However, it has not been shown for any other bird (Wada, 1924; Counter and Borg, 1979). In *Gallus,* Counter and Borg (1979) did demonstrate activation of the middle ear muscle during the bird's own vocalisations (Fig. 5.19), but this response was mediated by efferent signals rather than by acoustic self-stimulation.

Based on the little evidence available, several functions for the columellar muscle in birds have been proposed. The acoustic reflex in owls may be interpreted as a well-balanced tuning mechanism, possibly holding the sound intensity in the "best working range" of the inner ear receptors over a wider frequency range (Iljitschew, 1974). At high intensities, the reduction of sound energy transmitted to the inner ear can be regarded as a protective mechanism against acoustic over-stimulation from outside (Iljitschew, 1974) or during the bird's own vocalisation, as in *Gallus* (Counter and Borg, 1979). On the other hand, in the Common Starling (*Sturnus vulgaris*), the activation of the columellar muscle enhances the bandpass filter properties of the middle ear, benefiting its most sensitive range, which in most songbirds corresponds to the frequency range of

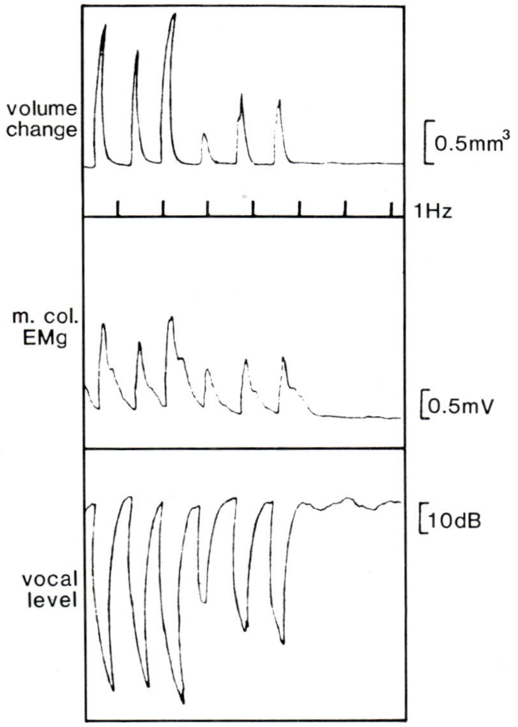

Fig. 5.19. In *Gallus* the columellar muscle is activated by efferent commands synchronously with the bird's own vocalisations. From bottom, the graphs show (1) the sound recorded during a chick's vocalisation, (2) the electromyogram of the columellar muscle and (3) the change in tympanic cavity volume which is one result of columellar muscle activity. From Counter and Borg (1979), courtesy of Almquist and Wiksell, Stockholm.

their own songs (Oeckinghaus and Schwartzkopff, 1983). These authors suggested that the mechanism may provide protection against impairment of hearing ability by low-frequency wind noises that occur during flight, and against masking of important sound signals by such noises.

The columellar muscle may well play different roles in different birds, and further comparative studies are required.

VI. Two-receiver systems

In Section II on the biophysical aspects, two possible strategies for sound localisation in birds have been advanced. Thus, birds may obtain directional cues by computing binaural intensity differences of pure tones or overall spectral differences. Alternatively, binaural onset time or ongoing phase differences may be used. These differences can be enhanced either by making each ear inherently directional, or by creating binaural asymmetries with respect to the receiver properties of the two ears. Both possibilities are realised in birds: in most cases, sound transmission via the interaural pathway enables each ear to act as a pressure difference receiver. Binaural asymmetries are especially developed in owls.

A. *Sound transmission through the interaural pathway*

1. *Anatomy*

Within the avian skull, several air cavities occur between the two ears. These result from connections between the right and left pharyngotympanic tubes and from the connections of the middle ear cavities with the surrounding air spaces. The cavities provide a potential pathway for the transmission of airborne sound from one middle ear to the other (Wada, 1924; Stellbogen, 1930; Schwartzkopff, 1952; Payne, 1971; Henson, 1974).

The interaural pathway may be formed from different components. In most birds, the main pathways are developed from the communication between the rostral air spaces in the orbitosphenoid bones and the pharyngotympanic tubes. However, the actual form is very variable, even in related genera. In the Boreal Owl (*Aegolius funereus*), Norberg (1978) described a communicating channel between the rostral air spaces about 8 mm in width and 4 mm in height, whereas the passage to the pharyngotympanic tube measured only about $0{\cdot}5 \times 2$ mm. In the Barn Owl (*Tyto alba*), the pharyngotympanic tube and the channel leading to the rostral air space join on each side, the common tube anastomosing with the corresponding tube of the other side. The result is an air chamber which occupies about a quarter of the entire skull volume (Payne, 1971). It communicates with each rostral air space via two large openings.

In *Coturnix,* the pathway appears to be produced mainly from the communication between the rostral air spaces (Hill *et al.*, 1980) (Fig. 5.20). The only obstacles for sound transmission are the bony trabeculae and spicules and the channels of the carotid arteries.

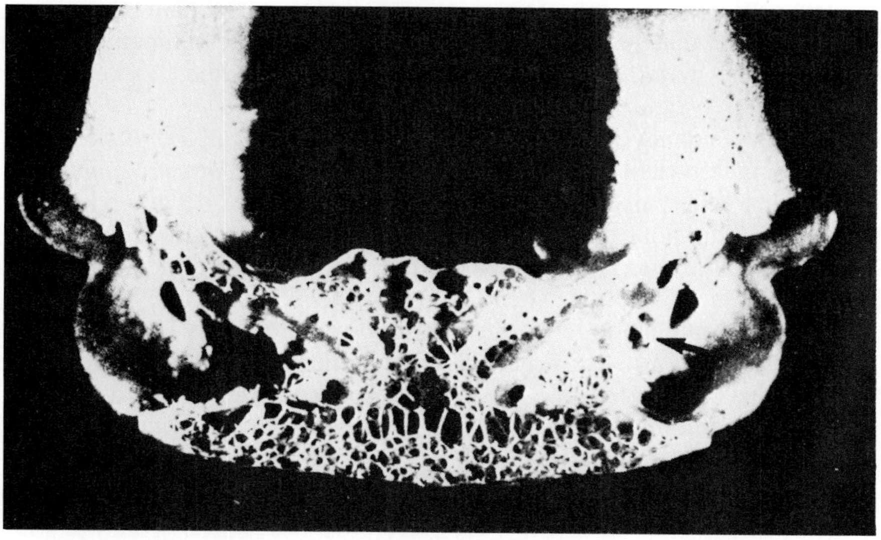

Fig. 5.20. Transverse section through the skull of the Japanese Quail (*Coturnix japonica*) showing the extreme trabeculation of the region between left and right middle ears which provides the interaural pathway. On the right the arrow points to the tympanic ring and the columella. From Lewis and Coles (1980), courtesy of Elsevier Biomedical Press, Amsterdam.

2. *Sound transmission characteristics*

Using long-term pressure fluctuations, Wada (1924) was the first to demonstrate experimentally the acoustic coupling of the two ears in birds. Interaural sound conduction has since been described by Schwartzkopff (1952) and Schwartzkopff and Bremond (1963), but only subsequent studies in *Gallus, Coturnix* and the Barn Owl (*Tyto alba*) have attacked its importance for directional hearing in birds.

In such a linked-receptor system, sound travelling through the head to reach the inner surface of the opposite tympanic membrane has to be transmitted across the ipsilateral membrane and the interaural cavity. Both these structures can contribute to an attenuation of sound energy, and this attenuation will be frequency-dependent. The degree of attenuation will be one major factor in determining whether the ears act as linked or as independent receiver systems. However, measurements of interaural sound transmission and the assignment of attenuation values to various structures must be made with great care. There are considerable differences between measuring transmission using monaural sound presentation, binaural presentation, and free-field presentation. The biophysics of the acoustic cavity may be different in each case, giving completely different values.

In *Gallus,* Rosowski and Saunders (1980) found that interaural attenuation was primarily due to the tympanic membrane for frequencies between 100 Hz and 5 kHz. They measured transmission losses of between 25 and 30 dB across the membrane. The transmission loss in the interaural cavity itself was found to be negligible in this frequency range. Blocking of the contralateral ear improved interaural transmission considerably, thus excluding bone transmission. This improvement could be the result of the creation of a closed, resonant acoustic cavity, the improvement due to resonance compensating for the tympanic membrane attenuation. The major question is the extent to which the blocked condition represents the free-field state.

A strong dependency for transmission on frequency was found by Hill *et al.* (1980) in their free-field studies of the Japanese Quail (*Coturnix japonica*). Attenuation due to the tympanic membrane decreased from 12 dB at 0·5 kHz to 0 dB at 1·6 kHz, and did not exceed 2 dB for frequencies up to 8 kHz (Fig. 5.21a). On the other hand, the interaural cavity produced only minimal attenuation at low frequencies: from 0·5 to 5 kHz attenuation remained within 5 dB, but for frequencies above 5 kHz attenuation increased sharply by about 15 to 20 dB/octave (Fig. 5.21b). Thus, the system as a whole, tympanic membrane plus interaural cavity, acts as a bandpass filter (Fig. 5.21c), the low-frequency limit being determined by the tympanic membrane, the properties of the air cavity producing the high-

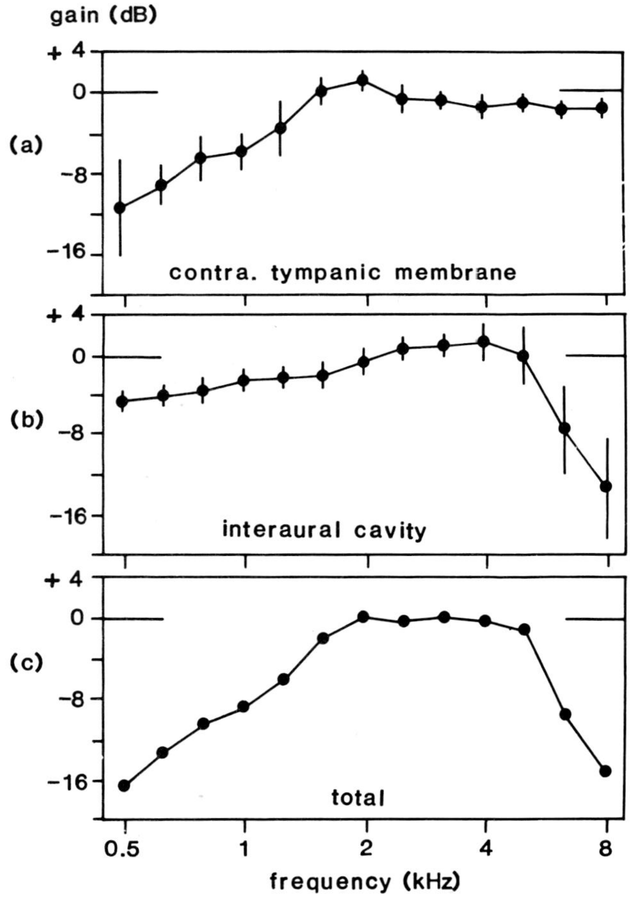

Fig. 5.21. Attenuation of sound passing through the interaural pathway between the two middle ears in the Japanese Quail (*Coturnix japonica*), determined from free-field experiments. The sound has to pass (a) the contralateral tympanic membrane, which acts as a high-pass filter, and (b) the interaural cavity, which acts as a low-pass filter. Together (c), the system acts as a bandpass filter with almost no sound attenuation in the frequency range between 2 and 4 kHz (which corresponds to the frequency range of a variety of biologically important sounds). After Hill *et al.* (1980), courtesy of the Company of Biologists, Ltd.

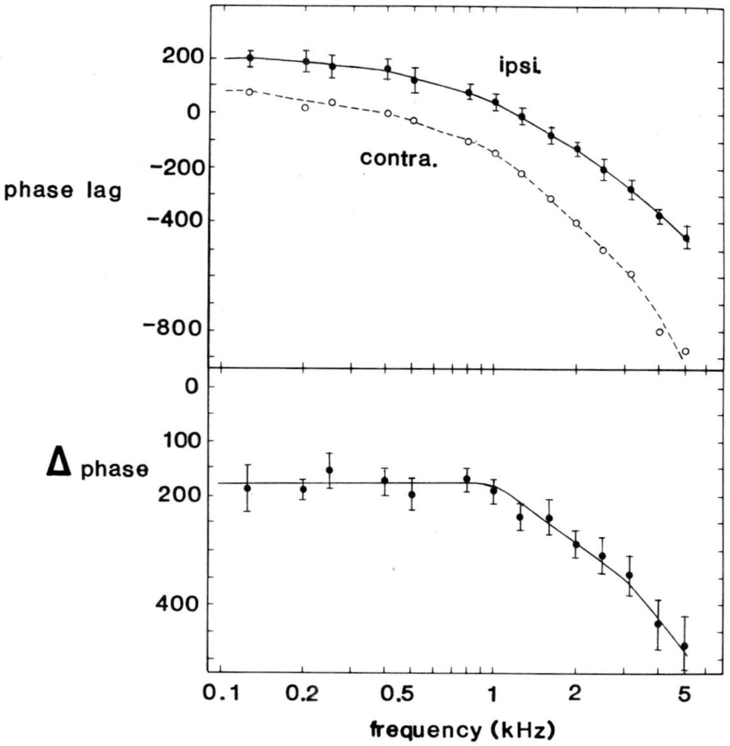

Fig. 5.22. The phase effects of interaural sound transmission. The upper part shows the phase lag of the CM potential of one ear with respect to the phase of the sound signals during monaural ipsilateral and contralateral stimulation. With the latter, the sound traversed the interaural pathway. The lower part shows the difference between the phase response for ipsilateral and contralateral stimulation. At low frequencies, the phase difference (Δphase) of 180° is due to the contralateral sound moving the tympanic membrane in the opposite direction; at higher frequencies, additional phase delays occur. From Rosowski and Saunders (1980), courtesy of Springer-Verlag, New York.

frequency cut-off. Similar results were obtained in the Barn Owl (*Tyto alba*) by Moiseff and Konishi (1981a). They characterised the interaural pathway as a low-pass filter with a high-frequency roll-off of about 35 dB/octave from 2 kHz, which is sharper than in the Japanese Quail. However, their monaural presentation technique may not clearly reflect the upper limit of air space transmission.

In addition to the sound energy, the phase of the signal is also affected by interaural transmission and is again dependent on frequency. One way of determining phase shifts is to record the cochlear microphonic potentials (CM) and compute the differences in phase between the CM and the sound at the ipsilateral and contralateral ears. Figure 5.22c shows data obtained in *Gallus* by Rosowski and Saunders (1980). The phase response of the CM for contralateral stimulation (i.e. via the interaural pathway) lags by 180° relative to that for ipsilateral stimulation, for frequencies below 0·8 kHz. This lag is due to the fact that the interaurally-conducted sound displaces the tympanic membrane of the recorded ear in the direction opposite to that which would have occurred with ipsilateral stimulation. Above 0·8 kHz, phase lag increased with frequency. In the Japanese Quail (*Coturnix japonica*), both the contralateral tympanic membrane and the interaural cavity were found to contribute to this high-frequency phase shift (Hill *et al.*, 1980).

3. *Implications for directional hearing*

Interaural sound transmission (at least in the frequency range up to 5 kHz or so) allows a considerable amount of sound energy to act on the inner surface of the tympanic membrane. This "internal" sound will interact (in pressure and phase) with the "external" sound arriving on the outer surface of each tympanic membrane. The ears are therefore coupled via the interaural pathway, and each tympanic membrane should act as an asymmetrical pressure difference receiver; that is, the response of each membrane will be the resultant (of the pressure and phase) of the internal and external sound energy (see Section II).

Since the phase of the sound entering each meatus will change with the angle of incidence of the sound, at any given frequency the phase difference across the tympanic membrane will also change with the angle of incidence. Assuming no attenuation during interaural transmission (internal and external sound pressures equal), the resultant tympanic membrane displacement will be the result of the transmembrane phase difference, which is correlated with the angle of incidence. Thus, providing the "acoustic pathways" (i.e. phase shifts) through the head and around the head are equal, minimal (up to 0°) phase difference may occur at some angle(s) at the ear contralateral to the sound source; at the ipsilateral ear, phase differences of up to 180° may occur (Michelsen, 1983; Lewis, 1983). It is important to emphasise, however, that under conditions of equal sound pressures, 180° phase differences can only double the effective force at the ipsilateral ear, that is, provide an increase of only 6 dB. In contrast, 0° phase difference at the contralateral ear can result in no tympanic membrane movement and therefore completely null the response of the ear.

The possibility that the bird ear acts as a pressure difference receiver was first considered by Schwartzkopff (1952). Subsequent investigations of the directional sensitivity of cochlear microphonic potentials under binaural stimulation conditions in *Gallus* and free-field conditions in the Japanese Quail (*Coturnix japonica*) have provided some experimental evidence in support of this hypothesis. Rosowski and Saunders (1980) showed that in *Gallus,* interaural transmission resulted in a binaural CM which was proportional to the pressure difference across the tympanic membrane. In the quail, Coles *et al.* (1980) measured the directional patterns of the CM through 360° azimuth. These directionality patterns were compared to the directional differences in sound intensity caused by diffraction and head shadow effects. Examples are shown in Fig. 5.23 for some frequencies. In these data it is clear that the CM directionality cannot be produced by head shadow effects. They are due to the differences in intensity and phase between the sound pressure acting on the outer and the inner surfaces of the tympanic membrane of the recorded ear as the sound source is rotated through 360°. Blocking the contralateral ear abolished the nulls and the CM directionality (Fig. 5.23). Thus, the marked directional changes in CM are primarily the result of the occurrence of "nulls"; these nulls represent spatial positions around the head for which the ear is relatively insensitive to sound.

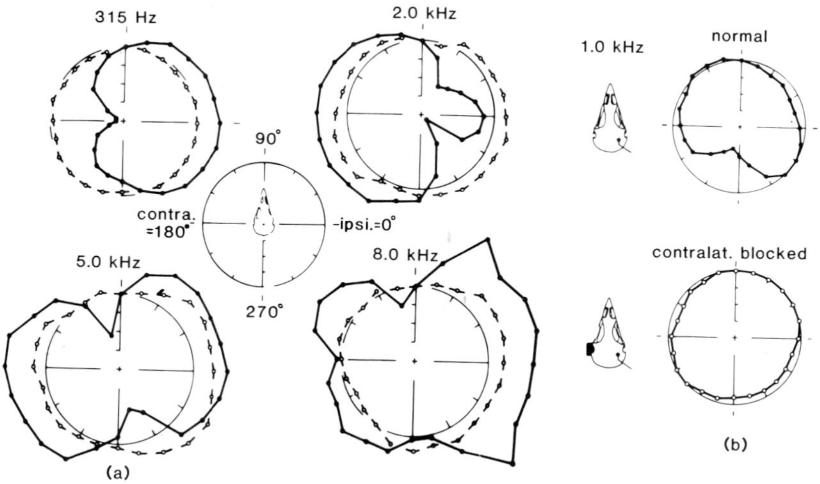

Fig. 5.23. Physiological evidence for interaural sound transmission and its effect on directional hearing. (a) Directional sensitivity patterns for the ear of the Japanese Quail (*Coturnix japonica*) at four different frequencies. The amplitude of CM potentials was measured while moving a sound source around the bird's head in azimuth. The directional patterns obtained are typical for pressure gradient (pressure difference) receivers. (b) The directional sensitivity of the quail's ear vanishes when the contralateral ear is acoustically blocked. Maximal interaural difference in CM amplitude decreases from 14 to 2 dB. From Coles *et al.* (1980), courtesy of the Company of Biologists, Ltd.

As can be expected from the acoustic transmission properties of the interaural pathway, CM directionality is highly frequency-dependent. Maximal directional effects occur at several frequency bands which are interspersed with bandwidths of directionally-insensitive frequencies. With increasing sound frequency, cardioid, figure-of-eight and multilobed directionality patterns develop (Fig. 5.23).

By functioning as pressure difference receivers, the chicken and quail ears have considerable inherent directionality and thus can provide monaural cues for localisation. However, these cues are ambiguous, as sensitivity changes in a single ear can be the result of a change in sound direction or distance, or both. Binaural comparisons of the directional sensitivities of both ears, on the other hand, can provide unequivocal directional cues. Coles *et al.* (1980) defined the directional index (DI) as the contrast in sensitivity between the ears (i.e. as a difference pattern), and from their results calculated the DI values for the quail auditory system. DI depends on angle of sound incidence and frequency but is independent of intensity. In the quail, the DI directional patterns changed from a single broad peak near the ipsilateral ear at low frequencies to sharp double peaks towards the front and the rear of the head at higher frequencies. Thus, there is a tendency for higher frequencies to generate best directional cues towards the front of the animal, with some possibility of front-to-rear errors caused by secondary peaks.

The very general occurrence of communicating middle ear cavities in birds suggests that the pressure difference principle may be a general system for sound localisation. However, the acoustic properties of the various interaural pathways cannot be predicted from anatomical measurements, and there is considerable need for comparative physiological studies. As mentioned earlier, interaural connections are very prominent in owls, which have the greatest degree of localisation acuity known (Knudsen, 1981). However, in owls, at appropriate predator–prey sound frequencies, sound transmission through the head appears to be of less significance than in other birds (Payne, 1971; Moiseff and Konishi, 1981a). Moiseff and Konishi used the thresholds of single auditory neurons in the magnocellular cochlear nucleus to calculate the effect of interaural transmission in the Barn Owl (*Tyto alba*). They concluded that "sound attenuation in the interaural pathway is too large to enable the two ears to function as coupled pressure-difference receivers at the sound pressure levels and for the frequency range (6–8 kHz) important for sound localisation by the owl." Thus, in the owls the solution of localisation problems at higher frequencies may have been approached in a different way, and this may be related to the need to localise prey in three-dimensional space. The use of directional cues derived from binaural asymmetry and the modification of facial feathers (as occurs also in some birds-of-prey) needs acoustic independence rather than the coupling of the two ears. This does not, however, preclude acoustic coupling in the low-frequency range under free-field conditions as may occur during species-specific communication. Little work has been done on this aspect of hearing in owls.

B. Bilateral ear asymmetry in owls

The Barn Owl is capable of locating a sound source in the dark with an error as little as 1 or 2° in both azimuth and elevation (Payne, 1962; Konishi, 1973). This remarkable ability, particularly in elevation, is not matched by any other terrestrial animal. It was Norberg (1977) who pointed out that in the field, the owl is usually well above the ground, so that a given error angle in elevation localisation will produce a greater error distance on the ground than will the same error angle in azimuth. He proposed, therefore, the existence of a particularly strong selection pressure for the improvement of localisation acuity in elevation. In several owl species, such an improvement has been achieved by the development, almost unique in vertebrates, of spatial binaural asymmetry. The occurrence of this asymmetry shows a high correlation with nocturnal activity, feeding on small rodents and shrews, and locating ability (Iljitschew, 1974). It is found in owl species of 9 genera of a total of 29. The asymmetry often results in left–right differences in the positions of the ear openings with respect to the horizontal plane and/or in the orientation of the external meatuses (see Section III). Ear asymmetry in owls has been morphologically described by a number of authors; a comprehensive review of the literature is given by Norberg (1977). However, it was only after biophysical measurements (Norberg, 1968; Payne, 1971; see also Iljitschew, 1974) and behavioural and physiological studies (Schwartzkopff, 1962; Payne, 1971; Konishi, 1973; Knudsen and Konishi, 1979; Knudsen, 1981) that an experimental basis for theories on localisation mechanisms in owls with binaural asymmetries was established.

1. Forms of binaural ear asymmetry in owls

Owls with good localisation acuity generally possess large, complex external ears. Bilateral asymmetry affects only the external ear; asymmetry in the middle and internal ears has not been reported for any species. In most cases the asymmetry involves only soft structures such as the feathers, the preaural skin flaps, and the external acoustic openings. but in four species (*Strix uralensis, Strix nebulosa, Aegolius funereus and Aegolius acadicus*), bilateral asymmetry has been described in the temporal parts of the skull.

In the Long-eared Owl (*Asio otus*) the external acoustic openings are located at different vertical levels on each side of the head, but the size and shape of the large ear slits are equal. Caudal to the slits, the external ear is divided by a more-or-less horizontal dermal septum. On the left side the region above this septum opens into the meatus, and the part below it opens into a blind cavity (the "diverticulum" of Pycraft, 1898) (Fig. 5.24). On the right side, the meatus opens below the septum and the blind cavity opens above the septum. Similar structures are found in *Asio flammeus* and *Asio capensis*. In the Barn Owl (*Tyto alba*), the ear openings in the skin are relatively small and about equal in size, but the opening on the left is located higher than that on the right (Konishi, 1973). Asymmetry is enhanced by the big, almost square preaural flap (*operculum auris*) which covers the opening of each ear in this species. The left flap is placed higher than the right one, and additionally, faces about 15° more dorsally (Payne, 1971) (Fig. 5.24). The asymmetry extends to the elaborate facial ruff which forms two troughs collecting sound for each ear (see Section III). The ruff on the left is directed slightly ventrally and the ruff on the right slightly dorsally, corresponding to the different positions of the acoustic openings and preaural flaps. A difference in the location of the acoustic opening is also found in the Bay Owl (*Phodilus badius*). This species lacks preaural skinfolds, but the cavities which

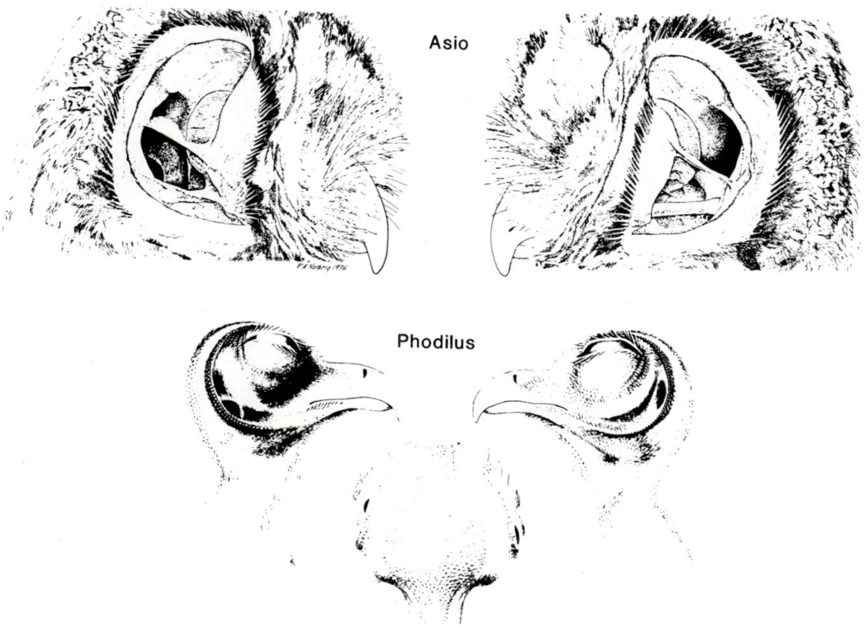

Fig. 5.24. Bilateral external ear asymmetry in owls. In the Bay Owl (*Phodilus badius*) the ear openings in the skin are located at different heights and are of different shapes. In the Long-eared Owl (*Asio otus*) the external ear slits are of similar size on both sides, but the external ear cavity is horizontally divided by a dermal septum; the meatus opens below this septum on the right and above it on the left. From Norberg (1977), courtesy of the Royal Society, London.

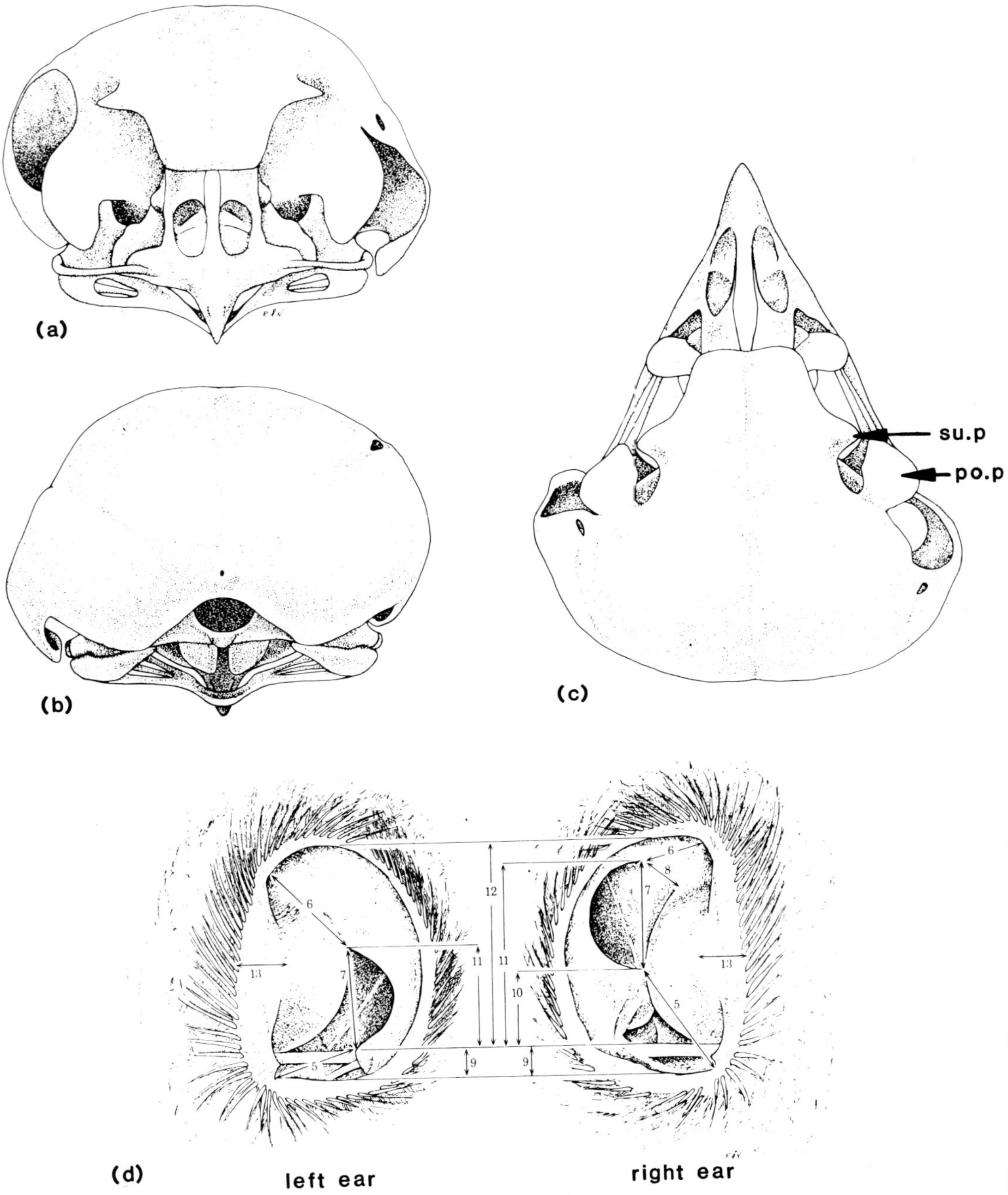

Fig. 5.25. Bilateral ear asymmetry in the Boreal Owl (*Aegolius funereus*). In this owl the asymmetry extends to skull structures mainly affecting the squamosooccipital wings. po.p., Postorbital process; su.p., supraorbital process. As in *Asio* (Fig. 5.24), the ear openings in the skin are placed at different heights and are of different shapes. Together the morphological asymmetries result in the axis through the centre of the tympanic membrane and the ear opening in the skull forming an angle of about 40° in elevation between the left and right ears. From Norberg (1978), courtesy of the Royal Society, London.

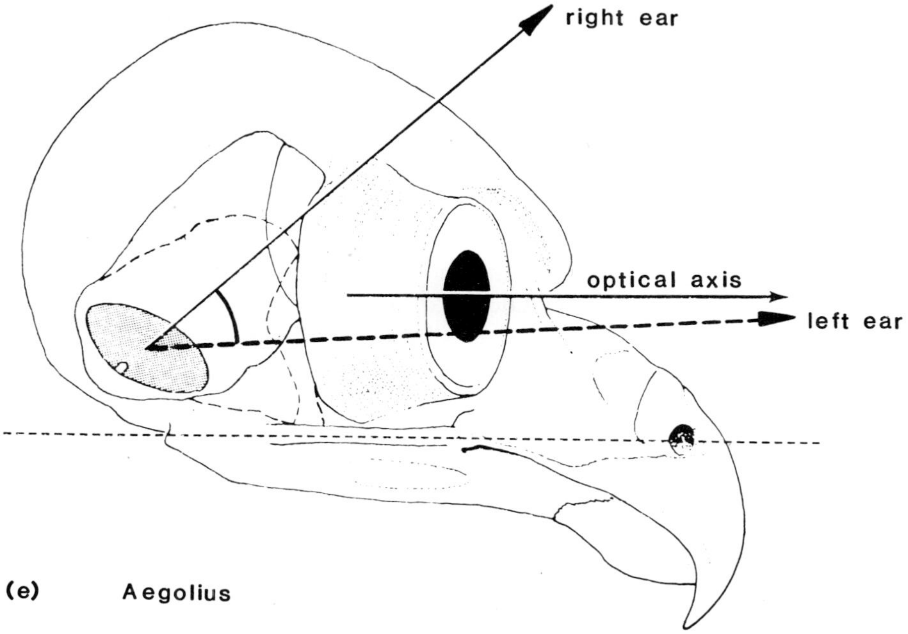

Fig. 5.25(e)

occur between the skin and skull differ in form between the left and the right sides (Pycraft, 1903) (Fig. 5.24).

In the Boreal Owl (*Aegolius funereus*) and the Saw-Whet Owl (*Aegolius acadicus*) the differences in the acoustic opening level and orientation are not basically due to soft structures but to skull asymmetry (Norber, 1977, 1978) (Fig. 5.25). In *Aegolius funereus,* the slit-like ear openings in the skin extend the height of the skull and are symmetrical (Fig. 5.4a). However, the lateral parts of the skull caudal to the postorbital process show extreme bilateral asymmetry, which is most pronounced in the squamoso-occipital wings (Fig. 5.25a–c). The heights of the ear apertures in the skull are only about half the height of the skin slits. The right bony opening is placed about 6 mm higher than the left (Fig. 5.25d). There is a blind cavity below the right ear aperture which is not found on the left side. The right opening faces more dorsally than the left, the vertical difference in their orientation being about 12° (Fig. 5.25e). Skull asymmetry also occurs in two other species, the Ural Owl (*Strix uralensis*) and the Great Gray Owl (*Strix nebulosa*), but it is not as pronounced as in *Aegolius*. It affects only the squamoso-occipital wings, the left wing reaching further rostrodorsally than the right.

It should be stressed again that bilateral asymmetry does not extend to the middle and inner ears in any of the owls investigated. Even in the *Aegolius* species, with their remarkable skull asymmetries, the shape, area and orientation of the tympanic membrane, as well as all the middle ear structures, are bilaterally symmetrical (Norberg, 1978). This is not as surprising as it seems at first: to extract the most useful information from the different sound transmission properties of the external ears, the stimulus-transforming apparatus and the receptor organs must be as perfectly matched as possible. In relation to this problem, Payne (1971) suggested that the prominence of the interaural pathway in owls may be related to the need for matching, by providing a quick pressure equalisation between the right and left middle ears. However, in what way this internal pressure equalisation is effected when an external differential pressure due to asymmetry has been created is not clear, nor is its purpose.

The variety of anatomical modifications leading to ear asymmetry in different owl species suggests its parallel evolution along at least a few lines. This idea was expressed by Payne (1971). Norberg (1977), after a comprehensive and careful comparative anatomical study, proposed a scheme for the evolution of ear asymmetry in owls along five independent lines. From acoustical measurements and functional considerations he concluded that even the slightest ear asymmetry may considerably enhance the acuity of spatial localisation. If the owl uses the method of minimising interaural differences, then during evolution changes in the auditory periphery would have been advantageous even without simultaneous changes in the neuronal networks computing the binaural information.

2. *Acoustic effects of binaural asymmetry*

Binaural directional cues such as interaural differences in arrival time and/or ongoing phase, pure-tone and/or overall intensity and spectral cues can, of course, be achieved by symmetrical ears. However, on or near the sagittal plane, where localisation by minimising interaural differences can be done most accurately (Schwartzkopff, 1962), they remain symmetrical (and therefore ambiguous) with respect to elevation. Bilateral ear asymmetry leads to differential reception of a sound signal by the left and right tympanic membranes, and this alters with changes in elevation. Binaural asymmetry is, in other words, correlated with the need for localisation in the vertical plane.

Norberg (1968) measured the sound transmission properties of the external ears of the Boreal Owl (*Aegolius funereus*), replacing the tympanic membranes, in a carefully-reconstructed head, with microphone probes. He found that both time and intensity parameters were affected by the ear asymmetry. The study concentrated on those directions of sound incidence which produced no interaural phase or intensity differences. With symmetrical ears all these directions were located in the median plane of the head. However, with the *Aegolius* head, the directions producing no interaural *phase* differences deviated to the left or right of the median plane with angles of elevation (Fig. 5.26). However, the directions producing no interaural *intensity* differences remained in the median plane for all elevation angles when the sound frequency was below about 8 kHz. Above this frequency the directional curves became asymmetrical (Fig. 5.26), maximum sensitivity being directed below the horizontal plane for the left ear and above it for the right. With respect to sound intensity, asymmetry in transmission properties is reflected not only in the directions producing minimal differences, but in the complete spatial sensitivity patterns of the ears. In Fig. 5.27, schematic representations of the spatial sensitivities of the left and right ears of the Long-eared Owl (*Asio otus*) are shown. It is clear that the spatial characteristics of the two ears are slightly different even at frequencies below 6 kHz. With increasing frequency, the patterns become more complex and the differences between left and right ears greater (Iljitschew, 1974). In the Barn Owl (*Tyto alba*), Payne

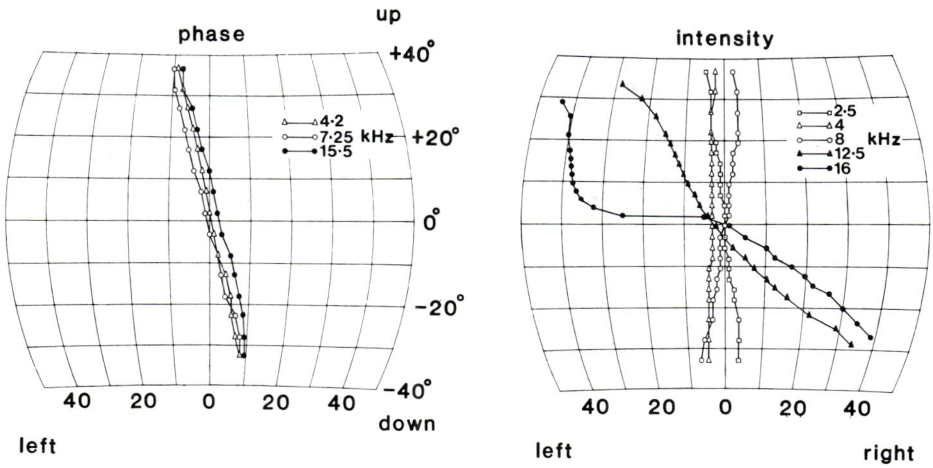

Fig. 5.26. Acoustic consequences of bilateral ear asymmetry in the Boreal Owl (*Aegolius funereus*). The diagrams show those directions in front of the owl which produce no interaural differences in phase or intensity. with perfectly symmetrical ears, these lines would always occur in the median saggital plane for all frequencies. In the Boreal Owl (see Fig. 5.25), the lines for no phase difference are shifted to the left ear above the horizontal plane and to the right ear below it. A similar effect occurs with the lines for no intensity differences, but in contrast to phase, it depends on and increases with frequency. From Norberg (1968), courtesy of the Royal Swedish Academy of Sciences.

(1971) measured the spatial characteristics of the two ears in greater detail. He also found that the complexity of the sensitivity patterns increased with frequency. For both ears, the regions of highest sensitivity were oriented to the front of the owl, parallel to the line of sight. At frequencies above 8 kHz, sharply-defined lows or "nulls" appeared and confined these regions of high sensitivity. The vertical orientations of the regions of both extreme high and extreme low sensitivity were different in the two ears, occurring about 10 to 15° higher in the right ear and thus corresponding to the morphological asymmetry.

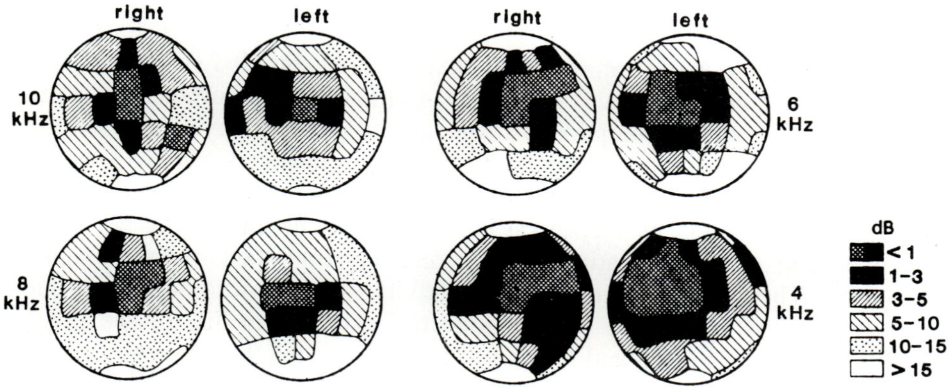

Fig. 5.27. Acoustic consequences of bilateral ear asymmetry in the Long-eared Owl (*Asio otus*) (see also Fig. 5.24). The diagrams show the differences in directional sensitivity patterns of the left and right ears. The diagrams represent the attenuation of sound measured at the site of the tympanic membrane with respect to the sound intensity applied externally. Complexity, and left–right differences, increase with frequency but are present even at 4 kHz. From Iljitschew (1974), courtesy of George Thieme, Leipzig.

Hence, using pure tones as stimuli, bilateral ear asymmetry has two main biophysical effects. First, in the median plane it results in a left–right shift in the position of equal sound phase which is dependent on elevation angle. Second, it produces different spatial sensitivity patterns in each ear, which are more pronounced at high (5–8 kHz) than at low frequencies and are also more pronounced near the line of sight of the owl. With complex sounds rather than pure tones, the different spatial characteristics of the two ears for each component frequency will result in differences in the resultant spectra at the two tympanic membranes. These spectral differences will themselves change with both azimuth and elevation, and thus provide spatial cues. Spectrally complex sounds and the presence of high frequencies do occur in the biotope: during movement or feeding, small rodents and shrews produce sounds rich in transients and with frequencies up to at least 20 kHz (Payne, 1971; Iljitschew, 1974).

3. *Theories on sound localisation in owls*

A first hypothesis for sound localisation in owls, based on theoretical considerations, was proposed by Pumphrey (1948). On the basis of their results Norberg (1968, 1978) and Payne (1971), have criticised and modified Pumphrey's ideas and proposed their own hypotheses for the Boreal Owl (*Aegolius funereus*) and the Barn Owl (*Tyto alba*), respectively. Both hypotheses are based on the behavioural observation that an owl faces a sound source before it attempts to strike at it. Norberg and Payne suggested that this head movement results in the minimising of binaural intensity differences and that comparing the intensities of different frequency bands of spectrally complex sounds provides cues for elevation. When the owl has moved its head so as to receive all the frequency components of a signal with equal intensities in both ears, it will be facing the sound source. However, Konishi (1973) showed that the head-facing movement of the Barn Owl was itself quite accurate, even when the sound signal was shorter than the duration of the head movement. Acuity is best for directions in front of the animal, but the owl must still be able to locate a sound source accurately when it arrives from other directions.

The behavioural and physiological data of Konishi (1973), Knudsen and Konishi (1979) and Knudsen (1981, 1983) confirm the importance of bilateral asymmetry and binaural comparison for localisation. They have also shown that for accurate localisation complex signals with frequencies above 5 kHz are needed. The increase in bilateral differences with respect to the vertical plane is similar to that reported by Payne (1971). Plugging one ear during behavioural experiments resulted in locating errors which corresponded to the anatomical asymmetry. The asymmetry in spatial sensitivity was shown to be mainly due to the asymmetrical orientation of the facial ruff rather than that of the preaural flaps. In general, Knudsen and Konishi (1979) concluded that the Barn Owl uses interaural comparison of the sound spectrum to determine both the elevation and the azimuth of a sound source. This cue is sufficient on its own to calculate elevation, but for an accurate calculation of azimuth, the interaural difference in signal onset time (rather than transient differences) is an essential second, independent cue. Schwartzkopff (1962), Norberg (1968, 1978) and Payne (1971) also noted that the method of localisation by intensity differences may be supported by interaural differences in signal onset time and/or phase. Interaural time differences, as measured in biophysical studies, are in the range up to 0·2 msec (Schwartzkopff, 1962; Norberg, 1968; Payne, 1971). In the central nervous system of the Barn Owls, time differences as short as 0·02 msec can be resolved (Moiseff and Konishi, 1981b). Thus, owls may use time differences for azimuthal localisation in a similar manner to mammals.

The variety of anatomical forms of ear asymmetry in owls and their independent evolution, and the variety of ecological specialisations, suggest that different localisation strategies may be employed by different species. In addition, very little is known about the role of controlled movements of ear structures during localisation. All investigators have stated that the exact positions of the feathers which form the facial ruff and of the ear flaps greatly influence the directional sensitivity pattern of the owl's external ear. Schwartzkopff (1962) suggested for the Long-eared Owl (*Asio otus*) that the "active" component of ear position control may be as important in localisation as the "passive" component given by the ear structures, including the asymmetry. The relative importance of the directional cues arising from external ear asymmetry and ear position control need to be investigated behaviourally, biophysically and physiologically for a number of different species before a full understanding of the mechanisms of sound localisation in the owl can be reached.

VII. Conclusion

"They are hidden, secret things," wrote Pumphrey (1948) when he compiled the body of knowledge then available about the structure and physiology of the ears of birds and found it not too extensive. Since then, detailed morphological investigations and an increasing interest in auditory biophysics, physiology and behaviour have greatly contributed to our knowledge of the form and function of the avian ear. These investigations have revealed the importance and the efficiency of hearing in birds which, at least in part, are due to the design of the external and middle ears. Although the basic construction of the external and middle ears in birds is more like their reptilian ancestors than is that of mammals, the design has evolved to an extent that the inner ear and central nervous system can perform recognition and localisation to a very high level. Highly-specialised structures are found in the external and middle ears of a number of bird species. In some cases, problems have been solved by similar specialisations which have evolved independently in different bird groups. In other cases, the same problems have resulted in different solutions, even in closely-related species, for example localisation in owls. The peripheral auditory systems of birds therefore provide instructive and illustrative material for the study of how structures are formed and modified to optimise their function in response to physical principles and ecophysiological demands.

References

Autrum, H. (1940). Das Richtungshören von *Locusta* und Versuch einer Hörtheorie fur Tympanalorgane des Locustidentyps. *Z. vergl. Physiol.* **28,** 326–352.

Beranek, L. L. (1954). "Acoustics." McGraw-Hill, New York.

Borg, E., Counter, S. A. and Rydquist, B. (1979). Contraction properties and functional morphology of the avian stapedius muscle. *Acta oto-lar.* **88,** 20–26.

Coles, R. B., Lewis, D. B., Hill, K. G., Hutchings, M. E. and Gower, D. M. (1980). Directional hearing in the Japanese Quail (*Coturnix coturnix japonica*). II. Cochlear physiology. *J. exp. Biol.* **86,** 153–170.

Counter, S. A. and Borg, E. (1979). Physiological activation of the stapedius muscle in *Gallus gallus*. *Acta oto-lar.* **88,** 13–19.

Crompton, A. W. (1953). The development of the chondrocranium of *Spheniscus demersus* with special reference to the columella auris of birds. *Acta zool. Stockh.* **34,** 71–146.

de Beer, G. R. (1937). "The Development of the Vertebrate Skull." Clarendon Press, Oxford.

Freye-Zumpfe, H. (1953). Befunde im Mittelohr der Vögel. *Wiss. Z. Martin-Luther-Univ. Halle Wittenb.* **2,** 445–461.

Gaudin, E. P. (1968). On the middle ear of birds. *Acta oto-lar.* **65,** 316–326.

Golubjewa, T. B. (1972). The reflex activity of the tympanal muscle in the owl *Asio otus*. *Zh. evol. Biol. Fiziol.* **8,** 173–181.

Henson, O. W. (1974). Comparative anatomy of the middle ear. *In* "Handbook of Sensory Physiology" (W. D. Keidel and W. D. Neff, Eds.), Vol. V. Springer-Verlag, Berlin and New York.

Hill, K. G., Lewis, D. B., Hutchings, M. E. and Coles, R. B. (1980). Directional hearing in the Japanese Quail (*Coturnix coturnix japonica*). I. Acoustic properties of the auditory system. *J. exp. Biol.* **86,** 135–151.

Iljitschew, W. D. (1974). Adaptionsökologische Parallelismen-mosaikartige Evolution. Das Hörsystem der Vögel als Objekt der funktionellen Morphologie. *Biol. Zbl.* **93,** 165–180.

Iljitschew, W. D. and Izwekowa, L. M. (1961). Some pecularities of the function of the auditory analyzer in birds. *Zool. Zh.* **40,** 1704–1714.

Kartaschew, N. and Iljitschew, W. D. (1964). Über das Gehörorgan der Alkenvögel. *J. Ornithol.* **105,** 113–136.

Khanna, S. M. and Tonndorf, J. (1972). Tympanic membrane vibration in cats studied by time-averaged holography. *J. acoust. Soc. Am.* **51,** 1904–1920.

Knudsen, E. I. (1980). Sound localisation in birds. *In* "Comparative Studies of Hearing in Vertebrates" (A. N. Popper and R. R. Fay, Eds.). Springer-Verlag, New York.

Knudsen, E. I. (1981). The hearing of the barn owl. *Sci. Am.* **245,** 82–91.

Knudsen, E. I. (1983). Space coding in the vertebrate auditory system. *In* "Bioacoustics" (D. B. Lewis, Ed.). Academic Press, London and New York.

Knudsen, E. I. and Konishi, M. (1979). Mechanisms of sound localization in the barn owl *(Tyto alba)*. *J. comp. Physiol.* **133,** 13–21.

Konishi, M. (1973). How the owl tracks its prey. *Am. Scient.* **61,** 414–424.

Krause, G. (1901). "Die Columella der Vogel (Columella Auris Avium). Ihr Bau und dessen Einfluss auf die Feinhörigkeit." Friedländer, Berlin.

Kuhn, A., Müller, C. M., Lepplsack, H. J. and Schwartzkopff, J. (1982). Heartrate conditioning used for determination of auditory threshold in the starling. *Naturwissenschaften* **69,** 245.

Lewis, D. B. (1983). Directional cues for auditory localisation. *In* "Bioacoustics" (D. B. Lewis, Ed.). Academic Press, London and New York.

Lewis, D. B. and Coles, R. (1980). Sound localisation in birds. *Trends NeuroSci., Pers. Ed.* **3,** 102–105.

Müller, H. J. (1960). Die Morphologie und Entwicklung des Craniums von *Rhea americana* Linné. *Z. wiss. Zool.* **165,** 221–319.

Michelsen, A. (1983). Biophysical basis of sound communication. *In* "Bioacoustics" (D. B. Lewis, Ed.). Academic Press, London and New York.

Møller, A. R. (1965). An experimental study of the acoustic impedance of the middle ear and its transmission properties. *Acta oto-lar.* **60,** 129–149.

Moiseff, A. and Konishi, M. (1981a). The owl's interaural pathway is not involved in sound localisation. *J. comp. Physiol.* **144,** 299–304.

Moiseff, A. and Konishi, M. (1981b). Neuronal and behavioural sensitivity to binaural time differences in the owl. *J. Neurosci.* **1,** 40–48.

Norberg, R. Å. (1968). Physical factors in directional hearing in *Aegolius funereus* (Strigiformes), with special reference to the significance of the asymmetry of the external ears. *Ark. Zool.* **20,** 181–204.

Norberg, R. Å. (1977). Occurrence and independent evolution of bilateral ear asymmetry in owls and implications on owl taxonomy. *Phil. Trans. R. Soc.* **280,** 376–408.

Norberg, R. Å. (1978). Skull asymmetry, ear structure and function, and auditory localization in Tengmalm's owl, *Aegolius funereus* (Linné). *Phil Trans. R. Soc.* **282B,** 325–410.

Oeckinghaus, H. (1978). Einfluss von Kontraktionen des Mittelohrmuskels auf die Entladungen von Hörnervenelementen beim Star. *Verh. dt. zool. Ges.* **71,** 191.

Oeckinghaus, H. and Schwartzkopff, J. (1975). Elektrische Aktivierung des Mittelohrmuskels beim Star. *Naturwissenschaften* **62,** 582.

Oeckinghaus, H. and Schwartzkopff, J. (1983). Electrical and acoustical activation of the middle ear muscle in a songbird. *J. comp. Physiol.* **150,** 61–67.

Payne, R. S. (1962). How the Barn Owl locates prey by hearing. *Living Bird* **1,** 151–159.

Payne, R. S. (1971). Acoustic location of prey by barn owls *(Tyto alba). J. exp. Biol.* **54,** 535–573.

Payne, R. S. and Drury, W. H. (1958). Marksman of the darkness. *Nat. Hist., N.Y.* **67,** 316–323.

Pohlman, A. G. (1921). The position and functional interpretation of the elastic ligaments in the middle ear region of *Gallus. J. Morph.* **35,** 229–262.

Portmann, A. (1950). Les organes des sens. *In* "Traité de Zoologie" (P. P. Grassé, Ed.), Vol. 15. Masson, Paris.

Pumphrey, R. J. (1948). The sense organs of birds. *Ibis* **90,** 171–199.

Pumphrey, R. J. (1961). Part II. Sensory organs: hearing. *In* "Biology and Comparative Physiology of Birds" (A. J. Marshall, Ed.), Vol. 2. Academic Press, New York and London.

Pycraft, W. P. (1898). A contribution to our knowledge of the morphology of owls. Part I. Pterylography. *Trans Linn. Soc. Lond., Zool.* **7,** 223–275.

Pycraft, W. P. (1903). On the pterylography of *Photodilus. Ibis* **45,** 36–48.

Pye, J. D. (1980). Echolocation signals and echoes in air. *In* "Animal Sonar Systems" (R. G. Busnel and J. F. Fish, Eds.). Plenum Press, New York.

Rosowski, J. J. and Saunders, J. C. (1980). Sound transmission through the avian interaural pathway. *J. comp. Physiol.* **136,** 183–190.

Saiff, E. I. (1974). The middle ear of the skull of birds: the Procellariiformes. *Zool. J. Linn. Soc.* **54,** 213–240.

Saiff, E. I. (1976). Anatomy of the middle ear region of the avian skull: Sphenisciformes. *Auk* **93,** 749–759.

Saiff, E. I. (1978). The middle ear of the skull of birds: Pelecaniformes and Ciconiiformes. *Zool. J. Linn. Soc.* **63,** 315–370.

Saiff, E. I. (1981). The middle ear of the skull of birds: the ostrich, *Struthio camelus. Zool. J. Linn. Soc.* **73,** 201–212.

Saunders, J. C. and Johnstone, B. M. (1972). A comparative analysis of middle ear function in non-mammalian vertebrates. *Acta oto-lar.* **73,** 353–361.

Schwartzkopff, J. (1949). Uber Sitz und Leistung von Gehör und Vibrationssinn bei Vögeln. *Z. vergl. Physiol.* **31,** 527–608.

Schwartzkopff, J. (1950). Beitrag zum Problem des Richtungshörens bei Vögeln. *Z. vergl. Physiol.* **32,** 319–327.

Schwartzkopff, J. (1952). Untersuchungen über die Arbeitsweise des Mittelohres und das Richtungshören der Singvögel unter Verwendung von Cochlea-Potentialen. *Z. vergl. Physiol.* **34,** 46–68.

Schwartzkopff, J. (1955). On the hearing of birds. *Auk* **72,** 340–347.

Schwartzkopff, J. (1957). Die Grössenverhaltnisse von Trommelfell, Columella-Fussplatte und Schnecke bei Vögeln verschiedenen Gewichts. *Z. Morph. Ökol. Tiere* **45,** 365–378.

Schwartzkopff, J. (1962). Zur Frage des Richtungshörens von Eulen (Striges). *Z. vergl. Physiol.* **45,** 570–580.

Schwartzkopff, J. (1973). Mechanoreception. *In* "Avian Biology" (D. S. Farner and J. R. King, Eds.), Vol. III. Academic Press, New York and London.

Schwartzkopff, J. and Bremond, J. C. (1963). A method for obtaining cochlear potentials in birds. *J. Physiol., Paris* **55,** 495–518.

Smith, G. (1904). The middle ear and columella of birds. *Q. J. Microsc. Sci.* **48,** 11–22.

Stellbogen, E. (1930). Über das äussere und mittlere Ohr des Waldkauzes *(Syrnium aluco,* L.*). Z. Morph. Ökol. Tiere* **19,** 686–731.

Wada, Y. (1924). Beiträge zur vergleichenden Physiologie des Gehörorgans. *Pflügers Arch. ges. Physiol.* **202,** 46–49.

Wever, E. G. and Lawrence, M. (1954). "Physiological Acoustics." Princeton University Press, Princeton.

6
Inner ear

CATHERINE A. SMITH

Department of Otolaryngology, Medical School, Oregon Health Sciences University, Portland, Oregon, U.S.A.

CONTENTS

I. Introduction

The avian inner ear (*auris interna*) has two functions: hearing and equilibration. Hearing takes place in the cochlear organ (*organum cochleare*), equilibration in the vestibular organ (*organum vestibulare*). The vestibular organ was developed early in marine vertebrates and has changed very little during the evolutionary process. Thus, the vestibular organ of birds is somewhat different from that of mammals but not remarkably so. On the other hand, the auditory receptor organs of the inner ear have distinctive forms in each of the classes of terrestrial vertebrates, birds being no exception. The avian cochlear organ is especially interesting, because it contains some structural features which are common to all birds thus far examined, and others which vary between species.

The inner ear is embedded within the skull and is composed of two parts, the membranous labyrinth and the bony labyrinth, the former being enclosed by the latter. The membranous labyrinth is filled with the fluid endolymph, the spaces between the membranous and bony labyrinths containing the fluid perilymph.

II. Bony labyrinth

The bony labyrinth (*labyrinthus osseus*) is formed by compact bone and for the most part is situated between the medial and lateral compact layers of the skull. Except for its most medial parts (the medial walls of the cochlea and the rostral semicircular canal), the bony labyrinth is surrounded by pneumatised bone. The bony labyrinth consists of three divisions: the bony semicircular canals caudally, the cochlea rostrally and the vestibule in the middle.

The vestibule (*vestibulum*) is the central part of the bony labyrinth and connects the cochlea rostrally with the semicircular canals caudally. It contains the utricle and saccule of the membranous labyrinth.

The form of the three bony semicircular canals (*canales semicirculares ossei*) shows only minor variations among vertebrates. In birds, the rostral or anterior semicircular canal (*canalis semicircularis rostralis*) is larger than the other two canals (Figs. 6.1, 2). Thus, the outside circumference of the rostral canal in the chicken is approximately 20 mm, whereas that of the horizontal or lateral semicircular canal (*canalis semicircularis horizontalis*) is approximately 12 mm. In humans, it is the caudal canal which is largest, but there is little difference between the three. One end of each semicircular canal is enlarged to form the ampulla (*ampulla ossea*). The caudal end of the rostral semicircular canal and the upper end of the caudal semicircular canal (*canalis semicircularis caudalis*) unite to form a common canal, the *crus osseum commune*. The bony semicircular canals contain the semicircular ducts of the membranous labyrinth.

The avian *cochlea* does not have the coiled snail shell form of the mammalian cochlea. Instead, it is a flattened tube which varies in length among species. In *Columba,* for example, the bony cochlea is approximately 5·3 mm in length as measured from its distal tip or apex (*apex cochleae*)* to the posterior lip of the vestibular window. In the Barn Owl (*Tyto alba*), in contrast, it is more than twice as long. The cochlea is almost straight in very small birds (including many passerines), but in *Columba* (Takasaka and Smith, 1971) and *Gallus* (Tanaka and Smith, 1978) it is slightly curved. The curvature is more pronounced in the longer cochlea of the Barn Owl (*Tyto alba*). There has been much speculation about the significance of the coiled form of the cochlea of mammals. One explanation is that it helps to conserve space. This seems logical when one considers the 36-mm-long cochlear duct of humans. Furthermore, even the smallest mammals have longer cochlear ducts than have birds of comparable sizes. Medium-sized birds such as *Columba* and *Gallus* have no difficulty accommodating a straight, 4- to 5-mm-long cochlea in their skulls. A longer structure might create difficulties, but among the 45 species studied by Schwartzkopff and Winter (Schwartzkopff, 1968), only 6 had cochlear ducts which were 7 mm or more in length. In addition to the space limitations for a long, straight structure, it seems probable that, as the basilar membrane lengthened in mammals and the frequency range increased, it became more practical from a standpoint of vascularisation and innervation to supply the organ from the central core of a coil.

There are two openings in the lateral wall of the bony labyrinth—the vestibular (oval) window (*fenestra vestibularis*) and the cochlear (round) window (*fenestra*

*The avian cochlea contains the *macula lagenae* at its apex.

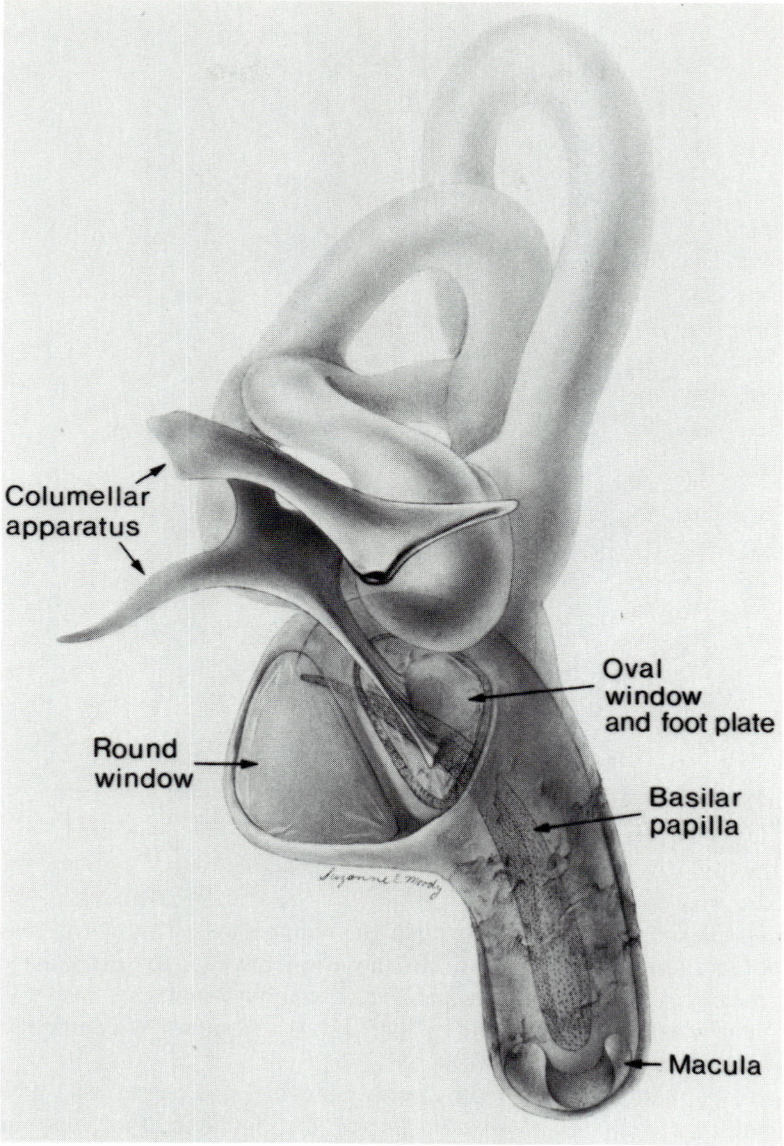

Fig. 6.1. Drawing of the bony labyrinth of *Gallus*. The oval (vestibular) window opens into the perilymphatic scala vestibuli of the cochlea. The round (cochlear) window opens into the scala tympani. The cochlea has been made transparent so that the basilar papilla and the macula of the lagena can be seen inside. From Tanaka and Smith (1978), courtesy of Alan R. Liss, Inc., New York.

cochlearis)—both of which are located at the base of the cochlea (*basis cochleae*) and which are perpendicular to each other (Fig. 6.1).Both windows open onto the middle ear space. The vestibular window is closed off by the footplate or base of the columella (*basis columellae*). The base is held in place by the annular ligament of the columella (*lig. annulare columellae*), which spans the space between the bony perimeters of the base and window. The cochlear window is larger and more rounded than the vestibular window and is located caudoventral (inferior) to

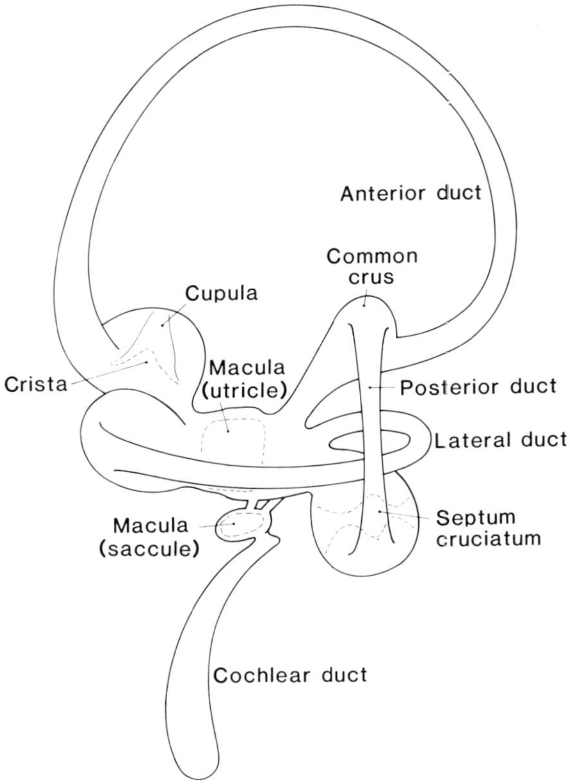

Fig. 6.2. Schematic drawing of a lateral view of the avian membranous labyrinth showing the cochlear duct, saccule, utricle and the anterior (rostral), posterior (caudal) and lateral (horizontal) semicircular ducts. Some of the neuroepithelial structures inside are depicted by dotted lines.

it. In *Columba* the long axis of the cochlear window is 2·5 mm, whereas that of the vestibular window is 2·3 mm. In the Barn Owl (*Tyto alba*) the cochlear window measures 3·5 × 2·0 mm and the vestibular window is 1·6 × 1·3 mm. The cochlear window is closed off by the fibrous secondary tympanic membrane (*membrana tympanica secundaria*).

The perilymphatic spaces of the cochlea are the *scala vestibuli,* which is in contact with the *tegmentum vasculosum* of the cochlear duct, and the *scala tympani,* which is in contact with the basilar membrane of the cochlear duct. The scala vestibuli terminates in the perilymphatic cisterna of the vestibule and is small in the bird. Schwartzkopf (1968) stated that the tegmentum vasculosum was in contact with the bony walls, implying that the scala vestibuli may be slit-like in some birds; this is not the case in all. The scala vestibuli is enlarged in the region of the cochlear window to form a cisterna (*cisterna scalae vestibuli*) at the proximal end of which it communicates with the perilymph-filled vestibule. Its distal end is called the *fossa scalae vestibuli.* The scala tympani terminates at the cochlear window, below which it forms a large recess (*recessus scalae tympani*). Schwartzkopf (1968) described two small passageways between the scala vestibuli and the scala tympani, a basal interscalar canal (*canalis interscalaris basalis*) at the base of the cochlea, and an apical interscalar canal (*canalis in-

terscalaris apicalis) at the apex of the cochlea. Neither the scala tympani nor the *scala vestibuli* persists distal to the basilar papilla, and the membranous lagena is indeed in contact with the otic capsule. This is somewhat different from the situation in the mammlian ear in which the perilymphatic scalae communicate widely at the helicotrema, which is at the apical tip of the cochlea. According to von Bekesy (1960b), such a helicotrema would be a disadvantage for a bird, since a large opening in a short cochlea would greatly reduce the ear's sensitivity.

A small perilymphatic canalicule (*canaliculus perilymphaticus*), which opens into the recess, connects the scala tympani with the subarachnoid space. The facial and vestibulocochlear nerves are transmitted through another opening in the internal acoustic fossa in the medial wall of the bony labyrinth.

The bony labyrinth protects the delicate structures within the inner ear from physical trauma. The pneumatised bone in the bird probably acts as a shock absorber and reduces the possibility of fractures to the otic capsule from head blows. In addition, the bony labyrinth acts as a barrier of low permeability which retains the large fluid pool of perilymph in place.

III. Membranous labyrinth

A. General characteristics

The membranous labyrinth (*labyrinthus membranaceus*) fits inside the bony labyrinth, and the two have a similar form and size (Figs. 6.1, 2). However, except for the ampullated ends of the semicircular ducts, the fit is not a tight one. The vestibular ducts as well as the cochlear duct are smaller than their bony counterparts, the space between the membranous and bony labyrinths being filled with perilymph. The vestibular membranous labyrinth is somewhat more complex than the outer form of the bony vestibule suggests, since this single, large, bony cavity contains the saccule, the utricle, the sacculocochlear duct, the utriculosaccular duct, the endolymphatic duct and the vestibular end of the cochlear duct. The membranous labyrinth of birds resembles that of mammals, in that it is a closed epithelial tube filled with endolymph and is divided into compartments which contain the neuroepithelia and other highly-specialised cell groups. The greatest histological differences between the inner ears of birds and mammals are found in the cochlear duct.

B. Inner ear fluids

Perilymph (*perilympha*) surrounds the membranous labyrinth and fills in the space between it and the bony labyrinth. It conducts the vibrations set up in the tympanic membrane and columella by airborne sound to the endolymph and basilar membrane. It is likely that it is also the source of nutrients to the cells of the basilar papilla, for there are no blood vessels within this structure. The vestibular structures, on the other hand, are well endowed with subepithelial capillary networks in both birds (Schmidt, 1964) and mammals (Smith, 1953).

There are no data for the composition of perilymph in birds. However, in mammals perilymph is similar to tissue fluid (Smith *et al.*, 1954). Peterson *et al.* (1978) have found that reptilian and mammlian perilymph have much the same composition. Thus, it seems probable that avian perilymph is also similar and that, in all cases, perilymph is a blood ultrafiltrate.

Endolymph (*endolympha*) fills the cavity of the membranous labyrinth and is most definitely not a blood ultrafiltrate. The chemical composition of endolymph in mammals has been shown to be exceedingly high in potassium and low in sodium (Smith *et al.*, 1954; Bosher and Warren, 1968). The only available data for the bird were reported by Johnstone *et al.* (1963), in the form of ratios, the potassium:sodium ratio in cochlear endolymph in *Columba* being 3. This is considerably less than the ratio of 114 found by Peterson *et al.* (1978) in the cat, but greater than the 0·1 ratio observed by these workers in the lizard.

A positive resting potential (relative to perilymph) can be recorded from cochlear endolymph, as first demonstrated by von Bekesy (1952) in the cat. The mammalian endocochlear potential is anoxia-sensitive and has a value close to +90 mV. In *Columba,* Schmidt (1963) found that the endocochlear potential is also anoxia-sensitive and that it ranged from +11 to 20 mV, with an average of +17 mV. In crocodiles, Schmidt and Fernandez (1962) found that the potential was not anoxia-sensitive and that it ranged from −1 to +5 mV (average +2 mV). This physiological difference between crocodiles and birds is interesting because the cochlear ducts of the two have remarkable histological similarities. Both have a distinctive tegmentum vasculosum which is probably analogous to the mammalian stria vascularis. These structures are generally conceded to be the source of the positive component of the endocochlear potential, as well as the high potassium of endolymph.

The endocochlear potential is closely related to sensory cell function, although its precise role is unknown. When the high positive endocochlear potential of mammals is reduced experimentally to a negative value, the cochlear microphonics and hearing are greatly depressed (Brummett *et al.*, 1977; Bosher, 1979). The avian cochlea apparently functions quite adequately with a lower endocochlear potential and with what may be a lower potassium concentration in the endolymph.

C. Cochlear organ

1. Cochlear duct

The cochlear duct (*ductus cochlearis*) is a finger-like tube which projects rostrally (anteriorly) from the saccule and is closed off at its distal end (Fig. 6.2). The length of the cochlear duct varies between species, ranging from 2·75 mm in small birds to 10 mm or more in the owls *Strix aluco* and *Bubo bubo* (Schwartzkopf, 1968). It is connected to the saccule by the sacculocochlear duct (*ductus sacculocochlearis*), which is much larger in some birds such as *Columba* than in mammals. In both *Columba* (Schmidt, 1964; Takasaka and Smith, 1971) and the Barn Owl (*Tyto alba*), the sacculocochlear duct carries strands of tegmentum vasculosum cells into the saccule.

Two sensory structures are present in the cochlear duct, the basilar papilla and

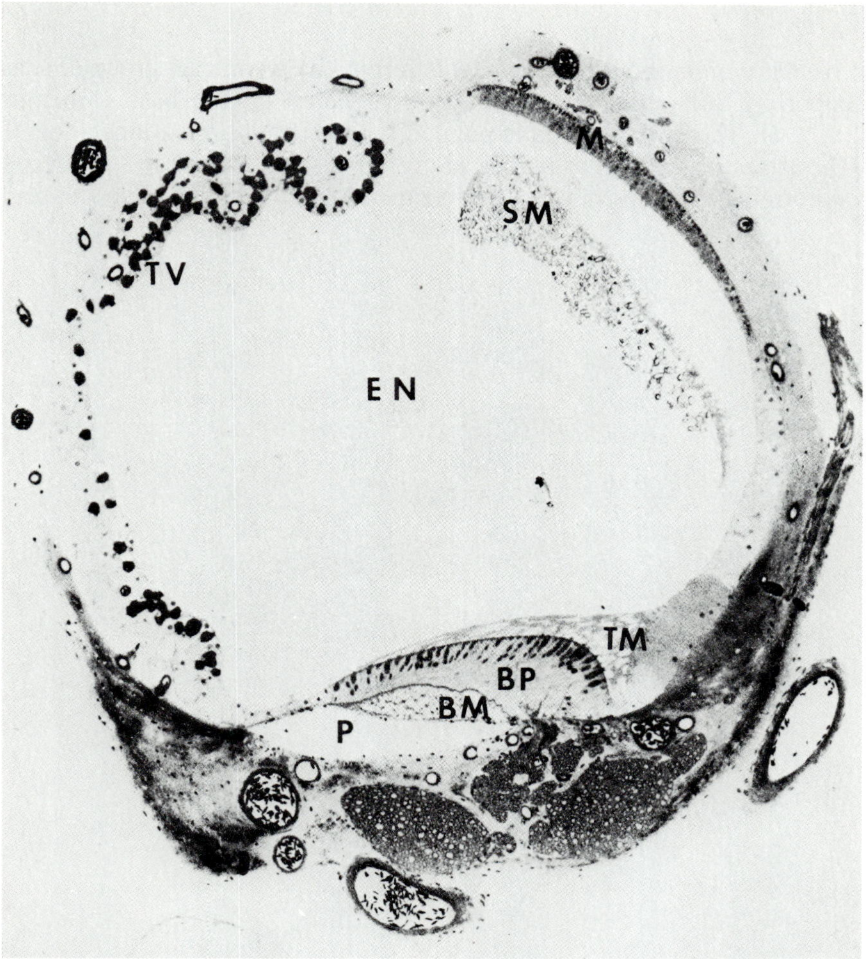

Fig. 6.3. Photomicrograph of a cross-section through the cochlear duct of *Gallus* near the distal end. The macula of the lagena (M) is visible at upper right. The statoconial membrane (SM) has been displaced in preparation. BM, Basilar membrane; BP, basilar papilla; EN, endolymph of scala media; P, perilymph of scala tympani; TM, tectorial membrane; TV, tegmentum vasculosum. ×113. From Tanaka and Smith (1978), courtesy of Alan R. Liss, Inc., New York.

the macula of the lagena (Figs. 6.1, 3). The basilar papilla rests on the basilar membrane and is continuous for almost the entire length of the duct. The distal end is occupied by the macula of the lagena. A thickened layer of specialised epithelium, the tegmentum vasculosum, closes off the duct above the papilla (Figs. 6.3, 4). The cavity of the cochlear duct is filled with endolymph and is surrounded by perilymph in the scala vestibuli and the scala tympani.

The distinctive characteristics of the avian cochlear duct include (a) a short, slightly-curved tubular form, (b) a basilar papilla composed of a compact mass of cells with a distinctive sensory cell organization, (c) a well-vascularised, specialised epithelial layer, the tegmentum vasculosum, which forms one wall of the duct, and (d) a statoconial macula at its distal tip.

2. *Basilar papilla*

The basilar papilla (*papilla basilaris*) is formed by a compact group of sensory and supporting cells, the *neuroepithelium,* situated on the basilar membrane (Figs. 6.3, 4). The proximal end is pointed but the distal tip is rounded off (Fig. 6.1). The edges of the basilar papilla are referred to as "inferior" and "superior," conforming to the position of the cochlea within the skull. The "superior"

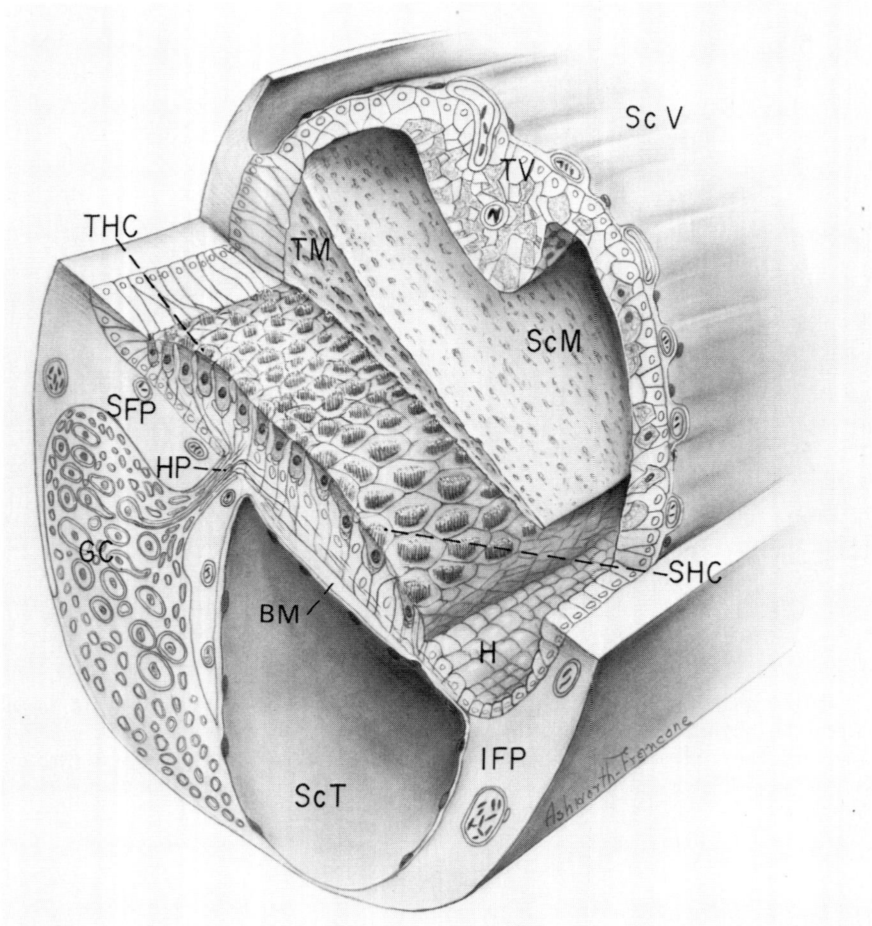

Fig. 6.4. Schematc drawing of a transverse section through the cochlear duct of *Columba.* The basilar papilla, composed of sensory and supporting cells, rests on the basilar membrane (BM). Short hair cells (SHC) are located on the free part of the basilar membrane and tall hair cells (THC) over the superior fibrocartilaginous plate (SFP). The tectorial membrane (TM), a part of which has been removed to display the surface of the papilla and the hair bundles on the sensory cells, is attached to epithelial cells on the superior wall of the cochlear duct and covers the entire sensory surface. The hyaline cells (H) are seen on the right side, stretching from the inferior edge of the papilla to the inferior fibrocartilaginous plate (IFP). The tegmentum vasculosum (TV) closes off the scala media (ScM) and separates it from the scala vestibuli (ScV), which has been left open in the drawing. The ganglion cells (GC) of the cochlear nerve are seen below the superior fibrocartilaginous plate at lower left. Their peripheral processes pierce the habenula perforata (HP) and are then distributed to the hair cells. ScT, Scala tympani. From Takasaka and Smith (1971), courtesy of Academic Press, New York and London.

side is comparable from a histological standpoint to the medial side of the mammalian basilar membrane. The length of the papilla varies in different species, being slightly over 2 mm in the small birds, 4·4 mm in *Gallus*, and 9·5–11·5 mm in the Barn Owl (*Tyto alba*). In some birds, such as *Columba* (Takasaka and Smith, 1971) and *Gallus* (Tanaka and Smith, 1978), the papilla gradually broadens towards the distal end, the width of the broad tip in *Gallus* measuring 0·3 mm. In the Budgerigar (*Melopsittacus undulatus*) and the Barn Owl (*Tyto alba*), however, the change in width of the papilla is not so gradual.

(a) Sensory cells. The sensory receptors are hair cells similar in general form to all auditory-vestibular sensory cells in vertebrates. Their apical ends are filled with many microfilaments (forming the so-called cuticular plate) from which a bundle of cilia protrudes (Fig. 6.5). Most of these are stereocilia containing actin filaments (Flock and Cheung, 1977). However, one of the long cilia situated to one side of the bundle is a kinocilium with microtubules, although the microtubular structure and dynien arm* structure is irregular and atypical (Tanaka and Smith, 1978). Thus, it seems unlikely that this kinocilium is motile. The cilia are arranged in a stepwise pattern, the longest ones being inserted into the tectorial membrane. Takasaka and Smith (1971) found that the kinocilium in each hair cell of the basilar papilla of *Columba* is located on the same side of each hair bundle (Figs. 6.4–6). If it is assumed that the hair bundle is polarized morphologically by the kinocilium, then the polarization direction is constant throughout the pigeon's papilla. In *Gallus* the morphological polarization is fairly uniform for most of the papilla, but at the distal tip a small number of hair cells have the opposite polarity (Tanaka and Smith, 1978). Kinocilia are absent from many of the hair cell bundles in *Gallus* (Fig. 6.7) (Tanaka and Smith, 1978). In comparison with the regular position of the kinocilium in the avian basilar papilla, kinociliar polarization is quite irregular in the small basilar papillae of reptiles (Miller, 1973), as well as being highly variable in all vertebrates in the maculae of the vestibular organ. Kinocilia are totally absent from the mature mammalian cochlea.

In the past, great emphasis has been placed on the possible role of the kinocilium in sensory cell activation. Lowenstein and Wersall (1959) and Flock (1965, 1971) demonstrated in the lateral line organ of fish that when the hair bundle was moved in the direction of the kinocilium, hair cell activation occurred; when it was moved in the opposite direction, inhibition took place. Nevertheless, some doubts regarding kinociliar importance have always been present because of their absence in the mammalian cochlea. Definitive studies by Hudspeth and Jacobs (1979) have provided evidence that the kinocilium plays little or no active part in hair cell depolarization or hyperpolarization but that the stereocilia do mediate hair cell transduction. However, they confirmed the earlier studies regarding the relationship of the directional movement of the hair bundle to hair cell excitation and inhibition. It may well be that the important feature in determining whether the sensory cell is activated or inhibited resides in the arrangement of tall and short stereocilia.

The second feature common to hair cells is the presence of nerve endings, generally located on their basal ends. There are two types of nerve terminals,

*The dynien arms are the inner arms of the microtubular doublets and the locus of the majority of ciliar and flagellar ATPase.

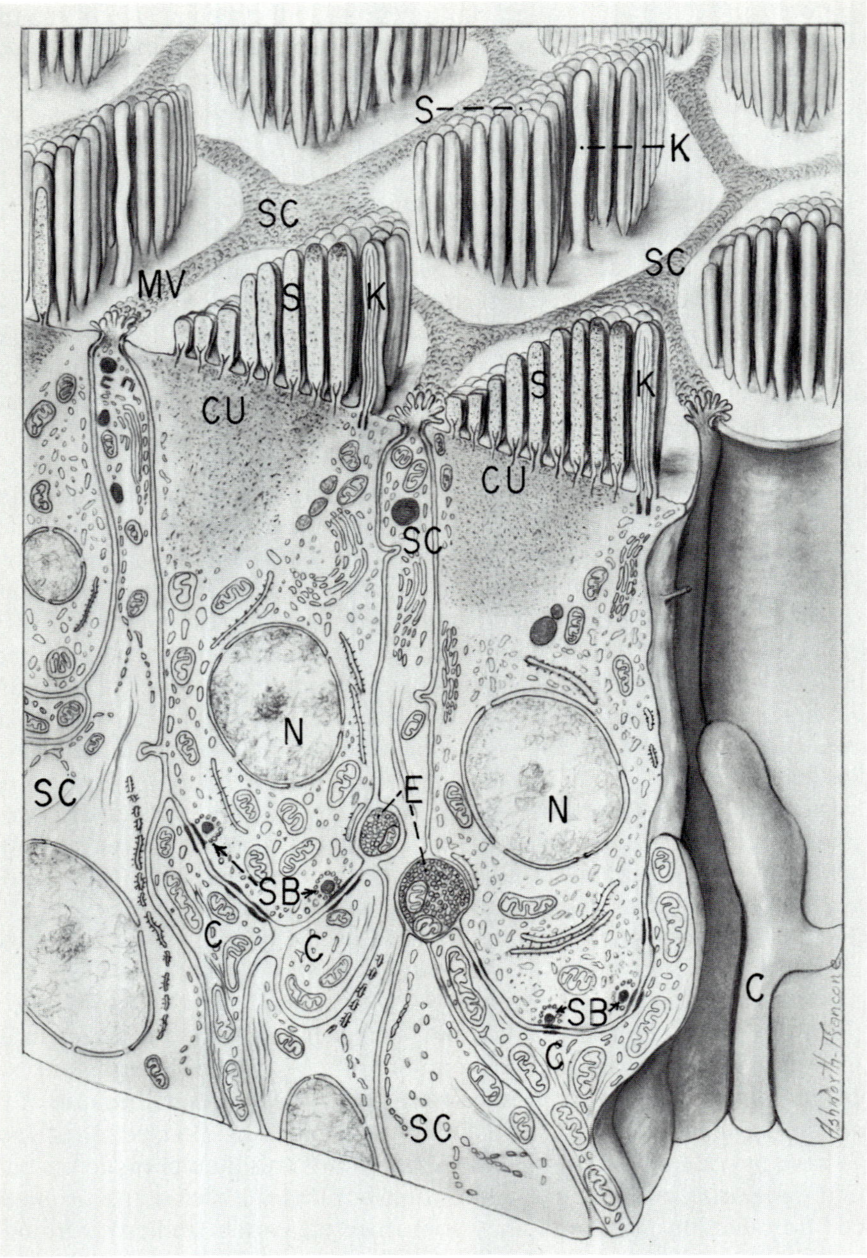

Fig. 6.5. Drawing showing tall hair cells and supporting cells (SC) in the basilar papilla of *Columba*. Three hair cells are sectioned longitudinally to demonstrate their intracellular structure. At right, whole cells have been removed to show the contours of the sensory cells and nerve endings. The microvillous tufts (MV) on the supporting cell surfaces form thick angular borders about each sensory cell. Each hair bundle is composed of a large number of stereocilia (S), which have a stepwise increment in height and one long kinocilium (K). Cochlear nerve endings (C) are finger-like, with one or two processes (two cells at right) or club-like, with membrane thickenings and synaptic balls (SB). Efferent terminals are small (E). CU, Cuticular plate; N, nucleus. From Takasaka and Smith (1971), courtesy of Academic Press, New York and London.

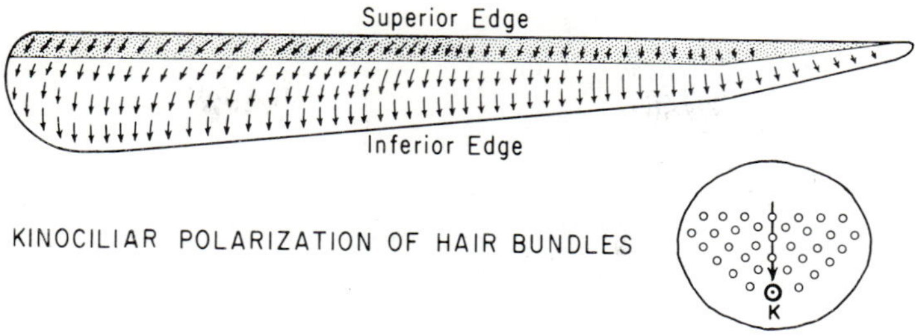

Fig. 6.6. Diagram showing the morphological (kinociliar) polarization on the hair cell bundles in the basilar papilla of *Columba*. The drawing of the hair bundle on the hair cell lower right explains the arrows. Most cells are polarized in a direction parallel to the transverse axis (or pointing to the inferior edge), but some on the dotted area (above the fibrocartilaginous plate) have a more diagonal polarization. From Takasaka and Smith (1971), courtesy of Academic Press, New York and London.

afferent and efferent. The afferent fibres (or cochlear nerve dendrites) have terminals of quite variable shape (Fig. 6.5). Some take the form of small boutons, some of shallow calyces and others of long, finger-like processes with multiple synapses. Membrane thickenings as well as synaptic ball structures inside the hair cells are associated with the afferent synapse. The efferent nerve endings, which have their source of origin in the brain stem (Boord, 1961), are filled with small, round vesicles and terminate as small boutons on the tall, intermediate and lenticular hair cells but form large calyces on the typical short hair cells.

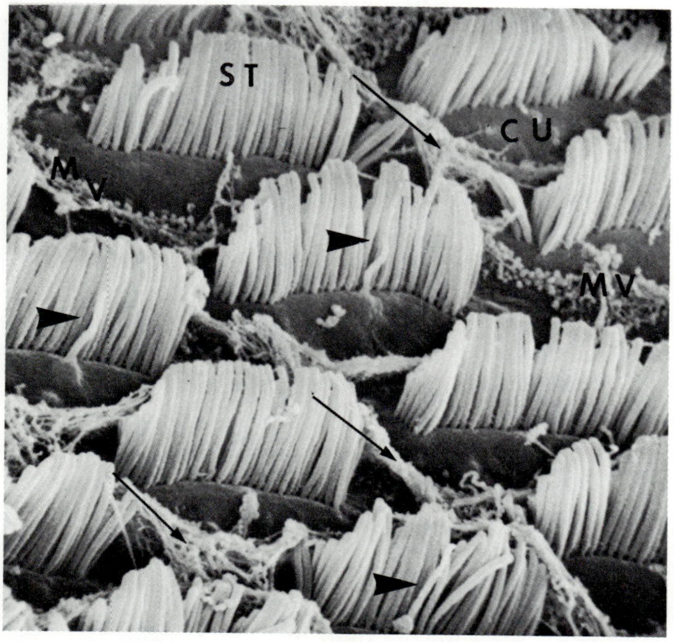

Fig. 6.7. Scanning electron micrograph of the hair bundles on the tall hair cells of the basilar papilla of *Gallus*. Only three bundles have kinocilia (arrowheads). Some remnants of the tectorial membrane (arrows) are still attached to the supporting cell microvilli (MV). CU, Cuticular plate; ST, stereocilia. ×5000. From Tanaka and Smith (1978), courtesy of Alan R. Liss, Inc., New York.

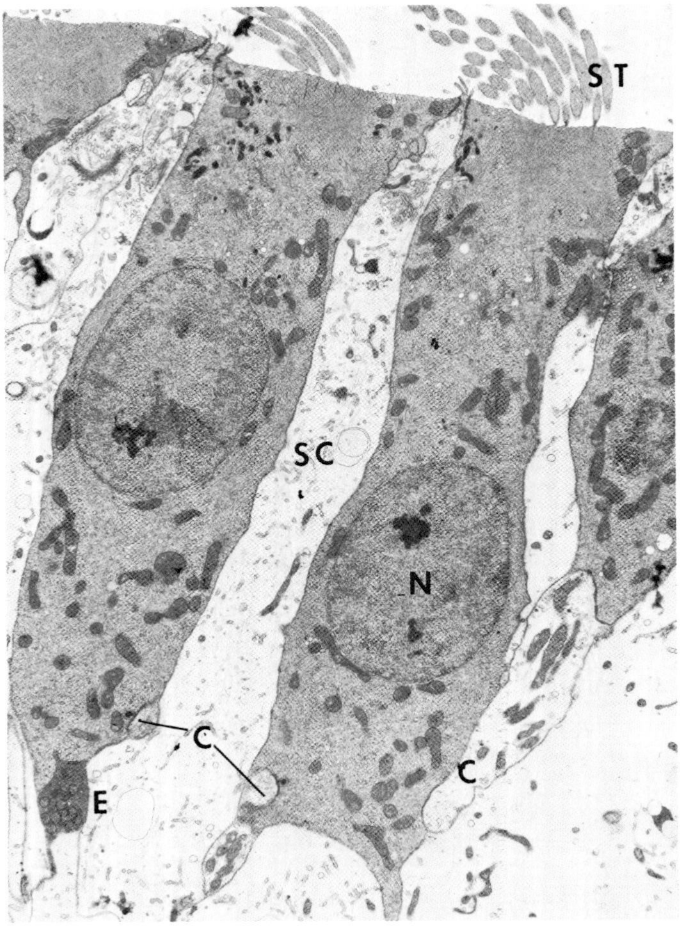

Fig. 6.8. Transmission electron micrograph of tall hair cells in the basilar papilla of *Gallus*. Cochlear sensory nerve endings (C) are visible on all three hair cells. A small efferent bouton (E) is present only on the cell on the left. N, Nucleus; ST, stereocilia; SC, supporting cell. ×5000 From Tanaka and Smith (1978), courtesy of Alan R. Liss, Inc., New York.

Three major types of sensory cells have been recognised: tall, intermediate and short hair cells. Included with the short hair cells are some that are atypical: the hemispherical cell and the lenticular cell.

The tall hair cells are columnar cells with large cochlear synapses and a few efferent boutons (Figs. 6.4, 5, 8, 9a). In both *Columba* and *Gallus,* they are located along almost the entire length of the papilla at the superior edge (close to the attachment of the tectorial membrane) and completely fill the distal rounded end (Fig. 6.11). They are absent from the proximal tip of the papilla.

The typical short hair cells have short bodies, greatly-enlarged cuticular plates and large efferent nerve terminals at their basal ends (Figs. 6.4, 9b, 10a). They are present in large numbers on the free basilar membrane—that is, the part of the membrane not supported by cartilage—and in *Columba* extend from the proximal tip for about two-thirds of the length of the membrane. In *Gallus,* they almost reach the distal end of the membrane (Fig. 6.11).

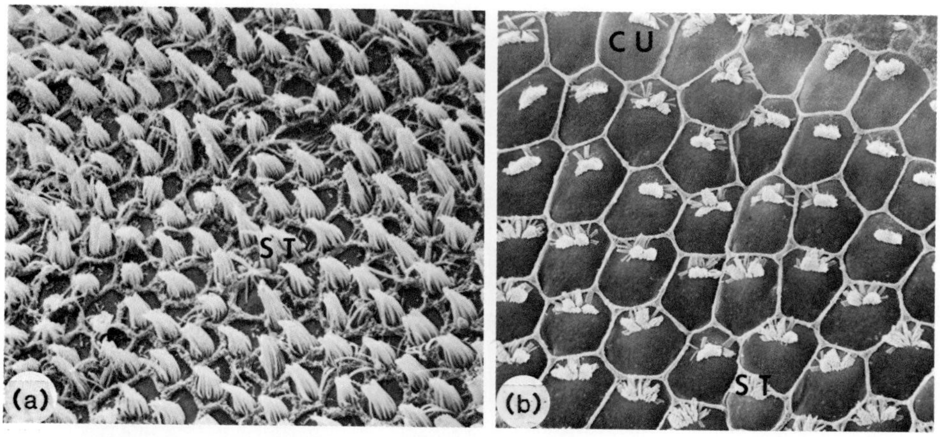

Fig. 6.9. Scanning electron micrographs of the surface of the basilar papilla of *Gallus*. (a) Tall hair cells. ST, Stereocilia. ×1000. (b) Short hair cells. CU, Cuticular plate; ST, stereocilia. ×800.

The intermediate cells, as their name suggests, are intermediate in form to the tall and short cells. A moderate number of these cells were found in *Columba* by Takasaka and Smith (1971) and in the Barn Owl (*Tyto alba*) by Smith *et al.* (1985), where they are situated between the groups of tall and short cells (Fig. 6.11). Tanaka and Smith (1978), however, could not identify them in *Gallus*.

The different types of sensory cell are not mixed together but are arranged in groups which are restricted to certain sites on the basilar membrane (Fig. 6.11). Furthermore, this site seems to be species-specific. Thus, Smith and Tanaka (1981) found an interesting difference in sensory cell distribution between the House Sparrow (*Passer domesticus*), the Budgerigar (*Melopsittacus undulatus*) and the Pine Siskin (*Carduelis pinus*). Although all three types of cell were found in each species, the cell type present at the distal end was different: intermediate cells occurring in *Carduelis,* many tall and a few short cells in *Passer* and a large number of short cells in *Melopsittacus*. When it is recalled that each type of cell is not only distinctive in form but likewise has a distinct pattern of innervation, the species specificity in distribution assumes an added relevance. However, more information on different individuals and species is necessary before the importance of the distribution can be determined.

The distribution of the sensory cells in the Barn Owl (*Tyto alba*) is different from that of any of the species discussed previously. A few lenticular cells were found in *Gallus* among the short cells at the proximal tip of the basilar papilla (Fig. 6.10b) (Tanaka and Smith, 1978). These cells are numerous in the Barn Owl, many being present on the proximal 3 mm of the papilla (Smith *et al.*, 1985). The lenticular cell is classed with the short cells because of its flattened, non-columnar shape and because its apical end is broad (Fig. 6.10b). It is unique in that only a part of the surface exposed to the endolymph has a cuticular plate, and the plate plus stereocilia are on the opposite side of the cell from that in the more common short cell.

Studies on *Columba* led Takasaka and Smith (1971) to hypothesise that tall and short hair cells could be compared to mammalian inner and outer hair cells,

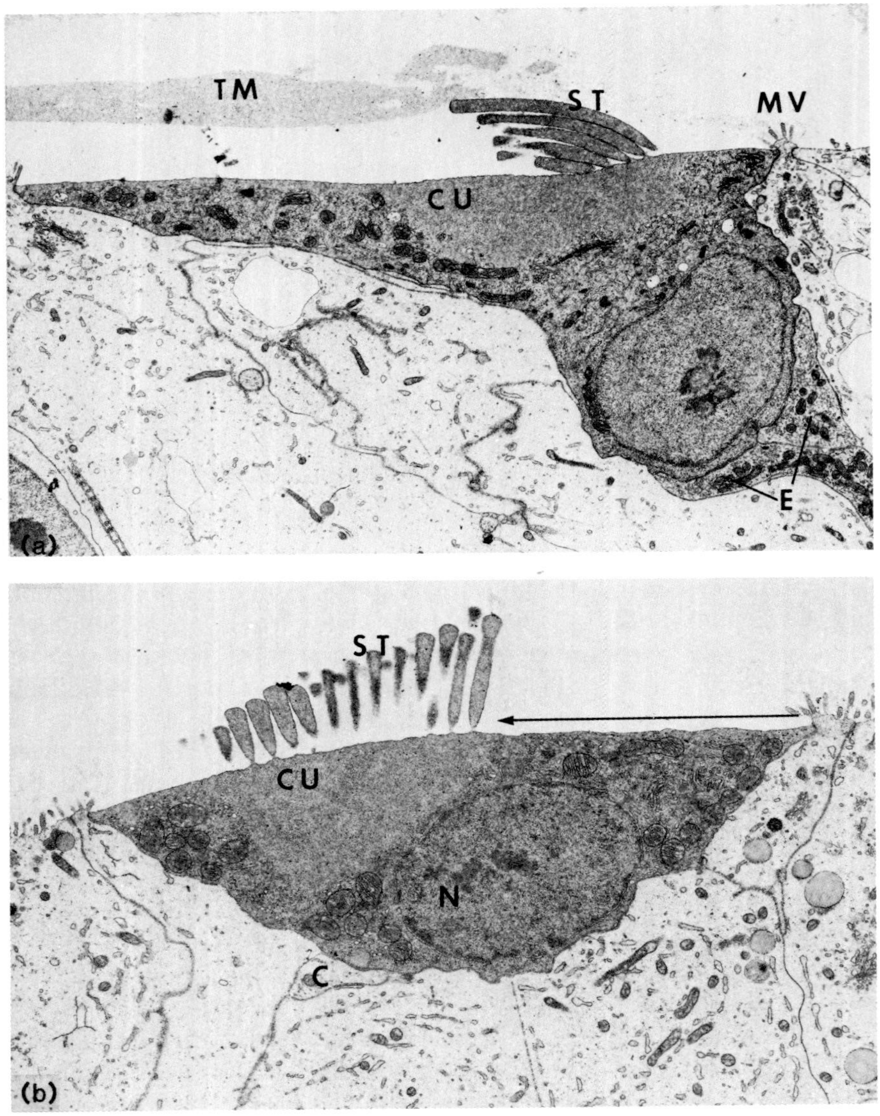

Fig. 6.10. Transmission electron micrographs of short hair cells in the basilar papilla of *Gallus*. (a) A typical short hair cell with its large apical end. The extended part of the cell on the left contains a thin extension of the cuticular plate (CU) which is mainly situated on the right. The basal end is enclosed by two large efferent nerve endings (E). The tallest cilia (ST) are attached to the tectorial membrane (TM), which extends in thin strands to attachment sites on the microvilli (MV) of the supporting cells. ×5800. (b) One of the atypical short hair cells, the lenticular cell. The cuticular plate (CU) and hair bundle (ST) are on the opposite side of the cell to that in (a). C, Cochlear sensory nerve ending; N, nucleus. ×10,300. From Tanaka and Smith (1978), courtesy of Alan R. Liss, Inc., New York.

respectively, on the basis of their location and pattern of innervation. Such comparisons, however, may have minimal value because avian and mammalian auditory organs are quite different in structure and at present there are still many unresolved questions about relative functions of inner and outer hair cells in the organ of Corti. Greater profit might be derived from a consideration of the

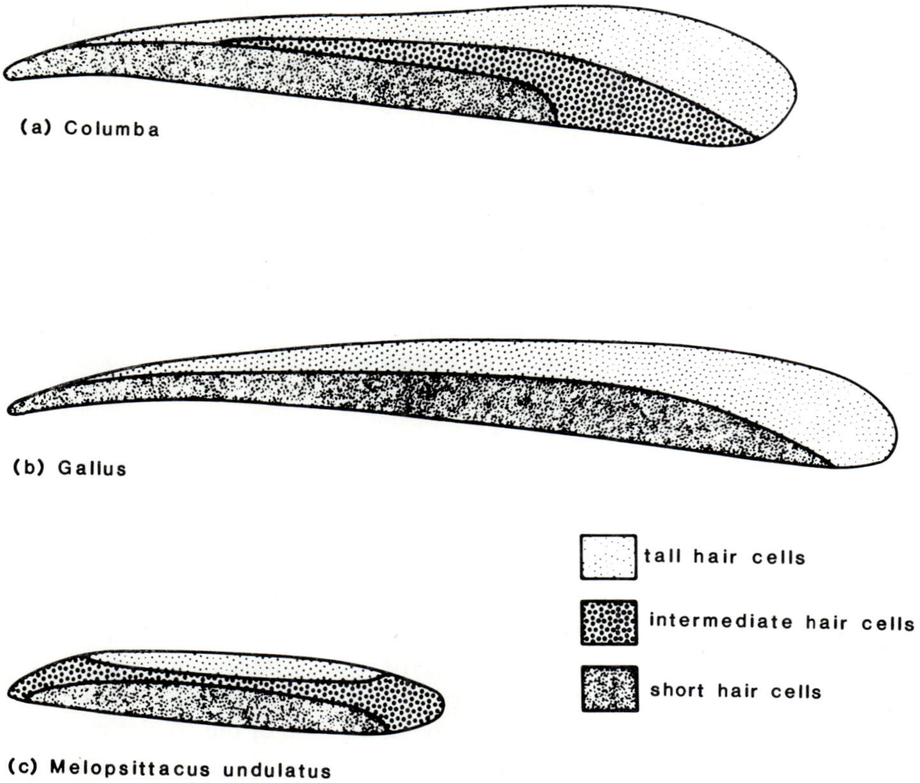

Fig. 6.11. Drawings showing the locations of the tall, intermediate and short hair cells on the basilar papilla of (a) *Columba*, (b) *Gallus* and (c) the Budgerigar (*Melopsittacus undulatus*). All are drawn to the same scale.

reptilian basilar papilla which, like that in birds, is a conglomerate of hair cells and supporting cells. Amongst the interesting data yielded by the long-term studies by Wever (1978) on the reptilian ear is the remarkable variability to be found even in a single papilla. The observations by Mulroy (1974) on the Alligator Lizard (*Gerrhonotus multicarinatus*) illustrate this feature quite well. The papilla of the Alligator Lizard is small (0·4 mm) and has a total of 160 cells. The hair cells at the proximal end have short stereocilia, have unidirectional kinociliar polarization, and are innervated by both afferent and efferent nerve fibres. A well-defined tectorial membrane covers the hair cells. The hair cells at the distal end, in contrast, have long cilia and bidirectional kinociliar polarization, but no organised tectorial membrane and no efferent nerve endings. The auditory receptor cell potentials and the unit activity in the afferent nerves, along with the movements of the basilar membrane, were recorded in the Alligator Lizard by Weiss *et al.* (1974, 1976, 1978). Their results led them to the conclusion that there was frequency selectivity as well as a tonotopic organisation of cochlear nerve fibres, but that this was determined by something in addition to or other than the motion of the basilar membrane. Perhaps small birds employ selective hair cell arrangements and innervation features as a similar analytical tactic.

(b) Supporting cells. The supporting cells are tall, columnar cells which reach from the basilar membrane to the surface of the papilla (Figs. 6.4, 5, 7, 8, 10).

They envelop the nerve fibres and the basal ends of the hair cells, then continue up in between the sensory cells to the surface. A number of supporting cells contribute to the envelope encircling each hair cell. The apical part of the supporting cell, between the cuticular plates of the sensory cells, is attenuated and surmounted by a thick tuft of short microvilli. These latter are attachment sites for the tectorial membrane.

There are no blood vessels below the neuroepithelium on the freely-moveable basilar membrane. A capillary network capable of supplying the needs of an active sensory organ would hardly be desirable on a vibrating membrane. It would result in an inconsistent, uncontrollable load due to the vagaries of the blood circulation in capillaries. The vibration itself could result in capillary stasis at a critical period in cell metabolism. This avascularity is in marked contrast to the vestibular maculae and cristae, which are non-vibrating structures and overlie dense capillary networks.

3. *Basilar membrane*

There are two extracellular fibrous membranes within the cochlea (the basilar membrane and the tectorial membrane), both of which are important accessories in the transduction process. The basilar membrane (*membrana basilaris*) is attached to the superior and inferior fibrocartilaginous plates in the walls of the cochlea and is suspended over the perilymph of the scala tympani (Figs. 6.3, 4, 12a). The basilar papilla covers most of the upper side of the basilar membrane. However, the neuroepithelium does not entirely coincide with that part of the basilar membrane suspended over the perilymph of the scala tympani (Figs. 6.4, 12a). In fact, in *Columba,* the tail hair cells occupy the area of the basilar papilla over the superior fibrocartilaginous plate and only a small part of the free basilar membrane. This is not too different a situation from that present in the basal part of the mammalian organ of Corti, where the inner hair cells are located over bone, whereas the tunnel rods and outer hair cells are situated on the free basilar membrane.

The basilar membrane fibres are oriented transversely to the long axis of the cochlear duct and are packed together to form a well-defined layer just beneath the epithelium. A thicker mass of more loosely-arranged fibres (the loose fibrous mass) usually protrudes into the perilymph (Fig. 6.12a). The loose fibrous mass is common to the basilar membrane of all species so far examined, including *Columba, Gallus, Melopsittacus undulatus, Passer domesticus, Carduelis pinus* and *Tyto alba.* In addition, a dense, fibrous mass has been found in the proximal part of the basilar membrane of the Barn Owl (*Tyto alba*) but not in any of the other birds listed before (Smith *et al.,* 1985).

Measurements made on the basilar membrane of *Gallus* showed that its width increases in a regular manner from 0·07 mm at the proximal tip to a maximum of 0·395 mm at the distal end just before it becomes rounded off. It seems probable that a similar increase in width also occurs in *Columba.* However, in the Budgerigar (*Melopsittacus undulatus*) the short (2·3 mm in length) basilar membrane shows little change in width over the distal 1·5 mm, that is, over 65% of its total length.

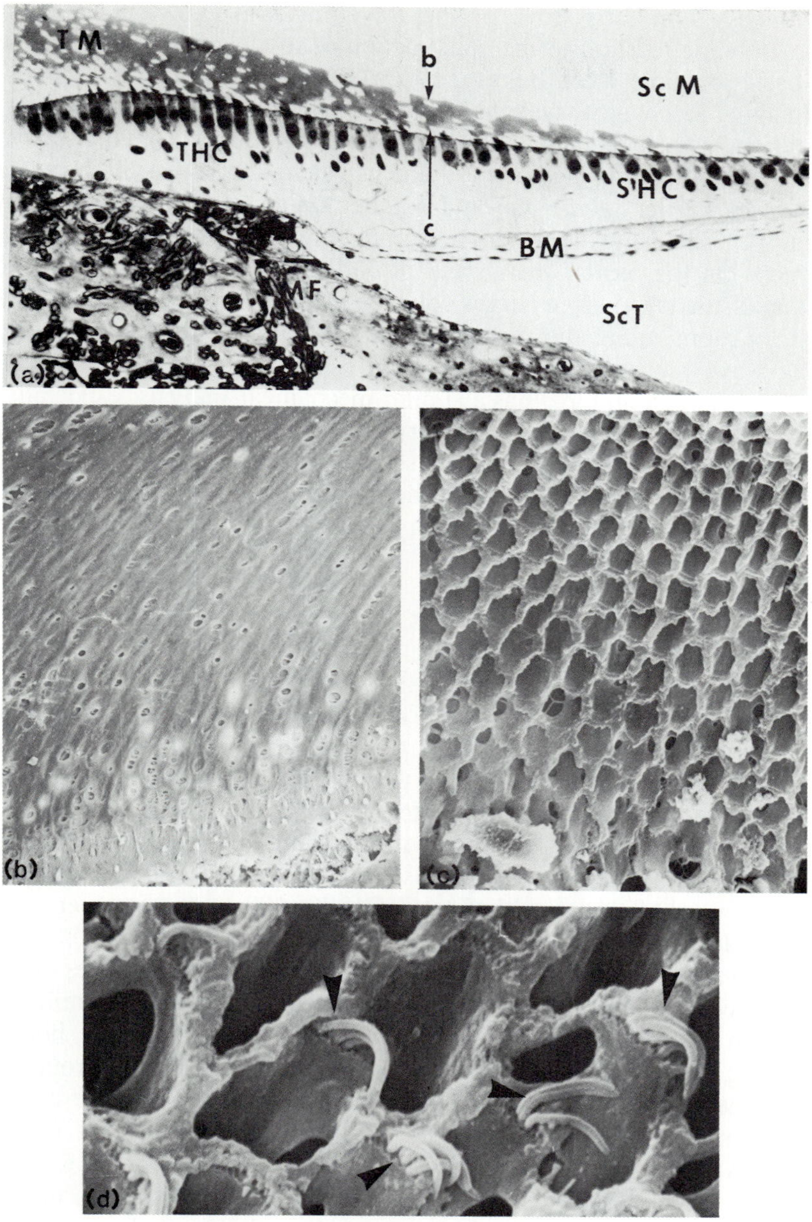

Fig. 6.12. The structure of the tectorial membrane and its relationship to the cilia. (a) Photomicrograph of the basilar papilla of *Columba* showing the insertion of the cilia of the tall hair cells (THC) and short hair cells (SHC) into the tectorial membrane (TM). BM, Basilar membrane; MF, myelinated nerve fibres; ScM, scala media; ScT, scala tympani. Arrows marked b and c refer to the surfaces viewed in (b) and (c). ×290. From Takasaka and Smith (1971), courtesy of Academic Press, New York and London. (b) Scanning electron micrograph of the upper surface, arrow marked b in (a), of the tectorial membrane of *Gallus*. ×300. (c) Scanning electron micrograph of the lower surface, arrow marked c in (a), of the tectorial membrane of *Gallus* showing the honeycomb structure. ×1500. (d) Scanning electron micrograph of the lower surface in (c) at higher magnification showing cilia (arrowheads), which have been disrupted from their hair cells during dissection but still remain attached to the edges of some of the cavities. ×5000. (b)–(d) From Tanaka and Smith (1975), courtesy of Annals Publishing Co., St. Louis.

Von Bekesy's investigations of the travelling wave movements along the basilar membrane led him to the construction of models from which he derived several basic principles (von Bekesy, 1960a). One of these was that "in continuously changing mechanical systems, there is an energy flow from the stiffer to the softer section." This results in a travelling wave that moves from the stapes to the apical end in the mammalian cochlea. The wavelength is longer in the stiffer part (the narrower basilar membrane and smaller organ of Corti at the base) and shorter at the softer part (the wider basilar membrane at the apex). From his direct observations on the basilar membrane of *Gallus,* von Bekesy (1960b) demonstrated that distinct resonance curves could be measured between 0·1 and 3 kHz at different locations along the membrane, giving support for the existence of a travelling wave pattern in the ear of *Gallus*. The question arises whether the shorter cochlea of small birds has a similar mechanical analysis, especially when physical properties seem to change very little over a large part of the basilar membrane. Undoubtedly, the peaked sensitivity curve and limited frequency range of most birds are reflections of their short cochleas. The Barn Owl (*Tyto alba*) provides an exception, with a long basilar membrane and a wide frequency range.

4. Tectorial membrane

The tectorial membrane (*membrana tectoria*) is the other of the two extracellular fibrous membranes (Figs. 6.3, 4, 12). It originates from the columnar epithelial cells on the superior side of the papilla, spreads out in a continuous sheet across the papilla, and attaches to the simple cuboidal epithelium at the inferior edge of the papilla. In addition, it is anchored by fibrous extensions from its undersurface to all the supporting cells of the papilla (Figs. 6.7, 10). The membrane is composed of innumerable fine fibrils (70–100 Å diameter), embedded in a mucopolysaccharide gel.

The relationship of the hair cells to the tectorial membrane becomes clear when it is examined by scanning electron microscopy (Tanaka and Smith, 1975). Viewed from the undersurface, the membrane has a characteristic honeycomb structure (Fig. 6.12c, d). The cilia of the hair cells project as a group into the cavities of the honeycomb, the longest cilia being inserted into the membrane. This insertion is sufficiently strong that after fixation some cilia retain their attachment even though they have broken away from the hair cells. Cavities and channels other than those of the honeycomb abound in the membrane in *Columba* (Hoshino, 1973) and *Gallus* (Tanaka and Smith, 1975), and this has led to the assumption that the tectorial membrane is fully permeable to endolymph.

Although the tectorial membrane of birds is considerably thicker than its mammalian counterpart, it undoubtedly has the same function. Vibrations of the columellar base in the vestibular window create wave movements in the perilymph of the scala vestibuli which are transmitted through the scala media to the basilar membrane. In some birds, vibrations may be transmitted directly to the tegmentum vasculosum (Schwartzkopff, 1968). It is usually assumed that the basilar membrane and the basilar papilla move as one. As the tectorial membrane has a firm attachment at one side, a shearing movement of the membrane (and the cilia embedded in it) would be created across the papilla, resulting in depolarization of the hair cells and nerve stimulation. The precise mechanism by which this occurs (deformation at the sides of the stereociliar membrane or at the constricted, pedestal-like base, etc.) is obscure at present. Nevertheless, Hudspeth and Corey (1977) have clearly demonstrated that when the stereocilia are displaced in one direction, depolarization occurs; when displaced in the other direction, the hair cell is hyperpolarized.

5. *Tegmentum vasculosum*

The cochlear duct is closed off above the papilla by the *tegmentum vasculosum* (Figs. 6.3, 4). The highly-specialised tegmental cells are separated from the hyaline cells (Held, 1926) on the inferior side of the basilar membrane, as well as from the columnar cells (which are the sources of the tectorial membrane) on the superior edge by a small number of cuboidal cells.

The tegmentum vasculosum is composed of two major types of cell which will be called "dark" cells and "light" cells for lack of a more definitive name (Figs. 6.3, 4, 13, 15a). The dark cell is flask-shaped with a short, thick neck and an enlarged round base, the nucleus usually being located just below the neck. The remainder of the bulbous base is packed with mitochondria, the large number of which give the cell a basophilic property. The cell membrane of the basal part of the cell is tremendously enlarged by infoldings which enclose the mitochondria. The apical end of the dark cell has coarse, short, irregular microvilli which project into the endolymph. The light cell also faces the endolymph, but its apical surface

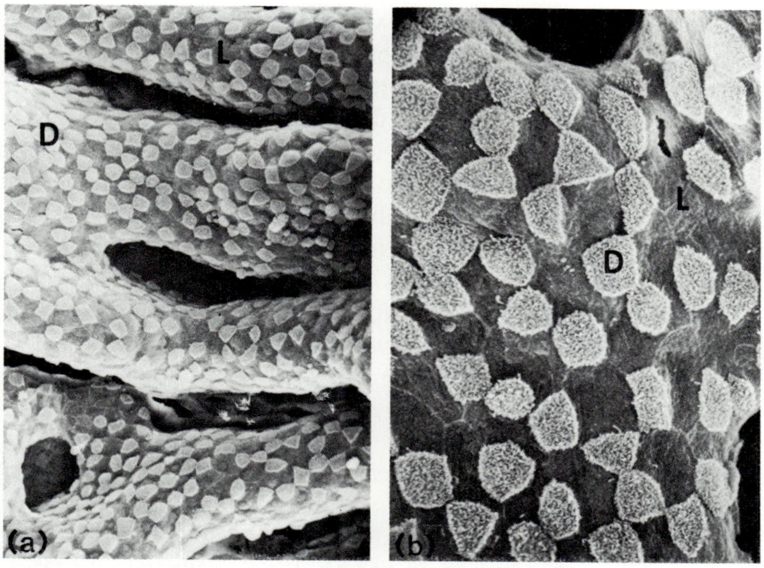

Fig. 6.13. Scanning electron micrographs of the surface facing the endolymph of the tegmentum vasculosum of *Gallus*. (See arrow in Fig. 6.15.) (a) Cords of light (L) and dark (D) cells. The numerous microvilli on the dark cells give them the "lighter" appearance. ×300. (b) Higher magnification showing that the dark cells (D) have many microvilli on their apical ends, whereas the light cells (L) are smooth. × 1000. Courtesy of Dr. Katsuhiko Tanaka.

is fairly smooth. The light cells separate the dark cells from each other at the free surface of the tegmentum.

The tegmentum vasculosum is well vascularised. Schmidt (1964) showed that arterioles supplying the tegmentum enter at the sides (superior and inferior), whereas blood is drained by venules situated at the top of the cochlear duct. The close relationship which is present in mammals between marginal and intermediate cell processes and capillaries in the stria vascularis does not seem to be duplicated in the avian tegmentum vasculosum. The stria vascularis of mammals is the source of the endocochlear potential and maintains the high potassium concentration of endolymph. We may assume that the tegmentum vasculosum serves the same function in birds.

The numerous cells of the tegmentum vasculosum are clustered together in cords which arch across the scala media (Figs. 6.4, 13). According to Schwartzkopff (1968), these cords fill most of the cochlear duct. Although this may be the case for some birds, such as the Barn Owl (*Tyto alba*), which has an exceedingly thick tegmentum, it is not the arrangement found in the adult chicken (Tanaka and Smith, 1978) or in *Columba*. In young domestic fowl, however, the tegmentum is coarse and thick. Schmidt (1964) showed that in *Columba* the cords of cells do not extend over the roof of the scala media in the distal half of the cochlear duct but are restricted to the sides. He also found that the tegmental cells in *Columba* continue through the sacculocochlear duct into the saccule. The stria vascularis of mammals, in contrast, is strictly a cochlear structure.

6. *Innervation*

Terminals of the cochlear part of the vestibulocochlear nerve have been found on all the hair cells. The form and synaptic area of the endings vary according to the type of hair cell, as described in a previous section. The peripheral processes of the cochlear nerve have been studied in *Columba* by Boord (1969) and Takasaka and Smith (1971). They found that the axons lose their myelin sheaths immediately before passing through the openings in the basilar membrane (the habenula perforata) and then take an almost straight course across the papilla, little branching occurring before the terminal arborization. There is no information regarding the ratio of nerve fibres to the different types of cells, or even whether they are innervated by separate nerve fibres. However, Boord (1964) found that there were 5136 cochlear nerve fibres in *Columba,* a species in which Goodley and Boord (1966) counted an average of 10,400 cochlear hair cells.

The cochlear ganglion (*ganglion cochleare*) is composed of bipolar neurons which are clustered together beneath the habenula perforata of the basilar membrane. Boord and Rasmussen (1963) clearly defined the central projections of the cochlear nerve by use of experimental lesion methods in *Columba*. The axons exit through the internal auditory fossa and bifurcate as they enter the medulla. The lateral (ascending) branches end in an orderly fashion (proximal to distal) in the angular cochlear nucleus. The medial (internal) branches terminate in an equally orderly manner in the magnocellular cochlear nucleus.

Konishi (1970) found that physiological responses measured by means of exploring microelectrodes inserted into the magnocellular cochlear nucleus and the angular nucleus showed an orderly transition of high to low frequencies across each nucleus. He interpreted this as a reflection of the travelling wave moving from a relatively rigid part of the basilar membrane to a more flexible portion. This is probably correct for some birds. The basilar membrane and papilla of *Gallus,* for example, do have width and cellular increases from the proximal to distal parts, and evidence for a travelling wave pattern has been provided by von Bekesy (1960b). But the basilar membranes of some of the smaller birds not only are exceedingly short but do not show great changes over most of their lengths. It seems probable, therefore, that some factors other than (or in addition to) the travelling wave must be active in small songbirds, as has been discussed previously.

Fibres from another neural tract also terminate on the hair cells. These fibres make up the "efferent cochlear bundle," so called because the bundle has its origin in the brain stem (Boord, 1961), although the location of the cells has not been precisely defined. Schwarz *et al.* (1981) have suggested they may be in the vicinity of the lateral reticular paragigantocellular nucleus. Boord (1961) cut the decussation of the efferent cochlear bundle in the floor of the fourth ventricle between the genua of the facial nerves and followed the degenerating neurons into the cochlea. He found degenerated efferent fibres along the vestibular nerve, in the vestibulocochlear anastomosis, and finally in the cochlear nerve to the basilar papilla. The distribution of the nerve fibres within the papilla was studied by Takasaka and Smith (1971) using the acetylcholinesterase technique. They found that the nerves branched extensively between the supporting cells before terminating on the sensory cells and that the size of the efferent nerve endings varied with the specific cell type. The short hair cells have large efferent calyces on their basal ends, whereas the tall hair cells have only one or more small boutons.

The precise role of the efferent olivocochlear tract in hearing is still obscure, although it has been shown many times that experimental stimulation of the tract in the brain stem of mammals results in a slight increase in the cochlear microphonics and a marked decrease in the cochlear nerve action potential (AP) produced in response to sound. The study by Desmedt and Delwaide (1965) on *Columba* gave convincing evidence that faradic stimulation of the crossed efferent tract between the facial genua in the floor of the fourth ventricle produced similar results, but with some interesting differences. First, the magnitude of neural (AP) inhibition was considerably less than found in the cat using the same techniques (Desmedt, 1962). Second, the potentiation of the cochlear microphonic potential in the pigeon was greater relative to neural inhibition, as well as being greater in magnitude than ever seen in the cat. They proposed that these differences could be the result of major differences in synaptic organisation in the two vertebrates, that is, extensive hair cell synapses and relatively few axodendritic synapses in the pigeon *versus* relatively few hair cell synapses and numerous axodendritic synapses in the cat. It is true that the pigeon efferent fibres make very few axodendritic contacts on cochlear nerve fibres beneath the hair cells (Smith, 1968), whereas in mammals these are numerous, especially beneath inner hair cells. However, if the innervation of mammalian outer hair cells is compared with avian short hair cells, there is not too much difference, since both have very large efferent synaptic areas. It may be that single-unit recordings from the bird would yield more informative data. However, such data have not been helpful in resolving the dichotomies of the mammalian efferents. We know too little about avian peripheral innervation patterns at present to formulate really valid theories. Further study on birds might be useful in solving mammalian problems, since the abundance of axosomatic contacts and lack of axodendritic synapses seem to point toward a more simple system. At any rate, it seems clear that both the avian and the mammalian cochleas are under the same sort of centrally-modulated control.

7. *Cochlear function and hearing*

There have been a number of good studies on peripheral auditory function, as well as on hearing in birds. For the most part, these have indicated that hearing in birds is not quite as good as in humans, but otherwise not too different within certain frequency limits. As one would expect from the short avian basilar membrane, sensitivity is limited to a narrow frequency range. In addition, the sensitive peak for most birds is located at 1 to 5 kHz (Fig. 16.14).

Pierson and Dallos (1976) re-examined the cochlear potentials in *Columba* and found that the hair cell cochlear microphonics showed properties similar to those found in mammals. The cochlear microphonic isopotential function was well matched to middle ear function below 4 kHz. Also, cochlear microphonic input–output function (relationship of the cochlear microphonic amplitude to sounds of regularly-increasing intensities) was linear, similar to that of mammals.

The study by Sachs *et al.* (1974) on single nerve fibres in *Columba* indicated that responses were similar to those of the cat, except that spontaneous neural

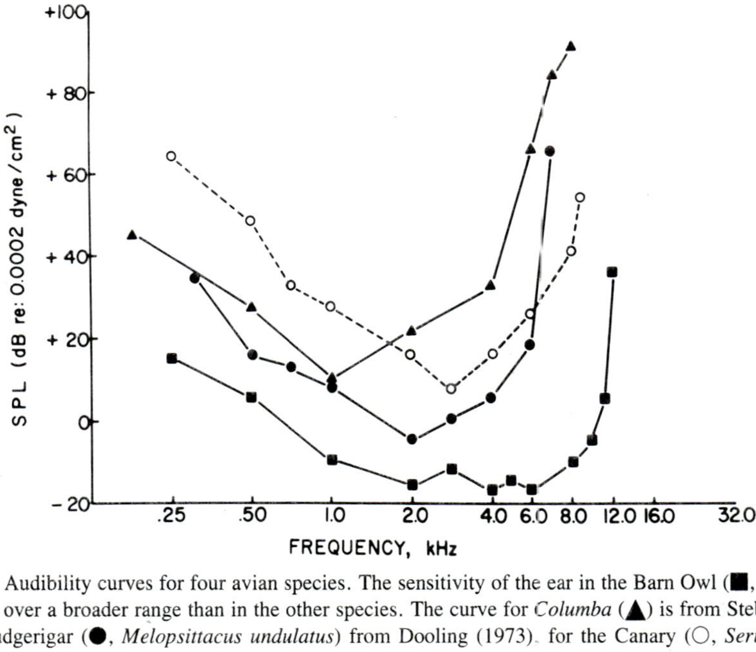

Fig. 6.14. Audibility curves for four avian species. The sensitivity of the ear in the Barn Owl (■, *Tyto alba*) is very good over a broader range than in the other species. The curve for *Columba* (▲) is from Stebbins (1970), for the Budgerigar (●, *Melopsittacus undulatus*) from Dooling (1973), for the Canary (○, *Serinus canaria*) from Dooling *et al.* (1971), and for the Barn Owl (*Tyto alba*) from Konishi (1973). From Smith (1981), courtesy of Springer-Verlag, Heidelberg.

activity was higher. Konishi (1970) explored responses evoked by sound in the brain stem cochlear nuclei (the magnocellular and angular cochlear nuclei) from 10 different songbirds. He found there was a definite tonotopic organisation in both nuclei but that this was better organised and the range wider in the angular nucleus. In the magnocellular nucleus, the distribution was more skewed towards the most sensitive range. The largest number of units (and the most sensitive) were clustered around 1 to 5 kHz. Some birds, however, showed more responses in the lower end of the frequency spectrum (*Passer domesticus* at 2 to 3 kHz, for example) and others in the higher frequencies (*Junco hyemalis,* at 3 to 7 kHz). He also found that the peak of the sensitivity curve was correlated with the frequency spectrum of the bird's song.

It is not surprising that physiological data from the periphery are rather similar in birds and mammals. The hair cells are dependent on basilar membrane and tectorial membrane movements for their activation, and once this mechanism is set in operation, the hair cells undoubtedly respond in a similar fashion in whatever organ or animal they may be located. The findings of Boord and Rasmussen (1963) that nerve fibres from proximal, central and distal parts of the cochlea have orderly projections onto the cochlear nuclei, when considered along with the association of organised frequency responses from the nuclei (Konishi, 1970), give good evidence that there is a frequency analysis in the bird's cochlea. The intriguing question that still remains is precisely how this is accomplished.

Audibility curves have been obtained from a number of species, and these demonstrate that hearing fits in well with physiological data, except that sensitivity seems to be more sharply centered (Fig. 6.14). Only the Barn Owl (*Tyto alba*) shows good sensitivity over a wide range, but both its ear and hearing are remarkable exceptions among birds thus far examined.

The excellent chapter "Auditory Perception in Birds" by Dooling (1983) explored many psychological aspects of avian hearing in considerable detail, and some of these features are summarised below:

1. *Frequency discrimination* is only about one-half to one-third as sensitive in birds as in humans over a narrow range of 1 to 4 kHz.
2. *Intensity discrimination.* Although it had been suggested by Pumphrey (1961) that birds might have better powers of intensity discrimination than mammals, later data from Dooling and Saunders (1975), Sinnott *et al.* (1976) and Dooling (1983) indicate this is not the case.
3. *Temporal discrimination.* There is no evidence available to suggest that birds are significantly better than other vertebrates at detecting changes in the temporal aspects of an acoustic signal.

All in all, it appears that birds hear very well over a limited frequency range and that this is all that is necessary for their propagation. The exceptions that have been studied so far are the large nocturnal owls. Of these, the Barn Owl (*Tyto alba*) has been the subject of intensive investigation by Konishi and his associates (Konishi, 1973; Quine and Konishi, 1974; Knudsen, 1981). This species has good sensitivity over a wide frequency range (Fig. 6.14). Its hearing is as good or better than the cat, from 1 to 7 kHz. The high-frequency cut-off is sharp above 8 kHz (Konishi, 1973), and frequency discrimination is close to that of mammals (Quine and Konishi, 1974). Quine and Konishi argued that frequency discrimination might be even better than indicated because their owls had to perform the difficult task of frequency discrimination by memory and the test involved much longer time intervals than usually used in determining $\Delta F/F$. The good sensitivity of the Barn Owl over a wide frequency range is undoubtedly correlated with its long basilar membrane and probably with a larger number of nerve fibres, although we have no data at present regarding the latter. The good frequency discrimination over 3·5 to 10 kHz may also be a reflection of the longer membrane. Nevertheless, good frequency discrimination over a range of 1 to 4 kHz is possessed by the Budgerigar (*Melopsittacus undulatus*) (Dooling and Saunders, 1975), which has a short basilar membrane, only 2 to 3 mm in length, and one wonders how this small bird accomplishes such a feat.

8. *Macula of the lagena*

At the slightly-expanded blind end of the cochlear duct, the lagena, there is a second *neuroepithelium,* the macula of the lagena (*macula lagenae*) (Fig. 6.1). There are still some questions regarding the function of this structure, but it seems to be primarily a vestibular organ. The hair cells are covered by a gelatinous statoconial membrane (*membrana statoconiorum*) in which are embedded small calcified particles, the *statoconia* (Figs. 6.3, 15, 18b). The broad, ribbon-like shape of the macula lagenae is contoured to fit inside the rounded lagena, each end being curled up but not forming a complete circlet. The lagenar macula of *Columba* is approximately 1·5 mm long and has an area of 0·93 mm^2 (Rosenhall, 1970). A statoconial macula is found also in the cochlear duct of reptiles, but in mammals it is present only in monotremes. The structure of the lagenar macula is similar to that of other statoconial maculae present in the vestibule. It is composed of a group of sensory cells and supporting cells (Fig. 6.15). The supporting cells rest on the basal lamina, with their nuclei lined up in the basal parts of the cells. The supporting cells of the lagena are shorter than the supporting cells in the basilar papilla, with the result that the lagenar macula is not as thick as the auditory organ (Fig. 6.3). Beneath the neuroepithelium is a layer of connective tissue which contains many small blood vessels and myelinated nerve fibres.

The sensory cells are tall columnar cells. In the mammalian vestibular organ, Wersall (1956) was able to identify two types of hair cell on the basis of their innervation. The type I hair cell has a deep chalice (or cup) that envelops all the cell except the hair-bearing extremity. The terminal chalice of a single nerve fibre may contain more than one hair cell. The type II hair cell has small, knob-like nerve endings (Fig. 6.16). In birds, Jorgensen and Andersen (1973) found that most of the chalices in the lagena had one to four hair cells, but that some contained as many as seven. Thus, summation of sensory cell activity could readily occur in nerves which envelop many hair cells. Furthermore, the type I nerve dendrites are of larger diameter than the nerve fibres terminating in small knobs on the type II cells, so that transmission from the type I cells to the brain stem nuclei would be faster than from the type II cells. Birds thus have available not only an extra macula but also a fair number of what appear to be low-threshold, fast-conducting vestibular nerve units, enabling them to control their

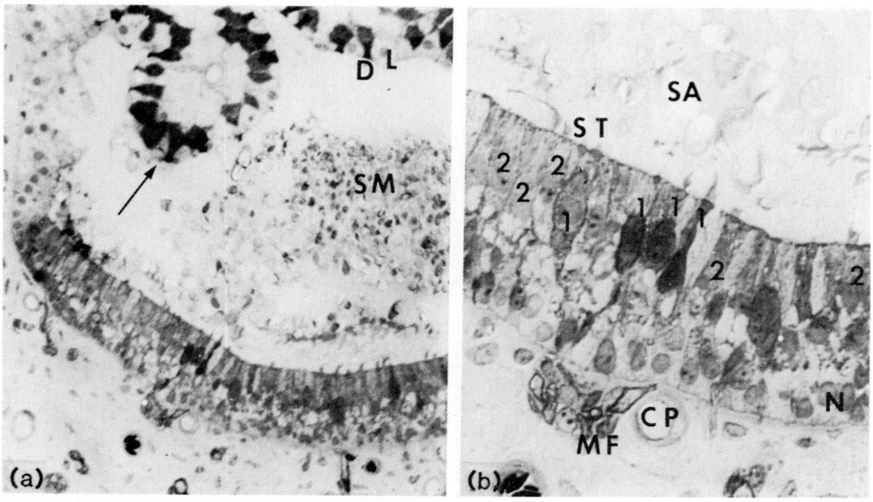

Fig. 6.15. Photomicrographs of a cross-section through half the macula lagenae of a 17-day-old domestic fowl. (a) The neuroepithelium crosses the lower part of the figure. The tegmentum vasculosum with its light (L) and dark (D) cells is visible above. The arrow indicates the surface seen in Fig. 6.13. The statoconial membrane (SM) and the statoconia are above the hair cells. ×290. (b) Higher magnification of the neuroepithelium in (a). 1, Type I cells; 2, type II cells; CP, capillary; MF, myelinated nerve fibres; N, nuclei of supporting cells; SA, statoconia; ST, hair bundle. ×740.

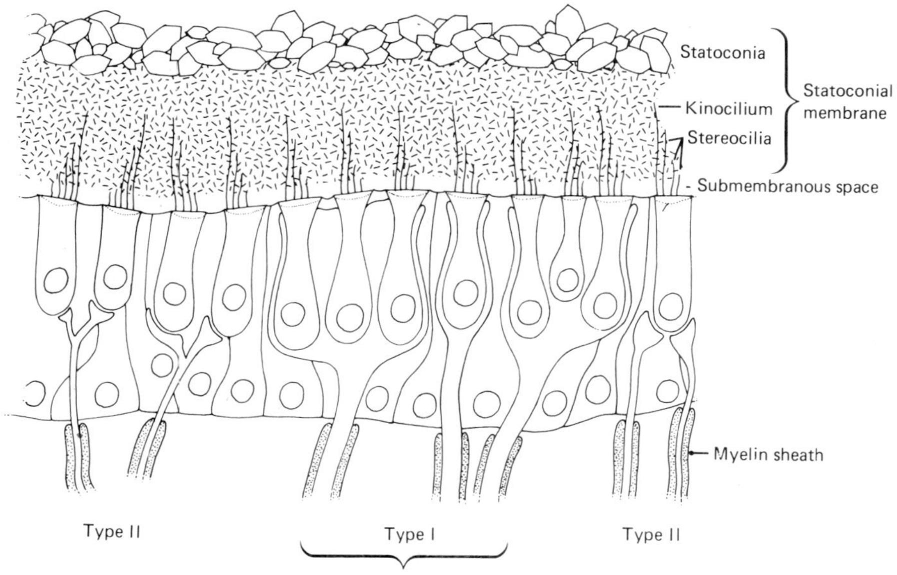

Fig. 6.16. Schematic drawing of a macula based in part on Jorgensen and Andersen (1973). The afferent innervation of the type I and type II hair cells is shown. The efferent innervation has not been included.

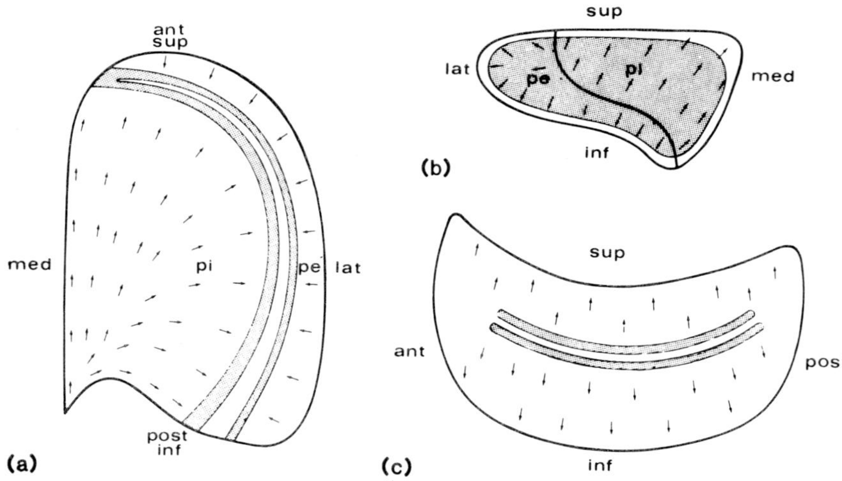

Fig. 6.17. Drawings showing the form and shape of the three maculae in *Columba*. The shaded areas indicate the location of the type I hair cells. They are located in stripes (striolae) in the utricle (a) and the lagena (c). There is no obvious striola in the saccule (b). Arrows show the kinociliar polariztion of the hair cells. pe, *Pars externa*; pi, *pars interna*. From Rosenhall (1970), courtesy of Springer-Verlag, New York.

flight in space. Rosenhall (1970) found that most of the hair cells in the lagenar macula of *Columba* were type II, the type I hair cells being limited to two narrow stripes in a central area called the ‘‘striola,’’* which covers about 20% of the macula (Fig. 6.17c). Both he and Jorgensen and Andersen (1973) studied the kinociliar polarization and concluded that the direction was opposed on opposite sides of the striola.

The sensory cells of the lagena are somewhat taller and the hair bundles are longer than those on the basilar papilla (Fig. 6.15). The cilia project up into the gelatinous membrane. In mammals, vestibular cilia generally do not reach the statoconial surface of the membrane, and it may be assumed that the same is the case for birds. Rosenhall (1970) found that the statoconial membrane did not cover the entire macula in *Columba,* the inferior side and the curled-up edges being free of the gel. The superior side was covered by the gel but without statoconia. However, it is well known that the gelatinous membranes in the inner ear are subject to considerable shrinkage during fixation and dehydration, and it is difficult to assess the exact amount of distortion that has occurred in fixed material.

Boord and Rasmussen (1963) traced the central terminations of the lagenar nerve in *Columba* using nerve transection methods. Their results show that the majority of fibres terminate in the dorsolateral, dorsomedial and superior vestibular nuclei. A few fibres could be traced to the cerebellar nuclei and to the reticular formation. A number of lagenar fibres were also traced to the auditory nuclei including the ventral part of the angular nucleus and the ventrolateral part of magnocellular nucleus. Further studies by Boord and Karten (1974) have

*It was found by Werner (1933) that the macular surfaces were not entirely uniform and that there was a central striola (or stripe) on most maculae, where the statoconial drift was thinner and the cells beneath different from those in the remainder of the macula. Several studies on birds have shown that the type I cells predominate in the striolae of their maculae (Rosenhall, 1970; Jorgensen and Andersen, 1973).

confirmed these projections to the vestibular nuclei as well as adding details regarding specific areas of termination. One concludes that the statoconial organ in the cochlear duct is a part of the vestibular system and is concerned for the most part with equilibration. The curved shape of the lagenar macula would seem to represent a valuable addition to the flatter macula sacculi and macula utriculi in gravitational assessments. The function of those neurons which terminate in the auditory nuclei is puzzling. Obviously, more experimental work on the function of the lagenar macula in birds is necessary.

D. Vestibular organ

The vestibular labyrinth (Fig. 6.2) consists of the saccule, utricle and semicircular ducts. The saccule is joined to the cochlea by the sacculocochlear duct; this is a short, wide tube, filled with endolymph, which transmits in its wall strands of the tegmentum vasculosum from the cochlea to the saccule. The utricle and saccule are connected by the short utriculosaccular duct (*ductus utriculosaccularis*). Both the utricle and the saccule contain statoconial organs. The three enlarged ends (ampullae) of the semicircular ducts are found on the medial and lateral sides of the body of the utricle. Each semicircular duct makes a half circle and then terminates in the utricle. The horizontal duct ends alone but the rostral and caudal ducts join to form the *crus membranaceum commune*.

The vestibular labyrinth contains six separate sensory structures: one in the saccule (macula sacculi), two in the utricle (macula utriculi and crista neglecta), and one in each of the three ampullae of the semicircular ducts (crista ampullaris). The function of the vestibular organ as a whole is that of equilibration, and the sensory structures are responsive to angular and linear acceleration and positional changes. It is generally considered that it is the semicircular ducts which are sensitive to angular acceleration, whereas the statoconial organs are responsive to linear acceleration and to gravity. Nevertheless, the function of the different organs is not totally separated, and according to some authors (Bard, 1968) the evidence points to interaction between the semicircular ducts and the statoconial organs, at least under some circumstances. The function of the saccule in higher vertebrates has never been clearly defined, and the macula of the utricle seems to be the more important statoconial organ.

1. Saccule

The saccule (*sacculus*) is small and located in a bony recess on the vestibular wall caudal to the vestibular window. The tegmentum vasculosum forms one wall of the saccule in the Barn Owl (*Tyto alba*). It is also a prominent, well-vascularised saccular structure in *Columba* (Schmidt, 1963). The macula occupies the other wall and is attached to the bone. Arising caudally from the saccule and adjacent to the utriculosaccular duct is the endolymphatic duct (*ductus endolymphaticus*). It traverses the vestibular aqueduct to end in an enlargement, the endolymphatic sac (*saccus endolymphaticus*) within the cranial cavity (Baird, 1974).

The *macula sacculi* is small (slightly more than 1 mm long in *Tyto alba*) and oval in shape. It is composed of a compact group of hair cells and supporting cells located over a connective tissue base through which the capillaries and myelinated nerve fibers course (Fig. 6.16). The statoconial membrane (*membrana statoconiorum*) overlies the hair cells and has the shape of a rice grain.

Rosenhall (1970) found that in *Columba* the macula sacculi measured 1 mm in length and had a mean surface area of 0·46 mm^2. No striola was obvious, and most of the hair cells in the central part of the macula were of type I. Type II cells were much fewer in number and were located mainly at the thin peripheral border (Fig. 6.17b). The cells were polarized in opposite directions on each side of an S-shaped curve along the long axis of the structure (Fig. 6.17b). Jorgensen and Andersen (1973) found a similar pattern in seven other avian species. They also noted that there were more type I hair cells per nerve chalice in the macula sacculi than in the macula of the lagena or utricle. In *Gallus*, for example, chalices containing four to eight cells were numerous in the saccule, whereas chalices with one to four cells were prevalent in the lagena.

The microscopic structure of the vestibular hair cells in all higher vertebrates is very similar (see Smith and Rasmussen, 1968; Lim, 1973; and Smith and Tanaka, 1975, for accounts of the fine structure of the vestibular hair cells in mammals). Each hair bundle has a long kinocilium which is in a central notch in the row of

Fig. 6.18. (a) Scanning electron micrograph of hair cells in the macula of the utricle of *Columba*. K, Kinocilium; ST, stereocilia. ×4100. Courtesy of Dr. David Lim. (b) Scanning electron micrograph of statoconia from *Columba*. ×1600. From Lim (1974), courtesy of S. Karger, Basel.

the tallest stereocilia (Fig. 6.18). Macular kinocilia in birds are of moderate length (Rosenhall, 1970), and not nearly as long as some in the cristae of the semicircular ducts. The cilia seem to be more tightly bound to each other than in the cochlea, and they remain clustered together in a tight bundle even after the statoconial membrane has been removed (Lim and Lane, 1969; Rosenhall, 1970). Lim and Lane (1969) found that the stereocilia were graduated in two directions, that is, from side-to-side as well as from bottom-to-top (Fig. 6.18).

2. *Utricle*

The utricle (*utriculus*) is a wide, membranous tube which reaches from the caudal ampulla on one side of the vestibular cavity to the rostral and horizontal ampullae on the other side, a distance of slightly more than 3 mm in the Barn Owl (*Tyto alba*). Money *et al.* (1974) found that the long diameter of the utricle in the Red-tailed Hawk (*Buteo jamaicensis*) was 2·1 mm, whereas in the Budgerigar (*Melopsittacus undulatus*) it was only 0·58 mm. The utricle expands near the lateral ampulla to accommodate the macula.

The *macula utriculi* is larger than that of the saccule. The measurements by Rosenhall (1970) on the macula utriculi of *Columba* revealed that it had a mean surface area of 1·29 mm^2 and that it contained 25,000 sensory cells in contrast to 7200 cells in the saccule. The striola was formed by a double row of cells, mostly of type I (Fig. 6.17), close to the rounded edge. Kinociliar polarization was in opposite directions medial and lateral to the striola (Fig. 6.17a).

Jorgensen and Andersen (1973) studied the macula utriculi in a number of species and found only slight variations in its structure. In most species, including, for example, *Phalacrocorax carbo, Ardea cinerea, Buteo buteo, Gallus, Columba, Melopsittacus undulatus, Strix aluco, Pica pica, Sturnus vulgaris* and *Passer domesticus,* there was a trizonal arrangement of the cells in the striola (such as shown in Fig. 6.17a). The striola of the Mute Swan (*Cygnus olor*) was different in that it had a single broad zone of type I cells, which contained a few scattered type II cells.

When the maculae of the guinea pig are compared with those of *Columba* (which has a cranial cavity only slightly smaller than that of the guinea pig), some interesting points emerge. Table I shows that the macula sacculi in each is approximately the same size and has a similar number of sensory cells. However, the macula utriculi of *Columba* is more than twice the size of that of the guinea pig and has almost three times the number of sensory cells. The pigeon, thus, has essentially twice as many sensory cells in the two maculae of the vestibule as has the guinea pig. These figures become even more impressive when one recalls that Jorgensen and Andersen (1973) found that there are large numbers of what is assumed to be an especially sensitive receptor cell unit, in the form of a single, large nerve chalice enclosing numerous sensory cells. Similar receptor cell units are present in mammals but do not contain as many cells. There appears to be nothing reported in mammals comparable to the findings of Jorgensen and Andersen (1973) in the saccule of the Great Cormorant (*Phalacrocorax carbo*), in which 60% of the endings on type I hair cells involved three to six cells, or in the saccule of *Gallus* in which over 90% of the endings involved three to nine cells.

TABLE I. **A comparison of the area and hair cell number in each macula in the inner ear of the guinea pig (*Cavia cobaya*) and domestic pigeon (*Columba livia*).**

	Area of macula	Number of sensory cells	References
Cavia cobaya			Lindeman (1969)
Macula sacculi	0.495 mm^2	7,560	
Macula utriculi	0.541 mm^2	9,260	
Total number of macular hair cells		16,820	
Columba livia			Rosenhall (1970)
Macula sacculi	0.46 mm^2	7,200	
Macula utriculi	1.29 mm^2	25,200	
Macula lagenae	0.93 mm^2	15,400	
Total number of macular hair cells		47,800	

When the lagenar cells are considered along with the other maculae (Table I), one concludes that the bird has a remarkable gravitational input, even if it is shown in the future that those lagenar fibres terminating in the auditory nuclei (Boord and Rasmussen, 1963) have some other function. Boord and Karten (1974) have suggested that since the boundaries of the pigeon's cochlear nuclei are not precisely defined, the sub-divisions to which the lagenar fibres project should perhaps be considered to be unique to the lagenar nerve and not part of the cochlear nucleus at all.

3. *Statoconial membrane*

The statoconial membranes are similar in all three maculae. An extracellular mucopolysaccharide gel with a fibrous component covers the macula, and the sensory hair bundles project up into it. The gel is highly hydrated, and the usual fixatives and dehydrants do not preserve its normal structure. Only the fine fibrous network, greatly shrunken, remains. Nevertheless, some features are generally accepted as representing the form of the membrane in life (Fig. 6.16). Just above the macular surface there appears to be a "submembranous" space where the fibrous component is more dispersed and poorly organised. The hair bundles traverse the space and project up into the more condensed fibrous part of the membrane. It is probably across this submembranous space that the critical movements of the statoconial membrane occur.

The *statoconia* are piled on top of the condensed part and, at least in mammals, are embedded in a fine fibrous meshwork (Smith and Tanaka, 1975). Lim and Lane (1969) and Lim (1974) found that the statoconia in *Columba* (Fig. 6.18b) are cylindrical in shape and pointed at both ends, similar to those in mammals. In mammals, the statoconia vary considerably in length, ranging from 1 to 20 μm. They are composed primarily of calcium carbonate (Iurato and dePetris, 1967).

Rosenhall (1970) observed that the statoconia in the ear of *Columba* do not extend to the periphery of the maculae, although the gel component of the membrane, without statoconia, continues somewhat further, particularly in the utricle. Substantial fibrous attachments to areas outside the hair cells of the utricle were found by Money and Correia (1972) in the Snowy Owl (*Nyctea scandiaca*) and the Great Horned Owl (*Bubo virginianus*), and by Money *et al.* (1974) in the Budgerigar (*Melopsittacus undulatas*) and the Red-tailed Hawk (*Buteo jamaicensis*). Some particular adaptations for holding the statoconial membrane in place would seem of particular importance for birds, because of their aerial acrobatics. Money *et al.* (1974) noted that the extramacular attachments seemed finer in the lagena and the saccule. The tegmentum vasculosum may help to hold the statoconia in place in the lagena and saccule. However, the saccule is so small that statoconial dislodgement would be most unlikely.

4. *Semicircular ducts*

The rostral, caudal and horizontal semicircular ducts (*ductus semicircularis rostralis, ductus semicircularis caudalis, ductus semicircularis horizontalis*) are thin-walled membranous tubes which arise from the utricle (Fig. 6.2). One end of each duct is enlarged, forming the membranous ampulla (*ampulla membranacea*), which contains a neuroepithlium, the *crista ampullaris*. The ducts, which are much smaller in diameter than the ampullae from which they emerge (Fig. 6.19), form approximately two-thirds of a circle and end at the side of the utricle opposite to that on which their ampullae are located. The horizontal duct terminates between the common crus and the caudal ampulla.The two vertical ducts (the rostral and caudal) join above the vestibule to form the common membranous crus (*crus membranaceum commune*), which opens into the utricle close to the macula. The rostral duct is longer than the caudal and horizontal ducts. In the Snowy Owl (*Nyctea scandiaca*) and the Great Horned Owl (*Bubo virginianus*), for example, Money and Correia (1972) observed that the rostral duct was 32 mm long, whereas the caudal and horizontal ducts were only 23 mm in length. The radii of the ducts also vary. Money and Correia (1972) found in the owls that the horizontal duct had a radius of 0·17 mm, the caudal duct a radius of 0·13 mm, and the rostral duct a radius of 0·24 mm. For measurements in other species, see Jones and Spells (1963). It is difficult to know whether the species differences that exist in the overall size of the ducts have a real functional importance or whether they are more related to differences in head and body size.

The *neuroepithelium* is composed of a mass of supporting and sensory cells arranged on the top and sides of a ridge or crest which has its long axis perpendicular to that of the ampulla. The hair bundles of these sensory cells are taller than those in the maculae and the kinocilia are even longer. In *Columba,* Landolt *et al.* (1975) found that the hair bundle lengths in the crista ranged from 28 to 48 μm. This is similar to measurements obtained in the cristae of the guinea pig (35 to 45 μm) by Smith and Tanaka (1975). The hairs of the sensory cells penetrate up into the overlying gelatinous membrane, the cupula, which fills the space between the crista and the roof of the ampulla.

The horizontal ducts of the left and right ears are in the same plane. The vertical ducts in each ear are in opposing planes, about 45° to each other. Thus, the rostral duct of one ear and the caudal duct of the opposite ear are in the same plane and are synergistic. It was demonstrated by Lowenstein and Wersall (1959) that in mammals the kinocilia are on the same side of all the sensory hair cell bundles within a single crista, but that the horizontal and vertical ducts of the same ear had opposite polarizations. Wersall *et al.* (1965) later found that this polarization was common to other vertebrates including the bird (*Gallus*). Specifically, the hair bundles in the horizontal duct are oriented so that their kinocilia face the utricle. The kinocilia on the cristae of the rostral and caudal ampullae face the duct end. When the head is rotated in a plane which, for example, coincides with caudal duct of one ear and the rostral duct of the opposite ear, the hair cells on one crista would be activated and those on the crista of the opposite ear inhibited due to opposing endolymph flow.

The cristae in the vertical ducts of birds have a peculiarity, the *septum cruciatum,* which divides the neuroepithelium into two parts. The septum cruciatum is also found in reptiles but only rarely in mammals. Igarashi and Yoshinobu (1966) studied the structure in several vertebrate classes and described it as a mound or eminence rather than a septum. "Septum" would imply a complete division of the ampulla, which is not the case. The structure is formed by a fibrous core covered by an epithelium composed of cuboidal and "dark" cells. The long

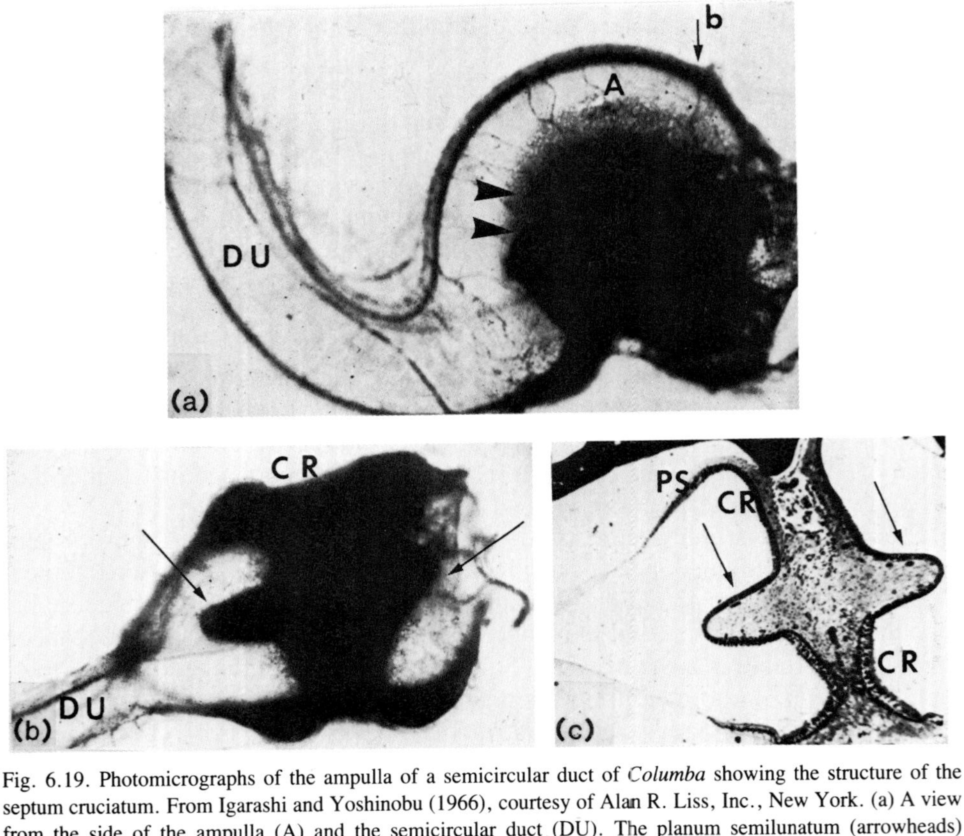

Fig. 6.19. Photomicrographs of the ampulla of a semicircular duct of *Columba* showing the structure of the septum cruciatum. From Igarashi and Yoshinobu (1966), courtesy of Alan R. Liss, Inc., New York. (a) A view from the side of the ampulla (A) and the semicircular duct (DU). The planum semilunatum (arrowheads) obscures the crista. The arrow marked b refers to the plane of the view in (b) ×40. (b) The ampulla viewed from above. The mound of the septum cruciatum (arrows) on both sides of the crista (CR) is clearly visible. Note that it is longer on the semicircular duct (DU) side. ×40. (c) A section of the ampulla through the septum cruciatum (arrows) in the same plane as in (b). The two neuroepithelial "saddles" of the crista (CR) are clearly visible. PS, Planum semilunatum. ×50.

axis of the mound is perpendicular to the crista, hence the term *cruciatum* (Fig. 6.19). Igarashi and Yoshinobu found that the neuroepithelium was thus divided into two parts, each with its own nerve fibres and cupula, and that structure was present only in the vertical ducts and never in the horizontal duct. Landolt *et al.* (1975) observed that the septum cruciatum of *Columba* formed two separate cristae each with a distinctive saddle shape. The two central horns of the saddles (adjacent to the eminence) were structures not present in the lateral crista. These authors suggested that the hair cells located on the slopes of the central horns could provide an additional input (either quantitative or qualitative) to the vestibular nuclei.

Specialised cells are located on the sides of the ampulla. They form a half-moon collar of uniformly-staining cells around the ends of the crista and are called the *planum semilunatum*. Dohlman (1964) demonstrated in *Columba* that the cells were active in secreting the mucopolysaccharides for the cupula. He also described other kinds of specialised cells on the slopes of the crista and in the perimacular region of the utricle. Some of these groups (other than the *planum semilunatum*) have alternating light and dark cells which resemble the cells in the *tegmentum vasculosum*. Dohlman showed that these other specialised cells were active in secretion and absorption, and he believed they were responsible for maintenance of vestibular endolymph. Extrasensory cells with similar characteristics are also present in the mammalian vestibular labyrinth and are assumed to have a similar function (Smith, 1970). One difference between the vestibular cell groups described previously and those in the *tegmentum vasculosum* and *stria vascularis* is that the latter structures have dense capillary networks, whereas the vestibular dark cells are not associated with capillaries. One would assume, therefore, that there was a lesser degree of activity in the vestibular dark cells than in the tegmentum vasculosum.

5. *Cupula*

The *cupula* overlies the crista and reaches to the roof and sides of the ampullar dome. It is a mucopolysaccharide fibrous gel similar to the tectorial and otoconial membranes. As with the others, it is difficult to preserve its normal hydrated form. According to Money *et al.* (1974), the cupula has attachments to the roof and sides of the avian ampulla.

The fibrous content of the cupula is more dispersed than in the other gel membranes, and Iurato and dePetris (1967) found that the fibres themselves were finer. That the fibres display a degree of organisation has been known for years. The hair bundles project up into small longitudinal channels in the cupula where the fibres are fewer in number. Furthermore, a definite subcupular space is present, also containing fewer fibres, just above the neuroepithelial surface.

The membranous labyrinth is filled with endolymph, and in the ampullae it is the endolymph movement which is the initial event in sensory cell activation. The communicating openings between the semicircular ducts and utricle are much wider than those between the utricule, the saccule and the cochlear duct, so that movements of endolymph between the utricle and the semicircular ducts could readily take place if no obstructions were present. However, it was Steinhausen

(1933) and Dohlman (1935) who demonstrated that the cupula is one impediment to the flow of endolymph. They found that the cupula divided the ampulla into two parts and that fluid pressure from one side or the other caused the cupula, as well as the hair cell cilia embedded in it, to move in a "swinging door" fashion. This implies rather large movements, and other investigators have argued that a "sliding cupula" or "drum membrane" are better descriptive terms. A second impeding factor which has frequently been discussed (Dohlman, 1981) resides in the size of the ampulla in relation to its duct. Dohlman emphasised that the cross-sectional area of the ampulla was always greater than that of the duct and that this would result in a reduction of cupular deflection.

It has become clear that the critical movements of the hair bundle are very small. In mammals extrapolated basilar membrane displacement at threshold is approximately 10^{-3} Å (Dallos, 1973). In the bullfrog, Hudspeth and Jacobs (1979) demonstrated that deflections of the hair bundle over a range greater than 0·5 μm resulted in saturation of the transducer. It seems apparent that the larger cupular movements visualised in earlier years probably occur only under excessive stimulation. Physiological stimulation probably takes place in the subcupular space, and this may be caused by fluid pressure changes, as was hypothesised by Dohlman (1981) in a comprehensive re-evaluation of his earlier observations. Alternately, minute cupular shifts across the surface of the neuroepithelium would be equally possible and just as effective.

6. *Crista neglecta*

The *crista neglecta* is a small neuroepithelial papilla found in the utricle close to the caudal ampulla. It is innervated by a small branch of the nerve which supplies the caudal semicircular duct. The crista neglecta was studied by Jorgensen (1972) in a large number of birds. He found that both types I and II hair cells were present but that there was no striola and no organised arrangement of the cells. The kinociliar polarization was the same as that in the caudal duct. Although Jorgensen did not describe a cupula, Igarashi and Lansdon (1964) observed that in the cat the structure sometimes has cupula-like strands over it. The function of the crista neglecta is obscure. It has a macular or papillary form, but its location near the crista ampullaris, as well as the fact that its nerve is derived from that of the caudal ampulla and that its kinociliar polarization is the same as that in the caudal ampulla, leads one to the assumption that the crista neglecta and the crista ampullaris of the caudal duct must have a common function. Since it is on the leading side of the crista ampullaris, its nerve fibres are closer to the vestibular nuclei so that a "priming" activity might be set up in the brain stem. However, the number of hair cells and nerve fibres in the crista neglecta are few. Until the central terminations of its nerve fibres have been defined, speculation does not seem warranted.

7. *Vestibular endolymph*

No information is available regarding the vestibular endolymph of birds, but it is probably not too different from that of the guinea pig. Vestibular endolymph in mammals is known to have a high concentration of potassium (Smith *et al.*, 1954). The ratio of potassium (144 mEq/liter) to sodium (15·8 mEq/liter) is 9:1, that is, lower than in cochlear endolymph.

The direct current potentials of vestibular endolymph have been measured in both birds and mammals. In the guinea pig, Smith *et al.* (1958) found the endovestibular potential to be +4 mV in the utricle and +1 mV in the saccule. Eldredge *et al.* (1961) recorded −1 mV from the ampullae of the semicircular ducts. The studies by Schmidt (1963) on *Columba* revealed that the endovestibular potentials were also lower than those in the cochlea, but slightly higher than recordings from the same location in guinea pigs. Measurements from the ampullae of the pigeons averaged +4 mV and from the utricle about +12 mV. No recordings were made from the saccule (which would have been technically difficult due to both its small size and the presence of the tegmentum vasculosum). It would be reasonable to assume that the saccular potential would be close to that of the cochlear duct in view of the presence in the saccule of the tegmentum vasculosum and the large diameter of the sacculocochlear duct.

Acknowledgements

The author is grateful to Dr. Mark Konishi for supplying the ears of the Barn Owls (*Tyto alba*), to Dr. Makato Igarashi for permission to use the micrographs in Fig. 6.19, to Dr. David Lim for furnishing the micrographs in Fig. 6.18, to Dr. Katsuhiko Tanaka for furnishing the micrographs in Fig. 6.13, to Dr. Ulf Rosenhall for permission to use the drawings in Fig. 6.17, and to Dr. Nancy Russell for critically reading the manuscript.

References

Baird, I. L. (1974). Anatomical features of the inner ear in submammalian vertebrates. *In* "Handbook of Sensory Physiology" (H. Autrum, R. Jung, W. Loewenstein, D. McKay and H. Teuber, Eds.). Springer-Verlag, Berlin and New York.

Bard. P. (1968). Postural coordination and locomotion and their central control. *In* "Medical Physiology" (V. B. Mountcastle, Ed.), Vol. 2. C. V. Mosby, St. Louis.

Boord, R. L. (1961). The efferent cochlear bundle in the caiman and pigeon. *Exp. Neurol.* **3,** 225–239.

Boord, R. L. (1964). The number and diameter of myelinated fibers in the statoacoustic nerve of the adult pigeon. *Am. Zool.* **4,** 101.

Boord, R. L. (1969). The anatomy of the avian auditory system. *Ann. N.Y. Acad. Sci.* **167,** 186–198.

Boord, R. L. and Karten, H. J. (1974). The distribution of primary lagenar fibers within the vestibular nuclear complex of the pigeon. *Brain Behav. Evol.* **10,** 228–235.

Boord, R. L. and Rasmussen, G. L. (1963). Projection of the cochlear and lagenar nerves on the cochlear nuclei of the pigeon. *J. comp. Neurol.* **120,** 463–475.

Bosher, S. K. (1979). The nature of the negative endocochlear potentials produced by anoxia and ethacrynic acid in the rat and the guinea pig. *J. Physiol., Lond.* **293,** 329–345.

Bosher, S. K. and Warren, R. (1968). Observations on the electrochemistry of the cochlear endolymph of the rat: a quantitative study of its electrical potential and ionic composition as determined by means of flame spectrophotometry. *Proc. R. Soc. B.* **171,** 227–247.

Brummett, R., Smith, C., Ueno, Y., Cameron, S. and Richter, R. (1977). The delayed effects of ethacrynic acid on the stria vascularis of the guinea pig. *Acta oto-lar.* **83,** 98–112.

Dallos, P. (1973). "The Auditory Periphery." Academic Press, New York and London.

Desmedt, J. E. (1962). Auditory evoked potentials from cochlea to cortex as influenced by activation of the efferent olivo-cochlear bundle. *J. acoust. Soc. Am.* **34,** 1478–1496.

Desmedt, J. E. and Delwaide, P. J. (1965). Functional properties of the efferent cochlear bundle of the pigeon revealed by stereotaxic stimulation. *Exp. Neurol.* **11,** 1–26.
Dohlman, G. F. (1935). Some practical and theoretical points in labyrinthology. *Proc. R. Soc. Med.* **28,** 1371–1384.
Dohlman, G. F. (1964). Secretion and absorption of endolymph. *Ann. Otol. Rhinol. Lar.* **73,** 708–723.
Dohlman, G. F. (1981). Critical review of the concept of cupula function. *Acta oto-lar., Suppl.* No. 376, 1–30.
Dooling, R. J. (1973). Behavioral audiometry with the parakeet *Melopsittacus undulatus. J. acoust. Soc. Am.* **53,** 1757–1758.
Dooling, R. J. (1983). Auditory perception in birds. *In* "Acoustical Communication in Birds" (D. E. Kroodsma and E. H. Miller, Eds.), Vol. I. Academic Press, New York and London.
Dooling, R. J. and Saunders, J. C. (1975). Hearing in the parakeet (*Melopsittacus undulatus*). Absolute thresholds, critical ratios, frequency difference limens, and vocalizations. *J. comp. Physiol.* **88,** 1–20.
Dooling, R. J., Mulligan, J. A. and Miller, J. D. (1971). Auditory sensitivity and song spectrum of the Common Canary *(Serinus canarius). J. acoust. Soc. Am.* **50,** 700–709.
Eldredge, D. H., Smith, C. A., Davis, H. and Gannon, R. P. (1961). The electrical polarization of the semicircular canals (guinea pig). *Ann. Otol. Rhinol. Lar.* **70,** 1024–1037.
Flock, A. (1965). Electron microscopic and electrophysiological studies on the lateral line canal organ. *Acta oto-lar., Suppl.* No. 199.
Flock, A. (1971). Sensory transduction in hair cells. *In* "Handbook of Sensory Physiology" (W. F. Loewenstein, Ed.), Vol. I. Springer-Verlag, Berlin and New York.
Flock, A. and Cheung, H. C. (1977). Actin filaments in sensory hairs of inner ear receptor cells. *J. Cell Biol.* **75,** 339–343.
Goodley, L. B. and Boord, R. L. (1966). Quantitative analysis of the hair cells of the auditory papilla of the pigeon. *Am. Zool.* **6,** 542.
Held, H. (1926). Die Cochlea der Säuger und der Vögel, ihre Entwicklung and ihre Bau. *In* "Handbuch der Normalen and Pathologischen Physiologie mit Beruksichtigung der Experimentellen Pharmakologie" (A. Bethe, G. von Bergmann, G. Embden and A. Ellinger, Eds.). Springer-Verlag, Berlin and New York.
Hoshino, T. (1973). Tectorial membrane and otolithic membrane of pigeon inner ear. A scanning electron microscope study. *Archvm. histol. jap.* **35,** 141–152.
Hudspeth, A. J. and Corey, D. P. (1977). Sensitivity, polarity and conductance change in the response of vertebrate hair cells to controlled mechanical stimuli. *Proc. natn. Acad. Sci. U.S.A.* **74,** 2407–2411.
Hudspeth, A. J. and Jacobs, R. (1979). Stereocilia mediate transduction in vertebrate hair cells. *Proc. natn. Acad. Sci. U.S.A.* **76,** 1506–1509.
Igarashi, M. and Lansdon, S. (1964). "Redefinition of the Macula Neglecta in Mammals," Res. Rep. No. 97. U.S. Naval School of Aviation Medicine. NASA Order No. R-93.
Igarashi, M. and Yoshinobu, T. (1966). Comparative observations on the eminentia cruciata in birds and mammals. *Anat. Rec.* **155,** 269–277.
Iurato, S. and dePetris, S. (1967). Otolithic membranes and cupulae. *In* "Submicroscopic Structure of the Inner Ear" (S. Iurato, Ed.). Pergamon Press, New York.
Johnstone, C. G., Schmidt, R. S. and Johnstone, B. M. (1963). Sodium and potassium in vertebrate cochlear endolymph as determined by flame microspectrophotometry. *Comp. Biochem. Physiol.* **9,** 335–341.
Jones, G. M. and Spells, K. E. (1963). Theoretical and comparative study of the functional dependence of the semicircular canal upon its physical dimensions. *Proc. R. Soc. B.* **157,** 403–419.
Jorgensen, J. M. (1972). The avian *neglecta Retzii. Acta zool., Stockh.* **53,** 155–163.
Jorgensen, J. M. and Andersen, T. (1973). On the structure of the avian maculae. *Acta zool., Stockh.* **54,** 121–130.
Knudsen, E. I. (1981). The hearing of the Barn Owl. *Sci. Am.* **245,** 112–125.
Konishi, M. (1970). Comparative neurophysiological studies of hearing and vocalizations in songbirds. *Z. vergl. Physiol.* **66,** 257–272.
Konishi, M. (1973). How the owl tracks its prey. *Am. Scient.* **61,** 414–424.
Landolt, J. P., Correia, M. J., Young, E. R., Cardin, R. P. S. and Sweet, R. C. (1975). A scanning electron microscopic study of the morphology and geometry of neural surfaces and structures associated with the vestibular apparatus. *J. comp. Neurol.* **159,** 257–288.
Lim, D. J. (1973). Ultrastructure of the otolithic membrane and the cupula. *Adv. oto-rhino-lar.* **19,** 35–49.
Lim, D. J. (1974). The statoconia of the non-mammalian species. *Brain Behav. Evol.* **10,** 37–51.
Lim, D. J. and Lane, W. C. (1969). Vestibular sensory epithelia. *Archs. Otolar.* **90,** 283–292.

Lindeman, H. (1969). Studies on the morphology of the sensory regions of the vestibular apparatus. *Adv. Anat. Embryol. Cell Biol.* **42,** 1–113.

Lowenstein, O. and Wersall, J. (1959). A functional interpretation of the electron microscopic structure of the sensory hairs in the cristae of the elasmobranch, *Raja clavata,* in terms of directional sensitivity. *Nature, Lond.* **184,** 1807–1808.

Miller, M. R. (1973). A scanning electron microscope study of the papilla basilaris of *Gekko gecko. Z. Zellforsch. mikrosk. Anat.* **136,** 307–328.

Money, P. and Correia, M. J. (1972). The vestibular system of the owl. *Comp. Biochem. Physiol.* **42A,** 353–358.

Money, K., Landolt, J. P., Correia, M. J. and Laufer, J. (1974). Anatomical and physiological investigations of the vestibular system of birds. *Brain Behav. Evol.* **10,** 212–227.

Mulroy, M. J. (1974). Cochlear anatomy of the alligator lizard. *Brain Behav. Evol.* **10,** 69–87.

Peterson, S. K., Frishkopf, L. S., Lechene, C., Oman, C. M. and Weiss, T. F. (1978). Element composition of inner ear lymphs in cats, lizards, and skates determined by electron probe microanalysis of liquid samples. *J. comp. Physiol.* **126,** 1–14.

Pierson, M. and Dallos, P. (1976). Re-examination of avian cochlear potentials. *Nature, Lond.* **262,** 599–601.

Pumphrey, R. J. (1961). Sensory organs. *In* ''Biology and Comparative Physiology of Birds'' (A. J. Marshall, Ed.), Vol. 2. Academic Press, New York and London.

Quine, D. B. and Konishi, M. (1974). Absolute frequency discrimination in the Barn Owl. *J. comp. Physiol.* **93,** 347–360.

Rosenhall, U. (1970). Some morphological principles of the vestibular maculae in birds. *Arch. klin. exp. Ohr., Nas. KehlkHeilk.* **197,** 154–182.

Sachs, M. B., Young, E. D. and Lewis, R. H. (1974). Discharge patterns of single fibers in the pigeon auditory nerve. *Brain Res.,* **70,** 431–447.

Schmidt, R. S. (1963). Types of endolymphatic potentials. *Comp. Biochem. Physiol.* **10,** 83–87.

Schmidt, R. S. (1964). Blood supply of pigeon inner ear. *J. comp. Neurol.* **123,** 187–204.

Schmidt, R. S. and Fernandez, C. (1962). Labyrinthine DC potentials in representative vertebrates. *J. cell. comp. Physiol.* **59,** 311–322.

Schwartzkopff, J. (1968). Structure and function of the ear and of the auditory brain areas in birds. *In* ''Hearing Mechanisms in Vertebrates'' (A. V. S. de Reuck and J. Knight, Eds.). Little, Brown, Boston.

Schwarz, I. E., Schwarz, D. W. F., Frederickson, J. M. and Landolf, J. P. (1981). Efferent vestibular neurons: a study employing retrograde tracer methods in the pigeon *(Columba livia). J. comp. Neurol.* **197,** 1–12.

Sinnott, J. M., Sachs, M. B. and Heinz, R. D. (1976). Differential sensitivity to frequency and intensity in songbirds. *J. acoust. Soc. Am.* **60,** S87.

Smith, C. A. (1953). The capillaries of the vestibular membranous labyrinth in the guinea pig. *Laryngoscope, St. Louis* **63,** 87–104.

Smith, C. A. (1968). Morphological features of axo-dendritic relationships between efferent and cochlear nerves in the cochlea of mammal and pigeon. *In* ''Structure and Function of Inhibitory Neuronal Mechanisms'' (C. von Euler, S. Skoglund and U. Soderberg, Eds.). Pergamon Press, New York.

Smith, C. A. (1970). Extrasensory cells of the vestibule. *In* ''Biochemical Mechanisms in Hearing and Deafness'' (M. Paparella, Ed.). Thomas, Springfield, Illinois.

Smith, C. A. (1981). Recent advances in structural correlates of auditory receptors. *In* ''Progress in Sensory Physiology'' (D. Ottoson, Ed.), Vol. 2. Springer-Verlag, New York.

Smith, C. A. and Rasmussen, G. L. (1968). Nerve endings in the maculae and cristae of the chinchilla vestibule, with a special reference to the efferents. *In* ''Third Symposium on the Role of the Vestibular Organs in Space Exploraton.'' *NASA, Spec. Publ. SP* **NASA SP-152,** 183–200.

Smith, C. A. and Tanaka, K. (1975). Some aspects of the structure of the vestibular apparatus. *In* ''The Vestibular System'' (R. Naunton, Ed.). Academic Press, New York and London.

Smith, C. A. and Tanaka, K. (1981). Further observation on the avian basilar papilla. *Abstract, ARO 4th Meet., Ass. Res. Otolar.,* No. 103, St. Petersburg Beach, Florida.

Smith, C. A., Lowry, O. H. and Wu, M.-L. (1954). The electrolytes of the labyrinthine fluids. *Laryngoscope, St. Louis* **64,** 141–153.

Smith, C. A., Davis, H., Deatherage, B. H. and Gessert, C. F. (1958). DC potentials of the membranous labyrinth. *Am. J. Physiol.* **193,** 203–206.

Smith, C. A., Konishi, M., and Schuff, N. (1985). Structure of the Barn Owl's (*Tyto alba*) inner ear. *Hearing Res.* (in press).

Stebbins, W. (1970). Studies of hearing and hearing loss in the monkey. *In* "Animal Psychophysics" (W. Stebbins, Ed.). Appleton-Century-Crofts, New York.

Steinhausen, W. (1933). Über den Nachweiss der Bewegung der Cupula in der intakten Bogengangsampulle, des Labyrinthes bei der naturlichen rotatorischen und calorischen Reizung. *Pflügers Arch. ges. Physiol.* **232,** 500–512.

Takasaka, T. and Smith, C. A. (1971). The structure and innervation of the pigeon's basilar papilla. *J. Ultrastruct. Res.* **35,** 20–65.

Tanaka, K. and Smith, C. A. (1975). Structure of the avian tectorial membrane. *Ann. Otol. Rhinol. Lar.* **84,** 287–297.

Tanaka, K. and Smith, C. A. (1978). Structure of the chicken's inner ear *Am. J. Anat.* **153,** 251–271.

von Bekesy, G. (1952). D.C. resting potentials inside the cochlear partition. *J. acoust. Soc. Am.* **24,** 72–76.

von Bekesy, G. (1960a). Experimental models of the cochlear with and without nerve supply. *In* "Neural Mechanisms of the Auditory and Vestibular Systems" (G. L. Rasmussen and W. F. Windle, Eds.). Thomas, Springfield, Illinois.

von Bekesy, G. (1960b). "Experiments in Hearing." McGraw-Hill, New York.

Weiss, T. F., Mulroy, M. J., and Altmann, D. W. (1974). Intracellular responses to acoustic clicks in the inner ear of the alligator lizard. *J. acoust. Soc. Am.* **55,** 606–619.

Weiss, T. F., Mulroy, M. J., Turner, R. B., and Pike, C. (1976). Tuning of single fibers in the cochlear nerve of the alligator lizard: relation to receptor morphology. *Brain Res.,* **115,** 71–90.

Weiss, T. F., Peake, W. T., Ling, A. and Holton, T. (1978). Which structures determine frequency selectivity and tonopic organization of vertebrate nerve fibers? Evidence from the alligator lizard. *In* "Evoked Electrical Activity in the Auditory Nervous System" (R. F. Naunton and C. Fernandez, Eds.). Academic Press, New York and London.

Werner, C. F. (1933). Die differentzierung der Maculae im Labyrinth insbesondere bei Säugetieren. *Z. Anat. EntwGesch.* **99,** 696–709.

Wersall, J. (1956). Studies on the structure and innervation of the sensory epithelium of the cristae ampullares in the guinea pig. *Acta oto-lar., Suppl.* No. 126.

Wersall, J., Flock, A. and Lundquist, P.-G. (1965). Structural basis for directional sensitivity in cochlear and vestibular sensory receptors. *Cold Spring Harbor Symp. Quant. Biol.* **30.**

Wever, G. (1978). "The Reptile Ear, its Structure and Function." Princeton University Press, Princeton.

7
Eye

GRAHAM R. MARTIN

Departments of Zoology and Comparative Physiology and Department of Extramural Studies, University of Birmingham, Birmingham, England

CONTENTS

I. Introduction

Optically, the avian eye is classified as a linear magnification system the function of which, like that of all other vertebrate eyes, is simply to produce a real image of objects and surfaces in the environment. This image of a three-dimensional world is produced upon a basically two-dimensional receptor surface (the retina) where information is extracted from the image. Both the nature of the image and the characteristics of the image-analysing mechanism, which includes higher centres of the central nervous system as well as the retina, determine the visual capacities of an animal.

Advances in visual science have increased in general terms our understanding of how the diversity of structures amongst vertebrate eyes may be interpreted as adaptations to the visual problems which different life-styles pose. Particularly important to this task have been advances in the following areas: (1) the analysis and description of the visual environment (e.g. Lythgoe, 1979), including the somewhat different description of the visual environment contained in Gibson's "ecological optics" (Gibson, 1961, 1979; Lee, 1980), (2) the theoretical analysis of the fundamental constraints on optical-imaging systems (e.g. Miller, 1979; Snyder, 1979; Land, 1981) and of the ways in which optical images can be analysed by receptor systems (Snyder *et al.*, 1977; Barlow, 1981), and (3) the visual capacities of vertebrates (e.g. Muntz, 1974).

Analysis of the diversity of eye structures as adaptations to the visual environment is not new, and the older summaries of vertebrate eye structure (Walls, 1942; Rochon-Duvigneaud, 1943; Duke-Elder, 1958) are typically framed within the assumptions of such adaptionist arguments. What the advances in visual science just listed have produced is a better understanding of the physical limitations and the complexity of the problems which the task of extracting information from the visual world actually presents. Concomitant with this progress has been an increased need for more thorough and detailed analyses of eye structure and visual capacities in birds. As in other areas of comparative anatomy and physiology, the descriptions of eye structure produced during the last century and in the early part of this century present intriguing and inspiring ideas to the sensory scientist of today. Texts such as those of Slonaker (1897), Wood (1917) and Franz (1934) contain many observations which are still of great value. However, there is often the need for much of this work to be confirmed and extended. For example Wood (1917) includes many detailed coloured illustrations of the ophthalmoscopic view of the fundus in a range of avian species. These reveal a diversity of fundic colour, numbers and positions of areas and foveae, and sizes and shapes of pecten oculi. However, Wood appears to have obtained his species for study somewhat randomly, and so there is little possibility of being able to discern amongst his data phylogenetic or ecological relationships which may be correlated with this diversity.

Thus, although this chapter draws for its comparative base on the work of early anatomists, interpreting the details of this work within a more modern framework is problematic. It is hoped that the chapter might inspire the continued collection of comparative data on avian eye structure and visual performance. These will provide important tests of the theoretical advances outlined earlier as well as further the understanding of many aspects of avian anatomy, physiology and behaviour.

Any division in the description of an organ as complex as the vertebrate eye must be arbitrary, and in this there is always the possibility that important functional links or correlations between certain structures will be minimised or overlooked. Here the avian eye is described principally by reference to its two primary functional divisions: (1) as an optical system which produces an image of the environment; and (2) as a sensory system which extracts information from that image.

II. The optical system

Included in this section are not only the main refractive components (cornea and lens) but also the accommodatory mechanisms and the pupil, and the size, shape and position of the eyes. Between them these features determine the extent of the visual environment which may be brought to a focussed image upon the retina. Eye size also appears to be related to the range of light levels over which the visual system can be considered to function adequately.

A. Eye size

It has often been remarked that the avian eye (*oculus*) is both relatively and absolute large (Walls, 1942; Duke-Elder, 1958; Meyer, 1977). The largest eye of any land vertebrate is reported to be that of the Ostrich (*Struthio camelus*) (body weight 63–104 kg; Cramp 1977), which has a bulbar axial length of 50 mm. The bulbar axis (*axis bulbi*) is a line passing through the centre of the cornea and lens to the retina. Even the much smaller Tawny Owl (*Strix aluco*) (body weight 450 g; personal observation) has an eye with an axial length of 28·5 mm, which is approximately 4·5 mm greater than that of the human eye (Hughes, 1977). Typically, birds' eyes are so large relative to the skull that they meet in the middle of the head and are usually separated by only a thin septum of bone (Fig. 7.1).

Optically the most important parameter of eye size is the anterior focal length or posterior nodal distance, since this determines the size of the retinal area over which the image of an object is spread. The longer the posterior nodal distance the larger the retinal image of an object. Miller (1979) has shown that a large posterior nodal distance is essential if the vertebrate eye is to achieve the maximum theoretical limit of visual resolution. This is because there exists a finite limit on the size, and hence packing density, of the retinal photoreceptors which sample the retinal image, and this will therefore determine the size of the smallest detail in the image which can be resolved. Clearly if the image of an object is spread over more photoreceptors, then the amount of detail which can be resolved in that image will increase. However, independent of any structural or metabolic limitations on eye size, this process cannot be continued *ad infinitum,* since the optical image will itself provide a limit to resolution because of diffraction effects and aberrations within the optical system. Thus, an eye any larger than that of the Ostrich (*Struthio camelus*) may have little adaptive value, since although the image will be larger, its quality may deteriorate.

It has also been argued (Martin, 1982) that a large posterior nodal distance is essential if an eye is to function adequately throughout the range of naturally occurring luminance levels. The large eye of the Tawny Owl (*Strix aluco*), and possibly other species of owl, has been interpreted on this basis. The eye of *Strix aluco* is perhaps best regarded not simply as a nocturnal eye, but that of an

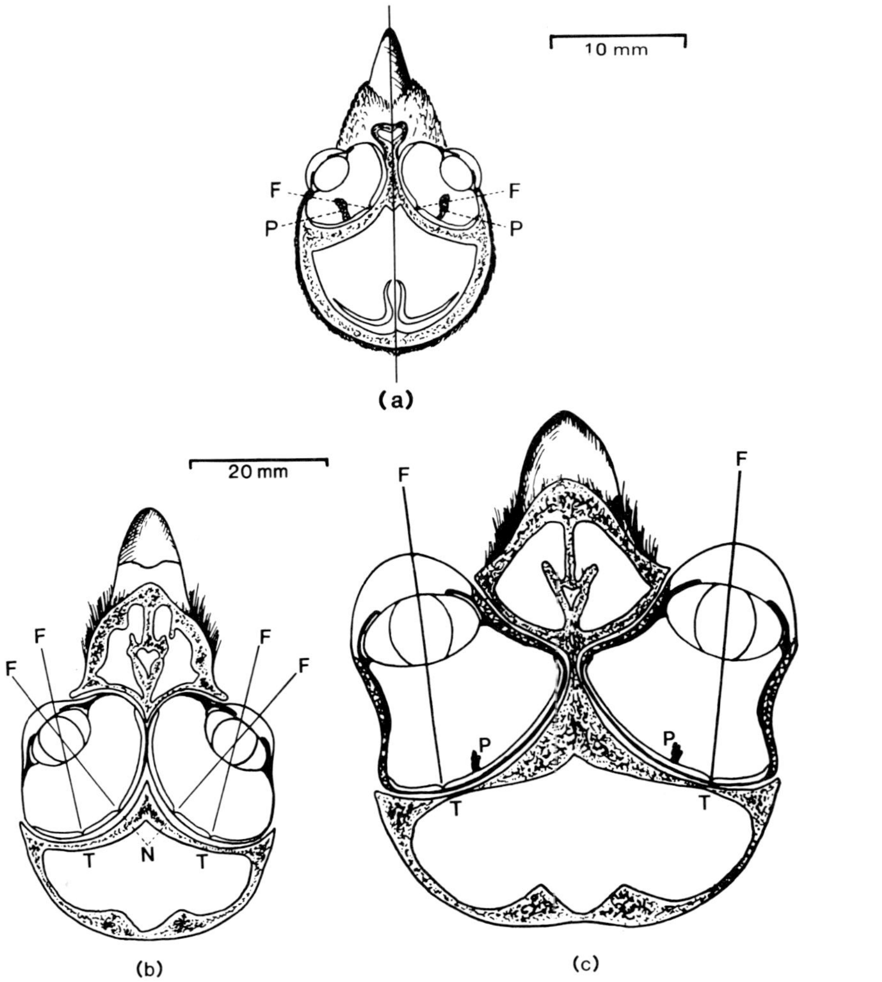

Fig. 7.1. Drawings of horizontal sections through the head of (a) the Black-capped Chickadee (*Parus atricapillus*), (b) the Broad-winged Hawk (*Buteo platypterus*) and (c) the Great Horned Owl (*Bubo virginianus*), showing the different shapes of eye found in birds and the position of eyes within the skull. Redrawn from Wood (1917) to a scale based on the average skull width in the collection of the British Museum (measurements provided by A. Knox). F, fovea; T, temporal fovea; N, nasal fovea; P, pecten.

arrhythmic species which functions adequately throughout the range of naturally occurring day-time, twilight and night-time light levels (Fig. 7.2).

Species such as warblers (Parulidae, Sylviidae) and finches (Fringillidae) with eyes of small bulbar axial length, that is, 6–8 mm (Walls, 1942; Rochon-Duvigneaud, 1943), may be restricted to their diurnal life style by the size of their eyes. Thus, it has been argued (Martin, 1983) that, because of the problems of optimally sampling the retinal image at various luminance levels (Snyder *et al.*, 1977), eyes of this size cannot function adequately over a wide luminance range, although they can function optimally within a narrow luminance range such as exists during day-time. It seems that as light levels fall, these birds must roost, since their eye design (attributable primarily to its size) prohibits the extraction of enough information from the visual environment to mediate flight safely under

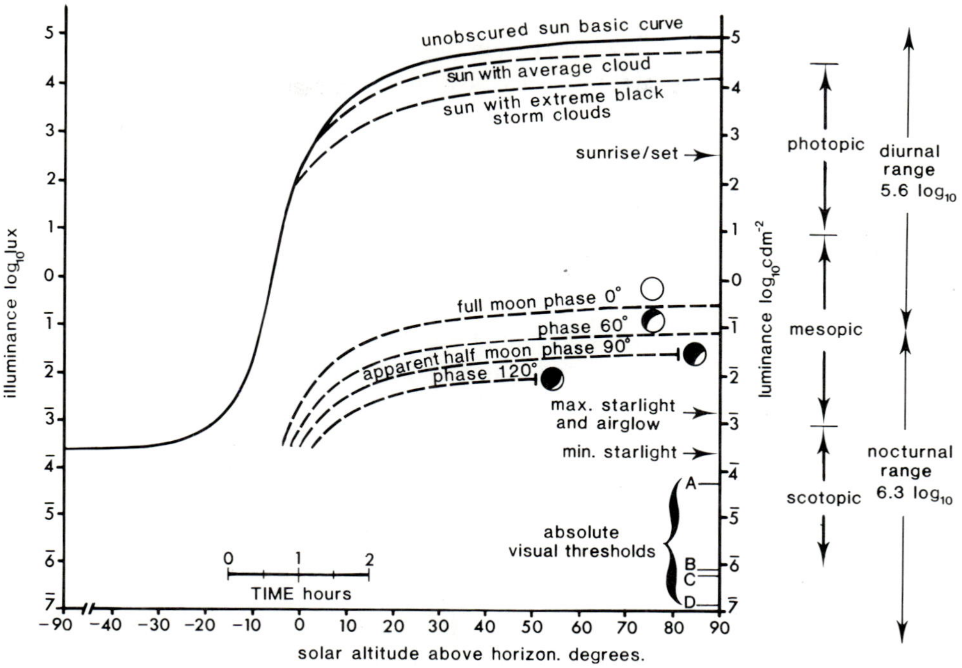

Fig. 7.2. The total range of naturally-occurring illumination levels experienced on earth. The luminance level which such illumination would produce on a highly-reflecting and diffusing surface such as white paper is also shown. Illumination levels due to the sun as a function of altitude, the moon as a function of altitude and phase, and starlight are shown, together with an indication of the maximum attenuation of light levels which may be produced by cloud cover. If account is taken of differences in the reflectance of different substrates, and attenuation by cloud and tree canopies, the total luminance range to which an eye can be naturally exposed (ignoring entry into caves and cavities) in temperate latitudes is approximately 12 $\log_{10}$ units ($\times 10^{12}$). This may be divided approximately equally between the naturally-occurring diurnal and nocturnal ranges and is shown here as the luminance levels of a grass or leaf litter substrate. The transition between day-time and night-time is defined as the sun passing 8° below the horizon. The basic curve presented here is for latitude 50° at the time of the summer solstice. At lower latitudes the transition from day- to night-time is more rapid, but the luminance levels experienced are approximately the same. At higher latitudes the transition from day- to night-time is slower, and the total range of luminance experienced on both a daily and annual basis is less than at lower latitudes. (Days are less bright and nights are less dark at higher latitudes.) At the pole there is no daily variation of luminance. The absolute visual thresholds are for (A) pigeon, (B) human, (C) Tawny Owl (*Strix aluco*) and (D) cat. Based in part on Martin (1982), courtesy of Springer-Verlag, Heidelberg.

low-light conditions. It is also worth noting that maximal visual resolution can only be achieved at the very highest naturally occurring light levels (Snyder *et al.*, 1977), and there is evidence (Fox *et al.*, 1976; Reymond and Wolfe, 1981; Hirsch, 1982) indicating that avian eyes designed to function close to the theoretical limit of resolution, such as those of the diurnal birds-of-prey (Falconiformes), may restrict the visually-guided behavior of the species to only high—that is, day-time (Fig. 7.2)—light levels.

The eyes of the nocturnal kiwis (Apterygidae) are described by Walls (1942, p. 650) as being "as degenerate as it is possible for an avian eye to be". This statement seems to be based on the relatively small size of the eye compared to body size, a feature also observed by other writers (Dawson, 1978). The bulbar axial length of the kiwi eye (≃8 mm) is similar to that of the majority of passeriforms, however. Possibly the kiwi eye is designed to function optimally, like that of the rat (Martin, 1982), at the low light levels which exist at night within their dense forest floor habitats. On theoretical grounds there appears to be no selective advantage in having a large eye if activity is restricted to low light levels such as these. Clearly the visual capacities and eye structure of species like kiwis and the Ostrich (*Struthio camelus*), which are representative of the two extremes in avian eye size, deserve close attention.

B. Eye shape

No avian eyes have a shape approximating the typically spherical form of the mammalian eye. Walls (1942) divided the shapes of the eyes found in birds into three broad categories: flat, globose and tubular (Fig. 7.1). The main differences between these eye categories lie in the relative proportions of their axial and equatorial diameters, and it is likely that a complete gradation exists, at least between the flat and globose types. Shape, however, is not independent of absolute size, the tubular shape being found only in species with the largest eyes including owls (Strigiformes) and some eagles (Falconiformes). The majority of avian species have a flat type of eye.

The visual function, if any, of these different shapes of eyes has not been determined. It seems likely that they result primarily from constraints imposed by the general form of the skull. Thus, the different shapes may simply be a solution to the problem of placing a relatively large eye within a small aviform skull. The shapes may also be regarded as the result of adaptations to reduce weight. For example, whilst the eye of the Tawny Owl (*Strix aluco*) has a similar axial length and similar optical properties to the human eye, its tubular shape encloses a smaller volume and hence weighs less than the human eye. Thus, the tubular shaped eye may be viewed as a sphere with sections trimmed away to facilitate both placement in the skull and reduction in the weight of the entire head.

Figure 7.1 shows that the basic structural differences underlying the eye shapes lie in the ciliary region between the retinal margin, *ora serrata*, close to the equator of the eye (the greatest circumference of the eyeball) and the conjunctival ring (*annulus conjunctivalis*) (Fig. 7.3), marking the junction of the corneal epithelium and the bulbar conjunctiva (*tunica conjunctiva bulbi*). This region is strengthened by a group of bones, the scleral ossicles (*ossicula sclerae*) (Fig. 7.3), which may number between 10 and 18 (the majority of species have 15). (Scleral ossicles are also a feature of reptilian, but not mammalian, eyes.) These bones, which are often pneumatised in large eyes, form a complete ring (*annulus ossicularis sclerae*) of overlapping plates (Fig. 7.4). The shape of the individual ossicles varies according to the shape of the eye, from flat or slightly convex in the flat type of eye, to concave in tubular eyes.

The posterior segment of the eye is always hemispherical (Figs. 7.1, 3). It is contained within the *sclera,* which consists of a cartilaginous lamina (*lamina cartilaginea sclerae*) and a fibrous layer, often as thick as the cartilage, external to it. The cartilaginous lamina attaches to the posterior portion of the scleral ossicles. Within it are areas of ossification (*ossicula posteriora sclerae*). In some species one of these areas of ossification forms a ring-shaped or horse-shoe-shaped bone around the optic nerve (*os nervi optici,* formerly Gemminger's ossicle). Although the existence of this bone was reported in 219 species by

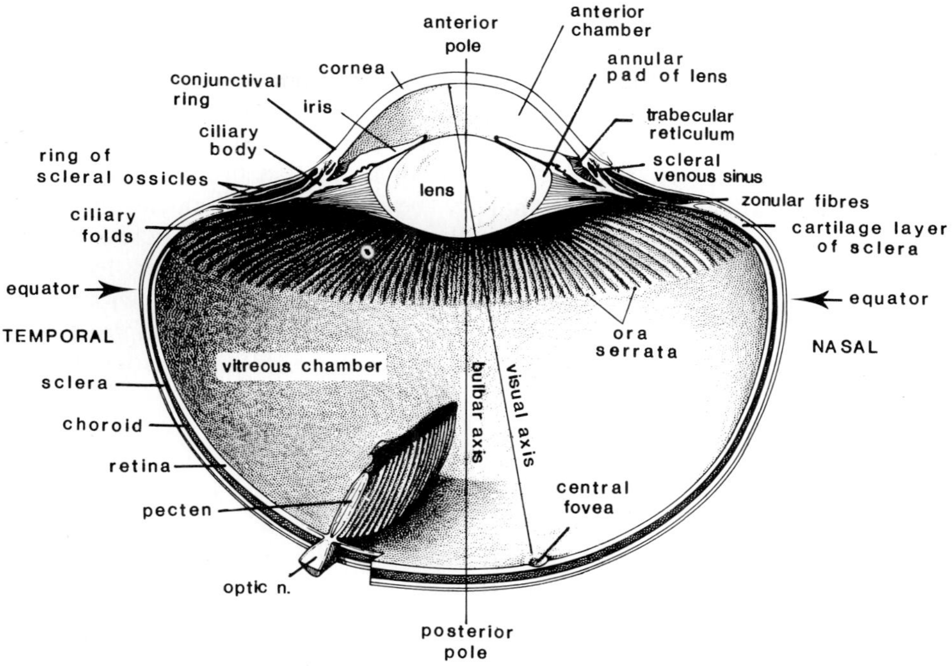

Fig. 7.3. Drawing of the eye of *Gallus* sectioned in the horizontal plane showing the interior of the ventral half of the bulb. From Evans (1979), courtesy of Academic Press, London and New York.

Tiemeier (1950), Duke-Elder (1958, p. 404) considers that "no satisfactory theory for its presence has been put forward".

Many, possibly all (Walls, 1942, p. 643), eyes of birds are asymmetrical structures. Typically there is a marked nasal–temporal asymmetry in the distance between the corneoscleral junction and the equator, but there is little (if any) asymmetry about a ventrodorsal axis. This "nasad-asymmetry" (Walls, 1942, p.

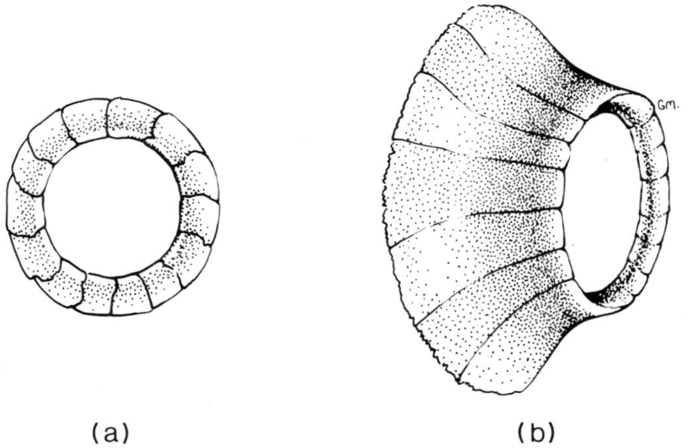

Fig. 7.4. Scleral ossicles. (a) The flat ossicles of *Gallus* are shown in anterior view. (b) The concave ossicles of the Tawny Owl (*Strix aluco*) came from the right eye and are shown in lateral view; in this species the medial ossicles are much shorter than the lateral ossicles. From Duke-Elder, Sir Stewart, editor: System of ophthalmology. Vol. I. The eye in evolution (by Duke-Elder, Sir Stewart), St. Louis, 1958, The C. V. Mosby Co.; after Franz.

301) is seen most clearly in Fig. 7.1c. It results in the retina being arranged asymmetrically about the bulbar axis (Fig. 7.3), and this is manifest in the complex asymmetry of the retinal visual field about the axis (see Section II,F). The function of this nasad-asymmetry was interpreted by Walls (1942, p. 300) as an adaptation which serves to enlarge the binocular field, and this conclusion has often been repeated. However, Walls offers no explanation of how such an enlargement is affected by the asymmetry, beyond a general conclusion that the bulbar axis is made to face more frontally. Analysis of the monocular and binocular visual fields of the Tawny Owl (*Strix aluco*) has, however, led Martin (1984) to propose that nasad-asymmetry serves to maximise the width of the cyclopean (or panoramic) visual field within the twin constraints imposed by the width of the visual field of each eye (the monocular field) and the width of the binocular field.

C. Cornea and lens

The avian eye, like that of all vertebrates, has just two principal refractive components: the *cornea,* which consists of a relatively simple convexo–concavo lens, and the more complex biconvex *lens.* Amongst the vertebrates there is considerable variation in the relative contributions of these two components to the total refractive power of the eye's optical system, and the avian eye appears to present no exception to this. Thus, Fig. 7.5 indicates that in the eye of the Tawny Owl (*Strix aluco*), the total refractive power (F_E) is divided approximately equally between the lens (F_L) and the cornea (F_C), whilst in the pigeon the refractive power of the cornea is over twice that of the lens. However, with such limited data it is not possible to discern any general rules which may account for differences in the ratio of lens:corneal refractive powers.

Descriptions of the complete schematic eyes are restricted to only three avian species (Fig. 7.5), although limited optical analyses have been conducted in three other species (Sivak, 1977; Sivak *et al.*, 1977). This lack of data may be partly explained by the need to conduct such investigations on live or freshly-dead material. Histological procedures are not suitable for such analyses (see Martin, 1983, for an outline of procedures), and hence the many illustrations of hemisected eyes which appear in, for example, Slonaker (1897) and Rochon-Duvigneaud (1943), are not suitable for detailed analyses of refractive properties, although they do indicate broad differences in the refractive components of the eyes.

Histologically the avian cornea is considered to be similar to that of other vertebrates. Within it can be distinguished the anterior epithelium (*epithelium anterius corneale*), the anterior limiting membrane (*lamina limitans anterior,* formerly Bowman's membrane), the substantia propria (*substantia propria cornealis*), the posterior limiting membrane (*lamina limitans posterior,* formerly Descemet's membrane) and the posterior epithelium (*epithelium posterius corneale*). It appears to be of uniform thickness ($\simeq$0.4 mm) with a refractive index relative to air in *Columba* of 1·337 (Gundlach *et al.*, 1945) and 1·369 in the Double-crested Cormorant (*Phalacrocorax auritus*) and Brown Pelican (*Pelicanus occidentalis*) (Sivak *et al.,* 1977). These values are within the range (1·337–1·39) found throughout the vertebrates (Sivak, 1978). The spectral transmission of the cornea

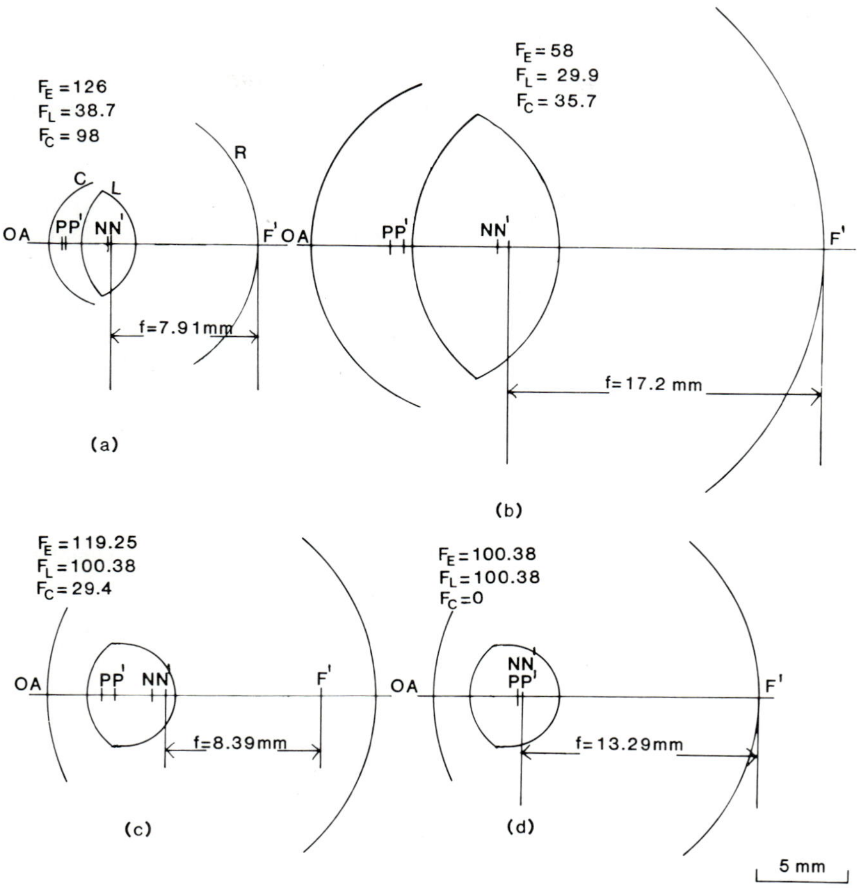

Fig. 7.5. Scaled drawings of the schematic (unaccommodated) eyes of (a) the pigeon, (b) the Tawny Owl (*Strix aluco*), (c) the Humboldt Penguin (*Spheniscus humboldti*) in air, and (d) the Humboldt Penguin in water. C, Cornea; L, lens; OA, optic axis; R, retina; P and P′, anterior and posterior principal points; N and N′, anterior and posterior nodal points; F′, second focal point; F_E, total refractive power of the eye; F_L, refractive power of the lens; F_C, refractive power of cornea (all powers are in dioptres, D); *f*, anterior focal length or posterior nodal distance.

of the pigeon extends into the near-ultraviolet region (Govardovskii and Zueva, 1977; Emmerton *et al.*, 1980).

The anterior chamber (*camera anterior bulbi*) between the cornea and lens (Fig. 7.3) contains the aqueous humour (*humor aquosus*), which is typically assigned a refractive index of 1·334. Thus, the relative refractive index between the cornea and the aqueous is very low, and hence the posterior surface (*facies posterior*) of the cornea has little refractive power compared with its anterior surface (*facies anterior*). Abelsdorff and Wesseley (1909) and Barany *et al.* (1957) reported that the anterior chamber of the Tawny Owl (*Strix aluco*) contains a highly-viscous, mucinous substance. This substance is relatively thick close to the cornea, and Barany *et al.* (1957) hypothesised that its presence reduced the metabolic load (lower turnover of aqueous humour) of the anterior chamber, which in this species has a relatively large volume. Whether such a substance occurs in the aqueous of other avian species with either large or small anterior chamber volumes is not known.

The avian lens may be divided into two principal components (Figs. 7.3, 6). The construction of the central body (*corpus centrale lentis*) is basically that of all vertebrate lenses, consisting of concentric layers of fibres (*fibrae lentis*) which decrease in their density from a central core and thus produce a gradient of refractive index. This central part is surrounded by the annular pad (*pulvinus annularis lentis,* sometimes referred to as the "Ringwülst"), a feature found only in the avian and reptilian lens (Walls, 1942, p. 276). In the annular pad the fibres are enlarged relative to those in the central body and are arranged radially rather than concentrically. The whole lens, which is typically very soft and pliable, is surrounded by the capsule (*capsula lentis*). Frequently between the body and the annular pad there is a vesicular space (*vesicula lentis*), which according to Duke-Elder (1958, p. 410) contains an albuminous fluid (*aqua vesiculae lentis*).

The function of the annular pad is disputed. Thus, Walls (1942, p. 277) argues that it is entirely without a direct optical role, since it is always shielded by the iris

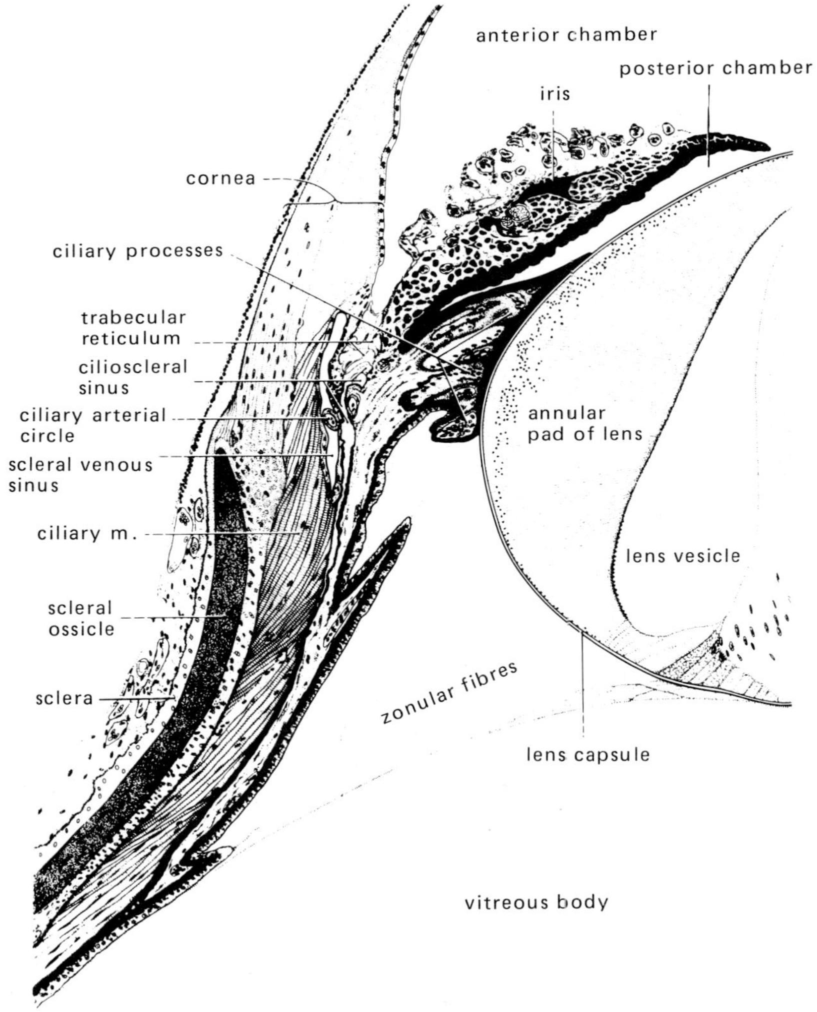

Fig. 7.6. Part of a meridional section of the eye of the Inca Dove (*Scardafella inca*) through the corneoscleral junction, ciliary region, iris and lens. From Evans (1979), courtesy of Academic Press, London and New York.

(although it does not seem improbable that light could enter it obliquely) but serves the optical system as part of the accommodative mechanism (see Section II,E). Slonaker (1918), however, argued that the pad had only a nutritive function, its epithelium secreting material into the vesicular space for subsequent use by the lens.

The relative size of the annular pad and central body of the lens appears to vary markedly between species (Rochon-Duvigneaud, 1943; Duke-Elder, 1958, p. 409), but it is not clear with which optical, metabolic or ecological feature this is correlated. Walls (1942) proposed that the size of the annular pad was related to the accommodative range of the lens. The central body of the lens is known to contain glycogen. However, while Rabaey (1963) proposed that this deposit had mainly an optical function in serving to increase the refractive index of the lens, Kaskel *et al.* (1975) (arguing from a broader comparative context, since they also found glycogen in the mammalian lens) proposed a primary metabolic role for it.

Marshall *et al.* (1973) presented some experimental evidence which indicated the presence of a refractive index gradient in the lens of the pigeon, but this has not been measured in detail in any avian species. Such a gradient is, however, commonly assumed to be present from the results of comparisons of the structural and optical properties of mammalian lenses (e.g. Nakao *et al.*, 1968; Philipson, 1969; Campbell and Hughes, 1981). The equivalent homogeneous (bulk) refractive index of the schematic eye models in birds (1·408 in the pigeon, Marshall *et al.*, 1973; 1·470 in *Strix aluco,* Martin, 1982; 1·5078 in *Spheniscus humboldti,* Martin and Yang, 1984) is similar to that found in other vertebrate eyes (see Table I of Hughes, 1977, for a summary of such data in species other than birds).

Both the size and shape of the avian lens show marked inter-specific variability, and this presumably is reflected in differences in the refractive power, accommodative range and visual fields of the eye. However, in the absence of sufficient detailed analyses of optical properties of individual eyes it would be premature to attempt to discern general functional properties of these lens types. Caution is necessary where lens shapes are described from an examination of fixed material. Also, the vertebrate lens is known to exhibit considerable intra-specific variation (Howcroft and Parker, 1977). Thus, while many examples of interesting and/or unusual shapes of avian lens have been described (Rochon-Duvigneaud, 1943; Duke-Elder, 1958), the functional explanations of these must await more detailed analyses of optical properties.

D. Iris and pupil

1. Iris

Pupil aperture is controlled by the *iris,* which in birds is highly mobile. The iris is very thin at its ciliary border (*margo ciliaris*), where it is attached to the ciliary body (*corpus ciliare*) at the iridocorneal angle (*angulus iridocornealis*) (Fig. 7.6). It thickens towards its midpoint, and thins again towards the pupillary border (*margo pupillaris*). The pupillary border is in contact with the anterior surface (*facies anterior*) of the lens. The margin of the pupil is folded (*plicae iridicae*). The anterior surface of the iris is lined by a single layer of flattened epithelial cells (*epithelium anterius iridis*). The posterior surface is lined by several layers of pigmented epithelial cells (*stratum pigmentum iridis*). Between the epithelial layers are connective tissue stroma (*stroma iridicum*) and two layers of muscle. The muscles of the iris, including the sphincter muscle of the pupil (*m. sphincter pupillae*) and the dilator muscle of the pupil (*m. dilator pupillae*), are generally considered to be striated. It is these striated muscles which are believed to be responsible for the rapidity of the pupillary response. Walls (1942, p. 647) reports the presence in a number of avian species of a second dilator system in which the muscle fibres are unstriated, and this has also been described by Oehme (1969) and Pilar and Vaughan (1971). These muscles of the iris are highly vascularised, and some inter-specific differences in the extent of the capillaries and veins have been reported (Mann, 1929; Duke-Elder, 1958). The blood supply to the iris is formed by the posterior ciliary artery; this divides at the margin of the iris to form an arterial circle (*circulus arteriosus iridicus*) which passes circumferentially within the iris.

The anterior surface (*facies anterior*) of the iris is typically heavily pigmented; both lipochromes and guanine are observed to be present, although the latter seems to be rare (Duke-Elder, 1958). Iris colour has been reviewed by Mann (1931), who described a wide range of colours including browns, yellows, blues, greens and reds. She also documented both sexual and seasonal variations in iris colour, as well as dietary influences. It seems likely that the specific coloration serves no optical function (although the presence of any pigment cells will improve the opacity of the iris), but instead acts as part of the animal's display plumage (with which it often strikingly contrasts) or contributes towards the cryptic coloration of the bird.

In columbiform species the iris possesses a collection of reflective cells known as iridocytes (*iridocyti*) forming a tapetum lucidum (*tapetum lucidum iridicum*) (Chiasson and Ferris, 1968). In the superficial layer of the iris the cells are scattered and have large, reflecting platelets. In the deep layer the cells form a relatively discrete iridocyte body (*corpus iridocytorum*) and have smaller platelets.

2. *Pupil*

There are four principal functions which a pupil (*pupilla*) of varying aperture may perform within the avian eye. Thus, it can (1) prepare the eye for dark adaptation (Woodhouse and Campbell, 1975), (2) protect the eye from photic and thermal damage (Marshall *et al.*, 1973), (3) improve image quality (Miller, 1979), and (4) function as an accommodative mechanism (Levy and Sivak, 1980). The latter will be dealt with in Section II,E,2,c.

As in similar optical systems (e.g. a camera lens), the avian pupil aperture is situated between the refractive components of the system (Fig. 7.3) rather than external to it, and hence it is necessary to distinguish between the real pupil and the entrance and exit pupils. The latter two are the pupils as seen when magnified by the eye's optic elements. The entrance pupil (the real pupil magnified by the cornea) is the most important, since it determines retinal illumination. It is the pupil seen when looking into the eye. The size of the entrance pupil relative to the real pupil will vary between eyes depending on the refractive power of their corneas.

(a) Pupil shape. The avian pupil is generally spherical, but there are reports of slightly ellipsoid pupils (Walls, 1942), and Wetmore (1919) described the eye of the Black Skimmer (*Rhynchops niger*) as having a slit pupil. Slit pupils are found in many vertebrate groups (see Table 6 of Walls, 1942); they may function to protect the retina of nocturnal or crepuscular forms when basking in bright sunlight, the differences in the musculature of the iris possibly permitting a slit pupil to be closed to a smaller aperture than a circular pupil. A slit pupil may also serve as a stenopaic aperture, improving (again because of its very small size) the depth of focus of the eye (Martin, 1983). Such a role for a slit pupil has been proposed in the Bearded Seal (*Phoca barbata*), among other mammalian species (Suthers, 1978). The Black Skimmer (*Rhynchops niger*) is reported to feed primarily during crepuscular or nocturnal conditions and spends the day on sand banks under bright sunlight (Fry, 1978). The possession of a slit pupil in this species therefore may be in accord with the hypothesis of protection from photic damage. However, data on acuity and depth of focus in the eye are not available. It is an intriguing problem as to why the three species of skimmer which comprise the Rhynchopidae are apparently the only birds which have a slit pupil.

(b) Pupil size and retinal illumination. The brightness of the retinal image of an extended source of light is inversely related to the square of the *f*-number of the eye (*f*-number = focal length *f*/pupil aperture *A*). The smaller the *f*-number the relatively brighter the image. Comparison of minimum *f*-number (maximum pupil diameter) in the pigeon and the Tawny Owl (*Strix aluco*) as representative avian species of diurnal and nocturnal–crepuscular habit show that maximum image brightness in these two species is very similar, and indeed varies little from the maximum image brightness in humans and other mammals (Martin, 1983). This finding is, however, in line with a more extensive survey of image brightness amongst vertebrate eyes (see Fig. 10 of Hughes, 1977), which demonstrated that across a wide range of species (including forms considered nocturnal and diurnal in habit), maximum image brightness varies by little more than sixfold. Thus, large pupil diameter and low *f*-number may perhaps not be considered as the primary adaptations of the nocturnal avian eye, as has previously been supposed (Walls, 1942) (see Section II,A).

The difference between maximum and minimum pupil size in vertebrate eyes with circular pupils produces a relatively small change in image brightness [$\simeq$16-fold (1.2 $\log_{10}$ units) in humans (Reeves, 1918), pigeon (Marshall *et al.*, 1973) and the penguin, *Spheniscus humboldti* (Martin and Young, 1984)] compared with the total range of luminance levels to which an eye can be exposed in the natural environment ($\simeq 7 \times 10^{11}$-fold; 11.8 $\log_{10}$ units) (Fig. 7.2), and it is thus clearly impossible for the pupil aperture to function to equalise image brightness at all natural luminances. However, within a particular habitat natural luminance levels change slowly, and even when clouds quickly cover the sun, luminance levels change only by a maximum of 10-fold (Fig. 7.2); moreover, when moving from an open habitat to below a tree canopy the maximum change in luminance level will only occasionally be as high as 100-fold (Martin, 1982). Thus, it is likely that pupil variation is sufficient to equalise retinal illuminance approximately, as ambient luminance levels change at their most rapid natural rate, thus obviating the need for retinal dark adaptation under these conditions.

Not at variance with these observations is the proposal that a variable pupil aperture may function to keep the eye, even under bright-light conditions, in a partially dark-adapted state (Woodhouse and Campbell, 1975). This may be of adaptive value, since the retinal mechanisms of dark adaptation are relatively slow, and a rapid diminution of light levels results in a decrease of visual sensitivity for a number of minutes if the pupil is not constricted before the luminance change.

(c) Pupil size and image quality. The pupil of any optical system is often employed to improve the quality of the image. This is achieved by what has been termed "optical surgery" (Weale, 1974), that is, simply removing rays which would pass peripherally through the optical system, since it is usually peripheral rays which are primarily responsible for the various aberrations to which all optical systems are prone. This function of the pupil in a vertebrate eye has been demonstrated convincingly in humans, where it has been shown that the quality of optical image is high as long as the pupil remains small. Above a pupil diameter of 2·4 mm, image quality in the human eye begins to deteriorate because increasingly-aberrant rays now form part of the image (Campbell and Gubisch, 1966). However, it is theoretically well understood (Miller, 1979) that for a perfect optical system the larger the pupil aperture the finer the detail which can be resolved in the image which it produces. Thus, different optical systems may become limited in their optical performance at different pupil sizes, and in the case of one bird, the Congo Snake Eagle (*Dryotriorchis spectabilis*), evidence has been presented (Shlaer, 1972) indicating that optical performance is superior to that of the human eye and that image quality is limited at a pupil diameter of 6·25 mm. The data of Miller (1979) on pupil size under bright-light conditions in eight species of diurnal birds-of-prey (Falconiformes) also indicate that image quality in all of these species becomes limited at a larger pupil diameter than in humans. In the human eye, steady-state pupil aperture does increase with falling light levels (Reeves, 1920), and this is probably related to the quality of the eye's optics. As luminance levels fall, acuity decreases, and so the poorer-quality image which is concomitant with a larger pupil aperture can be tolerated. Indeed, Denton (1956) and Campbell and Gregory (1960) have argued (based on work in the gekko and human eye, respectively) in favour of such an hypothesis, and it is possible that a similar relationship also applies for the same reasons in the avian eye.

It is clear from the previous sections that the function of the pupil, and the factors which may determine its size under a given set of conditions in any one species, are complex. It is of key importance in determining the quality of the retinal image as well as determining its brightness. This complexity is further increased in the avian eye by evidence which also shows the pupil to have a key role in accommodation, at least in certain species (see Section II,E).

E. Accommodation

An accommodative mechanism is essential for the adequate functioning of any vertebrate eye. Such a mechanism allows the eye to bring successively into clear focus objects at varying distances. Only if the eye is very small or the pupil can be reduced to the size of a pin-hole can objects at a wide range of distances from the eye be considered to be simultaneously in focus, and this is only achieved at the expense of resolution and sensitivity. The mechanism of accommodation in the avian eye has been the subject of some controversy, most recently reviewed by Meyer (1977). Both the lens and the cornea have been implicated, much of the debate centering on the actual mechanism involved in lenticular accommodation.

1. Accommodative range

Much of the data on the accommodative range appear to be conflicting. In pigeons, for example, it has been estimated at 8 to 12 D (Hess, 1910), 5 D (Levy and Sivak, 1980) and 17 D (Gundlach *et al.*, 1945). In owls (Strigiformes) the accommodative range is reported by Walls (1942), Rochon-Duvigneaud (1943) and Duke-Elder (1958) to be between 2 and 4 D, whilst a study by Murphy and Howland (1983) in 15 owl species found that it ranged from 0·6 to more than 10 D. Pumphrey (1948a) believed that the accommodative range of most birds is about 20 D. Accommodation of up to 50 D has been reported to occur in a number of aquatic species, such as the Great Cormorant (*Phalacrocorax carbo*) (Hess, 1910), the American Dipper (*Cinclus mexicanus*) (Goodge, 1960), the Hooded Merganser (*Mergus cucullatus*) and Redhead (*Aythya americana*) (Levy and Sivak, 1980). In these species it is argued that such an accommodative range is necessary to overcome the refractive error which would result in an aerial emmetropic eye when the refractive power of the cornea is negated by entry into water.

2. Accommodative mechanisms

(a) Musculature. All authors agree that accommodation in birds is mediated by the striated ciliary muscle (*musculus ciliaris*) (Fig. 7.6). However, there is conflicting evidence about how many sub-divisions of the muscle there are and about its mode of action. Whilst it is likely that some of the disagreement lies in genuine inter-specific differences, some may result from differences in anatomical technique (Meyer, 1977). The fibres of the ciliary muscle run meridionally and are attached to the inner surface of the sclera close to the posterior portion of the scleral ossicles. From the cartilage the muscle extends anteriorly across the width of the scleral ring to the corneoscleral junction, where it ends in close association with the venous sinus of the sclera (*sinus venosus sclerae*).

The ciliary muscle is described as consisting of from one to three separate muscles. The anterior portion, the anterior corneal muscle (*m. cornealis anterior,* formerly Crampton's muscle), originates on the anterior end of the scleral ring and inserts on the inner surface of the cornea at its margin. The posterior portion, the posterior corneal muscle (*m. cornealis posterior,* formerly Brücke's muscle), originates on the inner surface of the thin scleral sheet which forms the anchorage of the trabecular meshwork (*reticulum trabeculare*) connecting the base of the iris to the cornea, and inserts into the sclera opposite the posterior part of the ciliary body. A third group of muscle bundles, the radial fibres (*fibrae radiales,* formerly Müller's muscle), has been identified by some authorities. The muscle is sometimes described as a subdivision of either the anterior or the posterior muscle, but according to Rochon-Duvigneaud (1943) it consists of fibres which pass between the two muscles and should not be designated as a separate muscle. A fourth ciliary muscle was described by Lord (1956) in the Red-tailed Hawk (*Buteo jamaicensis*) but has not been recorded in any other species. The ciliary muscles attach to the lens capsule via the ciliary processes (*processus ciliares*), which are extensive and often fused. Marked inter-specific variations in the number and size of the muscles of the ciliary region have been recorded, and attempts have been made to correlate these with assumed differences in accommodative abilities. However, it seems that in no avian species has a thorough analysis of its ciliary muscles and powers of accommodation been conducted. On anatomical grounds the ciliary muscles could produce changes in the curvature of both the lens and the cornea, the latter being unique to the avian eye.

(b) Corneal accommodation. The anterior muscle, because of its insertion on the margin of the inner surface of the cornea, has been proposed as the principal muscle involved in corneal accommodation. Contraction of the muscle, it is argued, simply pulls the corneal margin inwards, thus producing an increase in the curvature of the cornea. The idea of corneal accommodation was first studied by Beer (1893) and Hess (1910, 1912) and subsequently investigated in the pigeon by Gundlach *et al.* (1945) and in the pigeon and Mallard (*Anas platyrhynchos*) by Levy and Sivak (1980). Using various drugs to induce contraction or relaxation of the ciliary muscles, Gundlach *et al.* (1945) recorded changes in corneal curvature amounting to a maximum of 17 D of accommodation. Both decreased and increased corneal curvatures were observed. Nicotine was reported as producing a consistent increase in corneal curvature equivalent to approximately 5 D of accommodation. However, Levy and Sivak (1980) replicated this study using nicotine sulphate, and although a mean change in accommodation of 5 D was recorded retinoscopically, it was not possible to detect any change in corneal curvature. They concluded therefore that accommodation in the pigeon and Mallard is entirely lenticular with no corneal component. Whether this finding applies to all avian species is not known. It is a problem which certainly warrants further investigation.

(c) Lenticular accommodation. Whilst there seems to be little dispute that lenticular accommodation occurs in birds, there is debate concerning the way in which the changes in lens curvature are brought about. The avian lens is always in firm contact with both the ciliary body and iris, each of which contains striated muscle. According to Meyer (1977), lenticular accommodation is achieved by an active squeezing of the lens, the posterior corneal muscle being the principal muscle involved. When this muscle contracts the ciliary body is thrust inwards and pushes the annular pad of the lens evenly along the extent of its equator. According to Walls (1942), the fibres of the annular pad act as columns which transmit the stress of the ciliary body directly to the central portion of the lens. Although inter-specific differences in the development of the annular pad have been correlated with presumed accommodatory abilities, no precise study of this relationship has been carried out (also see Section II,C). Forward movement of the lens during the deformation is prevented by the fibres (*fibrae zonulares*) of the ciliary zonule (*zonula ciliaris*) and the trabecular meshwork in the iridocorneal angle (Fig. 7.6). Slonaker (1918), however, proposed that in at least one species, the House Sparrow (*Passer domesticus*), deformation of the lens does not take place but instead the lens is moved forward as in the accommodative mechanism of fish.

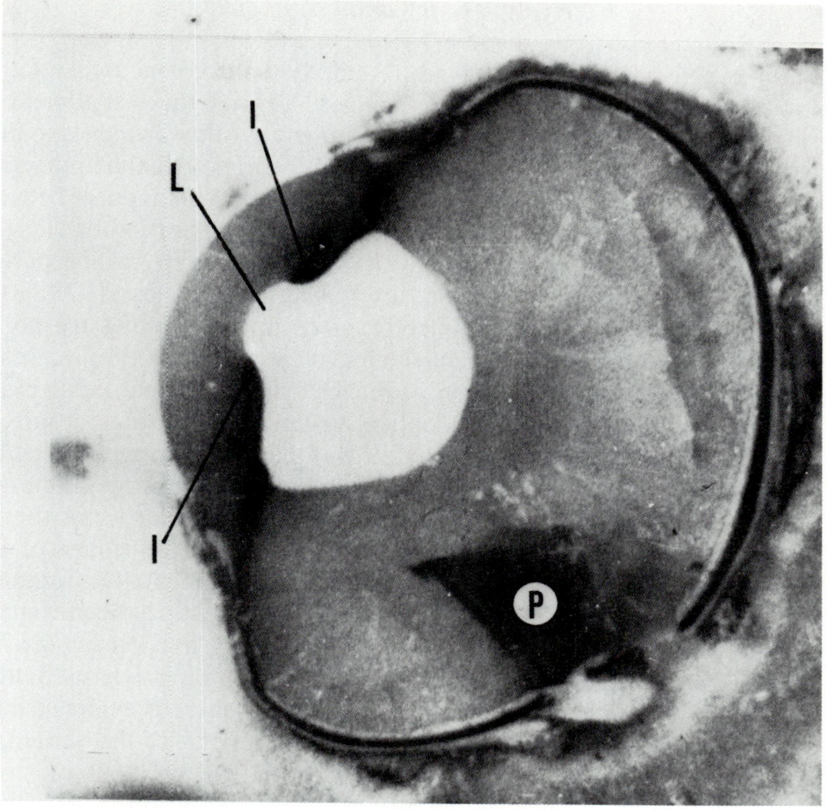

Fig. 7.7. Sagittal section of the accommodated eye of the Hooded Merganser (*Mergus cucullatus*). Nicotine sulphate was employed to stimulate accommodation and the eye frozen immediately after death. The lens (L) can be seen protruding through the pupil bounded by the iris (I). P, Pecten. From Levy and Sivak (1980), courtesy of Springer-Verlag, Heidelberg.

Hess (1910) proposed that in some birds in which accommodation is of a high magnitude (e.g. the Double-crested Cormorant, *Phalacrocorax auritus*), lenticular accommodation by the ciliary body is assisted by a further squeezing of the lens by the iris. Apparent confirmation of this squeezing of the lens and the resultant large increase in the curvature of its anterior surface has been demonstrated by Levy and Sivak (1980) in the Hooded Merganser (*Mergus cucullatus*) and the Redhead (*Aythya americana*). In these species the application of nicotine sulphate resulted in the eye accommodating by approximately 50 D, and examination of enucleated eyes frozen in the accommodated state showed that the central portion of the lens protruded through the pupil in some individuals (Fig. 7.7). They proposed that contraction of the iris sphincter forms a rigid disk onto which the ciliary muscle pushes the malleable lens and thus produces a highly-curved protuberance through the pupil. Clearly the curvature of the lens surface deformed in this manner will depend on the pressure applied to the lens and on the pupil diameter. Thus, under certain conditions pupil diameter in these species may be governed by accommodatory considerations in addition to those described in Section II,D. Whether deformation of the lens in this manner occurs naturally in these and other avian species, especially the non-aquatic forms, is not known.

F. Eye geometry and visual fields

The term eye geometry is used here to refer to both the geometrical relationship between the cornea and lens which principally defines the visual field of the individual eye (the monocular field), and the positions of the two eyes in the skull which determine how the monocular fields combine to produce the binocular and cyclopean (or panoramic) visual fields. It is necessary to distinguish between the optical and retinal visual fields. The former refers to the portion of visual space which is actually imaged by the eye, whilst the latter refers to that part of the visual field which is actually occupied by the retina (Hughes, 1977). It is the retinal field which is of particular interest, since this determines the portion of space in which visual control of behaviour may be possible. In all cases so far examined the retinal field is noticeably smaller than the optical field, indicating that full use is not made of the eye's optical system. The reasons for this are not clear but may result from a decrease in quality of the optical image formed by rays passing obliquely through the system. Such rays may pass through the annular pad, which, because of the different arrangement of lenticular fibres there compared with the lens core (see Section II,C), may produce an image of inferior quality. This, however, has not been demonstrated experimentally. Certainly, the portion of the optical field that is not served by the retina has been found to be larger on the nasal than on the temporal side (Martin, 1984; Martin and Young, 1984). This may reflect the need for better optical quality in the nasal field which subserves frontal vision (Pettigrew, 1979b). However, there is evidence that poor image quality in the periphery is not a necessary property of the vertebrate eye (Hughes and Vaney, 1978; Murphy and Howland, 1983).

The eyes of all birds are placed laterally within the skull. Figure 7.1 illustrates the typical relationships. No avian species has yet been described in which the bulbar axes face forwards, parallel to each other. Figure 7.8 shows that even in those species frequently described as having frontally-directed eyes, such as owls (Strigiformes), the optic axes may diverge by 55°. Detailed analyses of retinal visual fields seem to have been conducted in only three avian species (Fig. 7.8). These species show marked differences in all visual field parameters and pose intriguing problems for functional analyses. It has been shown that retinal field boundaries are defined by, or coincide with, external structures such as the bill (pigeons and penguins) and, in the case of the Tawny Owl (*Strix aluco*), with the feathering of the facial disk (Martin, 1984).

Despite extensive investigations of both the neurophysiology (e.g. Pettigrew, 1979b) and psychophysics (e.g. Blake and Fox, 1973) of binocular vision, the functional advantages of binocularity are disputed (Fox, 1979; Jones and Lee, 1981), and to date there is no clear understanding of the factors which may determine the extent of eye frontality and binocularity amongst vertebrates in general and birds in particular. Certainly the functional advantages of stereoscopic vision are in dispute, and the suggestion by Walls (1942) that there is a universal urge towards binocularity in order to acquire the benefits of stereopsis is in doubt. Hughes (1977) proposed that "the relative widths of the panoramic and binocular fields is determined by the interaction of separate pressures which are not universal in intensity but predominate only in certain life styles". Indeed he proposed that the animal's "praxic space", in which visually-controlled activities

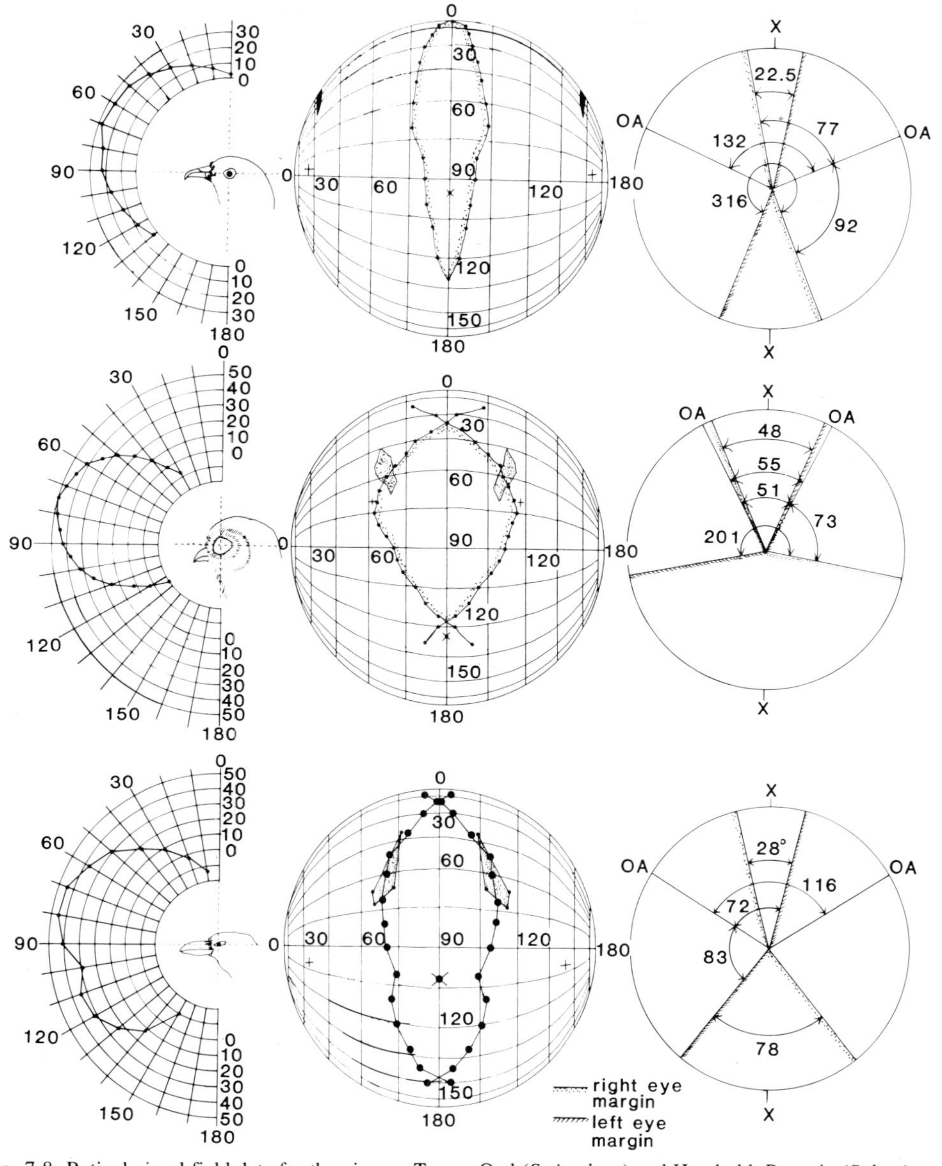

Fig. 7.8. Retinal visual field data for the pigeon, Tawny Owl (*Strix aluco*) and Humboldt Penguin (*Spheniscus humboldti*) (in air). In the left-hand column the mean width of the binocular field is shown as a function of elevation with the bird's head depicted in the correct orientation. The central column shows the position of the binocular field boundary presented in the form of an equatorial zenithal orthographic projection. In each projection the bird is facing the reader with the line of the bill pointing perpendicularly from the plane of the page at the centre of the projection. X, The projection of the bill tip; +, the projection of the optic axes. The projections of the pecten are shown shaded. The right-hand column shows diagrammatic presentations of the visual fields in the plane of the optic axis. The points X are in the bird's median sagittal plane. OA, Optic axis. All numerical values are in degrees.

involved in directed locomotion, prey capture, nest building and other manipulations of the environment take place, is the principal determinant of binocular field size. This supposes that visual fields are determined more by such factors as the possession of prehensile limbs and the diet of the animal, than by the consideration of fundamental problems of vision *per se,* from which general predictive relationships may be discerned.

It should be noted that an adequate functional analysis of visual field parameters would seem to require not only data on the actual retinal field sizes, but also data on the head position adopted by birds engaged in various tasks, as well as some cognizance of the extent and use of eye movements. All visual field data published to date refer to the static visual fields determined when the eyes are fixated on a target or when the animals are sedated. In both cases the eyes are regarded as having adopted their "primary" or resting position.

G. Eye movements

Figure 7.1 shows that the eyes of birds fit tightly within the skull. Compared with mammals, in no species of bird can the orbit be regarded as roomy. The result of this tight fit is that eye movements tend to be of smaller amplitude than in mammals and that the eye muscles are reduced to thin ribbons. Slonaker (1918) and Chard and Gundlach (1938) described in some detail the bulbar muscles (*musculi bulbi*) of the House Sparrow (*Passer domesticus*) and pigeon, respectively. Figure 7.9 shows the typical arrangement of the six bulbar muscles which include three opposing pairs, the dorsal and ventral straight muscles (*m. rectus dorsalis, m. rectus ventralis*), the medial and lateral straight muscles (*m. rectus medialis, m. rectus lateralis*), and the dorsal and ventral oblique muscles (*m. obliquus dorsalis, m. obliquus ventralis*).

The dorsal and ventral oblique muscles originate close to each other on the nasal side of the orbit and insert on the dorsal and ventral parts of the sclera. The pull of the dorsal oblique muscle is direct and, in contrast to mammals, is not transmitted through a pulley arrangement. The straight muscles originate from the caudal part of the orbit and insert on the anterior part of the sclera (medial straight muscle), the posterior part of the sclera (lateral straight muscle), the dorsal part of the sclera close to the dorsal oblique muscle (dorsal straight muscle) and the ventral part of the sclera close to the ventral oblique muscle (ventral straight muscle). In contrast to many reptiles and mammals, there is no retractor bulbi muscle.

The eye movements produced by these muscles in the restrained but alert pigeon have been categorised as being of four types (Nye, 1969), as follows:

1. *Impulses and oscillations.* These occur in short bursts at frequencies of about 30 Hz with an amplitude of 5° in the vertical direction and 3° horizontally. The oscillations last about 250 msec and decrease in amplitude. This movement is normally executed during an eye blink, and Nye (1969) suggested that it performs a polishing action on the cornea in conjunction with the nictitating membrane (see Section II,H).

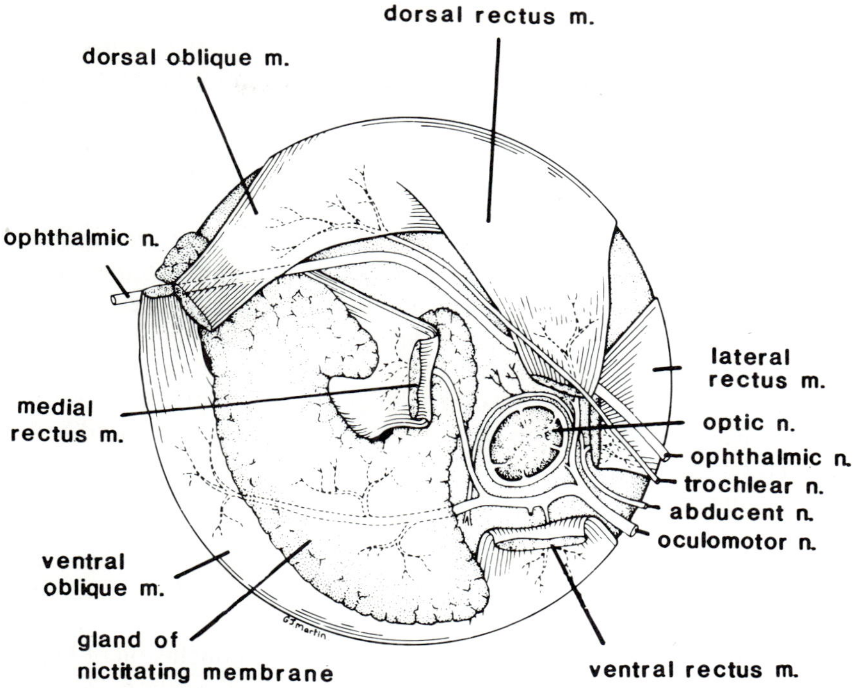

Fig. 7.9. Posterior view of the right eye of the House Sparrow (*Passer domesticus*) showing the four rectus muscles and the two oblique muscles, and their nerve supply. From Bubień-Waluszewska (1981), courtesy of Academic Press, London and New York.

2. *Drifts*. These are slow (1–5 deg/sec), with a total excursion of approximately 3–5 degrees; they are probably the eye movements exhibited when the animal is visually tracking a target.

3. *Flicks*. Nye (1969) found that these occurred rarely, with a maximum amplitude of between 3 and 5 degrees and a duration of about 20 msec. He interpreted them as occurring in response to sudden and unexpected motion of visual stimuli.

4. *Tremors*. These are of a very low amplitude (<12 seconds of arc) and are unlikely to serve any visual function.

Eye movements in the pigeon larger than those demonstrated by Nye (1969) have been reported by several workers. Thus, Goodson (1969) is quoted by Donovan (1978) to have recorded eye movements occurring in an arc of about 15°, while Bloch *et al.* (1981) reported that "saccadic" eye movements (probably equivalent to the "flicks" of Nye, 1969) could attain about 15° from the resting position.

The eye movements of only one other avian species seem to have been reported in detail. Following the frequently-cited statement by Walls (1942, p. 309) that the owl eye is completely immobile and that "it cannot be turned in the orbit even with a pair of pliers", Steinbach and Money (1973) and Steinbach *et al.* (1974) have shown that in the Great Horned Owl (*Bubo virginianus*), all the types of monocular eye movements found in other vertebrates are in fact present, but with a reduced amplitude. Thus, by tilting the head to varying degrees (up to a maximum of 90°) they showed that countertorsion eye movements to a maximum of 2·5 to 3·5° occur. Oscillations (polishing movements) of the kind reported in the pigeon were observed up to a maximum of 1·5°.

Unfortunately not all the functions of the different types of eye movement to be found amongst vertebrates are well understood. Ditchburn *et al.* (1959) have shown that the rapid, small-amplitude, saccadic deflections of the kind described before as occurring in the pigeon and the Great Horned Owl (*Bubo virginianus*) are probably essential for the maintenance of pattern vision in the retina. However, the function of the wide variety of pursuit movements to be found amongst vertebrates is not understood. Indeed Walls (1962) posed the question "Why have eye movements?" Framed within an evolutionary context his discussion led to the conclusion that the origin of eye movements lies in the need to keep an object fixed on the retina, not in the need to scan the surrounding environment. Since birds generally have a relatively light head with low inertia, and a highly-flexible neck, the need for eye movements to fixate moving objects precisely is much reduced compared with mammals. In birds it may be possible to move the whole head with comparable precision and speed to that which can be achieved by the movement of eyes of the same size in a mammal.

Nevertheless, it would seem ill-advised to accept the conclusion of Duke-Elder (1958) that eye movements in birds are either negligible or absent. Certainly the way in which eye movements may, for example, alter the extent of the binocular fields during the location of food by granivorous birds would seem worthy of investigation.

H. Nictitating membrane, lacrimal apparatus and eyelids

The structures discussed here are most appropriately considered as part of the optical system of the eye, since their primary function would seem to lie in the maintenance of image quality via simple protection and/or cleaning. A direct optical function, however, has also been suggested for the nictitating membrane.

1. Nictitating membrane

The nictitating membrane (*membrana nictitans*) is found in most amphibians, reptiles and birds. Detailed descriptions of it in birds are available for the House Sparrow (*Passer domesticus*) (Slonaker, 1918), the pigeon (Chard and Gundlach, 1938) and *Gallus* (Simić and Jablan-Plantic, 1959). Although the membrane is usually transparent or slightly opaque, in owls (Strigiformes) and dippers (Cinclidae) it is opaque and milky white. In all birds the membrane is elastic and forms a pouch in which the eye sits (Fig. 7.10). When retracted the membrane is barely visible on the nasal side of the orbit. Two striated muscles are responsible for its movements transversely across the front of the eye, the pyramidalis and quadratus muscles (*m. pyramidalis membranae nictitantis, m. quadratus membranae nictitantis*) (Fig. 7.10). The quadrate muscle at its insertion at the back of the eyeball

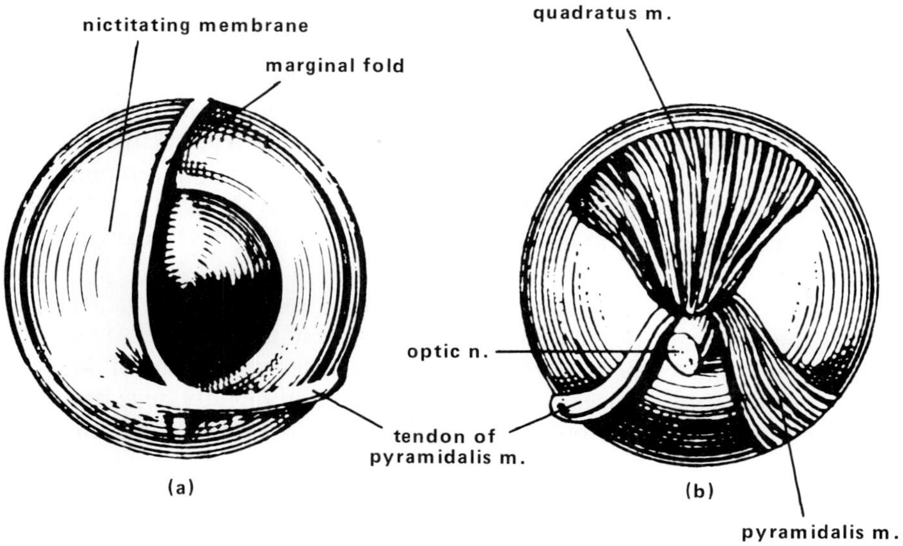

Fig. 7.10. The nictitating membrane of *Meleagris*. (a) Anterior view; (b) posterior view. From Duke-Elder, Sir Stewart, editor: System of ophthalmology, Vol. I. The eye in evolution (by Duke-Elder, Sir Stewart), St. Louis, 1958, the C. V. Mosby Co.; Bland-Sutton.

forms a fibrous sheath (*vagina fibrosa tendinis*) through which the tendon of the pyramidalis muscle (*tendo m. pyramidalis*) passes mediolaterally to insert on the ventral part of the free edge of the membrane. The dorsal part of the free edge of the membrane is fixed to the eyeball, however, and therefore the membrane moves like a pendulum as it passes backwards and forwards across the eye.

Although the nictitating membrane has no fibrous or cartilaginous structure, the free margin which sweeps across the cornea is composed of a relatively stiff band of connective tissue described by Slonaker (1918) and Chard and Gundlach (1938) as a plait-like structure, the marginal fold (*plica marginalis*). On its anterior surface the marginal fold is raised into a scoop- or hook-like projection, resembling in cross-section the barb of a fish hook. The fold in some strigiform species is pigmented brown. The posterior surface of the fold is smooth and fits closely over the corneal surface. As the membrane passes down over the anterior surface of the eye, the scoop-like projection is somewhat flattened and sweeps readily beneath the lids, which often close during the movement of the membrane. However, as the membrane is drawn back the tight contact with the surface of the cornea is lost and the scoop-like fold at the margin rubs against the lining of the lids. Its role in drawing fluids from the conjunctival space is discussed next.

2. *Lacrimal apparatus*

Two glands constitute the avian lacrimal apparatus (*apparatus lacrimalis*), the gland of the nictitating membrane (*glandula membranae nictitantis,* formerly Harder's gland) (Fig. 7.9) and the lacrimal gland (*glandula lacrimalis*). The movements of the nictitating membrane are both rapid and frequent and are lubricated by secretion from the gland of the membrane. This secretes directly into the space between the eyeball and the membrane. While the simple lubricating function of the secretions of this gland are not in doubt, its structure and function have received attention in recent years because of evidence that this gland is a source of antibody-producing cells. The relatively large size and accessibility of the gland of the membrane in birds has rendered it relatively easy to study. For details of the ultrastructure of the gland and its role in immunity, see for example the work of Burns (1975, 1976), Burns and Maxwell (1979), Maxwell and Burns (1979), Schramm (1980), Gupta and Maiti (1981) and Chapter 6, by Rose, in Vol. 2.

The gland of the nictitating membrane is large and irregular in shape and extends on the nasal side of the orbit from the region of the optic nerve anteriorly and beneath the insertions of the oculomotor muscles. The duct of the gland (*ductus gl. membranae nictitantis*) empties its secretion directly beneath the nictitating membrane in the dorsotemporal quadrant. Thus, as the nictitating membrane sweeps over the eye it is kept well supplied with secretion. The thin edge of the membrane polishes the cornea and removes any excess fluid from its surface. The scoop-like projection of the anterior surface of the marginal fold collects on its return movement any excess fluid which has accumulated between it and the inner surface of the eyelid. This fluid drains through the ostia of the lacrimal canaliculi (*ostia canaliculorum lacrimalium*) at the nasal commissure of the lids (*commissura nasalis palpebrarum*). It should be noted that a constant replenishment of the fluid covering the eye may be particularly important when birds are in flight to prevent drying out with a concomitant loss of optical quality. Certainly in laboratory preparations, constant moistening of the cornea is essential if optical quality is to be maintained.

Inter-specific differences in the size of the gland and the nature of its secretion are noted by Walls (1942, p. 425). For example, he reports that the gland in the Great Cormorant (*Phalacrocorax carbo*) is "very large" and suggests that this is related to the marine habits of this species, the oily secretion of the gland reducing the osmotic problems which might ensue when the cornea is immersed in water. However, systematic inter-specific data on the development of the gland and the nature and quality of its secretions seem to be lacking.

The lacrimal gland is situated in the ventrotemporal part of the orbit, its secretion emptying directly beneath the lower lid. There may be a single duct (*ductus gl. lacrimalis*) or multiple ducts (Slonaker, 1918). Walls (1942, p. 425) describes the inter-specific variation in the size of the gland as "inconsistent" and states that whilst the lacrimal gland is minute in one group of aquatic (marine) birds (penguins, Sphenisciformes), it is large in others (the freshwater dippers, Cinclidae). It is said to be entirely lacking in owls (Strigiformes). Clearly further comparative work on its development would be useful. The secretions from the gland of the nictitating membrane and the lacrimal gland are drained (in the pigeon at least) by two lacrimal canaliculi (*canaliculi lacrimales*), both of which are situated close to the nasal commissure of the lids. A groove, the "peripalpebral groove", was described by Slonaker (1918) and Chard and Gundlach (1938) as running from the opening of the canaliculus around the inner edge of the lower lid, with a smaller, but similar groove on the upper lid. They proposed that these grooves direct secretion to the canaliculi. The lacrimal canaliculi open via the nasolacrimal duct (*ductus nasolacrimalis*) into the lower portion of the nasal passage in such a position that secretions from the eye eventually drain into the mouth. Walls (1942) suggests that the lacrimal canaliculi are lacking in penguins (Sphenisciformes), but this observation does not appear to be based on anatomical evidence.

3. *Eyelids*

The dorsal and ventral eyelids (*palpebra dorsalis, palpebra ventralis*) are moveable and are controlled by three muscles. When open, the lids typically cover the whole of the anterior portion of the eye except the cornea. This gives the impression that the eyes of birds are smaller than they actually are. In general the upper lid is short and thick while the lower lid is longer, thinner and highly mobile. Closure of the eye is due mainly to the movement of the lower lid. The lid margins, which are thickened and segmented into a variable number of folds (*plica marginalis*), meet in a slight upwards curve above the pupil. The lids may be feathered as in the Ostrich (*Struthio camelus*), rheas (Rheidae) and owls (Strigiformes), or unfeathered as in pigeons (Columbidae), parrots (Psittacidae) and passeriform species.

Two types of blinking—"normal" and "protective"—have been described in the pigeon (Mowrer, 1937; Chard and Gundlach, 1938), but the status and generality of this distinction does not appear to have received further consideration. Eye closure in the pigeon is described as made up of three movements: (1) the nictitating membrane is pulled across the cornea followed by closure of the upper lid, (2) the lower lid pushes up to the middle of the pupil, and (3) the nictitating membrane moves back across the eye beneath the lower lid, the margin of which then rises to be level with the top of the pupil. How applicable this description is to birds generally is not known.

Clearly the lids and nictitating membrane, together with the secretions which empty into them, constitute an important and elaborate system for maintaining the optical quality of the cornea by removing foreign matter which may enter the eye, stopping it from drying out and preventing it from suffering osmotic disturbance when immersed in water. The nictitating membrane would seem to play an important part in this. It should be noted that movements of the nictitating membrane are also typically associated with oscillatory movements of the eyeball itself, which are interpreted by Nye (1969) as corneal "polishing" movements (see Section II, G).

4. *Refraction by the nictitating membrane*

A direct optical (refractive) function for the nictitating membrane in diving birds has been proposed. Ischreyt (1912) reported differences in the nictitating membranes of diving and non-diving ducks (Anatidae), noting especially that the membrane of the non-divers was less transparent than that of diving species. In some diving ducks, such as the Common Merganser (*Mergus merganser*) and the Common Loon (*Gavia immer*), a central, totally transparent window is present in the nictitating membrane. Ischreyt (1912) proposed that this membrane has a high refractive index, which would help compensate for the loss of refractive power when the cornea is negated by entry into water. Although this claim is often repeated, Sivak *et al.* (1978) have argued, using both anatomical data and optical analysis, that this function for the nictitating membrane is highly unlikely. In addition, personal observations on the pigeon and the Humboldt Penguin (*Spheniscus humboldti*) showed that the quality of the ophthalmoscopic view of the fundus deteriorates as the nictitating membrane sweeps across the cornea, suggesting that the optical quality of the apparently transparent membrane in these species is poor and that there is a slight opacity.

Thus, in all species it would seem safe to conclude that the nictitating membrane serves the optical system by ensuring the maintenance of the optical quality of the cornea, and not as a part of the actual refractive system.

III. The image-analysing system

Since the publication of the general descriptions of the avian visual system by Walls (1942), Rochon-Duvigneaud (1943) and Duke-Elder (1958), many advances have been made in understanding the structure and function of both the initial, ocular portion of the image-analysing system (the retina) and the ancillary structures which maintain its metabolism. There has also been progress in understanding the visual capacities of birds through the development of psychophysical techniques, although the value of such data in understanding eye structure is limited, since visual capacities reflect the processing of visual information by the higher centres of the central nervous system as well as its initial analysis within the retina. Many of these advances have been achieved in one species, the pigeon, and these have been reviewed and collated a number of times (Donovan, 1978; Granda and Maxwell, 1979; Emmerton, 1983). Comparative surveys of certain retinal structures are to be found in Crescitelli (1972), Lythgoe (1972), Rodieck (1973), Meyer (1977) and Gallego (1983). However a comprehensive review of the avian retina does not seem to be available. Even in the thorough survey of the vertebrate retina by Rodieck (1973), the data are collated according to structure rather than arranged taxonomically. Cajal's ''La rétine des Vertébrés'', first published in 1893 and provided in English translation as an appendix in Rodieck (1973), remains a valuable source of data on avian retinal structure.

A. *Retina*

The *retina* is divided into two major layers, the outer pigmented layer (*stratum pigmentosum retinae*) and the inner nervous layers (*strata nervosa retinae*).

Within the nervous layers four main divisions are acknowledged, and these are further subdivided into seven principal layers (Fig. 7.11):

1. The neuroepithelial layer (*stratum neuroepitheliale*) lies between the pigmented epithelium and the external limiting membrane. It contains the visual cell outer segments and the outer nuclear layer, which is composed of the visual cell nuclei.
2. The bipolar layer (*stratum bipolare*) is further subdivided as follows: (a) the outer plexiform layer (containing the visual cell synaptic terminals), (b) the inner

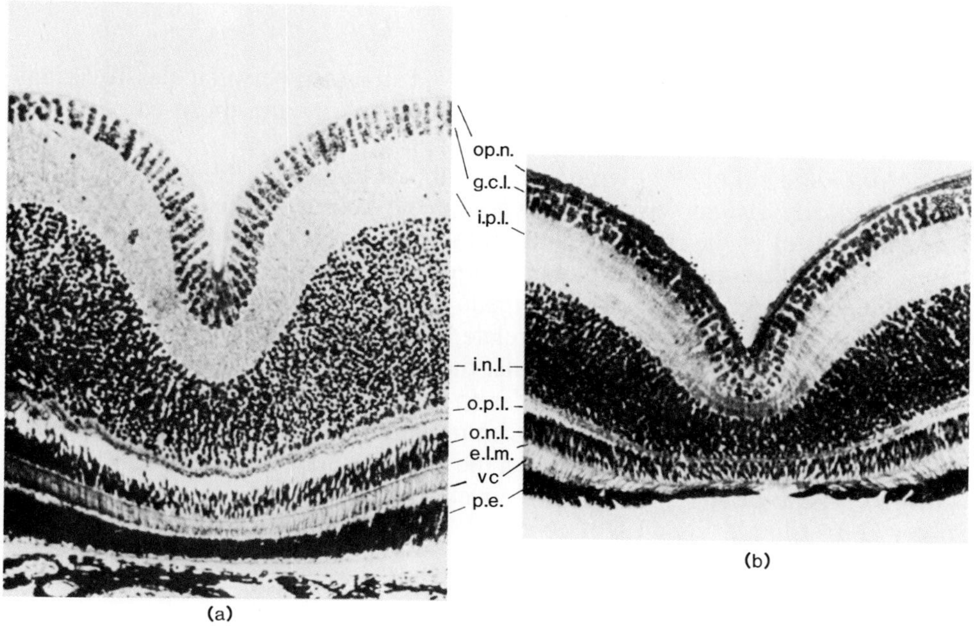

Fig. 7.11. Photomicrographs of sections through the fovea of (a) the White-capped Albatross (*Diomedea cauta*) (O'Day, 1940, courtesy of British Medical Association, London) and (b) the pigeon (Clarke and Whitteridge, 1976, courtesy of the Physiological Society), showing the principal layers of the retina. e.l.m., External limiting membrane; g.c.l., ganglion cell layer; i.n.l., inner nuclear layer; i.p.l., inner plexiform layer; o.n.l., outer nuclear layer; o.p.l., outer plexiform layer; op. n., optic nerve; p.e., pigmented epithelium; vc, visual cells.

nuclear layer (containing the nuclei of the horizontal, bipolar, amacrine and Müller cells) and (c) the inner plexiform layer.

3. The ganglion cell layer (*stratum ganglionicum*) contains the ganglion cell nuclei.

4. The optic fibre layer.

The external and internal limiting membranes are formed by the terminal bars of the fibres of adjacent Müller cells. These fibres pass radially through the retina and surround most neurons. Müller cells provide the primary mechanical support of the retina.

There have been many general histological analyses of the avian retina (see Duke-Elder, 1958, p. 419, for a list of 30 references spanning approximately a century), all of which attest to two principal conclusions.

1. Compared with that of most vertebrate groups, the avian retina is relatively thick. This increased thickness is principally due to the density and complexity of cells within the inner nuclear layers, and to the elaborate arborizations of these cells in the inner plexiform layer.

2. The retina is highly ordered. The principal cell layers are clearly distinguishable, and further concentric bands or strata are evident within these layers, especially the inner plexiform layer. Duke-Elder (1958, p. 419) remarked that "Histologically the retina of birds is the most beautiful and elaborate in its architecture in the animal kingdom; layers and sub-layers are clearly defined with each cell accurately in place".

Something of these two general features is evident in Figs. 7.11 and 17.

1. Nervous layers

(a) Neuroepithelial layer. It is necessary to distinguish between classifications of the visual cells which are based on cell morphology and those based on oil droplets and visual pigments.

(i) Morphological types of visual cell. Cajal (see Rodieck, 1973) described five morphologically-distinct visual cell types in the neuroepithelial layer (*stratum neuroepitheliale*) of the chicken retina: rods, upright (or straight) cones, oblique cones, and double cones. Upright and oblique cones are distinguished by the fact that the inner segment in the oblique type is displaced laterally. Double cones are composed of two separate but always related receptor cells. The principal cone of the pair is relatively tall and thin, whilst the accessory cone is short and broad. Using Golgi staining techniques similar to those of Cajal, Mariani and Leure-DuPree (1978) described the same five morphological types of visual cell in the pigeon (Fig. 7.12). They also indicated the strata in the outer plexiform layer in which the visual cells synapsed. In their electron microscope study Morris and Shorey (1967) and Morris (1970) observed, in addition to rods and double cones, three types of single cone in the chicken (Fig. 7.13), although the classification of one of these cell types was based not on gross morphological features but on differential staining of the oil droplets and variations in the cristae of the cell's mitochondria. They did not distinguish between straight and oblique cones. According to Meyer and May (1973), in the chicken double cones exceed single cones by a ratio of 2:1. Observations on the chicken (Morris, 1970) and Great Tit (*Parus major*) (Engstrom, 1958) indicate that the double-cone receptor cells form a regular mosaic. In *Parus* this pattern consists of a single cone surrounded by four pairs of double cones which form the sides of a square. These squares are arranged in rows. The morphology of the central cones is not known. Between these distinct groups both rods and single cones (also of unknown type) occur. As well as being present in the species just mentioned, double cones have also been observed in the Little Owl (*Athene noctua*), the Short-eared Owl (*Asio flammeus*), Bonelli's Eagle (*Hieraaetus fasciatus*), the Red Kite (*Milvus milvus*) (Gallego *et al.*, 1975a), the Common Quail (*Coturnix coturnix*) (Hazlett *et al.*, 1974) and several passeriforms (Rodieck, 1973). They are, however, reported to be absent in the Tawny Owl (*Strix aluco*) (Bowmaker and Martin, 1978), the Emu (*Dromaius novaehollandiae*), two species of tinamous (*Nothoprocta perdicaria, Nothoprocta cinerascens*) (Sillman *et al.*, 1981) and the Humboldt Penguin (*Spheniscus humboldti*) (Bowmaker and Martin, 1985).

It is widely believed that all single-cone receptors possess an oil droplet at the junction of the inner and outer segments (Fig. 7.13), and indeed this feature has been used in identification (e.g. Fite, 1973). However, cones without oil droplets have been observed in the Emu (*Dromaius novaehollandiae*), two species of tinamou (*Nothoprocta perdicaria, Nothoprocta cinerascens*) and the Red-tailed Hawk (*Buteo jamaicensis*) (Sillman *et al.*, 1981). Whether the accessory member of the double cones contains an oil droplet is also not clear. The available evidence on this in both pigeon and chicken retinas is conflicting (Cohen, 1963; Meyer and Cooper, 1966; Morris and Shorey, 1967; Meyer and May, 1973; Bowmaker, 1977; Bowmaker and Knowles, 1977; Mariani and Leure-DuPree, 1978).

The ultrastructure of the cones in the pigeon retina has been investigated by

Cohen (1963), who paid special attention to the structure of the outer segment. This appears to be basically similar to that of other vertebrates, although the membranous discs do not completely fill the segment (this has also been reported by Miller, 1979, in *Falco sparverius*) and some receptors contain three columns of discs, the significance of which has not been established. Whether they are a general feature of avian cones is also not known, although Cohen (personal communication) suggested that they could be the result of inbreeding amongst domestic stock and not a general feature of pigeon or avian outer segments.

There are no data on the possible functional differences between the straight and oblique types of single cone, although Cajal argued that the two populations would prove to have different properties. The function of the double cones has not been adequately explained. Cajal proposed that each cone of the pair is functionally independent, since he observed that the synaptic processes of the cones spread out in different strata of the outer plexiform layer. However, this was not confirmed by Mariani and Leure-DuPree (1978) in the pigeon (Fig. 7.12). Instead these authors showed that double cones are isolated as a pair from other cones by the processes of Müller cells (all single cones also appear to be isolated from each other by such processes). Furthermore, the outer segments of the cones within a pair are optically isolated from the outer segments of other cones by pigment granules contained within the cells of the pigment epithelium (single cones are similarly isolated from each other by pigment).

Engström (1958) proposed that the double cones of birds function in the colour vision mechanism, basing his argument on an analogy in function between the double cones of birds and fish. Although this role in colour vision seems to have been generally accepted, evidence in its support from studies of visual pigments and oil droplets is equivocal (see later). Delius *et al.* (1976) proposed that double cones may mediate the detection of the plane of polarization of light. Although they were unable to specify any particular mechanism, their argument has re-

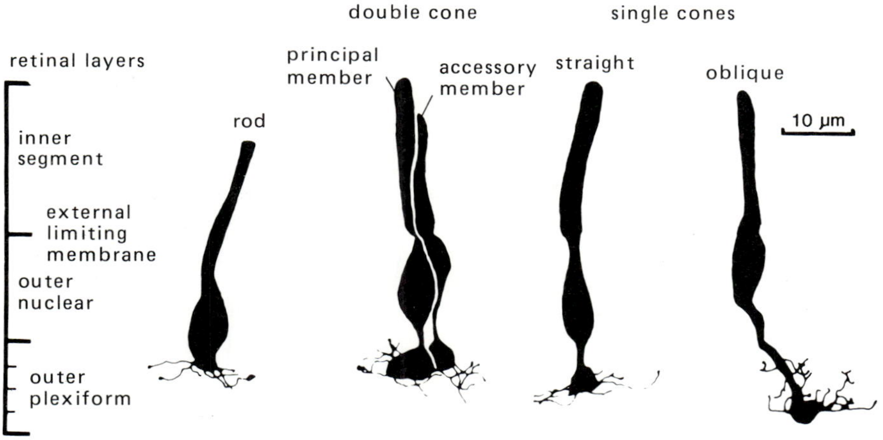

Fig. 7.12. Camera lucida drawings of photoreceptor types identified in vertically-sectioned Golgi-impregnated retinae of the pigeon by Mariani and Leure-DuPree (1978). The strata in the outer plexiform layer in which the arborizations of the cells synapse are indicated. Note that the outer segments of the cone receptors, which should be situated at the tips of the inner segments, had become detached and are therefore not shown in these drawings. Reproduced courtesy of Alan R. Liss, Inc., New York.

ceived support from Young and Martin (1984), who proposed a mechanism for polarization detection in the double cones based on an analysis of the optical properties of retinal oil droplets (see later).

(ii) Oil droplets. A single retinal oil droplet is found at the junction of the inner and outer segments of cone receptors in many vertebrate taxa, including all birds and some fish, amphibia, reptiles and mammals (monotremes and marsupials) (Muntz, 1972). In birds these spherical organelles are often brightly coloured due to the presence of dissolved carotenoid pigments of dietary origin (Meyer, 1971, 1977; Goldsmith *et al.*, 1984). The composition of the lipid forming the droplet has been described by Johnston and Hudson (1974, 1976).

Because of their position within individual cone receptors (Fig. 7.13), oil droplets have always been assigned a crucial role in avian vision. For discussions

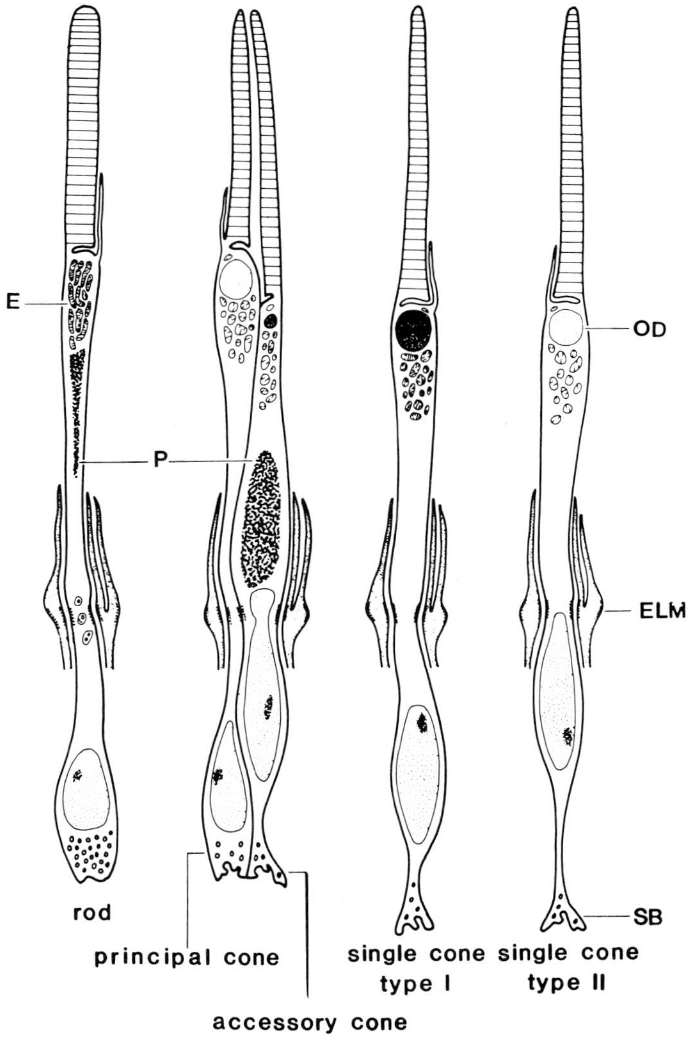

Fig. 7.13. Schematic drawings of five photoreceptor types identified in an electron microscope study of the retina of *Gallus* by Morris and Shorey (1967). E, Ellipsoid; OD, oil droplet; ELM, external limiting membrane; P, paraboloid; SB, synaptic body. Reproduced courtesy of Alan R. Liss, Inc., New York.

of the many theories of this role see Hahn (1916), Uchiyama (1930), Walls and Judd (1933), Walls (1942), Rochon-Duvigneaud (1943), Donner (1953, 1960), Duke-Elder (1958), Meyer *et al.*, (1965), Alpern (1968), King-Smith (1969), Bloch and Maturana (1971), Thompson (1971), Muntz (1972), Rodieck (1973), Sillman (1973), Baylor and Fettiplace, (1975), Wolbarsht (1976), Bowmaker (1977), Bowmaker and Knowles (1977), Bowmaker and Martin (1978, 1985), Mariani and Leure-DuPree (1978), Martin and Muntz (1978), Wallman (1979), Barlow (1981) and Young and Martin (1983). It has been argued that oil droplets are ideally placed to mediate chromatic filtering of incident light before it reaches the outer segment where the visual pigments are situated. At one time it was proposed (see Muntz, 1972, for a review) that colour vision in birds depends solely on such selective filtering and that only one type of visual pigment is present in the cone receptors. Evidence from visual pigment analysis suggests that this is not the case (see later). In addition it has been demonstrated in the Common Quail (*Coturnix coturnix*) that colour vision is little altered when all oil droplets are rendered colourless by dietary exclusion of carotenoids (Duecker and Schulze, 1977; Wallman, 1979). Non-visual functions including chemical storage (Hailman, 1976) and a general protective role against photo-oxidation (Kirschfeld, 1982) have also been proposed.

Many theories of oil droplet function have sought to explain the marked interspecific and intraretinal differences in oil droplet types. The aim has often been to explain apparent correlations between these features of droplet populations and general visual ecological problems experienced by different species in their natural habitats (see for example, Peiponen, 1964; Muntz, 1972). However, whilst many intriguing differences in oil droplet populations have been reported, the arguments put forward to explain these differences have not proved very satisfactory. This is because (1) there is a lack of understanding of how oil droplets actually function within individual receptors, and (2) there is difficulty in measuring the spectral filtering characteristics of the droplets. Since the droplets are small (diameter $<4\ \mu m$), highly refractile, and situated at different depths within the retina (see for example colour photographs illustrating this in Mariani and Leure-DuPree, 1978), characterisation of their spectral properties using the most highly-corrected microscope optics is not without error, and the same droplet may be made to appear a wide range of different colours by adopting different criteria for best focus in the microscope.

This problem has been overcome by the development of the technique of microspectrophotometry (Liebman, 1972). This measures the selective absorption of a narrow beam of light which is made to pass through the droplet. To date the technique has been applied to investigate the droplets of 26 avian species, including the pigeon (Bowmaker, 1977), chicken (Bowmaker and Knowles, 1977), Tawny Owl (*Strix aluco*) (Bowmaker and Martin, 1978), Humboldt Penguin (*Spheniscus humboldti*) (Bowmaker and Martin, 1985), two species of tinamou (*Nothoprocta cinerascens* and *Nothoprocta perdicaria*), the Emu (*Dromaius novaehollandiae*) and the Red-tailed Hawk (*Buteo jamaicensis*) (Sillman *et al.*, 1981), and three species of hummingbird (Trochilidae) and 12 species of Passeriformes (Goldsmith *et al.*, 1984). The overall conclusions to be drawn from these studies are that (1) coloured oil droplets in general act as "cut-off" filters, that is, absorbing a high proportion of light at short wavelengths and transmitting

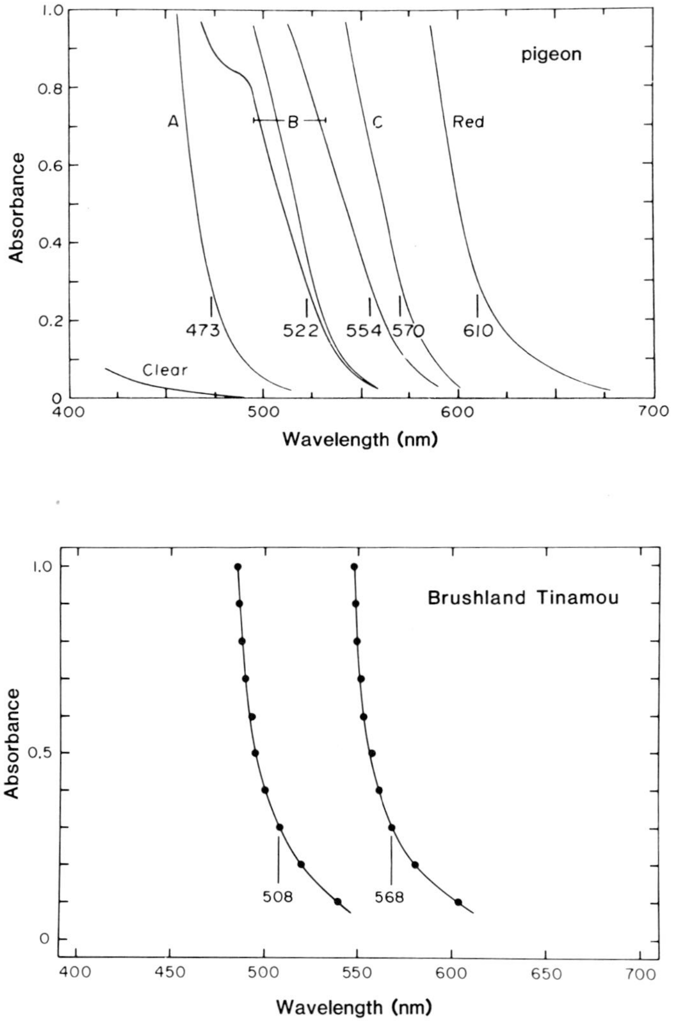

Fig. 7.14. Oil droplet absorbance spectra from the retina of the pigeon [Reprinted with permission from *Vision Res.* **17,** Bowmaker The visual pigments, oil droplets and spectral sensitivity of the pigeon. Copyright (1977), Pergamon Press, Ltd.] and the Brushland Tinamou (*Nothoprocta cinerascens*) (Sillman *et al.*, 1981, courtesy of Springer-Verlag, Heidelberg). In the pigeon the variable B type is found only in the principal member of the double cone. The variation in the absorbance of the B type occurs within individual retinae. The wavelengths shown are the 50% transmission points of the droplets. The oil droplet spectra of the pigeon are from the red oil droplet field.

all light at longer wavelengths (Fig. 7.14); (2) within the same retina, more than one type of oil droplet is usually found in the single cones; (3) within the same retina the oil droplets in the double cones are usually different (Fig. 7.14); (4) a colourless or transparent type of droplet occurs alongside the coloured droplets; and (5) the droplets show interspecific variations which may be classified into types.

Five types of oil droplet have been recognised in the pigeon and chicken, four in the owl, three in the penguin, and two in the Emu and tinamous. However, it has not been possible to correlate the microspectrophotometry data on the types of droplets with the two morphological types of single-cone receptors, although

Mariani and Leure-DuPree (1978) have argued that such a correlation might exist. They suggest that in the pigeon the straight cones may contain two types of oil droplet whilst the oblique cones contain an oil droplet of a different type. Thus, from an analysis of histological features and droplet types they proposed that five distinct functional receptor types occur in the pigeon retina. This picture is further complicated by the addition of data on the visual pigments which are found in combination with the oil droplet types (see later).

The question of oil droplet function is further complicated by analysis of their optical properties. Thus, Young and Martin (1984) have shown that oil droplets act within photoreceptors as converging lenses which may funnel light into the outer segment as well as produce spectral filtering. However, because of their small size an appreciable amount of light is diffracted around the droplet and is not subject to spectral filtering, while other light is scattered away from the receptor axis. Because the double cones are not optically isolated from each other, it has been argued that such scattered light passing from the oil droplet of the principal member of the double cone pair will pass transversely into the outer segment of the accessory member which lies adjacent to it. Since vertebrate photoreceptors are dichroic to transverse (but not longitudinal) illumination (Harosi and MacNicol, 1974), it has been hypothesised that this arrangement within the double cones could mediate the detection of the plane of polarization of incident light.

Note should also be taken of the so-called microdroplets (Pedler and Boyle, 1969) which are found throughout the inner segment of certain single-cone types in the pigeon retina. These are numerous and of very small diameter (<0.5 μm). Their function is unknown. Multiple oil droplets (up to a maximum of five) were reported in the accessory member of the double cones of quail (Hazlett *et al.*, 1974), but again their function is unknown.

(iii) Visual pigments. Data on the photopigments of the avian rod receptors are available for a wide range of species (Sillman, 1969; Lythgoe, 1972). All evidence suggests that rods contain a rhodopsin with the maximum absorbance (λ_{max}) in the range of 490 to 506 nm, the majority being in the range 500–506 nm. This range is quite comparable with that found in the rod receptors of other vertebrate groups, especially mammals.

Microspectrophotometric analysis of the visual pigments of cones has been conducted on the same species for which microspectrophotometric data on oil droplets are available. These have indicated the presence of three pigments in the cones of the pigeon, owl and penguin, two in the chicken, and one in the Emu and tinamou (Table I). It should be noted that due to sampling limitations of the microspectrophotometric technique, further pigments may be found in these species. Indeed there is electrophysiological, psychophysical and biochemical evidence (Graf and Norren, 1974; Norren, 1975; Govardovskii and Zueva, 1977; Fager and Fager, 1981) for one or two pigments with a maximum absorbance at short wavelengths (380–410 nm) in the pigeon and chicken retina which have not yet been found by the microspectrophotometric technique. What the microspectrophotometry results do indicate, however, is that visual pigments and oil droplets occur in specific combinations, and on this basis alone it can be argued that six different cone receptor types occur in the pigeon and chicken, five in the penguin, four in the owl and three in the Emu and tinamou. Whether these represent functionally-different receptor types, and how they relate to mor-

TABLE I. **Cone visual pigments identified in the avian retina.**[a]

Name	Pigment λ_{max} (nm)			
Chicken[b]				
Gallus gallus	(413[c], 417[d])	(449[d], 467[c])	497	569
Pigeon[e]				
Columba livia	(413[c])	460	515	567
Rook[f]				
Corvus frugilegus			497	565
Tawny Owl[g]				
Strix aluco		463	503	555
Aylesbury Duck[h]				
Anas platyrhynchos		449	500	567
Emu[i]				
Dromaius novaehollandiae				567
Chilean Tinamou[i]				
Nothoprocta perdicaria				566
Brushland Tinamou[i]				
Northoprocta cinerascens			498	564
Penjuin[j]				
Spheniscus humboldti	403	450		543

[a] All studies, except Govardovskii and Zueva (1977) and Fager and Fager (1981) employed microspectrophotometric techniques.
[b] Bowmaker and Knowles (1977).
[c] Govardovskii and Zueva (1977) (electrophysiological studies).
[d] Fager and Fager (1981) (pigment extracts).
[e] Bowmaker (1977).
[f] Bowmaker (1979).
[g] Bowmaker and Martin (1978).
[h] Jane and Bowmaker (1984).
[i] Sillman *et al.* (1981).
[j] Bowmaker and Martin (1985).

phological differences in the single cones, are not understood. However, one important result to emerge from these studies is that in the pigeon and chicken both members of the double cones contain the same visual pigment (which is also the same in both species, $\lambda_{max} \simeq 567$ nm), and if an oil droplet is present it is also identical in both members. Thus, both cones would appear to be functionally identical as far as their spectral response is concerned, and so the special function, if any, of the double-cone receptors in colour vision is unclear.

(b) Bipolar layer. The components, organisation and function of the remaining nervous layers of the retina are less well understood than the neuroepithelial layer, and Cajal's work still appears to provide the most thorough summary of these aspects of the avian retina.

Studies dealing specifically with the structure and response characteristics of the cells within the two plexiform layers and the inner nuclear layer of the avian retina are very few. This is because there is a general belief from past research (see for example Polyak, 1941; Stell, 1972; Rodieck, 1973; Gallego, 1983) that, for many cell types, the response characteristics and functions are similar across all vertebrate groups; therefore, physiologists interested in unravelling the complexity of the vertebrate retina have tended to choose for their studies species whose retinal cells are larger, and whose retina is more easily maintained as a physiological preparation, than the avian retina. Thus, much of the progress in this area has been in the study of reptiles and amphibia. The response properties of the various cells in the avian retina are assumed by analogy with those of these species. The avian retina has, however, not been completely neglected, and there have, for example, been studies of the horizontal neurons by Gallego and co-workers (Gallego *et al.*, 1975a,b; Gallego, 1976, 1982; Baron and Gallego, 1982) and by Mariani and Leure-DuPree (1977), of the bipolar neurons by Lockhart (1979), and of the amacrine neurons by Mariani (1982, 1983). In summarising work on the organisation of the outer plexiform layer (OPL) of the avian retina, Gallego (1983, p. 107) concluded that "although the photoreceptors and horizontal cells have been widely studied, as far as the bipolar cell is concerned, we have to rely mainly on Cajal's study which appears to be insufficient when the different bipolar types are studied with some modern techniques". Gallego has made a distinction between the organisation of the outer plexiform layer in diurnal and nocturnal birds. Diurnal birds show a stratified outer plexiform layer due to the distribution at three levels of the synaptic bodies of the photoreceptors (see Fig. 7.12), whilst nocturnal birds have their synaptic bodies lying at a single level. The functional significance of this difference is not known.

In the light of the limitations just mentioned it is perhaps sufficient to provide the following summary of cell types in the avian retina and refer the reader to the works of Rodieck (1973), Lockhart (1979), Gallego (1983) and Mariani (1983) for further details of the problems of interpretation. In the diurnal retina three concentric strata may be identified in the outer plexiform layer which correspond to the synaptic bodies of three morphological types of photoreceptor: (1) rods and double cones, (2) straight cones, and (3) oblique cones. These make synaptic connections with three different types of horizontal neurons (one type with a short axon and two axonless types; see Fig. 7.15) and an undetermined number of types (at least three) of bipolar neurons. The nuclei of the bipolar and horizontal neurons are found within the bipolar neuron layer, together with the nuclei of the

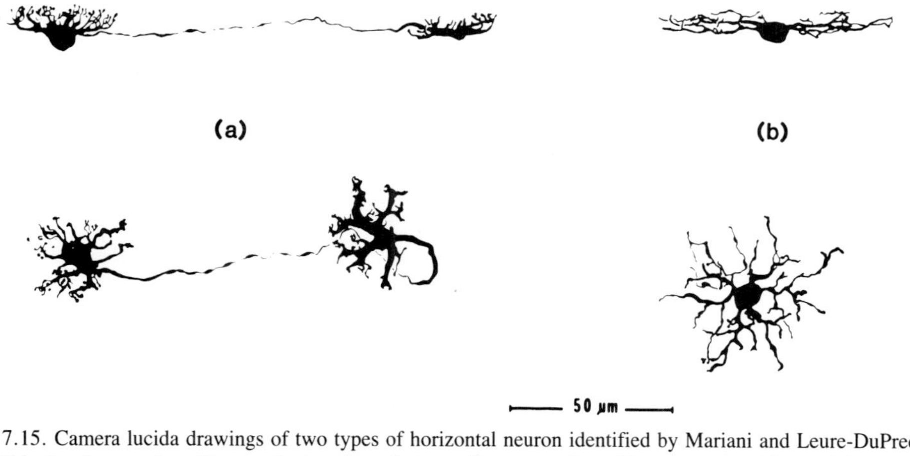

Fig. 7.15. Camera lucida drawings of two types of horizontal neuron identified by Mariani and Leure-DuPree (1977) in the pigeon retina. Top, vertical sections; bottom, flat preparations. Type 1 neurons (a) are of the short-axon type. The axon ends in the distal part of the outer plexiform layer and contacts both rods and cones, whilst the dendrites which arise from the cell body contact only cones. Type 2 neurons (b) are axonless, and their multi-branched dendrites terminate in the proximal part of the outer plexiform layer and contact only cones. Courtesy of Alan R. Liss, Inc., New York.

amacrine neurons and Müller cells (the last named being retinal gliocytes). Three types of amacrine neurons have been identified: (1) stratified amacrine neurons which lack an axon (at least five main strata being formed where their arborizations terminate in the inner plexiform layer), (2) diffuse amacrine neurons which are also axonless, and (3) amacrine neurons of association which have an intraretinal axon (Fig. 7.16). The Müller cell nuclei are found in a distinct stratum of the bipolar neuron layer, their fibres passing through the depth of the retina between the internal and external limiting membranes and surrounding all cell types, except possibly the amacrine neurons.

(c) Ganglion cell layer. In the retina of all vertebrates the ganglion cells are found to be morphologically heterogeneous, many types having been distinguished by such features as variations in somatic size, configuration and the strata in which their dendrites are distributed in the inner plexiform layer. In the

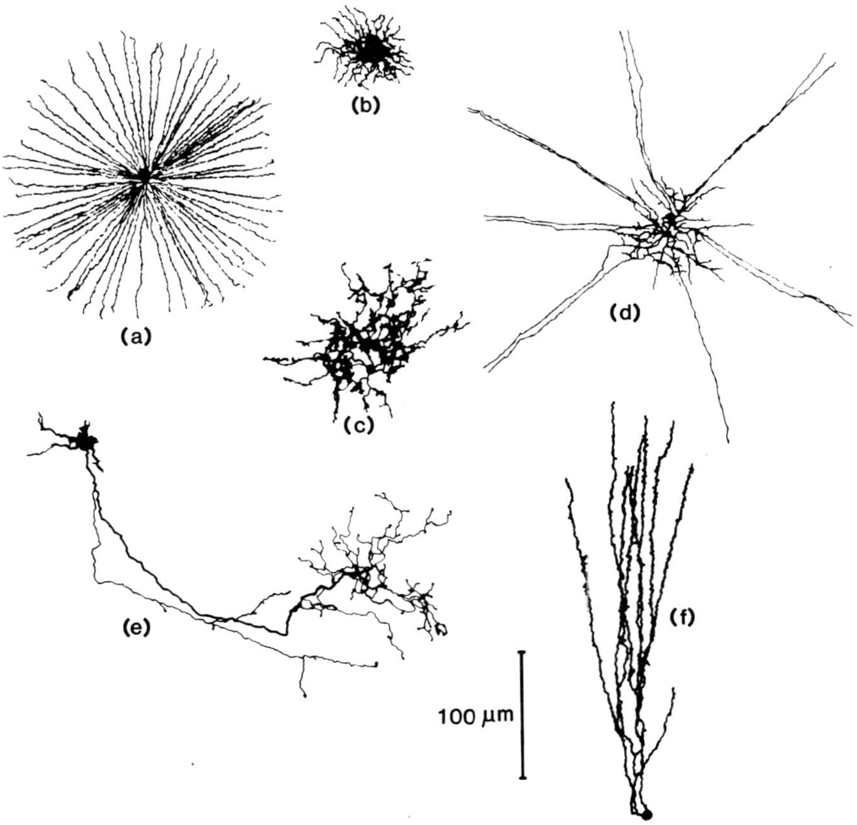

Fig. 7.16. Six examples of amacrine neurons identified in flat-mounted pigeon retinae using Golgi impregnation techniques by Mariani (1982, reprinted by permission from *Nature*, **298**, 654–655. Copyright © 1982 Macmillan Journals Limited; 1983). Types (a)–(d) lack an axon and are basically radially symmetrical. Type (e) is an example of an association amacrine neuron; it has an axon which separates the dendritic end and the axon terminal arborization. This type of cell could mediate the directional selectivity of ganglion cells found in the pigeon retina. Type (f) is an asymmetrical neuron the orientation of which is along the dorsoventral axis of the eye. Mariani (1983) proposed that neurons of this type could mediate responses of the verticality detector ganglion cells described by Maturana (1962). Courtesy of Springer-Verlag, Heidelberg, and Macmillan Journals Limited, London.

avian retina Cajal recognised two main types of ganglion cell: (1) monostratified cells, of which he found four forms in the chicken retina, and (2) polystratified cells, of which he recognised three forms. Cajal did not consider that he had identified all types of ganglion cell, and indeed a further so-called displaced ganglion cell was described in the avian retina by Dogiel (1888) as occurring, not in the ganglion cell layer proper, but along the inner margin of the bipolar neuron layer. The distribution of this "displaced" type of cell has been described by Hayes and Holden (1983). Fifteen different types of ganglion cell have been described in the mammalian retina (West and Dowling, 1972), and it is possible that just as many will be found in birds.

A number of attempts have been made in the pigeon to classify the ganglion cells by their response characteristics as measured electrophysiologically (Donner, 1953; Maturana, 1962; Maturana and Frenk, 1963; Pearlman and Hughes, 1976a,b; Holden, 1977a,b, 1979, 1981). Not surprisingly with such a complex group of cells, there has been considerable disagreement concerning their classification. According to their response to chromatic stimuli, Donner (1953) described three narrow-band chromatic units. Using mainly high-contrast achromatic stimuli, Maturana (1962) and Maturana and Frenk (1963) found evidence for six classes of ganglion cell, which respond selectively to relatively complex stimuli such as bright and dark stationary bars, straight edges and spots of light, convex edges, and moving edges and bars. Holden (1969, 1977a,b, 1981), using stimuli such as spots, annuli and slits, has described ganglion cells with directional and non-directional properties. The cells exhibit concentric receptive fields of small diameter with centres showing "on", "off" or "on–off" responses, and with antagonistic surrounds. All of these studies have been interpreted mainly in the context of the known response properties of retinal ganglion cell in other vertebrate groups, and a number of similarities and differences have been noted. It has not been possible to identify the morphological types of cells to which the response properties belong. However, there has been some progress in understanding which types of amacrine neuron may mediate certain of the complex response properties of the ganglion cells (Mariani, 1982, 1983) (Fig. 7.17).

(d) Centrifugal fibres. A remarkable and possibly unique feature of the avian retina (Ogden, 1968; Tripathi, 1974) is the presence of a system of centrifugal fibres which constitute an efferent pathway to the retina from the contralateral isthmo-optic nucleus of the mesencephalic *complexus isthmi*. The system has only been reported in detail in the pigeon and chicken. Cowan (1970) estimated that there are approximately 12,000 efferent fibres which terminate mainly along the proximal margin of the bipolar cell layer (Cowan and Powell, 1963), on the displaced ganglion cells and the amacrine neurons (Maturana and Frenk, 1965; Dowling and Cowan, 1966). A number of attempts have been made to elucidate the function of the centrifugal system. Pearlman and Hughes (1976a,b) investigated the effects of eliminating the activity of the efferent fibres on the responses of the retinal ganglion cells. They found that under these conditions the responsiveness of the ganglion cells decreased and the receptive fields shrank. Miles (1970, 1972a,b,c,d) and Rogers and Miles (1972) recorded from the ganglion cells while stimulating the efferent fibres and showed that the receptive fields expanded, with enhanced excitatory responses occurring in the central portion of the fields. It was concluded that the efferent fibres function to increase visual

sensitivity in general and that this may be important for improving performance on visual search tasks (such as detecting grain) in shadowed areas. However, Knipling (1978) showed that lesions of the centrifugal fibre tract produced no obvious deficits in near-field visual acuity when moderately bright stimuli were used, and so the general importance of the centrifugal system requires further investigation. Knowledge of the system and its characteristics in other avian species would be particularly interesting. Claims for centrifugal fibres in the mammalian retina appear to be controversial.

2. *Pigmented layer*

A pigmented layer of the retina (*stratum pigmentosum retinae*) is evidently present in all avian species. Its primary function is assumed (by analogy with that of other vertebrates) to be in the regeneration of visual pigments and the breakdown of membrane discs shed from the outer segments of the photoreceptors. The ciliary processes interdigitate with the photoreceptors. The pigment within the processes migrates towards the receptors when the retina is light-adapted; it retreats towards the tips of the cells (away from the receptors) during dark adaptation. During light adaptation the myoid region of the cone receptors shortens while that of the rods lengthens. The functional significance of such changes, which are well understood in most vertebrate groups (see Miller, 1979, for a review), is to ensure a complete transition between scotopic and photopic vision, and in light-adapted retinae to enhance the directional sensitivity of the cone receptors by protecting the outer segments from stray light. Such a removal of optical "cross-talk" is an essential requirement for the high spatial resolution of the avian retina during day-time (Miller, 1979).

In basic structure the avian choroid (*choroidea*) is similar to that of other vertebrates. It consists of a suprachoroid lamina (*lamina suprachoroidea*), a perichoroid space (*spatium perichoroideum*), a vascular lamina (*lamina vasculosa*), a choroidocapillary lamina (*lamina choroidocapillaris*) and a basal lamina (*lamina basalis*). According to Walls (1942, p. 644), the avian choroid is relatively thick compared to that of reptiles and mammals. It provides the primary blood supply to the retina. It is also heavily pigmented, the pigment acting to reduce reflections within the eye and also to prevent light entering through the wall. An apparently unique feature of the avian choroid is the presence of muscle fibres. First described by Wittich (1855), these striated and non-striated fibres pass radially through the thickness of the choroid. In the pigeon they are reported to be more numerous near the fovea (Kajikawa, 1923). Although the function of this muscle has not been established, Duke-Elder (1958, p. 405) proposed that it may regulate the amount of blood in the choroid. According to Walls (1942, p. 645), it could also form part of the accommodatory mechanism, the contraction and relaxation of the muscle possibly changing the position of the retinal surface in the manner of the fine-focus adjustment of a microscope.

No tapetum lucidum has been reported in the choroid of any avian species, although Nicol and Arnott (1974) have reported the presence of such a reflective structure in the pigment layer of the retina of goatsuckers (Caprimulgidae). These species are crepuscular in habit, and the adaptive advantage of such a structure, which serves to reflect light back towards the receptor outer segments after passing once through the retina, would seem clear. It is perhaps surprising in view of the widespread occurrence of tapeta in other vertebrate groups that such a structure has not been reported in any other nocturnal and crepuscular birds such as owls (Strigiformes) and the Oilbird (*Steatornis caripensis*). The tapetum of goatsuckers, like other retinal tapeta, produces a diffuse white reflection. It is composed of closely-packed lipid spheres (0·5 μm diameter) which occupy the processes of the pigment cells.

B. Retinal topography

The avian retina is far from homogeneous in its general morphology. As in other vertebrate groups, areas and foveae of various forms are present and there are marked intraretinal and inter-specific differences in the distribution of all types of retinal cell. Something of the general inhomogeneity of retinal structure can be readily appreciated in most birds simply by ophthalmoscopic examination of the fundus, although interpretation of the features which are seen is not particularly satisfactory. This diversity of structure was captured to a certain extent by Wood (1917) in a series of colour plates and line drawings depicting the retina in a range of avian species. Data and functional hypotheses on the areas and foveae of birds have been reviewed by Meyer (1977).

1. Retinal areas

An area is a local thickening of the retina which results from an increase in cell density in all the layers of the retina. The visual cells may also be longer and thinner than elsewhere in the retina. It is argued that an area is functionally a localised site of increased resolving power, that is a site which acts to extract from the retinal image the highest degree of spatial information which the image contains.

Areas are a common feature of the avian retina and may be of various shapes and size. Often more than one area is present, or an area may extend across a considerable length of the retina. There have been a number of studies of the inter-specific distribution of areas (Meyer, 1977), and it is argued that areas are most highly developed (i.e. represent the greatest local thickening and the highest density of visual and other retinal cells) amongst the diurnal birds-of-prey (Falconiformes). This correlates with arguments presented in Section II concerning the high degree of spatial information which is probably available in the retinal image of these species. There would seem to be no adaptive value in the possession by a small-eyed passerine species of an area as highly developed as that reported to occur in the retina of an eagle (Rochon-Duvigneaud, 1943). This is because the highest spatial frequency contained in the retinal image of the passerine is likely to be lower than that which the eagle area is capable of resolving.

Areas are reported to lie always above the dorsal end of the pecten. The most commonly occurring area, the central round area (*area centralis rotunda*), is located in the central region of the fundus (*fundus oculi,* the part of the interior of the eyeball around the posterior pole) above and to the nasal side of the pecten. Another area, the temporal area (*area temporalis*) is often located lateral to the central area. A third area, the central horizontal area (*area centralis horizontalis*), may extend as a horizontal band across the retina, sometimes joining the central and temporal areas. Functional explanations for the number and locations of these areas require an understanding of how they project within the visual field of the bird, and also where they project when the bird is engaged in specific tasks. Unfortunately, experimental data on these aspects are lacking. However, the general functional importance of areas in the avian eye can be appreciated by considering why areas occur at all. Since the area is a portion of retina capable of enhanced spatial resolution, it may be asked why the retinal thickening which defines the area does not extend over the whole retina? However, from the point of view of analysing spatial information within the retinal image, such a retina of generally high resolution would in fact be of little adaptive value.

The optimal extraction of spatial information from any image has been shown to require different-sized receptors at different spacings for each luminance and contrast level within the image (Snyder *et al.*, 1977). Only at the highest contrasts and luminance levels can spatial information close to the diffraction limit of, say, the eagle's eye (Shlaer, 1972) be extracted from the retinal image, and this requires small, densely-packed receptors. As luminance levels decrease, so a different size and density of visual receptor is required for the optimal extraction of spatial information. On theoretical grounds (Snyder, 1979) it is argued that it is not possible to have high spatial resolution at low light levels no matter how fine the receptor spacing. In fact at low light levels, small, closely-spaced receptors will be unable to detect the presence in the retinal image of even large objects. This can only be achieved with large receptors widely spread.

Thus, a retinal area will only have functional value at the highest levels of luminance and contrast, where it will be capable of extracting a high degree of spatial detail from the image. As light levels fall the adaptive value of the area will decrease, and parts of the retina in which receptors are larger and at a lower density will become optimal for the extraction of information from the retinal image. Therefore, it may be argued that an eye in which the retina had the structure of an area throughout would be restricted in its functioning only to the highest levels of luminance. Possibly some of the problems of receptor size and spacing required to give optimal extraction of spatial information over a wide luminance range can be overcome by a flexible system of neural summation of receptor outputs (possibly via the bipolar and/or ganglion cells; Snyder *et al.*, 1977) to provide different-sized "effective" receptors at different light levels.

2. *Foveae*

A central fovea (*fovea centralis*) is typically found within the central round area or the central horizontal area, and in many species a second fovea, the temporal fovea (*fovea temporalis*), occurs in the temporal area. Not more than two foveae have been reported in any species. The fovea takes the form of a generally circular depression of variable depth and size (Figs. 7.11, 17). It is produced by the radial displacement of the more internal layers of the retina, which results in a local

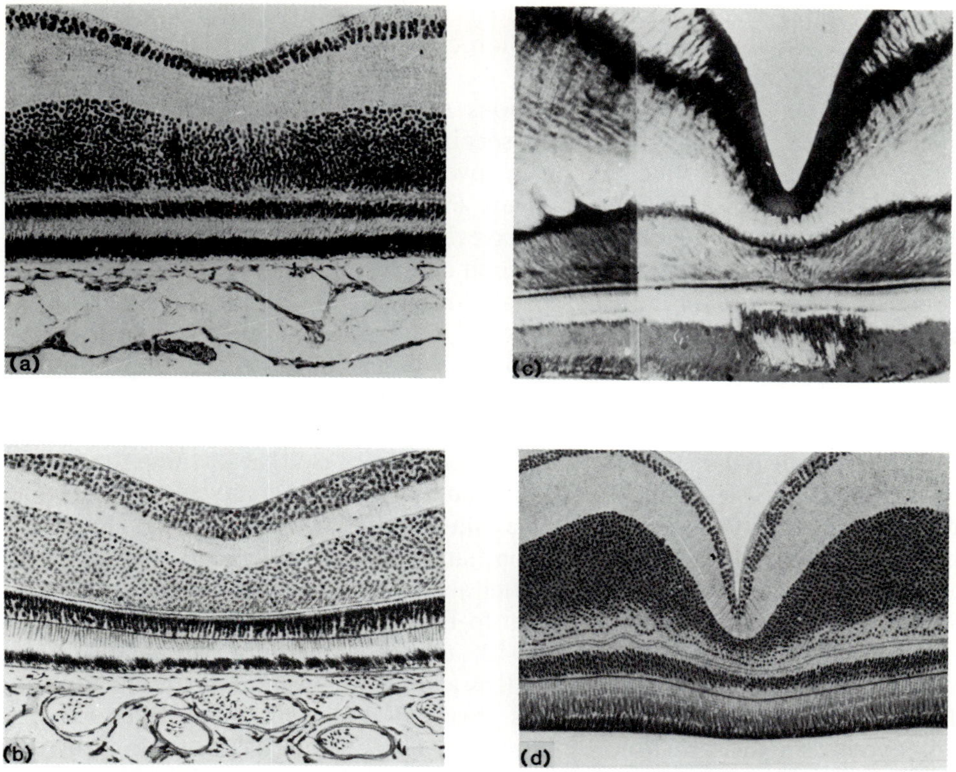

Fig. 7.17. Sections through the fovea of (a) the pigeon (Galifret, 1968, courtesy of Springer-Verlag, Heidelberg), (b) the Scops-Owl (*Otus scops*) (Rochon-Duvigneaud, 1943, courtesy of Masson Editeur, Paris), (c) a hawk (*Buteo*) (Snyder and Miller, 1978, reprinted by permission from *Nature* **275,** 121–129. Copyright © 1978 Macmillan Journals Limited.) and (d) the Least Tern (*Sterna albifrons*) (Rochon-Duvigneaud, 1943, courtesy of Masson Editeur, Paris).

thinning of the retina at the centre of the depression. In birds, however, there does not appear to be a displacement of the ganglion cell layer which is reported to occur in mammalian foveae (Hughes, 1977, p. 695). Two main foveal shapes are recognised: (1) *concaviclivate,* in which the depression is relatively broad and shallow with a concave or flat bottom (Fig. 7.17a), and (2) *convexiclivate,* in which the depression is deep and narrow with convex sides (Figs. 7.11, 17d). Examples of these types are found throughout fish, reptiles and mammals, as well as in birds. It should be noted that both varieties of fovea have been recorded in the pigeon (Figs. 7.11b, 17a).

Within the foveal region the length of the outer segments of the receptors is often increased, and this may lead to a so-called fovea externa in which the scleral surface of the retina is slightly convex (Figs. 7.11, 17d). The visual cells of the fovea are usually only cones, although the fovea of the Great Horned Owl (*Bubo virginianus*) contains both rods and cones (Fite, 1973). Visual cell density is greater in the fovea than elsewhere in the area, and the ratio of photoreceptors to ganglion cells in the fovea is also reported to reach its lowest value (see Fite,

1973, in *Bubo virginianus*). According to Rochon-Duvigneaud (1943) and Pumphrey (1948a), the receptor:ganglion cell ratio may reach unity within the fovea of some diurnal raptors.

(a) Function. Two principal questions regarding function have been asked about the fovea. The first concerns its specific visual function as a retinal feature within the vertebrates. Since the avian fovea is often considered to be the most highly developed of all vertebrate foveae (Walls, 1942; Pumphrey, 1948b), details of avian foveal characteristics have been important in formulating these general functional hypotheses. The second question is concerned with functional arguments to account for the inter-specific differences in the form and position of foveae within individual avian retinae.

(i) Direct improvement of visual acuity. This foveal function is supported by the observation that in the fovea (a) photoreceptor densities typically reach their highest value, (b) receptor: ganglion cell ratios are typically at their lowest, and (c) psychophysical data (in humans) show that acuity is highest within the fovea. Thus, a number of explanations have been proposed whereby the shape of the fovea is seen as an adaptation which maximises visual resolving power. However, this assumption is not without question, since Blough (1973) failed to find any decrease in visual acuity in pigeons after ablation of their foveae. The results of Uhlrich *et al.* (1982) also lend support to Blough's finding.

According to Walls (1937, 1940, 1942), the foveal margins, particularly of the convexiclivate type, act like a convex lens and locally magnify the retinal image. He suggested that the magnification is greater in the deep, more convex foveae. This hypothesis was criticised by Pumphrey (1948b) but has been revived in a modified form by Snyder and Miller (1978) and Williams and McIntyre (1980), who have argued that the small spherical depression in the very bottom of a deep fovea may act like the negative lens component of a telephoto lens system and thereby cause a localised increase in the eye's focal length, thus producing an enlarged image where receptor densities are highest. However, concaviclivate foveae could not function in this manner.

According to Wolff (1940), the fovea is simply a portion of retina where overlying layers are displaced in order to reduce the supposed deterioration of image quality which is produced by passage through the retinal tissue. Whilst this argument has been criticised by Weale (1966), an apparent deterioration in image quality as a result of passage through the retinal tissue has been demonstrated by Ohzu *et al.* (1972). Hughes (1977, p. 696) pointed out that a 1:1 receptor:ganglion cell ratio will produce a large increase in retinal thickness, and that if these extra cells are not displaced the deterioration of the image will be even greater than that determined by Ohzu *et al.* (1972). Whilst this function of the fovea may be important, it does not necessarily explain the various foveal shapes, although they could be the result of simple mechanical considerations resulting from the displacement of the retinal layers. According to Walls (1942), the shallow concaviclivate foveae are simply degenerate relative to the convexiclivate forms. However, the widespread occurrence of the former shape would seem to rule out such a simple explanation.

(ii) A fixation device. Although Pumphrey (1948b) also considered that the fovea caused localised refraction of light, he argued that this made the fovea particularly sensitive to displacement of the retinal image and that as a result the

fovea has two related functions, "the maintenance of accurate fixation and the sensitive appreciation of angular movement of fixed objects". Since high resolution requires accurate fixation, the fovea may fulfill this requirement for eye stability, which becomes more important with increased focal length of the eye (Martin, 1983), and thus a simple relationship would be predicted between foveal form and eye size.

(iii) A focus indicator. The hypothesised focus-indicating function would also seem to apply only to the convexiclivate foveae and concentrates on localised distortion of the retinal image produced in the spherical depression in the bottom of the foveal pit. Harkness and Bennett-Clark (1978) argued that this distortion is capable of yielding unambiguous information on focus error. The fovea therefore may function to improve resolution by providing precise information for the eye's accommodatory mechanism.

(iv) Detection of polarized light. Kreithen and Keeton (1974) concentrated on the way the surface of the foveal sides may reflect rather than refract light. They argued that the fovea could act as a radial analyser and respond differentially to planes of light polarization. It is claimed that light reflected from the foveal surface will be able to enter receptor outer segments perpendicular to their axes and that the dichroic property of the outer segments to such illumination (Harosi and MacNicol, 1974) will permit the detection of the plane of light polarization. However, this argument fails to take account of the directional sensitivity of receptors and the pigment screen which envelops cone receptor outer segments (see Section III,A).

Which of the functions just mentioned may actually apply to the avian foveae is difficult to assess. It is possible that more than one of these roles may be fulfilled simultaneously. For example, image quality could be improved in foveae of both types simply by removal of overlying tissue. As argued in Sections II,A and II,D, high spatial resolution is only possible in large eyes, and hence it may be predicted that the additional functions which a fovea may serve in enhancing resolving power could become more important as eye size increases. An elaborate device for the precise control of eye stability or focus indication may be of little adaptive advantage in a smaller (or nocturnal) eye in which resolution is inherently limited. Such a prediction is supported by the data of Fite and Rosenfield-Wessels (1975), who found a negative correlation (in nine avian species) between eye size and the width of the opening of the fovea.

3. *Inter-specific differences in the retinal position of areas and foveae*

The description and classification of the forms of areas and foveae have a long history. The first extensive summary was that of Slonaker (1897); subsequently important surveys have been carried out by Wood (1917), Rochon-Duvigneaud (1943), Duke-Elder (1958) and Meyer (1977). Meyer recognised eight different arrangements of areas and foveae in the avian retina (Fig. 7.18). It is probably sufficient to provide here only a brief summary of Meyer's review and urge the reader to study his account in detail. It should be noted that it is always assumed that areas and foveae are retinal structures the primary function of which is the enhancement of visual acuity.

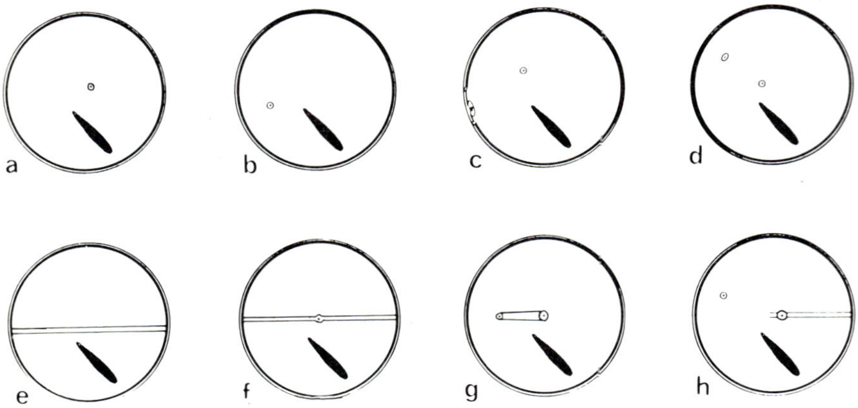

Fig. 7.18. Examples of the eight different arrangements of areas and foveae known to occur in the avian retina (data collated by Meyer, 1977). Each diagram shows the fundus of the right eye with the pecten (elongated black oval) drawn conventionally to indicate its position rather than its shape or size. (a) Steller's Jay (*Cyanocitta stelleri*); (b) an owl (Strigiformes); (c) a kingfisher (*Alcedo*); (d) American Bittern (*Botaurus lentiginosus*); (e) Manx Shearwater (*Puffinus puffinus*) and Northern Fulmar (*Fulmaris glacialis*); (f) American Coot (*Fulica americana*); (g) American Kestrel (*Falco sparverius*); (h) Common Tern (*Sterna hirundo*). From Meyer (1977), courtesy of Springer-Verlag, Heidelberg.

Both a central area and a central fovea are characteristic of most birds where they subserve monocular vision. Their specific function in the lateral visual field is not understood. A solitary temporal area and fovea are found in owls (Strigiformes), in which they project forward into the binocular field (Pettigrew, 1979a). When both central and temporal areas are present, they each contain a fovea. These bifoveate retinae are characteristic of many birds and are considered to be correlated with the habit of "feeding on the wing" as, for example, in some passerines, hummingbirds (Trochilidae) and kingfishers (Alcedinidae). The lateral foveae of these species would seem to be involved in binocular vision, but their projections have not been determined experimentally.

A linear area is found in many avian species. Although often regarded as unique to birds, such an area is known in reptiles (Brown, 1969). Furthermore, Hughes (1977) has presented evidence that, although linear areas are absent in mammals, linear bands of increased ganglion cell density are not uncommon. Duijm (1958) and Pearson (1972) tabulate the occurrence of such areas in birds. In general, linear areas seem to be associated with birds which either inhabit open spaces, or seek their nourishment from the ground and/or swim. According to Duijm (1958), when the bird is at rest this area is generally maintained in a horizontal position similar to that occupied by the semicircular canals. Many functions have been attributed to the linear areas. These include the minimisation of eye movements (Walls, 1942), movement detection (Pumphrey, 1961), spatial orientation (Diujm, 1958) and the localisation of the visual horizon (Pennycuick, 1960). Usually linear areas are associated with at least one fovea. The occurrence of a linear area associated with central and lateral foveae is regarded as an arrangement unique to birds. It is characteristic of diurnal birds-of-prey (Falconiformes), in which it is presumed to subserve both a high degree of visual acuity and stereoscopic vision. It should be noted that none of these functional explanations is without criticism. In very few species have the actual visual projections of areas and foveae been determined, and only one study (Diujm, 1958) has attempted to investigate at what angle the head is held. Also, the studies are unsystematic in that apparently no attempts have been made to test functional hypotheses by collecting data guided by the principles of phylogenetic relatedness or ecological convergence. Certainly these patterns of retinal topography are very intriguing, but their functional interpretation is subject to criticism.

4. *Oil droplets*

Considerable inter-specific differences in the colours of retinal oil droplets and in the patterns of their intraretinal distribution have been recorded. These data have been summarised by Muntz (1972), who proposed a number of functional explanations for the patterns. However, in view of the problems of detailed characterisation of oil droplets by light microscopy (on which all data on droplet distributions are based) and of the subtle differences in the spectral sensitivity of receptors containing the same oil droplet type (see Section III,A), functional interpretations of intraretinal and inter-specific differences in droplet populations are problematic.

(a) Intraretinal distribution. The most clearly defined intraretinal pattern of oil droplet distribution so far described is that of the pigeon (Galifret, 1968). In this bird the differences in oil droplet populations are so marked that sclerad illumination of the excised eye reveals that the fundus is divided into two distinct regions, the red and yellow oil droplet fields (Fig. 7.19a). These differences in the colour of the illuminated retina are due to a marked difference in their oil droplet population. The red field is sometimes referred to as the "pecking field", since its visual projection would seem to encompass an area around the bill, while the yellow field views the sky. However, the precise visual projections of these fields have not been determined. Not only does the relative proportion of cone receptors containing red oil droplets differ between the two fields, but also the absorption properties of droplets within the two fields differ (Bowmaker, 1977). It has been shown psychophysically that the spectral sensitivity of these two retinal areas differ (Martin and Muntz, 1978, 1979; Wortel *et al.*, 1984), and it has been proposed (Bowmaker, 1977) that differences in colour vision exist between these two parts of the retina.

A differentiation of the fundus on the basis of its oil droplet population into two clearly defined regions has not been reported in any species other than the pigeon. However, Peiponen (1964) presented data on a diversity of species which showed intraretinal differences in oil droplet populations. He reported that in many bird species the proportion of red droplets is greater in the dorsal part of the retina than in the ventral part, as in the pigeon. The phylogenetic extent of these intraretinal differences is not known and would clearly be helpful in evaluating functional hypotheses. According to Muntz (1972), the increased number of red oil droplets in the dorsal part of the retina is simply associated with better colour vision in that part of the retina, which generally looks downwards towards the ground or into vegetation. Walls and Judd (1933) suggested that red droplets may be associated with the reduction of "glare" and "dazzle". However, Muntz (1972) pointed out that red-filters may not be particularly appropriate for this role, which can quite

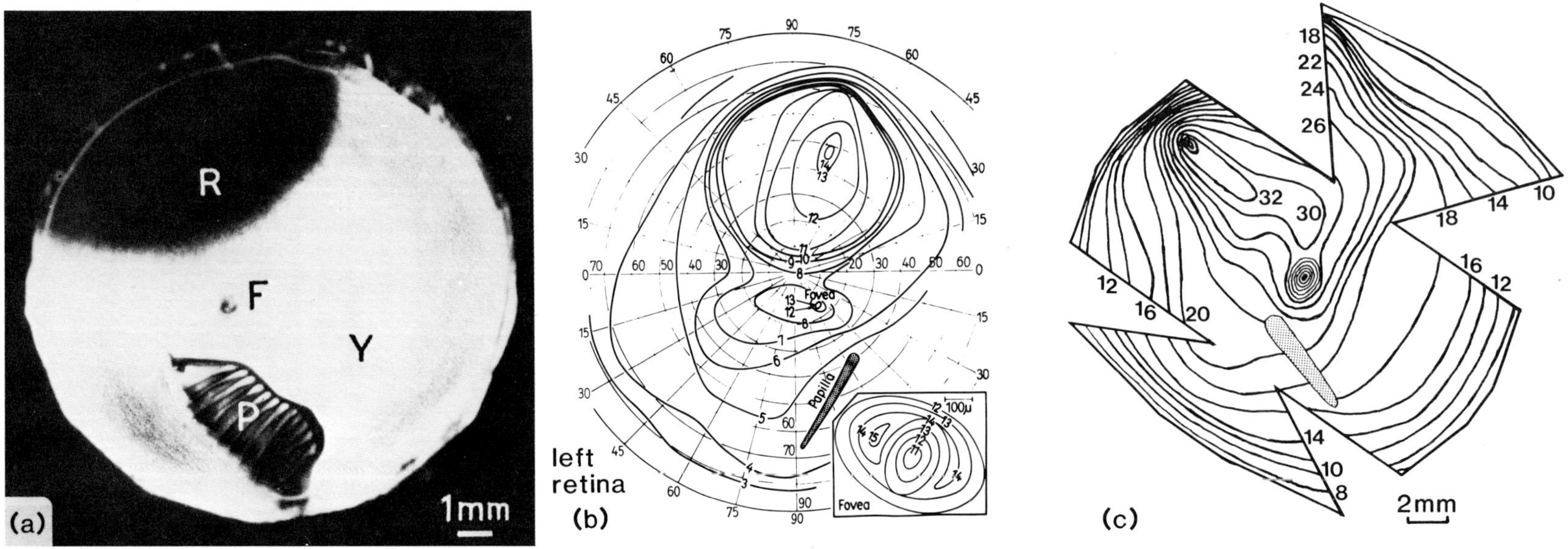

Fig. 7.19. The retina of the pigeon. (a) The fundus of the right eye photographed using transcleral illumination to reveal the red (R) and yellow (Y) oil droplet fields; also shown are the pecten (P) and the fovea (F). From Galifret (1968), courtesy of Springer-Verlag, Heidelberg. (b) Isodensity contour map of cells in sample areas (50 × 50 μm) of the bipolar cell layer of the left eye. The position of the pecten and fovea are indicated. Inset shows cell densities within the fovea. From Galifret (1968), courtesy of Springer-Verlag, Heidelberg. (c) Isodensity contour map of ganglion cells within the right eye based on a flat-mount preparation. The numbers ($\times 10^3$) indicate the density of nuclei per square millimetre of fixed retina. The pecten is indicated by stippling, and the fovea is just above it and to the right. From Bingelli and Paule (1969), courtesy of Alan R. Liss, Inc., New York.

effectively be fulfilled by a simple neutral filter. This can also be achieved by reduction of image brightness using the pupil, although this may also lead to a reduction in spatial resolution (see Section II,D).

(b) Inter-specific distribution. Despite the problems mentioned before of collecting reliable data on oil droplet colour, it is worth considering briefly the recorded inter-specific differences in droplet distribution which presumably reflect differences in the spectral sensitivity of cone receptor populations. Often the retinal position from which samples of droplet populations were taken in various species is not recorded, and in view of the known intraretinal differences recorded in other species, this severely limits the value of the data. Most data are attributable to Krause (1894), Erhard (1924), Walls and Judd (1933), Strother (1963) and Peiponen (1964). Despite the problems of interpretation, Muntz (1972) considered that the following conclusions were warranted, although it would seem reasonable to suggest that all of the species mentioned should be subjected to a more thorough examination.

1. In nocturnal birds, the proportion of coloured droplets is reduced compared with that in most diurnal species. However, coloured droplets are not absent in owls (Bowmaker and Martin, 1978), as previously suggested by Erhard (1924).
2. Swallows (Hirundinidae) and swifts (Apodiformes) have few red and orange droplets.
3. Birds that fly over water and need to see through the surface have a high proportion of red droplets. This seems to apply in such diverse groups as kingfishers (Alcedinidae), gulls and terns (Laridae), and the Gannet (*Morus bassanus*). However, birds that live on water but apparently do not need to see through the surface (e.g. the Shag, *Phalacrocorax aristotelis,* and the Manx Shearwater, *Puffinus puffinus*) do not have many red-coloured droplets.

In the absence of more thorough data on these species and similar data on others, it would seem premature to propose functional explanations beyond those outlined already to account for intraretinal differences in oil droplet populations.

5. *Ganglion cells and cells of the bipolar cell layer*

Galifret (1968), Bingelli and Paule (1969), and Yazulla (1974) have provided evidence of intraretinal differences in the distribution of ganglion cells and other neurons within the retina of the pigeon. Their results are shown in Fig. 7.19b and c. The two maps are not directly comparable because of differences in the projections used, but both indicate a high degree of differentiation within the retina including the presence of a second area in the red oil droplet field. It is interesting to note that the red–yellow oil droplet field differentiation at the receptor level does not appear to be correlated with differences in the distribution of cell nuclei within the inner retinal layers. In addition, Yazulla (1974) demonstrated that amacrine and bipolar synaptic densities do not parallel the cellular densities of amacrine and bipolar neurons across the retina. Why this is the case, and what its functional implications are, have not been established.

C. Pecten

The pecten (*pecten oculi*) is a vascular body projecting from the retina into the vitreous body (*corpus vitreum*) at the point of exit of the optic nerve (*nervus opticus*) (Figs. 7.3, 7). There is marked inter-specific variation in its size and form, and this has given rise to many theories about its function. The pecten consists exclusively of capillaries, extravascular pigmented stromal cells and a superficial covering membrane. It lacks both muscle and nervous tissue.

The pecten is most appropriately considered as part of the image-analysing system, since the consensus view is that its primary function is the nutrition of the inner layers of the retina, which in birds are avascular. However, this function is not accepted by all workers. In certain species it is possible that the pecten may also serve some secondary function, but this has not been verified experimentally. The structure and function of the pecten has been thoroughly reviewed by Meyer (1977), and the following brief account is based on this survey.

1. *Blood supply*

The arterial supply of the pecten is completely separate from that of the choroid. At the base of the pecten, each artery of the pecten (*a. pectinis oculi*), a branch of the ophthalmotemporal artery, forms a basilar artery which travels the length of the pecten giving rise to an arteriole for each pleat of the pecten. The arterioles open into an extensive capillary network. This is drained by venules which run parallel to the arterioles and open into a basilar vein at the base of the pecten. The basilar vein is drained by the vein of the pecten (*v. pectinis oculi*), a tributary of the ophthalmotemporal vein. (For further details of the blood supply see Chapter 5, by West *et al.*, in Vol. 2.)

2. *Types of pecten*

Three main morphological types of pecten are recognised: conical, vaned and pleated (Figs. 7.20, 21). Only one type (pleated) has been the subject of detailed histological investigations using light and electron microscopy.

(a) Conical pecten. The conical type of pecten is recognised only in kiwis (Apterygidae) (Lindsay-Johnson, 1901; Wood, 1917) and is considered to resemble the conus papillaris of reptiles. The optic disc in kiwis is circular (in other avian species it is linear), and the heavily pigmented, cone-shaped pecten arises from it. This type of pecten extends into the vitreous almost as far as the lens. Vanes and pleats are absent. Its status has been disputed. Walls (1942) considered it to be another indication of the general degenerative state of the kiwi eye (but see Section II,A), although Lindsay-Johnson (1901) proposed it to be intermediate between the pecten-like structures of various mammals and the pleated avian pecten. It is surprising that the conical pecten does not appear to have received further investigation since the description of it by Wood (1917).

(b) Vaned pecten. Little investigation has been done on the vaned type of pecten, which is reported to be present in the Ostrich (*Struthio camelus*), rheas

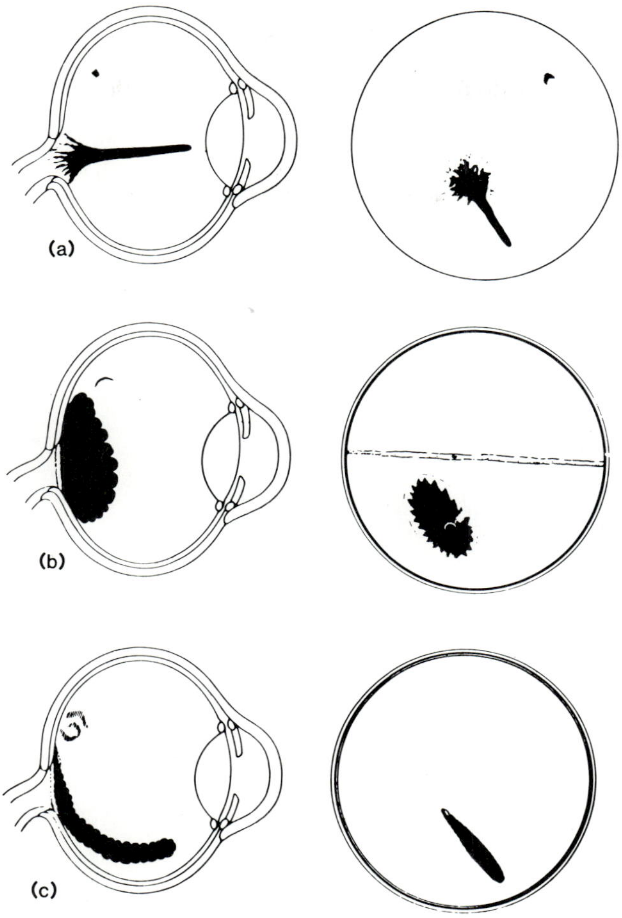

Fig. 7.20. Diagrammatic representations of the three morphological types of pecten. The drawings on the left are schematic sections through the eyeball and indicate the general shape and the extent to which the pecten extends into the posterior chamber towards the lens. The drawings on the right depict an ophthalmoscopic view of the fundus. (a) Conical pecten in the kiwi (*Apteryx*); (b) vaned pecten in the Ostrich (*Struthio camelus*); (c) pleated pecten in the Wood Pigeon (*Columba palumbus*). From Meyer (1977), courtesy of Springer-Verlag, Heidelberg.

(Rheidae) and tinamous (Tinamidae). Clearly more comparative data would be useful. It consists of a central vertical lamina from which arises a series of lateral vanes (Fig. 7.21).

(c) Pleated pecten. The pleated type of pecten is thought to occur in all the remaining avian orders. Meyer (1977) describes it as arising from the linear optic disc as a simple accordion-pleated lamina, which in most species is held in place at its free (apical) end by a more heavily pigmented crest or bridge (*pons pectinis*) of tissue. Inter-specific differences in all aspects of its gross morphology have been recorded, such as length, width, height, shape, number and arrangement of pleats, and extent of pigmentation, and this work has been analysed by Meyer (1977). Many of the studies have tried to correlate the gross structure of the pecten with other ocular features, for example the number of scleral ossicles and the diameter of the cornea. Meyer concludes that the available evidence indicates

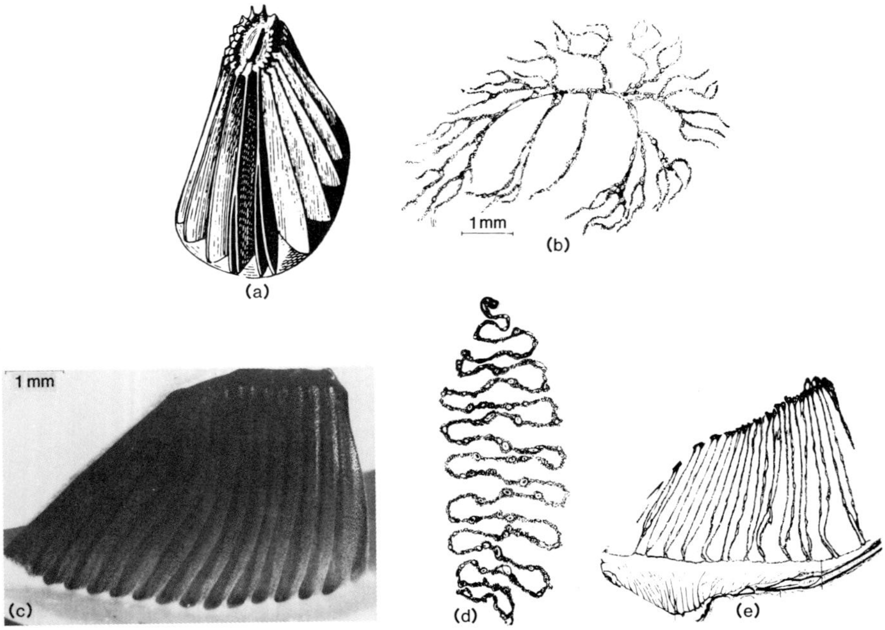

Fig. 7.21. The vaned and pleated types of pecten. (a) Drawing of the general appearance of the vaned pecten of the Ostrich (*Struthio camelus*) From Duke-Elder, Sir Stewart, editor: System of ophthalmology. Vol. I. The eye in evolution (by Duke-Elder, Sir Stewart), St. Louis, 1958, The C. V. Mosby Co.; and (b) transverse horizontal section near the base of the vaned pecten of the Greater Rhea (*Rhea americana*) (Rochon-Duvigneaud, 1943, courtesy of Masson Editeur, Paris). In (c)–(e) is shown the pleated type of pecten. (c) Lateral view, the pigeon (from Meyer, 1977, courtesy of Springer-Verlag, Heidelberg); (d) horizontal section, the Tawny Eagle (*Aquila rapax*); (e) longitudinal section, the Gannet (*Morus bassanus*) (Rochon-Duvigneaud, 1943, courtesy of Masson Editeur, Paris).

that, "active, diurnal birds with high acuity and monocular rather than binocular perception, tend to possess a larger and more pleated pecten, nocturnal species with poor vision a smaller one of simpler morphology".

3. *Function*

Not surprisingly for a structure which exhibits such great inter-specific diversity in its morphology, many functions have been proposed for the pecten. Meyer (1977) concludes from an analysis of the various hypotheses that "Much indirect evidence has been obtained from microscopic (light and electron), histochemical and experimental investigations, but more conclusive proof is still required to determine the essential functions of the pecten". Most of the functions discussed (Wingstrand and Munk, 1965, listed 30) involve the blood supply.

(a) Functions associated with the vascular supply. The most widely-accepted function of the pecten is that it acts as a supplementary nutritive organ, providing oxygen and nutrient for the inner portion of the retina. Its richness in blood vessels, particularly capillaries, its pleated arrangement resulting in an increased surface area, and its position close to the retina, all support this view. However, the results of experimental studies designed specifically to test this role have not

been unequivocal. Wingstrand and Munk (1965), for example, described how elimination of the arterial supply to the pecten subsequently led to the complete degeneration of the pecten as well as to degenerative changes in the vitreal layers of the retina. In contrast, direct destruction of the pecten has been claimed to have no visible effect on retinal histology (Francois and Neetens, 1974; Brach, 1975). Brach (1975) proposed that the pecten is involved in the regulation of intraocular pH. Kajikawa (1923) and Bacsich and Gellert (1935), on the other hand, considered that the pecten was primarily a heat exchange device and functioned to protect the eye against cooling. A role in heat transfer was also suggested by Griffin (1953), who, however, thought it worked in the opposite direction, removing excess heat from the eye.

(b) The pecten shadow. The inter-specific diversity in the gross structure of the pecten had led to proposals on its function related to its size, shape and position in the eye. Many of these hypotheses are ingenious and intriguing, but none have been supported by adequate experimentation. It is of course possible that in certain species the form and position of the pecten may have been secondarily

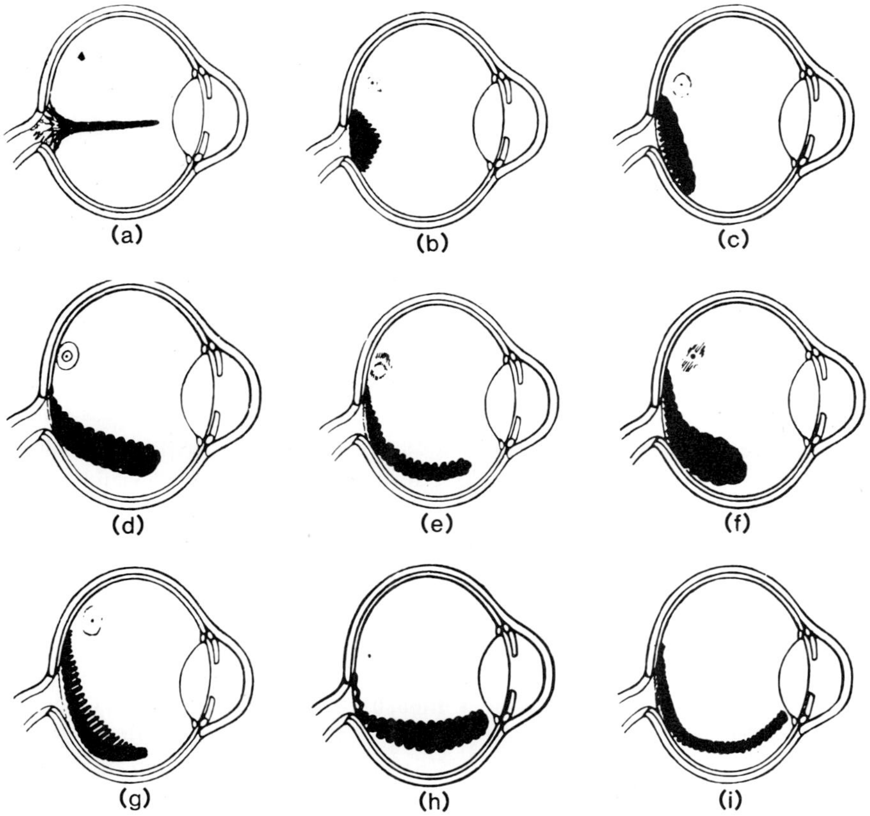

Fig. 7.22. Types of pecten. Schematic drawings of vertical sections through the eye showing the marked differences in shape and size of the pecten in various species. The diagrams are not drawn to scale. (a) A kiwi (*Apteryx*); (b) Common Kestrel (*Falco tinnunculus*); (c) Secretary Bird (*Sagittarius serpentarius*); (d) Herring Gull (*Larus argentatus*); (e) Wood Pigeon (*Columba palumbus*); (f) Greater Rhea (*Rhea americana*); (g) Laughing Kookaburra (*Dacelo novaeguineae*); (h) Barn Swallow (*Hirundo rustica*); and (i) Blue Jay (*Cyanocitta cristata*). From Duke-Elder, Sir Stewart, editor: System of ophthalmology. Vol. I. The eye in evolution (by Duke-Elder, Sir Stewart), St. Louis, 1958, the C. V. Mosby Co.; after Casey Wood.

adapted to subserve another function beyond a general nutritive one. A number of investigations have been concerned with the shadow which the pecten might throw on the retina, although Mann (1924) maintained that under normal viewing conditions the pecten shadow falls mainly along its own base and slightly to each side, the result being a small enlargement of the blind spot in the dorsal half of the visual field. However, in view of the diversity of shapes shown in Fig. 7.22 this conclusion may not apply to all species. Among the functions which have been ascribed to the shadow of the pecten are the following: (1) increased sensitivity for the perception of movement (Menner, 1938; Crozier and Wolf, 1944a,b), (2) improvement of visual acuity (Thomson, 1928, 1929), (3) a sextant to measure the angle of elevation of the sun (Griffin, 1952, 1953; Pettigrew, 1978), and (4) an intraocular shade against the glare of the sun (Barlow and Ostwald, 1972). Whilst all of these hypotheses have been subject to varying degrees of criticism, none has been adequately tested.

IV. Visual capacities

The visual capacities of birds have been studied using both electrophysiological and behavioural psychophysical techniques. The interpretation of such studies in themselves is problematic and would require more space than is available here to be fully discussed (see the analyses by Stebbins, 1970; Muntz, 1974; Hughes, 1977; Granda and Maxwell, 1979). Correlating visual capacities with morphological features of the eye is particularly difficult. This is because psychophysically determined visual capacities reflect processing of visual information by the whole visual system, not just the eye, and in the case of behavioural studies the processing of that information in the mediation of behavioural responses. It is obvious that the eye must limit the information available for processing by higher neural centres, and so in a very general sense all visual capacities are a function of eye structure. The following discussion is therefore brief and indicates only those areas where reasonably specific statements about visual capacities and their relation to avian eye structure can be made. More is known about the visual capacities of birds than is indicated here. The visual capacities of the pigeon have been reviewed in detail by Donovan (1978) and Emmerton (1983a).

A. Absolute visual sensitivity

Absolute visual sensitivity has been measured in the Tawny Owl (*Strix aluco*) (Martin, 1977), the pigeon (Blough, 1955) and two species of duck (Anatidae) (Hersloff *et al.*, 1974; Wells *et al.*, 1975). The data have shown that the owl is approximately 100 times more sensitive than the pigeon. Comparison of retinal image illumination in the owl and pigeon indicates that this difference is mainly due to retinal mechanisms, not to differences in optical properties of the eye (Martin, 1982). It also suggests that the owl retina has attained (along with the retina of mammals) the ultimate limit on visual sensitivity dictated by the quantal nature of light; that is, it is unlikely that any avian species will be found to have a retina the sensitivity of which is greater than that of owls.

B. Scotopic spectral sensitivity

This is generally regarded as reflecting the spectral sensitivity of the rod receptors (Muntz, 1974) and has been measured in only two avian species for which rod visual pigment data are available, the Tawny Owl (*Strix aluco*) (Martin, 1977; Bowmaker and Martin, 1978) and the pigeon (Donner, 1953; Blough, 1955; Ikeda, 1965; Alexandridis, 1967; Granda and Yazulla, 1971; Graf and Norren, 1974; Bowmaker, 1977). As in other vertebrates, good agreement is found between scotopic spectral sensitivity and the percentage of light absorbed at different wavelengths by the rod visual pigment.

C. Photopic spectral sensitivity

Photopic spectral sensitivity is thought to reflect the combined activity of the cone receptors only. Convincing correlations between known cone receptor sensitivities and photopic spectral sensitivity data are difficult to achieve because of the degree of latitude in the models employed. However, Bowmaker (1977) and Bowmaker and Knowles (1977) have had some success in modelling the observed spectral sensitivity curves with data on the known sensitivities of the cone receptor populations in the pigeon (Granit, 1942; Donner, 1953; Blough, 1957; Ikeda, 1965; Graf, 1969; Blough *et al.*, 1972) and chicken (Honigman, 1921; Armington and Thiede, 1956). However, Bowmaker (1977) was not able to account adequately, in terms of known differences in cone populations, for the observed differences in the spectral sensitivities of the red and yellow oil droplet fields of the pigeon (Martin and Muntz, 1978, 1979) or for the photopic spectral sensitivity of the Tawny Owl (*Strix aluco*) (Martin and Gordon, 1974a,b; Martin *et al.*, 1975).

D. Colour vision

Colour vision in vertebrates is regarded as principally a function of the retinal cones. Since all diurnal birds examined to date have such receptors, Muntz (1974) concluded that "no one doubts that they also have well developed colour vision". It also now seems clear that avian species regarded as nocturnal, such as the owls, have functional cone receptors (Bowmaker and Martin, 1978) and colour vision (Martin, 1974). Wavelength discrimination has been studied in the pigeon by Hamilton and Coleman (1933), Riggs *et al.* (1972), Wright (1972b, 1979), Jitsumori (1978), and Emmerton and Delius (1980). Attempts to account for these results in terms of the sensitivity of retinal receptors have, however, not been particularly successful.

E. Ultraviolet sensitivity

Visual sensitivity in pigeons to wavelengths within the near-ultraviolet region of the spectrum (minimum wavelength ~360 nm) was first demonstrated by

Wright (1972a), and later confirmed by Kreithen and Eisner (1978), Delius and Emmerton (1979), Emmerton (1983b), and Emmerton and Remy (1983); also wavelength discrimination in this spectral region was shown by Emmerton and Delius (1980). There is, however, debate about the relative sensitivity of pigeons in the ultraviolet region (Emmerton and Remy, 1983). Ultraviolet sensitivity has since been demonstrated in three species of hummingbird (Trochilidae) (Goldsmith, 1980; Goldsmith *et al.*, 1981) and in the Mallard (*Anas platyrhynchos*) (Parrish *et al.*, 1981). So far, a visual pigment to mediate such sensitivity has not been directly identified in these species, but there is circumstantial evidence for one (see Section III,A and Table I). Also, a cone pigment the sensitivity of which extends into the ultraviolet region with a maximum at 406 nm has been identified in the Humboldt Penguin (*Spheniscus humboldti*), but visual sensitivity in the ultraviolet region has not been demonstrated psychophysically (Bowmaker and Martin, 1985).

F. *Polarization sensitivity*

The ability of pigeons to detect the plane of polarization of light has been described by Kreithen and Keeton (1974) and Delius *et al.* (1976). There is also limited experimental evidence for polarization sensitivity in passerine species (Abel, 1982). The underlying mechanism is not known. Kreithen and Keeton (1974) proposed that the fovea may be involved (see Section III,C), whilst Delius *et al.* (1976) and Young and Martin (1984) have proposed that the double cones may mediate this sensitivity (see Section III,A).

G. *Visual resolution*

Generally, minimum separable acuity, involving the perception of black and white bars of high contrast, has been used to measure visual resolution in birds. Typically a narrow range of luminances has been employed. The species studied include the pigeon (Hamilton and Goldstein, 1933; Chard, 1939; Blough, 1971, 1973; Hodos *et al.*, 1976; Hodos and Leibowitz, 1977; Bloch and Martinoya, 1978; Uhlrich *et al.*, 1982), various passeriforms (Donner, 1951), two species of owl (Strigiformes) (Fite, 1973; Martin and Gordon, 1974a) and the American Kestrel (*Falco sparverius*) (Fox *et al.*, 1976). More detailed investigations, using stimuli varying in contrast as well as luminance and spatial frequency, have also been carried out in the pigeon (Nye, 1968) and *Falco sparverius* (Hirsch, 1982). In only a few species has acuity been measured over a wide range of luminances (e.g. Fite, 1973).

There have been many attempts to account for measures of visual resolution in terms of eye structure, most notably in terms of the maximum spatial frequencies available in the retinal image and/or in terms of the density of visual cells in conjunction with the focal length of the eye (see Muntz, 1974; Miller, 1979, for reviews). Unfortunately there have been few such analyses in birds. By far the most thorough investigation involving birds was that of Hirsch (1982), who was able to account for the measured acuity of *Falco sparverius* in terms of receptor density, eye focal length, and the hypothesised telephoto lens system of the convexiclivate fovea (Section III,C).

It is worth noting that Walls (1942), Rochon-Duvigneaud (1943) and Duke-Elder (1958) have proposed that visual acuity in diurnal birds-of-prey is many times higher than that of humans. However, this has not been confirmed experimentally in any species. Although Fox *et al.* (1976) reported that acuity in *Falco sparverius* exceeded that of humans by a factor of 2·6 times, this study may be criticised on methodological grounds. Furthermore, Hirsch (1982) reported that the minimum separable acuity in this species was very similar to that of humans under identical viewing conditions. However, the focal length of the eye of *Falco sparverius* is less than that of humans. Clearly acuity data in the larger-eyed accipitriol species would be of interest.

References

Abel, K. P. (1982). Skylight polarization patterns at dusk influence migratory orientation in birds. *Nature, Lond.* **299,** 550–551.

Abelsdorff, G. and Wessely, K. (1909). Vergleichend physiologie Untersuchungen über den Flüssigkeitswechsel des Auges in der Wirbeltierreihe. *Arch. Augenheilk.* **64,** 62–125.

Alexandridis, E. (1967). Pupillographische Untersuchung der Netzhautempfindlichkeit des Taubenauges. *Albrecht v. Graefes Arch. Ophthal.* **172,** 139–151.

Alpern, M. (1968). Distal mechanisms of vertebrate colour vision. *A. Rev. Physiol.* **30,** 279–318.

Armington, J. C. and Thiede, F. C. (1956). Electroretinal demonstration of a Purkinje shift in the chicken eye. *Am. J. Physiol.* **186,** 258–262.

Bacsich, P. and Gellert, A. (1935). Beiträge zur Kenntnis der Struktur und Funktion des Pectens im Vogelauge. *Albrecht v. Graefes Arch. Ophthal.* **133,** 448–460.

Barany, E., Berggren, L. and Vrabec, F. (1957). The mucinous layer covering the corneal endothelium in the owl *Strix aluco. Br. J. Opthal.* **41,** 25–30.

Barlow, H. B. (1981). Critical limiting factors in the design of the eye and visual cortex. *Proc. R. Soc. B.* **212,** 1–34.

Barlow, H. B. and Ostwald, T. J. (1972). Pecten of the pigeon's eye as an interocular shade. *Nature, Lond.* **236,** 88–90.

Baron, M. and Gallego, A. (1982). The horizontal cells of the falconidae retina. *In* "Structure of the Eye" (J. G. Hollyfield, Ed.), 4th edn. Elsevier, Amsterdam.

Baylor, D. A. and Fettiplace, R. (1975). Light path and photon capture in turtle photoreceptors. *J. Physiol., Lond.* **248,** 433–464.

Beer, T. (1893). Studien über die Accomodation des Vögelauges. *Pflügers Arch. ges. Physiol.* **53,** 175–237.

Bingelli, R. L. and Paule, W. J. (1969). The pigeon retina: quantitative aspects of the optic nerve and ganglion cell layer. *J. Comp. Neurol.* **137,** 1–18.

Blake, R. and Fox, R. (1973). The psychophysical enquiry into binocular summation. *Perception Psychophys.* **14,** 161–185.

Bloch, S. and Martinoya, C. (1978). Comparison of frontal and lateral visual acuity of the pigeon, under behavioural fixation. *J. Physiol., Lond.* **278,** 49P.

Bloch, S. and Maturana, H. R. (1971). Oil droplet distribution and colour discrimination in the pigeon. *Nature, Lond.* **234,** 284–285.

Bloch, S., Martinoya, C. and Rivaud, S. (1981). Eye movements in pigeons: participation in binocular fixation and visual pursuit. *J. Physiol., Lond.* **320,** 29P.

Blough, D. S. (1955). Method for tracing dark adaptation in the pigeon. *Science, Wash., D.C.* **121,** 703–704.

Blough, D. S. (1957). Spectral sensitivity in the pigeon. *J. opt. Soc. Am.* **47,** 827–833.

Blough, P. M. (1971). The visual acuity of the pigeon for distant targets. *J. exp. Analysis Behav.* **15,** 57–67.

Blough, P. M. (1973). Visual acuity in the pigeon II: effects of target distance and retinal lesions. *J. exp. Analysis Behav.* **20,** 333–343.

Blough, P. M., Riggs, L. A. and Schafer, K. L. (1972). Photopic spectral sensitivity determined electroretinographically for the pigeon eye. *Vision Res.* **12,** 477–485.

Bowmaker, J. K. (1977). The visual pigments, oil droplets and spectral sensitivity of the pigeon. *Vision Res.* **17,** 1129–1138.

Bowmaker, J. K. and Knowles, A. (1977). The visual pigments and oil droplets of the chicken retina. *Vision Res.* **17,** 755–764.

Bowmaker, J. K. and Martin, G. R. (1978). Visual pigments and colour vision in a nocturnal bird, *Strix aluco* (Tawny Owl). *Vision Res.* **18,** 1125–1130.

Bowmaker, J. K. and Martin, G. R. (1985). Visual pigments and oil droplets in the penguin, *Spheniscus humboldti*. *J. Comp. Physiol.* **156,** 71–77.

Brach, V. (1975). The effect of intraocular ablation of the pecten oculi of the chicken. *Invest. Ophthal.* **14,** 166–168.

Brown, K. T. (1969). A linear area centralis extending across the turtle retina and stabilized to the horizon by non-visual cues. *Vision Res.* **9,** 1053–1062.

Bubién-Waluszewska, A. (1981). The cranial nerves. *In* "Form and Function in Birds" (A. S. King and J. McLelland, Eds.), Vol. 2. Academic Press, London and New York.

Burns, R. B. (1975). Plasma cells in the avian Harderian gland and the morphology of the gland in the rook. *Can. J. Zool.* **53,** 1258–1269.

Burns, R. B. (1976). The structure of the lachrymal glands of the domestic fowl and the duck. *Res. vet. Sci.* **21,** 292–299.

Burns, R. B. and Maxwell, M. H. (1979). The structure of the Harderian and lacrimal gland ducts of the turkey, fowl and duck. A light microscope study. *J. Anat.* **128,** 285–292.

Campbell, F. W. and Gregory, A. H. (1960). Effect of size of pupil on visual acuity. *Nature, Lond.* **187,** 1121–1123.

Campbell, F. W. and Gubisch, R. W. (1966). Optical quality of the human eye. *J. Physiol., Lond.* **186,** 558–578.

Campbell, M. C. W. and Hughes, A. (1981). An analytic, gradient index schematic lens and eye for the rat which predicts aberrations for finite pupils. *Vision Res.* **21,** 1129–1148.

Chard, R. D. (1939). Visual acuity in the pigeon. *J. exp. Psychol.* **24,** 588–608.

Chard, R. D. and Gundlach, R. H. (1938). The structure of the eye of the homing pigeon. *J. comp. Psychol.* **25,** 249–272.

Chiasson, R. B. and Ferris, W. R. (1968). The iris and associated structures of the Inca Dove (*Scardafella inca*). *Am. Zool.* **8,** 818.

Clarke, P. G. H. and Whitteridge, D. (1976). The projection of the retina, including the 'red area' on to the optic tectum of the pigeon. *Q. J. exp. Physiol.* **61,** 351–358.

Cohen, A. I. (1963). The fine structure of the visual receptors of the pigeon. *Exp. Eye Res.* **2,** 88–97.

Cowan, W. M. (1970). Centrifugal fibres to the avian retina. *Br. med. Bull.* **26,** 112–118.

Cowan, W. M. and Powell, T. P. S. (1963). Centrifugal fibres in the avian visual system. *Proc. R. Soc. B.* **158,** 232–252.

Cramp, S. (1977). "The Birds of the Western Palearctic", Vol. 1. Oxford University Press, London.

Crescitelli, F. (1972). The visual cells and visual pigments of the vertebrate eye. *In* "Handbook of Sensory Physiology" (H. J. A. Dartnell, Ed.), Vol. VII/1. Springer-Verlag, New York and Berlin.

Crozier, W. J. and Wolf, E. (1944a). Theory and measurement of visual mechanisms. x. Modification of the flicker response contour, and the significance of the avian pecten. *J. gen. Physiol.* **27,** 287–313.

Crozier, W. J. and Wolf, E. (1944b). Flicker response contours for the sparrow, and the theory of the avian pecten. *J. gen. Physiol.* **27,** 315–324.

Dawsons, E. W. (1978). Apterygiformes. *In* "Bird Families of the World" (C. J. O. Harrison, Ed.). Elsevier-Phaidon, Oxford.

Delius, J., and Emmerton, J. (1979). Visual performance of pigeons. *In* "Neural Mechanisms of Behaviour in the Pigeon" (A. M. Granda and J. H. Maxwell, Eds.) Plenum Press, New York and London.

Delius, J., Perchard, R. and Emmerton, J. (1976). Polarized light discrimination by pigeons and an electroretinographic correlate. *J. comp. physiol. Psychol.* **90,** 560–571.

Denton, E. J. (1956). The responses of the pupil of the gekko to external light stimulus. *J. gen. Physiol.* **40,** 201–218.

Ditchburn, R. W., Fender, D. H. and Mayne, S. (1959). Vision with controlled movements of the retinal image. *J. Physiol., Lond.* **145,** 98–107.

Dogiel, A. S. (1888). Ueber die nervösen Elemente in der Retina der Ganoiden, Reptilien, Vögel und Säugetiere. *Anat. Anz.* **3,** 133–143.

Donner, K. O. (1951). The visual acuity of some passerine birds. *Acta zool. fenn.* **66,** 1–40.

Donner, K. O. (1953). The spectral sensitivity of the pigeon's retinal elements. *J. Physiol. Lond.* **122,** 524–537.

Donner, K. O. (1960). On the effect of the coloured oil droplets on the spectral sensitivity of the avian retina. *Proc. 12th Int. Ornithol. Congr., Helsinki,* **1,** 167–172.

Donovan, W. J. (1978). Structure and function of the pigeon visual system. *Physiol. Psychol.* **6,** 403–437.

Dowling, J. E. and Cowan, W. M. (1966). An electronmicroscopic study of normal and degenerating centrifugal fiber terminals in the pigeon retina. *Z. Zellforsch. mikrosc. Anat.* **71,** 14–28.

Duecker, G. and Schulze, I. (1977). Color vision and color preferences in Japanese Quail (*Coturnix coturnix japonica*) with colourless droplets. *J. comp. physiol. Psychol.* **91,** 1110–1117.

Duijm, M. (1958). On the position of a ribbon-like central area in the eyes of some birds. *Arch. néerl. Zool.* **13,** 128–145.

Duke-Elder, S. (1958). "System of Ophthalmology", Vol. 1. Henry Kimpton, London.

Emmerton, J. (1983a). Functional morphology of the visual system. *In* "Physiology and Behaviour of the Pigeon" (M. Abs, Ed.). Academic Press, London and New York.

Emmerton, J. (1983b). Pattern discrimination in the near-ultraviolet by pigeons. *Perception Psychophys.* **34,** 555–559.

Emmerton, J. and Delius, J. D. (1980). Wavelength discrimination in the 'visible' and ultraviolet spectrum by pigeons. *J. Comp. Physiol.* **141,** 47–52.

Emmerton, J. and Remy, M. (1983). The pigeon's sensitivity to ultraviolet and "visible" light. *Experientia* **39,** 1161–1163.

Emmerton, J., Schwemer, S., Muth, I. and Schlecht, P. (1980). Spectral transmission of the ocular media of the pigeon. *Invest. Ophthal. Visual Sci.* **19,** 1382–1387.

Engström, K. (1958). On the cone mosaic in the retina of *Parus major*. *Acta zool., Stockh.* **39,** 65–69.

Erhard, H. (1924). Messende Untersuchungen über den Farbensinn der Vögel. Zugleich ein Beitrag zur Lehre von der Schmuckfarben. *Zool. Jb., Abt. allg. Zool. Physiol.* **41,** 489–552.

Evans, H. E. (1979). Organa sensoria. *In* "Nomina Anatomica Avium" (J. J. Baumel, A. S. King, A. M. Lucas, J. E. Breazile and H. E. Evans, Eds.). Academic Press, London and New York.

Fager, L. Y. and Fager, R. S. (1981). Chicken blue and chicken violet, short wavelength sensitive pigments. *Vision Res.* **21,** 581–586.

Fite, K. V. (1973). Anatomical and behavioural correlates of visual acuity in the Great Horned Owl. *Vision Res.* **13,** 219–230.

Fite, K. V. and Rosenfield-Wessels, S. (1975). A comparative study of deep avian foveas. *Brain Behav. Evol.* **12,** 97–115.

Fox, R. (1979). Binocularity and stereopsis in the evolution of vertebrate vision. *In* "Frontiers in Visual Science" (S. J. Cool and E. L. Smith, Eds.). Springer-Verlag, Berlin and New York.

Fox, R., Lehmkuhle, S. W. and Westendorf, D. H. (1976). Falcon visual acuity. *Science, Wash., D.C.* **192,** 263–265.

Francois, J. and Neetens, A. (1974). Comparative anatomy of the vascular supply of the eye in vertebrates. *In* "The Eye" (H. Davson and L. T. Graham, Eds.). Academic Press, New York and London.

Franz, V. (1934). Höhere Sinnesorgane. 1. Vergleichende Anatomie des Wirbeltierauges. *In* "Handbuch der vergleichenden Anatomie der Wirbeltiere" (L. Bolk, E. Göppert, E. Kallius and W. Lubosch, Eds.), Vol. 2. Urban and Schwarzenberg, Berlin.

Fry, C. H. (1978). Charadriiformes, skimmers. *In* "Bird Families of the World" (C. J. O. Harrison, Ed.). Elsevier-Phaidon, Oxford.

Galifret, Y. (1968). Les diverse aires fonctionelles de la rétine du pigeon. *Z. Zellforsch. mikrosk. Anat.* **86,** 535–545.

Gallego, A. (1976). Comparative anatomy of horizontal cells in the vertebrate retina: mammals and birds. *In* "Neural Principles in Vision" (F. Zetter and R. Weiler, Eds.). Springer-Verlag, Berlin and New York.

Gallego, A. (1982). Organization of the outer plexiform layer of the tetrapoda retina: horizontal cells of mammals and avian retina. *In* "Structure of the Eye" (J. G. Hollyfield, Ed.), Vol. 4. Elsevier, Amsterdam.

Gallego, A. (1983). Organization of the outer plexiform layer of the tetrapoda retina. *In* "Progress in Sensory Physiology" (D. Ottoson, Ed.), Vol. 4. Springer-Verlag, Berlin and New York.

Gallego, A., Baron, M. and Gayoso, M. (1975a). Organization of the outer plexiform layer of the diurnal and nocturnal bird retina. *Vision Res.* **15,** 1027–1028.

Gallego, A., Baron, M. and Gayoso, M. (1975b). Horizontal cells of the avian retina. *Vision Res.* **15,** 1029–1030.

Gibson, J. J. (1961). Ecological optics. *Vision Res.* **1,** 253–262.

Gibson, J. J. (1979). "The Ecological Approach to Visual Perception". Houghton Mifflin, Boston.

Goldsmith, T. H. (1980). Hummingbirds see near ultraviolet light. *Science, Wash., D.C.* **207,** 786–788.

Goldsmith, T. H., Collins, J. S. and Perlman, D. L. (1981). A wavelength discrimination function for the hummingbird *Archilochus alexandri*. *J. comp. Physiol.* **143,** 103–110.

Goldsmith, T. H., Collins, J. S. and Licht, S. (1984). The cone oil droplets of avian retinas. *Vision Res.* **24,** 1661–1671.

Goodge, W. R. (1960). Adaptations for amphibious vision in the Dipper (*Cinclus mexicanus*). *J. Morph.* **107,** 79–91.

Goodson, J. E. (1969). Optokinetic responses of the pigeon. Ph.D. Thesis, George Washington University.

Govardovskii, V. I. and Zueva, L. V. (1977). Visual pigments of chicken and pigeon. *Vision Res.* **17,** 537–543.

Graf, V. A. (1969). A spectral luminosity function in the pigeon determined by flicker photometry. *Psychon. Sci.* **17,** 282–283.

Graf, V. A. and Norren, D. V. (1974). A blue sensitive mechanism in the pigeon retina: λ max 400 nm. *Vision Res.* **14,** 1203–1209.

Granda, A. M. and Maxwell, J. H. (1979). "Neural Mechanisms of Behaviour in the Pigeon". Plenum Press, New York.

Granda, A. M. and Yazulla, S. (1971). The spectral sensitivity of single units in the nucleus rotundus of the pigeon. *J. gen. Physiol.* **57,** 363–384.

Granit, R. (1942). The photopic spectrum of the pigeon. *Acta physiol. scand.* **4,** 118–124.

Griffin, D. R. (1952). Bird navigation. *Biol. Rev.* **27,** 359–393.

Griffin, D. R. (1953). Sensory physiology and the orientation of animals. *Am. Scient.* **41,** 209–244.

Gundlach, R. H., Chard, R. D. and Skahen, J. R. (1945). The mechanism of accommodation in pigeons. *J. comp. Psychol.* **38,** 27–42.

Gupta, S. K. and Maiti, R. B. (1981). Action of prolactin on the Harderian gland of the chicken. *Gen. comp. Endocrinol.* **45,** 289–292.

Hahn, E. (1916). Über den Farbensinn der Tagvogel und die Zapfenölkugeln. *Z. wiss. Zool.* **116,** 1–42.

Hailman, J. P. (1976). Oil droplets in the eyes of adult anuran amphibians: a comparative survey. *J. Morph.* **148,** 453–468.

Hamilton, W. F. and Coleman, T. B. (1933). Trichromatic vision in the pigeon as illustrated by the spectral hue discrimination curve. *J. comp. Psychol.* **15,** 183–191.

Hamilton, W. F. and Goldstein, J. L. (1933). Visual acuity and accommodation in the pigeon. *J. comp. Psychol.* **15,** 193–197.

Harkness, L. and Bennett-Clark, H. C. (1978). The deep fovea as a focus indicator. *Nature, Lond.* **272,** 814–816.

Harosi, F. I. and MacNicol, E. F. (1974). Visual pigments of goldfish cones: spectral properties and dichroism. *J. gen. Physiol.* **63,** 279–304.

Hayes, B. P. and Holden, A. L. (1983). The distribution of displaced ganglion cells in the retina of the pigeon. *Exp. Brain Res.* **49,** 181–188.

Hazlett, L. D., Meyer, D. B. and Susan, S. R. (1974). Visual cell ultrastructure in the Japanese Quail (*Coturnix coturnix japonica*). *Anat. Rec.* **178,** 371–372.

Hersloff, L., Lehner, P. N., Bolen, E. G. and Rylander, M. K. (1974). Visual sensitivity in the Black-bellied Tree Duck (*Dendrocygna autumnalis*), a crepuscular species. *J. comp. Physiol. Psychol.* **86,** 486–492.

Hess, C. (1910). Die Akkommodation bei Tauchervogeln. *Arch. vergl. Ophthal.* **1,** 153–164.

Hess, C. (1912). Vergleichende Physiologie des Gesichtssinnes. *In* "Handbuch der vergleichende Physiologie" (H. Winterstein, Ed.), Vol. 4. Fischer, Jena.

Hirsch, J. (1982). Falcon visual sensitivity to grating contrasts. *Nature, Lond.* **300,** 57–58.

Hodos, W. and Leibowitz, R. W. (1977). Near-field visual acuity of pigeons: effects of scotopic adaptation and wavelength. *Vision Res.* **17,** 463–467.

Hodos, W., Leibowitz, R. W. and Bonbright, J. C. (1976). Near-field visual acuity of pigeons: effects of head location and stimulus luminance. *J. exp. Analysis Behav.* **25,** 129–141.

Holden, A. L. (1969). Receptive properties of retinal cells and tectal cells in the pigeon. *J. Physiol., Lond.* **201,** 56P–57P.

Holden, A. L. (1977a). Concentric receptive field of the pigeon ganglion cells. *Vision Res.* **17,** 545–554.

Holden, A. L. (1977b). Responses of directional ganglion cells in the pigeon retina. *J. Physiol., Lond.* **270,** 253–269.

Holden, A. L. (1979). Non-directional and directional ganglion cells in the pigeon retina. *Vision Res.* **19,** 321–326.

Holden, A. L. (1981). Classifying and comparing retinal ganglion cells. *Brain Behav. Evol.* **18,** 188–193.

Honigman, H. (1921). Untersuchungen über Lichtempfindlichkeit und Adaptierung der Vogelanges. *Pflügers Arch. ges. Physiol.* **189,** 1–72.

Howcroft, M. J. and Parker, J. A. (1977). Aspheric curvature of the human lens. *Vision Res.* **17,** 1217–1223.

Hughes, A. (1977). The topography of vision in mammals of contrasting life styles: comparative optics and retinal organisation. *In* "Handbook of Sensory Physiology" (F. Crescitelli, Ed.), Vol. 7/5. Springer-Verlag, Berlin and New York.

Hughes, A. and Vaney, D. I. (1978). The refractive state of the rabbit eye: variation with eccentricity and correction of oblique astigmatism. *Vision Res.* **18,** 1351–1355.

Ikeda, H. (1965). The spectral sensitivity of the pigeon (*Columba livia*). *Vision Res.* **5,** 19–36.

Ischreyt, G. (1912). Zur vergleichen der Morphologie des Entenauges. *Arch. vergl. Ophthal.* **3,** 39–76.

Jane, S. D. and Bowmaker, J. K. (1984). A microspectrophotometric study of the visual pigments and oil droplets found in the retina of the Aylesbury duck *Anas platyrhynchos domesticus*. *Vision Res.* **24,** 1701.

Jitsumori, M. (1978). Wavelength discrimination function derived from post-discrimination gradients in the pigeon. *Jap. psychol. Res.* **20,** 18–28.

Johnston, D. and Hudson, R. A. (1974). Phospholipids of the cone-rich chicken retina and its photoreceptor outersegment membranes. *Biochim. biophys. Acta* **369,** 269–277.

Johnston, D. and Hudson, R. A. (1976). Isolation and composition of the carotenoid-containing oil droplets from cone photoreceptors. *Biochim. biophys. Acta* **424,** 235–245.

Jones, R. K. and Lee, D. N. (1981). Why two eyes are better than one: the two views of binocular vision. *J. exp. Psychol. H. P. P.* **7,** 30–40.

Kajikawa, J. (1923). Beitrage zur Anatomie und Physiologie des Vogelauges. *Albrecht v. Graefes Arch. Ophthal.* **112,** 260–346.

Kaskel, D., Schedtler, C. M., Hockwin, O. and Ciolo, S. (1975). Localization of glycogen in lenses of different species, demonstrated by histochemical methods. *Ophthalmic Res.* **7,** 65–72.

King-Smith, P. E. (1969). Absorption spectra and function of the coloured oil drops in the pigeon retina. *Vision Res.* **9,** 1391–1399.

Kirschfeld, K. (1982). Carotenoid pigments: their possible role in protecting against photooxidation in eyes and photoreceptor cells. *Proc. R. Soc. B.* **216,** 71–85.

Knipling, R. R. (1978). No deficit in near-field visual acuity of pigeons after transection of the isthmo-optic tract. *Physiol. Behav.* **21,** 813–816.

Krause, W. (1894). Die Retina. V. Die Retina der Vögel. *Int. Mschr. Anat. Physiol.* **11,** 1–66.

Kreithen, M. L. and Eisner, T. (1978). Ultraviolet light detection by the homing pigeon. *Nature, Lond.* **272,** 347–348.

Kreithen, M. L. and Keeton, W. T. (1974). Detection of polarized light by the homing pigeon, *Columba livia*. *J. comp. Physiol.* **89,** 83–92.

Land, M. F. (1981). Optics and vision in invertebrates. *In* "Handbook of Sensory Physiology" (H. Autrum, Ed.), Vol. VII/6B. Springer-Verlag, Berlin and New York.

Lee, D. N. (1980). The optic flow field: the foundation of vision. *Phil. Trans. R. Soc. B.* **290,** 169–179.

Levy, B. and Sivak, J. G. (1980). Mechanisms of accommodation in the bird eye. *J. comp. Physiol.* **137,** 267–272.

Liebman, P. A. (1972). Microspectrophotometry of photoreceptors. *In* "Handbook of Sensory Physiology" (H. J. A. Dartnall, Ed.), Vol. VII/1. Springer-Verlag, Berlin and New York.

Lindsay-Johnson, E. (1901). Contributions to the comparative anatomy of the mammalian eye, chiefly based on ophthalmoscopic examination. *Phil. Trans. R. Soc. B.* **194,** 1–82.

Lockhart, M. (1979). Quantitative morphological investigations of retinal cells in the pigeon. A golgi, light microscope study. *In* "Neural Mechanisms of Behaviour in the Pigeon" (A. M. Granda and J. H. Maxwell, Eds.). Plenum Press, New York and London.

Lord, R. D. (1956). A comparative study of the eyes of some falconiform and passeriform birds. *Am. Midl. Nat.* **56,** 325–344.

Lythgoe, J. N. (1972). The adaptation of visual pigments to the photic environment. *In* "Handbook of Sensory Physiology" (H. J. A. Dartnall, Ed.), Vol. VII/1. Springer-Verlag, New York and Berlin.

Lythgoe, J. N. (1979). "The Ecology of Vision". Clarendon Press, Oxford.

Mann, I. C. (1924). The function of the pecten. *Br. J. Ophthal.* **8,** 209–226.

Mann, I. C. (1929). Some observations on the vascularization of the vertebrate eye. *Trans. Ophthal. Soc. U.K.* **49,** 353–392.

Mann, I. C. (1931). Iris pattern in vertebrates. *Trans. Zool. Soc. Lond.* **21,** 355–412.

Mariani, A. P. (1982). Association amacrine cells could mediate directional selectivity in pigeon retina. *Nature, Lond.* **298,** 654–655.

Mariani, A. P. (1983). A morphological basis for verticality detectors in the pigeon retina: asymmetric amacrine cells. *Naturwissenchaften* **70,** s.368.

Mariani, A. P. and Leure-DuPree, A. E. (1977). Horizontal cells of the pigeon retina. *J. comp. Neurol.* **175,** 13–26.

Mariani, A. P. and Leure-DuPree, A. E. (1978). Photoreceptors and oil droplet colors in the red area of the pigeon retina. *J. comp. Neurol.* **182,** 821–837.

Marshall, J., Mellerio, J. and Palmer, D. A. (1973). A schematic eye for the pigeon. *Vision Res.* **13,** 2449–2453.

Martin, G. R. (1974). Color vision in the Tawny Owl (*Strix aluco*). *J. comp. physiol. Psychol.* **86,** 133–141.

Martin, G. R. (1977). Absolute visual threshold and scotopic spectral sensitivity in the Tawny Owl, *Strix aluco. Nature, Lond.* **268,** 636–638.

Martin, G. R. (1982). An owl's eye: schematic optics and visual performance in *Strix aluco* L. *J. comp. Physiol.* **145,** 341–349.

Martin, G. R. (1983). Schematic eye models in vertebrates. *In* "Progress in Sensory Physiology" (D. Ottoson, Ed.), Vol. 4. Springer-Verlag, Berlin and New York.

Martin, G. R. (1984). The visual fields of the Tawny Owl, *Strix aluco* L. *Vision Res.* **24,** 1739–1751.

Martin, G. R. and Gordon, I. E. (1974a). Visual acuity in the Tawny Owl (*Strix aluco*). *Vision Res.* **14,** 1393–1397.

Martin, G. R. and Gordon, I. E. (1974b). Increment threshold spectral sensitivity in the Tawny Owl (*Strix aluco*). *Vision Res.* **14,** 615–621.

Martin, G. R. and Muntz, W. R. A. (1978). Spectral sensitivity of the red and yellow oil droplet fields of the pigeon (*Columba livia*). *Nature, Lond.* **274,** 620–621.

Martin, G. R. and Muntz, W. R. A. (1979). Retinal oil droplets and vision in the pigeon (*Columba livia*). *In* "Neural Mechanisms of Behaviour in the Pigeon" (A. M. Granda and J. H. Maxwell, Eds.). Plenum Press, New York and London.

Martin, G. R. and Young, S. R. (1983). The retinal binocular field of the pigeon (*Columba livia:* English racing homer). *Vision Res.* **23,** 911–915.

Martin, G. R. and Young, S. R. (1984). The eye of the Humboldt Penguin (*Spheniscus humboldti*): Visual fields and schematic optics. *Proc. R. Soc., B* **223,** 197–222.

Martin, G. R., Gordon, I. E. and Cadle, D. R. (1975). Electroretinographically determined spectral sensitivity in the Tawny Owl (*Strix aluco*). *J. comp. physiol. Psychol.* **89,** 72–78.

Maturana, H. R. (1962). Functional organization of the pigeon retina. *Int. Congr. Ser.—Excerpta Med.* No. **49,** 170–178.

Maturana, H. R. and Frenk, S. (1963). Directional movement and horizontal edge detectors in the pigeon retina. *Science, Wash. D.C.* **142,** 977–979.

Maturana, H. R. and Frenk, S. (1965). Synaptic connections of centrifugal fibres in the pigeon retina. *Science, Wash. D.C.* **150,** 359–361.

Maxwell, M. H. and Burns, R. B. (1979). The ultrastructure of the epithelium of the Harderian and lacrimal glands of the turkey, fowl and duck. *J. Anat.* **128,** 445–459.

Menner, E. (1938). Die Bedeutung des Pecten im Auges des Vogels für die Wahrnehnung von Bewegangen, nebst Bemerkungen über seine Ontogenie und Histologie. *Zool. Jb., Abt. allg. Zool. Physiol.* **58,** 481–538.

Meyer, D. B. (1971). The effect of dietary carotenoid deprivation on avian retinal oil droplets. *Ophthalmic Res.* **2,** 104–109.

Meyer, D. B. (1977). The avian eye and its adaptations. *In* "Handbook of Sensory Physiology" (F. Crescitelli, Ed.), Vol. VII/5. Springer-Verlag, Berlin and New York.

Meyer, D. B. and Cooper, T. G. (1966). The visual cells of the chicken as revealed by phase contrast microscopy. *Am. J. Anat.* **118,** 723–724.

Meyer, D. B. and May, H. C. (1973). The topographical distribution of rods and cones in the adult chicken retina. *Exp. Eye Res.* **17,** 347–355.

Meyer, D. B., Cooper, T. G. and Gernez, C. (1965). Retinal oil droplets. *In* "The Structure of the Eye" (J. W. Rohen, Ed.). Schattauer, Stuttgart.

Miles, F. A. (1970). Centrifugal effects in the avian retina. *Science, Wash., D.C.* **170,** 992–995.

Miles, F. A. (1972a). Centrifugal control of the avian retina. I. Receptive field properties of retinal ganglion cells. *Brain Res., Osaka* **48,** 65–92.

Miles, F. A. (1972b). Centrifugal control of the avian retina. II. Receptive field properties of cells in the isthmo-optic nucleus. *Brain Res., Osaka* **48,** 93–113.

Miles, F. A. (1972c). Centrifugal control of the avian retina. III. Effects of electrical stimulation of the isthmo-optic tract on the receptive properties of retinal ganglion cells. *Brain Res., Osaka* **48,** 115–129.

Miles, F. A. (1972d). Centrifugal control of the avian retina. IV. Effects of reversible cold block of the isthmo-optic tract on the receptive field properties of cells in the retina and isthmo-optic nucleus. *Brain Res., Osaka* **48,** 131–145.

Miller, W. H. (1979). Ocular optical filtering. *In* "Handbook of Sensory Physiology" (H. Autrum, Ed.), Vol. VII/6A. Springer-Verlag, Berlin and New York.

Morris, V. B. (1970). Symmetry in a receptor mosaic demonstrated in the chick from frequencies, spacing and arrangement of the types of retinal receptor. *J. comp. Neurol.* **140,** 359–398.

Morris, V. B. and Shorey, C. D. (1967). An electromicroscope study of types of receptor in the chick retina. *J. comp. Neurol.* **129,** 313–340.

Mowrer, O. H. (1937). A comparison of the visual function of nictitation and blinking. *J. comp. Psychol.* **15,** 75–93.

Muntz, W. R. A. (1972). Inert absorbing and reflecting pigments. *In* "Handbook of Sensory Physiology" (H. J. A. Dartnall, Ed.). Vol. VII/1. Springer-Verlag, Berlin and New York.

Muntz, W. R. A. (1974). Comparative aspects in behavioural studies of vertebrate vision. *In* "The Eye" (H. Davson, Ed.). Vol. 6. Academic Press, New York and London.

Murphy, C. J. and Howland, H. C. (1983). Owl eyes: accommodation, corneal curvature and refractive state. *J. comp. Physiol.* **151,** 277–284.

Nakao, S., Fujimoto, S., Nagata, R. and Iwata, K. (1968). Model of refractive index distribution in the rabbit crystalline lens. *J. opt. Soc. Am.* **58,** 1125–1130.

Nicol, J. A. C. and Arnott, H. J. (1974). Tapeta lucida in the eyes of goatsuckers (Caprimulgidae). *Proc. R. Soc. B.* **187,** 349–352.

Norren, D. V. (1975). Two short wavelength sensitive systems in pigeon, chicken and daw. *Vision Res.* **15,** 1164–1166.

Nye, P. W. (1968). The binocular acuity of the pigeon measured in terms of the modulation transfer function. *Vision Res.* **8,** 1041–1053.

Nye, P. W. (1969). The monocular eye movements of the pigeon. *Vision Res.* **9,** 133–144.

O'Day, K. (1940). The fundus and fovea centralis of the albatross (*Diomedea cauta cauta* Gould). *Br. J. Ophthal.* **24,** 201–207.

Oehme, H. (1969). Der Bewegungsapparat der Vogeliris (Eine vergleichende morphologisch-funktionelle Untersuchung). *Zool. Jahrb. Abt. Anat. Ontog. Tiere* **86,** 96–128.

Ogden, T. E. (1968). On the function of efferent retinal fibres. *In* "Structure and Function of Inhibitory Neuronal Mechanisms" (C. von Euler, S. Skoglund and U. Soderberg, Eds.). Pergamon Press, New York.

Ohzu, H., Enoch, J. M. and O'Hair, J. C. (1972). Optical modulation by the isolated retina and retinal receptors. *Vision Res.* **12,** 231–244.

Parrish, J., Benjamin, R. and Smith, R. (1981). Near ultraviolet light reception in the Mallard. *Auk* **98,** 627–628.

Pearlman, A. L. and Hughes, C. P. (1976a). Functional role of efferents to the avian retina. I.: analysis of retinal ganglion cell receptive fields. *J. comp. Neurol.* **166,** 111–122.

Pearlman, A. L. and Hughes, C. P. (1976b). Functional role of efferents to the avian retina. II. Effects of reversible cooling of the isthmo-optic nucleus. *J. comp. Neurol.* **166,** 123–132.

Pearson, R. (1972). "The Avian Brain". Academic Press, London and New York.

Pedler, C. and Boyle, M. (1969). Multiple oil droplets in the photoreceptors of the pigeon. *Vision Res.* **9,** 525–528.

Peiponen, V. A. (1964). Zur Bedeutung der Ölkugeln im Farbensehen der Sauropsiden. *Ann. zool. Fenn.* **1,** 281–302.

Pennycuick, C. J. (1960). The physical basis of astro-navigation in birds: theoretical considerations. *J. exp. Biol.* **37,** 573–593.

Pettigrew, J. D. (1978). A role for the avian pecten oculi in orientation to the sun? *In* "Animal Migration, Navigation and Homing" (K. Schmidt-Koenig and W. T. Keeton, Eds.). Springer-Verlag, Berlin and New York.

Pettigrew, J. D. (1979a). Binocular visual processing in the owl's telencephalon. *Proc. R. Soc. B.* **204,** 435–454.

Pettigrew, J. D. (1979b). Comparison of retinotopic organization of the visual wulst in nocturnal and diurnal raptors, with a note on the evolution of frontal vision. *In* "Frontiers of Visual Science" (S. J. Cool and E. L. Smith, Eds.). Springer-Verlag, Berlin and New York.

Philipson, B. (1969). Distribution of protein within the normal rat lens. *Invest. Ophthal. Visual Sci.* **8,** 258–269.

Pilar, G. and Vaughan, P. C. (1971). Ultrastructure and contractions of the pigeon iris striated muscle. *J. Physiol., Lond.* **219,** 253–266.

Polyak, S. (1941). "The Retina". University of Chicago Press, Chicago.

Pumphrey, R. J. (1948a). The sense organs of birds. *Ibis* **90,** 171–199.

Pumphrey, R. J. (1948b). The theory of the fovea. *J. exp. Biol.* **25,** 299–312.

Pumphrey, R. J. (1961). Sensory organs: vision. *In* "Biology and Comparative Physiology of Birds" (A. J. Marshall, Ed.), Vol. 2. Academic Press, New York and London.

Rabaey, M. (1963). Glycogen in the lens of birds' eyes. *Nature, Lond.* **198,** 206–207.

Reeves, P. (1918). Rate of pupillary dilation and contraction. *Psychol. Rev.* **25,** 330–340.

Reeves, P. (1920). The response of the average pupil to various intensities of light. *J. opt. Soc. Am.* **4,** 35–43.

Reymond, L. and Wolfe, J. (1981). Behavioural determination of the contrast sensitivity function of the eagle *Aquila audax. Vision Res.* **21,** 263–271.

Riggs, L. A., Blough, P. M. and Schafer, K. L. (1972). Electrical responses of the pigeon eye to changes in wavelength of the stimulating light. *Vision Res.* **12,** 981–991.

Rochon-Duvigneaud, A. (1943). "Les Yeux et la Vision des Vertébrés". Masson, Paris.

Rodieck, R. W. (1973). "The Vertebrate Retina: Principles of Structure and Function". W. J. Freeman, San Francisco.

Rogers, L. J. and Miles, F. A. (1972). Centrifugal control of the avian retina. V. Effects of lesions of the isthmo-optic nucleus on behaviour. *Brain Res., Osaka* **48,** 147–156.

Schramm, U. (1980). Lymphoid cells in the Harderian gland of birds. An electron microscopical study. *Cell Tiss. Res.* **205,** 85–94.

Shlaer, R. (1972). An eagle's eye: quality of retinal image. *Science, Wash., D.C.* **176,** 920–922.

Sillman, A. J. (1969). The visual pigments of several species of birds. *Vision Res.* **9,** 1063–1077.

Sillman, A. J. (1973). Avian vision. *In* "Avian Biology" (D. S. Farner, J. R. King and K. C. Parkes, Eds.), Vol. III. Academic Press, New York and London.

Sillman, A. J., Bolnick, D. A., Haynes, L. W., Walter, A. E. and Loew, E. R. (1981). Microspectrophotometry of the photoreceptors of palaeognathos birds—the emu and tinamou. *J. comp. Physiol.* **144,** 271–276.

Simić, V. and Jablan-Plantic, O. (1959). Morphologie Beitrag über den Mechanismus der dritten Augenlids bei den Hausvögeln. *Anat. Anz.* **106,** 76–85.

Sivak, J. G. (1977). The role of a flat cornea in the amphibious behaviour of the Blackfoot Penguin (*Spheniscus demersus*). *Can. J. Zool.* **54,** 1341–1345.

Sivak, J. G. (1978). A survey of vertebrate strategies for vision in air and water. *In* "Sensory Ecology: Review and Perspectives" (M. A. Ali, Ed.). Plenum Press, New York.

Sivak, J. G., Lincer, J. L. and Bobier, W. (1977). Amphibious visual optics of the eyes of the Double-crested Cormorant (*Phalacrocorax auritus*) and the Brown Pelican (*Pelecanus occidentalis*). *Can. J. Zool.* **55,** 782–788.

Sivak, J. G., Bobier, W. R. and Levy, B. (1978). The refractive significance of the nictitating membrane of the bird eye. *J. comp. Physiol.* **125,** 335–339.

Slonaker, J. R. (1897). A comparative study of the area of acute vision in vertebrates. *J. Morph.* **13,** 445–493.

Slonaker, J. R. (1918). A physiological study of the anatomy of the eye and its accessory parts of the English Sparrow (*Passer domesticus*). *J. Morph.* **31,** 351–359.

Snyder, A. W. (1979). Physics of vision in compound eyes. *In* "Handbook of Sensory Physiology" (H. Autrum, Ed.), Vol. VII/6A. Springer-Verlag, Berlin and New York.

Snyder, A. W. and Miller, W. H. (1978). Telephoto lens system of falconiform eyes. *Nature, Lond.* **275,** 127–129.

Snyder, A. W., Laughlin, S. B. and Stavenga, D. G. (1977). Information capacity of eyes. *Vision Res.* **17,** 1163–1175.

Stebbins, W. C., Ed. (1970). "Animal Psychophysics". Appleton-Century-Crofts, New York.

Steinbach, M. J. and Money, K. E. (1973). Eye movements of the owl. *Vision Res.* **13,** 889–891.

Steinbach, M. J., Angus, R. G. and Money, K. E. (1974). Torsional eye movements of the owl. *Vision Res.* **14,** 745–746.

Stell, W. K. (1972). The morphological organization of the vertebrate retina. *In* "Handbook of Sensory Physiology" (N. G. F. Fuortes, Ed.), Vol. VII/2. Springer-Verlag, Berlin and New York.

Strother, G. K. (1963). Absorption spectra of retinal oil globules in turkey, turtle and pigeon. *Exp. Cell Res.* **29,** 349–355.

Suthers, R. A. (1978). Sensory ecology of birds. *In* "Sensory Ecology: Review and Perspectives" (M. A. Ali, Ed.). Plenum Press, New York.

Thompson, G. (1971). The photopic spectral sensitivity of gulls measured by electroretinographic and pupilometric methods. *Vision Res.* **11,** 719–731.

Thomson, A. (1928). The riddle of the pecten with suggestions as to its use. *Trans. Ophthal. Soc. U.K.* **48,** 293–331.

Thomson, A. (1929). The pecten, considered from an environmental point of view. *Ibis* **12,** 608–639.

Tiemeier, O. (1950). The os opticus of birds. *J. Morph.* **86,** 25–46.

Tripathi, R. C. (1974). Comparative physiology and anatomy of the outflow system. *In* "The Eye" (H. Davson and L. T. Graham, Eds.), Vol. 5. Academic Press, New York and London.

Uchiyama, T. (1930). Beiträge zur Morphologie des Lipoidstoffwechels. III. Über die Öltropfen und Glykogengebilde in den Sehzellen der Netzhaut des Haushuhns. *Virchows Arch. path. Anat. Physiol.* **27,** 631–641.

Uhlrich, D. J., Blough, P. M. and Blough, D. S. (1982). The pigeon's distant visual acuity as a function of viewing angle. *Vision Res.* **22,** 429–431.

Wallman, J. (1979). Role of the retinal oil droplets in the color vision of Japanese quail. *In* "Neural Mechanisms of Behaviour in the Pigeon" (A. M. Granda and J. H. Maxwell, Eds.). Plenum Press, New York.

Walls, G. L. (1937). Significance of the foveal depression. *Albrecht v. Graefes Arch. Ophthal.* **18,** 912–919.

Walls, G. L. (1940). Postscript on image expansion by the foveal clivus. *Albrecht v. Graefes Arch. Ophthal.* **23,** 831–832.

Walls, G. L. (1942). "The Vertebrate Eye and its Adaptive Radiation". Cranbrook Institute of Science, Bloomfield Hills, Michigan.

Walls, G. L. (1962). The evolutionary history of eye movements. *Vision Res.* **2,** 69–80.

Walls, G. L. and Judd, H. D. (1933). The intraocular filters of vertebrates. *Br. J. Ophthal.* **17,** 641–675; 705–725.

Weale, R. A. (1966). Why does the human retina possess a fovea? *Nature, Lond.* **212,** 255–256.

Weale, R. A. (1974). Natural history of optics. *In* "The Eye" (H. Davson and L. T. Graham, Eds.), Vol. 6. Academic Press, New York and London.

Wells, M. C., Lehner, P. N., Bolen, E. G. and Rylander, M. K. (1975). Comparison of scotopic sensitivity in diurnal (*Anas platyrhynchos*) and crepuscular (*Dendrocygna autumnalis*) ducks. *J. comp. physiol. Psychol.* **88,** 940–944.

West, R. W. and Dowling, J. E. (1972). Synapses on to different morphological types of retinal ganglion cells. *Science, Wash., D.C.* **178,** 510–512.

Wetmore, A. (1919). Note on the eye of the Black Skimmer. *Proc. biol. Soc. Wash.* **32,** 195.

Williams, D. S. and McIntyre, P. (1980). The principal eyes of a jumping spider have a telephoto component. *Nature, Lond.* **288,** 578–580.

Wingstrand, K. G. and Munk, O. (1965). The pecten oculi of the pigeon with particular regard to its function. *Biol. Skr. Dan. Vid. Selsk.* **14,** 1–64.

Wittich, V. (1855). Vergleichend histologische Mittheilungen. *Albrecht v. Graefes Arch. Ophthal.* **2,** 124–132.

Wolbarsht, M. L. (1976). The function of intraocular colour filters. *Fedn. Proc. Fedn. Am. Socs. exp. Biol.* **35,** 44–50.

Wolff, E. (1940). "The Anatomy of the Eye in its Orbit". Lewis, London.

Wood, C. A. (1917). "The Fundus Oculi of Birds especially as viewed by the Ophthalmoscope". Lakeside Press, Chicago.

Woodhouse, J. M. and Campbell, F. W. (1975). The role of the pupil light reflex in aiding adaptation to the dark. *Vision Res.* **15,** 649–653.

Wortel, J. F., Wubbels, R. J. and Nuboer, J. F. W. (1984). Photopic spectral sensitivity of the red and the yellow field of the pigeon retina. *Vision Res.* **24,** 1107–1113.

Wright, A. A. (1972a). The influence of ultraviolet radiation on the pigeon's color discriminations. *J. exp. Analysis Behav.* **17,** 325–337.

Wright, A. A. (1972b). Psychometric psychophysical hue discrimination functions for the pigeon. *Vision Res.* **12,** 1447–1464.

Wright, A. A. (1979). Color-vision psychophysics: a comparison of pigeon and human. *In* "Neural Mechanisms of Behaviour in the Pigeon" (A. M. Granda and J. H. Maxwell, Eds.). Plenum Press, New York.

Yazulla, S. (1974). Intraretinal differentiation in the synaptic organization of the inner plexiform layer of the pigeon retina. *J. comp. Neurol.* **153,** 309–324.

Young, S. R. and Martin, G. R. (1984). Optics of retinal oil droplets. A model of light collection and polarization detection in the avian retina. *Vision Res.* **24,** 129–137.

8

Structure and function of avian somatosensory receptors

K.-M. GOTTSCHALDT*

Department of Neurobiology, Max Planck Institute for Biophysical Chemistry, Göttingen, Federal Republic of Germany

CONTENTS

I. Introduction

As in all vertebrates, it is the function of somatic receptors in birds to provide information about the physical conditions of the internal and external environment. The somatosensory receptors in the skin have to detect mechanical, thermal or noxious stimuli coming from the outside world and to transform them into neural signals conducted via afferent nerve fibres to the central nervous system. Such exteroceptors are essential for enabling an animal to behave appropriately under changing environmental conditions. Other somatic receptors provide information about the internal conditions within the limbs, the cardiovascular and respiratory systems, and the viscera, including mechanical, thermal, chemical

*Deceased. Requests for offprints should go to C. Vahle-Hinz, Physiology Institute, University of Würzburg, D-8700 Würzburg, Federal Republic of Germany.

and noxious factors. These enteroceptors often subserve autonomous regulatory systems operating outside the animal's conscious sensation and experience.

A primary goal in somatosensory receptor research has been to order the multitude of receptors, and several classifications have been proposed which, however, were based purely on either anatomical or electro-physiological investigations. Thus, morphological and functional classification schemes first reflected a rather parochial approach, and it was not until interdisciplinary barriers were gradually overcome that correlations between distinct morphological and functional receptor types could be established.

In the early days of purely anatomical receptor research it was believed that the different types of somatic receptors were nothing but variations of a single basic receptor structure. Krause (1881), for instance, insisted that all receptors were varieties of his endbulbs, whereas Merkel (1878, 1880) was convinced that his cells (Tastzellen) were the fundamental structural elements of cutaneous receptors. However, others (Key and Retzius, 1876; Ruffini, 1905) recognised the existence of essentially different morphological types of somatic receptors. This view was shared by Botezat (1906) who had investigated receptors in the dermis of the beak of birds*, but he also postulated that in the different morphological receptor types the sensory nerve fibre would still terminate in a uniform manner. Today, the existence of different receptor structures is firmly established, although different morphological criteria have been applied for their classification (Quilliam, 1966; Andres, 1969, 1974; Munger, 1971a,b, 1977; Andres and von Düring, 1973; King *et al.*, 1974; Halata, 1975; Chouchkov, 1978; Malinovský and Páč, 1982).

On an electro-physiological basis, somatic receptors or their afferent nerve fibres have been classified according to (a) their specific sensitivity to different stimulus modalities, such as mechanical, thermal, chemical or noxious stimuli, (b) the rate of adaptation of the afferent discharges recorded in single nerve fibres, (c) the conduction velocities of these nerve fibres and (d) the specific combination of excitability and response characteristics of individual afferent units consisting of a receptor and its afferent nerve fibre (Burgess and Perl, 1973; Iggo, 1974; Bannister, 1976; Horch *et al.*, 1977).

After correlations between different morphological receptor types and different

*Berkhoudt (1980) reported that the structure of the beak of the Mallard (*Anas platyrhynchos*) is the same as that of the White Pekin duck as described by Lucas and Stettenheim (1972, p. 23). The rhamphotheca is defined by Lucas (1979, p. 13) in the *Nomina Anatomica Avium* as "the horny covering of the beak". The anseriform rhamphotheca of both the upper and the lower beak possesses two horny components. (1) The nail, at the rostral tip, is a plate of *hard* horny tissue. (2) The remainder of the horny tissue (and hence the great majority) of the upper and lower rhamphothecae consists of a thin, flexible layer of soft leathery tissue characterised by soft (membranous) keratin; unfortunately there is no *anatomical* name for this soft component of the rhamphotheca. In this chapter the hard horny component (1) at the tip of the rhamphotheca will be called the *nail;* the horny component (2) will be designated as the *soft part of the rhamphotheca*. Both the nail and the soft part of the rhamphotheca are, of course, epidermal derivatives, and therefore both of them are intimately associated with the underlying dermis; this relationship is important, because it is the *dermis* that carries the neuronal receptors of the beak. It would be convenient to have a term to cover both the horny rhamphotheca *and* its underlying dermis together. In the literature on avian mechanoreceptors the terms "beak skin", "ordinary beak skin", and "beak skin proper" have been used for this purpose. However, it is anatomically somewhat unsatisfactory to regard beak and skin as identical structures, and furthermore the term "beak skin" is not always readily understood by ornithologists. Therefore, in this chapter formal anatomical terms will be used in full, including for example rhamphothecal dermis and nail dermis.

electro-physiological response types had been established and the fine structure of the receptors had been studied with the electron microscope, it became possible to search for specific structure–function relationships in individual receptor types. It has been proposed that the functional properties of mammalian mechanoreceptors are determined not only by the biophysical transducer characteristics of the sensory nerve terminals but also by the specific way in which the latter are installed in the different structural components forming the receptors. Thus, similar response properties should result from similar microstructural relationships between the sensory nerve endings and their surrounding tissues; this principle should hold even if the receptors as a whole have such dissimilar gross morphologies as, for instance, the straight lanceolate nerve endings in the facial sinus hair follicles and the Ruffini nerve endings in the skin or tendons of the cat (Iggo and Gottschaldt, 1974). The purpose of such considerations was to identify the various categories of structure–function relationships in the different types of receptors. From these it should be possible to establish a classification for somatic receptors that not only takes into account both morphological and physiological criteria but also understands these as causal expressions of the different types of biophysical processes taking place in the different receptor types. This approach will be extended in the present chapter to the avian somatosensory receptors in the hope that it also may help in future studies to predict the functional properties of a morphologically identified receptor, or, vice versa, to forecast some of its fine-structural characteristics when only the physiological response properties are known.

Unfortunately, most of our knowledge about the relationships between structure and function in vertebrate somatic receptors has been derived from mammals and is almost exclusively restricted to those receptors which have well-developed terminal structures and large-diameter myelinated afferent nerve fibres. However, although in birds some receptors with quite unique structural features have evolved, it is believed that the same biophysical laws apply to them as to their mammalian counterparts. Furthermore, in birds, electro-physiological studies allowing correlations with morphological and especially electron microscopic findings have generally been restricted to receptors in the beak. Therefore, this chapter will deal primarily with the receptors in the beak and what we know, or believe we know, about their form and function; it is hoped that at least the principal insights gained from the study of beak receptors can be transferred to all other somatosensory receptors in the rest of the bird's body.

II. General morphological characteristics of somatic receptors

Somatosensory receptors are composed of different structural elements which together form a functional unit, namely (a) the sensory nerve terminal with its specific transducer function, (b) the afferent nerve fibre which transmits action potentials, and (c) the more-or-less differentiated non-nervous tissue elements adjoining the nerve terminal.

When analysed by morphological criteria alone, somatosensory receptors have been classified according to the particular specialisations of the non-nervous tissue elements occurring at and around the terminal sensory nerve fibre, the presence or absence of myelination of the afferent nerve fibres, or simply the location of the receptors (Poláček, 1966; Andres, 1969, 1974; Munger, 1971a,b, 1977; Andres and von Düring, 1973; Halata, 1975; Saxod, 1978a; Chouchkov, 1978; Malinovský and Páč, 1982). Since the functional properties of a receptor are determined first by the biophysical properties of the receptor membrane, classification schemes based solely on morphological criteria inevitably grouped together receptors which actually have quite different functional properties. For instance, intraepidermal nerve endings supposedly responding specifically to either thermal or mechanical stimuli were placed in the same receptor group. Whatever their function, however, as Andres (1966) first pointed out, the nerve terminals in all the different morphological kinds of somatic receptors have certain ultrastructural features in common: there are always abundant mitochondria, the axoplasm is differentiated to a receptor matrix, and the receptor membrane stains much more conspicuously than the membrane of the conductile axon indicating different properties of these two regions. The receptor matrix is characterised by a finely-grained and filamentous ground substance containing many clear and occasional dense-cored vesicles, microfilaments, microtubules and various other cell organelles (e.g. see Section IV,A,1). Many receptor terminals contain glycogen granules, but this is not a consistent finding even in the same type of receptor. Mechano-receptive nerve endings of myelinated afferent axons are in addition characterised by finger-like processes which project into the surrounding tissue (Andres, 1969, 1974; Andres and von Düring, 1973; Halata, 1975; Chouchkov, 1976, 1978).

As far as is known, unmyelinated and thin myelinated sensory nerve fibres all terminate as "free nerve endings", that is, the receptors are ensheathed only by Schwann cell processes. Such free nerve endings are found in the skin, muscles, joint ligaments, tendons and inner organs. In mammals they comprise nociceptors, thermoreceptors, mechanoreceptors and chemoreceptors, but in birds little is known about their specific function or morphology. Another large group of somatosensory receptors is supplied by myelinated nerve fibres of a diameter larger than 2 or 3 μm. They all seem to be mechanoreceptors in which the nerve endings are associated with more-or-less elaborate non-nervous tissue elements. Such mechanoreceptors occur in the skin, muscles, joints, ligaments, tendons and intestines and belong to both exteroreceptors and enteroreceptors.

III. Physiological classification of somatosensory receptors

Functionally, somatosensory receptors can be classified according to the specific kind of stimulus to which they are most responsive. Thus, receptors responding to a mechanical, thermal, nociceptive or chemical stimulus modality can be distinguished. Each of these four main classes of receptors comprises sub-groups with a sub-modality sensitivity; that is, the receptors of a given sub-modality group are specifically sensitive to one particular kind of stimulus within a modality class. For instance, there is a cold sub-modality and a warm sub-modality in the temperature modality group. The meaning of the term modality, as used here, differs from that in some of the somatosensory literature where it often categorises receptor responses resulting from stimulation of, for instance,

TABLE I. **Stimulus modalities and sub-modalities.**

Stimulus modality	Stimulus sub-modality
Mechanical	Acceleration (vibration)
	Velocity
	Amplitude
Thermal	Cold
	Warm
Noxious	High-threshold mechanoreceptors
	Noxious heat
	Noxious cold
	Polymodal
Chemical	Biogenic amines
	pH
	Others

"deep" versus "superficial" tissues, even when all of these responses may originate only from mechanoreceptors. Thus, in the present chapter the term modality defines a distinct physical stimulus class, and the term sub-modality a sub-class of such stimuli as indicated in Table I. This table is based on electrophysiological studies in mammals (Burgess and Perl, 1973), but it can be assumed to be valid for birds although the existence of avian receptors for all the different stimulus sub-modalities remains to be established.

IV. Mechanoreceptors

The low-threshold mechanoreceptors studied in birds so far were all supplied by myelinated nerve fibres with a conduction velocity above 15 m/sec and a presumed diameter of more than 3 μm (Dorward, 1970a; Dorward and McIntyre, 1971; Necker, 1972, 1973, 1974a,b; Gregory, 1973; Leitner and Roumy, 1974a; Gottschaldt, 1974a; Necker and Reiner, 1980; Gottschaldt *et al.*, 1982). These mechanoreceptors belong to one modality class and comprise three functional sub-groups, each of which responds specifically to a particular physical aspect of a mechanical stimulus. Figure 8.1 illustrates responses of these three functional sub-modality types. Mathematically the physical stimulus component to which each of these three types of unit is specifically responsive corresponds to the second, first and "zeroth" derivative of the displacement amplitude over time. Generally, a receptor responding to a lower-order stimulus derivative will also respond to stimulus components corresponding to the next or the two next higher-order derivatives of the displacement. Thus, slowly-adapting receptors usually show, in addition to their obligatory amplitude response, a velocity response related to the movement part of the stimulus. Similarly, velocity-sensitive receptors normally respond also to acceleration components in mechanical stimuli. According to this rule all functional types of mechanoreceptors should also respond to the high acceleration components contained in high-frequency vibratory stimuli. In some situations, such as the mammalian sinus hair follicle, this has in

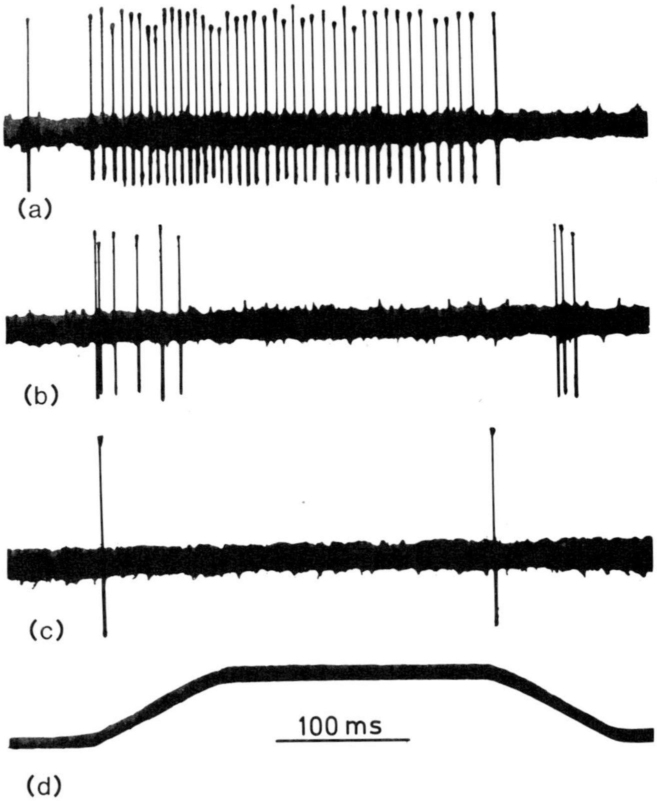

Fig. 8.1. Responses of three different types of mechano-sensitive units, recorded in one experiment from single nerve fibres of the ophthalmic nerve in the goose. In each case an identical stimulus (d) was used, that is, a ramp-shaped indentation of the rhamphotheca overlying the receptor's location. (a) A slowly-adapting amplitude-sensitive response; (b) a rapidly-adapting velocity-sensitive response; (c) a rapidly-adapting acceleration-sensitive response.

fact been demonstrated (Gottschaldt and Vahle-Hinz, 1981). However, usually the mechanical filter characteristics of the tissue embedding the receptors, or the fine structure of the receptors themselves, seem to prevent access of high-frequency vibratory stimulus components to the sensory nerve endings. Thus, slowly-adapting receptors and the velocity-sensitive rapidly-adapting receptors in the dermis of the beak of birds, for instance, are generally unresponsive to high-frequency vibratory stimuli (Gottschaldt, 1974a). Consequently the detection of such stimuli is left to a receptor type in which a structural specialisation has resulted in a particular sensitivity to high-frequency vibrations.

A. *Receptors detecting acceleration components in mechanical stimuli*

1. *Structure and function of Herbst corpuscles*

In 1848 Herbst discovered corpuscular nerve endings along the leg and wing bones of chickens and pigeons which resembled the somewhat larger mammalian

Vater–Pacinian corpuscles. For some time these Herbst corpuscles were thought to function as pressure or osmoreceptors, but it is now established that they are mechanoreceptors responding to the acceleration components in vibratory stimuli (Skoglund, 1960; Dorward, 1970a; Dorward and McIntyre, 1971; Gregory, 1973; Leitner and Roumy, 1974a; Gottschaldt, 1974a; Gottschaldt *et al.*, 1982), as shown in Fig. 8.2. Responses may be elicited with vibratory stimuli between 50 and at least 2000 Hz, and a one-to-one discharge can be maintained at frequencies between 100 and 1000 Hz. However, Herbst corpuscles are incapable of coding either the velocity or the amplitude of mechanical displacements. The structure of Herbst corpuscles has been well studied with the light microscope, especially by Carrière (1882; for reviews see Malinovský, 1967; Berkhoudt, 1980). Only the higher resolution of electron microscopy has revealed further morphological details (Rodriguez-Perez and de Iraldi, 1961; Quilliam, 1966; Munger, 1966, 1971a,b; Andersen and Nafstad, 1968; Saxod, 1968, 1970a, 1973a, 1978a; Poláček, 1969; Goglia, 1969a; Andres, 1969, 1974; Cobb and Bennett, 1970; Nafstad and Andersen, 1970; Halata, 1971a; Chouchkov, 1973, 1978; Abraham, 1978; Malinovský and Páč, 1978, 1980; Halata and Munger, 1980; Gottschaldt *et al.*, 1982).

Despite a considerable variability in the size and shape of Herbst corpuscles located in different sites of the bird's body and in different species (Poláček *et al.*, 1966; Malinovský, 1967; Malinovský and Zemanek, 1969, 1971), they always

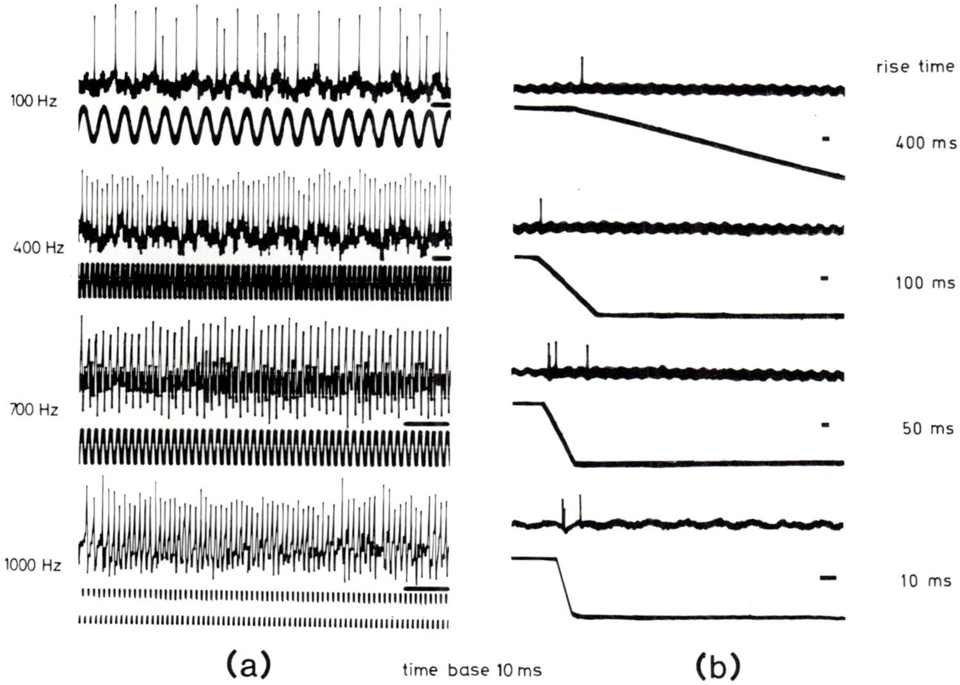

Fig. 8.2. Responses of Herbst corpuscle afferents. (a) One-to-one responses to vibratory stimuli at 100 to 1000 Hz as indicated on the left. (Note occasional double responses to a given vibratory cycle at 100 Hz.) (b) Responses to trapezoidal stimuli of constant amplitude attained in the different rise times shown on the right. Discharges are related to the "on" acceleration or the "off" deceleration beginning or ending the movement part of the ramp stimulus, but no consistent relationship between discharge frequency and movement velocity can be established (Gottschaldt, 1974a).

possess a characteristic arrangement of nervous and non-nervous structural components, the sensory nerve ending in the centre of the corpuscle being enclosed by cells which form a lamellar system around it. In addition, this lamellar system is usually surrounded by a wide space, which in turn is demarcated externally by a distinct capsule and contains a circularly-arranged system of collagen fibres (Figs. 8.3a, 7a). Thus, by far the greatest part of a Herbst corpuscle's diameter is made up by these auxiliary receptor structures, the specific kind and arrangement of which seem to determine the distinctive transducer property of this receptor (Gottschaldt *et al.*, 1982).

The auxiliary structures in the Herbst corpuscle are extensions or specialisations of the non-nervous constituents of its afferent nerve fibre. (1) The *outer capsule* is derived from the perineurium and is formed by four to eight layers of perineural cells. They are limited by a basal lamina against (2) the *outer zone* (also called the subcapsular space), an enlargement of the endoneural space, which contains the collagen fibre system and surrounds the lamellated structure of (3) the *inner core* formed by specialised Schwann cells.

The fibre system in the outer zone is a coiled two-dimensional network of collagen fibrils expelled from flattened endoneural fibrocytes, rather than the perforated lamellar system suggested by Saxod (1978a). It is wound around the inner core in many loops, becoming more tightly packed towards the centre of the receptor (Figs. 8.3a, 7a). The collagen fibrils fuse with the basal lamina of the outer perineural capsule cells as well as of the inner core cells; in this way the inner core with the sensory nerve ending becomes three-dimensionally suspended, via the collagen fibre system, from the outer capsule. The spaces between the collagen fibrils are filled by an apparently mucous fluid.

The perineural capsule and subcapsular space are sometimes developed sparingly; the receptors then constitute the "simple sensory corpuscles" in the skin of pigeons and domestic fowl (Malinovský and Páč, 1978). The sensory nerve ending, however, always remains enclosed by the lamellar system of the inner core. Such simplified versions of the Herbst corpuscle seem to correspond to the mammalian Golgi–Mazzoni corpuscles, which may be considered as miniature forms of Pacinian corpuscles and also share their functional properties (Sakada and Aida, 1971; Iggo and Gottschaldt, 1974).

The afferent nerve fibre enters the Herbst corpuscle at one pole and runs tortuously towards the inner core where the myelin sheath ends. There is always at least one node of Ranvier within each corpuscle. Occasionally the fibre branches into two or more terminal fibres before or after the myelin ends. The sensory nerve terminal usually is oval in transverse section and ends as a round or club-shaped enlargement. It is completely embedded in the inner core, which is formed by 8 to 30 pairs of Schwann cells lined up in two opposing rows along each pole of the oval profile of the receptor terminal. These Schwann cells possess a peculiar structure and polarity. The nuclei are located eccentrically at one side of the cell, so that they bulge from the inner core into the space of the outer zone (Figs. 8.3, 7); the bulk of the cytoplasm extends in numerous semicircular lamellar processes around the axon and towards the perikaryons of the opposite and neighbouring Schwann cells. The semicircular Schwann cell lamellae interdigitate with lamellae originating from the perikaryon of the opposite Schwann cell (Fig. 8.7b); consequently there is a cleft, lying along a line between the two nuclei, which houses the oval profile of the receptor axon (Fig. 8.3b).

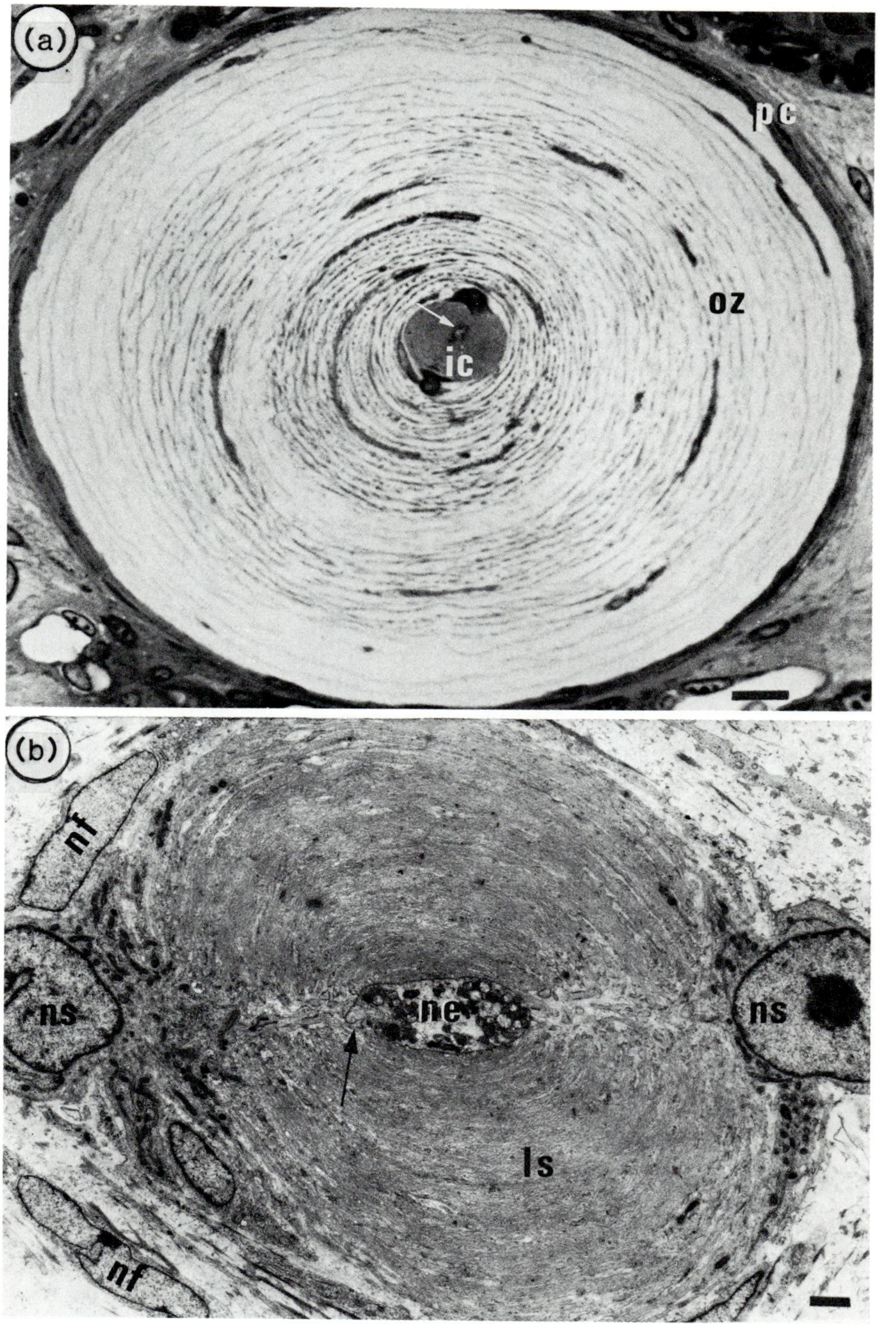

Fig. 8.3. Transverse sections through Herbst corpuscles of the goose. (a) Light micrograph from a semi-thin section. pc, Perineural capsule; oz, outer zone with coiled collagen fibres and scattered fibrocytes; ic, inner core with central sensory axon (arrow). ×850. Calibration bar: 10 μm. (b) Electron micrograph from a thin section showing the inner core enlarged. ns, Nuclei of two opposing Schwann cells; ls, interdigitating lamellae of these Schwann cells. Note "cleft" region connecting Schwann cell nuclei with the longer axis of the oval profile of the sensory nerve ending (ne). Arrow, blunt axon process of a transducer site; nf, nuclei of fibrocytes in the outer zone. ×5800. Calibration bar: 1 μm.

The number of, thickness of, and distance between the lamellae varies in Herbst corpuscles of different species according to the site of their location as well as the age of the animal. In Herbst corpuscles of the bill of the adult goose, up to 80 lamellae may cover each flat pole of the oval-shaped receptor axon. They are often only 50 nm thick and separated by interlamellar spaces measuring 15–20 nm in width. In newly-hatched goslings the inner core is larger in diameter, and the lamellae are thicker and separated by wider interlamellar spaces (Gottschaldt *et al.*, 1979). Therefore, during maturation the lamellar system becomes more tightly wound around the sensory axon as the inner core grows in length. At the same time the outer zone expands and the coiled collagen fibre network in it gains more and more loops. In the palate of the adult chicken, the structure of the inner core often resembles that seen in goslings; that is, it is larger in total diameter, and the individual lamellae are thicker and separated by wider spaces than in adult geese. Furthermore, the width of the outer zone relative to the diameter of the inner core is considerably smaller in Herbst corpuscles of the chicken than those of geese, and the overall shape of the corpuscles is frequently more longitudinal than oval (Fig. 8.4a,b). Saxod (1967) found Herbst corpuscles of the chicken type in the feathered skin of the duck, whereas Herbst corpuscles in the duck's bill resemble those in the goose's bill.

In a study of several species of songbirds, Ziswiler and Trnka (1972) distinguished four types of lamellated receptors, with typical Herbst corpuscles and typical Vater–Pacini corpuscles at opposite extremes of a morphological continuum. Without ultrastructural evidence and with inevitable fixation artifacts from the decalcification of the beaks, their claim for the existence of Vater–Pacini corpuscles in birds is doubtful, especially when the Pacini corpuscles were much smaller than the recognised Herbst corpuscles. However, their study reveals, along with those of others (Poláček *et al.*, 1966; Malinovsky, 1967; Malinovský and Zemanek, 1971; Krulis, 1978), a considerable variability in form and extension of the inner core and outer zone in Herbst corpuscles which may be partly species-specific and partly representative of local adaptations in different parts of the body.

As in all mechano-receptive nerve endings, the receptor axon in the Herbst corpuscle typically contains many mitochondria and various other organelles (multivesicular bodies, myelin and dense bodies, lysosomes). Occasionally dense-cored vesicles and groups of glycogen granules can be found scattered in the axoplasm of the sensory nerve ending. The latter is further characterised by the presence of distinct structural differentiations which have been called "transducer sites", because there is reason to believe that they correspond to the actual sites at which the mechano-electric transduction process takes place and the receptor potential is generated (Gottschaldt *et al.*, 1982).

The transducer site is characterised by differentiations of the axoplasm and the receptor membrane and by structural contacts between the latter and the surrounding auxiliary structures. Finger-like processes, up to 3 μm long and 100–250 nm in diameter, project from the axon into the lamellar system of the inner core (Figs. 8.5, 6, 7b). They emerge from a conical extension of the axon membrane under which the axoplasm has a fine-grained ground substance that lacks mitochondria or other larger organelles but contains many clear vesicles and some short tubular structures; both of the latter measure 30–90 nm in diameter, and also occur within

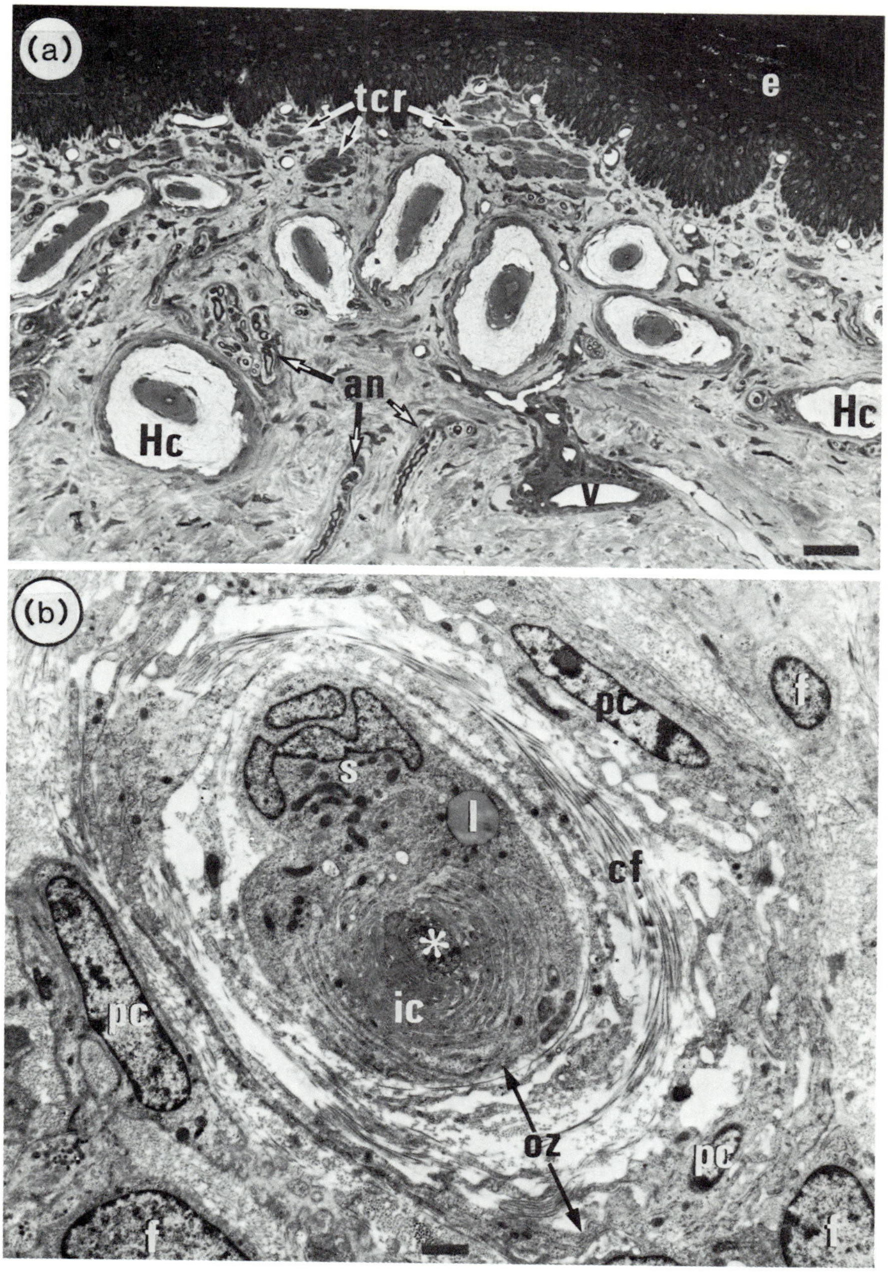

Fig. 8.4. (a) Herbst corpuscles (Hc) in the chicken palate. The corpuscles vary in size, and their inner cores are orientated irregularly with respect to the overlying epithelium (e). In addition to Herbst corpuscles, smaller and larger assemblies of terminal cell receptors (tcr) can be seen underneath the epithelium. v, Blood vessel; an, myelinated afferent nerve fibres. Semi-thin section. ×700. Calibration bar: 10 μm. (b) Electron micrograph of a small Herbst corpuscle or "simple encapsulated" receptor, in the chicken palate. The Schwann cell (s), processes of which contribute to the lamellar system of the inner core (ic), contains a large lipid droplet (l). Note the small extent of the outer zone (oz). cf, Collagen fibrils; asterisk, sensory nerve ending; pc, nuclei of perineural capsule cells; f, nuclei of surrounding connective tissue cells. ×5975. Calibration bar: 1 μm.

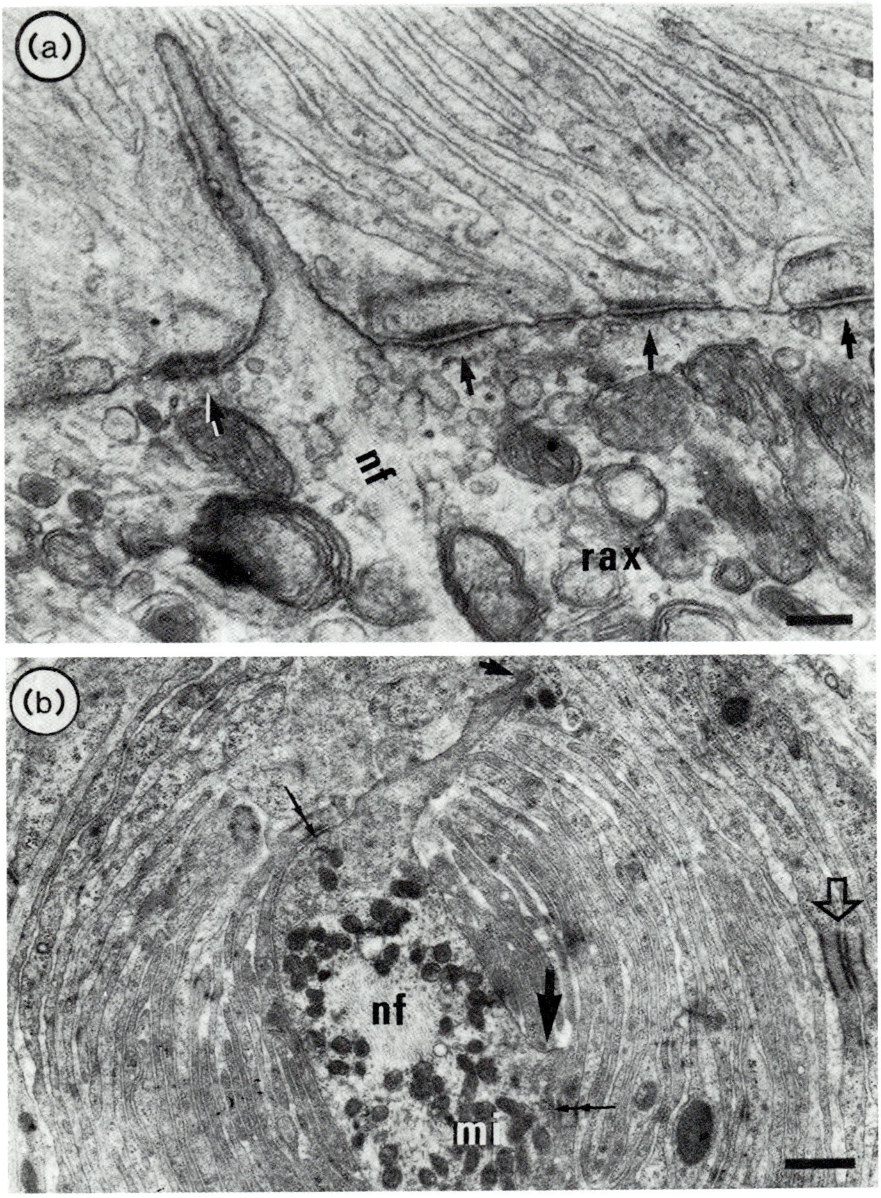

Fig. 8.5. (a) Transducer site at the terminal axon enlargement of a Herbst corpuscle of the goose beak. At the conical base of the finger-like axon process, mitochondria are typically absent, but clear vesicles are numerous. Neurofilaments (nf) extend into the axon process, which also contains some clear vesicles. Asymmetrical zonulae adherentes (arrows) surround the base of the axon process. rax, Receptor axon. ×46,750. Calibration bar: 200 nm. (b) Transducer site in a Herbst corpuscle of the chicken palate. The axon process projects into the cleft region of the inner core, thereby invaginating a Schwann cell lamella (small arrow). A cut-off axon process of another transducer site is marked by a large arrow. Double arrows, zonulae adherentes between membranes of receptor axon and Schwann cell lamellae. Note the large, symmetrical zonula adherens between adjacent Schwann cell lamellae (open arrow). The receptor axon contains a central bundle of neurofilaments (nf) surrounded by many mitochondria (mi). The latter are absent at the bases of the axon processes. ×19,800. Calibration bar: 0·5 μm.

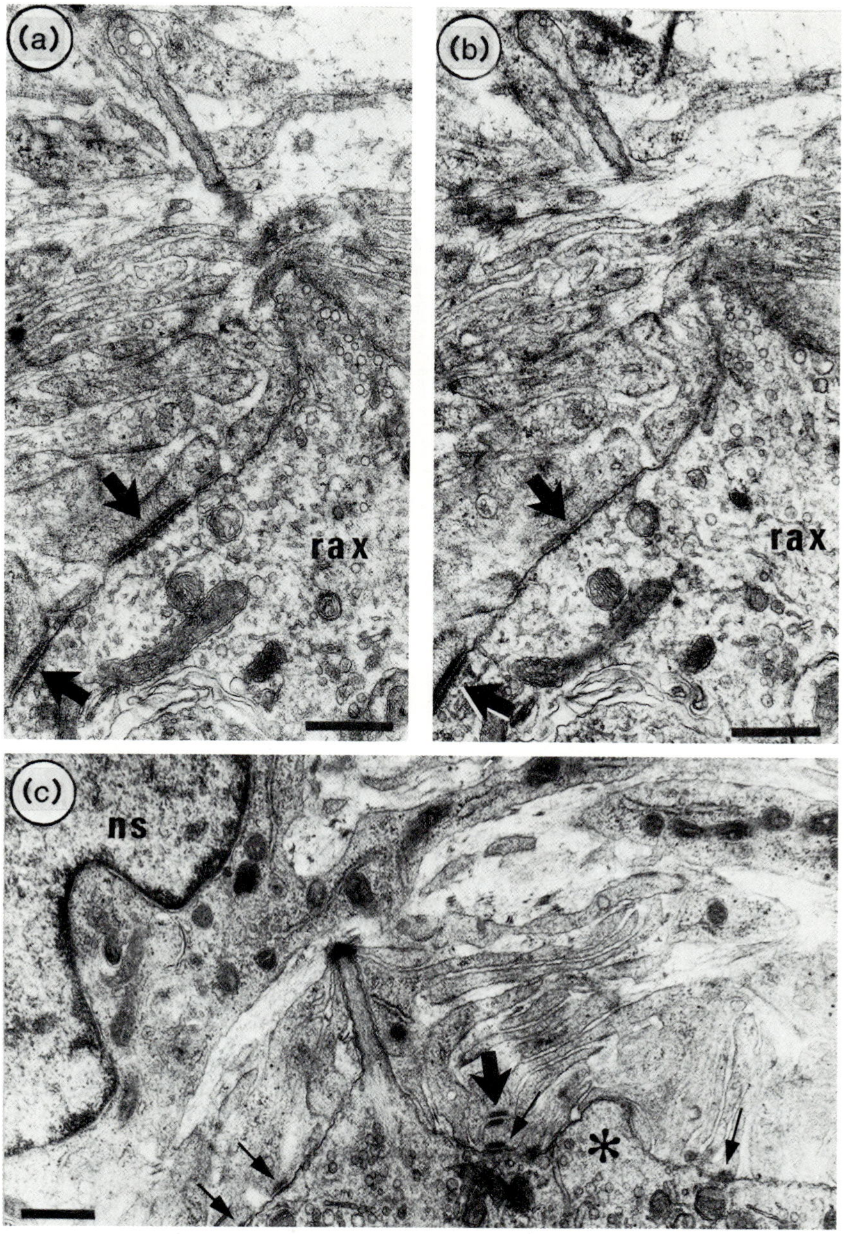

Fig. 8.6. Transducer sites at the terminal enlargement of the receptor axon in a goose Herbst corpuscle. (a) and (b) Consecutive serial sections. Note the axon process "diving" through lamellae of the inner core and extending almost into the outer zone. Clear vesicles are numerous in the axon process, and occur also in its club-like terminal enlargement. Note also the appearance and disappearance, respectively, of two zonulae adherentes (arrows) to the left of the axon process. ×26,000. Calibration bar: 0·5 μm. (c) Finger-like axon process pushing into and compressing lamellae of the inner core. The base of a second transducer site (asterisk) is seen to the right. Asymmetrical zonulae adherentes exist between the membranes of the receptor axon and the lamellae of the inner core (small arrows), whereas a symmetrical contact zone occurs between adjacent Schwann cell lamellae (large arrow). ns, Nucleus of Schwann cell. ×20,950. Calibration bar: 0·5 μm.

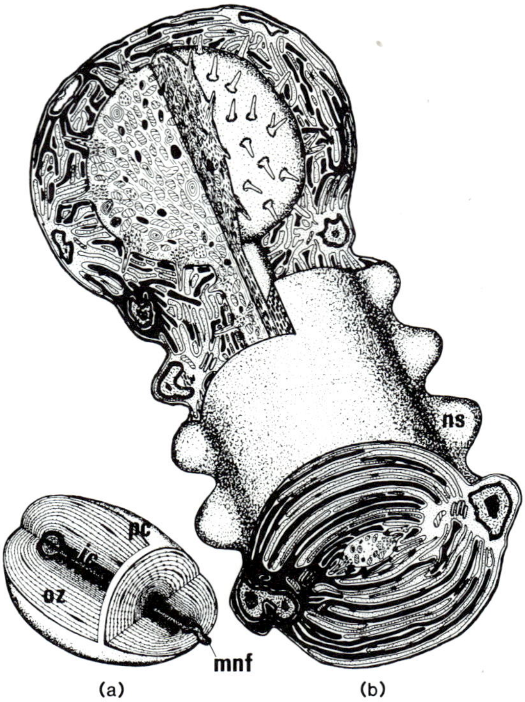

Fig. 8.7. Semischematic three-dimensional reconstruction of a Herbst corpuscle. (a) The gross structure as seen in light microscopy. pc, Outer perineural capsule; oz, outer zone with loops of the network of coiled collagen fibres; ic, inner core, with Schwann cell nuclei lined up along the central receptor axon; mnf, myelinated nerve fibre. (b) Fine structure of the terminal portion of the inner core. The latter is shown from the outside, and in transverse and longitudinal section. The two opposing Schwann cells which are cut in transverse section, as well as neighbouring Schwann cells shown in the longitudinal section, are distinguished by shading, one being black and the other stippled. The number of Schwann cell lamellae has been reduced. The sensory receptor axon with its terminal bulbous enlargement is shown in a surface view (on the right), in a horizontal longitudinal section (on the left) and in a vertical longitudinal section (in the centre). Finger-like axon processes project from the receptor membrane between the lamellae of the inner core. ns, Nuclei of Schwann cells bulging into the outer zone. From Gottschaldt *et al.* (1982).

the axon processes themselves (Figs. 8.5, 6). Whereas in badly-fixed tissue the axon processes look wrinkled, in better-preserved material they are usually straight and appear to be pushed out and kept stiff by numerous microfilaments which run longitudinally within the axon processes and seem to merge with the microfilament system in the central receptor axoplasm. Around the conical base of the axon processes the axon membrane is fastened to the cytolemma of the Schwann cell lamellae by asymmetrical desmosome-like zonulae adherentes. They form circular arrays of single contact points rather than continuous ridges, as becomes apparent in consecutive serial sections (Fig. 8.6a,b). Similar zonulae adherentes also connect neighbouring lamellae of the inner core, but these are always symmetrical (Figs. 8.5b, 6c).

The form, location and distribution of the transducer sites vary systematically along the receptor axon. They become more numerous towards the bulbous terminal enlargement of the receptor axon, where in a single section up to 10 axon processes project in all directions from its circumference. In the proximal part of the receptor axon the transducer sites occur only along the sharply-curved poles of its oval-shaped profile; they are usually shorter and thicker and extend into the cleft region of the inner core (Figs. 8.3b, 7b). In the distal part of the receptor axon, the axon processes become longer and thinner and usually project between the lamellae of the inner core (Figs. 8.5a, 6a,b). Occasionally, however, they project directly into the lamellae, squeezing them together (Fig. 8.6c). The axon processes appear to be actively pushed out of the receptor membrane with considerable force.

In young chickens Nafstad and Andersen (1970) observed a secondary innervation in some Herbst corpuscles which they assumed to have been derived from unmyelinated nerve fibres. In addition to the usual large central receptor axon, they illustrated small axon profiles between the inner-core lamellae. These axons contained clear and dense-cored vesicles and, like the normal sensory axons, were attached to the inner-core lamellae by asymmetrical desmosome-like (not synapse-like!) zonulae adherentes. Since in no other ultrastructural study have accessory nerve fibres been found in normal Herbst corpuscles, it is certain that their occurrence is exceptional, as was noted by Nafstad and Andersen. On the other hand, unmyelinated axons do appear in Herbst corpuscles re-innervated after crushing the afferent nerve (Chouchkov, 1979b). In fact, the central sensory axon in Nafstad and Andersen's micrograph is very dark and densely structured, like degenerating receptor axons in denervated Herbst corpuscles (Chouchkov, 1979a). Therefore, the occurrence of accessory nerve fibres in a Herbst corpuscle may indicate re-innervation following temporary denervation through the death of the original sensory neuron, as happens physiologically with many nerve cells in the course of normal development.

2. *Response properties of Herbst corpuscles as a consequence of their structure*

The morphology of the transducer sites in Herbst corpuscles suggests that the axon processes, fixed at their bases by the desmosome-like attachments, are deflected by displacements of the inner-core lamellae, but are returned to their original position by the inherent elasticity of their microfilaments as soon as the mechanical deformation of the lamellae ceases. Such displacements of the axon processes stretch the receptor membrane, especially in the region of the conical base of the axon processes, and this may activate mechanically-controlled membrane channels leading to an ionic current flow and thus to a receptor potential. There is electro-physiological evidence that the mammalian Pacinian corpuscle, with similarly structured transducer sites along its receptor axon (Nishi *et al.*, 1969; Andres, 1969; Chouchkov, 1971, 1978; Spencer and Schaumburg, 1973), has hot spots with transducer functions at the receptor membrane interspersed with conductile membrane areas (Loewenstein and Rathkamp, 1958; Sato and Ozeki, 1963; Ozeki and Sato, 1964). These hot spots seem to become more numerous towards the distal end of the receptor axon, where the number of morphological transducer sites increases. Whether or not the clear vesicles occurring inside the axon processes and at their bases play a rôle in the mechanoelectric transduction process is not known.

On the basis of electro-physiological evidence it has been postulated that in the different morphological receptor types the transducer properties of the mechanoreceptive nerve endings are uniform, and that in principle they would therefore be capable of generating receptor potentials—and hence discharges in the afferent nerve fibres—which correspond in their time course to the time course of the mechanical deformations effected on the receptor membrane (Gottschaldt and Vahle-Hinz, 1981; Gottschaldt *et al.*, 1982). Accordingly, all mechanoreceptor types should actually deliver slowly-adapting discharges in response to static displacement stimuli. However, as has been described already, Herbst corpuscles, and indeed all other varieties of receptors with lamellated auxiliary structures, do not respond to static displacements with ongoing discharges. This is because displacement (amplitude) as well as movement (velocity) components of mechanical stimuli may fail to reach the receptor axon due to the mechanical filter properties of the auxiliary receptor structures. Like the lamellar system surrounding the sensory nerve ending in Pacinian corpuscles (Hubbard, 1958; Loewenstein, 1971; Hunt, 1974), the coiled collagen fibre network in the outer zone of Herbst corpuscles as well as the Schwann cell lamellar system of their inner core seem to exert a high-pass filter function by which the velocity and amplitude components of mechanical stimuli become attenuated. Presumably such stimulus components succeed only in modifying the form of the auxiliary receptor structures, and in this way the stimulus energy is gradually dissipated between the outer capsule and the centre of the receptor. In contrast, acceleration components, especially those occurring in high-frequency vibratory stimuli, are transmitted much more easily by the auxiliary receptor structures and can therefore achieve the necessary brief deformations at the receptor membrane and particularly of the axon processes at the transducer sites. The mucous fluid filling the spaces between the collagen fibre network and between the lamellae of the inner core may play an important rôle in the particular mechanical filter characteristics of the auxiliary receptor structures in Herbst corpuscles.

It has been argued that the rapidly-adapting response properties of Pacinian afferents are due not only to the short duration of the receptor potential following a step-like stimulus, but also to a fast accommodation process at the nerve fibre itself by which the threshold for initiating propagated impulses is increased (Gray and Matthews, 1951; Mendelson and Loewenstein, 1964; Loewenstein and Mendelson, 1965; Loewenstein, 1971). Accommodation of the nerve fibre acting as a second filter mechanism may in fact increase the phasic character of afferent responses, as has been shown in crayfish stretch receptors (Nakajima and Onodera, 1969a,b). However, under *in vivo* conditions this mechanism should play only a minor rôle in Pacinian afferents, because otherwise it would collide with the well-documented continuous high-frequency responses of these fibres following high-frequency vibratory stimulation. In Herbst corpuscle afferents, at least, a substantial accommodation does not seem to take place as evidenced by the tonic responses of these fibres after cooling the receptor (see later and Gottschaldt *et al.*, 1982).

Whereas the general susceptibility of Herbst corpuscles exclusively to vibratory stimuli can be explained as before, some of the receptor's response characteristics may also be determined by the mechanical properties of the sensory axon itself, especially of the axon processes at the transducer sites. There is, for instance, a remarkable degree of phase locking in the one-to-one following responses to vibratory stimuli, even at the very highest frequencies possible, not only in Herbst corpuscles but also in other types of mechanoreceptors capable of following high-frequency vibratory stimuli (Gottschaldt and Vahle-Hinz, 1981). Such accurate phase locking requires that the receptor membrane or at least the axon processes at the transducer sites be able to oscillate at frequencies of 1000 Hz or more in a very precise manner, so that the successive oscillations do not slur into an average deflection of the axon processes. This necessitates well-developed elasticity alongside the receptor membrane, which seems to be supplied by the microfilaments extending into the axon processes. Moreover, there is no sign of temporal summation in the afferent responses to high-frequency vibratory stimuli, suggesting that it may also be missing in the successive oscillations of the receptor potential. This would indicate that the membrane channels, which enable the ionic flux to cause the receptor potential, operate very rapidly, and that their opening and closing may directly follow the changes in mechanical configuration at the receptor membrane.

An additional and/or alternative explanation for the lack of temporal summation in high-frequency vibratory responses may be that either the membrane deformation itself, or the inwards-directed Na^+ current resulting from it, may initiate an active mechanism counteracting the rise of the receptor potential. In fact, it has been shown in stretch receptor neurons of the crayfish that a hyperpolarizing potential is caused not only by a permeability increase for potassium ions but also by an electrogenic Na^+-pump (Nakajima and Takahashi, 1966). The operation of such mechanisms should not only cut short the time course of individual receptor potential oscillations and thus improve the capacity of frequency resolution, but also lead to a saturation of the receptor potential following large displacement stimuli; this indeed occurs in Pacinian corpuscles (Ozeki and Sato, 1964) and in slowly-adapting mechanoreceptors such as muscle spindles (Katz, 1950; Ottoson and Shepherd, 1971). The operation of an active pump mechanism at the sensory nerve endings would certainly require metabolic energy, and this may account for the striking abundance of mitochondria in the receptor terminals.

Another consequence of the special structural relationships between the receptor axon and the auxiliary structures in Herbst corpuscles appears to be a responsiveness to cold stimuli (Gottschaldt *et al.*, 1982). Many Herbst corpuscles in the dermis of the beak of geese generate a tonic afferent response on cooling, which reaches its maximal discharge frequency at temperatures around 25°C but ceases at still lower temperatures. In contrast, the vibratory sensitivity, decreases continuously as the temperature falls. These response characteristics of Herbst corpuscles have been explained by the effects that temperature changes may have either on the mechanically-elicited receptor potential itself, or on the auxiliary receptor structures. It is assumed that the auxiliary structures, and especially their collagen fibres, contract on cooling and that this leads to a static deformation of the receptor membrane or deflection of the axon processes. Consequently a continuous receptor potential is generated, resulting in a tonic afferent discharge (Gottschaldt *et al.*, 1982). However, not all Herbst corpuscles in the goose beak are temperature-sensitive, and apparently the mechanoreceptors in the feathered skin of pigeons also lack this property (Necker and Reiner, 1980). Perhaps only the particularly large Herbst corpuscles with their very elaborate auxiliary structures are temperature-sensitive. Anyway, the temperature sensitivity of Herbst corpuscles is probably non-specific, and not a specific transducer property of the receptor membrane which could be used to convey temperature information to the central nervous system (Gottschaldt *et al.*, 1982).

3. *Occurrence and function of Herbst corpuscles*

Herbst corpuscles are perhaps the most widely-distributed mechanoreceptors in the bird's body, occurring everywhere in the integument, along the large leg and wing bones, in tendons, muscles and joint capsules, and near the large blood vessels (Schildmacher, 1931; Stresemann, 1934; Stammer, 1961; Poláček *et al.*, 1966; Malinovský, 1967; Saxod, 1978a; Chouchkov, 1978; Halata and Munger, 1980). In general, they are situated wherever mechanical disturbances of tissue homeostasis may occur. Their location along blood vessels shows that although Herbst corpuscles always function as mechanoreceptors they do not always serve a tactile sensation. Older assumptions that Herbst corpuscles might also function as warmth receptors (Krause, 1881), osmoreceptors (Clara, 1925; Malinovský and Zemanek, 1971) or even as "pain" receptors (Stammer, 1961) are unlikely to be correct in view of the similar structural relationships between the sensory nerve ending and the auxiliary structures that are found in all varieties of Herbst corpuscles.

In feathered skin the receptors are located primarily at the base of the feather follicles (Stammer, 1961; Winkelmann and Myers, 1961; Ostmann *et al.*, 1963). These Herbst corpuscles, as well as the great numbers in the interosseous membrane between the fibula and tibiotarsus, have been interpreted from behavioural observations as vibration detectors enabling the bird to detect airborne and ground-borne vibrations after destruction of the inner ear (Schwartzkopff, 1949). The most functionally-versatile array of Herbst corpuscles occurs in the beak of birds. Here their number and location appear to be related to the manner in which the beak is used as a tactile exploratory organ during feeding. In charadriiforms, which use their bill for probing, the premaxillary and dental bones bear countless small cavities (lacunae) which are packed with Herbst corpuscles (Leydig, 1868; Clara, 1925; Bolze, 1968). Similar receptor-containing lacunae also occur in the distal parts of the beak bones of anseriforms (Quilliam, 1966; Gottschaldt and Lausmann, 1974). In addition, there are numerous receptors in the basal portion of the touch papillae of the bill tip organ (see Section VII) and in the rhamphothecal dermis. Berkhoudt (1980) showed in whole-mount preparations of the duck's rhamphotheca that Herbst corpuscles were located at regularly-spaced intervals. Their number increases towards the tip and the lateral ridges of the beak (Krogis, 1931; Gottschaldt and Lausmann, 1974; Krulis, 1978; Berkhoudt, 1980). In grain-feeding songbirds, especially the Fringillidae, some of which use their beak to cut seeds open and others to pinch them open, Herbst corpuscles, as well as other types of mechanoreceptors, are located at exactly those places in the beak which are mechanically involved in the seed-opening operations (Krulis, 1978). In the long, sticky tongue of the woodpecker, which is thrown out of the beak to catch prey, Herbst corpuscles are particularly numerous and attracted the attention of a royal investigator (Prince Ludwig Ferdinand of Bavaria, 1884).

B. Receptors detecting velocity components in mechanical stimuli

Included among this group of receptors are receptors in feathered skin, Grandry corpuscles and structurally-related "terminal cell" receptors.

The responses of mechanoreceptors which are specifically velocity-sensitive are always restricted to the movement part of mechanical stimuli (Fig. 8.1). The distinction of responses in rapidly-adapting velocity-sensitive units from those which are also rapidly adapting but arise from Herbst corpuscle afferents can become quite difficult; indeed the identification of a velocity-sensitive unit is often only possible by the demonstration of a quantitative stimulus–response relationship between afferent discharge frequency and stimulus velocity. Although rapidly-adapting mechanoreceptors have been reported in electro-physiological studies on the beak and feathered skin of pigeons, chickens, ducks and geese (Zeigler and Witkowsky, 1968; Dorward, 1970a; Necker, 1972, 1973, 1974a,b; Gregory, 1973; Roumy and Leitner, 1973; Leitner and Roumy, 1974a; Gottschaldt 1974a; Necker and Reiner, 1980; Gottschaldt *et al.*, 1982), there is still considerable uncertainty whether only one or more than one type of rapidly-adapting receptor exists in avian skin, and especially about the morphology of these receptors. Feathered skin will be considered first.

1. Receptors in feathered skin

Dorward (1970a) described the occurrence of at least two types of rapidly-adapting units, which were distinguishable from each other by their response and receptive field characteristics, and from Herbst corpuscle afferents by the lack of a vibratory sensitivity. The "down feather units" had low mechanical thresholds and large receptive fields including many down feather follicles. "Touch units", in contrast, had higher mechanical thresholds than down feather units, their responses adapted more slowly, and their receptive fields were always localised and occasionally discontinuous. The association of each type of unit with any particular, these light microscopic studies, using silver staining techniques, have apparently had slowly-adapting response characteristics and seemed to be associated with the bases of the large wing coverts. Thus, the touch units may well have represented more than one receptor type. The only other electro-physiological study on mechano-receptive afferents from the feathered skin (Necker and Reiner, 1980) did no more than merely mention the existence of rapidly-adapting receptors.

However fragmentary our physiological knowledge about the mechanoreceptors of feathered skin may be, there is sufficient evidence for the presence of more than one receptor type. This conclusion contrasts strongly with the results of morphological studies, which have revealed only Herbst corpuscles in feathered skin (Stammer, 1961; Winkelmann and Myers, 1961; Ostmann *et al.*, 1963). In particular, these light microscopic studies, using silver staining techniques, have reported the absence of nerve endings in the walls of the feather follicles, except for the transitional zone between rhamphotheca and feathered skin of the head (Stammer, 1961). Rapidly-adapting responses from this area have occasionally been recorded in single fibres of the ophthalmic nerve (K.-M. Gottschaldt, unpublished observations). They were apparently velocity-sensitive, had very low mechanical thresholds and possessed homogeneous receptive fields 5–30 mm in diameter. In contrast to the down feather units described by Dorward (1970a), all of the few units encountered also had an exquisite vibratory sensitivity, being readily entrained by directly touching the feathers with a vibrating tuning fork (440 Hz), whereas they did not respond to vibratory stimuli applied outside the receptive field from which the velocity-sensitive responses could be elicited. The receptors of these units may have been situated within the walls of the feather follicles. However, because of their great sensitivity and the density of the plumage it was not possible to localise them within the feather follicles themselves, for instance, by establishing threshold differences for their activation following isolated stimulation of the feathers or the intervening skin.

It is clear, then, that more electro-physiological as well as anatomical and especially ultrastructural studies are needed to clarify the mechano-sensitive innervation of the feathered skin. This is even more true for receptors responding to other than mechanical stimulus modalities.

2. *Physiology of velocity-sensitive receptors in the dermis of the beak*

Studies aimed at differentiating the rapidly-adapting units among the beak afferents into more than one type have yielded controversial results. In the duck, Gregory (1973) found only rapidly-adapting, low-threshold mechano-sensitive afferents, and because he was not convinced that two different response types existed he considered all units to be Herbst corpuscle afferents. However, Roumy and Leitner (1973) and Leitner and Roumy (1974a) were able to distinguish, in both chickens and ducks, two types of rapidly-adapting receptors on the basis of their vibration sensitivity or insensitivity. They also found, in chickens but not in ducks, a few slowly-adapting low-threshold mechano-sensitive units; in contrast, such units were frequently observed by Necker (1972, 1973, 1974a) in both pigeons and ducks, but unfortunately he did not attempt in turn to differentiate the rapidly-adapting units into different types.

In geese the picture is much more consistent, perhaps because there is only one study (Gottschaldt, 1974a). About 50% of all beak mechano-sensitive units, and about two-thirds of the rapidly-adapting units among them, were found to be velocity-sensitive and distinct from the velocity-insensitive but vibration-sensitive Herbst corpuscle afferents. The results of different procedures for identifying the receptors of the velocity-sensitive responses all pointed to the Grandry corpuscles, which were the only other receptors besides Herbst corpuscles that could be identified with light microscopic techniques in the dermis of the beak (Gottschaldt, 1974a,b; Gottschaldt and Lausmann, 1974). Therefore, the velocity-sensitive, rapidly-adapting units were called "Grandry units", although Grandry corpuscles may not be the only receptors in the beak of birds giving rise to these discharges.

Grandry units in the dermis of the soft part of the rhamphotheca of geese are not spontaneously active and have receptive fields varying in size from 1 to 100 mm^2. In large receptive fields, the afferent nerve fibre can often be activated from several isolated spots which may correspond to the location of different Grandry corpuscles supplied by the same afferent nerve fibre (Merkel, 1878; Dogiel, 1904; Andres and von Düring, 1973). The vibratory sensitivity of Grandry units is restricted to the low-frequency range, and only very rarely could they follow vibratory stimuli above 200 Hz with one-to-one discharges.

Usually, tangential movements causing a crumpling displacement of the rhamphotheca generate more vigorous discharges in Grandry units than perpendicular indentations. Responses in a single afferent nerve fibre to this latter kind of stimulation are illustrated in Fig. 8.8a–c. When a constant final amplitude of the displacement was used, the duration of the discharges decreased but their frequency increased with the movement velocity (Fig. 8.8a). If both the movement velocity and the displacement amplitude increased simultaneously (Fig. 8.8b), the duration of the discharges remained about constant but their frequency

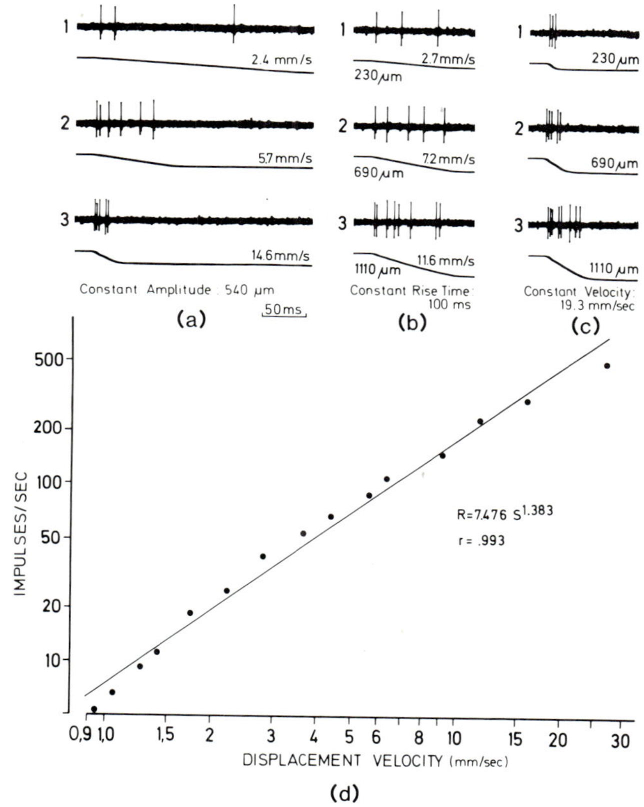

Fig. 8.8. Response profiles of a single Grandry unit in the dermis of the goose's beak. In (a) to (c) the original records are shown above the respective stimulus traces. (a) The velocity (as on the right of each of the three stimulus traces) of a trapezoidal movement, leading to a constant final displacement amplitude, was varied. (b) The rise time of the trapezoidal displacement was held constant but the amplitude was increased (as on the left), and this induced different movement velocities (as on the right). (c) The movement velocity was held constant, while the amplitude of the trapezoidal displacement was increased (as on the right). (d) The average discharge frequency of the phasic response is plotted against the displacement velocity. Stimulus mode: constant amplitude, as in (a). The regression line was calculated by the least-square method. The mathematical equation of the stimulus–response relationship is given on the right. R, Response measured in impulses per second; S, stimulus measured in millimetres per second; *r*, correlation coefficient.

still increased with the movement velocity, that is, inversely to the rise time of the displacement. If, in contrast, the displacement amplitude increased at a constant movement velocity (Fig. 8.8c), the duration of the response became longer with larger amplitudes but the discharge frequency remained about constant. Altogether this analysis shows that Grandry units respond in a quantitative manner only to the velocity and not to the amplitude component of mechanical displacements. As in all Grandry units, the stimulus–response relationship can be mathematically described by a power function (Fig. 8.8d).

Grandry units are not temperature-sensitive. The mechanical thresholds of individual units in the cere of the goose can vary over three orders of magnitude. The most sensitive receptors respond to von Frey hair stimulation of less than 5 mg, whereas others require stimulus intensities of up to 6.5 g. Also the velocity sensitivity of individual Grandry units (expressed by the slope of their stimulus–response functions) varied, with slope values between 0·37 and 1·04 (Gottschaldt, 1974a). These differences in threshold and velocity sensitivity have been accounted for by the varying size of individual Grandry corpuscles, which is much more conspicuous in the dermis of the beak of geese than ducks (Gottschaldt and Lausmann, 1974). It is significant that Gregory (1973) established in some units a quantitative relationship between the afferent discharge frequency and the stimulus velocity; this indicates that his population of rapidly-adapting afferent fibres must also have included Grandry units, despite his conclusion that all responses were derived only from Herbst corpuscles.

3. *Structure of Grandry corpuscles in the dermis of the beak*

It has already been indicated that Grandry corpuscles are not the only receptor structures in the beak of birds which give rise to rapidly-adapting, velocity-sensitive responses. In fact they represent only one particular configuration in a group of mechanoreceptors with distinct structural similarities but different response properties. Earlier controversies about the presence of these receptors in different avian species, their terminological designation, their morphological differences as well as similarities, and even their very existence, now appear reconcilable. The following description of the Grandry corpuscles, as the best-known avian example of these receptors, will illustrate their common structural characteristics.

Grandry corpuscles are generally believed to occur only in the beaks of waterfowl, but they have also been observed light microscopically in the beaks of grain-feeding songbirds (Ziswiler and Trnka, 1972; Krulis, 1978). The receptors lie in the dermis closely underneath the epithelium, but usually more superficially than Herbst corpuscles. A typical Grandry corpuscle (Fig. 8.9) consists of an orderly stack of specialised cells (here called Grandry cells) between which are sandwiched the disk-like terminals of an afferent nerve fibre, the whole complex being enclosed by Schwann cells. In the rhamphothecal dermis, Grandry cells and nerve terminals are always orientated approximately parallel to the surface of the overlying epidermis. The whole corpuscle is enclosed by a "capsule" of variable distinctness (Figs. 8.16, 17a,b).

Soon after Grandry corpuscles were first seen [probably by Leydig (1868) and

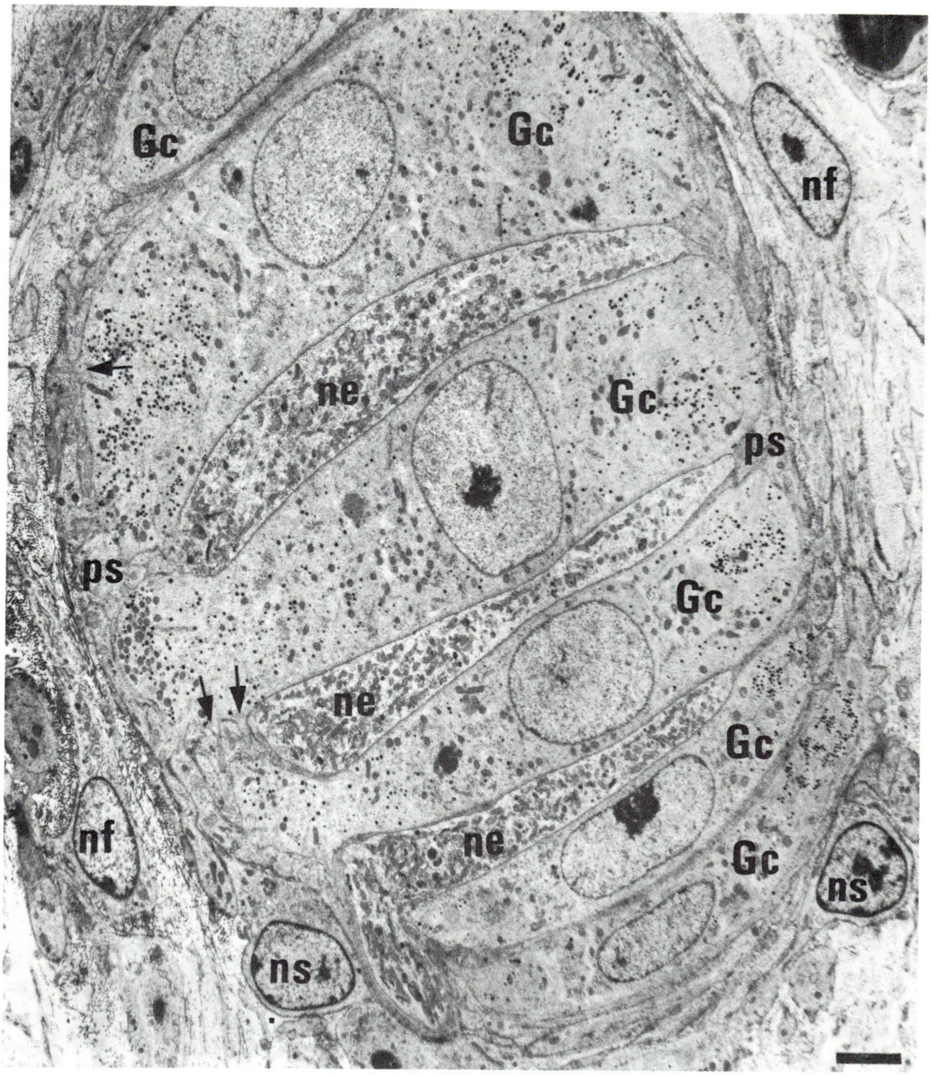

Fig. 8.9. Electron micrograph of a Grandry corpuscle in the beak of the goose. Gc, Grandry cells; ne, nerve endings; ns, nuclei of Schwann cells, processes (ps) of which enclose the stack of Grandry cells and interposed nerve endings; nf, nuclei of fibrocytes forming the outer capsule of the corpuscle. Characteristic electron-dense granules lie mainly in peripheral regions of the cytoplasm of the Grandry cells. Arrows point to microvilli of Grandry cells projecting into the Schwann cell enclosure. ×4175. Calibration bar: 2 μm.

then by Grandry (1869)], the special kind of cells associated with the afferent nerve endings was recognised as the main feature of these receptors; probably Hesse (1878) provided the first fairly accurate description of their light microscopic structure. Despite many subsequent anatomical studies (reviews by Malinovský, 1967; Berkhoudt, 1980), some authors denied the very existence of these receptors (as well as of others, such as Ruffini endings) and simply dismissed them as artifacts (Wedell *et al.*, 1955; Winkelmann, 1960; Winkelmann and Myers, 1961; Montagna and Parakkal, 1974).

Merkel (1875, 1878, 1880) believed that the specialised cells in Grandry corpuscles were identical with the intraepidermal "Tastzellen", which he had discovered in the snout skin of pigs and which now bear his name. He interpreted these cells as peripheral ganglion cells within which the afferent nerve fibre would terminate, although this had already been refuted (Key and Retzius, 1876; Ranvier, 1877; Hesse, 1878; Szymonowicz, 1897). A cytological similarity between mammalian Merkel cells and the specialised cells occurring in some avian cutaneous receptors, such as Grandry corpuscles, has always been assumed, and this has now been confirmed by ultrastructural studies.

Some authors proposed to call the specialised cells in Grandry corpuscles "Grandry cells" rather than Merkel cells, thus leaving open the possibility of a cytological, histological or functional distinction between Merkel and Grandry cells (Szymonowicz, 1897; Halata, 1971b; Munger, 1971a,b, 1977; Gottschaldt and Lausmann, 1974; Iggo and Andres, 1982). Nevertheless, a variety of other terms has been used for Grandry cells, such as periterminal, secretory, satellite, sensory or tactile cells. Moreover, where the specialised cells of birds are assembled into receptor configurations other than Grandry corpuscles, they usually have also been called Merkel cells, thus implying that avian "Merkel cells" are analogous to their mammalian counterparts but different from Grandry cells (Andersen and Nafstad, 1968; Nafstad, 1971a,b; Ziswiler and Trnka, 1972; Gottschaldt *et al.*, 1976; Krulis, 1978; Saxod, 1978a,b). In contrast, Idé and Munger (1978) proposed to name all specialised cells in avian mechanoreceptors "Grandry cells", and consequently called the receptors in the chicken toe skin "Grandry corpuscles", even though they were structurally unlike the typical Grandry corpuscles found in the beaks of waterfowl.

Seeking a way out of this confusion and also trying to avoid a terminological preconception about their function, Gottschaldt and Kräft (1978) proposed to call all variants of specialised cells occurring in vertebrate mechanoreceptors "terminal cells", indicating their location at the sensory nerve terminals. The traditional name "Merkel cell" should then be used only for their mammalian, intraepidermal variant and the term "Grandry cell" only for the particular variant occurring in the classical Grandry corpuscles. This terminology will also be used in the following sections of this chapter. [It should be noted, however, that to retain the terms Merkel cell and Grandry cell would contravene the *Nomina Anatomica* (1977) and the *Nomina Anatomica Avium* (1979), which require the abandonment of eponyms.]

Grandry cells measure 10–40 μm in diameter. Their number in individual Grandry corpuscles may vary from mostly 2 (Fig. 8.15a) up to 12, and there appear to be species differences (Gottschaldt and Lausmann, 1974). In ducks, most Grandry corpuscles contain two Grandry cells, with one nerve terminal sandwiched between them (Malinovský, 1967; Berkhoudt, 1980). However, the "formula of Ranvier" (1877), according to which the number of nerve endings should always be one less than the number of Grandry cells, does not always hold. The nerve endings occasionally outnumber the Grandry cells (Fig. 8.15a), the Grandry cells being sandwiched between the nerve endings. Moreover, in large corpuscles containing four or more Grandry cells (Fig. 8.9), nerve endings often occur only between the more centrally-located Grandry cells.

Electron microscopic studies have revealed that Grandry cells, as well as the terminal cells occurring in other receptor configurations, share a number of cytological features with mammalian Merkel cells (de Iraldi and Rodriguez-Perez, 1961; Andres, 1966, 1969, 1974; Quilliam, 1966; Munger, 1966, 1971a,b, 1977; Andersen and Nafstad, 1968; Saxod, 1968, 1970b, 1973a, 1978a,b; Goglia, 1969b; Halata, 1970, 1971b, 1975; Nafstad, 1971a,b; Andres and von Düring, 1973; Winkelmann and Breathnach, 1973; Chouchkov, 1974, 1978). These are dense-cored granules and microfilaments in the cytoplasm, microvilli projecting from the cell surface, and junctional complexes between the terminal cells and the adjoining nerve terminals.

The dense-cored granules are 60–200 nm in diameter. They are larger in foetal terminal cells but less electron-dense. Presumably deriving from the Golgi apparatus, the granules are membrane-bound, and often a clear space exists between the membrane and the dense core. Not only in the orderly-stacked Grandry cells but in other avian terminal cells, the granules are often assembled into small groups and occur predominantly in the peripheral parts of the cytoplasm. This is evident if the cell is cut through its centre including its nucleus (Figs. 8.9, 12a, 13a, 14b, 15a, 16, 18, 19, 20, 21b, 28b, 29, 30a); the granules appear to be homogeneously distributed in paranuclear sections only (Figs. 8.12b, 13a, 18, 20, 29a, 30b). Thus, in contrast to mammalian Merkel cells, the dense-cored granules are not concentrated in the cytoplasm adjoining the associated nerve endings. Indeed, the opposite is the case.

Another peculiarity of terminal cells is the occurrence of large, membrane-bound dense bodies which can measure up to 2 μm in size and then become visible even in light microscopy (Fig. 8.17a). These dense bodies are located predominantly in the paranuclear cytoplasm and resemble giant lysosomes. Usually the electron density of the smaller dense bodies is homogeneous (Fig. 8.10a), but the larger examples often contain one or several vacuoles of different size as well as small electron-dense granules (Fig. 8.10b). These latter granules correspond in size to the typical dense-cored granules of the terminal cells, and therefore the cytoplasmic dense-cored granules may become incorporated into the large dense bodies. If so, the dense-cored granules may have a life cycle entirely within the terminal cells and never be released from them, despite the general belief that they are "secretory" granules (Munger, 1965, 1966, 1971a,b, 1977; Andres, 1966, 1969, 1974; Iggo and Muir, 1969; Nafstad, 1971a,b; Halata, 1975; Saxod, 1978a; Chouchkov, 1978). In fact, despite intensive efforts (reviewed by Saxod, 1978a), the chemistry of the dense-cored granules is still unknown, and really conclusive evidence for their release from the terminal cells is lacking.

All terminal cells contain abundant microfilaments 6–8 nm in diameter. Presumably composed of actin molecules, the microfilaments form bundles which extend throughout the entire cytoplasm, thus building an elaborate cytoskeleton that was already recognised as the "chondriom of the Grandry corpuscle" in older silver impregnation studies (Dogiel, 1904; Lawrentjew, 1926). At the cell surface the microfilaments extend into a great number of microvilli; actually, the microvilli appear to be pushed straight out of the cell by bundles of microfilaments "growing" towards the cell surface (Figs. 8.11, 14a, 22a, 30). Microvilli may occur anywhere at the surface of terminal cells, but where the cells are stacked upon each other, as in Grandry corpuscles, the microvilli occur preferentially at their perimeter. Generally, the cell surface contacting the adjoining nerve terminals lacks microvilli, but occasionally a microvillus projects straight into a

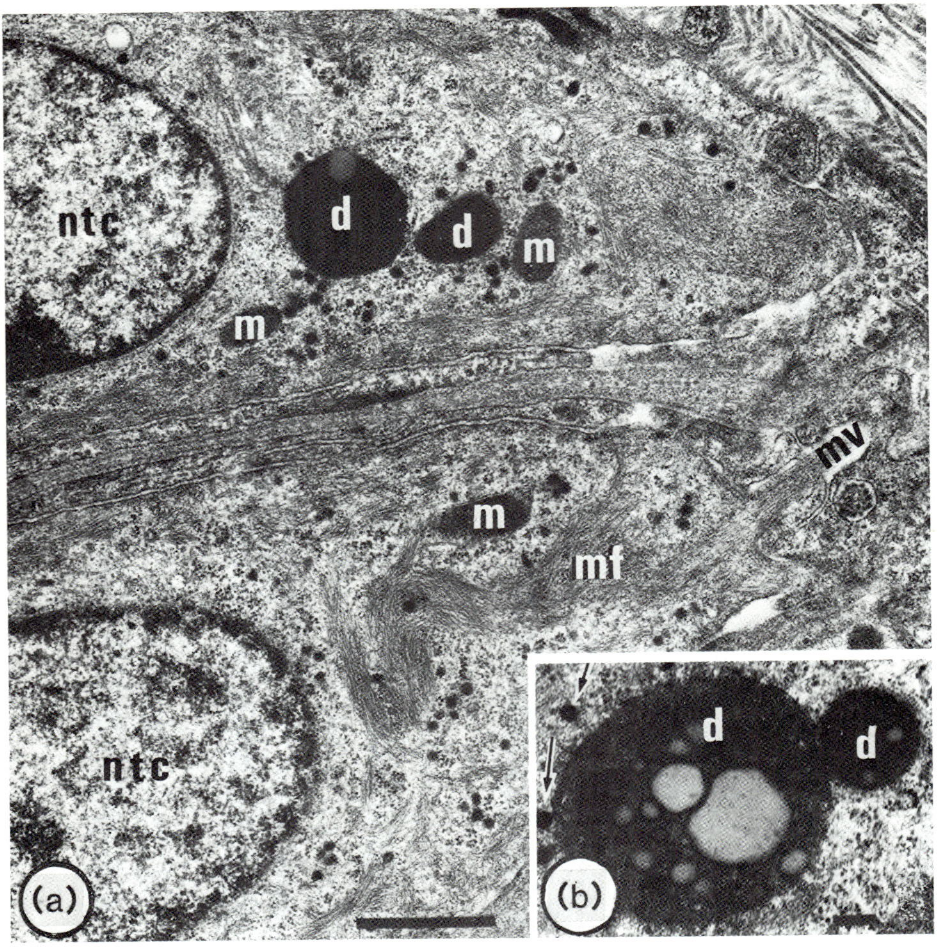

Fig. 8.10. Dense bodies in terminal cells of the chicken's palate. (a) The electron-dense bodies, d, are almost homogeneous. (b) The electron-dense bodies, d, contain several "vacuoles" and small electron-dense granules; small arrows point to dense-cored granules in the cytoplasm of the terminal cell. ntc, Nuclei of terminal cells; mf, microfilaments partly extending into a cut-off microvillus (mv); m, mitochondria. (a) ×18,170. Calibration bar: 1 μm; (b) ×28,135. Calibration bar: 200 nm.

nerve terminal (Fig. 8.11b). In the periphery of Grandry corpuscles, where the Grandry cells extend beyond the interposed nerve terminals, myriads of microvilli from neighbouring Grandry cells interdigitate with each other (Fig. 8.11a). The microvilli also project sidewards, thereby indenting the cell surface and sometimes even the nucleus of the surrounding Schwann cells, processes of which intermingle with the microvilli and fill out all gaps between them. Occasionally, small bundles of collagen fibrils can be seen in the conglomerate of microvilli and Schwann cell processes, sometimes in close proximity to the nerve endings (Fig. 8.11c).

The nuclei of terminal cells are round or oval and have a distinct nucleolus. They are distinguished from Schwann cell nuclei by their faint chromatin texture and a relatively slender fringe of chromatin along the nuclear membrane (Figs. 8.9, 12a, 14, 15, 18–20, 21b, 28b, 29, 30a). This is an additional criterion for

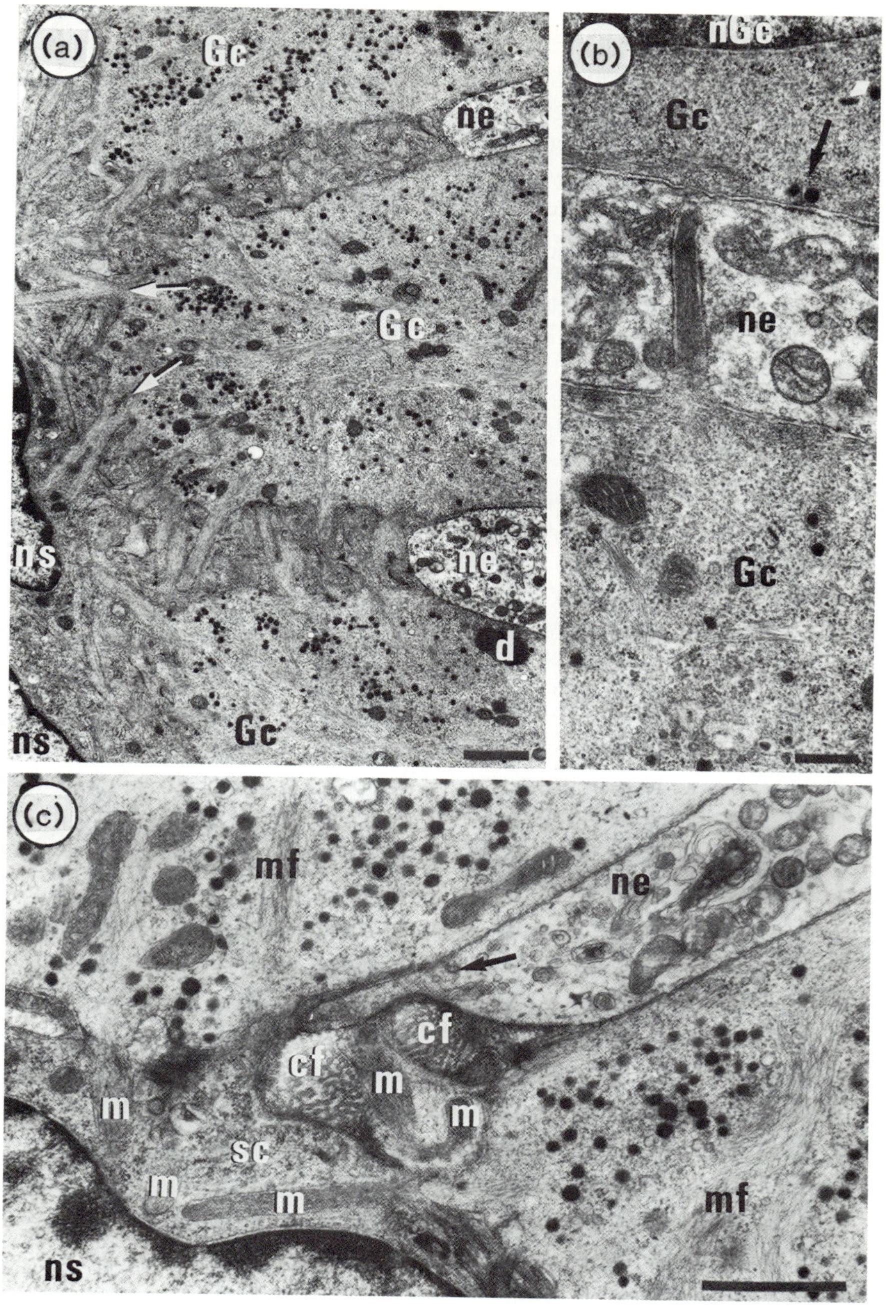

Fig. 8.11. Microvilli of Grandry cells. (a) Peripheral zone of a Grandry corpuscle, that is, the zone where the Grandry cells extend beyond the interposed nerve terminals. Microvilli of neighbouring Grandry cells (Gc) interdigitate with each other, and also project into the enclosing Schwann cells. ne, Nerve ending; ns, nucleus of Schwann cell; d, dense body. Arrows, branched microvilli. ×9285. Calibration bar: 1 μm. (b) Microvillus projecting into a nerve ending (ne) sandwiched between two Grandry cells (Gc); nGc, nucleus of Grandry cell. Arrow, two dense-cored granules lying near a faint contact zone between the nerve ending and the Grandry cell. ×46,140. Calibration bar: 200 nm. (c) Microvilli (m) in the peripheral zone of a Grandry corpuscle, projecting into the Schwann cell cytoplasm (sc) and also projecting between two small bundles of collagen fibrils (cf). The nerve ending (ne) has a transducer site with a finger-like axon process (arrow). mf, Microfilaments of Grandry cell; ns, nucleus of Schwann cell. ×21,100. Calibration bar: 1 μm.

identifying terminal cells and distinguishing them from Schwann cells; it is useful when both cell types are irregularly mixed as in terminal cell receptors of the chicken palate (Fig. 8.21b), or if dense-cored granules are not abundant in the terminal cells as in their early developmental stages. Near the nucleus of the terminal cells, typical centrioles may be seen (Figs. 8.15a, 21b, 28b, 29b), as well as a well-developed Golgi apparatus (Fig. 8.15a). The cytoplasm contains also many mitochondria and free ribosomes. In the paranuclear region there are usually some scattered ridges of rough endoplasmic reticulum. On several occasions Saxod (1968, 1970b, 1973a,b, 1978a) described "perinuclear organelles" in Grandry cells but illustrated them only by variants of the same picture. No other investigator has been able to confirm the existence of these structures. It is therefore certain that these "perinuclear organelles" do not exist in Grandry cells. They were probably an unusual configuration of the ordinary rough endoplasmic reticulum in Saxod's material. Thus, as suggested by Chouchkov (1978), there appear to be no special organelles which distinguish avian Grandry from mammalian Merkel cells.

Grandry corpuscles are innervated by a single myelinated nerve fibre which, however, may also supply other Grandry corpuscles (Merkel, 1878; Dogiel, 1904; Andres and von Düring, 1973; Saxod, 1978a). Where the Grandry corpuscle consists of only two Grandry cells, the nerve fibre ends with a single disk-like terminal between them. If the corpuscle contains additional Grandry cells, the pre-terminal nerve fibre branches into several endings which terminate in parallel between the Grandry cells. Alternatively, the afferent nerve fibre may not branch but instead form disk-like terminals in series, each one lying between two Grandry cells. Combinations of these variants may also occur (Berkhoudt, 1980).

The myelin sheath of the afferent nerve fibre either may reach the capsular zone of the Grandry corpuscle (Fig. 8.12b) or may end well before it, leaving an unmyelinated connecting segment up to 30μm long between the first node of Ranvier and the sensory terminals within the corpuscle (Fig. 8.12a). Not infrequently, the last part of the myelinated axon widens into a bulbous enlargement (Figs. 8.16, 17a) from which the unmyelinated pre-terminal axon, covered only by Schwann cell processes, emerges. Sometimes the pre-terminal axon enlarges just after it enters the corpuscle, at the branching of the terminal axon (Fig. 8.13a). Both types of enlargements contain many mitochondria and lysosomes which become particularly abundant within the terminal disks.

Whereas the interface between Grandry cells and nerve terminals is smooth, there are distinct structural differentiations at the edges of the nerve terminal disks which closely resemble the transducer sites in Herbst corpuscles. In particular, finger-like processes project from the receptor axon into the conglomerate of interdigitating microvilli, Schwann cell processes and scattered collagen fibrils which seal off the peripheral edges of the nerve terminal disks (Fig. 8.11c, 12b, 13c). The axon processes are 1–1·5 μm long, and contain longitudinally-orientated microfilaments (Fig. 8.13b) and both clear and dense-cored vesicles (Fig. 8.13c) 60–120 nm in diameter. Sometimes the axon processes are blunt or bulbous (Fig. 8.12b), containing a great number of small vesicles. The axon processes along the peripheral circumference of the nerve terminal disks give the latter the appearance of irregularly-toothed wheels, with the "teeth" sticking into the surrounding wreath of microvilli and Schwann cell processes derived from the auxiliary receptor structures.

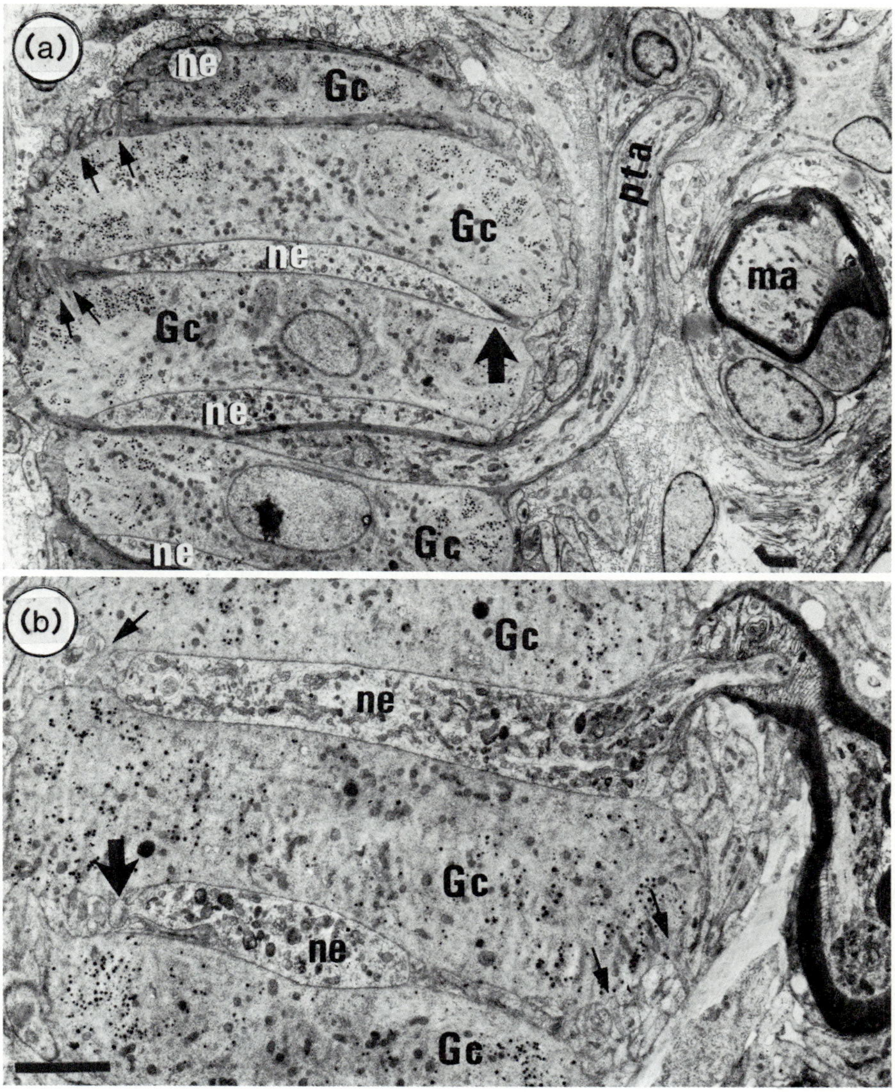

Fig. 8.12. Termination of sensory nerve fibres in Grandry corpuscles. (a) The myelin sheath ends at least 30 μm before the receptor; an unmyelinated pre-terminal axon (pta) connects the myelinated nerve fibre with the disk-like nerve terminals (ne). The pre-terminal axon is covered by Schwann cell processes. Serial sections established that the myelinated axon to the right (ma) did not innervate this particular Grandry corpuscle. The large arrow points to a transducer site on the nerve terminal; the small arrows indicate microvilli of Grandry cells (Gc). ×3175. Calibration bar: 2 μm. (b) The afferent nerve fibre is myelinated as far as the capsular zone, and then immediately continues into a terminal disk (ne). Large arrow, a bulbous axon process of a transducer site; small arrows, microvilli of Grandry cells (Gc). ×6500. Calibration bar: 2 μm.

The membranes of terminal cells and nerve endings are connected by special contact zones which have been particularly controversial (Saxod, 1978a). Some of these junctions have a morphological appearance reminiscent of synaptic contacts (Fig. 8.14a). These contact zones are between 200 and 600 nm wide and possess asymmetrical membrane thickenings formed by a fine filamentous material. Furthermore, a group of dense-cored granules adjoins the membrane of the

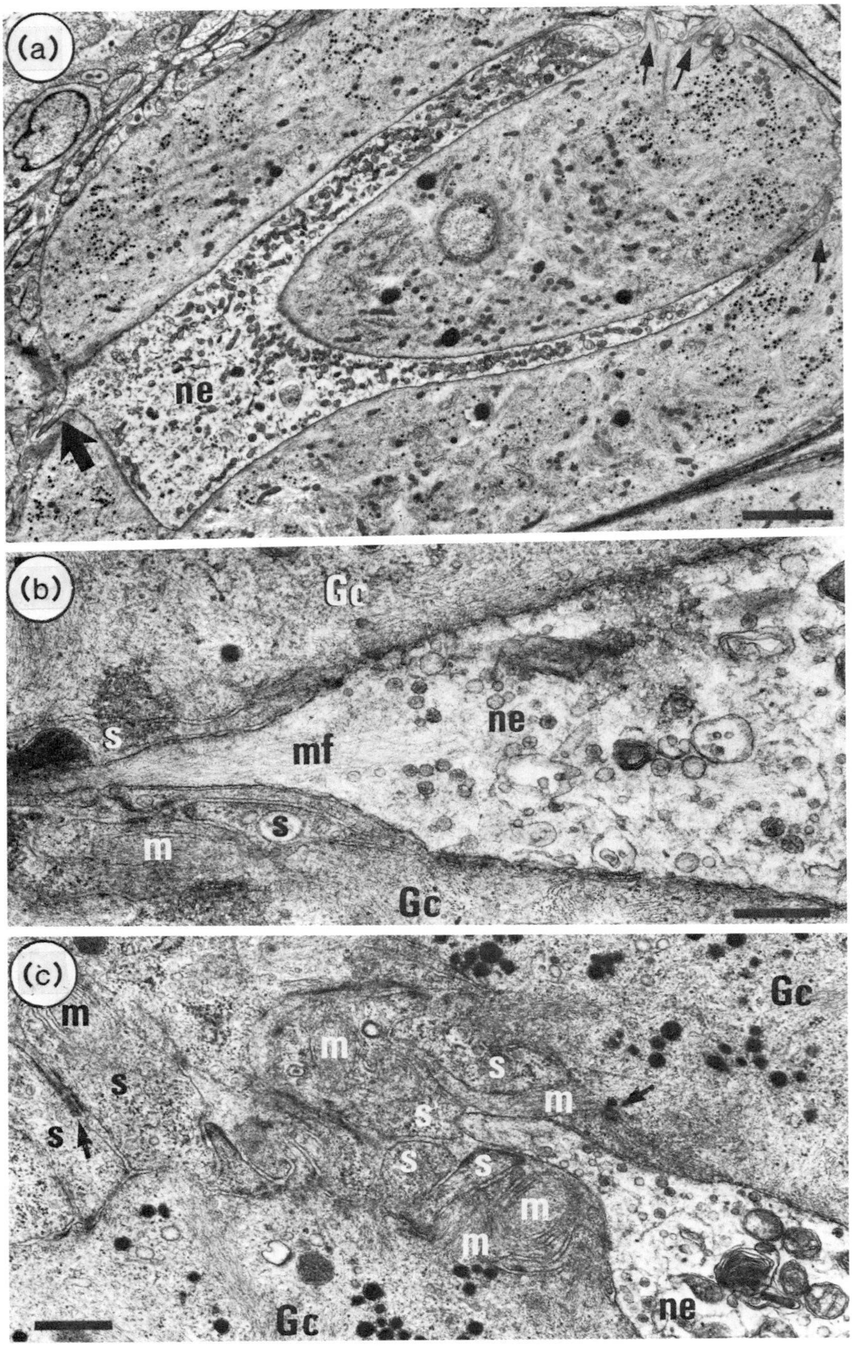

Fig. 8.13. Transducer sites on the nerve endings in Grandry corpuscles. (a) An axon process (large arrow) emerges from an enlargement of the nerve ending (ne), which gives rise to two nerve terminal disks. Small arrows, microvilli of Grandry cells. ×5360. Calibration bar: 2 μm. (b) An axon process containing longitudinally-orientated microfilaments (mf) projects between Schwann cell processes (s) and microvilli (m) of Grandry cells (Gc). Clear vesicles and several dense-cored vesicles are present in the nerve ending (ne) at the base of the axon process. ×28,175. Calibration bar: 0·5 μm. (c) An axon process projects into the conglomerate of Schwann cell processes (s) and microvilli (m), and contains clear and dense-cored vesicles. Small arrows, zonulae adherentes between neighbouring Schwann cell processes; Gc, Grandry cell. ×22,380. Calibration bar: 0·5 μm.

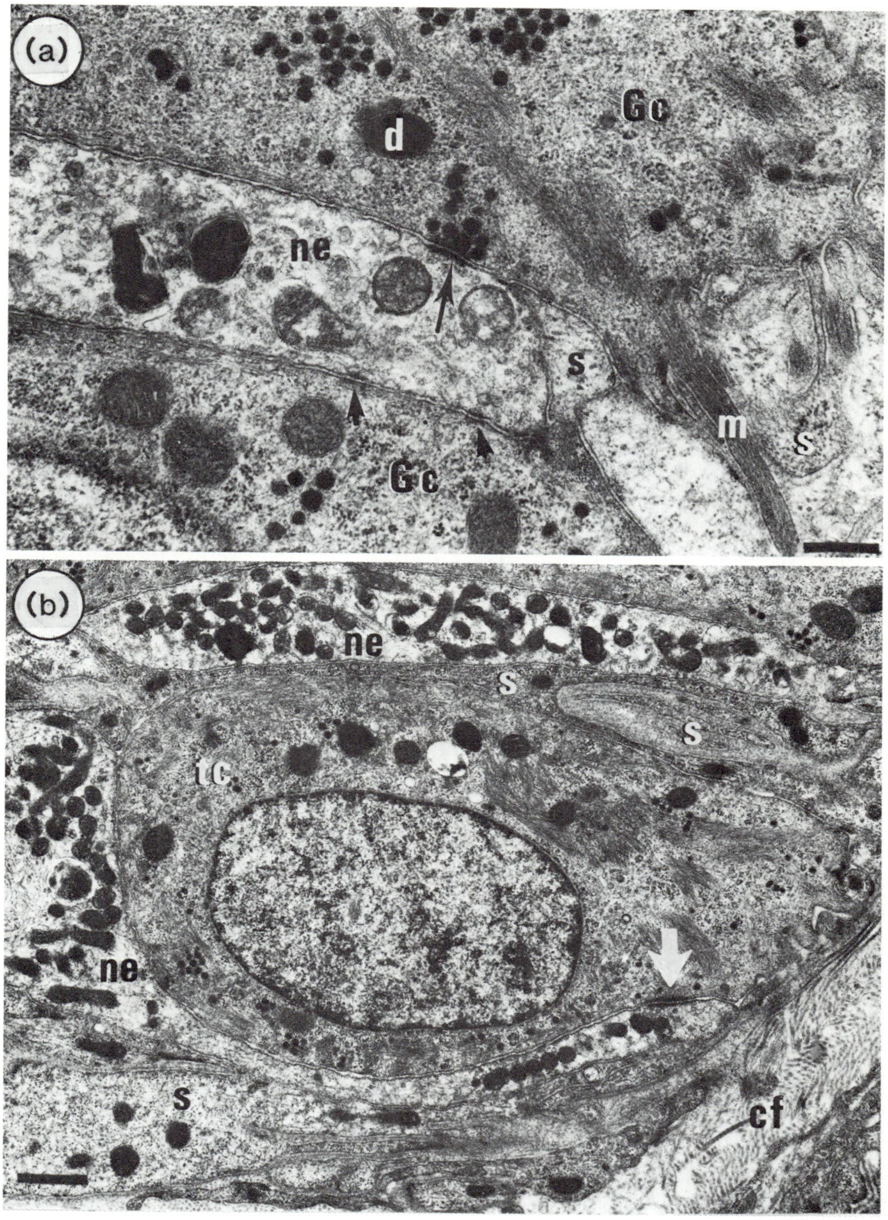

Fig. 8.14. Contact zones between terminal cells and adjoining nerve endings. (a) "Synapse-like" contact zone (large arrow) in a Grandry corpuscle of a newly-hatched gosling. At this age the Schwann cell processes (s) are still little elaborated, and therefore the microvillus (m) of the Grandry cell (Gc) projects into the intercellular connective tissue space. d, Dense body. Small arrows, contact points between the nerve ending (ne) and Grandry cell membranes. ×21,000. Calibration bar: 0·5 μm. (b) "Desmosome-like" contact zone (large arrow) in a terminal cell receptor of the adult chicken palate. Schwann cells (s) form lamellated processes enclosing the nerve endings (ne) and the terminal cell (tc), or separating them from each other. cf. Collagen fibrils. ×10,600. Calibration bar: 1μm.

terminal cell. Other contact zones correspond to desmosome-like zonulae adherentes, where the membrane thickenings are symmetrical and there is no accumulation of dense-cored granules at the side of the terminal cell (Fig. 8.14b). In both types of contact zone the intermembraneous cleft contains a faint filamentous material and the width of the cleft does not differ consistently from the intermembraneous cleft elsewhere.

It may well be that these two types of contact zone represent the extremes in a morphological continuum, since there are various transitional forms (Saxod, 1978a). Thus, asymmetrical junctions of the "synaptic" type can be found which lack the association with dense-cored granules, and there are also symmetrical junctions of the "desmosome" type where dense-cored granules are present. Occasionally, small groups of dense-cored granules are very close to the terminal cell membrane, but no specialisation of the two adjoining membranes can be recognised. In addition to the large contact zones, there are also small spot-like membrane contacts (Fig. 8.14a), and especially in Grandry corpuscles several of these can sometimes be seen lined up one beside the other in the peripheral parts of the interface between the Grandry cell and the nerve ending. Small desmosome-like zonulae adherentes also connect the membranes of terminal cells and surrounding Schwann cells, as well as neighbouring Schwann cells with each other. In the former case, dense-cored granules may be located in the vicinity of the desmosome-like contact zones.

The morphological variety of the contact zones, and especially their inconsistent association with dense-cored granules, obscure their functional significance. Whereas it seems certain that the "desmosome" type of contact mechanically stabilises the terminal cell–nerve interfaces, the possibility that granules or their contents are released from terminal cells at the "synaptic" type of membrane contact cannot be excluded. Nevertheless, the function of such "synapses" remains uncertain. The number of "synapse-like" contacts appears to be very variable even in the same morphological sub-group of terminal cell receptors. In Grandry corpuscles, for instance, these contacts were not found in several ultrastructural studies (de Iraldi and Rodriguez-Perez, 1961; Quilliam, 1966; Munger, 1966, 1971a,b; Saxod, 1968; Halata, 1971b; Chouchkov, 1974; Abraham, 1978). In our own material we found "synapse-like" contacts in some Grandry corpuscles and not in others. If such contacts did occur, the number of dense-cored granules associated with them varied considerably. In a single thin section, between 1 and rarely more than 12 granules were found at a "synapse-like" contact zone. From the size of the contact zones (200–600 nm), the number of dense-cored granules found at them in a single thin section, and the diameter of the granules (80–200 nm), it can be estimated that not more than 70 granules will occur at a single "synapse-like" contact zone and in most cases their number will be considerably smaller.

On the nerve terminal side of the contact zones, clear and dense-cored vesicles may be seen in the axoplasm. The occurrence of these vesicles is inconsistent, and they are not specifically associated with the contact zones because they can also be found elsewhere in the axoplasm. Moreover, the contact zones are mostly located at the peripheral parts of the interface between the terminal cell and the nerve ending, that is, near the previously described transducer sites of the nerve endings in which vesicles are particularly abundant. Since vesicles of the "synaptic type'' are common in all mechano-receptive nerve endings, the morphological picture does not seem to support the notion of a vesicle-related ''synaptic'' transport process operating from the nerve endings towards the terminal cells, as was discussed by Saxod (1978a) and Whitear and Lane (1981).

4. *Structure–function relationships in Grandry corpuscles*

The question of the functional nature of the ''synapse-like'' membrane contacts between terminal cells and nerve endings, as well as the controversy (Section IV,B,3) about the terminological designation of terminal cells in different types of the receptors, basically amount to the same problem, that is, the function of the terminal cells. The cardinal points are whether these cells are true sensory cells, that is, the actual mechano-electric transducer elements of the receptors, and if so, whether they may differ in their functional properties and thus give rise to different types of afferent responses. Despite the expression of more cautious views (Hesse, 1878; Szymonowicz, 1897; Munger, 1966, 1971a,b, 1977), it has been generally assumed since Merkel's day that terminal cells are sensory in function, and this view was reflected in such terms as ''sensory'' or ''tactile'' cells. This view gained support when electron microscopy revealed, not only in Merkel cells but also in Grandry and all other varieties of terminal cells, dense-cored granules in their cytoplasm and ''synapse-like'' junctions between these cells and the adjoining nerve endings (Andres, 1966, 1969; Andersen and Nafstad, 1968; Iggo and Muir, 1969; Halata, 1970; Saxod, 1970b, 1978a; Nafstad, 1971a,b; Winkelmann and Breathnach, 1973; Chen *et al.*, 1973; English, 1974).

On this morphological basis the concept emerged that Merkel cells, and similarly all other variants of terminal cells, were sensory cells located pre-synaptically to the adjoining nerve endings. It was particularly surmised that the microvilli of the terminal cells might function as sensors, and that their mechanical deformation would lead to the generation of a receptor potential in the terminal cells which, in turn, would activate the adjoining nerve endings via chemo-synaptic transmission at the ''synapse-like'' membrane contacts (Andres, 1966, 1969; Iggo and Muir, 1969; Hartschuh and Weihe, 1980). Since Merkel cells in mammalian skin are part of a slowly-adapting mechanoreceptor (Iggo, 1963; Tapper, 1965; Iggo and Muir, 1969; Munger *et al.*, 1971; Gottschaldt *et al.*, 1973) and because of the ultrastructural similarities between Merkel and Grandry cells, it was assumed that Grandry corpuscles would also function as slowly-adapting mechanoreceptors (Iggo, 1966). The argument was so tempting and authoritative that Leitner and Roumy (1974a) argued against their own experimental evidence (which really indicated that the non-vibration-sensitive, rapidly-adapting mechanoreceptors in the dermis of the duck's beak corresponded to Grandry corpuscles), because they expected these receptors to be slowly adapting.

The eventual demonstration that Grandry corpuscles are indeed rapidly-adapting mechanoreceptors (Gottschaldt, 1974a,b) was certainly thought to indicate a principal difference between Merkel and Grandry cells. At first, however, it was not regarded as conflicting with the notion that Grandry cells would function as sensory cells, and some electro-physiological findings appeared even to support this concept (Gottschaldt and Fakoya, 1975). Similarly, indirect electro-physiological evidence seemed to indicate the operation of a synaptic transmission process between Merkel cells and their adjoining nerve endings in mammalian type I slowly-adapting mechanoreceptors (Horch *et al.*, 1974). However, in agreement with earlier proposals (Kasprzak *et al.*, 1970), it has been shown in a combined electro-physiological and ultrastructural investigation that probably no chemo-synaptic transmission process operates between Merkel cells and their associated nerve endings in the generation of type I slowly-adapting responses deriving from the cat's sinus hair follicles (Gottschaldt and Vahle-Hinz, 1981, 1982). Consequently, Merkel cells are evidently not sensory cells, and therefore it was concluded that the trandsucer process must take place at the nerve endings themselves. A similar proposition had already been made on purely structural grounds for the avian Grandry corpuscles (Gottschaldt and Fakoya, 1977; Gottschaldt and Kräft, 1978), and most likely it is also true for all other kinds of terminal cell mechanoreceptors.

Curiously, it is again the morphology of terminal cells which, according to our present understanding of synaptic mechanisms, seems incompatible with the idea that a receptor potential is synaptically transmitted from Grandry cells to their adjoining nerve endings. Apart from their general paucity, the size of the "synapse-like" contact zones stands in striking contrast to the dimension of the Grandry cell–nerve ending interfaces (Fig. 8.15). The latter may be as large as 1200 μm^2, whereas the "synapse-like" contact zones measure only between 0·04 and 0·35 μm^2.

In view of the apparent mismatch between the total area of all "synapse-like" contact zones present at a given Grandry cell–nerve ending interface and the volume of the total nerve terminal disk, it is hard to imagine that a "synaptic" mechanism could sufficiently depolarize the nerve ending to elicit action potentials in the myelinated afferent nerve fibre, presumably at the first node of Ranvier. Moreover, in view of the small number of dense-cored granules usually found at the contact zones, the granules would have to contain a transmitter substance of unprecedented potency for depolarizing the nerve endings in order to enable the high discharge frequencies of up to 1200 impulses/sec recorded in some Grandry units (Gottschaldt, 1974a). Because the number of dense-cored granules at the "synapse-like" contact zones is usually small, these response characteristics could not even be explained by assuming that, following a strong stimulus, all granules were simultaneously released, apart from the fact that in such a case no granules would remain to enable responses to further stimuli following at short intervals. Fast replacement of the granules also is unlikely, because the vast majority of them lie far away from the "synapse-like" contact zones in parts of the Grandry cell's cytoplasm which are too remote to act as a storage pool of granules for the "synaptic" transmission.

In the absence of any further morphological basis for other mechanisms by which a receptor potential could be transmitted from Grandry cells to their adjoining nerve endings, the possibility that these cells are sensory in function can be virtually excluded. Consequently, Grandry cells, and almost certainly all other kinds of terminal cells associated with mechanoreceptors, are auxiliary receptor structures which do not generate receptor potentials, and therefore the mechano-electric transduction process must take place directly at the sensory nerve endings themselves.

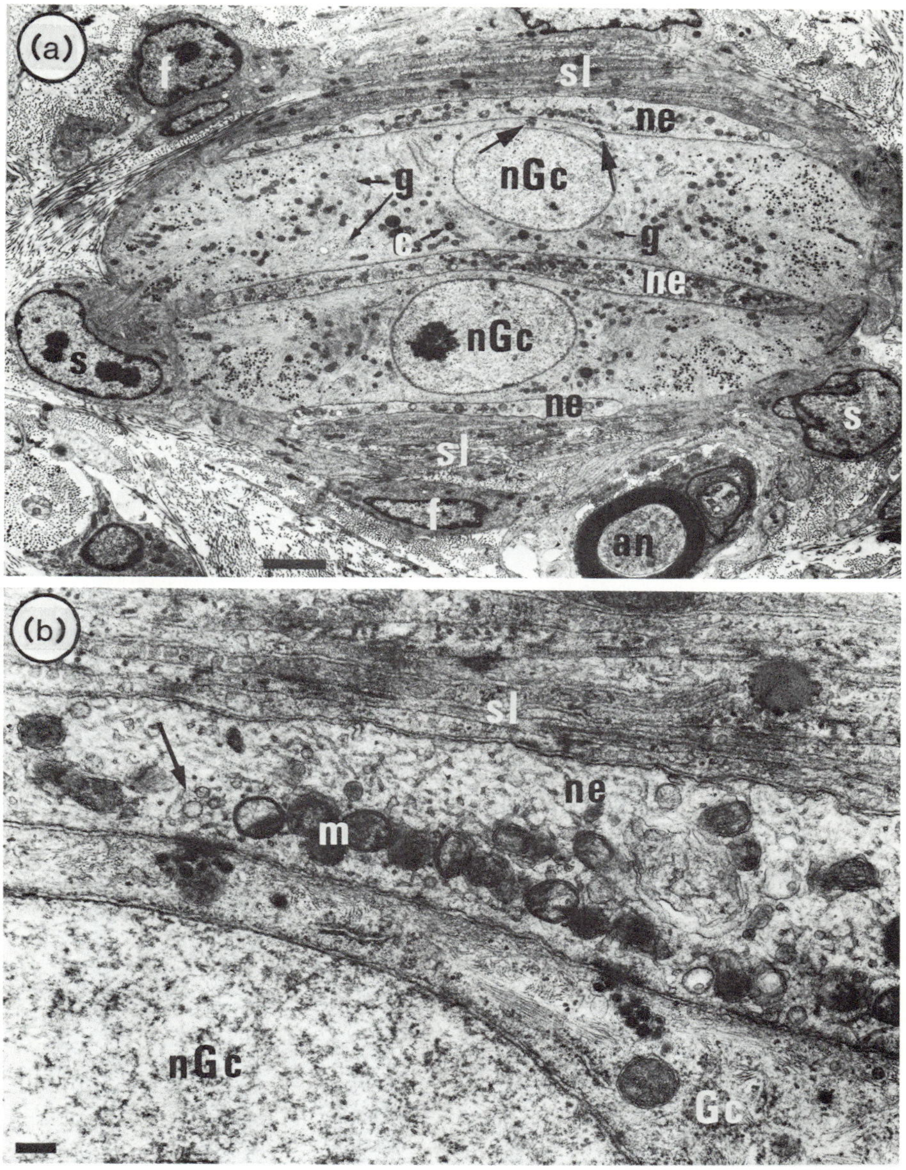

Fig. 8.15. (a) Two "synapse-like" contact zones (large arrows) in a Grandry corpuscle of the goose beak. Three nerve terminal disks (ne) adjoin two Grandry cells. Schwann cells (s) form lamellated processes (sl). f, Fibrocytes of outer capsule; an, myelinated afferent nerve fibre; g, Golgi apparatus; c, centriole. ×4200. Calibration bar: 2 μm. (b) The two contact zones in (a) shown at higher magnification. Gc, Grandry cell; nGc, nucleus of Grandry cell. Small arrow in nerve ending (ne) points to a group of clear vesicles; m, mitochondria. ×26,200. Calibration bar: 200 nm.

The discovery of transducer sites with axon processes around the perimeter of the flat nerve terminals in Grandry corpuscles reveals that the nerve endings are indeed equipped with essentially the same structural differentiations as the sensory nerve endings in other types of mechanoreceptors. This suggests that a mechanical deformation of the axon processes leads to the generation of a receptor potential essentially as for Herbst corpuscles (Section IV,A,2). However, a functional interpretation of the cytological characteristics of Grandry cells, and especially of the structural arrangement between them and the sensory nerve endings, suggests that the Grandry cells may act as mechanical interfaces modifying the time course of the effective deformation of the nerve endings. The structural relationships between the nerve endings with their transducer sites and the auxiliary receptor structures in Grandry corpuscles are illustrated in Fig. 8.16. In this schematic drawing some structural components possibly involved in structure–function relationships have deliberately been over-represented in size, and their real proportions should be inferred from the original electron micrographs.

The stack of Grandry cells and nerve endings is held in place by the enclosing Schwann cells and the peripheral capsule. The complex is mechanically stabilised further by desmosome-like zonulae adherentes between individual Schwann cell

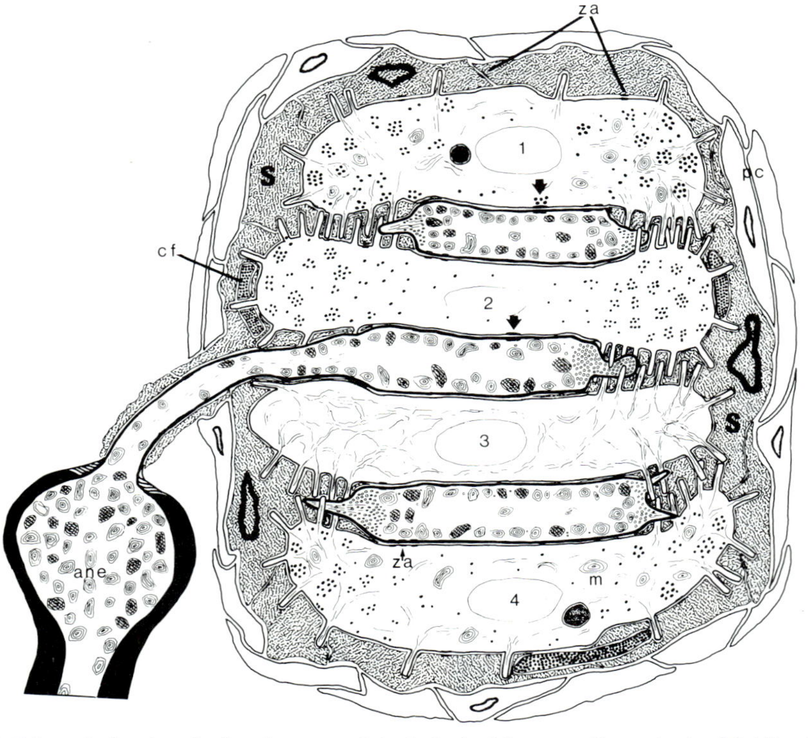

Fig. 8.16. Schematic drawing of a Grandry corpuscle in the beak of the goose. Some structural details relevant to structure–function relationships have been over-represented in size. The myelinated afferent nerve fibre forms a bulbous enlargement (ane), and then terminates in three disk-like nerve endings between four Grandry cells, the nuclei of which are numbered (1–4). In cell 2, only the distribution of the dense-cored granules is shown; in cell 3 only the cytoskeleton is shown. Microvilli of Grandry cells interdigitate with each other and with processes of the surrounding Schwann cells (s). Thus, a wreath of microvilli and Schwann cell processes is formed around the circumference of the nerve terminals, into which the axon processes of the transducer sites project. In addition to small zonulae adherentes (za), Grandry cells and nerve endings are connected by larger "synapse-like" contact zones associated with dense-cored granules (arrow in Grandry cell 1), as well as by larger desmosome-like contact zones (arrow in cell 2) which lack dense-cored granules. The Schwann cells surrounding the stack of Grandry cells and nerve terminal disks are also interconnected with each other and with the Grandry cells by zonulae adherentes. Bundles of collagen fibrils (cf) penetrate the Schwann cell sheath and may enclose microvilli of Grandry cells. A peripheral capsule (pc) encloses the receptor. Both the terminal enlargement of the myelinated nerve fibre and the nerve terminal disks are rich in mitochondria and other organelles.

processes, between Schwann cells and Grandry cells, and between the latter and the interposed nerve endings. Small bundles of collagen fibrils run between the Schwann cell processes and may contact the surface of the Grandry cells or even be pierced by their microvilli. It is important to note that, in contrast to many published electron micrographs, there are no gaps between the single cellular elements forming the corpuscle which would slacken their tight physical and mechanical connection.

The complex cytoskeleton within the Grandry cells, with its microfilament bundles pushing out against the cell surface and extending into the microvilli, should function as an elastic framework that keeps the cell membrane stretched out like an umbrella. Consequently the Grandry cells should possess not only a distinct mechanical load-bearing capacity but also considerable elastic properties. The latter seem to be especially evident, since the many microvilli seem to be kept straight and stiff by their microfilaments. Where the microvilli interdigitate with each other at the perimeter of the nerve terminal disks, they could act like the bristles of two brushes stacked face-to-face upon each other. Thus, a vertical compression of the corpuscle not only should push the microvilli further together, but, what seems more important, also shear them sidewards. In this way the amplitude of a mechanical deformation would be used up, and the deformation energy would be stored in the elastic elements of the cytoskeleton; a subsequent decompression of the corpuscle would free that energy and allow the microvilli to move back to their original position. Such movements of the microvilli should deflect the axon processes projecting between them, thus causing the generation of a receptor potential in the nerve endings.

Because the microvilli of a given Grandry cell are all connected with each other via the microfilaments of the cytoskeleton, a deformation of the cell may change the position of all its microvilli simultaneously. Conceivably, a displacement of only a few microvilli could induce, via the microfilament system, a displacement of all other microvilli. In this way a focal mechanical deformation would spread evenly across the Grandry cells and synchronously displace all axon processes projecting into the microvilli wreath. Thus, local receptor potentials arising at individual transducer sites could summate sufficiently to cross the threshold for the elicitation of propagated action potentials at the myelinated afferent nerve fibre. Different numbers of transducer sites at a single nerve terminal disk, as well as different numbers of disks deriving from a single afferent nerve fibre, may account for the broad range in both the mechanical threshold and the sensitivity of different Grandry units.

This model of structure and function relationships in Grandry corpuscles can account for at least some of the receptor's response properties. For instance, the inferred elastic properties of the Grandry cells would dampen high-frequency mechanical oscillations, especially when the cells are stacked upon each other. This would account for the poor vibratory sensitivity of Grandry corpuscles. The stacked arrangement of the Grandry cells, as well as the distinct relationship between their microvilli and the axon processes, may also account for the differing adaptation and transducer properties of Grandry corpuscles and mammalian Merkel cell receptors. [A model of the structure–function relationships in mammalian Merkel cell receptors has been discussed by Gottschaldt and Vahle-Hinz (1981).]

The strict movement sensitivity of Grandry units suggests that during the static phase of a mechanical stimulus the axon processes are not displaced but return to their original position. This may occur if the Grandry cells also have viscous properties and the cytoskeleton would passively give way to mechanical displacements, which is rather likely. A mechanical system with visco-elastic properties could equally well filter out high-frequency stimulus components and adjust to static stimulus components; on the other hand, the changes in displacement stimuli (movements or lower-frequency vibrations) would be least compensated and thus best transmitted to the sensory nerve endings. Under such conditions a receptor should respond qualitatively and quantitatively only to the movement part of the mechanical displacements and to its velocity. Alternatively, the lack of a static component in the responses of Grandry units could be explained by a fast accommodation of the spike-triggering zone at the myelinated nerve fibre. This assumption, however, would not account for the lack of an amplitude component in the movement responses of Grandry units, nor would it permit "off" responses at the return of trapezoidal displacements, as seen in Fig. 8.1. Another possible explanation of the response properties of Grandry units may be that the supposed actin in the microfilaments of the cytoskeleton responds actively to deformations of the Grandry cells to bring the microvilli back to their original position. Although this notion is speculative, such a mechanism would require energy and this is consistent with the abundance of mitochondria in Grandry cells.

The significance of the dense-cored granules remains obscure. If they are ever released from the terminal cells and not merely stored within them (for instance, to increase the viscosity of the cytoplasm), they may play a rôle in establishing and maintaining the contact with the sensory nerve terminals. At least for Merkel cells, a target or guiding function for the afferent nerve endings has been postulated (Scott *et al.*, 1981). For such a purpose a release of dense-cored granules could, in fact, take place at the "synapse-like" contact zones, but it would not then be directly involved in the mechano-electric transduction process. However, the presence of the contact zones must increase the mechanical stability of the interfaces between the terminal cell and the nerve ending, so that movements of the axon processes could more effectively stretch the receptor membrane.

5. *Other terminal cell receptors with presumed rapidly-adapting response properties*

(a) Terminal cell column receptors. Besides the Grandry corpuscles there is another configuration of terminal cell receptors in the beak of anseriform birds which could give rise to rapidly-adapting, velocity-sensitive responses. This receptor mingles with Grandry corpuscles in the distal parts of the touch papillae of the bill tip organ (Section VII) and was first called the "Merkel cell column receptor" (Gottschaldt *et al.*, 1976), but with new insight (Gottschaldt and Kräft, 1978) it will now be named the "terminal cell column receptor". The receptor is distinguished structurally by the columnar arrangement of its terminal cells and the lack of an outer capsular zone (Fig. 8.17b).

Typically, the myelin sheath of the afferent nerve fibre terminates some distance away from the receptor. A faint trace of the unmyelinated pre-terminal axon

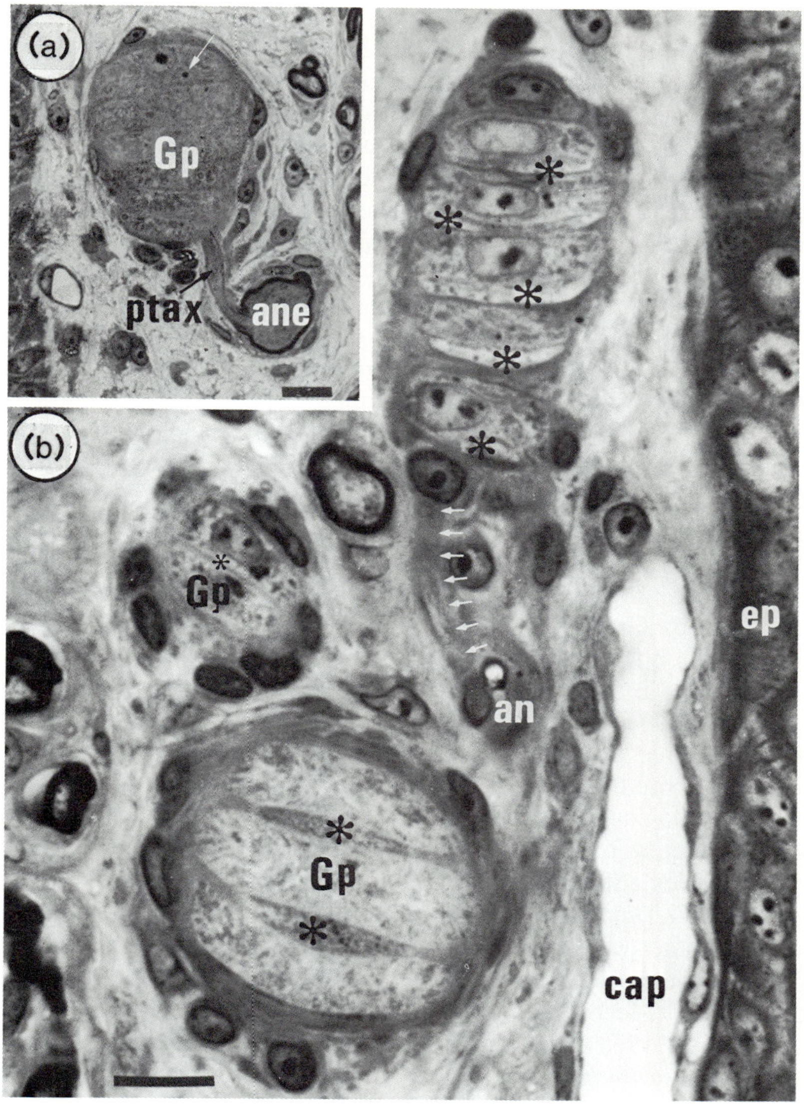

Fig. 8.17. Light micrographs of Grandry corpuscles and a terminal cell column receptor in touch papillae of the goose's bill tip organ. (a) Grandry corpuscle (Gp) consisting of six Grandry cells. The myelinated afferent nerve fibre shows a bulbous enlargement (ane) in its terminal portion. An unmyelinated pre-terminal axon (ptax) connects the myelinated axon with the nerve terminal disks in the receptor. The white arrow points to a large dense body. × 650. Calibration bar: 10 μm. (b) A large and a small Grandry corpuscle (Gp) alongside a terminal cell column receptor. The column receptor has six terminal cells, between which branches of the sensory nerve fibre (asterisks) can be recognised. The afferent nerve fibre (an) loses its myelinated sheath early. The arrows indicate the course of the unmyelinated pre-terminal axon, which is sheathed by Schwann cell processes. cap, Blood capillary; ep, epithelium of the touch papilla. ×1385. Calibration bar: 10 μm.

and its Schwann cell accompaniment is recognisable in Fig. 8.17b. Whereas the cytological characteristics of the terminal cells are in all respects identical to those of Grandry cells, their relationship to the sensory nerve terminals differs. The afferent nerve fibre does not terminate in disks but forms one or several flat bands meandering through the column of terminal cells (Figs. 8.19, 20a). Thus, the column receptors bear some resemblance to the human Meissner corpuscles (Cauna, 1966; Andres and von Düring, 1973), which also are rapidly adapting, velocity-sensitive mechanoreceptors (Knibestöll, 1973). Another difference from Grandry corpuscles is that the Schwann cells not only enclose the column of terminal cells but also send processes between them which often have a lamellated structure; they may separate neighbouring terminal cells and nerve endings from each other as well as terminal cells from nerve endings (Figs. 8.18–20).

There appears to be a morphological transition from typical Grandry corpuscles to typical terminal cell column receptors, and often a receptor combines structural characteristics of both receptor configurations. For instance, the receptor shown in Fig. 8.18 has the outer form of a large Grandry corpuscle with several flat, continuous and probably disk-like nerve terminals, but there are also profiles of smaller, more band-like nerve terminals; furthermore, Schwann cell processes are interposed in the stack (or column) of terminal cells and nerve endings. Similarly, it is difficult to decide whether the receptor in Fig. 8.20b represents two superimposed Grandry corpuscles separated by lamellated Schwann cell processes or a single terminal cell column receptor.

The morphological transition from Grandry corpuscles to terminal cell column receptors may explain their similar response properties. However, the finger-like axon processes of Grandry corpuscles are not common in column receptors. Nevertheless, the edges of the band-like nerve endings have the same characteristic axoplasmatic structure as the transducer sites of the nerve endings in Grandry corpuscles (Fig. 8.19). Possibly, therefore, a more diffuse stretching or wrenching of the receptor membrane leads to the generation of a receptor potential rather than a simultaneous displacement of many individual axon processes. The general mechanical properties of the terminal cells and their influence on the effective deformation of the nerve endings, however, should be similar in both Grandry corpuscles and terminal cell column receptors.

(b) "Diffuse" terminal cell receptors. In the palatine mucosa and toe skin of the chicken, another variety of terminal cell receptors is found. These have been called "subepithelial Merkel cell receptors" (Andersen and Nafstad, 1968; Nafstad, 1971a), "Merkel corpuscles" (Saxod, 1978b) or "Grandry corpuscles" (Idé and Munger, 1978), but perhaps the best name would be "diffuse" (diffusely-organised) terminal cell receptors because the aggregation of terminal cells into "corpuscles" is more-or-less lost. Instead, there are irregularly-shaped agglomerations of terminal cells, Schwann cells, fibrocytes and nerve endings, which lack an outer capsule and frequently form a virtually continuous receptor sheath underneath the epithelium (Fig. 8.21a). It is sometimes impossible to determine the boundaries of individual receptor complexes or of the territories of individual afferent nerve fibres. However, within these groups of terminal cells smaller complexes can often be distinguished which are separated from each other by connective tissue spaces, and sometimes these complexes resemble in structure small Grandry corpuscles or terminal cell column receptors. If the "diffuse" terminal cell receptors are located in the dermal papillae, where they find natural boundaries in the surrounding epidermal pegs, they gain a more orderly appearance, and then the arrangement of terminal cells, Schwann cells and nerve endings may again resemble that of Meissner corpuscles.

The terminal cells in the chicken are indistinguishable from Grandry cells in

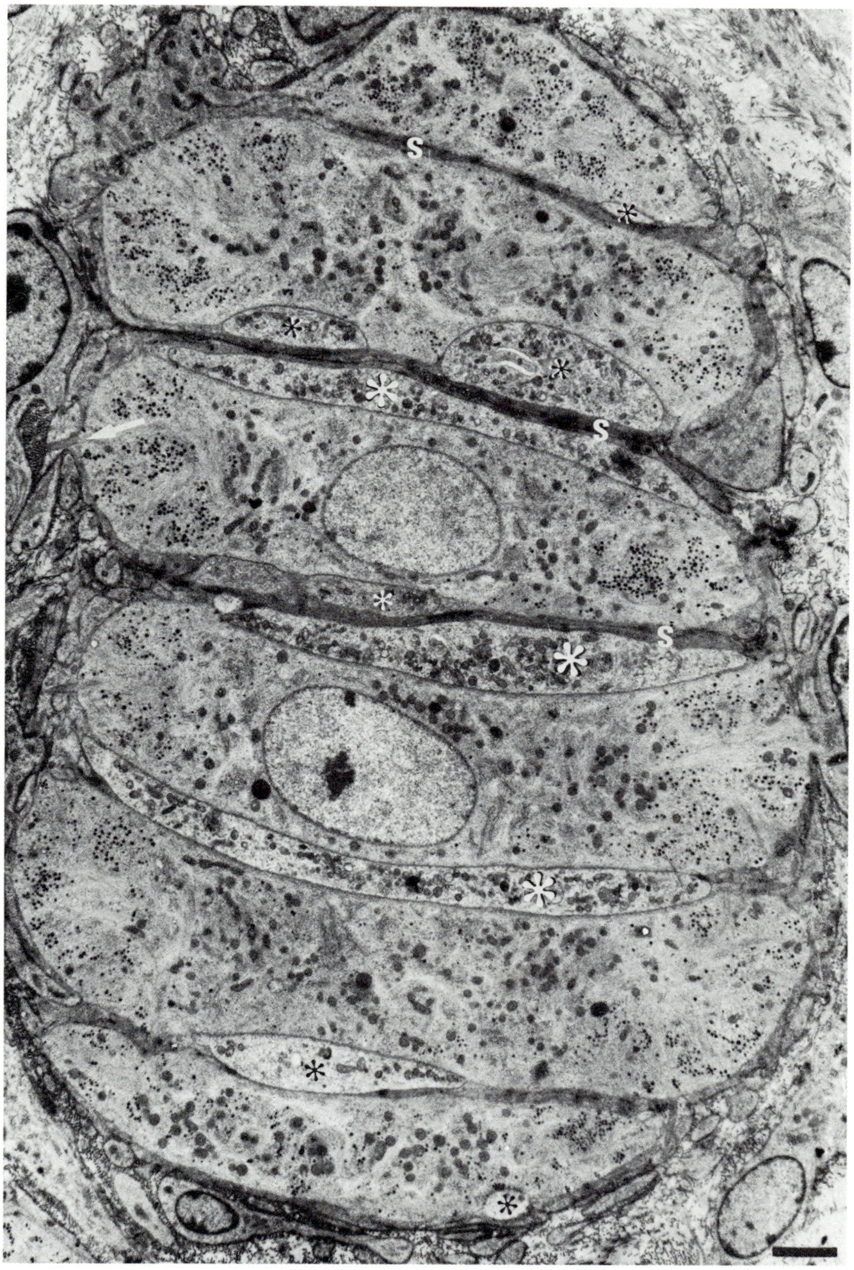

Fig. 8.18. Transitional form between Grandry corpuscle and terminal cell column receptor. The shape and arrangement of the terminal cells resemble those of Grandry cells. Some nerve terminals are wide and appear to be discoid (large white asterisks), whilst others are narrower and look more like transverse sections through band-like structures (small black asterisks). In the upper part of the receptor, Schwann cell processes (s) (lamellated at higher magnification) penetrate the receptor; these processes separate nerve endings from each other, and nerve endings from terminal cells. Arrow, a microvillus contacting a small bundle of collagen fibrils. ×4700. Calibration bar: 2 μm.

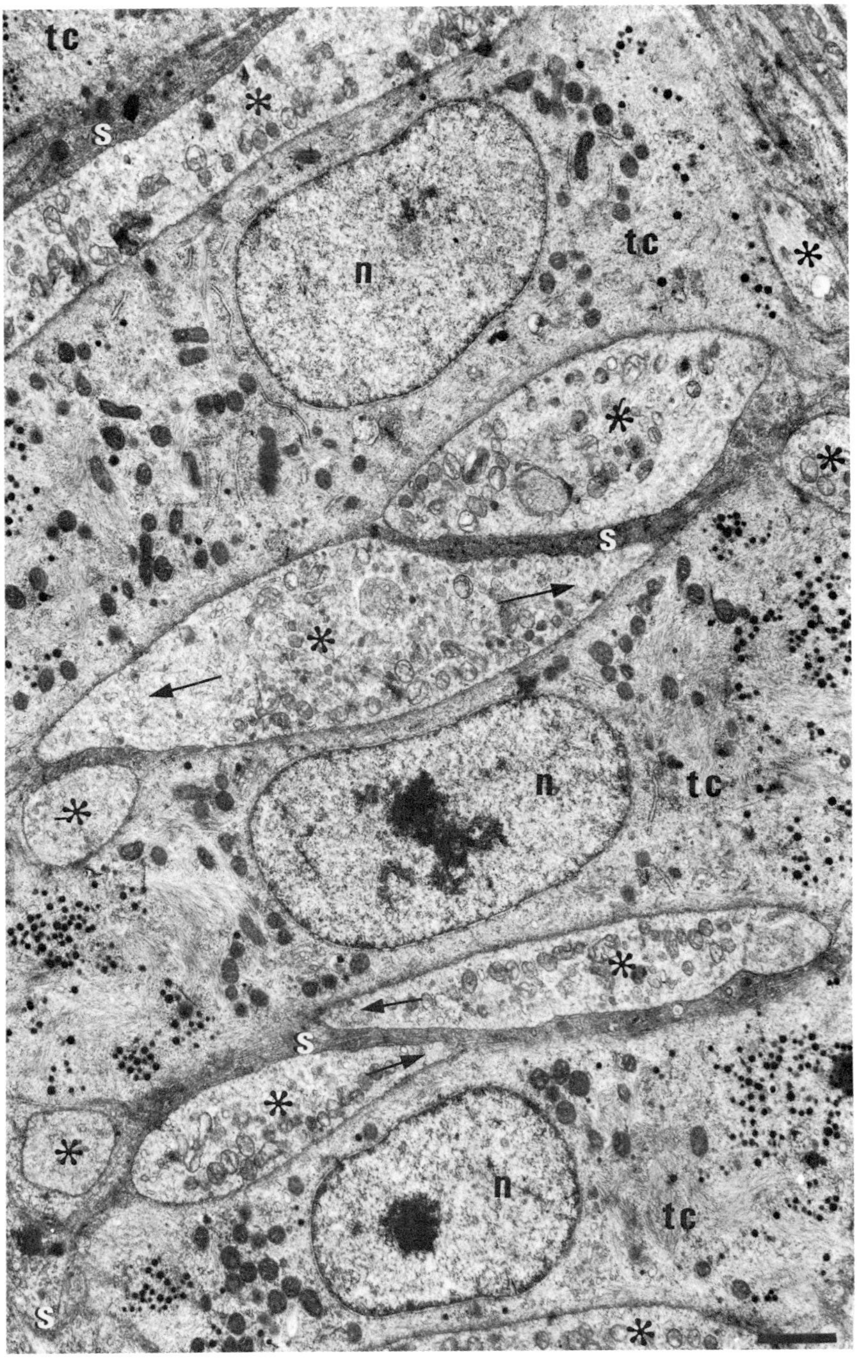

Fig. 8.19. Terminal cell column receptor. Several flat nerve endings (asterisks) course between the terminal cells (tc), which are cut in the plane of their nuclei (n). Lamellated Schwann cell processes (s) penetrate the column of terminal cells and nerve endings. The peripheral edges of the latter are extended (arrows) and contain small vesicles and fine granular material typical of the receptor matrix at a transducer site. Note distribution of dense-cored granules in the cytoplasm of the terminal cells. ×11,770. Calibration bar: 1 μm.

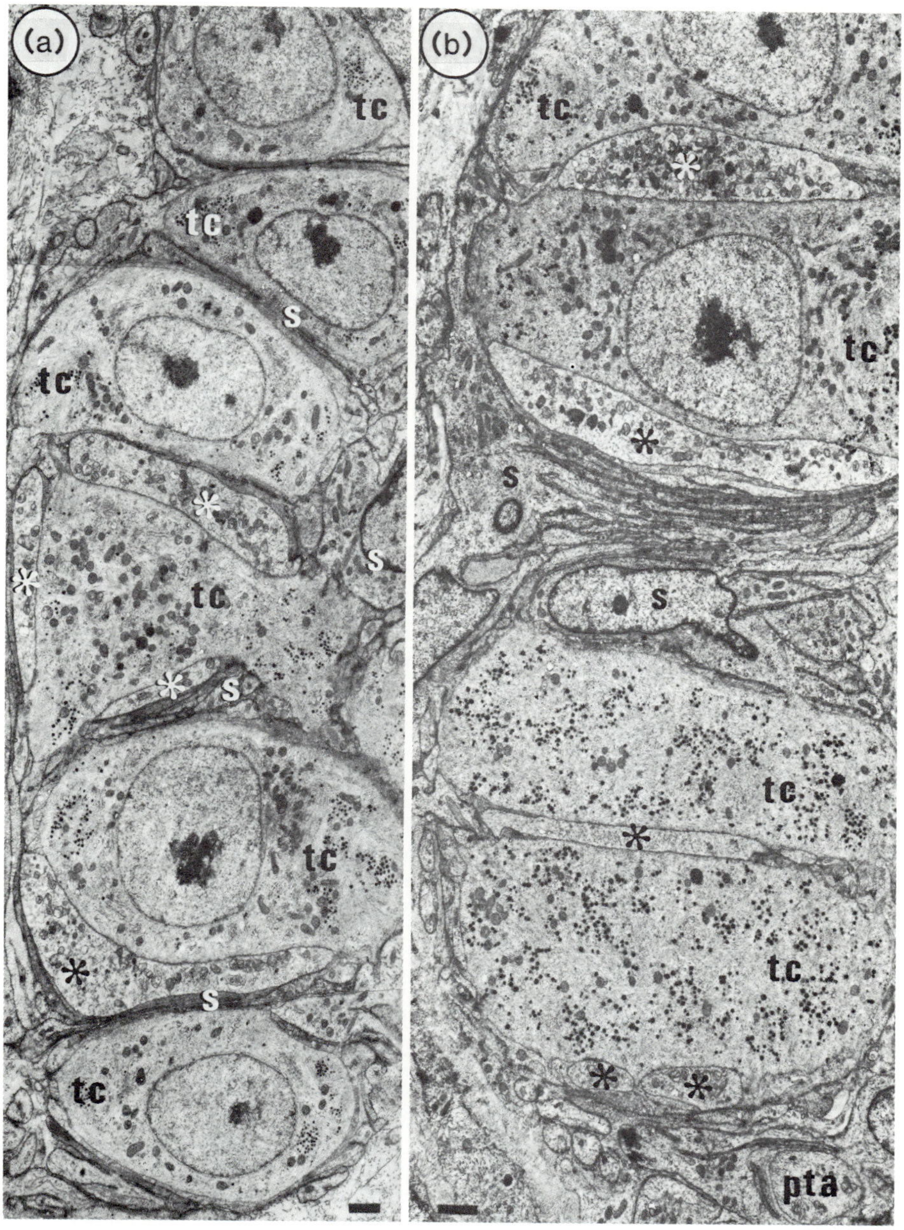

Fig. 8.20. (a) Terminal cell column receptor in its more typical appearance. The terminal cells (tc) are smaller than Grandry cells and are arranged in a column through which ribbon-like nerve endings (asterisks) wind their way. Schwann cell processes (s) are intercalated in the column of terminal cells. ×4500. Calibration bar: 1 μm. (b) Intermediate form between Grandry corpuscle and terminal cell column receptor. The upper pair of terminal cells (tc) is sectioned in the nuclear plane, the lower pair parallel to this plane. Note the difference in the distribution of dense-cored granules, in the upper and lower pairs of cells. Of the nerve ending between the lower two terminal cells only the lateral edge is visible in which there are no mitochondria. Schwann cells (s) with lamellated processes separate the upper and lower parts of the receptor. pta, Unmyelinated pre-terminal axon; asterisks, nerve endings. ×5570. Calibration bar: 1 μm.

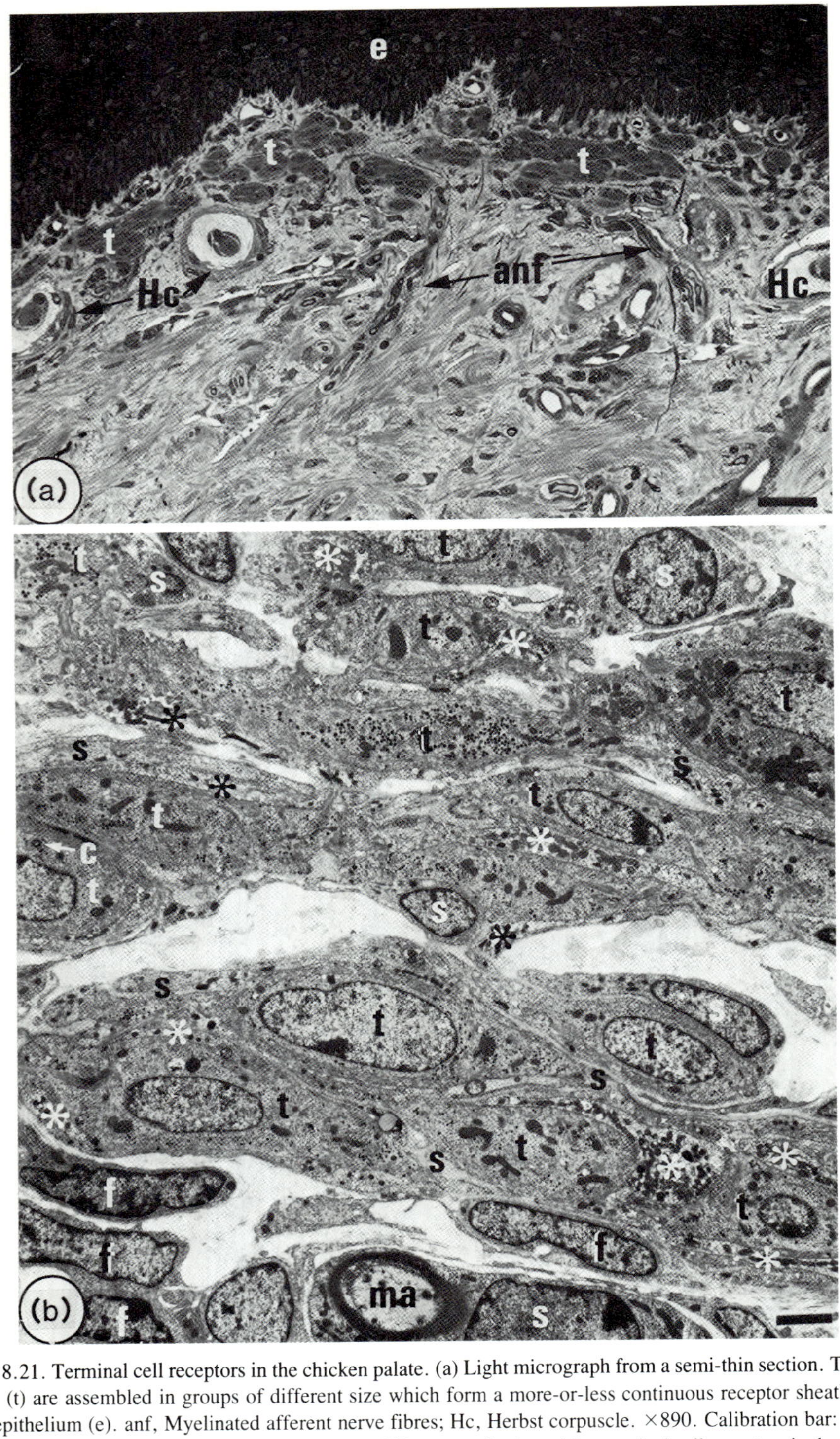

Fig. 8.21. Terminal cell receptors in the chicken palate. (a) Light micrograph from a semi-thin section. Terminal cells (t) are assembled in groups of different size which form a more-or-less continuous receptor sheath below the epithelium (e). anf, Myelinated afferent nerve fibres; Hc, Herbst corpuscle. ×890. Calibration bar: 10 μm. (b) Low-power electron micrograph showing the diffuse organisation of the terminal cell receptors in the chicken palate. Terminal cells (t) are mixed with Schwann cells (s) and their processes, and are contacted in an irregular manner by nerve endings (asterisks). Interspersed in the receptor complexes are connective tissue spaces containing collagen and elastic fibre elements. f, Fibrocyte; ma, myelinated afferent axon; c, centriole in terminal cell. ×4000. Calibration bar: 2 μm.

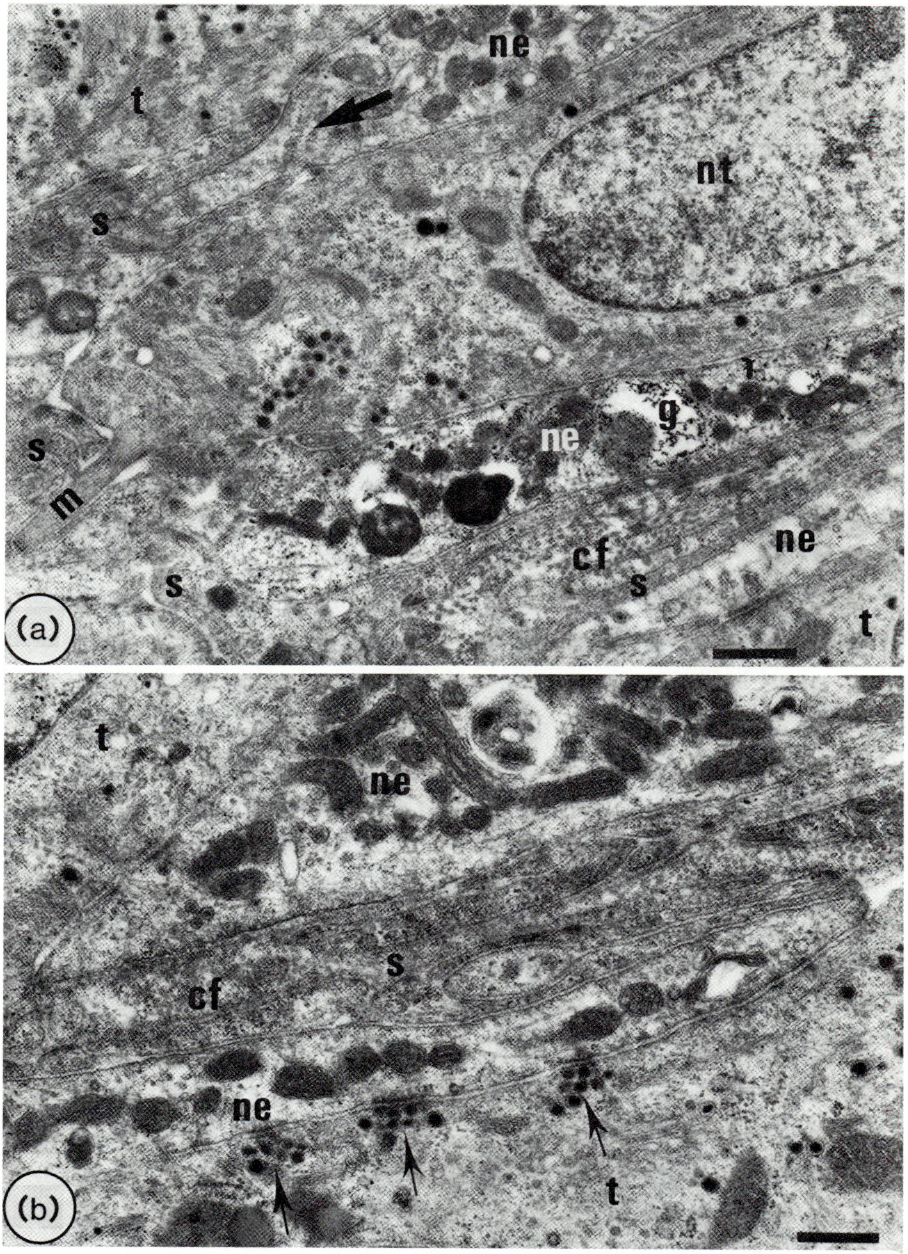

Fig. 8.22. (a) Transducer site with finger-like axon process (arrow) at a nerve ending (ne) of a terminal cell receptor in the chicken palate. t, Terminal cells; m, microvillus; nt, nucleus of terminal cell; s, Schwann cell with lamella-like processes; g, glycogen in nerve ending; cf, collagen fibrils. ×21,000. Calibration bar: 0·5 μm. (b) ''Synapse-like'' contact zones (arrows) between terminal cell (t) and nerve ending (ne). Schwann cell processes (s) with intervening collagen fibrils (cf) separate adjacent nerve endings. ×21,200. Calibration bar: 0·5 μm.

geese or ducks. Many microvilli project in all directions from the terminal cells. The latter are enclosed by Schwann cell processes, which often attain a lamellated configuration (Fig. 8.14b), but in general the Schwann cell investment of the terminal cells is much looser than that of Grandry cells. Occasionally, microvilli penetrate the Schwann cell sheath and project into the surrounding connective tissue spaces, where they come in contact with collagen fibrils.

Without a recognisable distinct pattern, the unmyelinated terminal branches of the afferent nerve fibre wander through the assembly of Schwann and terminal cells, contacting the latter at all possible sites (Fig. 8.21b). Thus, a systematic orientation of the interfaces between the terminal cells and their nerve endings parallel to the epidermal surface, which is so typical for Grandry corpuscles in the rhamphothecal dermis of geese, is not apparent in the "diffuse" terminal cell receptors. Differentiations of the receptor axoplasm typical of transducer sites are frequently encountered, and also finger-like axon processes have been found (Fig. 8.22a). Specialised contacts between terminal cells and nerve endings of both the "desmosomal" (Fig. 8.14b) and the "synaptic" type (Fig. 8.22b) are seen more frequently than in either Grandry corpuscles or terminal cell column receptors in the dermis of the beak of the goose. Perhaps the much looser organisation of the "diffuse" terminal cell receptors necessitates a greater number of adhesive contacts between the nerve endings and the terminal cells.

The assignment of rapidly-adapting, velocity-sensitive responses to the "diffuse" terminal cell receptors is still tentative. It is based firstly on the general structural similarities to other types of velocity-sensitive terminal cell receptors, and secondly it is inferred from electro-physiological findings. Roumy and Leitner (1973) described vibration-insensitive and vibration-sensitive rapidly-adapting units in the mucosal lining of the chicken's palate. The vibration-sensitive responses most likely originate in Herbst corpuscles, and since no other receptor structure has so far been identified in the chicken palate, it seems inevitable to conclude that the vibration-insensitive but velocity-sensitive responses are derived from the "diffuse" terminal cell receptors.

The limited knowledge about the gross and fine structure of these receptors and about their functional properties, makes it premature to speculate in detail about their structure–function relationships. However, in view of the preceding analysis (Section IV,B,4) of the function of terminal cells occurring in other morphological receptor types, it seems almost certain that in the chicken palate also the terminal cells are not sensory in function.

The developmental origin of terminal cells is an interesting problem. Saxod (1978b) described a so-called transitional cell, located seemingly between epithelium and dermis in the chicken palate, which showed cytological characteristics of both keratocytes and terminal cells. Transitional cells have also been seen in mammalian skin (Iggo and Muir, 1969; Nafstad, 1971b; English, 1974, 1977) and in the chicken palate by the present author (unpublished observation). However, it is not yet clear whether transitional and terminal cells are independent cells or a developmental sequence of cells. Furthermore, if the transitional cells should in fact cross the border between epidermis and dermis, as the morphological pictures suggest, it is still uncertain in which direction they migrate. Therefore, these transitional cells deserve further investigation.

C. Receptors detecting amplitude components in mechanical stimuli

The receptors in this group include Ruffini nerve endings, "branched" terminal cell receptors, Golgi tendon organs and muscle spindles.

1. Physiology of cutaneous receptors

The capability of a mechanoreceptor to measure amplitude components presupposes a quantitative relationship between afferent discharge frequency and stimulus amplitude. Because the amplitude corresponds to the "zeroth" derivative of the displacement against time, it has an inherent static characteristic, and therefore, ideal amplitude detectors must be able to respond to static stimulus conditions; that is, they must have slowly-adapting response properties. Reports about slowly-adapting mechanoreceptors in birds are few, and the literature is controversial about the occurrence of these receptors in avian skin.

In feathered skin Dorward (1970a) found only a few receptors with seemingly slowly-adapting responses, and no detailed analysis of their discharge properties was performed. In the beaks of ducks, Gregory (1973) and Leitner and Roumy (1974a) found no low-threshold slowly-adapting mechanoreceptors whatsoever. In contrast, Necker (1974a) encountered these receptors frequently but concentrated his investigations on their responses to thermal stimuli. Similarly, the occurrence of slowly-adapting mechano-receptive afferents in the trigeminal nerves of the pigeon was mentioned in several studies (Zeigler and Witkowsky, 1968; Necker, 1972, 1973, 1974a,b), but again no details about their quantitative responses to mechanical stimuli were communicated.

In the ophthalmic nerve of the goose, slowly-adapting units account for about 20% of the mechano-sensitive afferent fibres (Gottschaldt, 1974a; Gottschaldt *et al.*, 1982). The receptive fields of all these units are located in the tip of the bill, either underneath the horny plate of the nail or directly within the touch papillae of the bill tip organ (see Section VII). Many slowly-adapting receptors exhibit a directional sensitivity in the sense that a response is elicited only to an upwards or a downwards deflection of the nail or, alternatively, to an indentation of the dorsal or palatinal surface of the beak tip. Such directional sensitivity was also observed for slowly-adapting mechanoreceptors in the beak of pigeons and ducks (Necker, 1974a).

As in mammals, the responses of avian slowly-adapting mechanoreceptors consist of dynamic and static components related to the movement and static parts of displacement stimuli, respectively. Both the dynamic and the static responses are quantitatively related to the velocity and amplitude of the displacement stimuli; Fig. 8.23 illustrates an analysis of these stimulus–response relationships of a single slowly-adapting unit in the ophthalmic nerve of the goose. The receptor was activated by a displacement (indentation) of the nail covering the tip of the bill, and controlled mechanical stimuli were applied which had either constant amplitudes attained at different velocities (Fig. 8.23a), or increasing amplitudes attained at a constant movement velocity (Fig. 8.23b). Inspection of the original

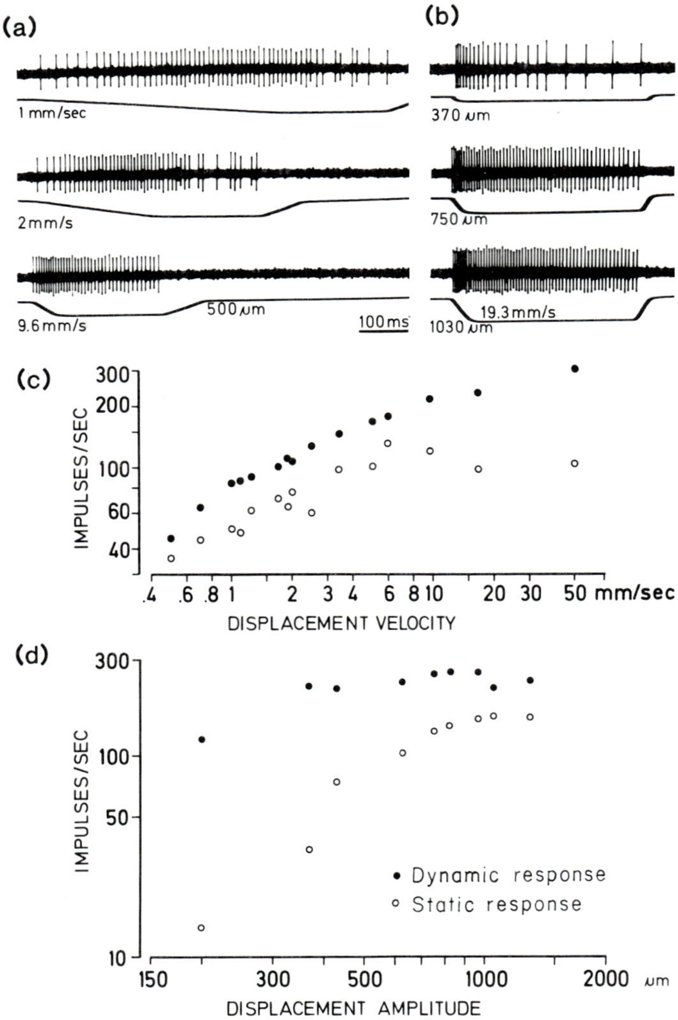

Fig. 8.23. Response profiles of a slowly-adapting unit in the ophthalmic nerve of the goose obtained by indenting the nail. (a) and (b) Original records with the appropriate stimulus traces below them. In (a) the velocity (on the left) of a constant-amplitude displacement was changed. In (b) the movement velocity was held constant and the displacement amplitude increased (as on the left). The average discharge frequency during the movement part (dynamic response) and during the static part (static response) of the trapezoidal displacement stimulus was plotted in (c) against the displacement velocity, and in (d) against the displacement amplitude.

records suggests a mutual interaction between the velocity and amplitude components of the stimulus in both the dynamic and the static response; this interaction is confirmed in the graphic display of the stimulus–response relationship on double logarithmic axes (Fig. 8.23c,d).

The nearly straight course of the dynamic response function (Fig. 8.23c) indicates a power function relationship between discharge frequency and movement velocity. The static response also increases with the movement velocity, especially if the duration of the static-stimulus phase is short with respect to its movement phase. This spread of the dynamic response onto the early static response has also been described for mammalian slowly-adapting mechanoreceptors (Werner and Mountcastle, 1965; Pubols and Pubols, 1976; Gottschaldt and Young, 1977). It is probably due partly to a delayed movement of the tissue intervening between the stimulator probe and the receptor, and also to an overlap between adapting dynamic and ongoing static response.

The effects of changing the stimulus amplitude at a constant movement velocity are recorded in Fig. 8.23b. The plot of the dynamic stimulus–response relationship under these conditions (Fig. 8.23d) shows that the discharge frequency already is high at small displacements (as it should be in view of the high movement velocity which was used), and that the discharge frequency increases further at higher amplitudes. This increase reveals an amplitude component in the dynamic response, which is added to the constant-velocity response as the displacement amplitude increases. Such an amplitude component can be seen directly in the upper record of Fig. 8.23a as a continuous increase of the discharge frequency during the rising phase of the displacement. However, the slope of the dynamic stimulus–response function in Fig. 8.23d is much lower than that in Fig. 8.23c, thus confirming the empirical finding of a higher dynamic than static sensitivity in all slowly-adapting units. The approximately linear increase of the static response curve in Fig. 8.23d (on double logarithmic axes) up to a displacement amplitude of about 800 μm shows that this response component also follows a power function relationship to the displacement amplitude. Stimuli of higher amplitudes exceeded the maximum possible displacement of the nail, which is limited by the underlying premaxillary bone, and therefore the static discharge frequency remained at a constant value. When allowance is made for the different scalings of the logarithmic axes, the slope of the static response curve in Fig. 8.23d is about eight times steeper than that in Fig. 8.23c; this reveals that the static response was more sensitive to changes of the displacement amplitude than of the velocity of the preceding movement.

Although the quantitative stimulus–response characteristics of all slowly-adapting receptors in the beak of the goose are in principle similar to those in the example illustrated in Fig. 8.23, there is a considerable variability amongst individual units in their relative dynamic and static sensitivity, as well as in their static discharge characteristics, the rate of adaptation, and the presence of a spontaneous activity in the absence of an intended stimulus. Fig. 8.24 shows discharges of five slowly-adapting units which had more regularly (Fig. 8.24a,b) and irregularly (Fig. 8.24c,d) spaced inter-spike intervals in their static responses and different dynamic sensitivities. The unit of Fig. 8.24d is unusual in that it showed a spontaneous activity and yet a predominantly dynamic response. The discharge shown in Fig. 8.24e derived from a receptor located in a single touch papilla of the bill tip organ of the upper beak and was characterised by a sequence of impulse doublets occurring at fairly regular intervals. Such discharge patterns have also been observed in afferents deriving from receptors located underneath the nail. Therefore, they are not unique to slowly-adapting receptors located in the touch papillae, and of these only few discharged with doublets as illustrated in Fig. 8.24e.

In the beak of pigeons and ducks, Necker (1974a) noted also slowly-adapting receptors discharging with regular and irregular inter-spike intervals. He proposed that the units with irregular discharges would correspond to the mammalian type I

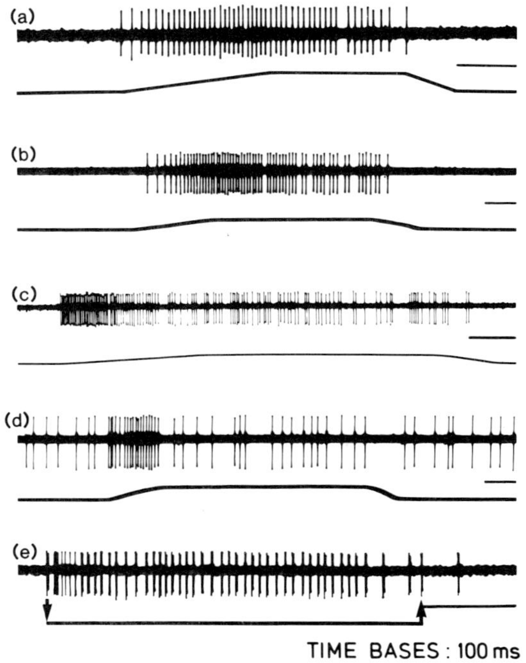

Fig. 8.24. Examples of different response characteristics of five slowly-adapting units in the ophthalmic nerve of the goose. The stimulus traces, representing the ramp-shaped indentations of the rhamphotheca overlying the receptors, are shown below the spike records. The response in (e) was elicited by an indentation of a single touch papilla of the upper bill tip organ, applied manually, during the time indicated by the bar with arrows, with a fine probe and under microscopic control. The time base is different for each record as indicated by the small bars.

receptors and those with regular discharges to type II receptors. However, the distinction between the two types of slowly-adapting receptors in mammals is based on a number of criteria of which the regularity of the static discharge is only one (Chambers *et al.,* 1972; Iggo, 1974). None of the other criteria have been taken into account by Necker (1974a), and when they were (Gottschaldt, 1974a), the existence of two types of slowly-adapting receptors in birds could not be substantiated. For instance, mammalian type I receptors are not spontaneously active, show a high dynamic sensitivity, are stretch-insensitive, and have an irregular inter-spike interval distribution (coefficient of variation always larger than 0·5) in their static discharge, which adapts continuously with at least two time constants. In addition, type I receptors are located intraepidermally and correspond to the "Merkel cell nerve endings" in both glabrous and hairy skin, as well as in sinus hair follicles (Iggo and Muir, 1969; Munger *et al.,* 1971; Gottschaldt *et al.,* 1973). In contrast, the slowly-adapting receptors in the beaks of birds which have an irregular discharge pattern are always located in the dermis, are often spontaneously active and show "silent periods" following a quick removal of a stimulus; moreover, their static response to a constant displacement of the nail may be maintained at a constant discharge frequency for a long time (Gottschaldt, 1974a). The same response properties were also apparent in the regularly-discharging, slowly-adapting units. Thus, the regularity of the static discharge does not correlate in a systematic manner with other response properties of the receptors, as it does in mammals. Therefore, the responses of all slowly-adapting receptors (at least in the ophthalmic nerve of the goose) have so many characteristics in common that they seem to derive from only one morphological type of receptor, although several variants of a basic receptor structure may exist.

As in mammalian skin, slowly-adapting receptors in avian skin may show a marked temperature sensitivity. This has been studied extensively in pigeons and ducks (Necker, 1972, 1973, 1974a,b), as well as in geese (Gottschaldt *et al.*, 1982). In most cases cooling the skin elicits a continuous discharge which reaches a maximal frequency at about 25°C and decreases again at lower temperatures. The responses to sudden temperature decreases show a dynamic component of variable distinctiveness. Conversely, warming the skin is followed by an increase of the discharge frequency if the starting temperature was below 25°C and a decrease if the starting temperature was above 25°C. Fast increases in the temperature lead to a transient cessation of activity, following which a discharge is resumed corresponding to the new adapted temperature. There is a considerable hysteresis in the responses to decreasing or increasing temperature changes. The temperature sensitivity of individual slowly-adapting receptors varies substantially, and in the goose only about two-thirds of the units showed any temperature sensitivity at all.

Because of the bell-shaped course of the temperature response curve at decreasing or increasing temperatures, and the presence of positive or negative dynamic responses following sudden temperature changes, the temperature responses of most slowly-adapting units in avian skin resemble those of specific cold receptors in mammalian skin (Hensel, 1973) and also those of mammalian slowly-adapting cutaneous receptors (Witt and Hensel, 1959; Iggo and Muir, 1969; Poulos and Lende, 1970a,b; Boldt and Biewald, 1971; Chambers *et al.*, 1972; Duclaux and Kenshalo, 1972; Booth and Hahn, 1974). However, in the beaks of pigeons, Necker (1974b) observed slowly-adapting mechanoreceptors which showed temperature responses resembling those of mammalian warmth receptors; that is, their activity decreased on cooling and increased on warming the skin. The basis for this unusual behavior is not known. It may be due to the presence of a distinct type of receptor in pigeons which is absent in the beaks of geese, or to a special embedding of a receptor (though common also to other birds) in the connective tissue of the pigeon beak. Unfortunately, the structure of slowly-adapting receptors in the beaks of pigeons is unknown.

The responses to mechanical stimuli are also influenced by temperature. In most cases an increase or decrease of the mechanical stimulus response is seen which parallels that of the pure temperature response of the receptor. It is significant that warming the receptive field may convert a slowly-adapting into a rapidly-adapting response (Gottschaldt *et al.*, 1982).

2. *Structure of cutaneous slowly-adapting receptors*

(a) Ruffini endings. Because of the similar response properties of slowly-adapting units originating from the goose beak and from mammalian type II receptors, it was proposed (Gottschaldt, 1974a) that the avian slowly-adapting units would derive from Ruffini endings as do their mammalian type II counter-

parts (Chambers *et al.*, 1972). Nerve endings of the Ruffini type have in fact been found at sites in the beak predicted for slowly-adapting receptors in electrophysiological experiments (Gottschaldt *et al.*, 1976, 1982).

Ruffini endings are associated with strands of collagen fibres which extend throughout the dermis underneath the nail at the beak tip in a direction parallel or oblique to the epidermis. The oblique strands apparently connect the tonofilaments of the epidermis with the Sharpey fibres of the premaxillary bone, and thus presumably fix the nail to the bony tip of the bill. The receptors are located close to the periosteum and are distinguished by a densely-packed system of collagen fibres (Fig. 8.25a,b); the latter are separated from the collagen fibres of the surrounding connective tissue by a loose transitional zone, which presumably allows some longitudinal displacement of the receptor relative to the connective tissue of the dermis. In contrast to mammalian Ruffini endings, a definite outer capsule is missing (Fig. 8.25).

On entering the receptor, the afferent nerve fibre loses its myelin sheath and then branches into fine nerve endings which meander between the collagen fibre strands of the receptor (Figs. 8.25, 26). Schwann cell processes enclose the nerve endings and also permeate the system of collagen fibres, sub-dividing it into smaller fascicles of collagen fibrils (Fig. 8.26a). These collagen fibrils may fuse with the basal lamina of Schwann cell processes. Occasionally the Schwann cell covering of the nerve endings is incomplete, and then collagen fibrils may make contact with the basal lamina of the nerve endings. The membranes of Schwann cell processes and nerve terminals are connected by small zonulae adherentes (Fig. 8.26b).

As in other mechanoreceptors, the axoplasm of the sensory nerve endings is characterised by abundant mitochondria and other organelles typifying the receptor matrix. Clear and dense-cored vesicles are frequently found. Along the nerve endings, specialised regions can be identified which correspond to the transducer sites of the other avian mechanoreceptors. Once again, the most conspicuous morphological characteristics of such transducer sites are finger-like axon processes which, in Ruffini endings come directly or indirectly into functional contact with the collagen fibre system of the receptor. Some axon processes penetrate the Schwann cell covering and basal lamina, and project directly among the surrounding collagen fibrils (Fig. 8.27a). More frequently, however, the axon processes remain enclosed by Schwann cell processes, which then establish a cellular bridge to the collagen fibrils (Fig. 8.27b). The axon processes contain longitudinally-oriented fine filaments which run into the axoplasm filling the conical bases of the axon processes. In this latter region there are scattered clear vesicles 50–100 nm in diameter, as well as microtubules. Mitochondria are absent in the axoplasm immediately surrounding the bases of the axon processes. Thus, in their general structure the transducer sites at the nerve terminals of Ruffini endings correspond to those in Herbst corpuscles. However, the axon processes are generally shorter than in Herbst corpuscles; also, specialised contact zones between receptor and Schwann cell membrane are usually absent in the immediate vicinity of the axon processes.

(b) Ruffini endings with terminal cell. Besides the more common variant of Ruffini nerve ending described already, a second variant occurs in conjunction with strands of collagen fibres traversing the dermis under the nail parallel to the

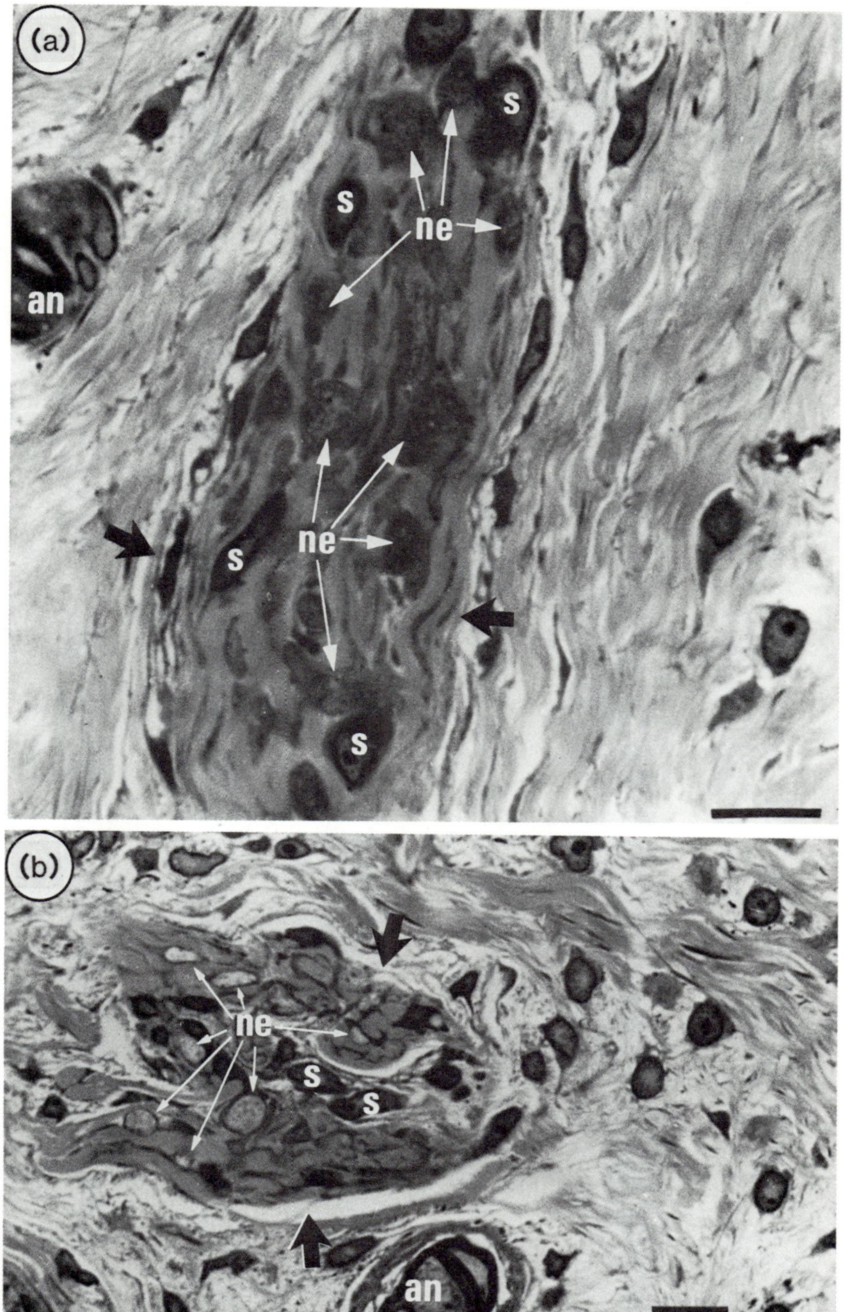

Fig. 8.25. Light micrographs of Ruffini endings in the dermis underlying the nail of the goose beak. (a) Longitudinal section. (b) Transverse section. The sensory nerve endings (ne) branch in a dense system of collagen fibres which is separated from the surrounding connective tissue by a loose transitional zone (large arrows). s, Nuclei of Schwann cells; an, myelinated afferent nerve fibres. (a) ×1624. Calibration bar: 10μm. (b) ×1120. Calibration bar: 10 μm.

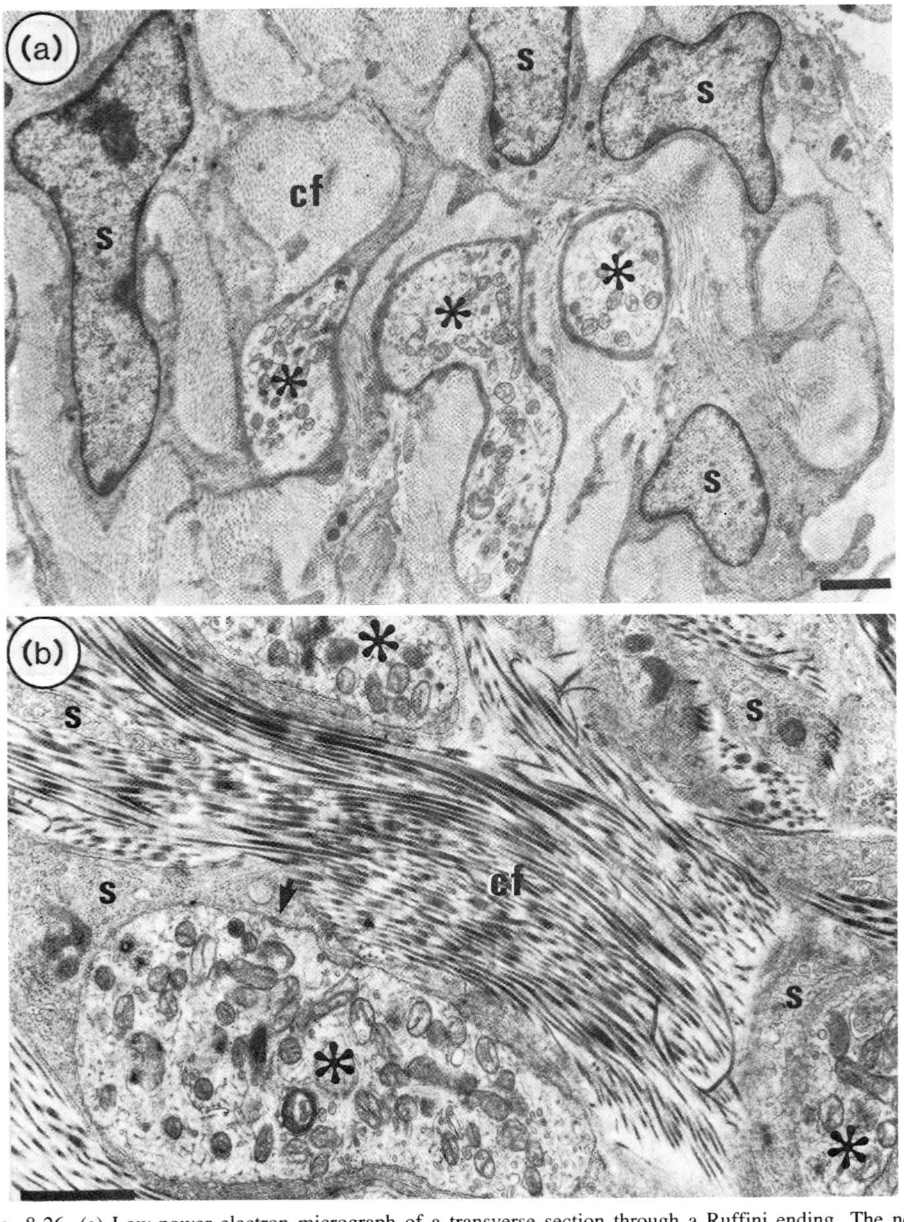

Fig. 8.26. (a) Low-power electron micrograph of a transverse section through a Ruffini ending. The nerve endings (asterisks) can be seen amidst numerous collagen fibrils (cf) sectioned transversely. Four nuclei of Schwann cells (s) are seen; their processes subdivide the collagen fibrils into smaller bundles, and also enclose the nerve endings. ×9229. Calibration bar: 1 μm. (b) Longitudinal section through a Ruffini ending, showing three nerve endings (asterisks) and several Schwann cell processes (s) at higher magnification. Note collagen fibrils (cf) fusing with the basal lamina of Schwann cell processes. Arrow, small zonula adherens. ×14,659. Calibration bar: 1 μm.

surface. These receptors lie about midway between the epidermis and periosteum and have been called ''Ruffini endings with terminal cell'' (Gottschaldt *et al.*, 1982) because of the association of a single terminal cell with the sensory nerve endings. This terminal cell usually is located shortly behind the first branching of the afferent axon, after the myelin sheath is lost. The terminal cell may be

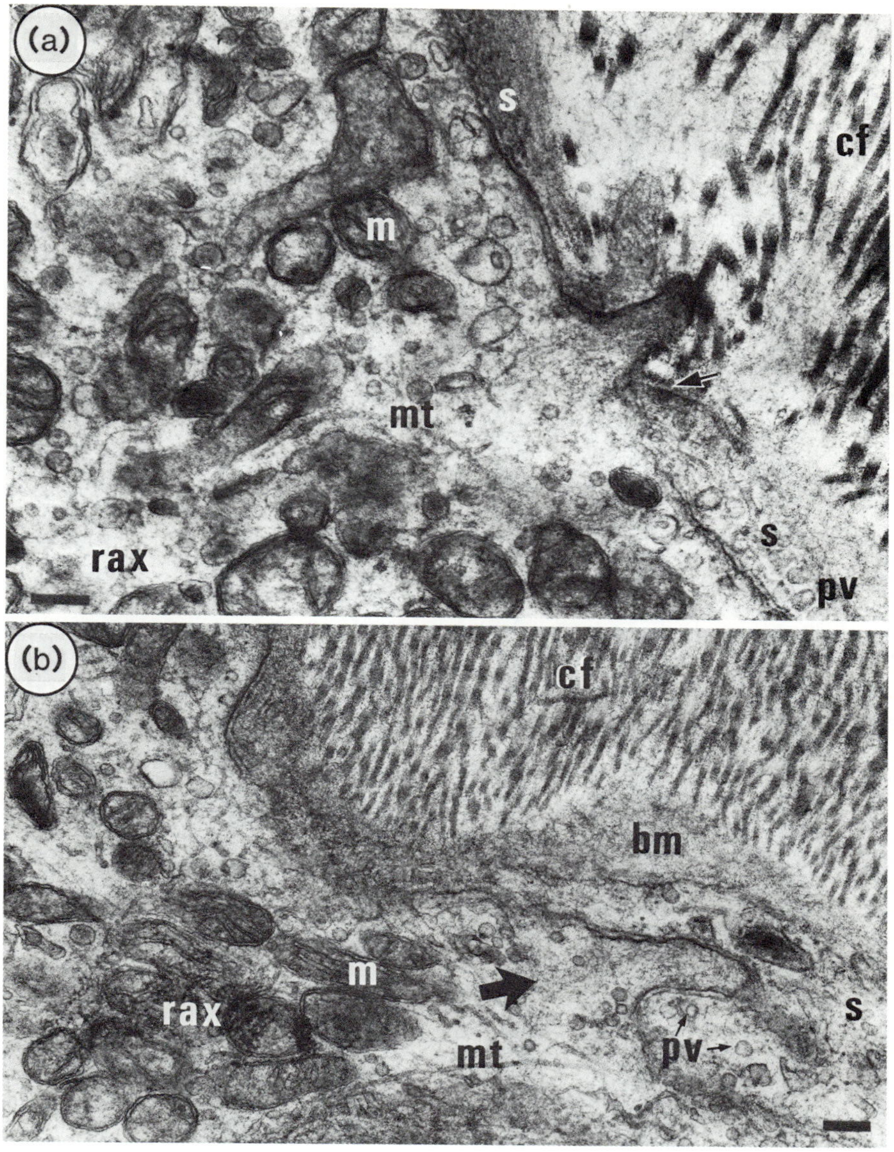

Fig. 8.27. Transducer sites in Ruffini endings. (a) An axon process of the receptor axon (rax) penetrates the Schwann cell sheath (s) and its limiting basal lamina (arrow), and comes into direct contact with collagen fibrils (cf). ×40,000. Calibration bar: 200 nm. (b) An axon process (arrow) of a receptor axon (rax) remains enclosed in a Schwann cell process (s). Collagen fibrils (cf) are in contact with the basal lamina (bm) of the Schwann cell processes. m, Mitochondria; mt, microtubules; pv, pinocytotic vesicles in Schwann cell process. ×34,760. Calibration bar: 200 nm.

contacted by several branches of the unmyelinated nerve endings (Fig. 8.28a) which also extend throughout the dense collagen fibre system of the receptor.

The terminal cell corresponds in all cytological details to the terminal cells found in other avian mechanoreceptors. Occasionally specialised junctional complexes of the ''synapse-like'' and the desmosome-like type can be observed (Fig. 8.28b). Both the nerve endings contacting the terminal cell as well as those coursing through the collagen fibre stroma of the receptor show the structural

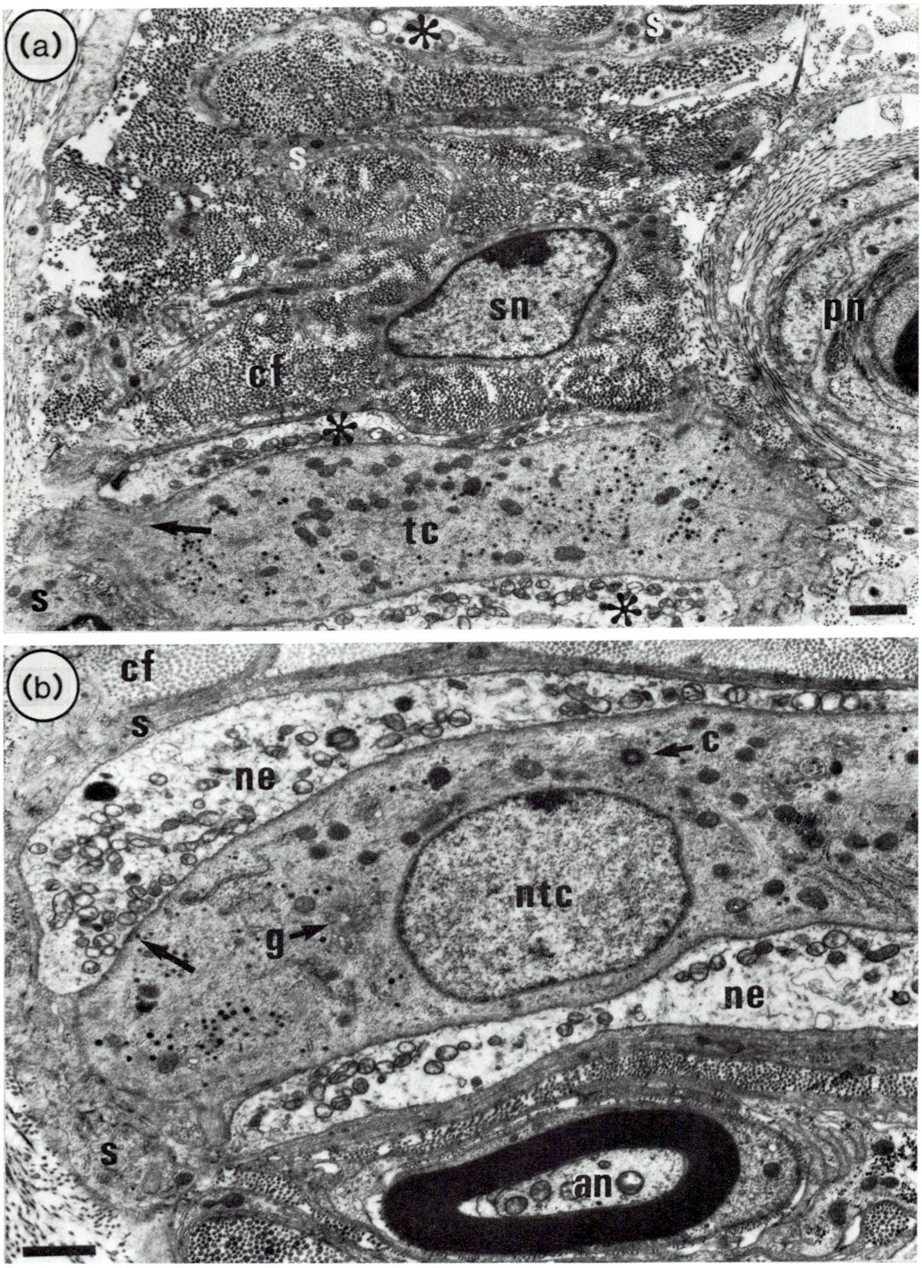

Fig. 8.28. Ruffini ending with terminal cell. (a) Nerve endings (asterisks) contact the terminal cell (tc) and branch in the densely-packed collagen fibrils (cf). sn, Schwann cell nucleus; s, Schwann cell processes; pn, perineurium of myelinated axon. Arrow, a microvillus in the terminal cell. ×8000. Calibration bar: 1 μm. (b) Terminal cell with nucleus (ntc), sandwiched by two nerve endings (ne). Long arrow, a specialised contact zone. g, Golgi apparatus; c, centriole; cf, collagen fibrils in transverse section; s, Schwann cell processes; an, myelinated afferent nerve fibre. ×10,000. Calibration bar: 1 μm.

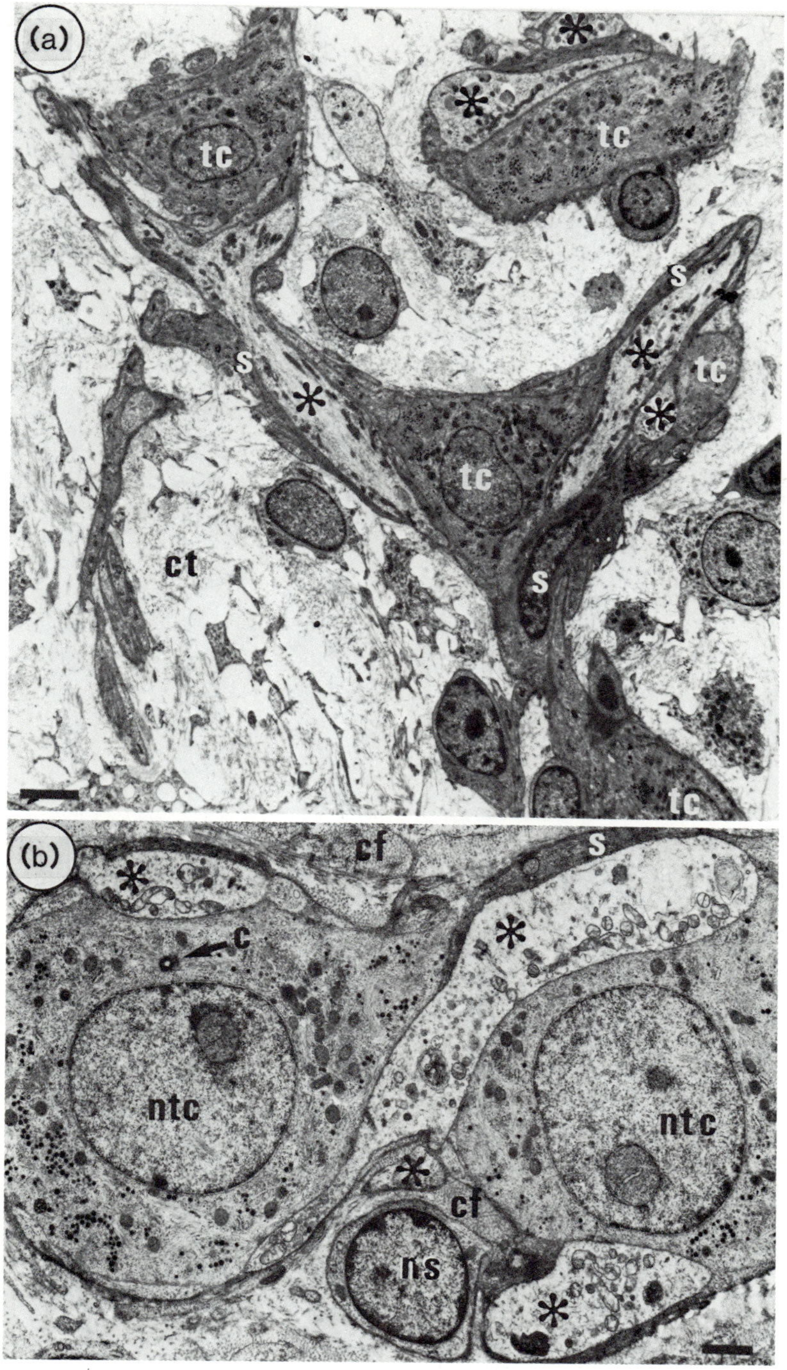

Fig. 8.29. (a) Low-power electron micrograph of a branched nerve ending with terminal cells in the distal part of a touch papilla of the goose beak. The nerve endings (asterisks) are in contact with several terminal cells (tc). The nerve endings are enclosed by Schwann cell processes (s) and branch in the connective tissue core (ct) of the touch papilla. ×4350. Calibration bar: 2 μm. (b) Two terminal cells in contact with three nerve endings (asterisks). Compare the chromatin structure of the terminal cell nuclei (ntc) with that of the Schwann cell (ns). c, Centriole; s, Schwann cell processes; cf, collagen fibrils. ×7070. Calibration bar: 1 μm.

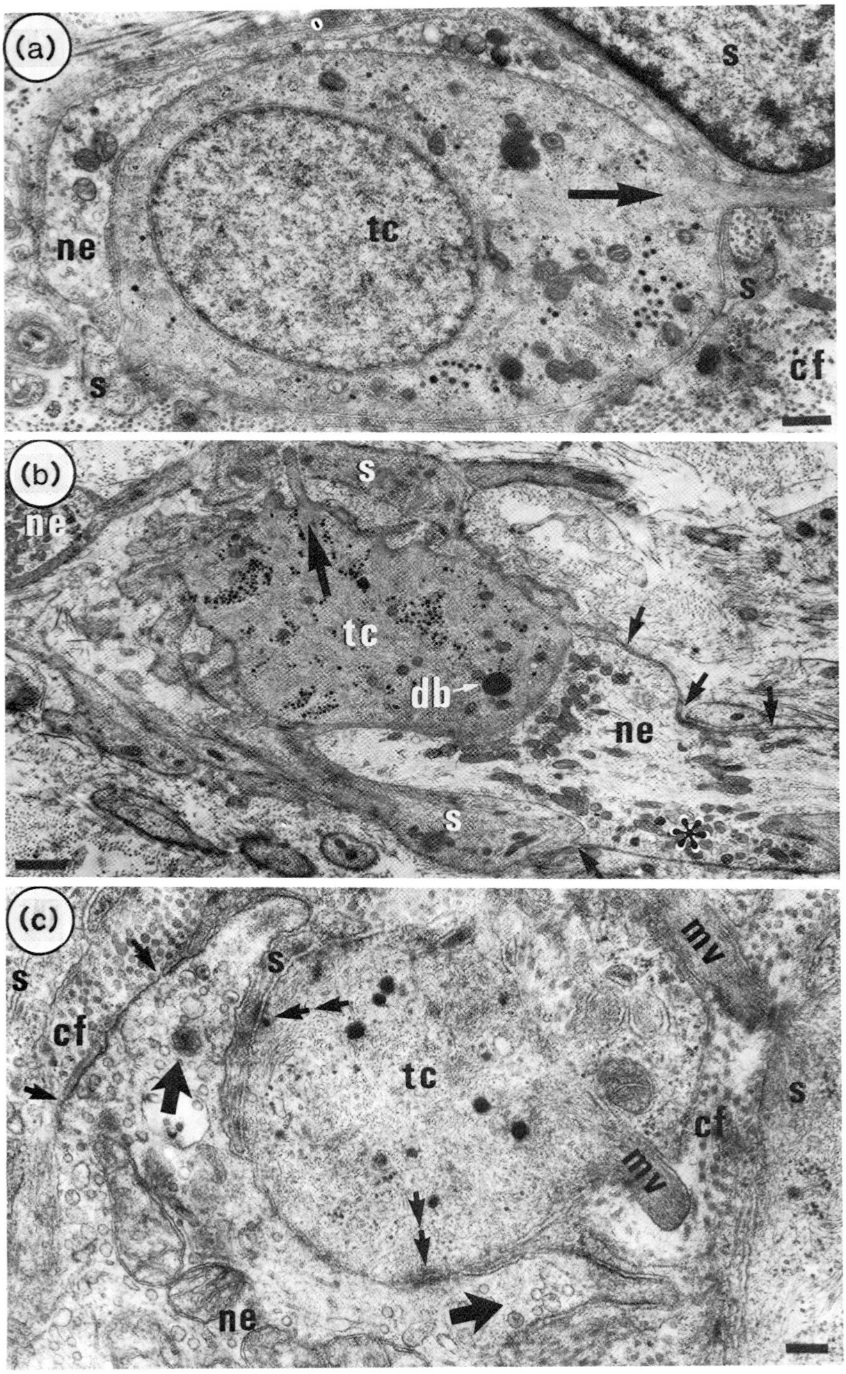

Fig. 8.30. Contacts between terminal cells and nerve endings, in branched nerve endings with terminal cells. (a) Nerve ending (ne) surrounds half of the terminal cell (tc). The latter has a microvillus (arrow) which penetrates between the processes of surrounding Schwann cells (s) and comes into contact with transversely-sectioned collagen fibrils (cf). Note the absence of dense-cored granules in the cytoplasm of the terminal cell in the zone which is apposed to the nerve ending. ×14,150. Calibration bar: 0·5 μm. (b) Nerve endings (ne), one seemingly making an ''en passant'' contact with a terminal cell (tc). A microvillus (large arrow) projects through a connective tissue gap into a Schwann cell process (s). The large nerve ending forms a ''snail foot'' (asterisk) which bears on either side a transducer site. Note incomplete Schwann cell enclosure (small arrows) of the nerve

characteristics of mechanoreceptor nerve endings, and also apparent transducer sites are present.

(c) Branched nerve endings with terminal cells. Although the existence of slowly-adapting mechanoreceptors with response properties similar to those of Ruffini endings could be established in the touch papillae of the bill tip organ (Section VII), by electro-physiological studies, no Ruffini endings have been found there. However, in the distal portions of the touch papillae there is a special type of terminal cell receptor which shares some morphological characteristics with Ruffini endings and, therefore, may well be the source of the slowly-adapting responses. This receptor is characterised by a profuse branching of the sensory nerve terminals (as in Ruffini receptors) and the occurrence of many isolated terminal cells which sit, like flowers on a climbing plant, on the branches of the nerve endings. Hence this receptor will be called a "branched nerve ending with terminal cells".

The branching unmyelinated nerve endings derive from a myelinated afferent nerve fibre. After the myelin sheath has ended, the axon continues for 10 to several hundred microns, and then branches into the actual sensory nerve endings. The unmyelinated pre-terminal axon is covered by Schwann cell processes, but no terminal cells occur along its course. Although the branching pattern of the sensory nerve endings resembles that in Ruffini endings, the formation of a special system of collagen fibres, in which the terminal arborization of the sensory axon would take place, is absent in branched nerve endings with terminal cells. Instead, the nerve endings, covered by Schwann cell processes, course through the collagen fibre system forming the connective tissue core of the touch papilla. The arborization of the nerve endings is strictly confined to the dermis, and branches of the receptor axon have never been seen to enter the epidermis.

The terminal cells occur along the course of the nerve endings and especially at their branching points (Fig. 8.29a). Occasionally two such cells can be seen in contact with the same as well as with different nerve endings (Fig. 8.29b), all presumably deriving from the same parent axon. Cytologically the terminal cells resemble those occurring in other receptor types. However, they are considerably smaller than, for instance, Grandry cells, rarely exceeding 10 μm in their largest diameter. Dense-cored granules are again abundant and also are typically concentrated in regions of the cell remote from the sensory nerve endings (Figs. 8.29, 30). As in other types of receptors, the terminal cells are covered by Schwann cell processes but their microvilli frequently extend through the Schwann cell covering and then contact the collagen fibrils of the surrounding connective tissue (Figs. 8.29a, 30a–c). As always, the cytoplasm of the terminal cells contains a cytoskeleton of microfilaments which extend into the numerous microvilli.

The sensory nerve endings may contact the terminal cells in various ways

ending. db, Dense body in the terminal cell. ×7500. Calibration bar: 1 μm. (c) Transducer sites with axonal processes (large arrows) of a nerve ending (ne) contacting a terminal cell (tc). The lower double arrow points to a zonula adherens connecting the membrane of the terminal cell with that of the nerve ending; the upper double arrow indicates a connection to a Schwann cell process. Small single arrows delineate a gap in the Schwann cell (s) covering of the nerve ending, allowing collagen fibrils (cf) to make direct contact with the receptor membrane. Microvilli (mv) of the terminal cell also are in contact with collagen fibrils. ×30,000. Calibration bar: 200 nm.

(Figs. 8.29, 30). Quite often a terminal cell appears to be contacted by several branches of a nerve ending or may be circularly enclosed by a single nerve ending. "Synapse-like" junctions between terminal cell and nerve ending have not been identified in our material, although hundreds of contacts between terminal cells and nerve endings were examined. However, a few desmosome-like zonulae adherentes were observed not only between the adjoining membranes of nerve endings and terminal cells but also between the latter and their enclosing Schwann cell processes (Fig. 8.30c). Where the nerve endings contact the terminal cells (Fig. 8.30c), and also along the course of their branches (Fig. 8.30b), axonal processes can be seen with the structural characteristics of transducer sites. The Schwann cell enclosure of the nerve endings is not always continuous, and then either the axon membrane, or the axon processes of transducer sites, make close contact with collagen fibrils. In such cases only the basal lamina intervenes between the axolemma and the collagen fibrils (Fig. 8.30c). Thus, in the arborization of the sensory nerve terminals in a collagen fibre stroma, and in the contacts of the nerve terminal membrane with collagen fibrils, the structure of the branched nerve endings with terminal cells resembles that of Ruffini endings.

3. *Structure–function relationships in cutaneous slowly-adapting receptors*

A prerequisite for the generation of slowly-adapting responses to maintained displacement stimuli is that the stimuli cause maintained deformations of the sensory receptor endings. In Herbst corpuscles (Section IV,A,2) this is prevented by the special arrangement, and the resulting mechanical high-pass filter function, of the auxiliary receptor structures. In Ruffini endings and all "stretch" receptors structurally related to them, it is the different morphology of the auxiliary receptor structures which actually enables static stimulus components to be transmitted to the sensory nerve terminals, thus making possible their maintained deformation.

As we have seen, in both the Ruffini endings with (Section IV,C,2,b) and without (Section IV,C,2,a) an associated terminal cell, as well as in branched nerve endings with terminal cells, there is an intimate association between sensory nerve endings (or their transducer sites) and collagen fibrils. The latter may be part of a special system of collagen fibres forming the supportive framework for the branching sensory nerve endings, as in Ruffini endings; alternatively, they may belong to the regular connective tissue of the dermis, as in branched nerve endings with terminal cells. In every case, however, a deformation of the skin surface will cause length changes in the collagen fibres, and since these fibres are only slightly expansile this should lead to deformations of the sensory receptor endings corresponding in time course to that of the applied stimulus. Although no direct structural connections seem to exist between the receptor membrane and the collagen fibres, a functional contact between them is established either by axon processes of transducer sites projecting among the collagen fibrils or by the fusing of the latter with the basal lamina of the nerve endings. Deformation of the receptor terminals may also be achieved via Schwann cell processes, which on one side are connected to the nerve endings by zonulae adherentes, and on the other side are attached to the connective tissue elements by the merging of collagen fibrils in the basal lamina of the Schwann cells.

As in other types of vertebrate mechanoreceptors, receptor potentials in avian slowly-adapting receptors very probably originate directly from mechanical deformations of the sensory nerve endings, and presumably result from the opening of mechanically-controlled ion channels. No evidence exists suggesting that the release of a chemical substance is a prerequisite for the generation of receptor potentials. However, different mechanisms or, specifically, different types of membrane channels with different kinetics may be responsible for the generation of dynamic and static response components, as has also been discussed for muscle spindles (Ottoson and Shepherd, 1971). Exactly what makes the mechano-electric transduction mechanism more sensitive for dynamic than for static stimulus components is little understood. Mechanical properties of the auxiliary receptor structures may partly account for the different dynamic and static sensitivity of the receptor, but it is also likely that different biophysical mechanisms operate at the receptor membrane and lead to the dynamic and static responses. The existence of two such independent biophysical mechanisms is further suggested by the presence of two time courses of adaptation for the dynamic and static response. In addition, in slowly-adapting mechanoreceptors of the sinus hair follicles in the cat, the mechanisms sustaining the dynamic response component apparently can be activated in isolation using high-frequency vibratory stimuli (Gottschaldt and Vahle-Hinz, 1981).

The functional significance of the terminal cells associated with the sensory nerve endings in the two variants of Ruffini endings, and especially in the branched nerve endings with terminal cells, is unclear. Since such cells seem not to be involved in the generation of receptor potentials in other terminal cell receptors, it is unlikely that they would do so in slowly-adapting receptors. However, the microvilli of the terminal cells frequently gain access to the stimulus-transmitting collagen fibre system, and hence the terminal cells may function as mechanical interfaces between the connective tissue structures and the sensory nerve endings. In addition, the strategic location of the terminal cells at the branching points of the sensory nerve endings may impose a distinct branching pattern. This could affect the propagation of receptor potentials, arising in individual branches of the sensory nerve endings, to the spike initiation zone of the afferent nerve fibre; this in turn may lead to the generation of the more regular or irregular discharge pattern observed in the avian slowly-adapting units.

Another consequence of the functional connection between the collagen fibres and the sensory nerve endings in Ruffini endings appears to be their temperature sensitivity. As for Herbst corpuscles (Section IV,A,2), it has been argued that cooling of the collagen fibres results in their thermal contraction, which is interpreted at the receptor membrane as a maintained mechanical stimulus and thus leads to a continuous afferent discharge (Gottschaldt *et al.*, 1982). Conversely, warming of the receptor results in a thermal expansion of the collagen fibres leading to a reduction and eventually complete abolition of the static response component following a mechanical stimulus. The finding that by warming the receptor the dynamic response can be obtained in isolation lends additional support for the assumption that different mechanisms underlie the generation of the dynamic and static response component. Besides the mechanical effects which temperature changes exert on the auxiliary receptor structures in Ruffini endings, they also affect the biophysical mechanisms sustaining the membrane potential and the mechano-electric transduction process; this can account for the typical bell-shaped response curve obtained with decreasing or increasing temperature variations across a temperature range of maximal thermal responsiveness (Gottschaldt *et al.*, 1982).

4. Slowly-adapting mechanoreceptors in the locomotory apparatus of birds

As in mammals, there are three types of slowly-adapting mechanoreceptors associated with the locomotory apparatus of birds. These are muscle spindles, Golgi tendon organs and the Ruffini endings occurring in articular capsules and joint ligaments.

(a) Physiology of avian muscle spindles and tendon organs. Responses from avian muscle spindles have been recorded in apparently only two electro-physiological studies, both of them in ducks. Manni *et al.* (1965) recorded slowly-adapting jaw muscle afferents with microelectrodes in the mesencephalic trigeminal nucleus, responding to either opening or closing the beak. Discharge pauses during electrically-induced contractions of the jaw muscles identified some of the units as muscle spindle afferents. A more detailed study was performed by Dorward (1970b), who investigated single units isolated from the sciatic nerve. Muscle spindle afferents showed dynamic–static responses to muscle stretch followed by a cessation of the discharge at stimulus end. During twitch contractions of extrafusal muscle fibres, spindle responses decreased or ceased altogether, whereas electrical stimulation of fusimotor fibres increased the discharge frequency in spindle afferents. Thus, like mammalian muscle spindles (Hunt, 1974), the avian counterparts also showed essentially an "in parallel" response behavior. However, there are also differences between mammals and birds. The conduction velocities of avian spindle afferents had a unimodal distribution corresponding to fibre diameters between 5 and 12 μm, and no evidence suggested the existence of two fibre populations comparable to the mammalian group Ia and II fibres innervating primary and secondary muscle spindles, respectively. Likewise, no separation in dynamic and static types of spindle responses could be established, but this problem may still receive a different solution (Adal and Chew Cheng, 1980b).

Dorward (1970b) also recorded responses derived from Golgi tendon organs. These units showed a typical "in series" behavior and were excited by both muscle stretch and active muscle contractions. Stimulation of fusimotor nerve fibres did not increase the discharge frequency in tendon organ afferents. The conduction velocities in these fibres also showed a homogeneous distribution corresponding to fibre diameters from 4 to 11 μm in diameter. Usually the threshold tensions required for maintained discharges in tendon organ afferents were characteristically high, but some units also showed unexpectedly low mechanical thresholds and were even spontaneously active when no tension was applied to the muscle.

(b) Structure of avian muscle spindles. Following a number of light microscopic studies (Huber and de Witt, 1897; Regaud and Favre, 1904; Tello, 1917, 1922; Germino and D'Albora, 1965; de Anda and Rebollo, 1967; Rebollo and de

Anda, 1967; Sağlam, 1968; deWet *et al.*, 1971; Maier and Eldred, 1971; Veggetti and Palmieri, 1971; Adal and Chew Cheng, 1980a), the fine structure of avian muscle spindles has been described by several investigators (James and Meek, 1973; Adal, 1973; Barker, 1974; Uehara *et al.*, 1976; Adal and Chew Cheng, 1980b). According to these reports, which do not agree in all details, avian muscle spindles measure between 700 μm and several millimetres in length, and 30–160 μm in their largest diameter. In small birds or in small muscles the spindles are generally smaller and more simple in structure than in large birds or in larger muscles. Their number per unit muscle volume varies in different muscles, small muscles having relatively more spindles than large ones. Also, the spindle density in corresponding muscles is higher in small than in large birds. On average, the spindle density is considerably higher in avian than in mammalian muscles. For instance, in neck muscles of two species of woodpecker the spindle density per unit muscle volume was 18 times higher than in the longissimus capitis muscle of humans, which is thought to have the highest spindle density in *homo sapiens* (Voss, 1963; Sağlam, 1968).

Each muscle spindle is composed of different structural elements, namely (1) intrafusal muscle fibres with their sheath cells, (2) sensory and motor nerve fibres with their endings, (3) connective tissue cells forming an inner capsule around the muscle fibres where the sensory endings are located, (4) blood vessels, (5) a free periaxial space around the muscle fibres, and (6) an outer capsule (Fig. 8.31).

The outer capsule is a continuation of the perineurium of the nerve fibres entering the spindle and consists of several layers of perineural cells each of which is limited by a basal lamina and has numerous pinocytotic vesicles. Collagen fibrils are scattered between the perineural cells. The fluid-filled periaxial space contains small blood vessels and is sub-divided by thin processes of septal cells, which are fibrocytes connecting the outer capsule with the inner capsule.

A peculiar specialisation of the inner capsule is found in the equatorial region of avian intrafusal muscle fibres. In addition to the endomysial sheath (a connective tissue embedding all muscle fibres), the sensory region is surrounded by an asymmetrical circular collagen fibre system that in transverse sections is shaped like a signet-ring pulled over the muscle fibre (Fig. 8.31). This collagen crescent is produced by fibrocytes forming the inner capsule around each muscle fibre and is usually widest where the sensory nerve endings terminate. Beyond the equatorial zone, the collagen crescent tapers rapidly and eventually merges with the ordinary endomysium of the muscle fibre.

The intrafusal muscle fibres extend between the two poles of the spindle. Along each muscle fibre it is possible to distinguish a mid-equatorial region comprising about 10% of the total muscle fibre length, two juxtaequatorial regions, and two polar regions. The number of muscle fibers per spindle varies between one and eight, about one-fifth of the spindles being monofibrillar. If several fibres are present, they lie close together at the spindle poles but diverge towards the equatorial region where they are separated from each other by inner capsule cells and the periaxial space. At the spindle poles the sarcolemma of the intrafusal muscle fibres attaches to the inner surface of the outer capsule, but some muscle fibres may also extend beyond the outer capsule poles. In such cases the sarcolemma of the intrafusal muscle fibres attaches directly to the endomysium of extrafusal muscle fibres, whereas normally the poles of the outer capsule are

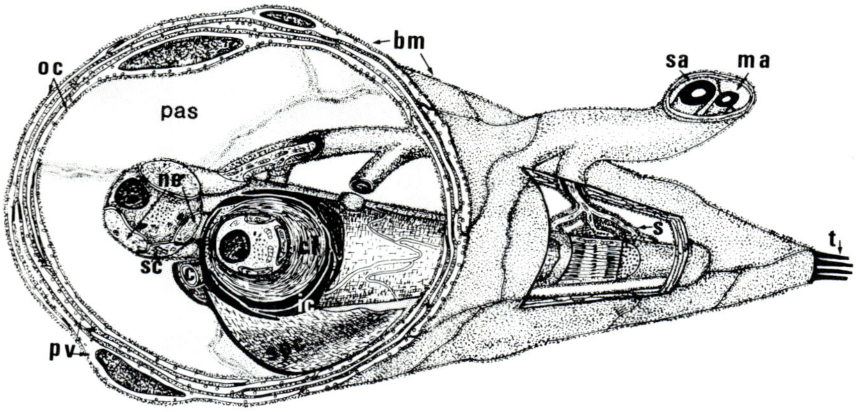

Fig. 8.31. Three-dimensional schematic view of an avian muscle spindle transected in the equatorial zone. The spindle contains two intrafusal muscle fibres, only the right of which is shown surrounded by the inner capsule (ic). This has been omitted in the left, slightly more enlarged, muscle fibre, in order to show the sensory terminals (ne) on its surface more clearly. A myelinated sensory axon (sa) and a smaller motor axon (ma) terminate in the equatorial and polar region of the spindle, respectively. The sheath of the entering nerve fibres blends with the outer capsule (oc) of the spindle. These perineural capsule cells contain many micropinocytotic vesicles (pv), and each cell is limited by a basal lamina (bm). A septal cell (spc) connects the outer and inner capsule, thereby subdividing the periaxial space (pas) into different compartments. The inner capsule is formed by fibrocytes enclosing the asymmetrical crescent of collagen fibres (cf) encircling the muscle fibre with its sensory endings. A small capillary (c) extends between the connective tissue cells. After entering the spindle, the sensory axon divides, loses the myelin sheath and forms several branches each terminating in bulbous enlargements embedded in the intrafusal muscle fibres. Desmosome-like attachment zones connect receptor and muscle fibre membranes. Also the satellite cells (sc) have such attachments to the nerve endings (ne). The latter and the satellite cells lie below the basal lamina of the muscle fibre. The terminals of the motor fibre, instead, lie above the basal lamina and are covered by Schwann cell processes (s). t, Tendon fibres of intrafusal muscle fibres leaving the spindle.

connected with the endomysium of the extrafusal muscle fibres. Because of this fixation of the spindle on the connective tissue components of the extrafusal muscles, stretch of the latter is transmitted also to the intrafusal muscle fibres.

In contrast to mammalian intrafusal muscle fibres, nuclear chain cannot be distinguished from nuclear bag fibres in birds, although the nuclei are most numerous in the equatorial region. However, from their diameter (de Anda and Rebollo, 1967; Sağlam, 1968), histochemical properties (Rebollo and de Anda, 1967) and ultrastructural characteristics (Adal, 1973), it appears that at least two types of intrafusal muscle fibre exist, but this has not been accepted by all investigators (Germino and D'Albora, 1965; James and Meek, 1973; Barker, 1974). The diameter of intrafusal muscle fibres near the equatorial zone varies between 5 and 11 μm, and decreases slightly towards the spindle poles. Myofilaments are absent or scarce in the equatorial zone but become increasingly densely packed towards the polar regions. Z-Lines can be seen regularly, but M-bands are not always clearly visible. The sarcotubular system is well developed, showing triads and terminal cisternae of the sarcoplasmic reticulum below the sarcolemma. Leptomeric fibrils occur in the polar regions of the muscle fibres, and many mitochondria are scattered in the sarcoplasm. The plasma membrane of the muscle fibres is coated by a basal lamina, which also extends across flat satellite cells lying on the surface of the muscle fibres.

Each muscle spindle is innervated by a single myelinated sensory nerve fibre which enters in the juxtaequatorial region. It may already have divided outside the spindle capsule and subdivides further within the periaxial space (Fig. 8.31). The axon branches terminate in unmyelinated endings in the equatorial region of the intrafusal muscle fibres. The nerve endings form numerous elongated or bulbous enlargements which are connected by thin axon segments and lie in indentations of the muscle fibre below its basal lamina. They are partly enclosed by processes of satellite cells. The axoplasm of the nerve terminals shows all the morphological ingredients of mechano-sensitive nerve endings. There are axial neurofilaments and microtubules, many mitochondria, various organelles, small vesicles and a receptor matrix underneath the receptor membrane. However, no transducer sites with an axon process have been observed. Instead, the receptor membrane is fixed to the muscle membrane by numerous zonulae adherentes, which also occur between the receptor membrane and the cytolemma of the satellite cells. A similar attachment of the nerve endings to the plasma membrane of the muscle fibres has been described in mammalian muscle spindles (von Düring and Andres, 1969). It appears clear that by means of such attachment zones a lengthening of the central part of the muscle fibre will stretch the receptor membrane. This, in turn, should cause the generation of a receptor potential, paralleling in time course the extension of the equatorial region.

An annulospiral form of nerve ending, as in mammals, is not observable in avian muscle spindles. Instead, the unmyelinated nerve terminals form a profuse fibre system weaving around and along the equatorial zone of the intrafusal muscle fibres. According to Adal and Chew Cheng (1980a), the sensory ending in thin spindles (diameter <80 μm) forms an elongated and narrow terminal system, whereas in large spindles the sensory ending is shorter and wider, and the nerve terminals show a more tortuous course and have more bulbous enlargements than those in thin spindles. In agreement with electro-physiological results (Dorward, 1970b), there are no secondary sensory endings, at least in duck muscle spindles (Adal and Chew Cheng, 1980a).

The fusimotor fibres enter the spindle either together with the sensory axon or independently at the polar region of the spindle. It is not quite clear whether more than one fusimotor fibre may innervate a given spindle. The motor fibres terminate in either "en grape" or "en plaque" endings in the polar region of the muscle fibres (Fig. 8.31). The terminals are rich in synaptic vesicles and lie above the basal lamina of the muscle fibres, covered by Schwann cell processes. There is no folding of the sub-synaptic muscle fibre membrane. In addition to the fusimotor fibre supply, several unmyelinated axons enter the spindle with the sensory axon. They seem to terminate among the cells of the inner capsules, but their function is unclear.

(c) Structure of Golgi tendon organs. Golgi tendon organs (Golgi, 1880) are slowly-adapting "stretch" receptors (Boyd, 1954), usually located at the musculotendinous junctions of skeletal muscles, but they can also be situated entirely within the tendon fibres. The receptors also occur in association with intermuscular and intramuscular tendons, such as aponeuroses or tendons of pennate muscles (Barker, 1974). Generally the receptors are fusiform in shape, but some may be bifid or trifid at one or both ends (Huber and de Witt, 1900; Barker, 1974). Although most of the morphological, and especially the ultrastructural

studies, on Golgi tendon organs were conducted on mammalian material (Merrillees, 1962; Goglia and Sklenská, 1969; Shimizu, 1971; Sklenská, 1972, 1973; Schoultz and Swett, 1972, 1974), there are also some electron microscopic investigations of avian tendon organs (Uehara *et al.*, 1976; Haiden and Awad, 1981) which reveal a close structural similarity of these receptors in mammals and birds.

Avian Golgi tendon organs are somewhat smaller than their mammalian counterparts, measuring on average about 450 μm in length and 60 μm in their largest diameter. The receptor complex is limited by a capsule that is again a continuation of the perineural sheath and similar in structure to that of muscle spindles or Herbst corpuscles. The interior of the receptor is filled with loosely-packed, longitudinally-orientated collagen fibres which are divided into bundles by processes of special septal cells. The latter derive from fibroblasts and lack a basal lamina. In ducks the cytoplasm of septal cells contains many lipid droplets (Haiden and Awad, 1981), whereas in finches these are apparently absent (Uehara *et al.*, 1976). Small blood vessels extend between the capsule and the septal cells. At the two ends of the receptor the capsule narrows and forms a tight collar around the collagen fibres, which extend beyond the capsule and merge with the endomysium of two or four muscle fibres at one end, and with collagen fibres of the extrafusal muscle tendon at the other end. In this way the collagen fibres within the receptor are placed "in series" with muscle and tendon fibres.

The sensory Ib nerve fibre measures in birds only 6–7 μm in diameter. After entering the capsule the axon divides several times and eventually terminates in many unmyelinated nerve endings meandering between the collagen fibres. The terminal arborization of the sensory nerve fibre extends throughout almost the entire inner capsular space of the receptor. The individual branches are sheathed by Schwann cell processes or at least covered by a basal lamina separating the receptor membrane from adjacent collagen fibrils. Within the larger bundles formed by the septal cells, Schwann cell processes demarcate smaller groups of collagen fibrils. Occasionally, a nerve ending and its enclosing Schwann cell process may completely encircle a small bundle of collagen fibrils. The axoplasm of the sensory nerve endings shares all the cytological characteristics already described for other mechano-sensitive nerve endings, except that transducer sites with axon processes have not yet been described in avian Golgi tendon organs.

In its special association between sensory nerve endings and collagen fibres, the structure of Golgi tendon organs is practically identical to that of Ruffini endings. Therefore, the same structure–function relationships should be valid in the two receptor types. For the mammalian Golgi tendon organ it has been suggested that tension applied to the muscle–tendon complex leads to a straightening of the collagen fibres inside the receptor capsule, and that this would result in a deformation of the sensory nerve endings coursing between them. The mechanical deformation of the nerve endings would then directly cause the generation of a receptor potential (Bridgman, 1968; Swett and Schoultz, 1975). Because of the structural similarity between mammalian and avian Golgi tendon organs, it can be predicted that the same structure–function relationships exist in these receptors. Indeed, they should be principally similar to those discussed for mammalian (Chambers *et al.*, 1972; Iggo and Gottschaldt, 1974) or avian (Gottschaldt *et al.*, 1982) Ruffini endings.

In mammals paciniform receptors may apparently be associated with Golgi tendon organs (Barker, 1974). This has not yet been established in birds, but it would not be a surprising observation. In mammals there is an unmyelinated nerve fibre supply to Golgi tendon organs and their adjoining muscle fibres, but this too has not been found in birds.

(d) Structure and function of Ruffini endings in articular capsules and joint ligaments. The joints of mammals are innervated by two functional kinds of low-threshold mechano-sensitive fibres (Skoglund, 1973), that is, rapidly-adapting units deriving from Pacinian corpuscles and slowly-adapting units from typical Ruffini endings (Boyd and Roberts, 1953; Boyd, 1954; Andrew, 1954). The latter type of receptor was first described by Sfameni (1902) in joint capsules and later by Gardner (1944). Joint innervation in different vertebrate species has been extensively reviewed at the light microscopic level by Poláček and his collaborators (Poláček, 1966; Poláček *et al.*, 1966; Sklenská, 1969).

The physiological properties of avian joint mechanoreceptors have not yet been determined. However, anatomically the presence of two receptor types—Herbst corpuscles and Ruffini endings—has been demonstrated, and these probably correspond in function to Pacinian corpuscles and Ruffini endings of mammalian joints, respectively (Poláček, 1966; Poláček *et al.*, 1966; Sklenská, 1969; Palmieri, 1969; Malinovský and Zemanek, 1970; Halata and Munger, 1980). Herbst corpuscles most certainly are rapidly-adapting mechanoreceptors, detecting fast transients and vibratory stimulus components in joint movements, whereas Ruffini endings are slowly-adapting mechanoreceptors capable of measuring static stimulus components such as rotation angles of the joints.

According to Poláček (1966), Poláček *et al.* (1966) and Sklenská (1969), the number of Ruffini endings, called "spray-like endings" by these authors, in avian joint capsules is low. Palmieri (1969) could not find these receptors at all, and Halata and Munger (1980) had difficulties in detecting them ultrastructurally. However, this latter study showed that Ruffini endings in the articular capsule of the pigeon are about 80 μm long and 30 μm in diameter, and are innervated by myelinated axons about 3 to 4 μm in diameter. The structure of the receptor is virtually identical to that of the Golgi tendon organ, except for its smaller size. The collagen fibres inside the receptor capsule are continuous with the collagen fibres of the articular capsule. The sensory nerve endings, sheathed by Schwann cell processes and/or a basal lamina, branch between the bundles of collagen fibrils, which are compartmentalised by Schwann cell processes and septal cells. Like Ruffini endings in the rhamphothecal dermis (Gottschaldt *et al.*, 1982), but perhaps unlike avian Golgi tendon organs (Haiden and Awad, 1981), the nerve terminals in Ruffini endings of avian joint capsules possess morphological transducer sites with axon processes. Thus, ultrastructurally the Ruffini endings in avian and mammalian joint capsules reveal the same morphological characteristics (Goglia and Sklenská, 1969; Sklenská, 1972, 1973; Halata, 1977; Halata and Munger, 1980), and it can therefore be assumed that they also share the same functional properties.

The relative paucity of Ruffini endings in avian joint capsules is surprising, since the total number of receptors is higher in avian than in mammalian joints (Poláček, 1966; Poláček *et al.*, 1966). However, most of these receptors are

Herbst corpuscles. Also the relative number of mechanoreceptors in different joints varies, apparently according to the functional requirements. Thus, the receptor supply of the elbow joint is far richer than that of the knee or hip joints. In addition to Herbst corpuscles and Ruffini endings, "free" nerve endings have also been found in avian joint capsules using silver impregnation techniques. The sensory nature, let alone their functional properties, as well as the fine structure of these "free" nerve endings are unknown.

V. Cutaneous thermoreceptors

As in mammals, the skin of birds participates in thermoregulation (Steen and Steen, 1965; Richards, 1970, 1971; Rautenberg, 1971; Necker, 1977). Presumably, therefore, avian skin, like mammalian skin, is supplied with receptors sensing external temperatures. In several studies temperature-sensitive primary afferent units have been recorded (Kitchell *et al.*, 1959; Necker, 1972, 1973; Gregory, 1973; Gottschaldt, 1974a; Leitner and Roumy, 1974b; Necker and Reiner, 1980), but the characterisation of their electro-physiological properties has been less complete than in mammals (Hensel, 1973). The main deficits are that a distinction between responses from temperature-sensitive mechanoreceptors and from specific thermoreceptors has not always been convincing, and that the conduction velocity of specifically temperature-sensitive afferent fibres has never been determined. For instance, in the duck beak Gregory (1973) and Leitner and Roumy (1974b) described temperature-sensitive responses in supposedly small-calibre afferent fibres (judged by the amplitudes of the action potentials); however, they failed to detect any slowly-adapting responses from mechanoreceptors, which are not only frequent in the trigeminal nerves but also show (Section IV,C,1) a distinct temperature sensitivity (Necker, 1974a; Gottschaldt *et al.*, 1982). Moreover, a precise determination of the receptive field characteristics of specific thermo-sensitive units, which are typically spot-like in mammalian skin (Hensel, 1973), has not been achieved in avian skin. Thus, although the existence of specific thermo-receptive afferents in avian skin is very probable, they have not been demonstrated unequivocally or characterised electro-physiologically.

Provided the temperature-sensitive units described in the literature included at least some specific thermo-receptive afferents, it appears that avian thermoreceptors have very similar response properties to their mammalian counterparts. Most of the thermo-sensitive units had cold fibre properties, but a few also behaved like mammalian warmth fibres. Cold fibres usually exhibited a slow spontaneous discharge at room temperature. They were active in a temperature range between 2 and 40°C and possessed static response maxima between 20 and 30°C. On sudden cooling most cold fibres responded with a transient increase of discharge, following which an impulse frequency was reached that corresponded to the static temperature response at the given adapted temperature. The dynamic sensitivity varied considerably between individual cold fibres, and some even lacked any appreciable dynamic responses (Necker, 1972). Where a dynamic response was present, it increased with the size and steepness of the temperature change and depended also on from where to where a temperature change occurred. Sudden

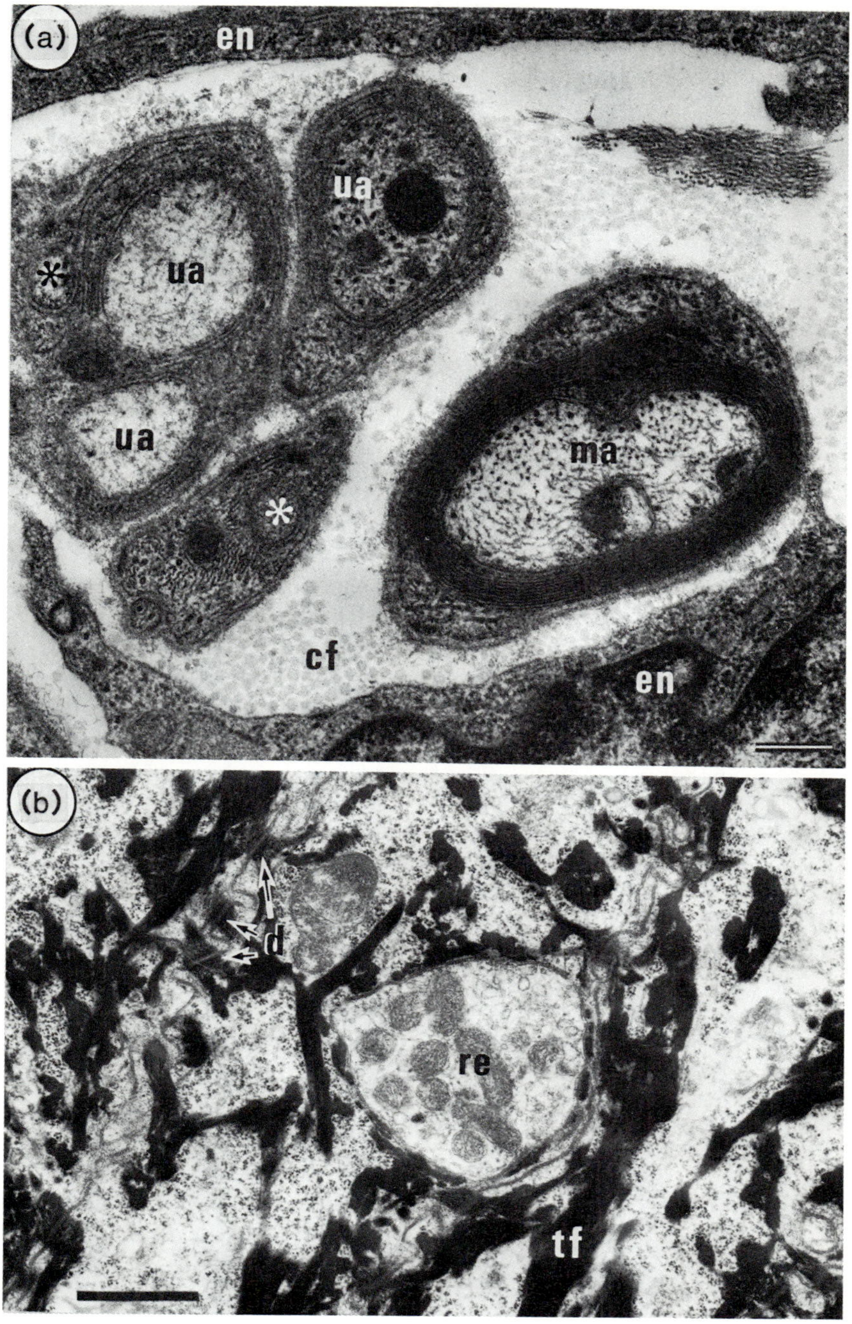

Fig. 8.32. (a) A group of small nerve fibres in the palate of the chicken possibly supplying thermoreceptors or nociceptors. ma, Myelinated axon (''A'' fibre); ua, unmyelinated axon (''C'' fibre); asterisks, very small unmyelinated axons; cf, collagen fibrils; en, endoneural fibrocytes. ×20,900. Calibration bar: 0·5 μm. (b) Intraepithelial receptor ending (re) in the epidermis of a touch papilla in the bill tip organ of a young gosling. tf, Tonofilaments of epidermal cell; d, desmosomes between epidermal cells. ×18,000. Calibration bar: 1 μm.

warming of the skin caused inverse responses in cold fibres, that is, the discharge decreased transiently or ceased altogether; a new discharge level was then attained which corresponded to the finally maintained adapted temperature.

A significant difference between mammalian and avian cold fibres is that the latter display much lower maximal discharge frequencies in both the dynamic and static responses. In mammals, dynamic cold fibre responses may reach up to 300 impulses/sec (Hensel, 1973), whereas in birds the discharge frequencies never exceeded 30 impulses/sec in the studies by Gregory (1973), Necker (1972, 1973) and Leitner and Roumy (1974b). Similarly, the maximal discharge frequencies of static responses lie around 20 impulses/sec in mammalian cold fibres but do not exceed 8 impulses/sec in birds. This low-frequency range in avian cold fibre responses indicates only a small information capacity.

Warmth-sensitive fibres so far have been recorded only in pigeons (Necker, 1972, 1973; Necker and Reiner, 1980). Such fibres also were usually active at room temperature, but in contrast to cold fibres, the discharge frequency decreased if the beak or the receptive field was cooled. A few fibres showed a progressive increase of their impulse activity on slowly warming the skin, reaching a maximum of about 65 impulses/sec at 44°C. Most units, however, behaved more like threshold detectors; that is, above a certain temperature the discharge was virtually switched on and then stayed constant even if the temperature of the skin was raised further. Warmth units of the first group also showed dynamic response properties, but no quantitative investigation has been performed. The generally higher discharge frequencies in static warmth than in cold fibre responses agree with findings in mammalian skin (Hensel, 1973), but warmth-sensitive fibres behaving as threshold detectors have not been observed in mammals.

The thermo-sensitive afferents probably belong to small myelinated or unmyelinated nerve fibres. Such fibres do, of course, occur, but in the palate of the chicken (Fig. 8.32a) their proportion of the afferent nerve fibres is surprisingly small. At least the cold receptors may correspond to the intraepithelial nerve endings which have been demonstrated by silver stains (Saxod, 1978a) and electron microscopy (Gottschaldt *et al.*, 1976; Fig. 8.32b). The abundance of mitochondria and the finely-grained axoplasm identify the intraepidermal nerve ending in Fig. 8.32b as a receptor terminal, similar in appearance to the subepidermal and intraepidermal nerve endings of mammalian cold receptors (Hensel *et al.*, 1974). The structure of warmth receptors is unknown, and it is not even certain whether they are located in the epidermis or the dermis.

VI. Cutaneous nociceptors

By cutting or pinching the soft sheath of the beak in anaesthetised geese, typical nociceptive responses may be elicited, such as an increase in blood pressure, heart beat or respiratory frequency; also, a tear flow is often initiated. The skin of birds therefore appears to be supplied with nociceptive afferents. In fact, a few fibres carrying presumably nociceptive responses have been recorded in several studies (Dorward, 1970a; Gregory, 1973; Leitner and Roumy, 1974b; Necker and Reiner, 1980; Gottschaldt *et al.*, 1982), from which it appears that avian cutaneous

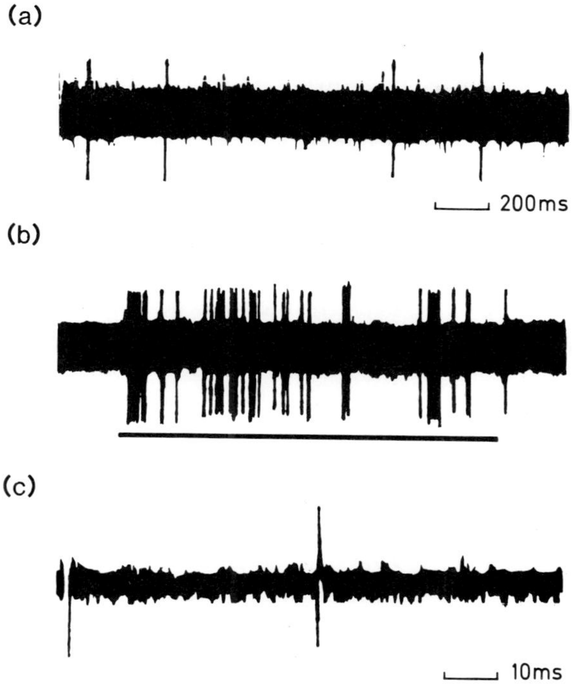

Fig. 8.33. Discharges in a nociceptive "C" fibre of the ophthalmic nerve in the goose. (a) "Spontaneous" activity following previous noxious stimulation. (b) Response to strong squeezing of the upper eyelid with serrated forceps. The bar below the spike record indicates the duration of the stimulus. Time base same as in (a). (c) Response to electrical stimulation of the nerve about 60 mm distal from the recording site.

nociceptors fall into functional categories similar to those in mammalian skin (Burgess and Perl, 1973). Most of them had properties of high-threshold mechanoreceptor afferents, but a few responded only to noxious heat stimuli (temperature threshold above 45°C) or were activated by strong cooling of the beak.

The receptive fields of nociceptive fibres were usually ill-defined, especially in units responding to noxious temperature stimuli. Before any damaging stimulus was applied, the units usually lacked an appreciable spontaneous activity. Judged by the spike amplitudes, all nociceptive afferents were of small calibre, but in only two instances was the conduction velocity measured directly (Gottschaldt *et al.*, 1982). One is illustrated in Fig. 8.33. The unit had a small receptive field at the upper eye lid of the goose and was at first recognised in the recorded nerve filament by electrical stimulation of the ophthalmic nerve. Originally the fibre was not spontaneously active, but after several strong pinchings of the eyelid a slow ongoing activity was recorded (Fig. 8.33a). Noxious mechanical stimuli elicited a slowly-adapting response (Fig. 8.33b) which increased in discharge frequency with stimulus strength. The response latency after electrical stimulation of the ophthalmic nerve amounted to 46 msec (Fig. 8.33c) at a conduction distance of approximately 60 mm, corresponding to an unmyelinated "C" fibre with a conduction velocity of 1·3 m/sec.

Nociceptors are present in both the beak and the feathered skin. They probably also occur in deep tissues, but electro-physiological evidence is lacking unless the "irritant receptors" described by Callanan *et al.* (1974) in the avian respiratory system are classified as chemically-sensitive nociceptors. The structure of nociceptive nerve endings is unknown. However, some of the intraepidermal nerve endings described previously (Section V) may belong to nociceptive afferents.

VII. Structure and function of the bill tip organ

In birds in general the tip of the beak is formed by hard horny tissue, thus enabling the bill to be used as a tool or weapon (see Chapter 3 by J. McLelland, in Vol. 1). At first glance, this design requirement appears to conflict with the important function of the beak as a tactile exploratory organ indispensable for food discrimination and plumage care. However, in many avian species the horny substance shelters a unique sense organ, which is perhaps the most sophisticated of the "complex sensory structures" (Iggo and Gottschaldt, 1974) occurring in vertebrate skin or its derivatives. In its functional significance this "bill tip organ" (Gottschaldt and Lausmann, 1974) may be considered as an avian equivalent of the mammalian sinus hair system or Eimer's organ in the mole.

The bill tip organ was first described in the beak of parrots (Goujon, 1869) and subsequently studied in a number of species (Ihlder, 1870; Jobert, 1872; Key and Retzius, 1876; Merkel, 1880). Despite several re-descriptions in the present century (Schildmacher, 1931; Lange, 1931; Stresemann, 1934; Lüdicke, 1940), the existence and functional significance of the bill tip organ gradually fell into oblivion. In fact, apart from the survey by J. McLelland (Chapter 3 in Vol. 1), it has not been mentioned in recent reference books dealing with the avian nervous system, and especially, the somatosensory component (Portmann, 1961; Sturkie, 1965; Bolton, 1971; Schwartzkopff, 1973; Pearson, 1976). Thus, the bill tip organ was virtually rediscovered by new light and electron microscopic studies, which also revealed further details about its morphology and innervation (Gottschaldt and Lausmann, 1974; Berkhoudt, 1976, 1980; Gottschaldt *et al.*, 1976), as well as the composition and growth mechanisms of the horny material enclosing the sensory structures (Menzel and Lüdicke, 1974).

The general structure of the bill tip organ appears to be similar in all avian species in which it is known to occur. Its presence or absence probably depends on the way the beak is employed during the feeding process. Birds which use their beaks primarily as a grain-pecking tool, such as pigeons or sparrows, have no bill tip organ. If, however, the beak is used in a more instrumental way to search for, catch, select or manipulate food—as, for instance, in parrots or waterfowl—a well-developed bill tip organ can be found. The bill tip organ of geese and ducks has been studied in most detail (Gottschaldt and Lausmann, 1974; Berkhoudt, 1976, 1980; Gottschaldt *et al.*, 1976) and will therefore serve to illustrate its structure.

In geese and ducks (Anatidae) the outer surfaces of the upper bill tip, and the outer and inner surfaces of the lower bill tip are covered by a horny plate shaped like a finger nail, that is, the nail or neb of the upper bill (*unguis maxillaris*) and of the lower bill (*unguis mandibularis*). In the upper bill the horny tissue of the nail also extends for some distance onto the palatine surface but without forming a

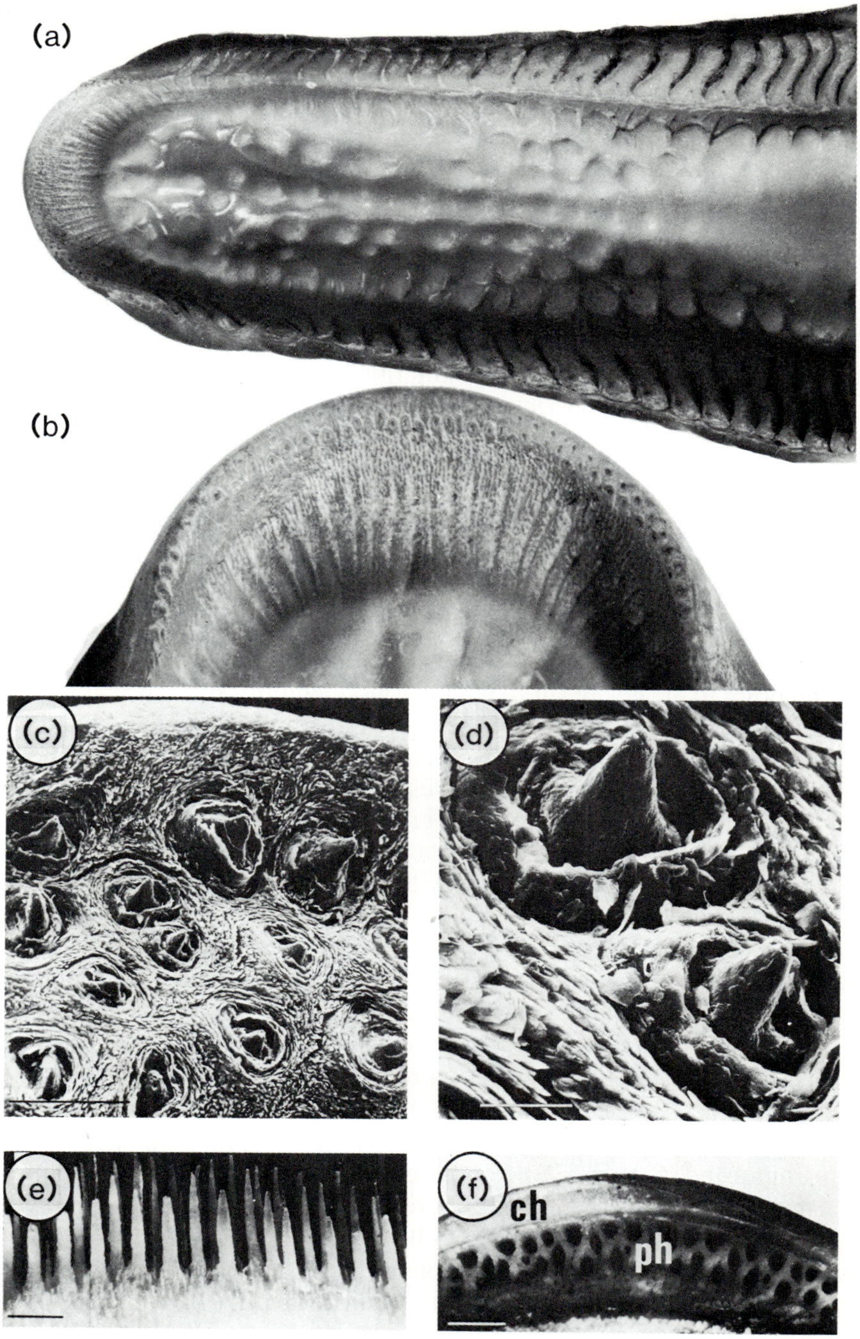

Fig. 8.34. Bill tip organ in the upper beak of the goose. (a) Palatine aspect of the upper beak. (b) Bill tip shown at higher magnification revealing several rows of small tubules opening along the underside of the curvature of the rostral rim of the upper beak. (c) and (d) Scanning electron micrographs showing the conical tips of the touch papillae protruding from the openings of the horny tubules. Calibration bar in (c) is 100 μm; in (d) 25 μm. (e) Isolated dermal cores of touch papillae, after removal of the horny rim of the nail. (f) The horny rim of the nail after it has been removed, showing the openings of the tubules of papillary horn which had previously contained the dermal cores exposed in (e). ch, Hard superficial horn of the nail; ph, soft papillary horn. Calibration bar in (e) and (f): 1 mm.

plate-like structure (Fig. 8.34a,b). Hidden underneath the hard outer horn of the surface of the nail, and indeed completely enclosed within horny tissue, are many cylindrical "touch papillae" which protrude from the deep dermis towards the rostral rim of the beak where their tips reach the free surface (Fig. 8.35). Thus, each touch papilla consists of a dermal core which, of course, arises from the dermis underlying the bill; however, it is covered by a tubular horny coat which is strictly epidermal in origin. This horny coat is relatively soft, and is known as papillary horn. In the upper bill of the goose the distal ends of the touch papillae can just be recognised at low magnification as tiny surface irregularities occurring in one or several rows along the underside of the curved rim (Fig. 8.34a). At higher magnification these surface irregularities can be resolved as minute crater-like structures from which project the distal tips of the touch papillae (Fig. 8.34b).

The horny tissue derives from two sources. The hard superficial horn of the nail is produced by the epidermis of the bill and grows towards the rim (cutting edge) of the bill tip (Menzel and Lüdicke, 1974). The softer papillary horn is produced by the keratinizing epidermis which covers the dermal core. Thus, each touch papilla has a small, soft, horny tubule around itself; this tubule is capped distally by a tiny cone, covered with keratinized epithelium (Fig. 8.35b), which protrudes from the opening on the surface of the nail (Fig. 8.34c,d) of a minute cylindrical channel (Fig. 8.35b). The cones are moveable in both the transverse and longitudinal directions. By incising (in fixed material) the horny substance at the base of the bill tip organ, the rostral rim of the nail can be pulled away and the arrays of isolated dermal cores of the touch papillae are then exposed (Fig. 8.34e), as are the openings of the corresponding tubules of papillary horn in which the dermal cores had previously been contained (Fig. 8.34f).

The number, size and shape of individual touch papillae varies between species, in the upper and lower beak, and with their site on the rim of the bill. In the upper bill tip organ of the goose, up to three rows of touch papillae occur in the central portion but only one row remains at the sides. Centrally, the papillae measure up to 2·5 mm in length and 100–200 μm in diameter. In the lateral parts of the bill tip organ the papillae become shorter (about 0·5 mm) and thicker (250 μm). In the lower bill tip organ of the goose there are up to five rows of papillae. Generally, the lower bill tip organ has more papillae than the one in the upper bill. For instance, in the domestic goose 100 and 180 papillae occur in the upper and lower bill tip organ, respectively. In the Mallard (*Anas platyrhynchos*) the corresponding numbers are only 33 and 180, because the nail of the upper bill is much smaller than in geese. In the chicken there are about 20 papillae in the lower but none in the upper bill. In the lower bill of parrots the papilla system is well developed (Goujon, 1869), but the number of papillae and their internal organisation are unknown (Lüdicke, 1940; Menzel and Lüdicke, 1974).

Each touch papilla is innervated by a bundle of between 30 and 100 myelinated axons and also by some unmyelinated nerve fibres. About 80% of the myelinated axons are larger than 3 μm in diameter, and terminate on different kinds of mechanoreceptors which are concentrated in the distal (peripheral, superficial) two-thirds of each papilla. In the proximal (central, deep) parts of the papillae only Herbst corpuscles are found, whereas in their distal parts there are only terminal cell receptors (Fig. 8.35a,b). The schematic drawing of a single papilla from a duck's bill tip organ in Fig. 8.35b shows only Grandry corpuscles in the

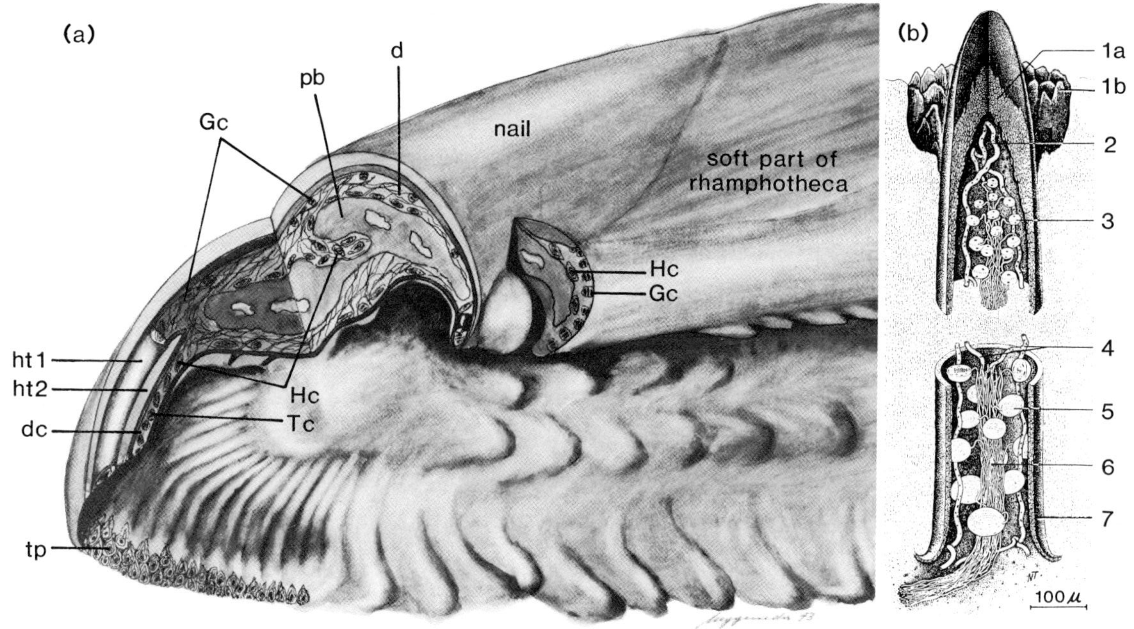

Fig. 8.35. (a) Schematic drawing of the bill tip organ in the upper beak of the goose. A segment has been cut out of the beak tip to show the course of the touch papillae in longitudinal section. A second segment has been cut out of the thickened ridge forming the left lateral edge of the beak. ht1, Outer surface of the horny tubule of a touch papilla. ht2 and dc belong to a touch papilla, which has been cut longitudinally; ht2 is the tubule of soft papillary horn of this touch papilla, and dc is the dermal core. tp, Conical tip of touch papilla; Gc, Grandry corpuscles; Hc, Herbst corpuscles; Tc, terminal cell receptors; pb, premaxillary bone; d, dermis. (b) Drawing of a single touch papilla embedded in the hard horn of the bill tip organ of the duck. 1a, Conical tip of the papilla, capped by keratinized epithelium; 1b, opening of the horny tubule, from which the conical tip of the papilla protrudes; 2, apical vascular loop; 3, Grandry corpuscle, shown as an example of a terminal cell receptor; 4, arteriovenous anastomoses; 5, Herbst corpuscle; 6, bundle of nerve fibres innervating the papilla; 7, wall of the tubule of soft papillary horn. From Berkhoudt (1980), by permission.

distal part of the papilla; however, as in the goose, terminal cell column receptors and branched nerve endings with terminal cells may also be present there. In the goose, the latter two receptor complexes have, in fact, been found only in the distal part of touch papillae but not in the rhamphothecal dermis. Although in the goose all four different types of mechanoreceptors are always present in a given papilla, their relative number varies considerably and does not seem to be related to either the size of the papilla or its place within the bill tip organ.

The Herbst corpuscles in the touch papillae have a preferential orientation with their longer axis lying parallel to the long axis of the papilla. However, Herbst corpuscles with other orientations can also be found. Therefore, it appears that the preferential orientation of the oval receptors is not an adaptation to functional requirements but a consequence of the limited space available in the papilla. In Grandry corpuscles, the orientation of the disc-like nerve endings relative to the epidermis differs from that in the rhamphothecal dermis. In the latter, the nerve endings always lie parallel to the epidermis, whereas in touch papillae they lie at right angles to it (Fig. 8.17b). Thus, the orientation of the nerve terminal discs is in general parallel to the surface from which mechanical stimuli are transmitted, that is, in case of the touch papillae, the distal ends of the papillae.

Microstimulation under microscopic control shows (Gottschaldt, unpublished observation) that in the ophthalmic nerve of the goose a single nerve fibre innervates only one touch papilla. A given receptor is activated by either longitudinal displacements of the touch papilla or transverse deflections of the cones protruding from the opening of the horny tubules. The mechanical thresholds of individual receptors is very low if the papilla is stimulated directly, whereas fairly strong stimuli are required for eliciting responses from neighbouring touch papillae or the surrounding horn.

The excellent mechanical insulation between neighbouring papillae should greatly increase the tactile discriminatory capacity of the bill tip organ. No less important, however, must be the extraordinary receptor density in this complex sensory structure. With up to 25 papilla openings per square millimetre of surface area, and at least 40 receptors in each papilla, more than 1000 receptors may be activated from 1 mm^2 of the rim of the goose's bill. This high receptor density in the bill tip organ is reflected in the disproportionally large representation area of the bill tip organ in the forebrain of geese and ducks (Gottschaldt *et al.*, 1980; Berkhoudt *et al.*, 1981).

In the total afferent outflow from the bill tip organ, all those mechanical stimulus parameters are encoded which may occur during tactile exploration of objects and surface textures. Thus, the Herbst corpuscles detect high velocity and vibratory stimulus components, the Grandry corpuscles and terminal cell column receptors encode the velocity components of mechanical stimuli, and the branched nerve endings with terminal cells presumably measure the static stimulus components. Probably there are additional receptors in the touch papillae for thermal and painful stimuli; electro-physiological evidence is lacking, but the nerve endings in the papilla epithelium (Gottschaldt *et al.*, 1976; Fig. 8.32b) offer an appropriate morphological basis for such receptors.

Touch papillae also possess a special vascular system (Gottschaldt *et al.*, 1976; Berkhoudt, 1980). Usually, one artery ascends in the centre of the papilla towards its tip, whereas the blood is drained off through a venous plexus located below the epithelium. The blood flow thus resembles the water flow in a fountain. There are numerous arteriovenous anastomoses (Fig. 8.35b) with many catecholaminergic nerve endings. This elaborate vascular system may have a twofold significance for the sensory function of the bill tip organ. Firstly, by adjusting the blood supply the internal turgor of the papillae can possibly be varied, and this should affect the sensitivity of transmission of mechanical stimuli through the connective tissue core of the papillae and to the different types of mechanoreceptors. Secondly, the bill tip must experience considerable temperature changes, which would affect the function of the receptors unless a thermal homeostasis can be maintained. Since the receptors are located inside the arteriovenous "fountain", temperature gradients from the outside to the inside may be considerably diminished by appropriate vasomotor control within the papillae.

VIII. Concluding remarks

Upon completion of this survey of the morphology and physiological properties of avian somatosensory receptors, one may take stock of the actual state of knowledge of these receptors and the principal insights which have been gained about their structure–function relationships. The greatest shortcoming undoubtedly is that adequate knowledge is entirely restricted to mechanoreceptors supplied by large-diameter myelinated afferent nerve fibres. Electro-physiologically, this is inevitable because the activity in large-diameter fibres can be studied much more easily than that in small ones. Morphologically, the reason is that large fibres usually terminate in more-or-less elaborate structures which can be recognised with both the light and electron microscope much more easily than the terminals of thin afferent nerve fibres. Electron microscopic studies have shown that the actual nerve endings in different morphological types of mechanoreceptors have so many ultrastructural characteristics in common that these features alone cannot explain their different functional properties. The most significant morphological criterion for determining the functional properties of a mechanoreceptor appears to be the way in which the sensory nerve endings are *directly connected* to non-nervous receptor structures deriving from both the nervous sheath cells and connective tissue elements. On the other hand, the size, shape or location of a receptor do not seem to be systematically related to its functional properties. Also, the individual components of the auxiliary receptor structures do not have the same significance for a receptor's function. For instance, in morphological receptor classifications much emphasis has been placed on whether a given receptor is encapsulated or not. It now appears that, with respect to function, the presence or absence of an outer capsule is perhaps the least significant structural component of a receptor. Thus, both Herbst corpuscles and, say, muscle spindles have a well-developed capsule, but their response properties are very different. Conversely, Ruffini endings in the dermis of the beak of geese lack a distinct capsular structure and yet have very similar response properties to the encapsulated Golgi tendon organs.

Unique to birds is the morphological variety of cutaneous receptors incorporating specialised terminal cells. Arguments have been put forward suggesting that the terminal cells are not directly involved in the mechano-electric transduction process of these receptors, and it is also certain that single types of terminal cell receptors may have quite different response properties. Nevertheless, the clarification of the precise rôle played by the terminal cells in the receptor function, of their cytological origin and, in particular, the chemical content of their mysterious dense-cored granules are major questions for future research. An even greater challenge is the need to elucidate the structure and function of somatosensory receptors supplied by thin myelinated and unmyelinated nerve fibres.

Peripheral receptors are the first link in a chain of neurons subserving somatic sensation. Compared with mammals, relatively little is known about the functional and structural organisation of the avian central somatosensory system (Necker, 1981), nor can we really appreciate what kind of skin sensation a bird may consciously experience. However, in several instances a relationship between the properties of cutaneous receptors and a distinct behavior can be suggested. For instance, mechanoreceptors associated with the breast feathers of siskins (Fringillidae), and perhaps also in other birds, function as an air-current sense organ in the control of flight behavior (Gewecke and Woike, 1978; Woike and Gewecke, 1978). When freely suspended in a harness, siskins not only initiate flight movements when air is blown against their breast feathers, but also adjust the thrust and the direction of the flight movements according to the strength and direction of the airstream.

Another example of interaction between mechanoreceptor function and a distinct behavior is found in geese and ducks. These animals perform fast mandibular movements in order to discriminate edible from non-edible items (Zweers *et al.*, 1977). The vast majority of mechanoreceptors in the oral cavity respond only to the first- or second-order derivative of mechanical displacements, that is, to the movement components of the stimuli. Since non-living food particles do not move, the birds have to move them themselves by straining water through the beak and by jaw movements, thereby providing the necessary stimulus components to which the mechanoreceptors can respond. Further details about the rôle of intraoral receptors in the control of feeding actions are discussed by H. Berkhoudt in Chapter 9 of this volume.

Finally, the horny material enclosing the bill tip organ grows continuously and would certainly over-grow the openings of the touch papillae unless it were constantly abraded. Some avian species, like parrots and other psittaciforms, achieve this by rubbing lower and upper beak against each other. Therefore, the functional significance of this typical behavior not only "consists in a modelling of the blunt end of the mandible" (Homberger and Ziswiler, 1972) but undoubtedly is also necessary to maintain the optimal function of the bill tip organ. Geese normally seem to abrade excessive horn substance by scratching the beak on the ground while grasping food. If they are kept imprisoned under conditions where this is prevented, they start scratching the beak tip on rough surfaces such as the brick walls of their cage and, in the course of some weeks, may even seriously damage the building (personal observation)!

Acknowledgements

The major part of the material, presented in this article for the first time, is derived from my Habilitation thesis (Göttingen, 1980). I would like to thank I.

Kräft, J. Lübke and J. Mick for their support and participation in this study. I am indebted to Miss A. Müller for expertly preparing the illustrations, to Mrs. B. Scheufler for patiently assembling the reference list, to Miss M. Knudson and G. Müller for attentive typing of the manuscript and to D. Leong and H. Kräft for helpful suggestions and critical reading. The encouraging support of C. Vahle-Hinz during all phases of this study was particularly valuable. The work was supported by the SFB 33 of the Deutsche Forschungsgemeinschaft.

Editors' note: We feel sure that Kay would have wished us to make known the expert scientific assistance which was generously given to us by Dr. C. Vahle-Hinz in resolving the complications of editing the chapter after the author's death.

References

Abraham, A. (1978). Ultrastrukturelle Untersuchungen an der Wachshaut von Enten unter besonderer Berücksichtigung der Grandryschen und Herbstschen Körper. *Z. mikrosk.-anat. Forsch.* **92,** 81–118.

Adal, M. N. (1973). The fine structure of the intrafusal muscle fibres of muscle spindles in the domestic fowl. *J. Anat.* **115,** 407–413.

Adal, M. N. and Chew Cheng, S. B. (1980a). The number, distribution and density of muscle spindles in two wing muscles of the domestic duck. *J. Anat.* **131,** 541–548.

Adal, M. N. and Chew Cheng, S. B. (1980b). The sensory ending of duck muscle spindles. *J. Anat.* **131,** 657–668.

Andersen, A. E. and Nafstad, P. H. J. (1968). An electron microscopic investigation of the sensory organs in the hard palate region of the hen. *Z. Zellforsch. mikrosk. Anat.* **91,** 391–401.

Andres, K. H. (1966). Über die Feinstruktur der Rezeptoren an Sinushaaren. *Z. Zellforsch. mikrosk. Anat.* **75,** 339–365.

Andres, K. H. (1969). Zur Ultrastruktur verschiedener Mechanorezeptoren von höheren Wirbeltieren. *Anat. Anz.* **124,** 551–565.

Andres, K. H. (1974). Morphological criteria for the differentiation of mechanoreceptors in vertebrates. *In* "Mechanoreception" (J. Schwartzkopff, Ed.). Proc. Rhein. Westf. Akad. Wiss., Vol. 53. Westdeutscher Verlag, Opladen.

Andres, K. H. and von Düring, M. (1973). Morphology of cutaneous receptors. *In* "Handbook of Sensory Physiology" (A. Iggo, Ed.), Vol. II. Springer-Verlag, Berlin and New York.

Andrew, B. L. (1954). The sensory innervation of the medial ligament of the knee joint. *J. Physiol., Lond.* **123,** 241–250.

Bannister, L. H. (1976). Sensory terminals of peripheral nerves. *In* "The Peripheral Nerve" (D. N. Landon, Ed.). Chapman and Hall, London.

Barker, D. (1974). The morphology of muscle receptors. *In* "Handbook of Sensory Physiology" (C. C. Hunt, Ed.), Vol. III/2. Springer-Verlag, Berlin and New York.

Bavaria, Ludwig Ferdinand of (1884). Über Endorgane der sensiblen Nerven in der Zunge der Spechte. *Sitzungsber. Kgl. bayer. Akad. Wiss., math.-physik. München* pp. 183–192.

Berkhoudt, H. (1976). The epidermal structure of the bill tip organ in ducks. *Netherl. J. Zool.* **26,** 561–566.

Berkhoudt, H. (1980). The morphology and distribution of cutaneous mechanoreceptors (Herbst and Grandry corpuscles) in bill and tongue of the mallard (*Anas platyrhynchos* L.). *Netherl. J. Zool.* **30,** 1–34.

Berkhoudt, H., Dubbeldam, J. L. and Zeilstra, S. (1981). Studies on the somatotopy of the trigeminal system in the mallard (*Anas platyrhynchos* L.) IV. Tactile representation in the nucleus basalis. *J. comp. Neurol.* **196,** 407–420.

Boldt, W. and Biewald, G.-A. (1971). Impulsmuster kutaner Mechanorezeptoren des Feldhamsters (*Cricetus cricetus* L.) bei verschiedenen Hauttemperaturen. *Zool. Jb., Abt. allg. Zool. Physiol.* **75,** 479–492.

Bolton, T. B. (1971). The structure of the nervous system. *In* "Physiology and Biochemistry of the Domestic Fowl" (D. J. Bell and B. M. Freeman, Eds.), Vol. 2. Academic Press, London and New York.

Bolze, G. (1968). Anordnung und Bau der Herbstschen Körperchen in Limicolenschnäbeln im Zusammenhang mit der Nahrungsfindung. *Zool. Anz.* **181,** 313–355.

Booth, C. S. and Hahn, J. F. (1974). Thermal and mechanical stimulation of type II receptors and field receptors in the cat. *Exp. Neurol.* **44,** 49–50.

Botezat, E. (1906). Die Nervenendapparate in den Mundteilen der Vögel und die einheitliche Endigungsweise der peripheren Nerven bei den Wirbeltieren. *Z. wiss. Zool.* **84,** 205–360.

Boyd, I. A. (1954). The histological structure of the receptors on the knee joint of the cat correlated with their physiological response. *J. Physiol., Lond.* **124,** 476–488.

Boyd, I. A. and Roberts, T. D. (1953). Proprioceptive discharges from stretch-receptors in the knee-joint of the cat. *J. Physiol., Lond.* **122,** 38–58.

Bridgman, C. F. (1968). The structure of tendon organs in the cat: A proposed mechanism for responding to muscle tension. *Anat. Rec.* **162,** 209–220.

Burgess, P. R. and Perl, E. R. (1973). Cutaneous mechanoreceptors and nociceptors. *In* "Handbook of Sensory Physiology" (A. Iggo, Ed.), Vol. II. Springer-Verlag, Berlin and New York.

Callanan, D., Dixon, M., Widdicombe, J. G. and Wise, J. C. M. (1974). Responses of geese to inhalation of irritant gases and injections of phenyl diguanide. *Resp. Physiol.* **22,** 157–166.

Carrière, J. (1882). Kurze Mitteilungen zur Kenntnis der Herbstschen und Grandryschen Körperchen in dem Schnabel der Ente. *Arch. mikrosk. Anat.* **21,** 146–164.

Cauna, N. (1966). Fine structure of the receptor organs and its probable functional significance. *In* "Touch, Heat and Pain" (A. V. S. de Reuck and J. Knight, Eds.). Churchill, London.

Chambers, M. R., Andres, K. H., von Düring, M. and Iggo, A. (1972). Structure and function of the slowly adapting type II mechanoreceptor in hairy skin. *Q. J. exp. Physiol.* **57,** 417–445.

Chen, S. Y., Gerson, S., and Meyer, J. (1973). The fusion of Merkel cell granules with a synaptic-like structure. *J. invest. Derm.* **61,** 290–292.

Chouchkov, Ch. (1971). Ultrastructure of Pacinian corpuscles in men and cats. *Z. mikrosk.-anat. Forsch.* **83,** 17–32.

Chouchkov, Ch. (1973). On the ultrastructure of the Herbst corpuscles in the bill skin of the Pekin duck. *C. R. Acad. Bulg. Sci.* **26,** 1705–1708.

Chouchkov, Ch. (1974). On the ultrastructure of the Grandry's corpuscles in the bill skin of the Pekin duck. *C. R. Acad. Bulg. Sci.* **27,** 129–132.

Chouchkov, Ch. (1976). Ultrastructural differences between the preterminal nerve fibres and their endings in the mechanoreceptors, with special reference to their degeneration and mode of uptake of horseradish peroxidase. *In* "Progress in Brain Research, Somatosensory and Visceral Receptor Mechanisms" (A. Iggo and O. B. Ilynski, Eds.). Elsevier, Amsterdam.

Chouchkov, Ch. (1978). Cutaneous Receptors. *Adv. Anat. Embryol. Cell Biol.* **54,** 1–62.

Chouchkov, Ch. (1979a). Degeneration and regeneration of some mechanoreceptors. An ultrastructural study. I. Ultrastructure of denervated Herbst corpuscles. *Z. mikrosk.-anat. Forsch.* **93,** 353–364.

Chouchkov, Ch. (1979b). Degeneration and regeneration of some mechanoreceptors. An ultrastructural study. III. Ultrastructure of reinnervated Herbst corpuscles. *Z. mikrosk.-anat. Forsch.* **93,** 425–432.

Clara, M. (1925). Über den Bau des Schnabels der Waldschnepfe. Beiträge zur Kenntnis der Lamellenkörperchen. *Z. mikrosk.-anat. Forsch.* **3,** 1–108.

Cobb, J. L. S. and Bennett, T. (1970). Herbst corpuscles in the smooth muscles in the wings of chicks. *Experientia* **26,** 768–769.

de Anda, G. and Rebollo, M. A. (1967). The neuromuscular spindles in the adult chicken. *Acta anat.* **67,** 437–451.

de Iraldi, A. P. and Rodriguez-Perez, A. P. (1961). Ultrastructura de los corpusculos de Grandry. *Trab. Inst. Cajal Invest. biol.* **53,** 185–188.

deWet, P. D., Farrell, P. R. and Fedde, M. R. (1971). Number and morphology of muscle spindles in the transversus abdominis muscle of the chicken. *Poult. Sci.* **50,** 1349–1357.

Dogiel, A. S. (1904). Über die Nervenendigungen in den Grandryschen und Herbstschen Körperchen in Zusammenhang mit der Frage der Neuronentheorie. *Anat. Anz.* **25,** 558–574.

Dorward, P. (1970a). Response patterns of cutaneous mechanoreceptors in domestic duck. *Comp. Biochem. Physiol.* **35,** 729–735.

Dorward, P. (1970b). Response characteristics of muscle afferents in the domestic duck. *J. Physiol., Lond.* **211,** 1–17.

Dorward, P. and McIntyre, A. K. (1971). Responses of vibration-sensitive receptors in the interosseus region of the duck's hind limb. *J. Physiol., Lond.* **219,** 77–87.

Duclaux, R. and Kenshalo, D. R. (1972). The temperature sensitivity of the type I slowly adapting mechanoreceptors in cats and monkeys. *J. Physiol., Lond.* **224,** 647–664.

English, K. B. (1974). Cell types in cutaneous type I mechanoreceptors (Haarscheiben) and their alterations with injury. *Am. J. Anat.* **141,** 105–126.

English, K. B. (1977). The ultrastructure of cutaneous type I mechanoreceptors (Haarscheiben) in cats following denervation. *J. comp. Neurol.* **172,** 137–163.

Gardner, E. (1944). The distribution and termination of nerves in the knee-joint of the cat. *J. comp. Neurol.* **80,** 11–32.

Germino, N. I. and D'Albora, H. (1965). Succinic-dehydrogenase activity in the neuromuscular spindles of the chick. *Experientia* **21,** 45–46.

Gewecke, M. and Woike, M. (1978). Breast feathers as an air-current sense organ for the control of flight behaviour in a songbird (*Carduelis spinus*). *Z. Tierphysiol.* **47,** 293–298.

Gogli, C. (1880). Sui nervi dei tendini delluomo e di un nuovo organo nervoso terminale musculo-tendineo. *Mem. R. Acad. Sci. Torino* **32,** 359–385.

Goglia, C. (1969a). Ricerche elettron-microscopiche ed istochimiche sopra i corpuscoli di Herbst del becco dell'anitra prima e dopo stimulazione meccanica e prima e dopo resezione delle fibre sensitive. *Acta med. rom.* **7,** 35–58.

Goglia, C. (1969b). Ricerche elettron-microscopiche ed istochimiche sopra i corpuscoli di Grandry del becco dell'anitra prima e dopo alcuni interventi sperimentali. *Acta med. rom.* **7,** 59–77.

Goglia, C. and Sklenská, A. (1969). Ricerche ultrastructurali sopra i corpuscoli di Ruffini delle capsule articolari nel coniglio. *Quaderni Anat. pratica* **25,** 14–27.

Gottschaldt, K.-M. (1974a). The physiological basis of tactile sensibility in the beak of geese. *J. comp. Physiol.* **95,** 29–47.

Gottschaldt, K.-M. (1974b). Mechanoreceptors in the beaks of birds. *In* "Mechanoreception" (J. Schwartzkopff, Ed.). Proc. Rhein. Westf. Akad. Wiss. Vol. 53. Westdeutscher Verlag, Opladen.

Gottschaldt, K.-M. and Fakoya, A. (1975). Responses from a mechanoreceptor after antidromic stimulation of its afferent nerve fibre. *Neurosci. Abstr.* **1,** 133.

Gottschaldt, K.-M. and Fakoya, A. (1977). Backfiring in trigeminal afferent nerve fibres of the goose elicited by antidromic stimulation. *In* "Pain in the Trigeminal Region" (D. J. Anderson and B. Matthews, Eds.). Elsevier, North-Holland, Amsterdam.

Gottschaldt, K.-M. and Kräft, I. (1978). A new hypothesis about the receptor mechanisms in avian terminal cell receptors. *J. Physiol., Lond.* **284,** 67–68.

Gottschaldt, K.-M. and Lausmann, S. (1974). The peripheral morphological basis of tactile sensibility in the beak of geese. *Cell. Tiss. Res.* **153,** 477–496.

Gottschaldt, K.-M. and Vahle-Hinz, C. (1981). Merkel cell receptors: structure and transducer function. *Science, Wash., D.C.* **214,** 183–186.

Gottschaldt, K.-M. and Vahle-Hinz, C. (1982). Evidence against transmitter function of met-enkephalin and chemosynaptic impulse generation in "Merkel cell" mechanoreceptors. *Exp. Brain Res.* **45,** 459–463.

Gottschaldt, K.-M. and Young, D. W. (1977). Properties of different functional types of neurones in the cat's rostral trigeminal nuclei responding to sinus hair stimulation. *J. Physiol., Lond.* **272,** 57–84.

Gottschaldt, K.-M., Iggo, A. and Young, D. W. (1973). Functional characteristics of mechanoreceptors in sinus hair follicles of the cat. *J. Physiol., Lond.* **235,** 287–315.

Gottschaldt, K.-M., Andres, K. H. and von Düring, M. (1976). Fine structure and function of the bill tip organ in geese. *Neurosci. Abstr.* **2,** 1912.

Gottschaldt, K.-M., Mick, J. and Kräft, I. (1979). A morphological investigation of the quantitative and qualitative ontogenetic development of the tactile receptor system in the beak of geese. *Pflügers Arch. ges. Physiol.* **382,** Suppl., R51.

Gottschaldt, K.-M., Veenman, C. L., and Steindler, D. A. (1980). Projection of tactile beak afferents into the forebrain of the goose. *Proc. Int. Union Physiol. Sci.* **14,** 441

Gottschaldt, K.-M., Fruhstorfer, H., Schmidt, W. and Kräft, I. (1982). Thermosensitivity and its possible fine-structural basis in mechanoreceptors in the beak skin of geese. *J. comp. Neurol.* **205,** 219–245.

Goujon, E. (1869). Sur un appareil de corpuscles tactiles situé dans le bec des perroquets. *J. Anat. Physiol.* **6,** 449–455.

Grandry, M. (1869). Recherches sur les corpuscules de Pacini. *J. Anat. Physiol.* **6,** 390–395.

Gray, J. A. B. and Matthews, P. B. C. (1951). A comparison of the adaptation of the Pacinian corpuscle with the accommodation of its own axon. *J. Physiol., Lond.* **114,** 454–464.

Gregory, J. E. (1973). An electrophysiological investigation of the receptor apparatus in the duck's bill. *J. Physiol., Lond.* **229,** 151–164.

Haiden, G. J. and Awad, E. A. (1981). The ultrastructure of the avian Golgi tendon organ. *Anat. Rec.* **200,** 153–161.

Halata, Z. (1970). Zu den Nervenendigungen (Merkelsche Endigungen) in der haarlosen Nasenhaut der Katze. *Z. Zellforsch. mikrosk. Anat.* **106,** 51–60.

Halata, Z. (1971a). Die Ultrastruktur der Lamellenkörperchen bei Wasservögeln (Herbstsche Endigungen). *Acta anat.* **80,** 362–371.

Halata, Z. (1971b). Ultrastructure of Grandry nerve endings in the beak skin of some aquatic birds. *Folia morph., Prague* **3,** 225–232.

Halata, Z. (1975). The mechanoreceptors of the mammalian skin. Ultrastructure and morphological classification. *Adv. Anat. Embryol. Cell Biol.* **50,** 1–77.

Halata, Z. (1977). The ultrastructure of the sensory nerve endings in the articular capsule of the knee joint of the domestic cat (Ruffini corpuscles and Pacinian corpuscles). *J. Anat.* **124,** 717–729.

Halata, Z. and Munger, B. L. (1980). The ultrastructure of Ruffini and Herbst corpuscles in the articular capsule of the domestic pigeon. *Anat. Rec.* **198,** 681–692.

Hartschuh, W. and Weihe, E. (1980). Fine structural analysis of the synaptic junction of Merkel cell–axon complexes. *J. invest. Derm.* **75,** 159–165.

Hensel, H. (1973). Cutaneous thermoreceptors. *In* "Handbook of Sensory Physiology" (A. Iggo, Ed.), Vol. II. Springer-Verlag, Berlin and New York.

Hensel, H., Andres, K. H. and von Düring, M. (1974). Structure and function of cold receptors. *Pflügers Arch. ges. Physiol.* **352,** 1–10.

Herbst, G. (1848). "Die Pacinischen Körper und ihre Bedeutung. Ein Beitrag zur Kenntnis der Nervenprimitivfasern." Vandenhoeck and Ruprecht, Göttingen.

Hesse, F. (1878). Über die Tastkugeln des Entenschnabels. *Arch. Anat. Entwgesch.* pp. 288–318.

Homberger, D. G. and Ziswiler, V. (1972). Funktionell-morphologische Untersuchungen am Schnabel von Papageien. *Rev. Suisse Zool.* **79,** 1038–1048.

Horch, K. W., Whitehorn, D. and Burgess, P. R. (1974). Impulse generation in type I cutaneous mechanoreceptors. *J. Neurophysiol.* **37,** 367–381.

Horch, K. W., Tuckett, R. P. and Burgess, P. R. (1977). A key to the classification of cutaneous mechanoreceptors. *J. invest. Derm.* **69,** 75–82.

Hubbard, S. J. (1958). A study of rapid mechanical events in a mechanoreceptor. *J. Physiol., Lond.* **141,** 198–218.

Huber, G. C. and de Witt, L. M. (1897). A contribution on the motor nerve endings and on the nerve endings in the muscle spindles. *J. comp. Neurol.* **7,** 169–230.

Huber, G. C. and de Witt, L. M. (1900). A contribution on the nerve terminations in neuro-tendinous end-organs. *J. comp. Neurol.* **10,** 159–208.

Hunt, C. C. (1974). The Pacinian corpuscle. *In* "The Peripheral Nervous System" (J. I. Hubbard, Ed.). Plenum Press, New York and London.

Idé, C. and Munger, B. L. (1978). A cytological study of Grandry corpuscle development in chicken toe skin. *J. comp. Neurol.* **179,** 301–324.

Iggo, A. (1963). New specific sensory structures in hairy skin. *Acta Neuroveg., Wien* **24,** 175–180.

Iggo, A. (1966). Discussion remark on page 115 to the paper of T. A. Quilliam: Unit design and array patterns in receptor organs. *In* "Touch, Heat and Pain" (A. V. S. De Reuck and J. Knight, Eds.). Churchill, London.

Iggo, A. (1974). Cutaneous receptors. *In* "The Peripheral Nervous System" (J. I. Hubbard, Ed.). Plenum Press, New York and London.

Iggo, A. and Andres, K. H. (1982). Morphology of cutaneous receptors. *A. Rev. Neurosci.* **5,** 1–33.

Iggo, A. and Gottschaldt, K.-M. (1974). Cutaneous mechanoreceptors in simple and in complex sensory structures. *In* "Mechanoreception" (J. Schwartzkopff, Ed.). Proc. Rhein. Westf. Akad. Wiss., Vol. 53. Westdeutscher Verlag, Opladen.

Iggo, A. and Muir, A. R. (1969). The structure and function of a slowly adapting touch corpuscle in hairy skin. *J. Physiol., Lond.* **200,** 736–796.

Ihlder, H. (1870). Die Nervenendigung in der Vogelzunge. *Arch. Anat. Physiol. wiss. Med.* 238–250.

James, N. T. and Meek, G. A. (1973). An electron microscopical study of avian muscle spindles. *J. ultrastruct. Res.* **43,** 193–204.

Jobert, M. (1872). Recherches sur la structure intime du bec de la spatule. *C. R. Hebd. Séanc. Acad. Sci., Paris* **75,** 1780–1782.

Kasprzak, H., Tapper, D. N. and Craig, P. H. (1970). Functional development of the tactile pad receptor system. *Exp. Neurol.* **26,** 439–446.

Katz, B. (1950). Depolarization of sensory terminals and the initiation of impulses in the muscle spindle. *J. Physiol., Lond.* **111,** 261–282.

Key, A. and Retzius, G. (1876). "Studien in der Anatomie des Nervensystems und des Bindegewebes", 2. Hälfte. Samson and Wallin, Stockholm.

King, A. S., McLelland, J., Cook, R. D., King, D. Z. and Walsh, C. (1974). The ultrastructure of afferent nerve endings in the avian lung. *Resp. Physiol.* **22,** 21–40.

Kitchell, R. L., Strom, L. and Zotterman, Y. (1959). Electrophysiological studies of thermal and taste reception in chickens and pigeons. *Acta physiol. scand.* **46,** 133–151.

Knibestöll, M. (1973). Stimulus–response functions of rapidly adapting mechanoreceptors in the human glabrous skin area. *J. Physiol., Lond.* **232,** 427–452.

Krause, W. (1881). Die Nervenendigungen innerhalb der terminalen Körperchen. *Arch. mikrosk. Anat.* **19,** 53–136.

Krogis, A. (1931). On the topography of Herbst and Grandry's corpuscles in the adult and embryonic duck bill. *Acta zool., Stockh.* **12,** 241–276.

Krulis, V. (1978). Struktur und Verteilung von Tastrezeptoren im Schnabel-Zungenbereich von Singvögeln, im besonderen der Fringillidae. *Revue Suisse Zool.* **85,** 385–447.

Lange, B. (1931). Integument der Sauropsiden. *In* "Handbuch der vergleichenden Anatomie der Wirbeltiere" (L. Bolk, E. Goppert, E. Kallius and W. Lubosch, Eds.) Vol. II. Urban and Schwarzenberg, Berlin.

Lawrentjew, B. J. (1926). Über das Chondriom der Grandryschen Körperchen. *Z. mikrosk.-anat. Forsch.* **6,** 241–255.

Leitner, L. M. and Roumy, M. (1974a). Mechanosensitive units in the upper bill and in the tongue of the domestic duck. *Pflügers Arch. ges. Physiol.* **346,** 141–150.

Leitner, L. M. and Roumy, M. (1974b). Thermosensitive units in the tongue and in the skin of the duck's bill. *Pflügers Arch. ges. Physiol.* **346,** 151–155.

Leydig, F. (1868). Über den Bau, insbesondere der Vaterschen Körper, des Schnabels der Schnepfe. *Arch. mikrosk. Anat.* **4,** 195–205.

Loewenstein, W. R. (1971). Mechano-electric transduction in the Pacinian corpuscle. Initiation of sensory impulses in mechanoreceptors. *In* "Handbook of Sensory Physiology, I. Principles of Receptor Physiology" (W. R. Loewenstein, Ed.), Vol. I. Springer-Verlag, Berlin and New York.

Loewenstein, W. R. and Mendelson, M. (1965). Components of receptor adaptation in a Pacinian corpuscle. *J. Physiol., Lond.* **177,** 377–397.

Loewenstein, W. R. and Rathkamp, R. (1958). The sites for mechanoelectric conversion in a Pacinian corpuscle. *J. gen. Physiol.* **41,** 1245–1265.

Lucas, A. M. (1979). *Integumentum commune. In* "Nomina Anatomica Avium" (J. J. Baumel, A. S. King, A. M. Lucas, J. E. Breazile and H. E. Evans, Eds.). Academic Press, London and New York.

Lucas, A. M. and Stettenheim, P. R. (1972). "Avian Anatomy. Integument", Agriculture Handbook 362, U.S. Department of Agriculture, Washington, D.C.

Lüdicke, M. (1940). Aufbau and Abnutzung der Hornzähne und Hornwülste des Vogelschnabels. *Z. Morph. Ökol. Tiere* **37,** 155–201.

Maier, A. and Eldred, E. (1971). Comparisons in the structure of avian muscle spindles. *J. comp. Neurol.* **143,** 25–40.

Malinovský, L. (1967). Die Nervenendkörperchen in der Haut von Vögeln und ihre Variabilität. *Z. mikrosk.-anat. Forsch.* **77,** 279–303.

Malinovský, L. and Páč, L. (1978). Ultrastructure of simple sensory corpuscles in the domestic pigeon. *Folia morph., Prague* **26,** 170–171.

Malinovský, L. and Páč, L. (1980). Ultrastructure of Herbst corpuscules from the beak skin of the pigeon. *Z. mikrosk.-anat. Forsch.* **94,** 292–304.

Malinovský, L. and Páč, L. (1982). Morphology of sensory corpuscles in mammals. *Acta Fac. med. Univ. Brunensis* **79.**

Malinovský, L. and Zemanek, R. (1969). Sensory corpuscles in the beak skin of the domestic pigeon. *Folia morph., Prague* **17,** 241–250.

Malinovský, L. and Zemanek, R. (1970). Sensory nerve endings in the joint-capsule of the large limb joints in the domestic hen (*Gallus domesticus*) and the rook (*Corvus frugilegus*). *Folia morph., Prague* **18,** 206–212.

Malinovský, L. and Zemanek, R. (1971). Sensory innervation of the skin and mucosa of some parts of the head in the domestic fowl. *Folia morph., Prague* **19,** 18–23.

Manni, E., Bortolami, R. and Azzena, G. B. (1965). Jaw muscle proprioception and mesencephalic trigeminal cells in birds. *Exp. Neurol.* **12,** 320–328.

Mendelson, M. and Loewenstein, W. R. (1964). Mechanisms of receptor adaptation. *Science, Wash., D.C.* **144,** 554–555.

Menzel, R. and Lüdicke, M. (1974). Funktionell-anatomische und autoradiographische Untersuchungen am Schnabelhorn von Papageien (Psittaci). *Zool. Jb., Abt. Anat. Ontog. Tiere* **93,** 175–218.

Merkel, F. (1875). Tastzellen und Tastkörperchen bei den Haustieren und beim Menschen. *Arch. mikrosk. Anat.* **11,** 636–652.

Merkel, F. (1878). Die Tastzellen der Ente. *Arch. mikrosk. Anat.* **15,** 415–427.

Merkel, F. (1880). "Über die Endigungen der sensiblen Nerven in der Haut der Wirbeltiere". Fues's Verlag, Leipzig.

Merrillees, N. C. R. (1962). Some observations on the fine structure of a Golgi tendon organ in the rat. *In* "Symposium on Muscle Receptors" (D. Barker, Ed.). University Press, Hong Kong.

Montagna, W. and Parakkal, P. F. (1974). "The Structure and Function of the Skin". Academic Press, New York and London.

Munger, B. L. (1965). The intraepidermal innervation of the snout skin of the opossum. A light and an electron microscopic study, with observations on the nature of Merkel's Tastzellen. *J. Cell Biol.* **26,** 79–96.

Munger, B. L. (1966). The ultrastructure of Herbst and Grandry corpuscles. *Anat. Rec.* **154,** 392–394.

Munger, B. L. (1971a). The comparative ultrastructure of slowly and rapidly adapting mechanoreceptors. *In* "Oral–Facial Sensory and Motor Mechanisms" (R. Dubner and Y. Kawamura, Eds.). Appleton-Century-Crofts, New York.

Munger, B. L. (1971b). Patterns of organization of peripheral receptors. *In* "Handbook of Sensory Physiology. Principles of Receptor Physiology" (W. R. Loewenstein, Ed.), Vol. I. Springer-Verlag, Berlin and New York.

Munger, B. L. (1977). Neural-epithelial interactions in sensory receptors. *J. invest. Derm.* **69,** 27–40.

Munger, B. L., Pubols, L. M. and Pubols, B. H. (1971). The Merkel rete papilla—a slowly adapting sensory receptor in mammalian glabrous skin. *Brain Res.* **29,** 47–61.

Nafstad, P. H. J. (1971a). Comparative ultrastructural study on Merkel cells and dermal basal cells in poultry (Gallus domesticus). *Z. Zellforsch. mikrosk. Anat.* **116,** 342–348.

Nafstad, P. H. J. (1971b). On the ultrastructure of neuro-epithelial interactions. The dermal innervations in the snout of the pig. *Z. Zellforsch. mikrosk. Anat.* **122,** 528–537.

Nafstad, P. H. J. and Andersen, A. E. (1970). Ultrastructural investigation of the innervation of the Herbst corpuscle. *Z. Zellforsch. mikrosk. Anat.* **103,** 109–114.

Nakajima, S. and Onodera, K. (1969a). Membrane properties of the stretch receptor neurones of crayfish with particular reference to mechanisms of sensory adaptation. *J. Physiol., Lond.* **200,** 161–185.

Nakajima, S. and Onodera, K. (1969b). Adaptation of the generator potential in the crayfish stretch receptors under constant length and constant tension. *J. Physiol., Lond.* **200,** 187–204.

Nakajima, S. and Takahashi, K. (1966). Post-tetanic hyperpolarization and electrogenic Na pump in stretch receptor neurone of crayfish. *J. Physiol., Lond.* **187,** 105–127.

Necker, R. (1972). Response of trigeminal ganglion neurons to thermal stimulation of the beak in pigeons. *J. comp. Physiol.* **78,** 307–314.

Necker, R. (1973). Temperature sensitivity of thermoreceptors and mechanoreceptors on the beak of pigeons. *J. comp. Physiol.* **87,** 379–391.

Necker, R. (1974a). Dependence of mechanoreceptor activity on skin temperature in sauropsids. II. Pigeon and duck. *J. comp. Physiol.* **92,** 75–83.

Necker, R. (1974b). Temperature sensitivity of slowly adapting mechanoreceptors in the beaks of pigeons. *In* "Mechanoreception" (J. Schwartzkopff, Ed.). Proc. Rhein. Westf. Akad. Wiss., Vol. 53. Westdeutscher Verlag, Opladen.

Necker, R. (1977). Thermal sensitivity of different skin areas in pigeons. *J. comp. Physiol.* **116,** 239–246.

Necker, R. (1981). Physiology and behavior of the pigeon somatosensory system. *In* "Progress in Sensory Physiology" (D. Ottoson, Ed.). Springer-Verlag, Berlin and New York.

Necker, R. and Reiner, W. (1980). Temperature-sensitive mechanoreceptors, thermoreceptors and heat nociceptors in the feathered skin of pigeons. *J. comp. Physiol.* **135,** 201–207.

Nishi, K., Oura, C. and Pallie. W. (1969). Fine structure of Pacinian corpuscles in the mesentery of the cat. *J. Cell Biol.* **43,** 539–552.

Ostmann, O. W., Ringer, R. K. and Tetzlaff, M. (1963). The anatomy of the feather follicle and its immediate surroundings. *Poult. Sci.* **42,** 958–969.

Ottoson, D. and Shepherd, G. M. (1971). Transducer properties and integrative mechanisms in the frog's muscle spindle. *In* "Handbook of Sensory Physiology" (W. R. Loewenstein, Ed.), Vol. I. Springer-Verlag, Berlin and New York.

Ozeki, M. and Sato, M. (1964). Initiation of impulses at the non-myelinated nerve terminal in Pacinian corpuscles. *J. Physiol., Lond.* **170,** 167–185.

Palmieri, G. (1969). Sensory nervous expansions in duck and chicken joints. *Riv. biol.* **62,** 345–350.

Pearson, R. (1976). "The Avian Brain". Academic Press, London and New York.

Poláček, P. (1966). Receptors of the joint, their structure, variability and classification. *Acta Fac. med. Univ. Brunensis* **23,** 1–107.

Poláček, P. (1969). Die Ultrastruktur des Herbstschen Körperchens im Vergleich mit dem Vater-Pacinischen Körperchen (vorläufige Mitteilung). *Sb. Ved. Pr. Lek. Fak. Univ. Karlovy Hradci* **12,** 417–426.

Poláček, P., Sklenská, A. and Malinovský, L. (1966). Contribution to the problem of joint receptors in birds. *Folia morph., Prague* **14,** 33–42.

Portmann, A. (1961). Sensory organs: skin, taste and olfaction. *In* "Biology and Comparative Physiology of Birds" (A. J. Marshall, Ed.), Vol. II. Academic Press, New York and London.

Poulos, D. A. and Lende, R. (1970a). Response of trigeminal ganglion neurons to thermal stimulation of oral–facial regions. I. Steady state response. *J. Neurophysiol.* **33,** 508–517.

Poulos, D. A. and Lende, R. (1970b). Response of trigeminal ganglion neurons to thermal stimulation of oral–facial regions. II. Temperature change response. *J. Neurophysiol.* **33,** 518–526.

Pubols, B. H. and Pubols, L. M. (1976). Coding of mechanical stimulus velocity and indentation depth by squirrel monkey and raccoon glabrous skin mechanoreceptors. *J. Neurophysiol.* **39,** 773–787.

Quilliam, T. A. (1966). Unit design and array patterns in receptor organs. *In* "Touch, Heat and Pain" (A. V. S. De Reuck and J. Knight, Eds.). Churchill, London.

Ranvier, L. (1877). De la terminaison des nerfs dans les corpuscules du tact. *C. R. Hebd. Séanc. Acad. Sci., Paris* **85,** 1020–1023.

Rautenberg, W. (1971). The influence of the skin temperature on the thermoregulatory system of pigeons. *J. Physiol., Paris* **63,** 396–398.

Rebollo, M. A. and de Anda, G. (1967). The neuromuscular spindles in the adult chicken. *Acta anat.* **67,** 595–608.

Regaud, C. and Favre, M. (1904). Les terminaisons nerveuses et les organes nerveux sensitifs de l'appareil locomoteur. I. Les terminaisons nerveuses et les organes nerveux sensitifs des muscles striés squelettaux. *Rev. gen. Histol.* **1,** 1–140.

Richards, S. A. (1970). The role of hypothalamic temperature in the control of panting in the chicken exposed to heat. *J. Physiol., Lond.* **211,** 341–358.

Richards, S. A. (1971). The significance of changes in the temperature of the skin and body core of the chicken in the regulation of heat loss. *J. Physiol., Lond.* **216,** 1–10.

Rodriguez-Perez, A. P. and de Iraldi, P. A. (1961). Algunos datos acerca de la microscopia electronica de los corpusculos de Herbst. *Trab. Inst. Cajal Invest. biol.* **53,** 190–195.

Roumy, M. and Leitner, L. M. (1973). Activités afférentes provenant du bec supérieur de la poule domestique. *C. R. Hebd. Séanc. Acad. Sci., Paris Sér. D.* **277,** 1791–1794.

Ruffini, A. (1905). Les expansions nerveuses de la peau chez l'homme et quelques autre mammifères. *Rev. gen. Histol.* **1,** 420–540.

Sağlam, M. (1968). Morphologische und quantitative Untersuchung über die Muskelspindeln in der Nackenmuskulatur (M. biventer cervicis, M. rectus capitis dorsalis und M. rectus capitis lateralis) des Bunt- und Blutspechtes. *Acta anat.* **69,** 87–104.

Sakada, S. and Aida, H. (1971). Electrophysiological studies of Golgi–Mazzoni corpuscles in the periosteum of the cat facial bones. *Bull. Tokyo Dental Coll.* **12,** 255–272.

Sato, M. and Ozeki, M. (1963). Response of the non-myelinated nerve terminal in Pacinian corpuscles to mechanical and antidromic stimulation and the effect of procaine, choline and cooling. *Jap. J. Physiol.* **13,** 564–582.

Saxod, R. (1967). Histogénèse des corpuscules sensoriels cutanés chez le poulet et le canard. *Archs. Anat. microsc. Morph. exp.* **56,** 153–166.

Saxod, R. (1968). Ultrastructure des corpuscules sensoriels cutanés de Herbst et de Grandry chez le canard. *Archs. Anat. microsc. Morph. exp.* **57,** 379–400.

Saxod, R. (1970a). Etude au microscope électronique de l'histogénèse du corpuscule sensoriel cutane de Herbst chez le canard. *J. ultrastruct. Res.* **33,** 463–492.

Saxod, R. (1970b). Etude au microscope électronique de l'histogénèse du corpuscule sensoriel cutané de Grandry chez le canard. *J. ultrastruct. Res.* **32,** 477–496.

Saxod, R. (1973a). Organization ultrastructurale des corpuscules sensoriels cutanés des oiseaux. *Ann. Centre Univ. Savoie* **1,** 79–98.

Saxod, R. (1973b). Les organites périnucléaires du corpuscule sensoriel cutané de Grandry. Organization ultrastructurale et formation. *Tiss. Cell* **5,** 269–280.

Saxod, R. (1978a). Development of cutaneous sensory receptors in birds. *In* "Handbook of Sensory Physiology" (M. Jacobson, Ed.), Vol. IX. Springer-Verlag, Berlin and New York.
Saxod, R. (1978b). Ultrastructure of Merkel corpuscles and so-called "Transitional Cells" in the white leghorn chicken. *Am. J. Anat.* **151,** 453–474.
Schildmacher, H. (1931). Untersuchung über die Funktion der Herbstschen Körperchen. *J. Ornithol.* **79,** 374–415.
Schoultz, T. W. and Swett, J. E. (1972). The fine structure of Golgi tendon organs. *J. Neurocytol.* **1,** 1–26.
Schoultz, T. W. and Swett, J. E. (1974). Ultrastructural organisation of the sensory fibers innervating the Golgi tendon organ. *Anat. Rec.* **179,** 147–162.
Schwartzkopff, J. (1949). Über Sitz und Leistung von Gehör und Vibrationssinn bei Vögeln. *Z. vergl. Physiol.* **31,** 527–608.
Schwartzkopff, J. (1973). Mechanoreception. *In* "Avian Biology" (D. S. Farner and J. R. King, Eds.), Vol. III. Academic Press, New York and London.
Scott, S. A., Cooper, E. and Diamond, J. (1981). Merkel cells as targets of the mechanosensory nerves in salamander skin. *Proc. R. Soc. B.* **211,** 455–470.
Sfameni, A. (1902). Recherches anatomiques sur l'existence des nerfs et sur leur mode de se terminer dans le tissu adipeux, dans le perioste, dans le perichondre et dans le tissus qui reforcent les articulations. *Arch. ital. Biol.* **38,** 49–101.
Shimizu, M. (1971). Ultrastructure of Golgi's tendon spindle of mouse. (In Japanese.) *Yonago Igaku Zasshi* **22,** 285–296.
Sklenská, A. (1969). Über Unterschiede in Häufigkeit, Verteilung und Struktur der Rezeptoren in einigen grossen Gelenken der Extremitäten von Wirbeltieren. *Z. mikrosk.-anat. Forsch.* **80,** 249–259.
Sklenská, A. (1972). Contribution to the ultrastructure of the Golgi tendon organ. *Folia Morph., Prague* **20,** 195–197.
Sklenská, A. (1973). Die Ultrastruktur des Golgischen Sehnenorgans bei der Katze. *Acta anat.* **86,** 205–221.
Skoglund, S. (1960). Properties of Pacinian corpuscles of ulnar and tibial location in cat and fowl. *Acta physiol. scand.* **50,** 385–386.
Skoglund, S. (1973). Joint receptors and kinaesthesis. *In* "Handbook of Sensory Physiology" (A. Iggo, Ed.), Vol. II. Springer-Verlag, Berlin and New York.
Spencer, P. S. and Schaumburg, H. H. (1973). An ultrastructural study of the inner core of the Pacinian corpuscle. *J. Neurocytol.* **2,** 217–237.
Stammer, A. (1961). Die Nervenendorgane der Vogelhaut. *Acta biol., Szeged* **7,** 115–131.
Steen, I. and Steen, J. B. (1965). The importance of the legs in the thermoregulation of birds. *Acta physiol. scand.* **63,** 285–291.
Stresemann, E. (1934). Sauropsida: Aves. *In* "Handbuch der Zoologie" (W. G. Kükenthal and T. Krumbach, Eds.), Vol. VII/2. de Gruyter, Berlin.
Sturkie, P. D. (1965). "Avian Physiology", 2nd edn. Cornell University Press, Ithaca, New York.
Swett, J. E. and Schoultz, I. W. (1975). Mechanical transduction in the Golgi tendon organ: A hypothesis. *Arch. ital. Biol.* **113,** 374–382.
Szymonowicz, L. (1897). Über den Bau und die Entwicklung der Nervenendigungen im Entenschnabel. *Arch. mikrosk. Anat.* **48,** 329–358.
Tapper, D. N. (1965). Stimulus–response relationship in the cutaneous slowly adapting mechanoreceptors in hairy skin of the cat. *Exp. Neurol.* **13,** 364–385.
Tello, J. F. (1917). Genesis de la terminaciones nerviosas motrices y sensitivas. I. En el sistema locomotor de los vertebrados superiores. Histogenesis muscular. *Trab. Lab. Invest. biol. Univ. Madrid* **15,** 101–199.
Tello, J. F. (1922). Die Entstehung der motorischen und sensiblen Nervenendigungen. I. In dem lokomotorischen System der höheren Wirbeltiere. *Z. Anat. Entwgesch.* **64,** 348–440.
Uehara, Y., Campbell, G. R. and Burnstock, G. (1976). "Muscle and its Innervation. An Atlas of Fine Structure". Arnold, London.
Veggetti, A. and Palmieri, G. (1971). Sulla distribuzione, frequenza e morfologia dei fusi neuromusculari nei muscoli masticator di alcuni uccelli domestici. *Riv. biol.* **64,** 215–239.
von Düring, M. and Andres, K. H. (1969). Zur Feinstruktur der Muskelspindel von Mammalia. *Anat. Anz.* **124,** 566–573.
Voss, H. (1963). Untersuchungen über die absolute und relative Zahl der Muskelspindeln in weiteren Muskelgruppen (Mm. Scaleni und Rückenmuskeln) des Menschen. *Anat. Anz.* **112,** 276–279.
Wedell, G., Palmer, E. and Pallie, W. (1955). Nerve endings in mammalian skin. *Biol. Rev.* **30,** 159–195.

Werner, G. and Mountcastle, V. B. (1965). Neural activity in mechanoreceptive cutaneous afferents: Stimulus–response relations, Weber functions, and information transmission. *J. Neurophysiol.* **28,** 359–397.

Whitear, M. and Lane, E. B. (1981). Fine structure of Merkel cells in lampreys. *Cell Tiss. Res.* **220,** 139–151.

Winkelmann, R. K. (1960). Similarities in cutaneous nerve endorgans. *In* ''Advances in Biology of Skin'' (W. Montagna, Ed.), Vol. I. Pergamon Press, New York.

Winkelmann, R. K. and Breathnach, M. S. (1973). The Merkel cell. *J. invest. Derm.* **60,** 2–15.

Winkelmann, R. K. and Myers, T. T. (1961). The histochemistry and morphology of the cutaneous sensory end-organs of the chicken. *J. comp. Neurol.* **117,** 27–35.

Witt, I. and Hensel, H. (1959). Afferente Impulse aus der Extremitätenhaut der Katze bei thermischer und mechanischer Reizung. *Pflügers Arch. ges. Physiol.* **268,** 583–596.

Woike, M. and Gewecke, M. (1978). Kinematic and aerodynamic parameters in tethered flying siskins (*Cardelius spinus,* Passeres, Aves). *J. comp. Physiol.* **127,** 123–129.

Zeigler, H. P. and Witkowsky, P. (1968). The main sensory trigeminal nucleus in the pigeon: A single unit analysis. *J. comp. Neurol.* **134,** 255–264.

Ziswiler, V. and Trnka, V. (1972). Tastkörperchen im Schlundbereich der Vögel. *Rev. Suisse Zool.* **79,** 307–318.

Zweers, G. A., Gerritsen, A. F. C. and von Kranenburg-Voogd, P. J. (1977). Mechanics of feeding of the mallard (*Anas platyrhynchos* L.: Aves, Anseriformes). The lingual apparatus and suction-pressure pump mechanism of straining. *In* ''Contributions to Vertebrate Evolution'' (M. K. Hecht and F. S. Szaly, Eds.), Vol. 3. S. Karger, Basel.

9

Structure and function of avian taste receptors

HERMAN BERKHOUDT

Department of Morphology, Zoological Laboratory, Rijksuniversiteit Leiden, Leiden, The Netherlands

CONTENTS

I. Introduction

In the vertebrate nervous system many different end-structures, acting as receptors for both extrinsic and intrinsic stimuli, have evolved. Birds also possess many specialised receptors, including those located in the perioral skin and buccal cavity, which serve to screen their daily food and water intake.

Food may be selected on the basis of a number of features such as size, shape, colour and texture; in addition the chemical composition (or taste) of both food and water may be tested by gustatory and probably also olfactory organs. As most of the physical qualities are dealt with in this volume in other chapters on special sense organs, only taste will be considered here.

There is much experimental evidence that most birds have an acute sense of taste. Kare and Rogers (1976) outlined the possible functions of that sense as follows: it encourages the ingestion of nutrients, it discriminates between the available items of food, and possibly enables those that are toxic to be avoided. Perhaps because of economic factors, the positive response to gustatory discrimination, that is food preference and selection, has been the subject of many studies. Yet, in the natural environment the negative response, that is rejection, may be of more importance for the bird itself. The biological benefit of avoiding harmful food is self-evident.

As in all vertebrates, taste impulses in birds are generated in taste buds. A taste bud is a barrel or flask-shaped group of fusiform cells which differ markedly from the other epithelial cells of the surrounding oral mucosa (Graziadei, 1974). Taste buds were demonstrated much later in birds than in most other vertebrates, the first published description being that by Botezat (1904). The relatively small number of modern studies possibly reflects the technical difficulties in demonstrating their presence. The number of taste buds reportedly present in birds is very low in comparison with other vertebrates. However, this relatively low number appears to be an under-estimate. In addition to the inconspicuousness of avian taste buds, the emphasis of most investigators on the role of the tongue in gustation has diverted attention from other structures as possible locations for taste buds.

II. Avian taste receptors

Most reviews of avian taste agree that (1) very few species of birds have been studied, (2) these few species are poorly equipped with taste organs, and (3) the limited numbers of organs are confined to the caudal regions of the tongue and to the pharyngeal floor (Kare, 1971; Gentle, 1975a; Kare and Rogers, 1976; Wenzel, 1973, 1983). This is because observations on lingual taste buds predominate among the few modern studies of avian taste buds (Moore and Elliott, 1946; Lindenmaier and Kare, 1959; Duncan, 1960; Gentle, 1971b; Nalavade and Varute, 1977; Dmitrieva, 1981; Wenzel, 1983). As pointed out by Berkhoudt (1977), these results conflict with earlier data, published in German at the beginning of this century (Botezat, 1904, 1906; Bath, 1906a), as well as with other subsequent studies, all of which claim a wider distribution and higher numbers, sometimes in the same species (Saito, 1966b; Warner *et al.*, 1967; Berkhoudt, 1977). To date, little effort has been made to integrate all available information into an overall picture of avian taste organs. One of the objects of the present review is to present such an integration.

A. *Distribution of avian taste buds*

In the past the distribution and numbers of avian taste buds have been studied from serial sections only. Few studies exist where all sections were analysed in an effort to give either total counts or a graphic reconstruction of the distribution of the taste buds (Saito, 1966b; Moore and Elliott, 1946, respectively). Simple staining techniques aiming at visualising the taste bud pore have now been developed (Berkhoudt, 1977; Brouwer and Wiersma, 1978). These techniques give quick and accurate data both on distribution and densities, and can be followed by

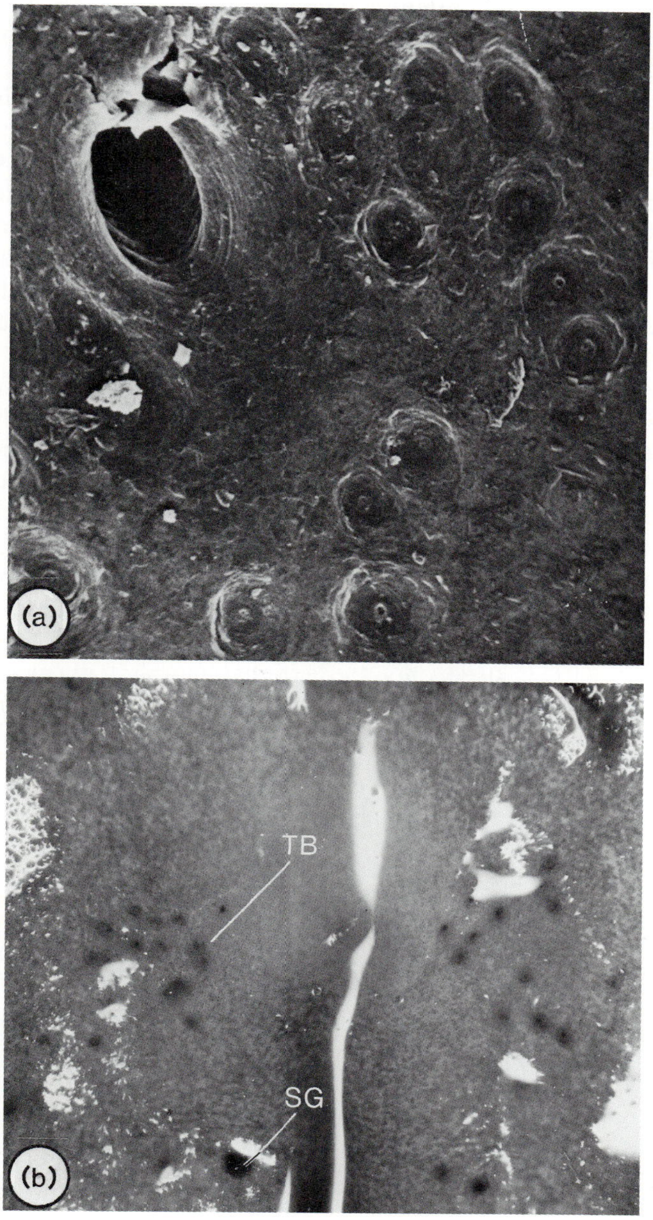

Fig. 9.1. Additional fields of taste buds located in the bill tip region of the Mallard (*Anas platyrhynchos*). (a) Scanning electron micrograph (SEM) showing a cluster of taste buds located in the palate, circumferentially around the monostomatic opening of the maxillary salivary glands. ×70. From Berkhoudt (1977). (b) Field of solitary taste buds located in the mucosa of the oral floor just caudal to the border of the mandibular nail. TB, Taste buds; SG, salivary glands. Pontamine sky-blue staining. ×14.

either a more detailed scanning electron microscopie (SEM) analysis or serial sectioning. Using a combination of techniques we were able to demonstrate in our laboratory that the description of Bath (1906a) of the distribution of taste buds in the Mallard (*Anas platyrhynchos*) is accurate, though incomplete. In the bill tip region, just caudal to the bill tip organ (see p. 446 of Chapter 8, by K.-M. Gottschaldt, of this volume), additional new fields of taste buds were discovered (Fig. 9.1a,b). This discovery aids an understanding of the discriminative capacities of this bird. In their "classical" discrimination experiment Zweers and Wouterlood (1973; see also Zweers *et al.*, 1977) offered trained Mallards an equal number of peas and plasticine imitations of the same size, shape and colour, hidden in a tray under wet sand. Within each session the birds were able to retrieve and swallow all the 50 peas, leaving the fake peas "untouched". This discrimination occurred between the beak tips under the sand. As no taste buds were reported from this region by Bath (1906a), it was presumed that discrimination was purely tactile and mediated by the bill tip organ (Berkhoudt, 1976). The discovery of the new rostral areas of taste buds necessitates a re-evaluation of these experimental data.

Avian taste buds can be distributed either free in the oral mucosa or in close association with the openings of the salivary glands. Botezat (1904, 1906) used the term solitary and glandular buds, respectively, to discriminate between these types. In the Mallard, solitary taste buds occur in the rostral part of the lower bill and in the caudal part of the upper bill. Glandular buds open directly into a duct of a salivary gland (Botezat 1904, 1906) and occur in many birds, including the chicken and pigeon. The distance between the taste bud pore and the opening of the salivary duct discriminates between the two types. Hence, taste buds that are oriented around salivary ducts are not simply glandular buds (Berkhoudt, 1977). Kurosawa *et al.* (1983) used only the external morphology as a criterion to distinguish two types of taste buds in the chicken, their far more abundant type I

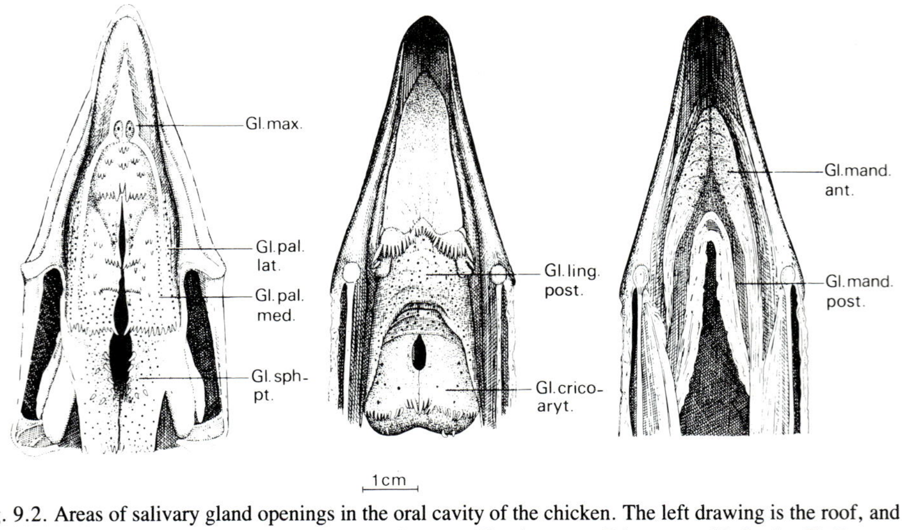

Fig. 9.2. Areas of salivary gland openings in the oral cavity of the chicken. The left drawing is the roof, and the two drawings on the right are the floor of the oropharynx. Gl. cricoaryt., Cricoarytenoid glands; Gl. ling. post., caudal lingual glands; Gl. mand. ant., post., rostral and caudal mandibular glands; Gl. max., maxillary gland; Gl. pal. lat., med., lateral and medial palatine glands; Gl. sph-pt., sphenopterygoid glands. ×0·72.

TABLE I. **Number of taste buds observed in the regions where the salivary glands open in the domestic fowl and pigeon.**

Area of salivary gland [see Saito (1966a) and Zweers (1982a)]	Number of taste buds observed[a]							
	Chicken (*Gallus domesticus*)					Pigeon (*Columba livia*)		
	Lindenmaier and Kare (1959)	Saito (1966b) $N = 5$	Gentle (1971b)	van Prooije (1978) $N = 17$	Kurosawa *et al.* (1983)	Bath (1906a)	Moore and Elliott (1946) $N = 6$	van Kan (1979) $N = 2$
Upper jaw								
gl. maxillaris monostomatica	—		—	18.1	25		—	
gl. palatinae lateralis	—	196.5 average	—	51.8	42		—	
gl. palatinae medialis	—		—	17.1	25		—	4
gl. sphenopterygoidea	—		—	0.4	20		—	
Lower jaw								
gl. mandibularis rostralis	—	84.4	—[b]	27.2	38	50–75	—	50
gl. mandibularis caudalis	—	—	—	0.8	12		—	—
gl. ling. rostralis (= *inferior*)	—	12.0	—	4.6	—		—	8
Tongue								
gl. ling. caudalis	5–12 chick	63.0	8–15	3.7	48		37 average	12
gl. cricoarytenoideae	24 (3-month cockerel)	4.8	—	0.7	3		25–59 range	54.5

[a] Ganchrow and Ganchrow (1985) counted an average of 316 taste buds in one-day-old chickens of which 69% were distributed in the palate, 29% in the in the mandibular mucosa, and 2% in the rostral part of the tongue.

[b] M. J. Gentle (personal communication) found an average of 56 taste buds in serial sections from a total of seven birds.

being elongated and pear-shaped (closely resembling Bath's type I; see Fig. 9.5), and their type II having a forked configuration. It is not clear from their English text if these types correspond with the just mentioned solitary and glandular buds respectively.

Since many investigators are, for a variety of reasons, especially interested in the chicken and pigeon, data collected in our laboratory on the distribution of their taste buds will be included here. As the location of taste buds interferes with the numerous small secretory openings of the salivary glands, no attempt has been made to illustrate this distribution, as was done for the Mallard (Berkhoudt, 1977). Instead, Fig. 9.2 shows where the salivary glands of the chicken open into the oral cavity and pharynx, and Table I summarises the number of taste buds associated with these openings of the salivary glands. The demarcation of these areas is based on the descriptions of Saito (1966a) for the chicken and Zweers (1982a) for the pigeon. Figure 9.14b illustrates the distribution of taste buds on the floor of the oropharynx of the chicken. The distribution of taste buds on the floor of the oropharynx of the pigeon is virtually identical to that shown in Fig. 9.14b (see Berkhoudt, 1984). The preceding accounts of the distribution of taste buds agree closely with the descriptions of Botezat (1904, 1906) for birds in general, combined with that of Bath (1906a), taste buds being invariably situated at the base of the tongue, on the floor of the oropharynx and in the palate (cf. also Krause, 1922; Ariëns Kappers *et al.*, 1936).

As already stressed by Bath (1906a), the region of the rostral mandibular glands was one of the most dense areas for taste buds in both the chicken and pigeon. In the chicken even higher relative densities occur around the monostomatic openings of the maxillary glands and along the lateral palatine glands. In the upper jaw most taste buds are associated with the secretory openings of the salivary ducts in both the pigeon and the chicken, but in the lower jaw the two species differ. In the chicken, taste buds are again orientated in a narrow ring around the openings (Fig. 9.3a–c), sometimes so close together that a cluster of taste buds looks like a single bud perforated by the salivary duct. Apparently this led Botezat (1904, 1906) to his concept of "perforated" glandular buds, a misinterpretation which he rectified in his 1910 article (Botezat, 1910). In the pigeon, however, the rostral mandibular glands possess relatively few openings (Zweers, 1982a), and these are located between the plicae of the oral mucosa on both sides of the tongue. Taste buds here, functionally quite understandable, are not only found circumferentially around the openings of the salivary ducts but also on the dorsum of the plicae (Fig. 9.4a,b).

The data on the distribution of taste buds in the pigeon and chicken can be summarised as follows. For the pigeon the data from our laboratory fit those of Bath (1906a), for both total numbers and distribution. The very accurate observations of Moore and Elliott (1946) cover only a limited part of our data (see also Wenzel, 1983), because these authors were interested only in the evolution of taste on the vertebrate tongue. For the chicken it is clear that Saito (1966b) gives complete data on numbers and distribution; the probable reason his work has not been accepted earlier is that he published it in Japanese. Since its translation for this author's 1977 work, his view is now more commonly accepted (see Gentle, 1971a, 1975a; and cf. Gentle, 1978). The difference between our findings and Saito's data could reflect strain differences (M. J. Gentle, personal communication), but age and differences in the techniques used to establish their presence seem to be more likely reasons. Ontogenetic differences were already stressed for the chicken by Saito (1966b), who found, like Lindenmaier and Kare (1959), almost double the total numbers in adults as compared with 1-day-old hatchlings (5–12 in the 1-day-old chick, 24 in the 3-month-old cockerel: Lindenmaier and Kare, 1959; 12 in the tongue of 1-day-old chicks, 25 in the tongue of adults: Saito, 1966b). In contrast to the chicken, a reverse trend was reported by Botezat (1910) for the House Sparrow (*Passer domesticus*) and by Duncan (1960) for the Bullfinch (*Pyrrhula pyrrhula*); this trend coincides with that in mammals, where

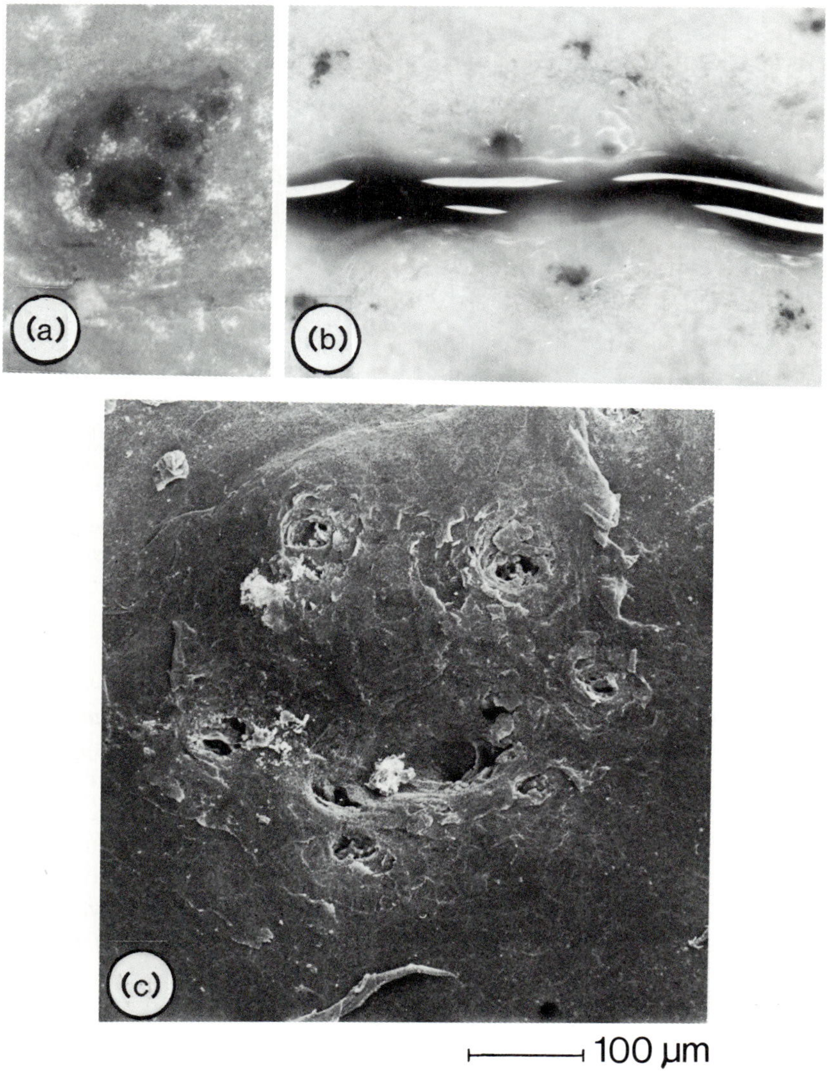

Fig. 9.3. (a) Typical ring-like arrangement of a group of taste buds around the opening of a rostral mandibular salivary gland in the chicken. Pontamine sky-blue stain. ×54. (b) A series of such openings of rostral salivary glands, and their associated buds. Pontamine sky-blue stain. ×12·6. (c) Scanning electron micrograph of the same preparation as in (a), showing that more taste bud pores can be distinguished than by using the Pontamine sky-blue technique. ×153.

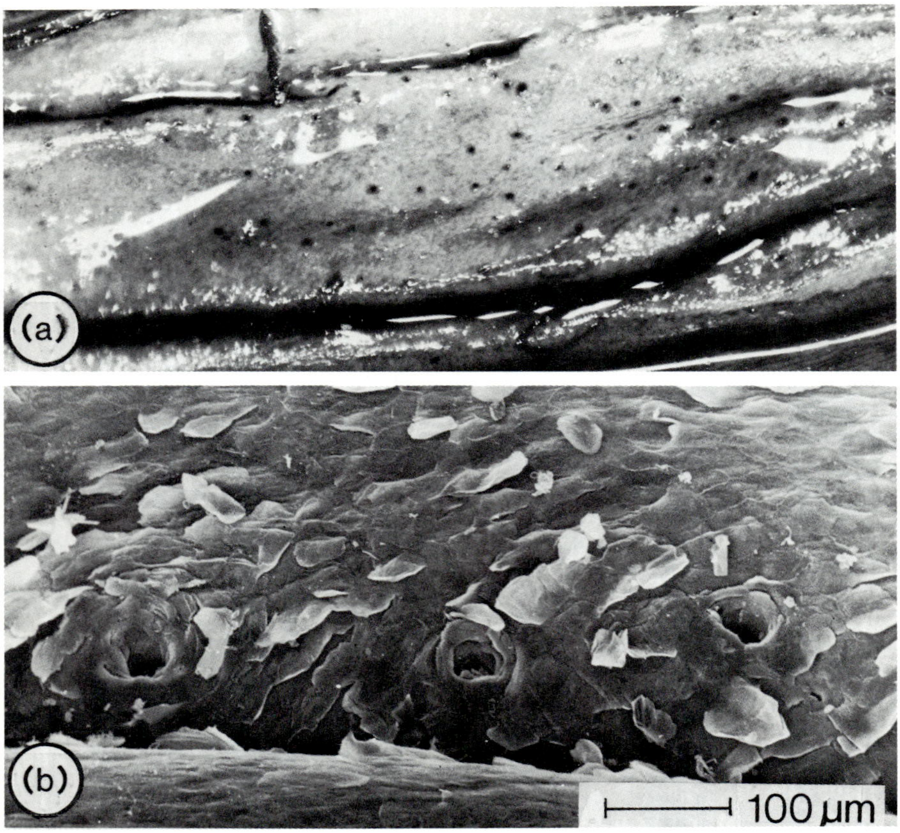

Fig. 9.4. (a) Taste buds (about 50) located on top of the plicae of the mandibular mucosa of the pigeon (*Columba livia*). Pontamine sky-blue staining. ×14. (b) Detail from (a) showing three single taste bud pores not associated with any salivary gland opening. SEM, ×170.

decreasing numbers of taste buds are again reported during ontogeny (Lalonde and Eglitis, 1961). The fact that van Prooije (1978) was unable to find the double numbers in adults may be explained by differences in counting techniques. In Saito's serial sections it is possible to count the same taste bud more than once in successive sections, thus leading to an overestimate of the total number of taste buds. In our technique of rapid surface staining, taste buds which have their pores inside the openings of the ducts of the salivary glands (a situation which is found to exist in serial sections) may be overlooked, and this may lead to an underestimate of the total numbers in adults (see Fig. 9.3a,c).

This section ends with a table of the maximum numbers reported in several avian species (Table II). It should be noted that the number of 30 to 70 tastebuds in the chicken attributed to Bath (1906a,b) by Lindenmaier and Kare (1959), Engelmann (1950), Saito (1966b), Gentle (1971b, 1975a) and Nalavade and Varute (1977) could not be found in either his thesis (Bath, 1906b) or in his 1906 article (Bath, 1906a), and should be excluded from the literature.

TABLE II. **Total number of taste buds reported in various birds.**

Bird species	Number of taste buds	Reference
Pink-headed Turtle Dove (*Streptopelia roseogrisea*)	54	Gentle (1975a)
Blue Tit (*Parus caeruleus*)	24	Gentle (1975a)
Bullfinch (*Pyrrhula pyrrhula*)	46	Duncan (1960)
Chicken	125; 250–350	van Prooije (1978); Saito (1966b)
Mallard (*Anas platyrhynchos*)	375	Berkhoudt (1977)
"Parrot" ("*Palaeornis rufirostris*")	300–400	Bath (1906a)
Pigeon	50–75	Bath (1906a); van Kan (1979)
Japanese Quail (*Coturnix japonica*)	62	Warner *et al.* (1967)
Common Starling (*Sturnus vulgaris*)	200	Bath (1906a)

B. Form and internal structure of avian taste buds

Primarily on the basis of shape and internal morphology, Bath (1906a) divided the taste buds in his material (19 species of birds) into three types (Fig. 9.5). Type I, the most common, is ovoid and has a central core of sensory and supporting (or sustentacular) cells, surrounded by follicular (or sheath) cells. Bath found this type in 13 of his 19 species, comprising mostly songbirds (oscines) and including the chicken and pigeon. Type II, a narrower and more elongated type which is restricted to ducks, flamingos and the Oystercatcher (*Haematopus ostralegus*), has the same cellular composition. Type III, a round type resembling that of mammals and amongst birds restricted to parrots, lacks the follicular cells.

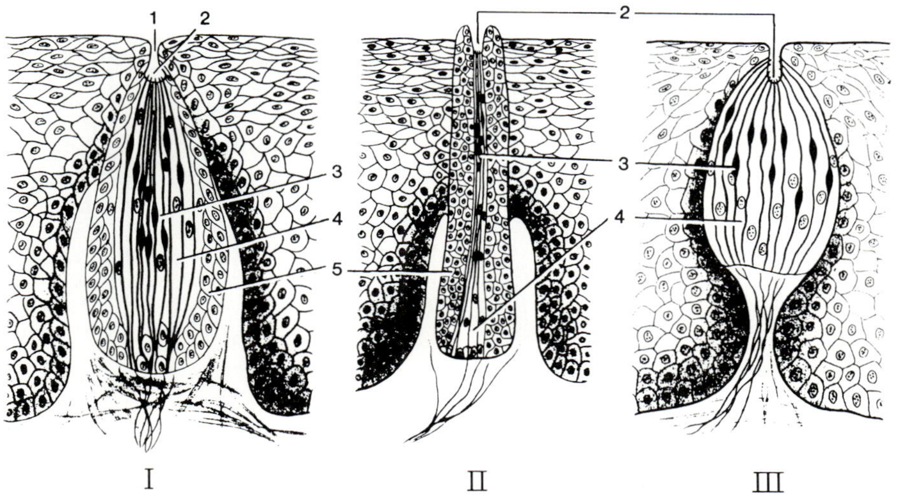

Fig. 9.5. Schematic representation of the three different types of taste bud distinguished by Bath (1906a). Type I occurred in most songbirds (oscines) and also the chicken and pigeon, type II in ducks and flamingos, type III in parrots. 1, Outer pore; 2, inner pore; 3, sensory cell; 4, supporting cell; 5, follicular cell. From Bath (1906a).

Bath's description and illustrations of the morphology of avian taste buds appear to be unrivalled (see Figs. 9.6a, 7a). From Bath's account and our own observations, the general composition of avian taste buds can be summarised as follows. Within the ovoid or flask-shaped taste bud a spindle-shaped (or in *Anas* a more cylinder-shaped) group of elongated fusiform sensory cells lies perpendicular to the mucosal surface. In some of our preparations the distinction between neuroepithelial and supporting cells described by Botezat (1904, 1906) and Bath (1906a) could not be made, although darker spindle-shaped cells were found surrounded by pale-staining polygonal cells, especially in Epon-embedded material. Many authors have stated that such a distinction is not valid, because both cells are transitional, representing different stages of the same basic cell type (Greschik, 1917, following Kolmer, 1910; de Lorenzo, 1963; Scalzi, 1967; Graziadei, 1969). Especially for the chicken taste bud, there is a debate as to how many different cell types can be distinguished. Lindenmaier and Kare (1959) and Gentle (1971b) found only a single cell type, whilst Saito found two, both sensory and supporting. In our preparations of chicken taste buds, we have found the same cell types as those described by Bath (1906a), that is sensory, supporting, follicular and basal cells (Fig. 9.8a,b). One feature not discussed by Bath (1906a,b) was the complex way in which the sensory cells can communicate with the oral cavity. Bath (1906a) illustrated for the chicken a pore similar to that found in the pigeon (see Fig. 9.6a), but both Saito (1966b), Gentle (1971b) and Kurosawa *et al.* (1983) described a lumen in the taste bud, comparable to that described in the

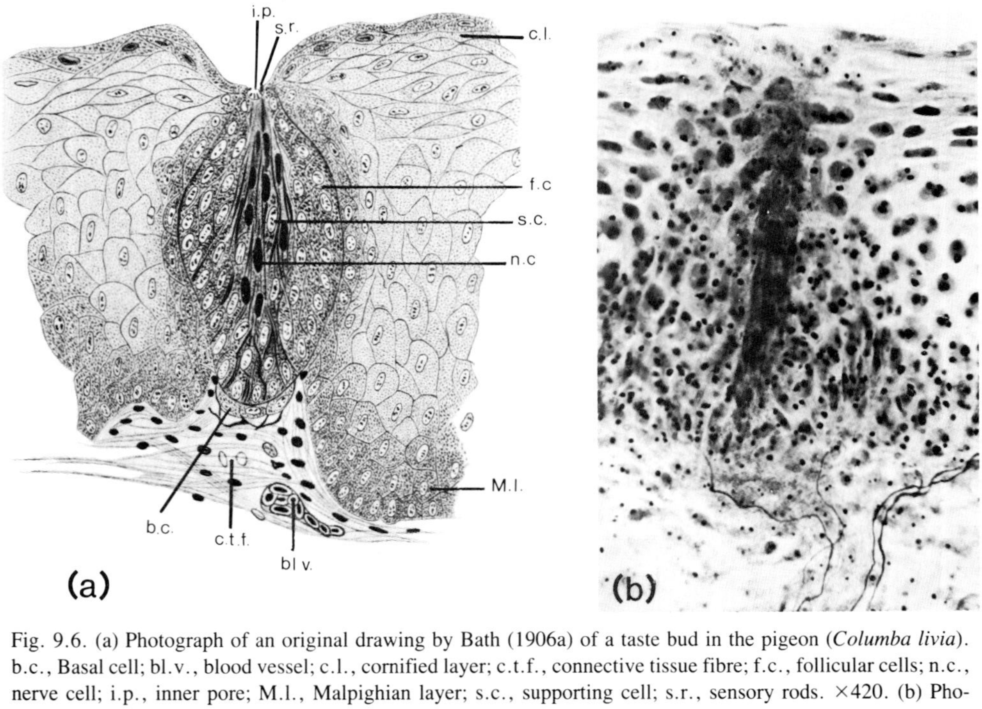

Fig. 9.6. (a) Photograph of an original drawing by Bath (1906a) of a taste bud in the pigeon (*Columba livia*). b.c., Basal cell; bl.v., blood vessel; c.l., cornified layer; c.t.f., connective tissue fibre; f.c., follicular cells; n.c., nerve cell; i.p., inner pore; M.l., Malpighian layer; s.c., supporting cell; s.r., sensory rods. ×420. (b) Photomicrograph of a taste bud in the mandibular mucosa of the pigeon (*Columba livia*). Silver stain of Sevier and Munger (1965). ×400.

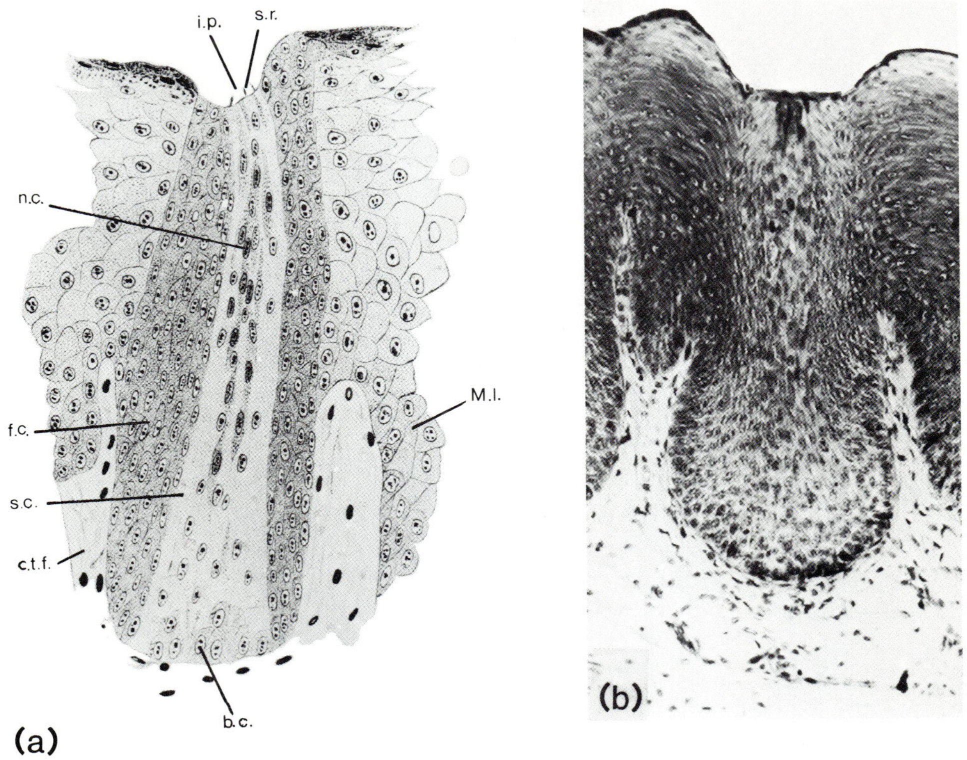

Fig. 9.7. (a) Photograph of an original drawing by Bath (1906a) of a taste bud in the mallard (*Anas platyrhynchos*). For abbreviations see Fig. 9.6a. ×420. (b) Photomicrograph of a taste bud in the rostral maxillary mucosa of the mallard (*Anas platyrhynchos*). Bath's stain. ×200. From Berkhoudt (1977).

quail by Warner *et al.* (1967), which communicates via meandering tubes with the oral cavity (Saito, 1966b). Kurosawa *et al.* (1983) refer to this communicating system as the "taste canal". The variability observed in the chicken taste bud goes far beyond the schematic figure of type I taste buds given by Bath (1906a). The variability of parrot taste buds was stressed earlier by Greschik (1917). Dmitrieva (1981), in the first EM study of lingual taste buds in the pigeon, distinguished the same four types of cells (Fig. 9.9) as those described by Bath (1906a); sensory, supporting, basal and follicular cells (she termed the latter perigemmal). Her recognition of both sensory and supporting cells is consistent with the structure of taste receptors in lower vertebrates (Graziadei, 1969; Graziadei and de Han, 1971; von Düring and Andres, 1976). A paper on the light- and electron-microscopy of the taste buds of the chicken was published by Kurosawa *et al.* (1983). They distinguished at least three types of cell: (1) light cells with typical afferent synaptic contacts with axon terminals, (2) dark cells with extensive axonal contacts but no synapses, and (3) flattened cells characterized by numerous filaments and free ribosomes. These three cells were interpreted respectively as gustatory cells, supporting cells and peripheral cells. Dmitrieva (1981) again emphasised that the greatest number of taste buds in the pigeon is in the

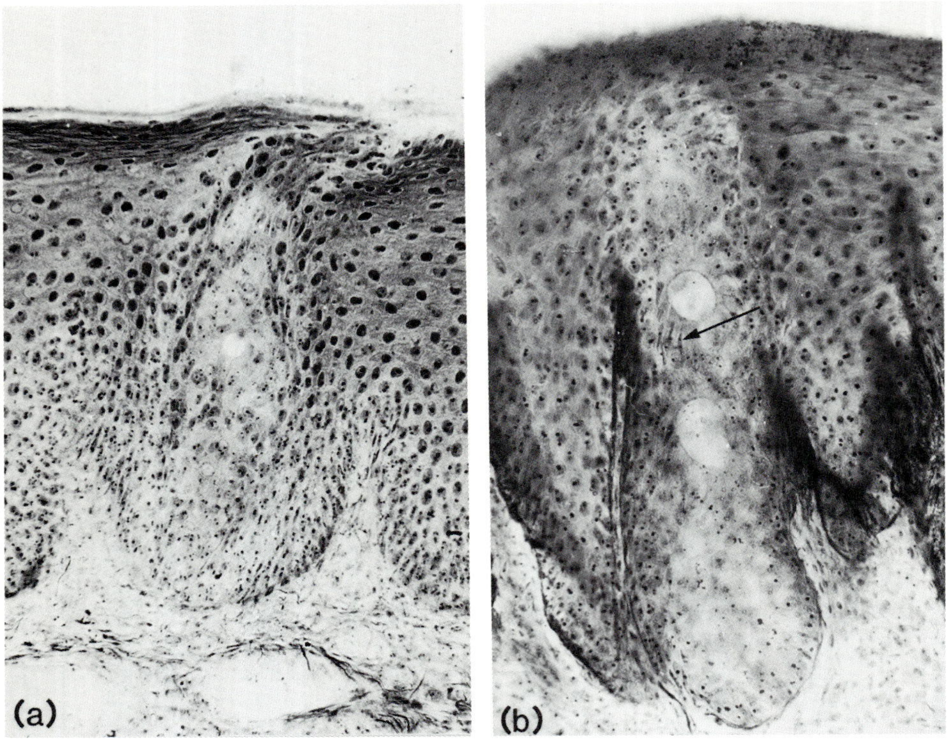

Fig. 9.8. Photomicrographs of taste buds from the maxillary region in the chicken. In (b) note lumen (empty spaces) inside the buds, and sensory cells (arrow). ×240.

region of the base of the tongue (see p. 468) around the openings of the salivary ducts; this agrees with the findings of Moore and Elliot (1946) but differs from our observations (see Wenzel, 1983). Each receptor cell possesses a large microvillar process at its apex (Fig. 9.9), which measures $3{\cdot}0 \times 1{\cdot}5$ μm. The smooth endoplasmic reticulum is located in the apical part of the receptor cell and is less developed than in the taste buds of other vertebrates. There are two centrioles in the apical part of the receptor cell. The supporting cell is characterised by a greater electron density of the cytoplasm, secretory granules in the apical part, and a well-developed smooth endoplasmic reticulum. The perigemmal cells (follicular cells) form several layers around the taste bud. Mitosis was observed in the follicular cells and occasionally in the supporting cells. There are three to five basal cells in the basal part of the taste bud in the pigeon. Basal cell cytoplasm contains small mitochondria, smooth and rough endoplasmic reticulum, and a large nucleus.

There are only a few reports on the histochemistry of avian taste buds. Gentle (1971b) and Pevzner (1964; see also Vinnikov, 1974) found a high level of acetylcholinesterase activity in the synaptic region. The mucosubstances in the taste pore region were shown by Nalavade and Varute (1977) to be neutral mucosubstances, comparable to those found in the rabbit by Scalzi (1967).

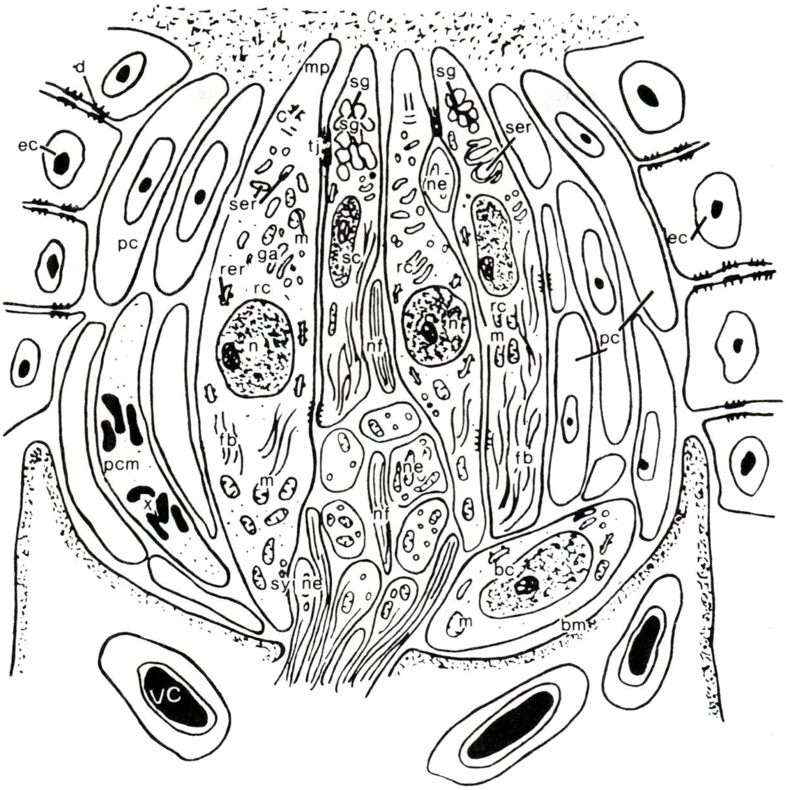

Fig. 9.9. Schematic drawing based on EM study of lingual taste buds in the pigeon (*Columba livia*) by Dmitrieva (1981). Reproduced with permission. bc, Basal cell; bm, basal lamina; c, centriole; ec, epithelial cells; d, desmosome; fb, fibrillar bundle; ga, Golgi apparatus; m, mitochondria; mp, microvillar process; n, nucleus; ne, nerve ending; nf, nerve fibres; pc, perigemmal cells; pcm, perigemmal cells in mitosis; rc, receptor cells; rer, rough endoplasmic reticulum; sc, supporting cell; ser, smooth endoplasmic reticulum; sg, secretory granules; sy, synaptic contact; tj, tight junction; vc, blood capillary; x, chromosome.

C. Innervation of avian taste buds

Botezat followed his first description of avian taste buds in 1904 by a second article in which he not only illustrated these organs but also provided a detailed description of their intrinsic innervation (Botezat, 1906). Bath (1906a) also attempted a description of avian taste bud innervation, but with less success (see Fig. 9.6a). Since at that time the reticular continuity theory was still dominant over the neuron theory, Botezat distinguished three different neuronal "nets" (Fig. 9.10). Firstly, a subgemmal plexus formed a cup under the taste bud; secondly, a perigemmal plexus formed a network around it; and finally, fibres of the intragemmal (or pericellular) plexus surrounded the sensory cells. To various extents the different neuronal nets indicated by Botezat are shown in the photomicrographs of taste buds in different bird species (Figs. 9.6b, 11a,b). Krol and Dubbeldam (1979) also reported a subgemmal plexus in the Mallard (as other authors did for other species: Gentle, 1971b; Pevzner, 1964) and showed that

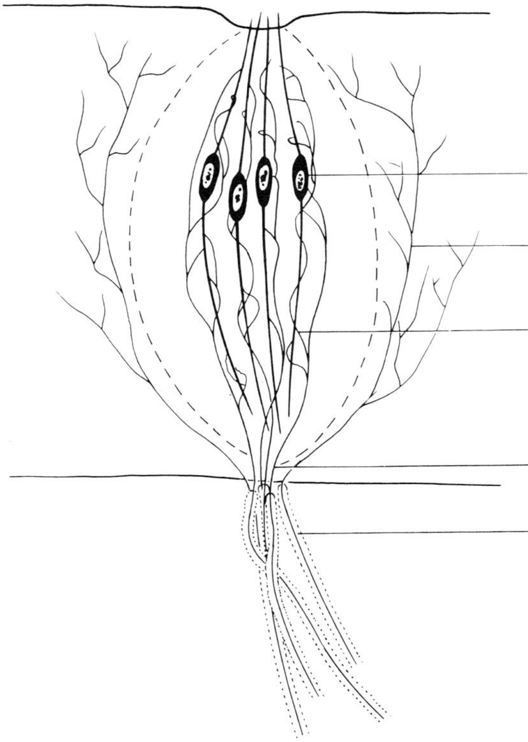

Fig. 9.10. Schematic figure based on Botezat (1906) showing nerve fibres of three different neuronal "nets". Subgemmal fibres form a plexus under the taste bud, perigemmal fibres form a plexus around the taste bud, and intragemmal fibres form a plexus between the sensory cells of the taste bud.

normal taste buds in *Anas* contain 20–30 intragemmal fibres. After nerve transection they found that most taste buds were devoid of axons and rapid degeneration occurred in a few days, a common finding also in other birds (Gentle, 1971b) and in many other vertebrates (Graziadei, 1969).

As many investigators only reported taste buds in the tongue, the lingual branch (*n. lingualis*) of the glossopharyngeal nerve (*n. glossopharyngeus*) was considered the only route along which gustatory impulses were transmitted to the brain stem (Kitchell *et al.*, 1959; Duncan, 1960; Halpern, 1962; Kadono *et al.*, 1966; Landolt, 1970). This idea apparently was so strong that Gentle (1971a) tried to deprive chickens of oral taste by bilateral transection of the lingual branch of the glossopharyngeal nerve. For tasting behaviour proper, no significant changes between surgically-altered birds and controls were observed, although aphagia and reduced rates of pecking and intake did result. The presence of extralingual chemoreceptors, that is chemoreceptors outside the innervation field of the glossopharyngeal nerve (Gentle, 1978), gives a sounder explanation of the latter result than do the postulated post-ingestional factors which were earlier thought responsible. The aphagia, and the reduced intake and reduced pecking rates, are better explained by the purely tactile deprivation of the tongue base which is inevitably coupled with cutting the lingual nerve. These are identical to the deficits reported after trigeminal nerve transection in the pigeon (Zeigler, 1975; Zeigler *et al.*,

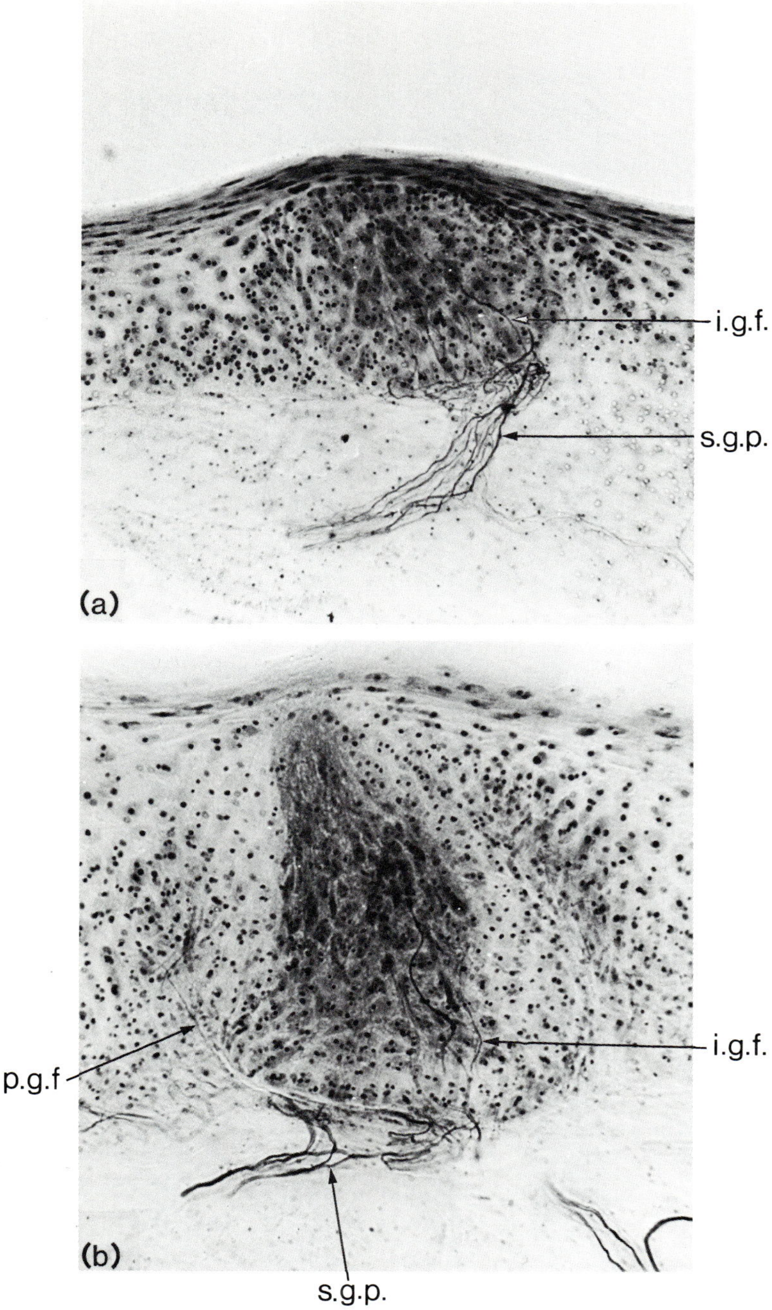

Fig. 9.11. Photomicrographs showing innervation of taste buds located in the oral mucosa of the lower jaw of the Carrion Crow (*Corvus corone*). Note the subgemmal plexus (s.g.p.) under the taste buds, and intragemmal fibres (i.g.f.) entering the bud between the sensory cells. p.g.f., Perigemmal fibres. (a) ×225. (b) ×360.

1975; see also Zeigler, 1976). Exactly which cranial nerve is responsible for the innervation of the different areas of taste buds was clarified by the work of Krol and Dubbeldam (1979). By selectively interrupting the palatine branch (*n. palatinus*) of the facial nerve (*n. facialis*), they were able to show that this nerve, travelling in the ventral premaxillary ramus (*r. premaxillaris ventralis*) of the medial ramus of the ophthalmic branch of the trigeminal nerve, innervates the rostral areas of taste buds in the upper bill of the Mallard. A supposed innervation of the area of taste buds in the lower jaw by fibres of the chorda tympani nerve, travelling in the *ramus sublingualis* of the trigeminal nerve, was postulated by Berkhoudt (1980). This is supported by earlier data (Cords, 1904; Chang Ho Yu, 1964; see also Bubién-Waluszewska, 1981). It is also supported by the electrophysiological findings of Gentle (1984), who was able to record gustatory activity from the chorda tympani nerve in the chicken following chemical stimulation of the taste buds associated with the rostral mandibular glands.

D. Central gustatory pathways in birds

On the central projection of gustatory information in birds comparatively few studies exist, in contrast to the situation in mammals (Norgren, 1977; Pfaffman, 1978). On the basis of degeneration studies, Dubbeldam *et al.* (1976) and Dubbeldam (1984), respectively, showed that the central representation of the chorda tympani nerve in the Mallard and pigeon is located in two distinct areas (both of which are subnuclei of the nucleus of the solitary tract, but not yet recognised by the *Nomina Anatomica Avium*): (1) in the *nucleus sensorius nervi facialis* (sVIId of Dubbeldam *et al.*, 1976), lying on the dorsum of the descending spinal tract of the trigeminal nerve (*tractus spinalis n. trigemini*), and (2) in the *nucleus ventrolateralis rostralis* (nVla of Dubbeldam, who also uses "anterior" instead of *rostralis*). For the glossopharyngeal nerve the main terminal fields, apart from sending a component to the main sensory nucleus of the trigeminal nerve (*nuc. sensorius principalis n. trigemini*) and to the descending trigeminal nuclei (*nuc. tractus spinalis n. trigemini*), were restricted to the rostral pole of the central rostral component of the nucleus of the solitary tract, and again to the same *nucleus ventrolateralis rostralis*. The fact that both the chorda tympani and the lingual nerve of IX converge on the same nucleus (i.e. the *nucleus ventrolateralis rostralis*) was the reason Dubbeldam (1977) proposed that this nucleus was involved in the processing of taste information. This tentative idea, elaborated in more detail by Dubbeldam *et al.* (1979), was supported by electro-physiological evidence demonstrated in the same area in the chicken by Gentle (1979). A further projection on an hypothesised secondary gustatory nucleus situated in the "stratum cellulare externum" in the caudoventral region of the hypothalamus, parallelling the situation in lower vertebrates (cf. Gentle, 1975b), was inferred by Gentle (1975a). However, in this case taste influence was measured indirectly by recording of the oral aversive behaviour following taste stimulation after diencephalic lesions. Earlier stimulation studies involving the same diencephalic areas (Gentle, 1973) showed that the oral behaviour which was elicited, such as exaggerated movements of the tongue and beak, closely resembled that obtained by stimulation of the *archistriatum*, the *tractus fronto-archistriaticus*, and the *tractus occipitomesencephalicus* (Putkonen, 1967). This last tract, described anatomically by Zeier and Karten (1971, 1973) has since been found to end on premotor structures in the *nuc. reticularis pars parvocellularis* (Rpc), a well-circumscribed area in the reticular formation associated with the initiation of jaw movements both in the pigeon (Berkhoudt *et al.*, 1982) and the Mallard (Arends and Dubbeldam, 1982). Therefore, the changes in oral behaviour observed by Gentle (1975a) may be more complex than simply attributable to gustatory hyposensitivity alone. It may be obscured by sensorimotor deficits resulting from intervention in two adjacent systems: (1) the ascending sensory quintofrontal tract and (2) the already-mentioned descending occipitomesencephalic tract. The first damage will result in sensorimotor deficits comparable to those described for the pigeon (Zeigler and Karten, 1973b; see also Zeigler, 1976). The second disruption will give mainly motor deficits like those described by Levine and Zeigler (1981) again in the pigeon. The prolonged aphagia observed in one-third of the experimental group suggests that this last proposition may be entirely realistic (Wright, 1968, 1975). One final problem which still needs further study is why not even one control bird, of which seven had lesions in the *tractus occipitomesencephalicus,* had no deficits (for instance the abolition of the oral aversion reaction), whilst six of the control birds showed a reduced rejection of quinine after the lesion. This probably cannot be attributed solely to species differences between the pigeon and chicken.

III. Avian taste behaviour

The early studies of avian taste perception were written in German, and again, just as for the early literature on taste bud anatomy, they are poorly represented in most subsequent works (Gentle, 1975a; Kare and Rogers, 1976). Rensch and Neunzig (1925) determined sensitivity thresholds for 60 species of birds to single chemical stimuli (salt, sour, bitter and sweet), mixed with the drinking water in a one-bottle test. The data for bitter stimuli are hard to compare with later results, as the stimulus not only involved gustatory but also visual and probably olfactory clues. Engelmann (1934) gave the results of two-bottle tests offering a wider range of chemicals in the same four classical taste modalities, adding saccharin and glycerin, in the pigeon, duck and domestic fowl. Among these three species the general acuity of taste discrimination diminished in the order pigeon, duck, chicken, but species differences for the various test stimuli were noted. For the chicken Engelmann (1937) distinguished four groups of taste reactions ranging from preference to aversion. Group I was palatable and preferred over pure water; natural sugars were in this group. Group II ($MgCl_2$ and glycerin) showed preference over groups III and IV and was only offensive in higher concentrations. Group III (NaCl) was preferred over group IV, and again was only offensive in high concentration. Group IV (acetic and hydrochloric acid) was offensive in low concentrations.

For taste behaviour in other birds, species as well as individual differences exist. The differences in taste behaviour for the four main taste qualities will now be summarised, the earlier German data being added to the excellent reviews by Gentle (1975a) and Kare and Rogers (1976) of the modern work on taste behaviour in birds.

A. Reactions to sweet stimuli

Both Rensch and Neunzig (1925) and Engelmann (1934, 1937, 1960) reported a modest preference in many species for natural sugars mixed with the drinking water. Similar preferences are shown by many other birds like the Bobwhite (*Colinus virginianus*) and Japanese Quail (*Coturnix japonica*) (Hamrum, 1953; Brindley, 1965; Brindley and Prior, 1968; Harriman and Milner, 1969), Great Tit (*Parus major*) (Warren and Vince, 1963), pigeon (Duncan, 1960), chicken (Gunther and Wagner, 1971; Gentle, 1972, 1975a), and parrots, the Budgerigar (*Melopsittacus undulatus*), and the Broad-tailed Hummingbird (*Selasphorus platycercus*) (Kare and Rogers, 1976). In many other birds indifference was reported for many sugars, but only xylose was rejected (Kare and Medway, 1959; Kare, 1961, 1965; Kare and Rogers, 1976). Gentle (1975a), however, showed that the difference between preference and indifference arises from the fact that in short-term trials taste preference is measured, whereas in long-term trials post-ingestional feedback masks the initial preferences. The preference behaviour can be modified by caloric dilution of the diet (Kare and Ficken, 1963; Kare and Maller, 1967). Kare and Rogers (1976) raised the idea that sweet taste-active proteins, contained in many tropical fruits (Cagan, 1973), may aid in seed dispersal by birds. This suggestion may be supported by a note by Davison (1962), who showed that taste and not colour draws birds to berries and seeds. This contrasts with the observation by Engelmann (1957) who, for the fowl, found that the preference for different seeds is mediated by tactile instead of gustatory clues. Likewise Kare and Rogers (1976), citing work on other birds (Morris, 1955; Kear, 1960), stated that visual properties and the surface texture of the food have precedence over all other qualities in the bird's selection of food (see also Stevens and de Bont, 1980). M. J. Gentle (personal communication) noted for hummingbirds that the nectar of most flowers in Arizona is composed of sucrose, glucose and fructose (Percival, 1961). Although studies exist on tongue adaptations in these nectar-feeding birds (Bradley, 1971; Reed Hainsworth, 1973) and taste may be a factor in selection of nectar (Reed Hainsworth, 1973; Reed Hainsworth and Wolf, 1976), no studies are reported yet on the distribution and numbers of taste buds in any of these birds, except for Weymouth *et al.* (1964), who mentioned occasional taste buds on the caudal parts of the tongue.

B. Reactions to salt stimuli

Rensch and Neunzig (1925) determined sodium chloride thresholds in 60 different species of birds. The range observed was large, being very low in parrots (0·35%) and pigeons (0·5–0·9%) and extremely high in partridge (20%) and siskin species (37%). Engelmann (1934) also showed that pigeons rejected very low concentrations of sodium chloride (0·75%, 0·15 *M*), which were taken without hesitation by duck and fowl (thresholds 1·2%, 0·2 *M*; 1·5%, 0·25 *M*, respectively). Baby chicks were found by Engelmann (1950) to be more sensitive than adults, being about as sensitive as the pigeon (0·7%, 0·12 *M*). Similar values (0·9%, 0·15 *M*) were obtained by Pick and Kare (1962). Kare and Rogers (1976), citing Kare and Beily (1948), stated that chicks avoid higher concentrations of sodium chloride because they exceed the ability of the chick's kidney to handle it.

Kare and Rogers (1976) also listed the preferences of chicks for various concentrations of sodium salts and chloride metallic solutions, and could not relate the acceptability to any physical or chemical theory. Kare and Rogers (1976), citing Wood-Gush, stated that domestic fowl, held on a sodium- or calcium-deficient diet, develop a specific appetite that corrects for this deficiency. Later Hughes and Wood-Gush (1971) were able to substantiate this finding for calcium, but were unable to do so for sodium (Hughes and Wood-Gush, 1970). Many birds with salt glands will drink seawater, but the Herring Gull (*Larus argentatus*) will select pure water over salt solutions when given the choice (Harriman and Kare, 1966). As is already clear from the work of Rensch and Neunzig (1925), birds without salt glands can have a very wide tolerance (Bartholomew and Cade, 1958; Hamrum, 1953; Kare and Ficken, 1963; Warren and Vince, 1963). Mariotti and Fiore (1980), using operant-conditioning techniques, reported a discrimination threshold for the pigeon of 0·2% (0·034 *M*) sodium chloride, a factor four times lower than that reported with the usual two-bottle choice technique.

C. Reactions to sour stimuli

Rensch and Neunzig (1925) found that most of the birds they examined were very stenotolerant for acetic acid mixed with the drinking water, most thresholds being below 1% (0·17 *M*). Engelmann (1934) obtained similar results, the pigeon again being much more sensitive than the duck and domestic fowl (0·01 *M* in pigeon, versus 0·02 and 0·025 *M* in duck and fowl, respectively). In trials with hydrochloric acid, Engelmann (1950) again proved that the taste sensitivity in baby chicks was more pronounced than in adults (0·009 versus 0·025 *M*). Kare and Rogers (1976) reported that the Common Starling (*Sturnus vulgaris*) and Herring Gull (*Larus argentatus*) readily accept hydrochloric acid. They also cited Fuerst and Kare (1962), who concluded that chicks tolerate strong hydrochloric acid solutions, that is pH 2. Duncan (1960, 1964) showed once more that pigeons are surprisingly sensitive to acids, especially to hydrochloric acid. By contrast, Brindley (1965) and Brindley and Prior (1968) reported a preference for 0·05% HCl in the Bobwhite (*Colinus virginianus*) and Japanese Quail (*Coturnix japonica*). Gentle (1972), using both percentage preference and EEG arousal time, showed that the addition of glucose reduced the initial aversion of chicks to acid solutions in comparison with testing the two solutions separately. Kare and Rogers (1976) finally listed the percentages of daily intake in the chicken of different acids and bases at different pH levels.

D. Reactions to bitter stimuli

As already mentioned, the data of Rensch and Neunzig (1925) for bitter stimuli are hard to evaluate, as "Aufschwemmung von Aloë" is not a readily-reproduced stimulus. For quinine, Engelmann (1934) showed that the pigeon, which had been most sensitive in tests with salt and sour, seemed not to perceive the stimulus at all; in contrast, the duck and chicken showed a marked rejection at low concentrations. This contrasts with findings for the pigeon by Duncan (1960), who described a marked rejection at 0·1%. Gentle (1975b, 1976a) used quinine hydrochloride as a standard stimulus to test gustatory sensitivity following brain lesion in the chicken. He also found that quinine-HCl acceptability changed significantly on water deprivation (Gentle, 1976b). Sucrose octa-acetate (SOA), which tastes extremely bitter to humans, according to Kare and Rogers (1976) is taken readily by the Herring Gull and the chicken; the Bobwhite and Japanese Quail, however, showed aversive reactions (Brindley, 1965). Cane and Vince (1968) were able to demonstrate that Bobwhites gradually develop this ability to discriminate SOA, but it took much statistical manipulation to see this in a scatter of points around the line of 50% preference. Dimethylanthranilate was found highly offensive by a number of birds (Yang and Kare, 1968; Rogers, 1974). This compound is not only used as a repellent, but also to reduce intake in growing chicks and turkey poults (Kare and Rogers, 1976). Many insects and other arthropods secrete or contain highly-offensive, often bitter constituents, which can serve as a protection against predators (Yang and Kare, 1968; Brower, 1969). Because of their offensive taste, these compounds may help in teaching the bird to associate noxious effects with eating unpleasant or poisonous prey. By inducing illness, Wilcoxon *et al.* (1971) found that Bobwhites could associate a sour taste in their drinking water with illness, but that visual clues dominated over taste (see also Gillette *et al.*, 1980).

E. Reactions to other chemicals

Apart from the reactions to the four basic categories of taste, many reactions to more complex compounds and synthetic flavours have been reported (Kare *et al.*, 1957; Romoser *et al.*, 1958; Kare and Medway, 1959; Kare and Pick, 1960; Deyoe *et al.*, 1962). Many of these studies were undertaken for economic reasons, that is to study the effect on the rate of gain in body weight or feed conversion. It is clear from all studies that mixing flavours with the drinking water had more effect than food-addition (Kare and Pick, 1960; see also Gillette *et al.*, 1980). Very few studies deal with natural taste compounds. Van Heezik *et al.* (1983) were able to show that the Sanderling (*Calidris alba*) and Dunlin (*Calidris alpina*) could discriminate between jars filled with sand containing "taste" and "no-taste". Taste came from worms, which had been removed before the experiment. Similar results were obtained for two further species, the Purple Sandpiper (*Calidris maritima*) and Red Knot (*Calidris canutus*) (Gerritsen *et al.*, 1984). Engelmann (1957) tried to relate taste sensitivity to the total number of sensory cells present. The decrease of sensitivity to salts and acids in the order pigeon, duck, chicken was connected with the overall number of sensory cells by multiplying the number of taste buds by the number of sensory cells in each taste bud (based on Bath, 1906a). It is unfortunate that the numbers of taste buds he cited are inaccurate (for correct numbers see Table II).

IV. Electro-physiological studies of avian taste

Comparatively few studies exist, and only chickens and pigeons have been investigated (Kitchell *et al.*, 1959; Halpern, 1962; Kadono *et al.*, 1966; Landolt, 1970). All studies implicitly assume that taste buds occur only in the caudal region of the tongue, and that the lingual branch (*n. lingualis*) of the glossopharyngeal nerve is the only nerve in which taste responses are to be detected. Many of the results show a close correlation between electro-physiological and behavioural data. Both Kitchell *et al.* (1959) and Landolt (1970), for instance, were unable to record electro-physiological responses in the pigeon on quinine stimulation, a finding reported earlier from behavioural experiments by Engelmann (1934); in contrast, Duncan (1960) reported a rejection of this compound. Discrimination between $KHCO_3$ and $NaHCO_3$ based on operant conditioning, as observed in the pigeon by Mariotti and Fiore (1980), seems to be supported by the ranking of the K^+ after the Na^+ ion in what Landolt (1970) called the "ionic stimulation efficiency factor". For the chicken, Kadono *et al.* (1966) showed that the sensory threshold as established electro-physiologically is considerably lower than the preference threshold found in behavioural studies. They found the following thresholds: sodium chloride, 0·025 *M;* sucrose, 0·125 *M;* quinine-HCl, 0·01 *M*. Responses were obtained with 0·1 *M* acetic acid, 0·05 *M* citric acid, and 0·01 *M*. hydrochloric acid. The fact that the electro-physiological sensory threshold is much lower is in accordance with the finding of Halpern (1962), who showed that 0·0002 *M* of sucrose octa-acetate produces electro-physiological responses in the lingual nerve, even though it was accepted in preference tests (Kare, unpublished data). The single-unit recordings in the *nucleus ventrolateralis rostralis* of the chicken (Gentle, 1979) cannot be compared with the recordings from the lingual nerves, because the latter were obtained from either whole nerves or multi-unit preparations. However, the burst-like clusters in the firing rates observed by Landolt (1970) in the lingual nerve are also found in the *nucleus ventrolateralis rostralis.* No response to quinine-HCl was reported in the study by Gentle (1979), but this may be due to the small number of units sampled.

Another electro-physiological method of measuring taste sensitivity has been to look at arousal changes in the gross electrical activity of the brain surface, the EEG (Gentle, 1972, 1975a). This may provide a measure of sensitivity intermediate between the receptor threshold (as obtained in the peripheral nerve) and the behavioural preference threshold (Gentle, 1982). For 0·05 *M* acetic acid or hydrochloric acid, which did not give any behavioural effect, large increases in the arousal length were reported in the EEG (Gentle, 1975a, 1982). This is in accordance with the observations of Mariotti and Fiore (1980) using operant conditioning in the pigeon, who also found a much lower discrimination threshold than those reported from preference tests.

V. Feeding mechanics and the role of taste

The ability to formulate ideas about precisely which areas of taste buds are involved in screening the chemical composition of food and water during the successive phases of intake, requires an adequate analysis of the oral structures which are in contact with the food (or water) during each phase. This implies that the actions of the epidermal structures of the oropharynx have to be carefully analysed during the process of gathering, transport and deglutition (Zweers, 1982a). In only a small number of avian species has a functional morphological analysis been carried out in enough detail to serve this purpose. Zweers *et al.*

(1977) were able to do so for straining behaviour in the Mallard, and indeed the pathway of the food during straining accurately reflects the distribution of taste buds (Fig. 9.12). During pecking (diagram on right of Fig. 9.12), where transport is much more discontinuous (J. G. M. Kooloos, personal communication), the coincidence of the areas of taste buds and the areas of contact with the food is even more striking. Although no further study has yet been carried out on the tasting behaviour of this bird, it is evident that taste discrimination is hierarchically organised and that at different levels intake behaviour may be modified and interrupted. More data are available for the modification by taste of intake behaviour in pigeon and chicken. Before this modified behaviour is discussed, the sequence of events during the normal food intake of these birds should first be sketched. Both pigeons and chickens are extremely granivorous, using the pecking technique described by Zweers (1982a). Zeigler *et al.* (1975; see also Zeigler, 1976) illustrated the spatiotemporal organisation of pecking in the pigeon as did Hutchinson and Taylor (1962) for pecking in the chicken. Zeigler described only the first part of the pecking behaviour in detail (i.e. approach, beak opening and grasp), since his objective in studying the neuro-behavioural basis of visually-guided grasping behaviour in the pigeon was to establish a paradigm for visually-guided grasping in humans (H. P. Zeigler, personal communication). Therefore, as soon as the tongue becomes involved in transport of the seed, the rest of the

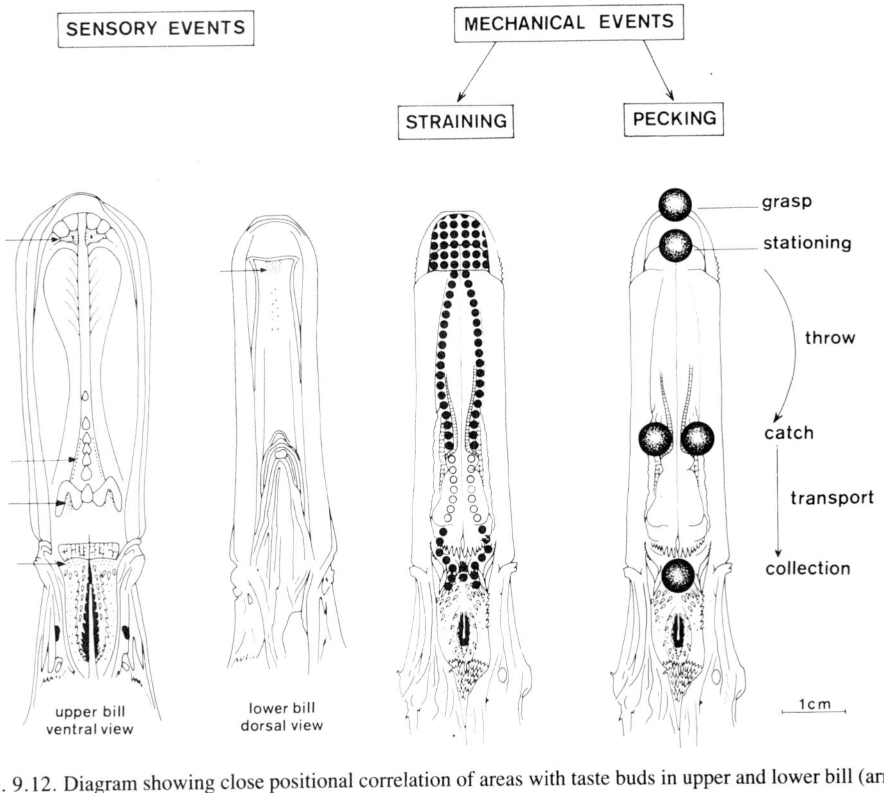

Fig. 9.12. Diagram showing close positional correlation of areas with taste buds in upper and lower bill (arrows) of the Mallard (*Anas platyrhynchos*) with the pathway of the food during straining and pecking. ×0·82. After Berkhoudt (1977, 1980).

behaviour is described relatively unspecifically as "mandibulation" and "swallowing". These last two behavioural sequences, showing intricate cooperation of head, beak and tongue actions, were further specified by Zweers (1982a,b), who gave a most detailed analysis of pecking in the pigeon using high-speed and X-ray cinematographic techniques. In addition to the familiar "catch-and-throw" transporting mechanism already known from the analysis by Zeigler (1976), he added an extra mechanism by which the small seeds are preferentially ingested, the "slide-and-glue" mechanism. Intermediate variants of the two techniques may occur (Zweers, 1982a,b). The mechanical events during pecking in the pigeon consist in principle of the following eight phases [based on Zweers (1982b) and Zeigler (1976)]:

1. *Preparatory phase,* consisting of an *arousal phase* resulting in visually-guided searching behaviour, and an *orienting phase* resulting in approach and fixation of the seed.
2. *Final approach* or *peck phase,* which is characterised by head descent and eye closure.
3. The *grasp phase,* where the seed is actually seized between the beak tips.
4. A *repositioning* or *stationing phase,* which occurs when particularly large seeds are not exactly in the preferred position to elicit the next throw phase for transport.
5. A *transport phase* (or mandibulation), in which small seeds are transported through the mouth by the "*slide-and-glue*" mechanism, consisting of tongue protraction during which the seed is glued to the tongue tip by saliva, followed by tongue retraction. Large seeds are transported by the more common "*catch-and-throw*" variant, where the tongue plays no direct role. Intermediate variants occur.
6. *Collection phase I,* where the food stays on the lingual base.
7. *Swallowing phase.*
8. *Collection phase II,* where the food stays in the rostral oesophagus.

For the mechanics of pecking in the chicken, Hutchinson and Taylor (1962) distinguished similar phases which were at that time called (1) interval between pecks, including fixation, (2) striking at the grain (same as peck phase), (3) prehension (same as grasp phase), (4) withdrawal (first throw or jerk), and (5) jerking the grain into the throat (same as second throw and catch). More data on deglutition in the fowl come from McLelland (1975, 1979), citing cineradiographic and cine-endoscopic observations reported in an abstract and doctoral thesis by White (1968, 1970, respectively). Glueing by saliva is again mentioned, but in a later stage than the tongue-tip glue in the pigeon. Also the description of the way the tongue moves the seed caudally differs from the description for the pigeon by Zweers (1982b). On the similarities and differences between the two species the following points can be made. The first is our observation that during the ingestion of small seeds the chicken uses the "tongue-tip glue" technique like the pigeon (Zweers, 1982b), and that no second throw occurs. Therefore, Plate I of Hutchinson and Taylor (1962), which shows pecking of small grains of wheat, illustrates either a next approach to a new seed or a complete withdrawal (i.e. second throw). This second throw, however, does occur if the chicken is fed large seeds, for example peas. To show the exact position of the pea during transport in

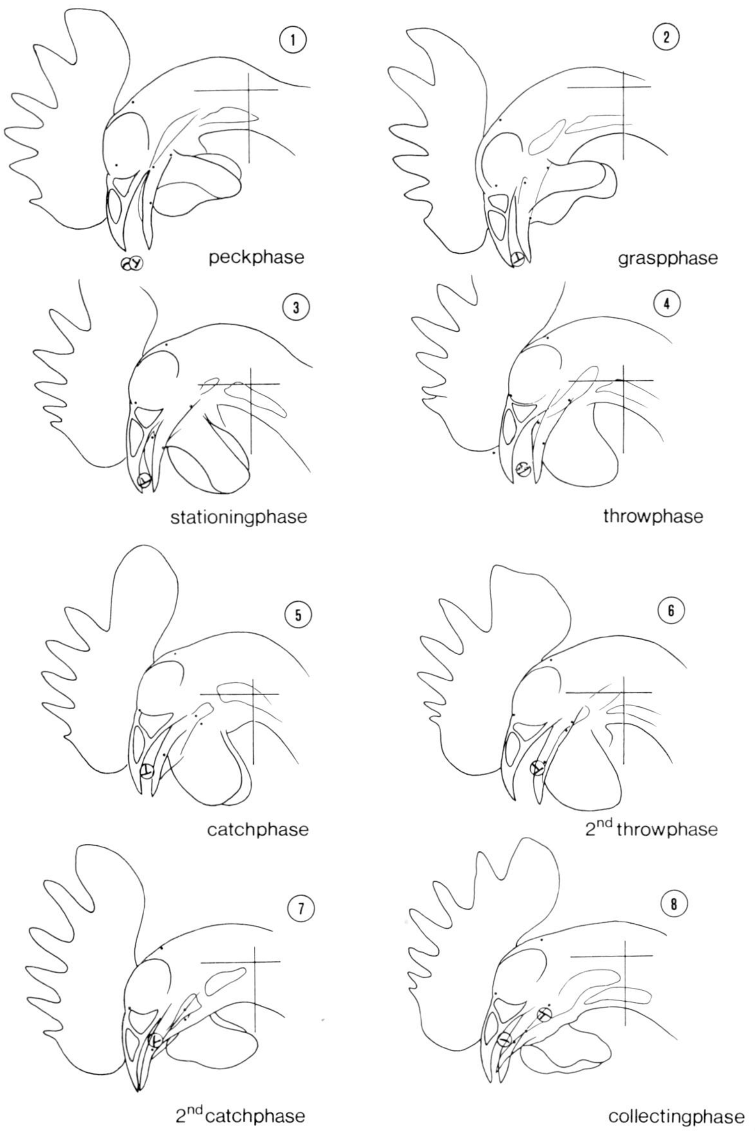

Fig. 9.13. Outline drawings based on a series of individual frames from X-ray film of pea-pecking in the chicken, showing the various phases of intake behaviour. Frames 1 to 7 are from the same pecking scene, but are not all adjacent frames. Frame 8 is from another scene. Based on unpublished cineradiographic data.

relation to the position of beak and tongue, see Fig. 9.13. The details of the mechanism of pecking can be disregarded here, since it is only necessary to know the position where the seed contacts the different epidermal structures of the oropharynx. For pea-pecking these positions can be summarised as in Fig. 9.14. As already stated, the pigeon adopts a similar procedure for large seeds (Zweers, 1982a,b). For small seeds, the contact with the rostral part of the palate will occur earlier and more frequently, in both the pigeon and the chicken (Hutchinson and Taylor, 1962; Zweers, 1982a,b). The caudally-pointing papillae on the palate prevent the seed being carried forwards during the next lingual protraction (White, 1968, 1970; Zweers, 1982a,b).

Zweers (1982a,b) was able to propose a so-called comparator model for the neurosensory control of food intake (see also Zeigler, 1976), combining the detailed results on the mechanics of pecking with information on the sensory organs. He postulated cerebral decision units where the actual incoming sensory input, registering the food properties, is compared with a pre-set stimulus expectancy value, which can be either innate or result from learning. When the criteria of this expectancy value have been satisfied, an adequately-timed, next mechanical unit will be recruited. It is tempting to speculate on such a model, and it may stimulate further lines of enquiry. However, one complication should be emphasised. In the forebrain alone there are at least four different neural structures which are candidates for such a role in mechanoreception: the *nucleus basalis,* the *neostriatum,* the *hyperstriatum ventrale,* and the *archistriatum* (Dubbeldam, 1984; Veenman, 1984; Wild *et al.,* 1984). For taste no data are yet available on any telencephalic control (Dubbeldam, 1984).

The role of taste in food intake in the pigeon will now be considered. My own observations have included a study of the effects of feeding pigeons with peas dipped in bitter (0·5 *M* $MgCl_2$) or acid (0.01 *M* hydrochloric acid) solutions. Bitter peas were taken without hesitation, but peas dipped in acid were rejected. In deprived pigeons confronted with these peas for the first time, the pea was almost swallowed whilst in collection phase I (phase 6, p. 485). When the bird "realised" that something was wrong, it stopped further intake, closed its eyes and, by vigorous head shaking and vomiting, forcefully ejected the unwanted food item. The next pea, without taste, was approached hesitantly, but it was swallowed normally. When in a next trial a new acid pea was given, the bird grasped it, stationed it (phase 4, p. 485) but immediately rejected it, this being the first level where taste can be screened, that is at the stationing level. From these experiments and our findings on the distribution of taste buds in the pigeon,

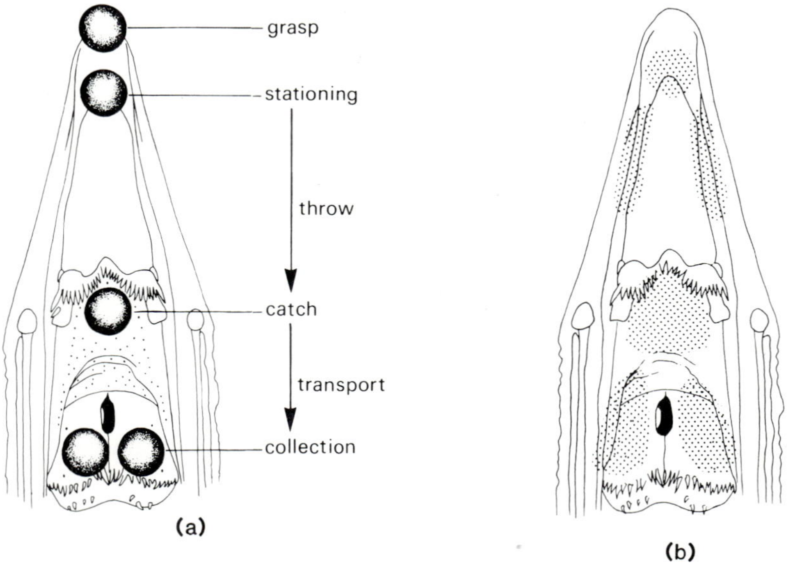

Fig. 9.14. Diagrams of the floor of the oropharynx of the chicken summarising (a) the positions of contact during the different phases of pecking of large seeds in the chicken, and (b) the coinciding distribution of taste buds (hatched areas). Based on van Prooije (1978); see also Table I.

Zweers (1982a,b) was able to give five possible positions in which taste may influence intake behaviour, coinciding with the phases in Fig. 9.14a. The positions occur (1) during stationing, (2) during contact with the rostral palate in the transport of small seeds, (3) during contact with the lingual base during the catch phase and subsequent lingual retraction phase, (4) during contact with the laryngeal mound (*mons laryngealis*) and/or (5) during contact with the caudal palate—both during the collection phase. From our experiments with peas dipped in various solutions, it is clear that screening of the taste of a food particle works as a safety mechanism, a warning device which in an unwanted situation results in a learned aversive reaction, a reaction which can be acquired within a few trials. That such a selection mechanism operates may be concluded from observation of food preferences in pigeons, in which the rapid adjustment to new seeds (Moon and Zeigler, 1979) may be mediated also by taste. For the chicken, too, aversion reactions to chemical stimulation as just shown for the pigeon have been reported, but in this case the taste substances were generally mixed with the drinking water instead of being added to the food (Gentle, 1976a, 1978; Gillette *et al.*, 1980). The sensory control of food intake was reviewed by Gentle (1979), who showed that oropharyngeal factors not only play a role in food selection but also provide an important positive reinforcement for the maintenance of feeding and drinking. Van Prooije (1978) was able to demonstrate the pathway followed by the food by presenting seeds dipped in a methylene blue solution, which left a clear blue track in the oropharynx. On the basis of her findings and the analysis of the X-ray film of pecking in the chicken (Fig. 9.13), we know that the positions of contact with the food (large seeds, peas) is as shown in Fig. 9.14a. These positions closely coincide with the areas of taste buds described by van Prooije (see Table I, and the hatched areas in Fig. 9.14b). The positions of contact with the food and the distribution of taste buds in the pigeon are so similar (Zweers, 1982b) that there is no need for an additional Figure for this species here (see Berkhoudt, 1984).

Oral aversion behaviour in the chicken was reviewed by Gentle (1982), who concluded that the oral behavioural repertoire to oral stimulation in the chicken consists of three main components: (1) beak–tongue movements, (2) head shake and (3) beak wipe. In oral aversion behaviour the recorded number of beak–tongue movements, head shakes and beak wipes proved to be proportional to both the types and concentrations of the chemicals tested (Gentle, 1978, 1982; Gentle and Harkin, 1979). The identity of the brain areas which control the different components of this oral behaviour is not completely clear (Gentle, 1981). Gentle stressed that the full repertoire of oral behaviour is also shown after stimulation with water; this fact may indicate that "beak–tongue movements" are too unspecified and probably underdefined as a behavioural entity, just like "mandibulation" was for the pigeon. During drinking in the Mallard, very specific movements, and also EMG patterns of beak and tongue muscles, coincide with the various sub-phases constituting drinking, that is, "immersion", "tipping-up" and "swallowing" (J. G. M. Kooloos, personal communication); this suggests that on close examination it may prove possible to distinguish "tasting" beak–tongue movements from "lapping" or "swallowing" sequences. This will necessitate a much more detailed analysis of both drinking and "tasting" behaviour in birds. The fact that "beak–tongue movements" persisted after forebrain lesions, whilst head shake and bill wipe were completely abolished (Gentle, 1981), probably may also be seen in this light. Responsiveness to chemical stimulation using Gentle's approach has been measured by Vince (1977) in the chicken embryo. Significant differences were found when the taste substances were mixed with egg fluid instead of distilled water. No preference or aversion was concluded from the changes in the different behavioural parameters following the various chemicals tested. One final point on oral aversion behaviour has to be made. Gentle and Dewar (1981) and Gentle *et al.* (1981) showed a significant decrease in the oral aversion reaction in chicks on a deficient diet for vitamin A and zinc, respectively. Probably this has more to do with increased friction, because of a lack of lubrication resulting from a loss of salivary secretion, than with gustatory effects.

Finally, two examples will be given from species in which one would not immediately think of gustatory reception as having any importance for feeding. The first is the great abundance of taste receptors in the Carrion Crow (*Corvus corone*) (Figs. 9.11a,b, 15). When given the same chemical stimulants (HCl and $MgCl_2$) as presented to our pigeons, no reaction to bitter stimuli occurred, but a profound reaction to sour stimuli was found. The many gustatory receptors in this species may be correlated with feeding on insects that might contain offensive or toxic constituents, as described for the Blue Jay (*Cyanocitta cristata*) by Brower (1969). Another possibility may be that assessment of the "quality" of carrion is a very tricky business, requiring careful testing before savouring. The second example is based on film observations of a Great-crested Grebe (*Podiceps cristatus*) which was saved from starvation during wintertime and fed on African

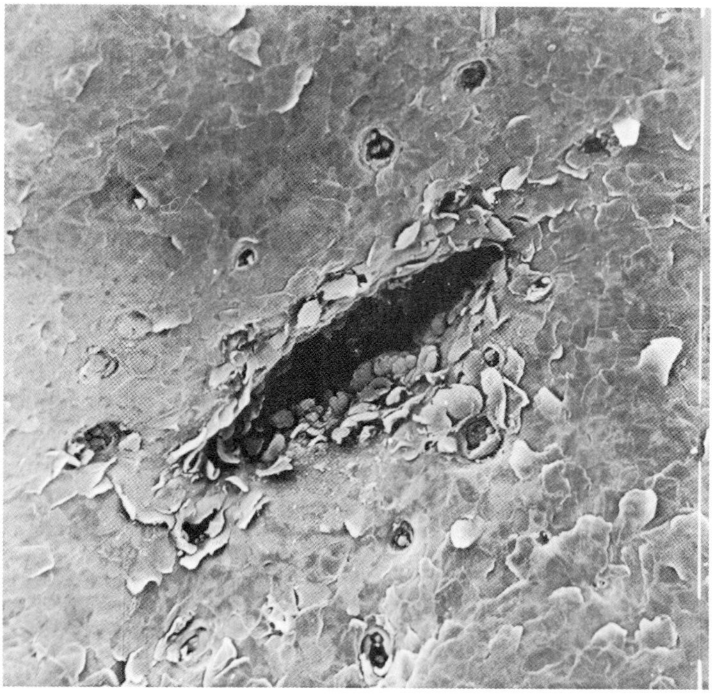

Fig. 9.15. SEM showing a cluster of taste buds located in the lower jaw, around the most rostral opening of the rostral mandibular salivary gland in the Carrion Crow (*Corvus corone*). ×35.

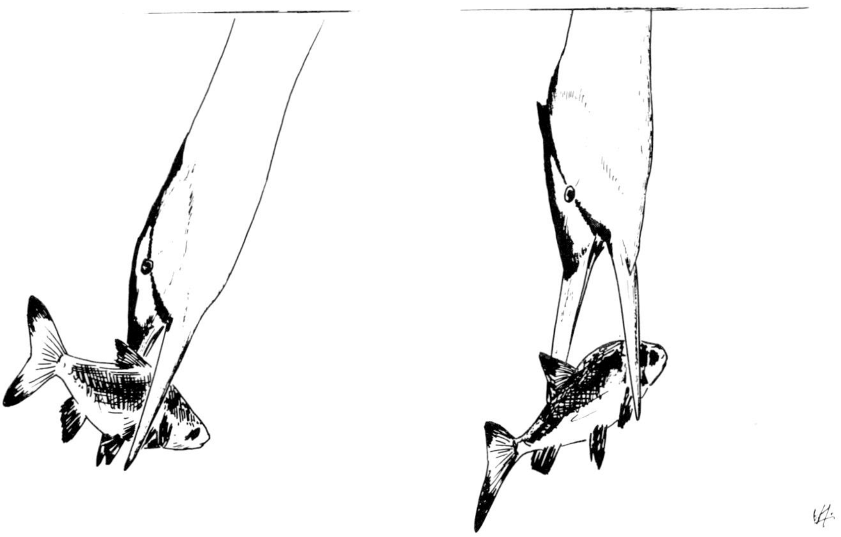

Fig. 9.16. Two individual cine frames from a Great-crested Grebe (*Podiceps cristatus*) rejecting an infected specimen of *Leuciscus leuciscus*, almost certainly on the basis of gustatory information. Note the position in the beak where the fish is being held during rejection; this position coincided with the site of taste buds in the mandibular region.

cichlids, which were kept in tanks in our laboratory. The bird had already swallowed a dozen of these *Haplochromis* species when we presented it with some leftover freshwater fishes (*Leuciscus* spp.). The bird darted three times to one of these fishes, oriented it head-forward, but refused to take it three times. The second refusal is shown in Fig. 9.16; the bird tested the fish at about one-third of the bill length and then released it. On inspection the refused specimen showed a severe infection of the slime layer of the epidermis. Later pontamine sky-blue staining and SEM inspection of preserved specimens of *Podiceps cristatus* showed that indeed taste buds were found in the mandibular region in the same position where rejection occurred. These two examples may show that much research on taste in birds still has to be done, especially in the normal environment and with more natural stimuli than the simple stimulants used in the laboratory to establish preference–aversion levels.

Acknowledgements

I gratefully acknowledge the discussions with my colleagues Dr. J. L. Dubbeldam, Drs. J. G. M. Kooloos and Dr. G. A. Zweers. Special thanks go to the head of the department, Professor Dr. P. Dullemeijer, and my former mentor Professor Dr. H. Ph. Zeigler, Department of Ornithology, American Museum of Natural History, New York, for their support and stimulation. I am greatly indebted to Dr. Judith Ganchrow, Hadassah Medical School, Jerusalem, Israel, for sending me her complete bibliography on avian taste, to Dr. Michael J. Gentle (Poultry Research Centre, Roslin, United Kingdom) for generously giving me access to his latest material, commenting on this manuscript, and many personal communications. Special thanks go to Dr. Raisa Pevzner and Dr. N. A. Dmitrieva from the Sechenov Institute of Evolutionary Physiology, Leningrad, for having allowed me to incorporate their EM data on lingual taste buds in the pigeon and for providing translations of their main findings from the Russian. In our laboratory the skilful microtechnical aid of Mr. C. Noorlander for the Epon-material and especially of Ms. Ardie Visser for semithin sections, silver staining and scanning electron microscopy is gratefully acknowledged. I thank the many students who contributed to this project, among them Ms. Annette van Prooije, Ms. Sylvia van Kan, Mr. Martin Corino, Mr. Peter van Noort and many others. Mr. J. Simons assisted in making the X-ray film of pea-pecking in the chicken. Dr. Melanie Stiassny carefully corrected the English text. The illustrations were made by our artists Mr. Cor Elzenga and Mr. Jan Herzberg. All photography was performed by Mr. Adrie 't Hooft. Ms. M. K. I. Vogelsang-Liem (of this department) and Mrs. M. Thompson (University of Liverpool) typed the manuscripts.

References

Arends, J. J. A. and Dubbeldam, J. L. (1982). Proprioceptive and exteroceptive afferents to the trigeminal and facial motor nuclei in the Mallard (*Anas platyrhynchos L.*) *J. comp. Neurol.* **209,** 313–329.

Ariëns Kappers, C. U., Huber, G. C. and Crosby, E. C. (1936). "The Comparative Anatomy of the Nervous System of Vertebrates Including Man", Vol. 1. (Reprinted by Hafner Publ. Co., New York, 1960.)

Bartholomew, G. A. and Cade, T. J. (1958). Effects of sodium chloride on the water consumption of house finches. *Physiol. Zool.* **31,** 304–310.

Bath, W. (1906a). Die Geschmacksorgane der Vögel und Krokodile. *Arch. für Biontologie* **1,** 5–47.

Bath, W. (1906b). Die Geschmacksorgane der Vögel und Krokodile. Thesis, University of Berlin.

Berkhoudt, H. (1976). The epidermal structure of the bill tip organ in ducks. *Netherl. J. Zool.* **26,** 561–566.

Berkhoudt, H. (1977). Tastebuds in the bill of the Mallard (*Anas platyrhynchos L.*). Their morphology, distribution and functional significance. *Netherl. J. Zool.* **27,** 310–331.

Berkhoudt, H. (1980). The morphology and distribution of cutaneous mechanoreceptors (Herbst and Grandry corpuscles) in bill and tongue of the mallard (*Anas platyrhynchos L.*). *Netherl. J. Zool.* **30,** 1–34.

Berkhoudt, H. (1984). The role of oral exteroceptive sense organs in avian feeding behaviour. *Proc. 1st Int. Symp. Vertebr. Morph., Giessen, 1983, Fortschritte der Zoologie* **30** (in press).

Berkhoudt, H., Klein, B. G. and Zeigler, H. P. (1982). Afferents to the trigeminal and facial motor nuclei in Pigeon (*Columba livia L.*): Central connections of jaw motoneurons. *J. comp. Neurol.* **209,** 301–312.

Botezat, E. (1904). Geschmacksorgane und andere nervöse Endapparate im Schnabel der Vögel (vorläufige Mitteilung). *Biol. Zbl.* **24,** 722–736.

Botezat, E. (1906). Die Nervenendapparate in den Mundteilen der Vögel und die einheitliche Endigungsweise der periphere Nerven bei den Wirbeltieren. *Z. wiss. Zool.* **84,** 205–360.

Botezat, E. (1910). Morphologie, Physiologie und phylogenetische Bedeutung der Geschmacksorgane der Vögel. *Anat. Anz.* **36,** 428–461.

Bradley, R. M. (1971). Tongue topography. *In* "Handbook of Sensory Physiology" (L. M. Beidler, Ed.), Vol. IV/2. Springer-Verlag, Berlin and New York.

Brindley, L. D. (1965). Taste discrimination in bobwhite and japanese quail. *Anim. Behav.* **13,** 507–512.

Brindley, L. D. and Prior, S. (1968). Effects of age on taste discrimination in the bobwhite quail. *Anim. Behav.* **16,** 304–307.

Brouwer, J. N. and Wiersma, A. (1978). Location of taste buds in intact taste papillae by a selective staining method. *Histochemistry* **58,** 145–151.

Brower, L. P. (1969). Ecological chemistry. *Sci. Am.* **220,** 22–29.

Bubién-Waluszewska, A. (1981). The cranial nerves. *In* "Form and Function in Birds" (A. S. King and J. McLelland, Eds.), Vol. 2. Academic Press, London and New York.

Cagan, R. H. (1973). Chemostimulatory protein: A new type of taste stimulus. *Science, Wash., D.C.* **181,** 32–35.

Cane, V. R. and Vince, M. A. (1968). Age and learning in quail. *Br. J. Psychol.* **59,** 37–46.
Chang Ho Yu (1964). Nervus facialis and nervus acusticus in the domestic fowl. (In Chinese.) *Tung Wu Hsueh Pao* **16,** 539–554.
Cords, E. (1904). Beiträge zur Lehre vom Kopfnervensystem der Vögel. *Anat. Hefte, Abt. 1.* **26,** 49–100.
Davison, V. E. (1962). Taste, not colour draws birds to berries and seeds. *Audubon Mag.* **64,** 346–350.
de Lorenzo, A. J. D. (1963). Studies on the ultrastructure and histophysiology in chemoreceptors. *In* "Olfaction and Taste" (Y. Zotterman, Ed.). Vol. I. Pergamon Press, New York.
Deyoe, C. W., Davies, R. E., Krishnan, R., Khaund, R. K. and Couch, J. R. (1962). Studies on the taste preference of the chick. *Poult. Sci.* **41,** 781–784.
Dmitrieva, N. A. (1981). Fine structural peculiarities of the taste buds of the pigeon *Columba livia.* (In Russian.) *Dokl. Akad. Nauk SSSR* **23,** 874–878.
Dubbeldam, J. L. (1977). "Touch" and "taste" within the brainstem of the mallard. *18th Dutch Fed. Meet., Leiden,* p. 185 (abstr.).
Dubbeldam, J. L. (1984). Afferent connections of nervus facialis and nervus glossopharyngeus in the pigeon (*Columba livia*) and their role in feeding behaviour. *Brain Behav. Evol.* **24,** 47–57.
Dubbeldam, J. L., Karten, H. J. and Menken, S. B. J. (1976). Central projections of the chorda tympani nerve in the mallard, *Anas platyrhynchos L. J. comp. Neurol.* **170,** 415–420.
Dubbeldam, J. L., Brus, E. R., Menken, S. B. J. and Zeilstra, S. (1979). The central projections of the glossopharyngeal and vagus ganglia in the mallard, *Anas platyrhynchos L. J. Comp. Neurol.* **183,** 149–168.
Duncan, C. J. (1960). The sense of taste in birds. *Ann. appl. Biol.* **48,** 409–414.
Duncan, C. J. (1964). The sense of taste in the feral pigeon: The response to acids. *Anim. Behav.* **12,** 77–83.
Engelmann, C. (1934). Versuche über den Geschmackssinn von Taube, Ente und Huhn. *Z. vergl. Physiol.* **20,** 626–645.
Engelmann, C. (1937). Vom Geschmackssinn des Huhns. *Forsch. Fortschr.* **13,** 425–426.
Engelmann, C. (1950). Über den Geschmackssinn des Huhnes IX. *Z. Tierpsychol.* **7,** 84–111.
Engelmann, C. (1957). "So leben Hühner, Tauben, Gänse." Neumann Verlag. Radebeul, East Germany.
Engelmann, C. (1960). Weitere Versuche über die Futterwahl des Wassergeflügels: über die Schmeckempfindlichkeit der Gänse. *Arch. Geflügelzucht Kleintierk.* **9,** 91–104.
Fuerst, F. F. and Kare, M. R. (1962). The influence of pH on fluid tolerance and preferences. *Poult. Sci.* **41,** 71–77.
Fuerst, F. F. and Kare, M. R. (1962). The influence of pH on fluid tolerance and preferences. *Poult. Sci.* **41,** 71–77.
Ganchrow, D. and Ganchrow, J. R. (1985). Number and distribution of taste buds in the oral cavity of hatchling chicks. *Physiol. Behav.* (in press).
Gentle, M. J. (1971a). Taste and its importance to the domestic chicken. *Br. Poult. Sci.* **12,** 77–86.
Gentle, M. J. (1971b). The lingual taste buds of *Gallus domesticus L.* Br. *Poult. Sci.* **12,** 245–248.
Gentle, M. J. (1972). Taste preference in the chicken, *Gallus domesticus L. Br. Poult. Sci.* **13,** 141–155.
Gentle, M. J. (1973). Diencephalic stimulation and mouth movements in the chicken. *Br. Poult. Sci.* **14,** 167–171.
Gentle, M. J. (1975a). Gustatory behaviour of the chicken and other birds. *In* "Neural and Endocrine Aspects of Behaviour in Birds" (P. Wright, P. G. Caryl and D. M. Vowles, Eds.). Elsevier, Amsterdam.
Gentle, M. J. (1975b). Gustatory hyposensitivity to quinine hydrochloride following diencephalic lesions in *Gallus domesticus. Physiol. Behav.* **14,** 265–270.
Gentle, M. J. (1976a). Using arousal changes in the EEG to indicate gustatory sensitivity following brain lesions in *Gallus domesticus. Br. Poult. Sci.* **17,** 151–156.
Gentle, M. J. (1976b). Quinine hydrochloride acceptability after water deprivation in *Gallus domesticus. Chem. Sens. Flav.* **2,** 121–128.
Gentle, M. J. (1978). Extra-lingual chemoreceptors in the chicken (*Gallus domesticus*). *Chem. Sens. Flav.* **3,** 325–329.
Gentle, M. J. (1979). Single unit responses from the solitary complex following oral stimulation in the chicken. *J. comp. Physiol.* **130,** 259–264.
Gentle, M. J. (1981). Oral behaviour following forebrain ablations in *Gallus domesticus. Behav. Processes* **6,** 109–120.
Gentle, M. J. (1982). Oral behaviour in response to oral stimulation in *Gallus domesticus. In* "Determination of Behaviour by Chemical Stimuli" (J. E. Steiner and J. R. Ganchrow, Eds.). IRL Press, London and Washington, D.C.
Gentle, M. J. (1984). The chorda tympani and taste in the chicken. *Experientia* **40,** 1253–1255.

Gentle, M. J. and Dewar, W. A. (1981). The effects of vitamin A deficiency on oral gustatory behaviour in chicks. *Br. Poult. Sci.* **22,** 275–279.

Gentle, M. J., Dewar, W. A. and Wight, P. A. L. (1981). The effects of zinc deficiency on oral behaviour and taste bud morphology in chicks. *Br. Poult. Sci.* **22,** 265–273.

Gentle, M. J. and Harkin, C. (1979). The effect of sweet stimuli on oral behaviour in the chicken. *Chem. Sens. Flav.* **4,** 183–190.

Gerritsen, A. F. C., van Heezik, Y. M. and Swennen, C. (1984). Chemoreception in two further *Calidris* species (*C. maritima* and *C. canutus*) with a comparison of the relative importance of chemoreception during foraging in *Calidris* species. *Netherl. J. Zool.* **33**(4), 485–496.

Gillette, K., Martin, G. M. and Bellingham, W. P. (1980). Differential use of food and water cues in the formation of conditioned aversion by domestic chicks (*Gallus gallus*). *J. exp. Psychol.* **6,** 99–111.

Graziadei, P. P. C. (1969). The ultrastructure of vertebrate tastebuds. *In* "Olfaction and Taste" (C. Pfaffman, Ed.), Vol. III. Rockefeller University Press, New York.

Graziadei, P. P. C. (1974). The olfactory and taste organs of vertebrates in a dynamic approach to the study of their morphology. *In* "Transduction Mechanisms in Chemoreception" (T. M. Poynder, Ed.). IRL Press, London and Washington, D.C.

Graziadei, P. P. C. and de Han, R. S. (1971). The ultrastructure of the frog's taste organs. *Acta anat.* **80,** 563–603.

Greschik, E. (1917). Geschmacksknospen auf der Zunge des Amazonenpapageis. *Anat. Anz.* **50,** 257–270.

Gunther, W. C. and Wagner, M. W. (1971). Preferences for natural and artificial sweeteners in heat-stressed chicks of different ages. *Proc. Ind. Acad. Sci.* **81,** 401–409.

Halpern, B. P. (1962). Gustatory nerve responses in the chicken. *Am. J. Physiol.* **203,** 541–544.

Hamrum, C. L. (1953). Experiments on the senses of taste and smell in the bobwhite quail (*Colinus virginianus virginianus*). *Am. Midl. Nat.* **49,** 872–877.

Harriman, A. E. and Kare, M. R. (1966). Aversion to saline solutions in starlings, purple grackles, and herring gulls. *Physiol. Zool.* **39,** 123–126.

Harriman, A. E. and Milner, J. S. (1969). Preference for sucrose solutions by Japanese quail (*Coturnix coturnix japonica*) in two-bottle drinking tests. *Am. Midl. Nat.* **81,** 575–578.

Hughes, B. O. and Wood-Gush, D. G. M. (1970). Investigations into specific appetites for sodium and thiamine in domestic fowls. *Physiol. Behav.* **6,** 331–339.

Hughes, B. O. and Wood-Gush, D. G. M. (1971). A specific appetite for calcium in domestic chickens. *Anim. Behav.* **19,** 490–499.

Hutchinson, J. C. D. and Taylor, W. W. (1962). Mechanics of pecking grain. *Proc. 12th World's Poultr. Congr.*, pp. 112–116.

Kadono, H., Okado, T. and Ohno, K. (1966). Neurophysiological studies of the sense of taste in the chicken. *Res. Bull. Fac. Agric. Gifu Univ.* **22,** 149–159.

Kare, M. R. (1961). Comparative aspects of taste. *In* "Physiological and Behavioural Aspects of Taste" (M. R. Kare and B. P. Halpern, Eds.). University of Chicago Press, Chicago.

Kare, M. R. (1965). The special senses. *In* "Avian Physiology" (P. D. Sturkie, Ed.). Baillière, Tindall and Cassell, London.

Kare, M. R. (1971). Comparative study of Taste. *In* "Handbook of Sensory Physiology" (L. M. Beidler, Ed.), Vol. IV/2. Springer-Verlag, Berlin and New York.

Kare, M. R. and Beily, J. (1948). The toxicity of sodium chloride and its relation to water intake in baby chicks. *Poult. Sci.* **27,** 751–758.

Kare, M. R. and Ficken, M. S. (1963). Comparative studies on the sense of taste. *In* "Olfaction and Taste" (Y. Zotterman, Ed.), Vol. I. Pergamon Press, New York.

Kare, M. R. and Maller, O. (1967). Taste and food intake in domesticated and jungle fowl. *J. Nutr.* **92,** 191–196.

Kare, M. R. and Medway, W. (1959). Discrimination between carbohydrates by the fowl. *Poult. Sci.* **38,** 1119–1127.

Kare, M. R. and Pick, H. L. (1960). The influence of the sense of taste on feed and fluid consumption. *Poult. Sci.* **39,** 697–705.

Kare, M. R. and Rogers, J. G. (1976). Sense organs. Taste. *In* "Avian Physiology" (P. D. Sturkie, Ed.). Springer-Verlag, Berlin and New York.

Kare, M. R., Black, R. and Allison, E. G. (1957). The sense of taste in the fowl. *Poult. Sci.* **36,** 129–138.

Kear, J. (1960). Food selection in certain finches with special reference to interspecific differences. Ph.D. Thesis, Cambridge University.

Kitchell, R. L., Ström, L. and Zotterman, Y. (1959). Electrophysiological studies of thermal and taste reception in chickens and pigeons. *Acta physiol. scand.* **46,** 133–151.

Kolmer, W. (1910). Über Strukturen im Epithel der Sinnesorgane. *Anat. Anz.* **36,** 281–299.

Krause, R. (1922). "Mikroskopische Anatomie der Wirbeltiere in Einzeldarstellungen: II. Vögel and Reptilien". de Gruyter, Berlin.

Krol, C. P. M. and Dubbeldam, J. L. (1979). On the innervation of tastebuds by n. facialis in the mallard, *Anas platyrhynchos L. Netherl. J. Zool.* **29,** 267–274.

Kurosawa, T., Niimura, S., Kusuhara, S. and Ishida, K. (1983). Morphological studies of tastebuds in chickens. *Jap. J. Zootech. Sci.* **54,** 502–510.

Lalonde, E. R. and Eglitis, J. A. (1961). Number and distribution of tastebuds on the epiglottis, pharynx, larynx, soft palate and uvula in a human newborn. *Anat. Rec.* **140,** 91–95.

Landolt, J. P. (1970). Neural properties of pigeon lingual chemoreceptors. *Physiol. Behav.* **5,** 1151–1160.

Levine, R. R. and Zeigler, H. P. (1981). Extratelencephalic pathways and feeding behaviour in the pigeon (*Columba livia*). *Brain Behav. Evol.* **19,** 56–92.

Lindenmaier, P. and Kare, M. R. (1959). The taste endorgans of the chicken. *Poult. Sci.* **38,** 545–550.

McLelland, J. (1975). Aves digestive system. *In* "Sisson and Grossman's The Anatomy of the Domestic Animals" (R. Getty, Ed.), Vol. 2. Saunders, Philadelphia.

McLelland, J. (1979). Digestive system. *In* "Form and Function in Birds" (A. S. King and J. McLelland, Eds.), Vol. 1. Academic Press, London and New York.

Mariotti, G. and Fiore, L. (1980). Operant conditioning studies of taste discrimination in the Pigeon (*Columba livia*). *Physiol. Behav.* **24,** 163–168.

Moon, R. D. and Zeigler, H. P. (1979). Food preferences in the pigeon (*Columba livia*). *Physiol. Behav.* **22,** 1171–1182.

Moore, R. A. and Elliott, R. (1946). Numerical and regional distribution of tastebuds on the tongue of the bird. *J. comp. Neurol.* **84,** 119–131.

Morris, D. (1955). The seed preferences of certain finches under controlled conditions. *Avic. Mag.* **61,** 271–287.

Nalavade, M. N. and Varute, A. T. (1977). Histochemical studies on mucins of vertebrate tongues II. Histochemical analysis of mucosubstances in lingual glands and tastebuds of some birds. *Acta histochem.* **60,** 18–32.

Norgren, R. (1977). A synopsis of gustatory neuroanatomy. *In* "Olfaction and Taste", (J. Le Magnen and P. McLeod, Eds.) Vol. VI. IRL, London and Washington, D.C.

Percival, M. S. (1961). Types of nectar in angiosperms. *New Phytol.* **60,** 235–281.

Pevzner, R. (1964). Distribution of acetyl cholinesterase activity in the tastebuds of vertebrates. (In Russian.) *Dokl. Akad. Nauk SSSR* **155,** 930–932.

Pfaffman, C. (1978). The vertebrate phylogeny, neural code and integrative processes of taste. *In* "Handbook of Perception. Vol. VI. Tasting and smelling" (E. Carterette and M. Friedman, Eds). Academic Press, New York and London.

Pick, H. and Kare, M. R. (1962). The effect of artificial cues on the measurement of taste preference in the chicken. *J. comp. physiol. Psych.* **55,** 342–345.

Putkonen, P. T. S. (1967). Electrical stimulation of the avian brain. *Ann. Acad. Sci. fenn. A5* **130,** 9–95.

Reed Hainsworth, F. (1973). On the tongue of a hummingbird: its role in the rate and energetics of feeding. *J. comp. Biochem. Physiol.* **46A,** 65–78.

Reed Hainsworth, F. and Wolf, L. L. (1976). Nectar characteristics and food selection by hummingbirds. *Oecologia, Berl.* **25,** 101–113.

Rensch, B. and Neunzig, R. (1925). Experimentelle Untersuchungen über den Geschmackssinn der Vögel II. *J. Ornithol.* **73,** 633–646.

Rogers, J. G. (1974). Responses of caged red winged blackbirds to two types of repellents. *J. Wildl. Management* **38,** 418–423.

Romoser, G. L., Bossard, E. H. and Combs, G. F. (1958). Studies on the use of certain flavours in the diet of chick. *Poult. Sci.* **37,** 631–633.

Saito, I. (1966a). Comparative anatomical studies of the oral organs of the poultry. IV. Macroscopical observation of the salivary glands. (In Japanese; English summary.) *Miyazaki Daigaku Nogakubu, Kenkyu Hokoku* **12,** 110–120.

Saito, I. (1966b). Comparative anatomical studies of the oral organs of the poultry V. Structure and distribution of tastebuds of the fowl. (In Japanese.) *Miyazaki Daigaku Nogakubu, Kenkyu Hokoku* **13,** 95–102.

Scalzi, H. A. (1967). The cytoarchitecture of gustatory receptors from the rabbit foliate papillae. *Z. Zellforsch. mikrosk. Anat.* **80,** 413–435.

Sevier, A. C. and Munger, B. L. (1965). A silver method for paraffin sections of neural tissue. *J. Neuropath. exp. Neurol.* **24,** 130–135.

Stevens, J. and de Bont, A. F. (1980). Choice by starlings among different cherry cultivars. *Agricultura, Heverlee* **28,** 421–436.

van Heezik, Y. M., Gerritsen, A. F. C. and Swennen, C. (1983). The influence of chemoreception on the foraging behaviour of two species of sandpiper, *Calidris alba* (Pallas) and *Calidris alpina* (L.). *Netherl. J. Sea Res.* **17,** 47–56.

van Kan, S. (1979). Touch and taste in the pigeon (*Columba livia domestica*). (In Dutch.) Internal report, Zoöl. Lab., Leiden.

van Prooije, A. (1978). The distribution, morphology and functional significance of tastebuds in the chicken (*Gallus domesticus L.*). (In Dutch.) Internal report, Zoöl. Lab., Leiden.

Veenman, C. L. (1984). The organisation of the nucleus basalis-neostriatum complex in the goose (*Anser anser L.*). Thesis, State University of Leiden.

Vince, M. A. (1977). Taste sensitivity in the embryo of the domestic fowl. *Anim. Behav.* **25,** 797–805.

Vinnikov, Y. A. (1974). Sensory reception. Cytology, molecular mechanisms and evolution. *In* "Molecular Biology, Biochemistry and Biophysics (A. Kleizeller, G. F. Springer and H. G. Wittmann, Eds.), Vol. 17. Springer-Verlag, Berlin and New York.

von Düring, M. and Andres, K. H. (1976). The ultrastructure of taste and touch receptors of the frog's taste organ. *Cell Tiss. Res.* **165,** 185–198.

Warner, R. L., McFarland, L. Z. and Wilson, W. O. (1967). Microanatomy of the upper digestive tract of the Japanese quail. *Am. J. vet. Res.* **28,** 1537–1548.

Warren, R. P. and Vince, M. A. (1963). Taste discrimination in the Great Tit (*Parus major*). *J. comp. physiol. Psychol.* **56,** 910–913.

Wenzel, B. M. (1973). Chemoreception. *In* "Avian Biology" (D. S. Farner and J. R. King Eds.), Vol. III. Academic Press, New York and London.

Wenzel, B. M. (1983). Chemical senses. *In* "Physiology and Behaviour of the Pigeon" (M. Abs, Ed.). Academic Press, London and New York.

Weymouth, R. D., Lasiewski, R. C. and Berger, A. J. (1964). The tongue apparatus in hummingbirds. *Acta anat.* **58,** 252–270.

White, S. S. (1968). Mechanisms involved in deglutition in *Gallus domesticus. J. Anat.* **104,** 177.

White, S. S. (1970). The larynx of Gallus domesticus. Ph.D. Thesis, University of Liverpool.

Wilcoxon, H. C., Dragoin, W. B. and Kral, P. A. (1971). Illness induced aversions in rat and quail, relative salience of visual and gustatory cues. *Science, Wash., D.C.* **171,** 826–828.

Wild, J. M., Arends, J. J. A. and Zeigler, H. P. (1984). A trigeminal sensorimotor circuit for pecking, grasping and feeding in the pigeon (*Columba livia*). *Brain Res.* **300,** 146–151.

Wright, P. (1968). Hypothalamic lesions and food/water intake in the Barbary dove. *Psychon. Sci.* **13,** 133–134.

Wright, P. (1975). The neural substrate of feeding behaviour in birds. *In* "Neural and Endocrine Aspects of Behaviour in Birds" (P. Wright, P. G. Caryl and D. M. Vowles, Eds.). Elsevier, Amsterdam.

Yang, R. S. H. and Kare, M. R. (1968). Taste response of a bird to constituents of arthropod defence secretions. *Ann. entomol. Soc. Am.* **61,** 781–782.

Zeier, H. and Karten, H. J. (1971). The archistriatum of the pigeon: organization of afferent and efferent connections. *Brain Res.* **31,** 313–326.

Zeier, H. and Karten, H. J. (1973). Connections of the anterior commissure in the pigeon *Columba livia L. J. comp. Neurol.* **150,** 201–216.

Zeigler, H. P. (1975). Trigeminal deafferentation and hunger in the pigeon (*Columba livia*). *J. comp. physiol. Psychol.* **89,** 827–844.

Zeigler, H. P. (1976). Feeding behaviour in pigeons. *In* "Advances in the Study of Behaviour" (J. S. Rosenblatt, R. A. Hinde, E. Shaw and C. Beer, Eds.), Vol. 7. Academic Press, London and New York.

Zeigler, H. P. and Karten, H. J. (1973a). Brain mechanisms and feeding behaviour in the pigeon (*Columba livia*). I. Quinto-frontal structures. *J. comp. Neurol.* **152,** 59–82.

Zeigler, H. P. and Karten, H. J. (1973b). Brain mechanisms and feeding behaviour in the pigeon (*Columba livia*). II. Analysis of feeding behaviour deficits after lesions of quinto-frontal structures *J. comp. Neurol.* **152,** 83–102.

Zeigler, H. P., Miller, M. and Levine, R. R. (1975). Trigeminal nerve and eating in the pigeon. *J. comp. physiol. Psychol.* **89,** 845–858.

Zweers, G. A. (1982a). The feeding system of the pigeon (*Columba livia L.*). *Adv. Anat. Embryol. Cell Biol.* **73,** 1–108.

Zweers, G. A. (1982b). Pecking of the Pigeon (*Columba livia*). *Behaviour* **81,** 173–230.

Zweers, G. A. and Wouterlood, F. G. (1973). Functional anatomy of the feeding apparatus of the mallard (*Anas platyrhynchos L.*). *Proc. 3rd Eur. Anat. Congr., Manchester,* pp. 88–89.

Zweers, G. A., Gerritsen, A. F. C. and van Kranenburg-Voogd, P. J. (1977). Mechanics of feeding in the mallard (*Anas platyrhynchos*). *In* ''Contributions to Vertebrate Evolution'' (M. K. Hecht and F. S. Szalay, Eds.), Vol. 3. S. Karger, Basel.

Index

Heavy type or ff. indicates pages where a subject is fully described. ff. means that more than one page is involved.

H

I

J

K

L

M

N

P

Q

R

S

U

V